BIOLOGY: THE FOUNDATIONS

SECOND EDITION

BIOLOGY
THE FOUNDATIONS

SECOND EDITION

STEPHEN L. WOLFE

University of California, Davis

Wadsworth Publishing Company
Belmont, California
A Division of Wadsworth, Inc.

Biology Editor: Jack C. Carey

Production Editor: Sally Schuman

Managing Designer: Mary Ellen Podgorski

Copy Editor: Don Yoder

Art Editor: Virginia Mickelson

Illustrations: Florence Fujimoto, Judith Lopez, Virginia Mickelson, Carla Simmons

Printed in the United States of America

3 4 5 6 7 8 9 10–87 86 85 84

ISBN 0-534-01169-1

Library of Congress Cataloging in Publication Data.

Wolfe, Stephen L., 1932–
  Biology, the foundations.

  Includes bibliographies and index.
  1. Biology. I. Title
QH308.2.W63 1983 574 82-6891
ISBN 0-534-01169-1   AACR2

Books in the Wadsworth Biology Series

Biology: The Unity and Diversity of Life, 2nd, Starr and Taggart

Energy and Environment: Four Energy Crises, Miller

Living in the Environment, 3rd, Miller

Replenish the Earth: A Primer in Human Ecology, Miller

Oceanography: An Introduction, 2nd, Ingmanson and Wallace

Biology books under the editorship of William A. Jensen, University of California, Berkeley

Biology, Jensen, Heinrich, Wake, Wake

Biology: The Foundations 2nd Wolfe

Botany: An Ecological Approach, Jensen and Salisbury

Plant Physiology, 2nd, Salisbury and Ross

Plant Physiology Laboratory Manual, Ross

Nonvascular Plants, Scagel et al.

Plant Diversity: An Evolutionary Approach, Scagel et al.

An Evolutionary Survey of the Plant Kingdom, Scagel et al.

Plants and the Ecosystem, 3rd, Billings

Plants and Civilization, 3rd, Baker

PREFACE

Like the first edition of *Biology: The Foundations,* the aim of this second edition is to lay a groundwork in the basic observations, hypotheses, and theories of biology while introducing the student to the philosophy and experimental approach used by biological scientists. To do this the book concentrates on five subjects considered to be the core of biology—cell biology, genetics, development, evolution, and ecology. These essential topics are supported by an introductory chapter on basic chemistry and a later chapter surveying the classification of living things. For those taking a single course, with no intention of pursuing further study in biology, the core subjects present the information needed to understand the fundamental conclusions of biology. For those intending to continue study in the field, the book provides the basic facts and concepts necessary for advanced study.

The facts and concepts discussed in the text are supported by key experiments from biological research showing how the topics were established as scientific truth and how the work was done. The status and direction of current research and some of the problems of interest to present-day biologists are also presented to show that biology is an active and rapidly developing science.

In this second edition the chapters describing cell biology have been thoroughly updated, and new chapters concentrating on energy and enzymes and the cell surface have been added. The balance between classic and modern genetics presented in the first edition has been retained. The development chapters have been expanded and updated, particularly on the subject of plant development. Although the chapter surveying the classification of living organisms has also been expanded, the concept and purpose of this chapter remain as a source of reference and review. The chapters on evolution have been expanded to include more emphasis on contemporary findings and the controversies presently sweeping this field. The ecology unit has been completely rewritten and updated

to reflect the contemporary approach to environmental biology.

As in the first edition, the core topics presented in each chapter are amplified by end-of-chapter supplements to make the book useful for teaching both majors and nonmajors. Some of the supplements, designed for students intending to major in biology, give additional information at more advanced levels. Other supplements provide relevant topics of general interest to both majors and nonmajors. By choosing the appropriate supplements in combination with the main chapters, the book can be tailored to meet the interests and aims of a variety of students and courses.

The second edition retains the study aids intended to make the book more useful and understandable to students. Key sentences, printed in boldface throughout each chapter, emphasize main points and direct the student's attention to major conclusions and concepts. Each chapter is also provided with study questions: these, in combination with the key sentences, facilitate review of the chapter topics. Information boxes give chapter overviews or present useful background information. Extensive diagrams and numerous photographs illustrate and amplify the material covered in the book. Finally, the key terms in the book are defined in a glossary. The language of the text is simple and direct, and every effort has been made to make the book clear to a beginning student in biology.

This edition has been reviewed exhaustively to make the book as accurate and pertinent as possible. To this end, no less than 36 reviewers commented on the manuscript at various stages during its completion. These include scientists with expert knowledge of the topics covered in the chapters and teachers with expertise and experience in biology instruction.

I am indebted to these reviewers, whose efforts greatly improved the accuracy, completeness, and coverage of this edition. I am also indebted to the scientists and authors who willingly and generously supplied photographs, diagrams, and tables for this and the first edition. Thanks are also due to Jack C. Carey, my subject editor at Wadsworth, whose enthusiastic support, cooperation, and suggestions were critical to the book, and to others at Wadsworth who provided their capable and professional assistance: Sally Schuman, production editor; Mary Ellen Podgorski, designer; Virginia Mickelson, art editor and illustrator; and illustrators Florence Fujimoto, Judith Lopez, and Carla Simmons.

Stephen L. Wolfe

EXTENSIVE MANUSCRIPT REVIEW

BRIEF CONTENTS

CONTENTS

Introduction

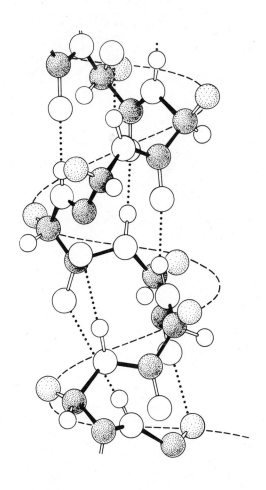

1

AN INTRODUCTION TO BIOLOGY AND THE SCIENTIFIC METHOD

We are all interested in life: what makes living organisms tick; how they grow, reproduce, and move; how they interact; why they take the forms they do; and how they and we got here in the first place. People have been trying to answer these questions as long as they have had the capacity to wonder about themselves and their surroundings. Until about 500 years ago, the answers were most often sought from religious and philosophical authorities rather than from direct observation of nature. In the fifteenth and sixteenth centuries, however, emphasis began to shift toward the observation of nature as the most reliable and productive way to study life and other natural phenomena. By the nineteenth century, techniques of observation and evaluation were refined into the **scientific method,** a procedure for observing, relating, and explaining the world around us. The scientific method has become established as the most effective procedure for investigating the world and the universe for one primary reason: simply because it works. It produces results that

---

The scientific method has become established as the primary method for explaining the world around us because *it works.*

---

have practical as well as intellectual value and form a reliable basis for further work and investigation.

HOW THE SCIENTIFIC METHOD WORKS

The aim of science is to explain as much of the universe as possible, to discover relationships in ourselves and our surroundings, and to draw general conclusions that can be applied with confidence to parts of the universe we have not seen. Often, these conclusions can also be used to foresee new relationships or predict the most likely course of future events.

The scientific method is based on careful observation and description of facts about the world. The observations may be as simple as counting the legs on a spider or as

---

The first step in the scientific method is careful observation and description of facts.

---

complex as determining the angles between the atoms of a molecule through x-ray diffraction and mathematical analysis. In either case, the facts are gathered as carefully and accurately as possible and the methods used are completely described so that the same facts can be observed by anyone who wishes to repeat the observations.

Once a solid base of carefully observed and described facts is established, a scientist attempts to explain them. This is done by forming a **hypothesis,** an educated guess about how the facts may be explained or related. Hypothesizing in science is as free and unrestricted as guessing in any field, with one exception: The hypothesis, to be scientific, must be testable by experimentation and further observation. The proposal that all insects have three pairs

---

To be scientific, a hypothesis must be testable by experiments.

---

of legs is an acceptable scientific hypothesis because it can be tested by collecting insects and counting their legs. In contrast, the hypothesis that humans have souls is not scientific because there is no conceivable way to construct a scientific test of its validity.

It is important to understand this distinction, because it lies at the very center of science and the scientific method. One philosopher who has dealt at length with the methods and approach of science is the Englishman Karl L. Popper. He points out that, to be valid as a scientific hypothesis, an idea must be capable of being *falsified*, or made untrue, by the results of experimental tests. From this point of view it is obvious that the proposal that all insects have three pairs of legs is scientific, because it can easily be falsified: all a scientist has to do to falsify the hypothesis is find an insect species that regularly has two pairs of legs (or less) or more than three pairs of legs. It is equally obvious that there is at present no experimental method available to determine whether humans have souls, and this hypothesis is therefore unscientific. The opposite hypothesis—that humans do not have souls—is also unscientific because, like the opposing idea, it cannot be experimentally tested or falsified. Therefore ideas that are testable fall within the domains of science and are scientific whereas other ideas that cannot, while possibly valid and true, are unscientific.

The next step in the scientific method is to *test* the hypothesis by experiment. The test must be carefully constructed to prevent bias by the investigator for or against the hypothesis. As part of the test, a deliberate search must be made for any possible exceptions to the hypothesis. If exceptions are found, or if new facts are discovered that contradict the hypothesis, it must be discarded or modified to fit the facts. Finally, any logical consequences of the hypothesis must be deduced and separately tested.

To understand how the scientific method might work in a specific example, consider a scientist observing the limbs of animals. He or she notes that humans, dogs, cats, monkeys, horses, cows, and rabbits have two pairs of limbs, one in front and one in back. From this the scientist hypothesizes that all mammals have four limbs arranged in two pairs, one anterior and one posterior. To test the hypothesis, counts are made of the limbs on other mammals such as bats, moles, bears, seals, and sea lions; these tests support the hypothesis. Then the scientist deduces that since whales are mammals, they must also have two pairs of limbs. A test of this deduction, however, reveals that whales have forelimbs but no hindlimbs. As a result, the hypothesis has been falsified and must be modified to state that "all mammals except whales have two pairs of limbs" or that "all mammals have two pairs of limbs or fewer." Further observation and testing reveal that members of only one other known group of mammals, the sea cows or manatees, have one pair of limbs. The hypothesis in its final form is then written to state that "all mammals except whales and sea cows have two pairs of limbs." The final hypothesis could be used to predict with confidence that any newly discovered mammals will have two pairs of limbs or fewer and that none having three pairs of limbs are likely to be found.

A hypothesis that withstands repeated experimental tests without exception is gradually accepted as scientific truth. This acceptance may take many years and involve repeated confirmation through experimentation. If exceptions are found at any time, the hypothesis is modified accordingly or rejected. If, after it is accepted as scientific truth, the hypothesis answers questions that apply widely to many fields in science and leads to the formation of other important hypotheses, it may be regarded as a scientific **theory.** Long-established theories of fundamental

---

A scientific theory, since it is extensively and completely supported by experimental evidence, is regarded as an established truth.

---

importance that are thoroughly supported by evidence are called **laws** of science.

Since scientific theories are supported by exhaustive experimentation, scientists usually regard them as established truths. Note that this use of the word *theory* is quite

---

The steps in the scientific method are (1) observation of facts, (2) formation of a hypothesis relating or explaining the observed facts, and (3) experimental tests of the hypothesis.

---

different from its meaning in everyday English. In common usage, it is used most often to label ideas that are merely speculative, as in the expression "Well, it's only a theory." This difference between the scientific and common usage of the word has led to much confusion. As a result many people fail to appreciate the extensive experimental evidence that supports scientific theories.

THE SCIENTIFIC METHOD IN BIOLOGY

Until the 1500s the intellectual world relied on the writings of the ancient Greeks as the primary guide for the study of nature. Although the Greeks had made beginnings in the study of life through direct observation, their methods were imperfect and many false impressions were recorded among the facts observed. Two events in the 1500s, both of them in the year 1543, revolutionized attitudes about nature and the most productive method for studying it. One was the publication of Nicolaus Copernicus's conclusion, based on the direct observation of the motions of the stars and planets, that the sun, not the earth, is at the center of the solar system. The second was the publication, by Andreas Vesalius, of his conclusions about the anatomy of the human body. Like Copernicus, Vesalius developed his ideas through direct observation—in this case by dissecting the human body. Both Copernicus's and Vesalius's conclusions directly contradicted the philosophical and religious teachings of the time. Since their conclusions were incontrovertible, however, they gradually became accepted as fact. With them came an acceptance of direct observation rather than reference to authority as the most practical and productive approach to the study of nature. The scientific method was born.

Physics and chemistry developed quickly in response to the new approach to study through direct observation. Complete application of the scientific method in biology lagged behind, however, primarily because of a persistent belief that living things are different from nonliving objects and cannot be studied by the same techniques. Underlying this belief was the concept that the organic substances of living organisms are intrinsically different from inanimate matter and can be produced only by the "vital force" of life. This view was shattered in 1828, when Friedrich Wöhler found crystals of an organic molecule, urea, among the chemicals produced when the inorganic substance ammonium cyanate was heated. Wöhler's synthesis of an organic molecule from an inorganic substance, and the many similar examples that followed, gradually established that the same elements occur in both living and inanimate objects and are governed by the same chemical and physical laws. The final foundation for the development of biology as a fully scientific field was laid by the publication, in 1859, of Charles Darwin's explanation of evolution. Darwin's hypothesis, now accepted as a scientific theory, explained how life, with all its apparent purpose and design, could have evolved through natural processes no different in character from those already investigated successfully in physics and chemistry. Through its logical extension, Darwin's hypothesis affirmed that life could have evolved from inanimate substances through inanimate processes—and, moreover, that life has no characteristics that cannot be studied by the same scientific method used with such success to study inanimate processes. From this point on, biology expanded rapidly as a completely scientific field with important and productive subdivisions such as botany, zoology, genetics, evolution, physiology, biochemistry, cell biology, ecology, and molecular biology.

Applying the scientific method may seem an almost impossible task reserved for intellectual giants of Darwin's stature. Although every age has its gifted individuals, the daily practice of science is much more attainable. At this level in biology, as in other sciences, the scientific method is rarely followed with the elegance suggested by the sequential steps of observation, hypothesis, and experimental testing. Probably most biologists work almost exclusively at the first level: observing and describing facts. They may rarely construct hypotheses consciously or carry out experiments to test them. These scientists describe the structures of biological objects, whether microorganisms, animals, plants, or molecules, or work out the details of biological processes. Their research is nevertheless a valuable contribution to the body of carefully described scientific fact.

The number of scientists who regularly construct and directly test hypotheses is much smaller. Even at this level, many biologists admit that luck or serendipity (the ability to find things not sought for) plays an important role in their work. They stumble on important discoveries while looking for something else or while carrying out an experiment to test another hypothesis. The scientific gift in this case is the ability to recognize a good thing when it is stumbled upon and then work back to the hypothesis that would have led to that finding. Of greatest importance, as many scientists stress, is constant involvement in laboratory or field research. In this way, the investigator is in a position to happen on findings that might present themselves unexpectedly.

Hypothesis making often takes the relatively simple form of modeling. One of the best examples comes from the scientific work leading to the discovery of the structure of the hereditary molecule called deoxyribonucleic acid

(DNA)—perhaps one of the greatest scientific accomplishments of all time. One problem to be solved was how molecular groups pair in the interior of the DNA molecule to hold the structure together. James D. Watson, who worked out DNA's structure with Francis H. C. Crick, hit on the solution by arranging models of the interior molecular groups cut from cardboard in various combinations on his desk top. Later, when other details of the structure were worked out, Watson and Crick built a scale model of the entire molecule, testing the fit of all the component atoms and their bonding distances and angles.

THE MOTIVES FOR BIOLOGICAL RESEARCH

The motivations of biologists toward scientific work are as complex as the forces driving any person to accomplish a goal. Curiosity about ourselves, our fellow creatures, or the physical objects of the world is a basic ingredient of scientific motivation. There is excitement in the discovery of information that no one has discovered and described before. There is also an element of play in science, an enjoyment in the manipulation of scientific ideas and apparatus and the chase toward a scientific goal. There are practical motivations too for scientific work—to cure disease, to invent work-saving devices, even to perfect the machines of war.

Ideally, scientists work selflessly to broaden the horizons of scientific truth without thought of personal gain. In practice, although a strain of altruism runs deep in almost every scientist, the usual drives and ambitions also motivate scientists, just as they do people in other occupations. A desire for personal fame, recognition, and the admiration and respect of others ranks high among the motivations of scientists. Many also work to achieve status and power as do other humans. In science, this amounts to the power to control large sums of money, extensive laboratories, and a staff of subordinate scientists and technicians.

None of this is meant to demean science or scientists. It is meant simply to stress that scientists are human, not superhuman, and that the goals and methodology of science are attainable by anyone interested in studying nature and life. Whatever the level of investigation or the motivation, the work of every scientist adds to the fund of human knowledge about ourselves and our world. And, for better or worse, scientific investigation—because it works—has provided a technology that has revolutionized the world and the quality of human life.

REPORTING THE RESULTS

The results of scientific investigation are reported in "papers" contributed by investigators to scientific journals. These journals are published by the thousands, usually on a monthly schedule; every university and college library devotes much of its funding and space to them. Scientists spend a major part of their time in an effort to keep up with the papers published in their own and related fields. Anyone familiar with scientific publication will verify that, because the number of journals and papers has increased prodigiously in recent years, the effort to keep up places ever greater demands on an investigator's time. The need for up-to-date information also requires scientists to attend national and international meetings in their fields, where findings can be exchanged without the delays inherent in publication. The information traded at meetings and published in journals is free for anyone to inspect and use.

The scientific facts, hypotheses, theories, and concepts described in a text like this are derived from the results reported in published papers—some recent and some classics dating to Darwin's time or earlier. This book summarizes the scientific information basic to all of biology in the most important and fundamental areas of contemporary work: cell biology, genetics, developmental biology, evolution, and ecology. As you read these chapters, remember that the information presented is not absolute, authoritarian, or dogmatic. It represents no more and no less than scientific hypotheses and theories supported by observations and experiments. All these observations, hypotheses, and theories are subject to change and modification if future research proves them to be false.

Questions

1. Why has the scientific method become the most widely used approach for discovering facts and relationships in the world around us?

2. Contrast the sources of information in the prescientific and postscientific world. Point out differences in the methods used for discovering facts and relationships.

3. Describe the steps in the scientific method.

4. What is a hypothesis?

5. What is the difference between a scientific and a nonscientific hypothesis?

6. How does a hypothesis become a theory? A law?

7. What is the difference between the common and scientific meanings of theory and law?

8. What events started the scientific revolution?

9. What was the significance of Wöhler's synthesis of urea?

10. Why was Darwin's theory of evolution important in the philosophy and practice of the scientific method in biology?

11. How are the discoveries of scientists announced?

12. What is the essential difference between the announcement of scientific information and the announcement of technological ideas developed by industry?

13. Do you think that some findings of scientists, such as those related to national security and defense, should be kept secret?

14. What ethical dilemmas might confront an investigator who makes a biological discovery of social importance—such as a cure for cancer or a procedure for altering the hereditary material of an individual?

15. Do you think that scientific research attempting to solve urgent human problems, such as finding a cure for cancer, is more important than discovering basic facts of biology, such as, for example, the research revealing the molecular structure of DNA? Why do you think so?

Suggestions for Further Reading

Asimov, I. 1964. *A Short History of Biology*. Natural History Press, New York.

Magner, L. N. 1979. *A History of Life Sciences*. Marcel Dekker, New York.

In its early years, biology was concerned primarily with the description of newly discovered species and the structures of animals and plants. This concentration on descriptive morphology began to change near the close of the nineteenth century, when new techniques for studying the chemistry of life were discovered. These techniques developed rapidly and were so successfully applied that by the mid-twentieth century much of biology was transformed into a biochemical and molecular science. As a result, it is impossible to understand the conclusions of modern biology without an introduction to chemistry and to the major types of molecules carrying out the activities of life.

ATOMIC STRUCTURE AND CHEMICAL BONDS

Although many of the molecules of living systems are highly complex, most of them fall into one or more of four distinct classes: (1) **carbohydrates,** (2) **lipids,** (3) **proteins,** and (4) **nucleic acids.** These four classes of organic molecules, along with water, form almost the entire substance of living organisms. Understanding the structure of these biological molecules, and the chemical bonds holding them together, will provide the basic information needed to follow the biochemical and molecular systems described in this book.

Atoms and Molecules

Molecules are formed from **atoms** linked together in definite numbers and ratios by chemical bonds. The atoms of a molecule may be the same, as in a molecule of oxygen (O_2) or hydrogen (H_2), or different, as in a molecule of water, which contains two hydrogen atoms linked to a single oxygen (H_2O).

An atom is the smallest unit possessing the chemical and physical properties of an element. Although nearly a

---

An atom is the smallest unit possessing the chemical and physical properties of an element.

---

hundred different kinds of atoms occur naturally on earth and link in various ways to form the molecules of both living and nonliving systems, all are basically similar in structure. Each atom consists of an atomic **nucleus** surrounded by one or more smaller, fast-moving particles, the **electrons** (Figure 2-1). Most of the space occupied by an atom contains the electrons; the nucleus represents only

2

A CHEMICAL BACKGROUND FOR BIOLOGY

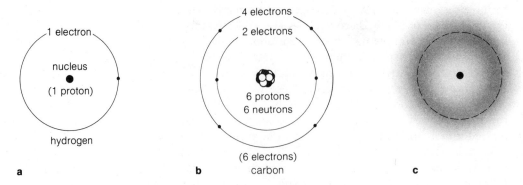

Figure 2-1 Atoms consist of a nucleus surrounded by fast-moving electrons. (**a**) In hydrogen, the simplest atom, the nucleus is surrounded by the orbital of a single electron. (**b**) Carbon, a more complex atom, has a nucleus surrounded by two successive shells of electrons. (**c**) The orbital (dotted line over region of deepest shade) represents the most probable locations for an electron to occupy; less probable locations are shown as regions of lighter shade.

about 1/10,000 of the total volume of an atom. The nucleus, however, is much heavier than all the electrons put together; it makes up more than 99 percent of the total mass of an atom.

The Atomic Nucleus All atomic nuclei contain one or more positively charged particles called **protons.** The number of protons in the nucleus of each type of atom is always the same. Hydrogen, the smallest atom, has a single proton in its nucleus; the nucleus of uranium, the heaviest naturally occurring atom, has 92 protons. Since the atoms of a given type always contain the same number of protons, this number, called the **atomic number,** specifically identifies an atom. Hydrogen, the simplest atom, with one proton in its nucleus, has the atomic number 1. Similarly, nitrogen with seven protons, and oxygen with eight, have atomic numbers of 7 and 8 respectively (Figure 2-2).

Protons are positively charged particles in an atomic nucleus. Neutrons are uncharged particles in an atomic nucleus.

Each type of atom, except hydrogen, also has a number of uncharged particles in its nucleus. These particles, called **neutrons,** occur in variable numbers approximately equal to the number of protons. Carbon, for example, in its most common form, has six protons and six neutrons in its nucleus. About 1 percent of naturally occurring carbon atoms have six protons and seven neutrons in their nuclei. Other carbon atoms, with six protons and eight neutrons, are found in natural carbons in even smaller proportions. These different forms of an atom, all with the same number of protons but varying numbers of neutrons, are called **isotopes.** The isotopes of an atom have essentially the same chemical properties, but differ in mass and other physical characteristics.[1]

A neutron and proton have the same mass. This mass is given an arbitrary value of 1, and atoms are assigned a **mass number** based on the total number of protons and neutrons in the atomic nucleus. (The mass of electrons is so small that they can be ignored in determinations of atomic mass.) The three isotopes of carbon with mass numbers of 12 (six protons plus six neutrons), 13 (six protons plus seven neutrons), and 14 (six protons plus eight neutrons) are identified as ^{12}C, ^{13}C, and ^{14}C.

Atomic isotopes have the same number of protons but differing numbers of neutrons.

Some isotopes of an atom are unstable and undergo radioactive decay. For example, the carbon isotope ^{14}C is unstable and slowly breaks down into an atom of nitrogen. In this breakdown, one neutron in the ^{14}C nucleus splits into a proton and an electron. The proton is retained in the nucleus, giving a new total of seven protons and seven neutrons, and the electron is ejected from the atom. When ejected, the electron is called a **beta particle:**

[1]Mass is defined as the tendency of a body or particle to resist an accelerating force. Consider the difference between kicking, or applying an accelerating force to, a soccer ball filled with either air or lead. The difference in mass would be immediately obvious.

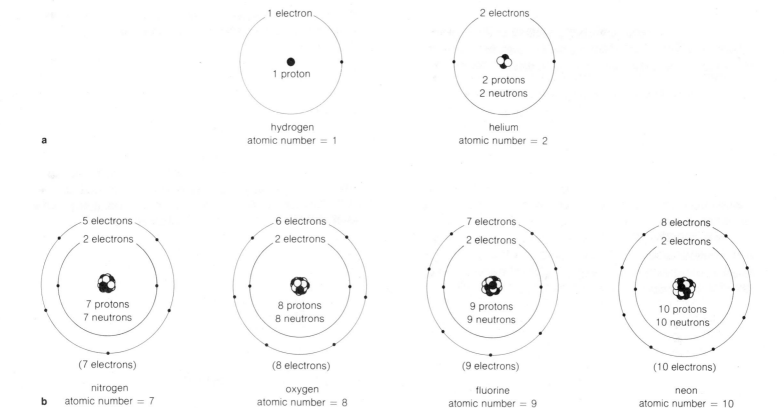

Figure 2-2 (**a**) Hydrogen and helium, with one and two protons respectively, have atomic numbers of 1 and 2; the atomic number is equal to the number of protons in the nucleus of the atom. (**b**) Atomic numbers of more complex atoms.

$$^{14}\text{C (6 protons + 8 neutrons)} \rightarrow$$
$$^{14}\text{N (7 protons + 7 neutrons) + 1 electron (beta particle)}$$

An unstable isotope of hydrogen, ^{3}H, called **tritium,** has one proton and two neutrons in its nucleus. It too undergoes radioactive decay with ejection of a beta particle from its nucleus. The electrons ejected as beta particles from ^{14}C and ^{3}H have considerable energy and can be detected by various means.

The radioactive isotopes of carbon and hydrogen, along with unstable isotopes of sulfur and phosphorus, among others, have been of great value in biological research because molecules containing them can be traced by their radioactivity as they go through chemical reactions in living organisms. A number of stable, nonradioactive isotopes, such as ^{15}N (heavy nitrogen), can be detected by their mass differences and have also proved to be valuable as tracers in biological experiments.

The Electrons of an Atom The electrons surrounding an atomic nucleus carry a negative charge and normally occur

Electrons are negatively charged particles that travel in orbitals around an atomic nucleus.

in numbers equal to the positively charged protons. Electrons, however, have a mass equivalent to only 1/1800 of the mass of a proton.

Electrons are in constant, rapid motion around the atomic nucleus. Until the 1920s, electrons were believed to follow definite paths in their movement around a nucleus, much as the planets follow a defined orbit around the sun. More accurate analysis of the behavior of electrons has shown that an orbiting electron may actually be found in almost any location, ranging from the immediate vicinity of an atomic nucleus to practically infinite space. In moving through these locations, an electron travels so fast that it can almost be regarded as being at all these points at the same time. If an electron could be tracked in its movements around the nucleus, however, it would be found to pass through some locations more frequently than others. These most probable locations surround the atomic nucleus in layers of different shapes called **orbitals.** In the hydrogen atom shown in Figure 2-1c, the orbital of the single electron moving around the nucleus is depicted as a shaded region. In this region, the most probable locations of the electron at any given time are shown as areas of deepest shade. Although either one or two electrons may occupy a given orbital, the most stable and balanced conditions are provided when an orbital is occupied by a *pair* of electrons.

The orbitals followed by electrons occur in successive layers or *shells* around an atomic nucleus. The innermost layer, the one closest to the nucleus, contains a maximum of two electrons and is thus filled by a single orbital. Hydrogen has one electron in this orbital; helium has two (see Figure 2-2a). Atoms with total numbers of electrons between three (lithium) and ten (neon) have two layers or shells of orbitals. The innermost layer contains two electrons in a single orbital. The second layer of orbitals, which may contain as many as eight electrons traveling in four orbitals, surrounds the nucleus at a greater distance. As this shell is filled, the atoms from lithium to neon are formed, including beryllium (with four electrons), boron (with five), carbon (six), nitrogen (seven), oxygen (eight), fluorine (nine), and neon (ten; see Figure 2-2b). Larger atoms have successive layers of orbitals that may contain more than eight electrons, up to a maximum of thirty-two electrons for any single shell. No matter what number of electrons occurs in the intermediate shells of large atoms, the most stable conditions are reached when the outermost shell contains a total of eight.

Electrons and the Chemical Activity of an Atom The chemical properties of an atom are determined largely by

---

The chemical properties of an atom are determined largely by the number of electrons in the outermost shell. Especially important is the difference between this number and the stable number of two for hydrogen and eight for all other atoms.

---

the number of electrons in the outermost shell. Most significant is the difference between this number and the stable number of either two (for atoms near helium in size) or eight (for larger atoms). Helium, with two electrons in its single shell, and atoms such as neon and argon, with eight electrons in their outer shells, are stable and essentially inert chemically. Atoms with outer shells containing

electrons near these numbers tend to gain or lose electrons to approximate these stable configurations. Sodium (Figure 2-3a), with two electrons in its first shell, eight in the second, and one in the third and outermost shell, readily loses its single outer electron to leave a stable second shell with eight electrons. Chlorine (Figure 2-3b), with seven electrons in its outermost shell, tends to attract an electron from another atom to attain the stable configuration of eight. Other atoms, such as oxygen or calcium, which differ from a stable configuration by two electrons, tend to gain or lose electrons in pairs. Atoms differing from the stable, eight-electron outer shell by more than two electrons are inclined to *share* electrons with other atoms rather than gain or lose electrons completely. This sharing pattern is characteristic of carbon, which has four electrons in its outer shell and thus falls at the midpoint between the tendency to gain or lose electrons. The relative tendency to gain, lose, or share electrons determines much of the chemical activity of an atom and underlies the formation of the chemical bonds holding the atoms of molecules together.

The most common atoms of living organisms are carbon (C), hydrogen (H), oxygen (O), and nitrogen (N).

The substances of living organisms are made up from the smaller atoms of nature—ranging from hydrogen, the smallest atom, to iodine, the largest found in any significant amount (Table 2-1). Most common are carbon, hydrogen, oxygen, and nitrogen, which together make up 95 percent or more of the matter of living organisms. Calcium, phosphorus, and sulfur are found in lesser quantities; the remaining atoms listed in Table 2-1 occur only in very small amounts.

The Formation of Chemical Bonds

Although some atoms, such as those of gold, can be found in free, uncombined form on earth, most are linked into molecules by chemical bonds. Most important of these linkages in biological molecules are **electrostatic bonds,** formed by electrical attractions between atoms, and **covalent bonds,** formed by electrons that are shared between atoms.

Electrostatic Bonds Electrostatic bonds form through the tendency of some atoms to gain or lose electrons to form stable outer shells. A bond of this type may form if sodium, which has a tendency to lose an electron, is brought together with chlorine, which has a tendency to gain an

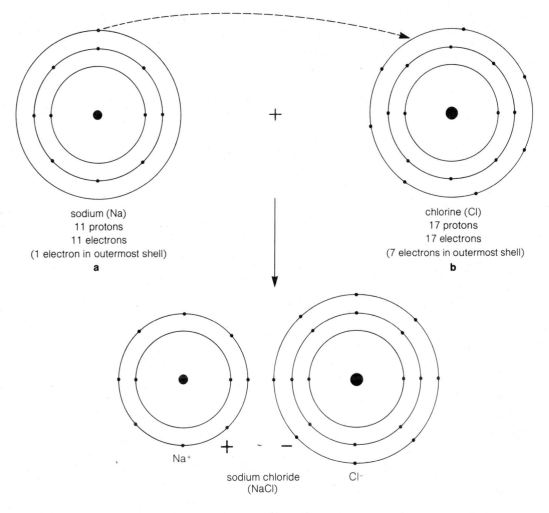

Figure 2-3 (**a**) Sodium, with one electron in its outermost shell and eight in its second shell, readily loses an electron to attain a stable state. (**b**) Chlorine, with seven electrons in its outer shell, readily gains an electron to attain the stable number of eight outer electrons. (**c**) Passage of an electron from sodium to chlorine creates the ions Na^+ and Cl^- (see text).

electron (Figure 2-3*a* and *b*). Under the correct conditions, the sodium atom readily loses an electron to the chlorine atom. After the transfer, the sodium atom, with eleven protons in its nucleus and only ten electrons in surrounding orbitals, carries a single positive charge. The chlorine, now with seventeen protons and eighteen electrons, carries a single negative charge. In this charged condition, the sodium and chlorine atoms are called **ions** and are identified as Na^+ and Cl^-. The difference in charge sets up a strong attraction that tends to hold the atoms together (Figure 2-3*c*). This attraction is termed an electrostatic or **ionic** bond.

Electrostatic bonds are readily broken and remade. In solid sodium chloride, Na^+ and Cl^- ions are held together by their opposite charges. If the sodium chloride is placed

Electrostatic bonds are formed by an attraction between ions of opposite charge.

in water, molecules of water take up positions between the Na^+ and Cl^- ions, greatly reducing the attraction of the ions for each other. The ions, each surrounded by a layer of water molecules, may then separate and diffuse through the solution as free ions. If the water molecules are re-

Table 2-1 Important Atoms in Living Organisms

Atom	Symbol	Atomic Number	Electrons in Outer Shell	Biological Function
Carbon	C	6	4	Basic atom of all organic compounds
Oxygen	O	8	6	Component of most biological molecules; final electron acceptor in many energy-yielding reactions
Hydrogen	H	1	1	Component of all biological molecules; H^+ ion important component of solutions
Nitrogen	N	7	5	Component of proteins, nucleic acids, and many other biological molecules
Sulfur	S	16	6	Component of many proteins
Phosphorus	P	15	5	Component of nucleic acids and molecules carrying chemical energy; found in many lipid molecules
Iron	Fe	26	2	Important in energy-yielding reactions; component of oxygen carriers in blood
Calcium	Ca	20	2	Found in bones and teeth; important in muscle contraction; found in cellular fluids and extracellular fluids; important in cell regulation and control
Potassium	K	19	1	Important in conduction of nerve impulses; activates many enzymes
Sodium	Na	11	1	Ion in solution in living matter
Chlorine	Cl	17	7	Ion in solution in living matter; required for photosynthesis
Magnesium	Mg	12	2	Part of molecules important in photosynthesis; important ion in many enzyme-catalyzed reactions
Copper	Cu	29	1	Important in photosynthesis and energy-yielding reactions
Iodine	I	53	7	Component of hormone produced by thyroid gland
Fluorine	F	9	7	Found in trace amounts in bones and teeth
Manganese	Mn	25	2	Found in trace amounts; required for photosynthesis
Zinc	Zn	30	2	Found in trace amounts; component of some enzymes
Molybdenum	Mo	42	1	Found in trace amounts; component of some enzymes
Boron	Bo	5	3	Essential for growth in higher plants

moved by evaporation, the Na^+ and Cl^- ions reassociate into a solid crystal as the solution dries. Water molecules are effective in causing the transition from bound to free ions because they have relatively positive and negative ends. The ends of the water molecules are attracted to the surface of the ions, surrounding them and thus reducing their effective charge.

Electrostatic bonds are common among the forces holding ions, atoms, and molecules together in biological systems. Because these bonds are easily made and broken under the influence of water, electrostatic bonds are frequently important in rapidly changing systems where molecules or ions are brought together briefly during their interactions.

Covalent Bonds Atoms such as carbon, with four electrons in their outer shells, have little tendency to gain or lose electrons completely. Instead, in forming chemical bonds these substances tend to share electrons to form stable outer shells. The hydrogen atom also shares its single electron readily to form a stable outer shell of two electrons.

Covalent bonds are formed through electron sharing between atoms to complete stable outer shells.

The sharing of electrons by two atoms of hydrogen to form a molecule of hydrogen H_2 (Figure 2-4) is the simplest example of the sharing process. Each hydrogen atom has a single electron occupying an orbital around its nucleus. If two hydrogen atoms approach closely enough, the single electron of each atom may join in a new, combined orbital that surrounds both nuclei. The new orbital, containing two electrons, satisfies the conditions for stability in atoms with a single shell of electrons. As a consequence, the hydrogen atoms tend to remain linked together, forming a stable molecule. This linking force, set up by shared electrons, is the covalent bond. Covalent bonds are relatively stable and can be broken only through the expenditure of considerable amounts of energy. In molecular diagrams, the covalent bond is designated by a pair of dots or a single line to represent the shared electrons:

$$H\cdot + \cdot H \rightarrow H{:}H \text{ or } H{-}H$$

Electrons are also shared in pairs to complete the outer shells of atoms larger than hydrogen. Carbon, with four unpaired outer electrons, forms four separate covalent bonds to complete its outermost shell. An example of this process, the formation of methane (CH_4) by electron sharing between one carbon and four hydrogen atoms, is shown in Figure 2-5.

Because carbon can form four separate covalent bonds, carbon atoms readily link together to form highly branched or chainlike structures. These carbon chains form the "backbone" of an almost unlimited variety of molecules. In the molecule shown in Figure 2-6, each covalent bond representing a pair of shared electrons is depicted as a single line between linked atoms. Note that each carbon atom is linked to its neighbors by a total of four covalent bonds.

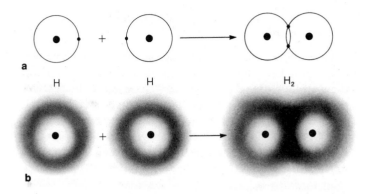

Figure 2-4 Electron sharing in the formation of covalent bonds. (**a**) Electrons are shared between two atoms of hydrogen to form H_2; the most probable locations of the electrons are shown as thin lines. (**b**) Here the most probable locations of electrons in H_2 are indicated by areas of deepest shade.

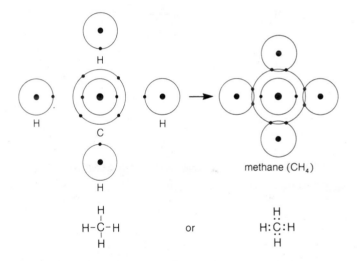

Figure 2-5 **Above:** Electron sharing between four hydrogen atoms and a carbon atom to form methane. **Below:** Two ways of diagramming covalent bonds.

The other atoms present in the molecule—nitrogen (N), oxygen (O), and hydrogen (H)—also form covalent linkages readily and are commonly found with carbon in biological molecules. In these linkages, nitrogen forms three covalent bonds, oxygen forms two, and hydrogen one. The double bond at the position marked by an arrow in Figure 2-6 indicates that two pairs of electrons are shared between the atoms involved in the linkage.

Unequal Electron Sharing, Polarity, and Hydrogen Bonds

Polarity Electrons are not always shared equally between two atoms held together by a covalent bond. In traveling their orbitals, the electrons may pass near one of the two atomic nuclei more frequently than the other. If this is the case, the atom retaining the electrons a greater percentage of the time tends to carry a relatively negative charge. The deprived atom becomes relatively positive. A covalent bond of this type has some of the properties of an electrostatic linkage and is said to be **polar.** If electrons are equally shared, so that there is no greater probability that they will be closer to one atom than another in the linkage, the covalent bond is **nonpolar.**

The covalent bonds linking two atoms of hydrogen to oxygen to form water are polar (Figure 2-7). The shared electrons in this molecule are more likely at any instant to be nearest the oxygen nucleus—leaving the hydrogen atoms with a relatively positive charge and giving the oxygen a relatively negative charge. This gives the entire molecule a polar character since the two hydrogen atoms

---

Polarity results from unequal electron sharing. Polar molecules have relatively positive and negative ends, and they associate most readily with polar or charged substances.

---

in water are located to one side of the oxygen (see Figure 2-7). Thus the whole molecule is polar in character and has relatively positive and negative ends.

Polar molecules attract and align themselves with other polar molecules and with ions. The polar environments created by this association tend to exclude nonpolar substances, which also prefer their own company. This tendency can be easily illustrated by mixing a polar substance, like water, with a nonpolar substance such as a fat or oil. No matter how vigorously the oil and water are shaken, they will quickly separate into polar and nonpolar environments when placed at rest. Because of their reaction to water, polar substances are often identified as **hydrophilic** (*hydro* = water; *philic* = preferring) and nonpolar substances as **hydrophobic** (*hydro* = water; *phobic* = avoiding).

Polar and nonpolar molecules are important in the organization of cells. The surface coating of cells is formed primarily from nonpolar molecules, but both the cell interior and the watery medium bathing the cell contain polar molecules. The nonpolar surface coat (a **membrane**) tends to exclude the polar molecules of both the inside and outside mediums and thus acts as a barrier to the free movement of substances to and from cells. The tendency of

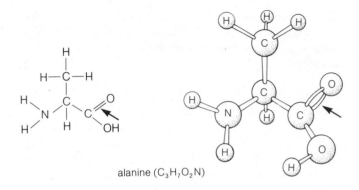

alanine ($C_3H_7O_2N$)

Figure 2-6 Alanine, an amino acid. Covalent bonds are indicated by solid lines in the structural formula (left) or rods in the model (right). Each line or rod represents a *pair* of shared electrons. Two pairs are shared in the double bond marked by the arrow.

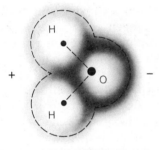

Figure 2-7 Polarity in the water molecule, created by unequal electron sharing in the covalent bonds between hydrogen and oxygen. The most probable locations of the shared electrons are indicated by regions of deepest shade. At any instant, the shared electrons are likely to be closer to the oxygen nucleus, making this end of the molecule relatively negative and the hydrogen end relatively positive.

polar and nonpolar groups to associate in mutually exclusive regions also provides part of the force folding many biological molecules into their three-dimensional forms or holding molecules together in complex molecular assemblies.

Hydrogen Bonds Hydrogen bonds also result from unequal electron sharing in covalent linkages. In the example given above, the hydrogen atoms in a water molecule are relatively positive because the electrons shared with oxygen tend at any instant to be nearer the oxygen nucleus. Because of this positive charge, a hydrogen nucleus in a water molecule is attracted to other atoms with a relatively

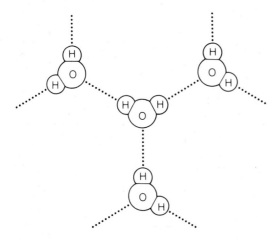

Figure 2-8 Hydrogen bonding (dotted lines) between water molecules. The bonds form a lattice that stabilizes water molecules in the liquid form.

Hydrogen bonds are formed by the attraction between a hydrogen nucleus, made relatively positive by unequal electron sharing, and a nearby atom that is relatively negative.

---

negative charge, such as the oxygen atom of an adjacent water molecule (Figure 2-8). This type of attraction, the **hydrogen bond,** is comparatively weak (about one-tenth as strong as a covalent bond) and is easily disturbed and broken. As shown in Figure 2-8, hydrogen bonds are frequently depicted by a dashed or dotted line between the hydrogen and adjacent atom involved in the linkage.

Hydrogen bonds are important in biological molecules because opportunities for hydrogen bonding are frequently so great that many such links may form. The individual bonds, although relatively weak, are collectively strong when numerous and lend stability to the three-dimensional structure of complex molecules such as nucleic acids and proteins. In these large biological molecules, hydrogen bonds form between a hydrogen nucleus linked covalently to nitrogen or oxygen (both of these atomic nuclei attract electrons more strongly than hydrogen) and another oxygen or nitrogen nucleus located nearby. Hydrogen bonds of this type are shown by the dotted lines in the protein molecule diagrammed in Figure 2-23.

Electrostatic linkages, polar and nonpolar covalent bonds, and hydrogen bonds are the primary forces that hold atoms together and arrange them in ordered patterns in biological molecules. Some of these forces, including electrostatic attractions, polar and nonpolar forces, and hydrogen bonds, also link molecules together into complex

---

The atoms of molecules are held together by electrostatic, covalent, and hydrogen bonds and by polar and nonpolar attractions.

---

assemblies. Particularly important among these molecular interactions is the tendency of nonpolar molecules or chemical groups to associate together when placed in a watery medium. The molecules of membranes, for example, are held together in a water-excluding environment by nonpolar associations, much as oil forms a separate, stable layer on water.

THE MAJOR MOLECULES OF LIVING ORGANISMS

Molecules containing carbon in combination with hydrogen and other atoms are defined as **organic** molecules. All other types are **inorganic.** Both molecular types are important as constituents of living organisms.

Inorganic Molecules

The Primary Inorganic Molecule: Water Water is the most abundant single molecule in living matter. Almost all other substances that make up living organisms are dissolved in water, and most of the chemical reactions of life take place in an aqueous environment. Water also enters directly into many of the reactions that build or break down complex molecules. In digestion, for example, one molecule of table sugar (sucrose) is broken into two smaller molecular units by the chemical addition of one molecule of water. (Reactions involving addition or removal of water are discussed in more detail in Information Box 2-1.)

Water also has the important property of changing temperature relatively slowly when heat is added or subtracted. This property helps organisms regulate their internal temperature and provides resistance to sudden, extreme temperature changes that could be fatal to life. The amount of heat required to raise one gram of water one degree Celsius (or centigrade) is called a **calorie.** It is an important measure of heat energy widely used in biology. (The Calorie familiar to dieters, the "large" C calorie or kilocalorie, equals 1000 of the small calories used in biology.)

Condensation and Hydrolysis

Many types of biological reactions in which molecular sub-units are assembled or disassembled involve addition or removal of water. As glucose and fructose are linked to-gether to form a disaccharide, for example, a molecule of water appears as an additional product (see Figure 2-10). In this type of assembly reaction, called a **condensation,** the components of water, H^+ and OH^-, are split from the reacting chemical groups of the combining molecules. (The atoms contributing to the formation of water in Figure 2-10 are shaded.) Most of the biological reactions in which complex molecules are assembled from smaller subunits are condensation reactions.

In the reverse reaction, chemical building blocks are broken off from a larger molecule with the *addition* of a molecule of water. This type of reaction, called a **hydrolysis,** is also common in biological systems:

The R in the chemical structures stands for a complex organic group that forms the rest of the molecule. Hydrolysis is a part of almost all biological reactions in which larger molecules are broken into their subunits. The reactions of digestion, in which proteins, fats, carbohydrates, and nucleic acids are broken into smaller units before being absorbed in the intestine, are typical hydrolysis reactions.

The strongly polar nature of water molecules gives water other properties of utmost significance for the survival of life. Because water molecules are small and highly polar, they tend to penetrate into the spaces between other charged and polar molecules in the solid form and form a coating that reduces the polar and electrostatic forces holding the solid together. As a consequence, the molecules of the solid separate readily and go into solution. Because of this property, water is a good solvent and dissolves a greater variety of substances than any other liquid.

Because of its polar structure, each water molecule can set up three hydrogen bonds with neighboring water molecules (Figure 2-8). These attractions establish a network or lattice that accounts for some of the unique properties of water. Because of the hydrogen bonds, water molecules tend to remain linked together and resist separation. This resistance makes water resist boiling and remain stable as a liquid until the relatively high temperature of 100°C is reached. Closely related substances, like H_2S, do not form an internal hydrogen-bond lattice in the liquid form and, as a consequence, boil at much lower temperatures. From the properties of H_2S and other molecules that are closely related to water in structure, it has been estimated that water would boil and form a gas at $-81°C$

without its hydrogen-bond lattice. It is difficult to imagine how life as we know it could exist under these conditions.

Another peculiarity of the hydrogen-bond lattice of water is that it separates significantly at the freezing point, opening spaces between water molecules as water freezes. As a result, water becomes lighter in density as it goes from a liquid to the frozen state. Most other substances increase in density during conversion from liquid to solid. This property of water allows ice to float at the top of lakes and streams, where it acts as an insulator to retard freezing of the water beneath. If ice became heavier than liquid water on freezing, it would continually form at the surface of lakes and streams in cold weather and sink to the bottom; eventually, the entire body of water would become solidly frozen. In colder regions, thawing in summer would occur only at the surface. Under these conditions, most forms of aquatic life would be unable to survive.

The polar nature of water molecules also gives water great internal **cohesion.** Because of polarity, water molecules in the liquid form set up a strong mutual attraction, producing a force that tends to pack them more closely together. This packing force has a significant effect at the water's surface, because the forces attracting the surface water molecules to underlying molecules are not balanced by forces exerted by the air molecules above the water. This unbalanced force pulls the surface layer even more tightly against the underlying water molecules, producing a tension that makes the surface semirigid. The combination of high surface tension and internal cohesion accounts for the fact that columns of water within microscopic tubes, as inside the trunks and stems of plants, can be raised to relatively great heights without separating. The high surface tension also permits small organisms, such as some insects and spiders, to walk on water.

Acids and Bases Most inorganic substances and many organic molecules act as either **acids** or **bases** in water solution. Acids and bases are substances that affect the concentration of hydrogen ions (H^+) and hydroxyl ions (OH^-) in water. Water always contains both ions because a proportion of the water molecules in the liquid form separate to produce H^+ and OH^- ions:

$$H_2O \rightarrow H^+ + OH^-$$

Acids are substances that, when dissolved in water, release additional hydrogen ions and thus increase the relative concentration of H^+ ions. Lemon juice and vinegar are common examples. Bases are substances that bind H^+ ions, or release additional hydroxyl ions, when dissolved in water and thereby reduce the relative concentration of H^+ ions. Caustic soda and ammonia are common bases.

The relative concentrations of H^+ and OH^- ions in a water solution determine the **acidity** of the solution. The degree of acidity greatly affects the chemical reactivity of many organic and inorganic substances dissolved in water.

Acidity is usually expressed as **pH** on a scale of 0 to 14. If the concentrations of H^+ and OH^- ions are equal, the solution is said to be **neutral** (pH 7). Acid solutions have greater concentrations of H^+ than OH^- ions and have pH values less than 7. Solutions with higher concentrations of OH^- ions are termed basic or alkaline and have pH greater than 7. The relative concentrations of H^+ and OH^- ions in living organisms are nearly equal, closely approximating pH 7. (For a quantitative description of the derivation of the pH scale, see Supplement 2-1.)

Other Essential Inorganic Molecules Oxygen is important in the final steps of the reaction sequences providing the primary sources of energy for the activities of life. Carbon dioxide, along with water, provides the raw material from which plants construct organic matter in photosynthesis. Plants absorb other inorganic substances necessary for their survival in the form of soluble inorganic salts. For example, nitrogen is absorbed in the form of nitrates of

various kinds; phosphorus and sulfur are absorbed in soluble phosphates and sulfates. Most of the metallic elements are absorbed in combination with nitrates, phosphates, and sulfates—as, for example, in magnesium, iron, and zinc sulfates. Animals also absorb essential elements such as iron, calcium, copper, iodine, cobalt, manganese, and phosphorus in the form of soluble inorganic salts.

Organic Molecules: An Overview *Reread 9/5/84*

Organic molecules are based on carbon and its interactions with other atoms, most frequently hydrogen, oxygen, and nitrogen. These atoms—carbon, hydrogen, oxygen, nitrogen—with lesser quantities of phosphorus and sulfur are the primary elements that combine to form the carbohydrates, lipids, proteins, and nucleic acids of living organisms.

Organic molecules are based on carbon and its interaction with hydrogen and other atoms.

Many of the substances in these molecular classes are large molecules with complex structures and activities. All, however, are based on different combinations of a few relatively simple organic subunits, primarily *organic acids, alcohols, aldehydes,* and *ketones*. The amino acids and the fatty acids, which are components of proteins and lipids respectively, are examples of organic acids that often appear as building blocks of biological molecules. These subunit molecules owe their acidic properties to the —COOH (*carboxyl*) group, which ionizes in solution to release an H^+ ion:

Alcohols, which serve as important building blocks of fats and many carbohydrates, have in common the reactive **alcoholic** group:

Linkages formed by the alcohols in building more complex substances are based primarily on the reactivity of the —OH segment of this group of atoms. Carbohydrates also contain a reactive *carbonyl* $\left(\diagup_{} C{=}O \right)$ group consisting of an oxygen double-bonded to a carbon of the chain. The carbonyl group occurs in either of two reactive groups. In one, called an **aldehyde,** the carbonyl group is carried on a terminal carbon of the chain:

$$-\overset{\displaystyle H}{\underset{\displaystyle |}{\overset{\displaystyle |}{C}}}-\overset{\displaystyle |}{C}{=}O$$

In the other reactive group, called a **ketone,** the oxygen is carried on a carbon at a point in the interior of the chain:

$$-\overset{|}{\underset{|}{C}}-\overset{|}{\underset{\displaystyle \|}{\underset{\displaystyle O}{C}}}-\overset{|}{\underset{|}{C}}-$$

These reactive groups and some other arrangements of atoms important in biological molecules are summarized in Table 2-2.

The Major Classes of Organic Molecules

Carbohydrates Carbohydrates are used as fuel substances by all cells. Long-chain molecules assembled from carbo-

---

Table 2-2 Important Reactive Groups of Organic Molecules

Radical	Structure	Reactivity
Alcohol	$-\overset{H}{\underset{|}{\overset{|}{C}}}-OH$	Reacts with organic acids to form many important biological substances; part of many carbohydrate molecules
Carboxyl (acid group)	$-C\overset{\displaystyle O}{\underset{\displaystyle OH}{\diagup\diagdown}}$	Ionizes in solution to release H^+ ions and thus acts as acid; part of amino acids, fatty acids
Carbonyl	$C{=}O$ in	
Aldehydes	$-\overset{H}{\underset{|}{\overset{|}{C}}}{=}O$	Important components of carbohydrates and intermediate compounds formed in synthesis and breakdown of carbohydrates
Ketones	$\overset{|}{\underset{|}{C}}{=}O$	
Amino	$-N\overset{\diagup\, H}{\diagdown\, H}$	Component of amino acids and other important biological molecules; can combine with H^+ ions in solution to produce an $-NH_3^+$ group, thus acting as base by reducing H^+ ion concentration
Phosphate	$-O-\overset{\displaystyle O^-}{\underset{\displaystyle \|}{\underset{\displaystyle O}{\overset{\displaystyle |}{P}}}}-O^-$	An acidic group; forms links between organic groups in complex biological molecules; enters in energy reactions as energy carrier
Sulfhydryl	$-S-H$	Reactive group on amino acid cysteine that enters into disulfide ($-S-S-$) linkages and stabilizes protein structure

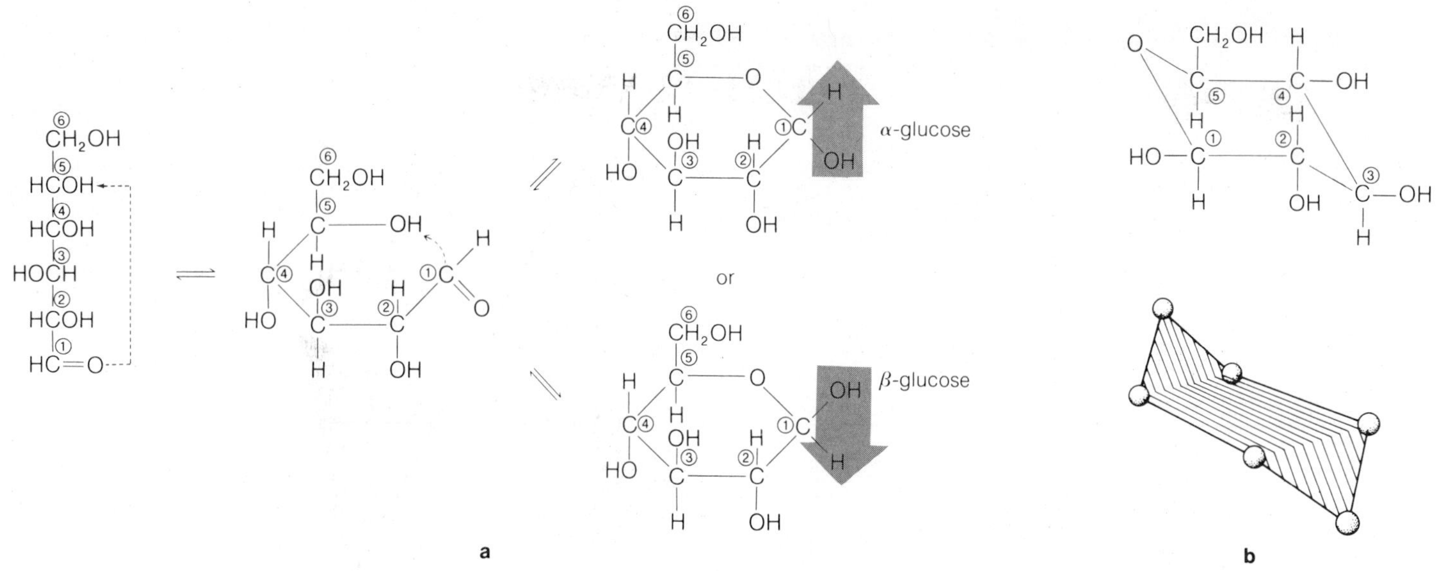

Figure 2-9 (a) Ring formation by glucose molecules in solution (see text). (b) The three-dimensional configuration of the glucose ring.

hydrate subunits are also important as structural molecules in some cells. The main constituent of plant cell walls, **cellulose,** is a long-chain molecule of this type. Carbohydrates are also stored in quantity as **starch** in plant cells or as **glycogen** in animal cells. Carbohydrate units also link to proteins to form **glycoproteins** (from the Greek *glykys* = sweet) and link to lipids to form **glycolipids.** Both these complex molecular types occur in quantity at cell surfaces, where they are active in recognizing stimuli reaching cells from the outside.

Basic Carbohydrate Structure Carbohydrates contain carbon, hydrogen, and oxygen in the approximate ratio 1C:2H:1O or (CH_2O). The basic building blocks of the carbohydrate

---

Carbohydrates are organic molecules containing carbon, hydrogen, and oxygen in the approximate ratio 1C:2H:1O or (CH_2O).

---

family are short carbon chains from three to seven carbons long. Each carbon atom in the chains forming the units, except one, carries an —OH (hydroxyl) group. The remaining carbon carries an oxygen attached by a double bond, forming a C=O (carbonyl) group (see Figure 2-9a). In most carbohydrates all the other available binding sites of the carbons are occupied by hydrogen atoms. The carbonyl oxygen may be at the end of the chain, forming an aldehyde $\left(\begin{array}{c} H \\ | \\ -C=O \end{array}\right)$ group, or in the interior of the chain,

forming a ketone $\left(\begin{array}{c} -C- \\ \| \\ O \end{array}\right)$ group (see Table 2-2).

One unit of carbohydrate of this type, with a chain of three to seven carbons linked to —OH groups and the single carbonyl oxygen, is known as a **monosaccharide.** Monosaccharides are named according to the number of carbons they contain. Of the various possibilities, **trioses** (three carbons), **pentoses** (five carbons), and **hexoses** (six carbons) are most abundant in nature. Table 2-3 lists some functions of the monosaccharides in living organisms.

Monosaccharides with five or more carbons can form stable ring structures in water solutions. Glucose, for example, occurs in either of two closely related ring structures formed by an interaction between the aldehyde $\left(\begin{array}{c} H \\ | \\ -C=O \end{array}\right)$ group at one end of the chain and the hydroxyl group at the next-to-last position at the other end of the chain (Figure 2-9a). The ring formed is not flat or in one plane in space, as suggested in Figure 2-9a, but commonly has the three-dimensional "chair" form shown in Figure 2-9b.

The two closely related ring forms of glucose differ only in the direction pointed by the hydrogen attached to the 1-carbon of the sugar (see shaded arrows in Figure 2-9a). The relatively small difference between these two forms, called alpha glucose and beta glucose, has great significance for the chemical properties of polysaccharides formed

Table 2-3 Carbohydrate Units Found in Nature

Number of Carbons	Type	Examples	Importance
3	Triose	Glyceraldehyde, dihydroxyacetone	Intermediates in energy-yielding reactions and photosynthesis
4	Tetrose	Erythrose	Intermediate in photosynthesis and energy-releasing reactions
5	Pentose	Ribose, deoxyribose, ribulose	Intermediates in photosynthesis; components of molecules carrying energy; components of informational nucleic acids DNA and RNA; structural molecules in cell walls of plants
6	Hexose	Glucose, fructose, galactose, mannose	Fuel substances; products of photosynthesis; building blocks of starches and cellulose
7	Heptose	Sedoheptulose	Intermediate in photosynthesis and energy-releasing reactions

from glucose units. Cellulose, for example, built up from glucose links in the beta form, is insoluble and cannot be digested as a food source by humans or most other animals. The starches, however, which contain glucose links in the alpha form, are soluble and easily digested. Other sugars with a general structural plan resembling glucose also have equivalent alpha and beta forms.

Linking Monosaccharides into Polysaccharides Monosaccharides may link together by twos to form **disaccharides** or in greater numbers to form **polysaccharides.** Some polysaccharides, such as starch, cellulose, and glycogen, are formed from monosaccharides linked end to end in either unbranched chains (as in cellulose) or chains with forks or branches (as in glycogen). Polysaccharide molecules commonly contain a single type of monosaccharide building block.

Linkage of the monosaccharides glucose and fructose to form the disaccharide sucrose illustrates the general pattern of the reactions forming disaccharides and polysaccharides (Figure 2-10). The reaction shown in Figure 2-10 also illustrates how water molecules frequently participate in reactions involving either assembly or disassembly of the building blocks of larger molecules (see Information Box 2-1). Cellulose, a polysaccharide (Figure 2-11a), is formed by a series of reactions of the same kind, linking glucose units into long chains. Plant starch molecules,

which also consist of long chains of glucose links, are closely similar (Figure 2-11b). Animal starch (glycogen) molecules resemble both cellulose and plant starch except that they are highly branched (Figure 2-11c).

Linkage of glucose units to form long chains in cellulose or starch molecules also illustrates a **polymerization** reaction—a common assembly mechanism forming large, complex molecules from repeats of a building block unit. In a polymerization reaction, the individual units, called **monomers,** join together like links in a chain to form the long, complex molecule called the **polymer.** Thus in cellulose formation the individual glucose molecules are the monomers of cellulose before they are joined together; the finished cellulose molecule is the polymer.

Much of the organic matter on earth consists of (1) carbohydrates in the form of monosaccharides, (2) polysaccharides containing glucose units as the sole building block, or (3) polysaccharides containing glucose in combination with other hexoses. Of the various polysaccharides containing glucose as the sole building block, starch and cellulose are the most abundant. Starchy foods such as potatoes and flour are diet staples over the entire world, and cellulose, in the woody tissues of plants, forms the basic supportive framework of trees, shrubs, and smaller plants. The woods used for construction of houses and furniture are about 50 percent cellulose; cotton is almost pure cellulose. It is estimated that cellulose alone makes up more than half the total organic matter on earth.

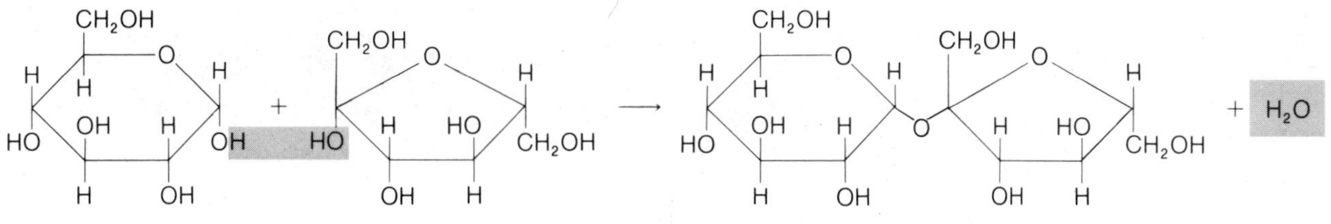

Figure 2-10 Combination of the monosaccharides glucose and fructose to form the disaccharide sucrose. The interaction, a typical condensation reaction (see Information Box 2-1), involves removal of the elements of a molecule of water from the interacting monosaccharides.

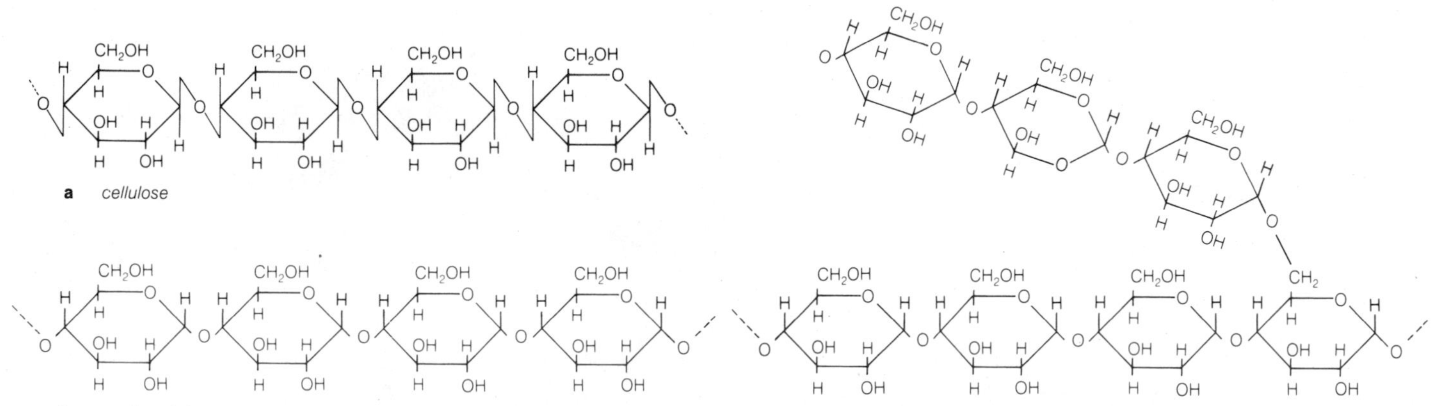

a cellulose

b amylose (plant starch)

c glycogen (animal starch)

Figure 2-11 (a) Cellulose, formed by end-to-end linkages of glucose molecules in the beta form. (b) Amylose, formed by end-to-end linkages of glucose molecules in the alpha form. (c) Glycogen, a branched molecule formed by glucose units joined by alpha linkages.

Lipids Lipids form a mixed group of biological molecules so diverse that they defy description in simple, all-inclusive terms. Perhaps the best definition is based on solubility. Lipids, according to this definition, are biological substances that dissolve more readily in nonpolar solvents such as acetone, ether, chloroform, and benzene than in

---

Lipids are biological substances that are more soluble in nonpolar solvents than in water.

---

water, a strongly polar solvent. These solubility properties reflect the presence of nonpolar groups that make up all or part of lipid molecules, giving them a strongly hydrophobic character. This characteristic is of extreme importance in the interactions of lipids in cells because lipids tend to set up water-excluding domains and barriers, as in

the cell membranes that form boundaries between and within cells. Apart from their structural and functional roles in membranes, lipids are important as an energy source and are stored and utilized in many cell types for this purpose. Three types of lipids—**neutral lipids, phospholipids,** and **steroids**—are especially important in biological systems and are found in quantity in different cells and tissues.

Neutral Lipids The neutral lipids, commonly found in living organisms as storage **fats** and **oils,** illustrate the basic molecular pattern of many cell lipids. The neutral lipids are so called because at the degrees of acidity present in

---

Neutral lipids serve as cellular fuel substances and provide protective coverings for cell surfaces.

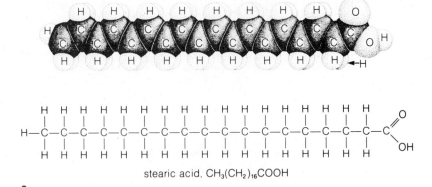

stearic acid, $CH_3(CH_2)_{16}COOH$

a

Figure 2-12 Two fatty acids. (**a**) Stearic acid, a saturated fatty acid. (**b**) Oleic acid, an unsaturated fatty acid. Note the "kink" introduced into the structure by the double bond.

oleic acid, $CH_3(CH_2)_7CH=CH(CH_2)_7COOH$

b

living cells they bear no charged groups. These lipids are formed by a combination between **fatty acids** and the alcohol *glycerol*.

A fatty-acid molecule consists of a long, unbranched chain of carbon atoms with attached hydrogens and other groups (Figure 2-12). A carboxyl (—COOH) group at one end of the chain gives the molecule its acidic properties. Almost all the fatty acids linked to glycerol in natural lipids have chains of carbon atoms in even numbers from 14 to 22; most have either 16 or 18 carbons. Within the chains, hydrogen atoms are bound to the carbon atoms forming the backbone of the molecules. If the maximum possible number of hydrogen atoms are bound to a fatty-acid chain, the chain is said to be **saturated** (Figure 2-12a). If hydrogen atoms are absent from adjacent carbon atoms in the interior of the chain, the fatty acid is said to be **unsaturated** (Figure 2-12b). At points where hydrogen atoms are missing from adjacent carbon atoms, the carbons

share a double instead of a single bond. This arrangement may occur at multiple points in the interior of the chain, forming the so-called **polyunsaturated** fatty acids. Unsaturated fatty acids have lower melting points than saturated fatty acids and are more abundant in living organisms. Neutral lipids containing unsaturated fatty acids are considered more desirable in our diet than fully saturated fatty acids. If eaten in large quantities, lipids containing saturated fats are thought by some authorities to contribute to high blood cholesterol levels and to increase the risk of heart and blood vessel disease.

Glycerol has three hydroxyl (—OH) sites at which fatty acids may attach (shaded in Figure 2-13). If a fatty acid binds to each of the three sites, the resulting compound is known as a **triglyceride**. The neutral lipids in living systems are primarily of this type.

In the formation of triglycerides, the three —OH sites of a glycerol molecule react with the carboxyl (—COOH)

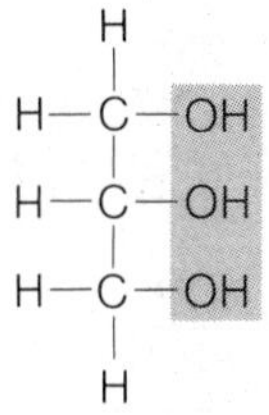

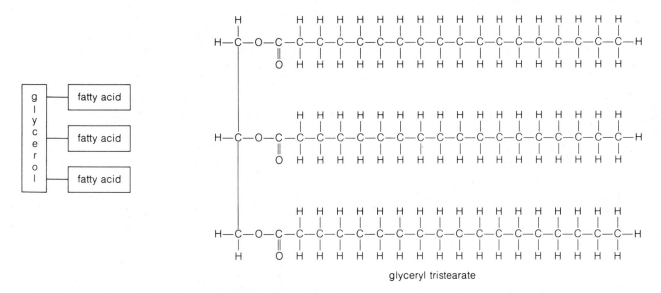

Figure 2-13 Glycerol. The shaded —OH groups enter into reactions with fatty acids to form lipid molecules of various kinds.

glyceryl tristearate

Figure 2-14 The basic structure of triglycerides (fats, oils, and some waxes), formed by interaction between glycerol and three fatty acids. The three fatty acids in the interaction may be the same or all different.

groups of three fatty acids (Figure 2-14). These interactions are typical condensation reactions (Information Box 2-1); for each linkage formed, one molecule of water is assembled as a by-product. The three fatty acids interacting with glycerol to form a triglyceride may be the same or all different; all possible combinations are found in nature.

Generally, the fluidity of triglycerides decreases as the length of their fatty-acid chains increases. Triglycerides that are liquid at biological temperatures are called oils; those that are semisolid or solid are called fats. Waxes, which contain very-long-chain fatty acids in combination with glycerol or other alcohols, are solid at biological temperatures. Waxes occur most frequently on the exterior surfaces of cells or cell walls, particularly in plants, where they form a protective covering that resists water loss and invasion by infective agents.

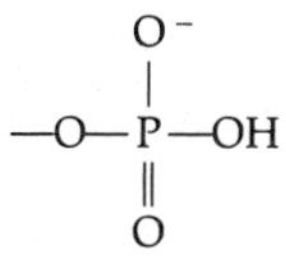

Phospholipids Triglycerides rarely occur as functional parts of cell structures such as membranes. Of much greater importance in this role are the **phospholipids,** a group of phosphate-containing molecules with structures basically similar to the triglycerides. In phospholipids, a phosphate group is substituted for one of the three fatty acids bound to glycerol (see Figure 2-15). In natural phospholipids, the phosphate group binds to either of the "outside" —OH groups on glycerol, never to the middle one. In membrane phospholipids, the phosphate group is linked in turn to one of a group of alcohols, usually containing one or more atoms of nitrogen. In these more complex phospholipids, the phosphate group thus forms a linking bridge between the glycerol and the alcohol (Figure 2-15).

Phospholipids form part of the basic framework of cell membranes.

Phospholipids have dual solubility properties because the end of the molecule containing the alcohol and phosphate group is polar and hydrophilic. The fatty-acid chains, in contrast, retain their strongly nonpolar and hydrophobic character. As a result, one end of the molecule is hydrophilic and the other hydrophobic.

When placed in solution, phospholipids take up arrangements that satisfy their dual solubility properties. For example, if introduced into the interface formed when a nonpolar solvent such as benzene is layered on water, phospholipids orient so that their nonpolar fatty-acid chains extend into the benzene and their polar phosphate–alcohol groups extend into the water (Figure 2-16a). Phospholipids placed under the surface of water (or any strongly polar substance) meet their dual affinities in an interesting and highly significant way: by forming layers just two molecules thick called **bilayers.** In the bilayers,

Phospholipids placed in water satisfy their dual solubility properties by forming bilayers.

the phospholipids orient so that the fatty-acid chains of the double layer associate together in a nonpolar, hydrophobic region in the interior of the layer. The phosphate–alcohol groups face the surrounding water (Figure 2-16b). Bilayers of this type form the basic structural framework of membranes (see Figures 4-3 and 4-4).

Steroids   The steroids comprise a class of lipids based on a complex framework of four interconnected rings of carbon atoms (Figure 2-17a). The various steroids differ in the position and number of double bonds linking the carbons in the rings and in the side groups attached to the rings. The most abundant group of steroids, the **sterols,** have a hydroxyl group linked at one end and a complex, nonpolar carbon chain at the opposite end of the ring structure (shaded groups in Figure 2-17b). Although the sterols are almost completely hydrophobic in character, the single hy-

droxyl group gives the end containing it a slightly polar or hydrophilic character. As a result, the sterols also tend to orient in arrangements that satisfy their dual solubility properties.

The sterol shown in Figure 2-17b and c, **cholesterol,** is an important part of the surface membranes of all animal cells; similar sterols occur in plant cell membranes. Cholesterol is also a factor in human heart and blood vessel disease. Deposits derived from cholesterol (a normal component of blood) gradually fill in the walls of blood vessels in some persons, where they cause serious complications by raising blood pressure and restricting the flow of blood to vital organs such as the heart.

Other steroids are important as **hormones** in animals. Although steroid hormones occur only in trace amounts in animal tissues, they have effects far out of proportion to their low concentration. The male and female sex hormones of humans and other animals are steroid hormones; so are the hormones of the adrenal cortex that regulate growth and other bodily functions. Some steroids occur as poisons in the venom of toads and other animals.

Proteins   Proteins are large, complex molecules that carry out three vital functions in living organisms: (1) as structural molecules, providing much of the framework of cells; (2) as **enzymes,** molecules that speed the rate of biological reactions; and (3) as molecules that provide movement

Proteins act as enzymes and structural molecules and provide movement to cells and cell parts.

to cells and cell parts. Proteins also function in less central roles, as in the protein-based hormones of animals (insulin is a prime example) and as antibodies.

The protein molecules carrying out these diverse functions are all basically similar in structure. All consist of one or more long, unbranched chains of subunits called **amino acids.** Although only 20 different amino acids link initially in different combinations to make proteins, these may be modified to other forms after synthesis, so that as many as 150 different amino acids may occur in natural proteins. For any given protein, however, the amino acid sequence shows little or no variation; this sequence determines the properties and chemical activity of a particular type of protein molecule. The total number of amino acids in different proteins may range from a minimum of 40 or so to giant molecules containing more than 50,000 amino acids.

The structure of proteins allows them to be made in an almost infinite variety of forms. At any point in the

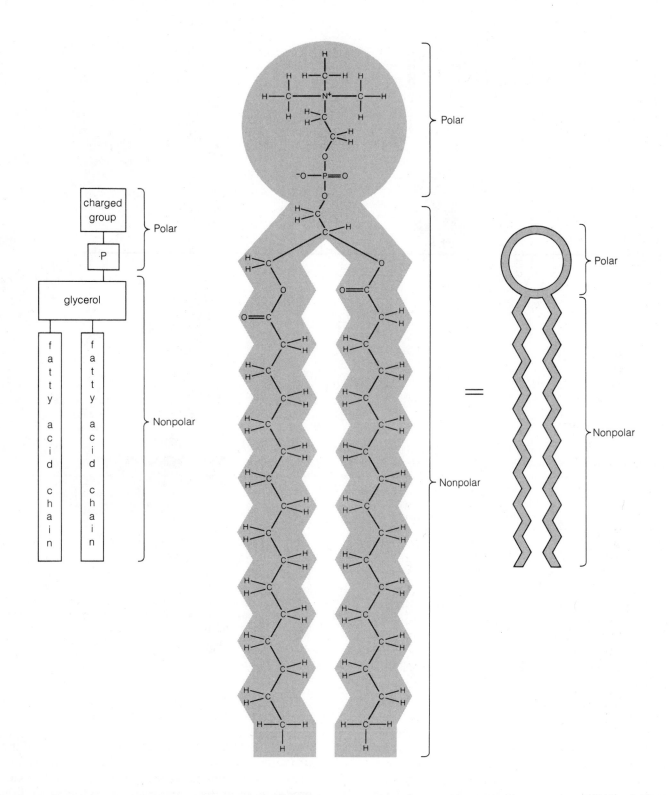

Figure 2-15 An important membrane phospholipid: phosphatidyl choline. Phospholipids are often diagrammed as a circle with two legs (at the right). The circle stands for the polar phosphate and charged group and the zigzag for the two long, nonpolar fatty-acid chains of the molecule.

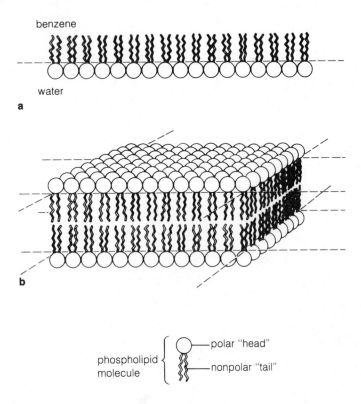

benzene

water

a

b

polar "head"

phospholipid molecule

nonpolar "tail"

Figure 2-16 (**a**) Phospholipid molecules introduced between the layers formed when benzene is poured on top of water orient with their nonpolar tails extending into the benzene and their polar heads facing the water. (**b**) Phospholipid molecules surrounded by water assemble into bilayers with their nonpolar tails associated together in the bilayer interior and their polar heads facing the surrounding water molecules.

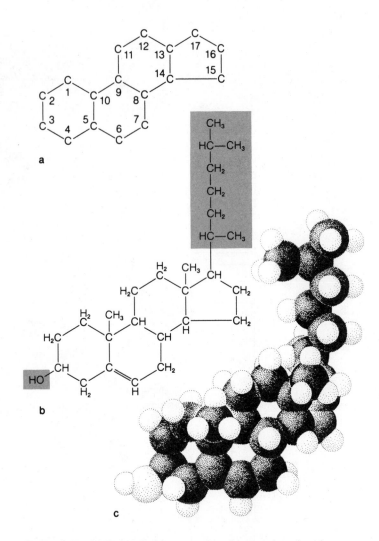

Figure 2-17 (**a**) The typical arrangement of four carbon rings in a steroid molecule. (**b**) and (**c**) A sterol, cholesterol. Sterols have a long, nonpolar side chain at one end of the molecule and a single —OH group at the other end (shaded). The —OH group makes its end of the cholesterol molecule slightly polar. The remaining parts of the molecule are nonpolar.

amino acid chain of a protein, any one of the 20 amino acids may be present in either modified or unmodified form. In even the smallest proteins, with roughly 50 amino acids, this allows 20^{50} different sequences to be made without modification of individual amino acids. This incomprehensibly huge number is equivalent to one unique protein for every gram of matter in the universe!

The Amino Acids Nineteen of the 20 amino acids initially assembled into proteins are based on the same structural plan (Figure 2-18). These amino acids have a central carbon atom, called the *alpha* (α) carbon, to which are attached, one on either side, an amino (—NH₂) group and a carboxyl (—COOH) group. One of the remaining bonds of the central carbon is linked to a hydrogen atom, giving the structure

$$NH_2-\underset{\underset{H}{|}}{\overset{|}{C}}-COOH$$

common to these amino acids. The fourth bond of the alpha carbon may be attached to any one of 19 different side chains. The chains range in complexity from a single hydrogen atom in the simplest amino acid, **glycine,** to long carbon chains or rings in other amino acids. Some of the more complex side chains contain oxygen, nitrogen, or sulfur atoms in addition to carbon and hydrogen. The re-

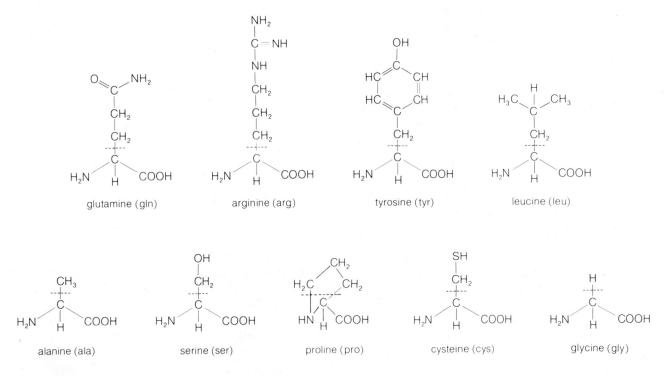

Figure 2-18 Representative amino acids. Side chains are shown above the dashed lines; the alpha carbon with its attached amino (—NH$_2$) and carboxyl (—COOH) group is depicted below. Only proline varies from this structural plan, which is common to nineteen of the twenty amino acids assembled into proteins inside cells. Proline, as shown, is closed into a ring form.

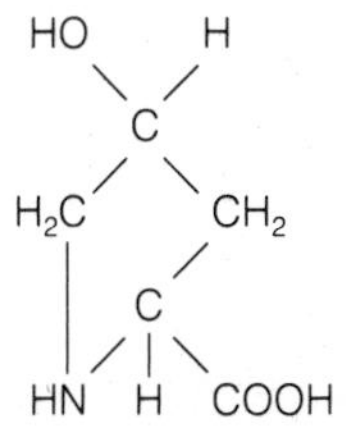

hydroxyproline

Figure 2-19 Hydroxyproline, a modified amino acid.

maining amino acid, **proline,** an exception to these general rules, is based on a ring structure including the central carbon atom (see Figure 2-18).

The types and configuration of the atoms in the side chain give each amino acid special properties, including differences in polarity, charge, and acidic or basic reaction. A number of amino acid side chains have groups capable of readily entering into reactions with atoms and molecules located outside the protein or elsewhere on the amino acid chain. These include amino (—NH$_2$), hydroxyl (—OH), carboxyl (—COOH), and sulfhydryl (—SH) groups.

Modified amino acids are common in natural proteins. The protein **collagen,** for example, a structural molecule that is abundant in animals, contains quantities of the amino acid **hydroxyproline** (Figure 2-19), a form modified from the original proline by addition of a hydroxyl group. These additional amino acids, as noted, are produced by chemical modifications of the original twenty amino acids after initial synthesis of a protein.

The Peptide Bond Amino acids can be readily linked together into chains of two or more units. The covalent bond linking the units together, the **peptide bond** or **peptide linkage,** is produced by a reaction between the amino group of one amino acid and the carboxyl group of a second amino acid (Figure 2-20). For each peptide bond formed, the equivalent of one molecule of water is released as a by-product.

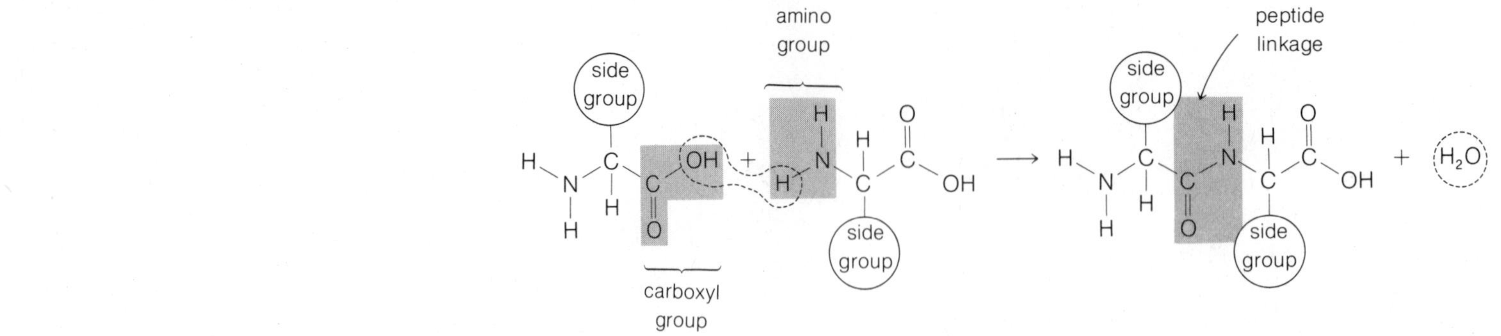

Figure 2-20 Reaction between two amino acids to form a dipeptide, bound together by a peptide bond (shaded).

The linkage of two amino acids into a short chain produces a *dipeptide*. Chains of more than two units are called **polypeptides.** The dipeptide shown in Figure 2-20 retains an amino group at one end and a carboxyl group at the other. Thus a dipeptide is similar to an amino acid and can enter into the formation of additional peptide linkages at either the carboxyl or the amino end. In nature, additional amino acids are added only to the carboxyl end of a growing polypeptide chain. The particular sequence in which amino acids occur in a finished chain is called the **primary structure** of a protein (see Information Box 2-2).

The Three-Dimensional Structure of Proteins   The sequence of amino acids describes only part of the structure of a protein molecule. The entire chain of amino acids folds and twists about itself into a three-dimensional arrangement that is as significant for the function of the protein as the sequence of amino acids itself. Some proteins, when folded in natural form, are essentially spherical or globular; others are long fibers. Some proteins, such as hemoglobin, the oxygen-carrying protein in the bloodstream of higher animals, are built up from several amino acid chains, folded and twisted into a composite three-dimensional structure.

Proteins are held in their globular or fibrous shapes by a number of bonds and attractions in addition to the peptide linkages between amino acids. Among these are

---

Proteins are made up from long chains of amino acids linked together by peptide bonds. They are held in their three-dimensional shape by hydrogen bonds, disulfide links, and polar and nonpolar attractions.

---

hydrogen bonds, electrostatic attractions between positively and negatively charged groups, associations between hydrophilic and hydrophobic amino acid side chains, and a covalent bond called a **disulfide (—S—S—) linkage.**

Although individual hydrogen bonds are relatively weak, many hydrogen bonds can be formed in proteins since each peptide linkage has hydrogen and oxygen atoms capable of forming them. The total effect of hydrogen bonding all along the chain is to restrict the twisting of the amino acid backbone and produce a limited number of ways in which a protein of given sequence may be folded. The greater the number of internal hydrogen bonds formed, the more inflexible the three-dimensional structure of the chain becomes.

Disulfide linkages (Figure 2-21) are produced by covalent bonding between the —SH (*sulfhydryl*) groups of cysteine amino acid side groups located at different points in the backbone chain. Where these links form, they anchor the amino acid chain of a protein in a folded position.

The net effect of hydrogen and disulfide bonds, electrostatic attractions, and hydrophilic and hydrophobic associations is to establish a unique three-dimensional shape for each protein with a particular sequence. This three-dimensional shape is called the **folding conformation** of a protein. The three-dimensional folding conformation of a relatively simple protein, *myoglobin*, is shown in Figure 2-22.

The folding conformation of a protein is flexible and changes readily in response to variations in temperature, pH, and the binding of other molecules, ions, or chemical groups. The resulting changes in shape, called **conformational changes,** are vital to the activities of proteins as enzymes and molecules providing cell movements. Extreme

Information Box 2-2

Protein Structure

The arrangement of acids and peptide chains in protein molecules is described in terms of their **primary, secondary, tertiary,** and **quaternary** structure:

Structural Level	Definition	Primary Bonds Holding Structure Together
Primary	Amino acid sequence of protein.	Covalent bonds of peptide linkages
Secondary	Folding or twisting of the amino acid chain into patterns such as the alpha helix, which are subparts of a protein molecule.	Hydrogen bonds
Tertiary	Complete three-dimensional folding pattern of a polypeptide chain. When proteins are formed from a single polypeptide chain, the tertiary structure describes the three-dimensional folding pattern of the entire molecule.	Hydrogen bonds, disulfide linkages, polar and nonpolar associations
Quaternary	The three-dimensional folding arrangement of a protein containing more than one polypeptide chain.	Same as tertiary structure

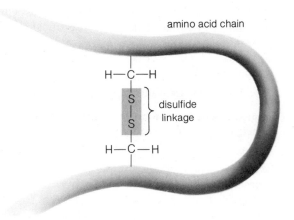

Figure 2-21 A disulfide linkage, formed by an interaction between sulfhydryl (—SH) groups on cysteine amino acids located in different parts of an amino acid chain. Disulfide linkages provide one of the bonds holding protein molecules in a three-dimensional folding conformation.

changes in temperature or pH seriously disturb the internal bonds and associations holding proteins together, causing the molecules to unfold or refold into random shapes. This effect, called **denaturation,** results in loss of the functional activity of the protein. For example, temperatures above about 50°C break most or all of the hydrogen bonds within proteins, causing them to unfold into random, extended shapes with total loss of functional activity. Denaturation of proteins at high temperatures accounts for much of the destructiveness of high fevers in humans and explains why almost no living organisms can tolerate temperatures above 45–55°C.

The Alpha Helix in Protein Structure The amino acid chain inside proteins often twists into a regular spiral called the **alpha helix** (Figure 2-23). The spiral, discovered by Linus Pauling and Robert Corey of the California Institute of Technology, is a relatively rigid structure containing 3.6

amino acids in each of its turns. The amino acids are placed so that the side chains extend outward from the spiral. The amino acid chain is held in the stable spiral structure by a regularly spaced series of hydrogen bonds formed between hydrogen and oxygen atoms along the chain.

The alpha-helical portions of protein molecules resist bending and deformation; where sharp bends are found in folded backbone chains the alpha helix gives way to a much less regular arrangement known as a **random coil.** Whether an alpha helix or a random coil forms in a region of the chain depends largely on the amino acids present. Some amino acids allow a stable helix to form; others reduce the stability or even interrupt the helix. Most significant of the latter kind of amino acid is proline, which has no hydrogen available for the formation of stabilizing hydrogen bonds. The unusual ring structure of proline also prevents the chain from twisting into the alpha helix at points containing this amino acid. The helix is therefore interrupted at any position occupied by proline. Other disturbances may be introduced by disulfide links. The myoglobin molecule diagrammed in Figure 2-22 shows the distribution of rigid alpha-helical spirals and the sharply bent, flexible regions containing random coils in this protein molecule.

Combinations Between Proteins and Other Substances By virtue of the different reactive groups carried on their amino acid side chains, proteins may bind to a variety of ions, chemical groups, and organic molecules within the cell. Inorganic ions—particularly sodium, potassium, chloride, calcium, magnesium, phosphate, copper, and iron ions—are often found in combination with enzymes. Vitamins also combine with proteins, particularly with enzymes active in the reactions capturing energy for cellular use. Proteins in combination with carbohydrates form glycoproteins; these complexes are frequently found at the surfaces of cells or in the gelatinous secretions of many cell types. Nonpolar amino acids are important in the hydrophobic associations between proteins and lipids that form lipoproteins; these complexes are basic structural units in membranes. Last but not least must be mentioned the electrostatic binding between proteins and nucleic acids to form the **nucleoproteins,** the major constituents of the systems storing and utilizing hereditary information in cells.

Nucleotides and the Nucleic Acids The two nucleic acids, **deoxyribonucleic acid (DNA)** and **ribonucleic acid (RNA),** are the informational molecules of all living organisms. Both of these polymer-type molecules are long structures built up from chains of repeating monomers or building

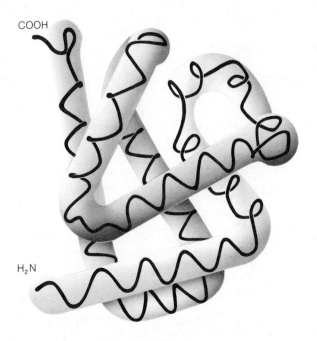

Figure 2-22 The three-dimensional folding of the amino acid chain in the protein myoglobin. The solid line shows the folding arrangement of the amino acid backbone of the molecule.

blocks: the **nucleotides.** The sequence of nucleotides in nucleic acid molecules forms a code that stores and transmits the cellular information required for cell growth and reproduction. Individual nucleotides also transfer units of energy or important reactants from one system to another in cells.

Nucleic acids store and transmit the information required for life.

Nucleotides serve both as informational units of the nucleic acids and as energy carriers in the cell.

The Nucleotides Nucleotides are the basic structural units of all these important substances. Each nucleotide (Figure 2-24) consists of (1) a nitrogen-containing base, (2) a five-carbon sugar, and (3) one or more phosphate groups. All these subunits are linked together into nucleotides by covalent bonds. The nitrogenous bases, called **pyrimidines** and **purines** (Figure 2-25), are ring-shaped molecules containing both carbon and nitrogen atoms. Three pyrimidine bases—uracil (U), thymine (T), and cytosine (C)—and two purine bases—adenine (A) and guanine (G)—are the most common bases in nucleotides.

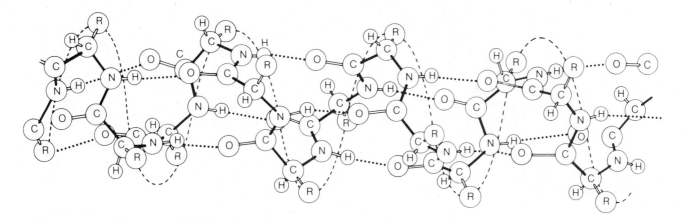

Figure 2-23 The alpha helix. The backbone of the amino acid chain is held in a spiral configuration (dashed line) by hydrogen bonds (short dotted lines) formed at regular intervals. The small spheres labeled "R" represent the side chains of amino acids in the alpha helix. Redrawn from "Proteins" by Paul Doty, *Scientific American 197*:173 (1957). Copyright © 1957 by Scientific American, Inc. All rights reserved.

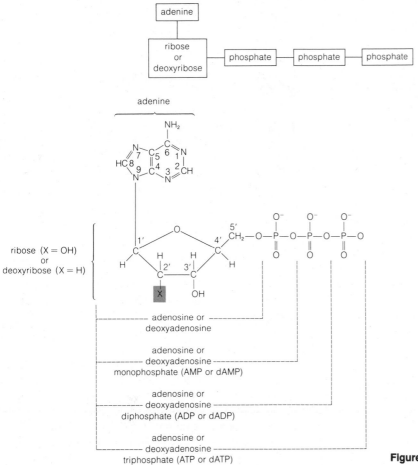

Figure 2-24 The structural plan of a nucleotide (see text).

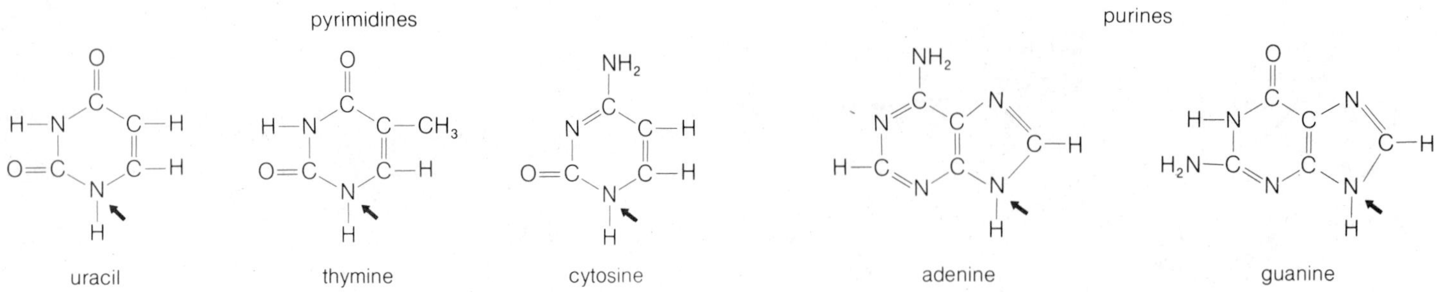

pyrimidines · purines

uracil · thymine · cytosine · adenine · guanine

Figure 2-25 The purine and pyrimidine bases of nucleic acids and nucleotides. The arrows indicate where the base links to ribose or deoxyribose sugars in the formation of nucleotides.

Information Box 2-3

Naming the Nucleotides

The term nucleo*tide* refers to a unit containing all three nucleotide subunits: a nitrogenous base, a five-carbon sugar, and one or more phosphates. Cells also contain units consisting of only the base and sugar without any phosphates. These units, called nucleo*sides*, are given names derived from the nitrogenous base present. The base–sugar complex containing adenine and ribose, for example, is called *adenosine;* if deoxyribose is the sugar in the complex, it is called *deoxyadenosine.* The remaining nucleosides are named as follows, using the convention that the suffix -*osine* is added when the base present is a purine and -*idine* is added when the base is a pyrimidine:

guanosine
deoxyguanosine } containing guanine

thymidine
deoxythymidine } containing thymine

cytidine
deoxycytidine } containing cytosine

uridine
deoxyuridine } containing uracil

The nucleo*tides* are then named by considering them as nucleo*sides* with added phosphates. The adenine–ribose nucleoside with one phosphate added, for example, is called *adenosine monophosphate* (AMP); with two phosphates it is called *adenosine diphosphate* (ADP); with three, *adenosine triphosphate* (ATP).

These nitrogenous bases link covalently in nucleotides to one of two 5-carbon sugars, either *ribose* or *deoxyribose*. The two sugars differ only in the chemical group bound to the carbon at the shaded position in Figure 2-24. Ribose has an —OH group at this position; deoxyribose has a single hydrogen. The ribose or deoxyribose sugars of nucleotides are also linked to a chain of 1-, 2-, or 3-phosphate groups, forming the mono-, di-, and triphosphates of the sugar–base unit as in the AMP, ADP, and ATP family of molecules shown in Figure 2-24. The two forms of the ATP nucleotide, one with the ribose and one with the deoxyri-

bose sugar, are written as ATP (with ribose) and dATP (with deoxyribose) to indicate which form is meant (see also Information Box 2-3).

DNA and RNA Nucleotides link together into long chains to form the two nucleic acids DNA and RNA (see Figure 8-3). The chain of nucleotides in DNA contains the sugar deoxyribose linked to one of the four bases adenine, thymine, guanine, or cytosine (A, T, G, C; see Figure 2-25). Each nucleotide in the chain making up RNA contains the

DNA is made up from chains of nucleotides containing the sugar deoxyribose and one of the four bases A, T, G, and C. RNA is made up from the sugar ribose and one of the four bases A, U, G, and C.

sugar ribose and one of the four bases adenine, uracil, guanine, or cytosine (A, U, G, C). Thus DNA and RNA differ only in the sugar present (ribose or deoxyribose) and in the presence of either uracil in RNA or thymine in DNA. As in the amino acid side groups of proteins, the nitrogenous bases of the nucleotides may be modified chemically to other forms after initial synthesis of a nucleic acid. These base modifications are especially common in some types of RNA molecules.

DNA exists in cells in the form of a **double helix** containing two nucleotide chains twisted around each other in a regular, double spiral (see Figure 8-4). The double helix is held together by hydrogen bonds and by hydrophobic associations between the nitrogenous bases of opposite chains in the spiral. RNA exists as single, rather than double, nucleotide chains in living cells. However, RNA molecules may fold back on themselves to form extensive regions with double-helical structure (see Figure 8-10). *Hybrid* double helices, containing one DNA and one RNA nucleotide chain wound into a double spiral, also occur as temporary intermediates during nucleic acid synthesis in living cells.

Questions

1. What are protons, neutrons, and electrons?

2. What determines the atomic number of an atom?

3. What is radioactive decay?

4. What are isotopes? Are all isotopes radioactive?

5. What is the difference between the electron orbitals and electron shells of an atom?

6. What determines the chemical activity of an atom?

7. What is an electrostatic bond? How are electrostatic bonds formed?

8. What is a covalent bond? How are covalent bonds formed? How are covalent bonds represented in molecular diagrams?

9. What are polar and nonpolar molecules? What condition produces polarity?

10. What do the terms *hydrophobic* and *hydrophilic* mean? How is the property of being hydrophobic or hydrophilic related to polarity?

11. How is polarity important in molecular and cellular structure?

12. Detergents allow particles of oil to mix intimately with water. From your reading in this chapter, how do you suppose detergents work?

13. What is a hydrogen bond? What conditions are necessary for the establishment of a hydrogen bond?

14. List the significant and unusual properties of water. How are these properties vital to the survival of life?

15. What is the difference between alpha glucose and beta glucose? Does this difference have any biological significance?

16. What are the primary differences between cellulose, plant starch, and animal starch (glycogen)?

17. What is the basic structural plan of a neutral lipid? Of a phospholipid? How do the solubility properties of neutral lipids and phospholipids differ?

18. What is a bilayer? How is bilayer formation related to the solubility properties of phospholipids?

19. What functions do proteins carry out in cells?

20. How do amino acids link together to form proteins? What determines the three-dimensional shape of a protein?

21. What are conformational changes in proteins? What is protein denaturation?

22. Diagram a carboxyl, hydroxyl, sulfhydryl, aldehyde, ketone, and phosphate group.

23. What is the difference between ATP and dATP?

24. How do DNA and RNA differ? In what ways are they the same?

25. What functions are carried out by the nucleic acids in living organisms?

Suggestions for Further Reading

Dickerson, R. E., and I. Geis. 1969. *The Structure and Action of Proteins.* Harper & Row, New York.

Dickerson, R. E., and I. Geis. 1976. *Chemistry, Matter, and the Universe.* W. A. Benjamin, Menlo Park, California.

White, E. H. 1964. *Chemical Background for the Biological Sciences.* Prentice-Hall, Englewood Cliffs, New Jersey.

SUPPLEMENT 2-1: ACIDITY AND PH

The term pH refers to the relative concentrations of hydrogen ions (H^+) and hydroxyl ions (OH^-) in a water medium. The number scale from 0 to 14 used to describe these concentrations is derived in the following way.

Water solutions always contain both H^+ and OH^- ions. At a temperature of 25°C the product of these concentrations has a constant value equal to 1×10^{-14}:

[Concentration of H^+] × [concentration of OH^-] = 1×10^{-14}

These concentrations are given in moles per liter. (One mole is equal to one molecular weight of a substance in grams.)

The pH of a solution is defined as the negative logarithm (to the base 10) of the concentration of H^+ ions in moles per liter, or

$$pH = -\log_{10} [H^+] \qquad (2\text{-}1)$$

where the brackets indicate concentration in moles per liter. If H^+ ions are present at a concentration of $0.0000001M$ ($1 \times 10^{-7}M$), for example, the $\log_{10}$ of this concentration is -7. The negative of the logarithm -7 is 7. Thus a water solution with H^+ ions at a concentration of $1 \times 10^{-7}M$ is said to have a pH of 7.

At pH 7 the concentration of H^+ and OH^- ions is equal:

$$(1 \times 10^{-7}M)(1 \times 10^{-7}M\ OH^-) = 1 \times 10^{-14}$$

At this concentration, the solution is said to be neutral. Solutions with pH higher than 7 have OH^- ions in excess and are basic or alkaline; solutions with pH less than 7 have H^+ ions in excess and are acid. Because pH values represent logs to the base 10 of the concentration, one pH unit represents a concentration difference of ten times. Thus a solution at pH 8 has ten times as many OH^- ions as a solution at pH 7. Cells typically have a pH at neutrality, or slightly on the basic side of neutrality, in the range 7.1 to 7.2.

The assembly of the carbohydrates, lipids, proteins, nucleic acids, and other organic molecules of living organisms—and the interactions between these molecules that amount to the activities of life—require a constant input of energy. The assembly and interaction of these molecules also require enzymes, the protein molecules that speed the rate of biological reactions at ordinary temperatures. These reactions are regulated as well as speeded by enzymes: the right enzymes are made in cells at the right place and time, and in the right numbers, to increase the rate of the biochemical interactions necessary for a given type of growth or life activity.

The enzyme-catalyzed reactions making up the activities of life are not unique to cells or the living condition. They can proceed outside cells, in a test tube, provided the necessary physical and chemical conditions are duplicated. Not even the enzymes normally catalyzing the reactions are necessary for the reactions to proceed. The primary role of enzymes is simply to increase the rate of reactions that would take place, however slowly, without the presence of an enzyme. Biochemical reactions have these characteristics because they must obey the chemical and physical laws operating everywhere in the universe, whether inside cells or in the nonliving surroundings.

To understand how enzymes increase the rate of biological reactions we must therefore understand the basic rules governing the probability and direction of chemical changes that take place throughout the universe. These rules, or laws, are principles drawn from **thermodynamics,** the science dealing with energy changes in all collections of matter.

3

ENERGY AND ENZYMES IN BIOLOGICAL REACTIONS

THE NATURE OF ENERGY AND ENERGY CHANGES

What is energy? Although everyone has an intuitive grasp of the meaning of the word, energy is difficult to define precisely. We cannot perceive energy itself; we can only observe its effects on physical objects in the environment. Because these effects often involve movement of physical objects—whether electrons, atoms, molecules, or larger pieces of matter—energy is often defined as "the capacity to do work." Although not entirely adequate, this definition is useful and for most purposes does not violate the truth.

There are two basic principles, called the **first and second laws of thermodynamics,** that describe the energy changes in reactions of all kinds and specify whether these changes are probable and will proceed on their own. Reactions that are probable are called *spontaneous* reactions in thermodynamics. This usage of the word *spontaneous* does not carry its more common suggestion that the change is instantaneous: spontaneous reactions, in thermodynamics, are simply reactions that will "go." They may take place at any rate from unmeasurably slow to practically instantaneous.

Spontaneous Reactions and the Laws of Thermodynamics

The first law of thermodynamics is a formal statement of the common observation that "you can't get something for nothing." The law was developed partly through studies of the many unsuccessful attempts to build perpetual motion machines. Machines of this type, once set in motion, would run indefinitely without any input of energy. All such machines, however, no matter how ingeniously they are constructed, eventually run down because of friction: some of the initial push of energy used to start the machine is lost as heat due to friction between the working parts. This heat flows from the machine into its surroundings and is lost. To keep the machine running, the lost energy must be constantly replaced by additional pushes of energy from the outside. Otherwise, the operator would be getting something for nothing—the "something" in this example being the energy needed to replace the energy lost as frictional heat.

Careful measurements of the heat flowing from such machines show that as the machines come to a stop, the total amount of heat lost due to friction is exactly equivalent to the amount of energy added in the initial push used to start them. These observations, and similar observations made in other energy changes, led to the formal statement of the first law of thermodynamics. This law affirms that in any process involving an energy change (as in the change from mechanical to heat energy in a so-called perpetual motion machine) the total amount of energy remains constant. In other words: Energy can neither be created nor destroyed in such changes. The first law

---

The first law of thermodynamics states that as changes take place in the universe, the total energy remains constant: Energy can neither be created nor destroyed.

---

means that the energy in the universe remains constant. Although it may be converted from one form to another, the total quantity remains the same.

It is useful to consider what total energy means when it applies to a collection of molecules. Part of the energy content of any collection of molecules at temperatures above absolute zero ($-273°C$) is reflected in their constant rotation, vibration, and movement from one place to another. This part of the total energy, reflected in molecular

---

Kinetic and potential energy make up the total energy content of a molecular system.

---

movement, is the **kinetic energy** of the collection of molecules. The second part depends on the energy contained in the arrangement of chemical bonds in the molecules. This second component is called **potential energy.**

The two energy components are interchangeable. An unlit match held at room temperature, for example, contains considerable potential energy in the form of the arrangement of chemical bonds in its wood and the phosphorus of its tip, but only moderate kinetic energy. Once struck, much of the potential energy is transformed into increases in the kinetic motion of the molecules of the match and its immediate surroundings—energy that we can detect as an increase in temperature. Some energy also flows from the burning match as heat and light.

What does the first law governing energy content have to do with the probability and direction taken by chemical reactions? From our own intuition, based on observations of ordinary events within our common experience, we would expect reactions to be spontaneous when the energy content of the system in its initial state is higher than in the final state. Take, for example, a system represented by a rock placed on the side of a steep hill. We know that the rock will roll spontaneously downhill until it comes to a stop at the bottom. At the initial state, when placed on the side of the hill, the rock has a quantity of potential energy equivalent to its distance from the center of the earth. On reaching its position of rest at the final state at the bottom of the hill, the rock is closer to the center of the earth and has a smaller amount of potential energy. The difference between the initial and final states is lost as heat energy to the surroundings. From this type of observation, we would therefore expect spontaneous chemical reactions to be those in which the total chemical energy of the products is smaller than that of the reactants. That is, we expect chemical reactions to proceed spontaneously to a state in which the molecules of the system have *minimum energy content*.

By itself the first law cannot be used to predict whether a chemical interaction is possible or to determine whether a reaction will release energy or require energy to take place. The additional information required to predict these characteristics of a reaction is supplied by the second law of thermodynamics. The second law too has been derived from everyday experience, but its meaning is more subtle and difficult to understand. The common observation that constant effort is required to keep anything in order is one basis for the second law. Things get out of place in our rooms, dust settles everywhere, tires go flat. When stated in the precise terms of thermodynamics, these observations affirm that as long as change of any kind takes place anywhere in the universe, the *change always increases the total disorder or randomness of the universe.* An increase of order in one part of the universe, such as

---

The second law of thermodynamics states that as changes take place anywhere in the universe, the total disorder (entropy) always increases.

---

the growth of a tree from a seed, can take place only at the expense of a greater reduction of order somewhere else in the universe. This change in order or randomness is called the **entropy** change in thermodynamics. In these terms the second law of thermodynamics states: As change takes place, the total entropy of the universe always increases.

The first and second laws together allow us to predict whether a reaction is possible and whether it will take place spontaneously: Reactions that proceed to a condition of minimum energy and maximum disorder (maximum entropy) are favored and will take place spontaneously—that is, without requiring an energy "push" from outside.

Spontaneously reacting systems generally proceed to a state of minimum energy content and maximum disorder (entropy).

In proceeding to completion, spontaneous reactions may give off energy to the surroundings. Energy given off in this way is called **free energy.** (The word *free* means simply that the energy is available to do work.) In living organisms, the primary work accomplished by free energy is the

---

Free energy is energy that is available to do work.

---

chemical work of assembling complex organic molecules from simple inorganic substances.

By themselves, reactions that assemble complex organic molecules from simple inorganic substances will not proceed spontaneously because the products of the reactions contain more energy and order than the reactants. These reactions proceed only when free energy is added from the outside at the expense of another reacting system

that releases free energy. This important principle is the basis for all life processes, including growth, reproduction, movement, and response to the environment. In these life processes chemical reactions that give off free energy are joined, or *coupled,* to reactions that require an input of energy. If the energy and entropy changes for the coupled reactions are added together, the requirements of the first and second laws are still met: the total changes are toward minimum energy content and maximum entropy.

ATP: The Cellular Coupling Agent

One of the most interesting things about the mechanisms coupling energy changes in living cells is that the primary coupling agent, the nucleotide ATP, is the same in all organisms from bacteria to humans. The basic structure of **ATP (adenosine triphosphate**; see Figure 3-1) consists of a nitrogenous base, *adenine,* linked to a five-carbon

---

ATP is the primary agent that couples reactions in all living organisms.

---

sugar, *ribose.* Also linked to the ribose sugar is a chain of three phosphate groups.

The part of ATP that makes it useful as a cellular coupling agent is the arrangement of the three phosphate groups in a row at one end of the molecule. One of the oxygens of each of these phosphate groups carries a strongly negative charge. In the ATP molecule, the three phosphate groups line up side by side in a position that brings the strongly negative charges close together. Since like charges repel, the close alignment of negative charges imposes an internal strain on the molecule. The strain is relieved—with the release of energy—if the phosphates are removed. The effect is much like compressing or releasing a spring. Adding phosphate groups, up to a limit of three, compresses the chemical spring and stores energy in the molecule. Removing them releases the spring and makes energy available for useful work. Usually, only the last of the three phosphates is added or removed, converting the molecule between ATP, with three phosphates, and ADP with two (Figure 3-1). (**ADP** is the compound **adenosine diphosphate**.) In some reactions, two of the phosphates are removed, producing **AMP (adenosine monophosphate**).

The reaction removing the terminal phosphate in the conversion of ATP to ADP is fully reversible:

$$ATP + H_2O \rightleftharpoons ADP + HPO_4^{2-} \qquad (3\text{-}1)$$

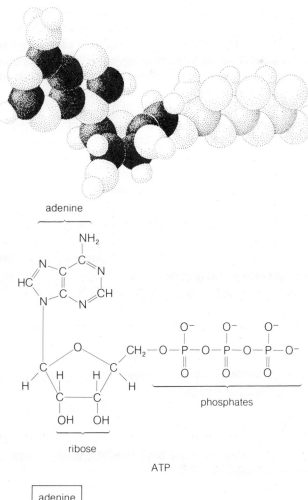

adenine

ribose

ATP

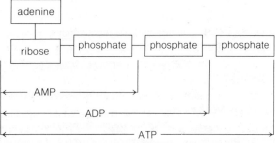

AMP

ADP

ATP

Figure 3-1 **Top:** The ATP molecule, adenosine triphosphate. **Bottom:** Sequential removal of the phosphate groups produces the closely related molecules adenosine diphosphate (ADP) and adenosine monophosphate (AMP).

In going to the right, the reaction releases a large quantity of free energy: for every mole of ATP converted to ADP, 7000 calories of energy is released.[1] (A calorie is an amount of energy equivalent to the heat required to raise 1 gram of water 1°C under standard conditions.) Running the reaction to the left, in the uphill direction, requires an *input* of 7000 calories for each mole of ATP produced.

Running the ATP reaction uphill is accomplished in cells by coupling ATP synthesis to other reactions that release free energy to produce an overall reaction that runs downhill. For example, the oxidation of glucose, a complex organic molecule broken down as a source of coupling energy by all living organisms, proceeds spontaneously if the glucose is simply ignited in the presence of oxygen:

$$C_6H_{12}O_6 + 6O_2 \rightarrow 6CO_2 + 6H_2O \qquad (3\text{-}2)$$

In the process, free energy amounting to 686,000 cal/mole is released from the reaction to the surroundings.

In cells, ADP plus an inorganic phosphate are added to this reaction, producing an overall sequence that synthesizes ATP:

$$C_6H_{12}O_6 + 6O_2 + 38ADP + 38HPO_4^{2-} \rightarrow$$

$$6CO_2 + 38ATP + 44H_2O \quad (3\text{-}3)$$

The reaction still proceeds spontaneously far to the right because—even with the ADP and phosphate added on the left and the ATP and H$_2$O on the right—the products contain considerably less energy and greater entropy than the reactants. Less free energy, however, amounting to 420,000 cal/mole, is released to the surroundings. The difference of 266,000 cal/mole between Reactions 3-2 and 3-3 represents potential energy captured in the ATP molecules synthesized as products in Reaction 3-3.

The ATP synthesized in coupled reactions is then used in other coupled sequences to drive reactions that, by themselves, would be required to go uphill and would therefore be very unlikely to proceed. The uphill reaction synthesizing glutamine from glutamic acid is

---

[1]A mole, the standard quantity used for making comparisons in chemical reactions, is approximately equal to 6 × 10^{23} molecules of a substance. The weight of a mole of a substance is determined from the sum of the atomic weights of all the atoms in a molecule of the substance in grams. A mole of water, for example, which contains two hydrogen atoms, each with an atomic weight of 1, and an oxygen atom, with an atomic weight of 16, weighs 2 + 16 = 18 grams.

$$\text{Glutamic acid} + NH_3 \rightarrow \text{glutamine} + H_2O \qquad (3\text{-}4)$$

This reaction requires an energy input of 3400 cal/mole to make it proceed to the right. In cells this uphill reaction is coupled to the breakdown of ATP to ADP and produces an overall downhill reaction that releases free energy and proceeds spontaneously to the right:

$$\text{Glutamic acid} + ATP + NH_3 \rightarrow$$
$$\text{glutamine} + ADP + HPO_4^{2-} \qquad (3\text{-}5)$$

Now the total products, glutamine, ADP, and phosphate, have less energy content and greater entropy than the total reactants, which include glutamic acid, ATP, and NH_3. Free energy amounting to 3900 cal/mole is released as each gram molecular weight of reactants is converted to products.

In coupled reactions, therefore, an uphill reaction requiring energy is coupled to the breakdown of ATP to produce an overall reaction that proceeds downhill, and the reacting system becomes thermodynamically possible. Almost all the chemical, electrical, and mechanical work of the cell—and all the other energy-requiring activities of growth, reproduction, movement, and responsiveness—are directly or indirectly coupled to the ATP $\rightarrow$ ADP conversion and made energetically favorable in this way (Figure 3-2).

THE ROLE OF ENZYMES

Enzymes and Enzymatic Catalysis

The laws of thermodynamics, as described up to this point, apply to all chemical reactions whether they occur inside cells or not. None can violate the rule that they must proceed to a level of minimum energy and maximum entropy. What effects do enzymes have, then, on the biochemical reactions taking place inside cells? The answer is

---

Enzymes increase the rate of spontaneous biological reactions. They cannot make a reaction proceed that will not take place spontaneously without an enzyme.

---

simply that enzymes greatly increase the *rate* at which spontaneous reactions take place. Enzymes cannot make a reaction go if it would not already proceed spontaneously without the enzyme.

Many reactions, although spontaneous, proceed so slowly that their rate at room temperature is essentially

unmeasurable. A good example is the conversion of glucose to CO_2 and H_2O:

$$C_6H_{12}O_6 + 6O_2 \rightarrow 6CO_2 + 6H_2O \qquad (3\text{-}6)$$

Although this reaction will proceed spontaneously to completion at room temperature if glucose is exposed to oxygen, the rate is so slow that it is essentially zero. If the correct enzymes are present, however, the reaction will quickly proceed to completion at room temperature.

Enzymes and the Energy of Activation

How do enzymes accomplish this feat? The basic mechanism depends on an alteration by enzymes of the **activation energy** required for a chemical reaction to occur. The activation energy may be visualized as an energy barrier over which the molecules in a system must be raised

---

Enzymes increase the rate of biological reactions by lowering the activation energy.

---

for reaction to take place (Figure 3-3a). This condition is analogous to a rock resting in a depression at the top of a hill (Figure 3-3b). As long as the rock is undisturbed, it will not spontaneously begin its travel downhill, even though the total "reaction," the progression of the rock downward, is energetically favorable. In this physical example the activation energy may be regarded as the effort that would be required to raise the rock over the lip of the depression holding it in place.

The requirement for energy of activation raises the question of why reactions proceed spontaneously at all. In other words, how can any molecules in a system go over the energy barrier? Where does the energy come from? Movement over the barrier is possible because, instead of being at rest as in the example of the rock at the top of a hill, molecules at temperatures above absolute zero are in constant motion. The amount of motion, or kinetic energy, is not the same in all of the molecules present, although the average energy is below the amount required for activation. Some molecules in the population are below this average, and some are above. The distribution occurs partly as a result of random collisions in which energy is gained and lost by individual molecules. Depending on the height of the energy barrier, these random collisions may raise a number of molecules to the energy level required for the reaction to proceed. This happens when molecules strike each other at exactly the right place with

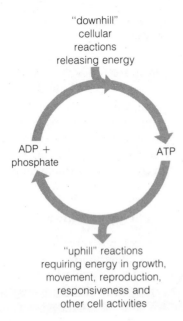

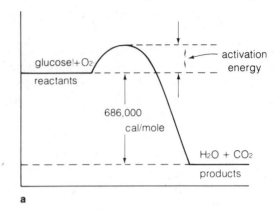

Figure 3-2 ATP and ADP cycle between energy-releasing and energy-requiring activities in the cell. Reactions that release energy are coupled to the synthesis of ATP from ADP and phosphate (top half of diagram). Reactions that require energy are made favorable by coupling them to the breakdown of ATP to ADP and phosphate, which releases sufficient energy for the reactions to proceed (bottom half of diagram). Some energy-requiring activities remove two phosphates from ATP, producing AMP and phosphate, which return to the energy-releasing mechanism to be converted back to ATP.

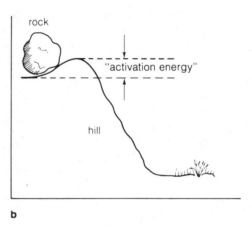

Figure 3-3 (**a**) The activation energy for the oxidation of glucose is an energy barrier over which glucose molecules must be raised before they can be oxidized to CO_2 and H_2O. (**b**) An analogous physical situation in which a rock is poised in a depression at the top of a hill. The rock will not move downward unless enough "activation energy" is added to raise it over the lip of the depression.

sufficient force for interaction to occur. A high energy barrier indicates that successful collisions of this type are not very likely to occur.

One way to increase the probability that molecules in the reacting system will pass over the energy barrier is to raise the temperature. At elevated temperatures, both the speed of travel of individual molecules and the frequency of collisions increase. This increase raises the probability that sufficiently forceful collisions, at the correct angle and place, will occur. Once large numbers of molecules begin to pass over the barrier, the reaction is frequently self-sustaining: sufficient heat energy is released by molecules converted to products to keep the reaction temperature high. Glucose may be ignited in this way by heating it over a flame. Once ignited, glucose will continue to burn in a self-sustaining reaction until conversion to products is complete.

For obvious reasons, ignition is not a satisfactory approach for pushing reacting molecules over an energy barrier in biological systems. Enzymes speed reactions by a different mechanism—by lowering the activation energy required for a reaction to proceed (Figure 3-4). Exactly how enzymes lower the activation energy is not completely understood. However, it is known that enzymes combine briefly with the reactant molecules during a reaction; the particular site on the surface of the enzyme molecule that combines with the reactants is called the **active site**.

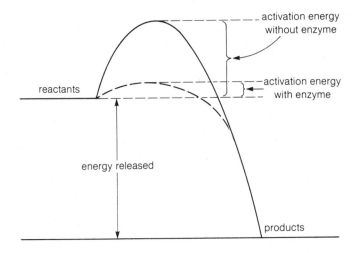

Figure 3-4 Enzymes increase the rate of a spontaneous reaction by reducing the activation energy (see text).

The Role of the Active Site Part of the mechanism lowering the energy of activation probably depends simply on the fact that, by virtue of their combination with the active site on the surface of the enzyme, reacting molecules are brought close together in a position favoring their collision and interaction. Combination with the active site may also align the molecules so that collisions between them are in the most favorable direction and place for reaction to occur. Moreover, binding of the reactants is believed to alter the arrangement of atoms in the substrate molecule, placing them under a strain that pushes the bonds holding them together out of alignment. This warping reduces the stability of the reacting molecules; once in this unstable state, they are more easily pushed by collisions in the direction of products. All these effects of the active site place the reactants in a state in which their interaction at ordinary temperatures is favored.

The Characteristics of Enzymes The enzyme molecules lowering the energy barrier of biological reactions have several important characteristics. All known enzymes are

---

All enzymes are proteins. While catalyzing a reaction, enzymes combine briefly with the reacting molecules.

---

proteins. In catalyzing biological reactions, enzymes combine only briefly with the reacting molecules and are released unchanged on completion of the reaction. As a result, a single enzyme molecule may cycle repeatedly through its reaction sequence, carrying many reactant molecules in succession over the reduced activation barrier. The number of reactions catalyzed by a single enzyme molecule, depending on the enzyme, may vary from 100 to more than 3 million per minute. The magnitude of these numbers means that only a relatively small number of enzyme molecules are required to catalyze a large quantity of reactants. Finally, enzymes are *specific* in their catalytic activity: usually, they are tailored to catalyze only a single

---

All enzymes are specific: they catalyze only a single reaction by combining with a single molecular type or closely related molecular group.

---

type of biochemical reaction and combine only with specific molecular types or closely related groups of molecules. The specific molecule or molecular group whose reaction is catalyzed is known as the **substrate** of the enzyme.

Thousands of different enzymes have been detected and described. Of these, several hundred have been purified to the extent that they can be crystallized and chemically described. These enzymes vary from relatively small molecules having about 100 amino acids to large complexes containing several polypeptide chains totaling thousands of amino acids. Many enzymes are linked in addition to an inorganic ion or nonprotein organic group that contributes to their catalytic function. These nonprotein groups, called **cofactors,** may be covalently linked to the enzyme or held by weaker linkages such as electrostatic attractions or hydrogen bonds.

The inorganic cofactors are all metallic ions, including iron, copper, magnesium, zinc, potassium, manganese, molybdenum, and cobalt. These ions, when present as part of the enzyme, contribute directly to the reduction of activation energy by the enzyme. That is, they form a functional part of the site on the enzyme that combines with substrates and speeds their conversion to products. The organic cofactors, called **coenzymes,** are all complex chemical groups containing segments derived from vitamins. (The biological role of vitamins, in fact, apparently involves forming parts of coenzymes of various kinds.) When present as cofactors, coenzymes act as *carriers* of chemical groups, atoms, or electrons removed from substrates during reactions. (A representative coenzyme is shown in Figure 6-13.)

Enzymes are named according to their substrates and the type of reaction they catalyze. Enzymes catalyzing the polymerization of nucleotides into DNA, for example, are called *DNA polymerases.* Certain enzymes, named before the current rules of nomenclature were agreed upon, are also known by their original or "trivial" names. The digestive enzymes *trypsin* and *pepsin* are examples of this group.

Factors Affecting Enzyme Activity

The effect of enzymes on biochemical reactions depends on the three-dimensional structure of the active site. As might be expected, exposing enzymes to conditions that seriously alter their folding conformation interferes with their ability to increase biological reaction rates.

Proteins are held in their final structure by interactions between their amino acid side groups (see p. 28). The pattern and strength of two types of these interactions,

---

Factors that cause enzymes to unfold interfere with their ability to increase the rate of biological reactions.

---

hydrogen bonding and electrostatic attraction, are highly sensitive to changes in temperature and pH of the medium surrounding the enzyme.

The Effects of Temperature Gradual rises in temperature increase the kinetic motion of both the amino acid chains and side groups forming an enzyme molecule and the molecules colliding with the enzyme in the surrounding solution. At elevated temperatures, these disturbances become strong enough to overcome the attraction of hydrogen bonds, which individually are relatively weak. As its internal hydrogen bonds are broken, an enzyme gradually unfolds and loses its normal three-dimensional folding conformation.

These changes affect enzymatic activity in a characteristic way (Figure 3-5). Initially, rises in temperature over the range from 0 to about 40°C increase the activity of most enzymes along the lines followed by all chemical reactions: each 10°C rise in temperature approximately doubles the reaction rate. This effect is due to increases in the force and frequency of collisions of enzymes and reactant molecules, reflecting the heightened kinetic motion of all molecules in the solution. The collisions at these temperatures are forceful enough to increase the reaction rate, but they do not seriously disturb enzyme structure. However, in most enzymatically catalyzed reactions, rates begin to fall off at temperatures above 40°C as collisions become violent enough to begin unfolding the enzyme; the drop in activity becomes steep at 55°C and falls to zero at 60°C. At the highest temperature, the disturbance in hydrogen bonding causes the enzyme to unfold into a completely inactive form, entirely counteracting the positive effects of increased kinetic motion at elevated temperatures.

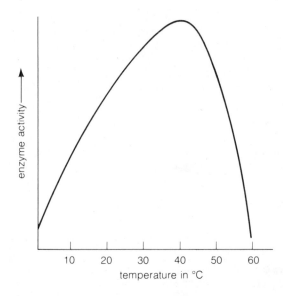

Figure 3-5 The effects of increasing temperature on enzymatic activity. Typically, the rate of the reaction reaches an optimum at 40 to 50°C. Once past the optimum, the reaction rate falls rapidly due to unfolding of the enzyme at elevated temperatures.

As a result of these two opposing effects of increased temperature, all enzymes exhibit an optimum activity representing the highest temperature at which reaction rates are increased with no significant disturbance of enzyme structure. For most enzymes, this optimum lies between 40 and 50°C. A few organisms, such as the bacteria living in hot springs, possess enzymes with structures so resistant to disturbances that they remain active at temperatures of 85°C or more.

The Effects of pH Changes in the pH of the surrounding medium affect enzyme structure and activity by altering the charge of groups carried on amino acid side chains. Particularly important among the affected groups are those, such as —COOH and —NH$_2$ groups, that are capable of releasing or accepting an H$^+$ ion and converting to a charged form. Each of these groups, depending on their location in a protein molecule, undergoes conversion from uncharged to charged form at a characteristic pH. Changes of this type affect the charge of groups holding enzymes in their final three-dimensional shape. These changes also alter the activity of charged groups that may be present in the active site of the enzyme.

As a result, changes in pH also affect enzyme activity in characteristic ways (Figure 3-6). Each enzyme has a pH optimum at which the correct charge is taken on by its amino acid side groups and the enzyme is most efficient in speeding the rate of its specific biochemical reaction. On either side of this optimum pH, the charges of these groups change; the resulting alterations in the folding pattern of the enzyme and the charges in the active site cause the rate of the reaction to drop off. The effects become more extreme as pH values depart from the optimum until the rate drops to zero. Some enzymes reach their optimum activity at an intermediate pH (Figure 3-6a); in most such cases, the optimum lies in the vicinity of neutrality, near pH 7. Others (Figure 3-6b and c) have their optimum catalytic activity at higher or lower pH; these enzymes drop in activity as the pH rises or falls from the optimum value. The digestive enzyme pepsin, for example, which speeds the breakdown of proteins being digested in the stomach, has its optimum activity at the acid pH characteristic of the stomach contents.

Enzyme Saturation and Inhibition   The fact that enzymes combine briefly with their substrate molecules to speed biological reactions has two important effects on the rate of a reaction being speeded by enzymatic activity. One effect is related to the concentration of substrate molecules. At very low concentrations of the substrate, the enzyme molecules do not collide frequently with the substrate molecules and the reaction proceeds slowly. As the concentration of substrate molecules increases, the reaction rate increases at first because collisions between the enzyme molecules and the reactants become more frequent (Figure 3-7). Further increases in substrate concentration continue to increase the reaction rate until the enzymes colliding with the substrate molecules are cycling through the reaction as fast as they can. (Each enzyme, as noted earlier in this chapter, has a maximum rate, the turnover number, at which it can combine with the reactants, speed the reaction, and release the products.) Since the enzymes are cycling as rapidly as they can, further increases in substrate concentration have no effect on the reaction rate. At this point, the enzyme is said to be *saturated*, and the reaction remains at the **saturation** level represented by the horizontal dotted line in Figure 3-7.

The characteristic saturation curve shown in Figure 3-7 provides a valuable biochemical tool for determining whether a given reaction is speeded by an enzyme in a biological system. To determine whether an enzyme is involved, the concentration of reactants is increased experimentally and the rate of the reaction is followed. If a point is reached at which further increases in reactants have no effect on the reaction rate, indications are good that an enzyme is involved. (Uncatalyzed reactions speed up almost indefinitely with increases in concentration of the reactants.)

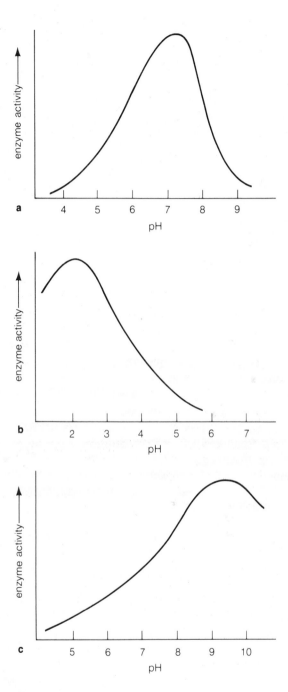

Figure 3-6   The effects of pH changes on enzyme activity. (**a**) Enzymes with maximum activity at neutral pH. (**b**) Enzymes with maximum activity at acid pH. (**c**) Enzymes with maximum activity at alkaline pH (see text).

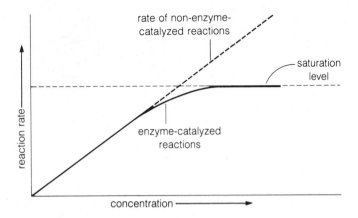

Figure 3-7 Saturation of an enzyme by increases in substrate concentration (see text).

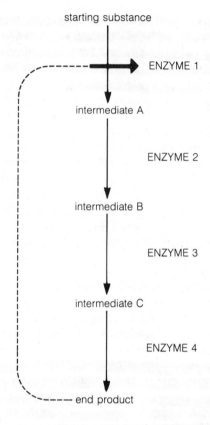

Figure 3-8 Feedback inhibition of an enzyme by a product of a biochemical pathway (see text).

The second effect of the combination of enzymes with their reactants is susceptibility to **inhibition** by nonreactant molecules that resemble the substrate molecules in structure. These molecules, by virtue of their resemblance to the substrate molecules, can combine with the reactive site of the enzyme. Since they cannot be converted to the products of the reaction normally catalyzed by the enzyme, they tend to remain attached to the active site without change and block access by the normal substrate. As a result, the rate of the reaction slows; if the concentration of the inhibitor becomes high enough, the reaction may stop completely.

Enzyme inhibition has both positive and negative effects on the activities of life. Many cellular systems use enzyme inhibition to control the rate of biochemical reactions. In certain complex pathways involving a series of enzymes that carry out sequential steps in the manufacture of a required substance such as an amino acid, for example, a product made by an enzyme at the end of the pathway acts as an inhibitor of an enzyme at the beginning of the pathway (Figure 3-8). As the end product accumulates and becomes abundant in the cell, it shuts off its own pathway so that it is not made in excess. Self-regulation through inhibition of a synthetic pathway by an end product in this way is called *feedback inhibition*. The combination is reversible, so that, if the concentration of the inhibitor falls in the solution surrounding the enzyme, the inhibitor is released and the enzyme takes up its normal activity again.

The negative effects of enzyme inhibition occur through the action of foreign substances that combine with en-

zymes, often irreversibly, and block their normal activity. Many poisons fall into this category. For example, the action of cyanide as a poison depends on its ability to act as an inhibitor of an enzyme important in the utilization of oxygen in cellular respiration.

Many examples of enzymes and enzymatic activity are presented throughout this book. In many respects, the story of cellular life is the story of enzymes—so much so that cells have been described simply as "bags of enzymes." While this is obviously an oversimplification, it is true that each of the thousands of individual biochemical reactions coordinated to achieve cellular life is speeded by its own enzyme, tailored by evolution to fit its normal substrates with specificity approaching perfection. This specificity permits cells to control their activities with precision and sensitivity by regulating which enzymes are present in the various cellular compartments at different times.

Questions

1. State the first and second laws of thermodynamics. Give examples in your own experience that illustrate the operation of the laws.

2. What does the word *spontaneous* mean in thermodynamics?

3. What does *energy content* mean with reference to molecules?

4. What is entropy?

5. How do energy content and entropy interact in determining whether chemical reactions will proceed to completion?

6. How do biological organisms run reactions that require energy? What does *coupling* mean?

7. It has sometimes been claimed that organisms can violate the second law of thermodynamics because they become more complex as they develop from a fertilized egg or seed. Why is this statement incorrect?

8. Which of the following reactions is spontaneous? Why?
   a. Fertilized egg → adult animal
   b. Fertilized egg + raw materials (glucose, fats, etc.) → adult animal + waste products

9. How does ATP act as the agent coupling uphill biological reactions to other reactions that release energy?

10. Draw the structures of ATP, ADP, and AMP.

11. What are calories? Moles? Cal/mole?

12. What effects do enzymes have on spontaneous reactions?

13. What is activation energy? How do enzymes affect activation energy?

14. What are the characteristics of enzymes?

15. Define cofactors and coenzymes.

16. What is the active site of an enzyme?

17. How do changes in temperature affect enzyme-catalyzed reactions?

18. How do changes in pH affect enzyme-catalyzed reactions?

19. What is enzyme saturation? Enzyme inhibition? Feedback inhibition?

Suggestions for Further Reading

Dickerson, R. E., and I. Geis. 1969. *The Structure and Action of Proteins*. Harper & Row, New York.

Koshland, D. E., Jr. 1973. "Protein Shape and Biological Control." *Scientific American* 229:52–64.

Miller, G. Tyler, Jr. 1971. *Energetics, Kinetics and Life*. Wadsworth, Belmont, California.

Cells and Cell Growth

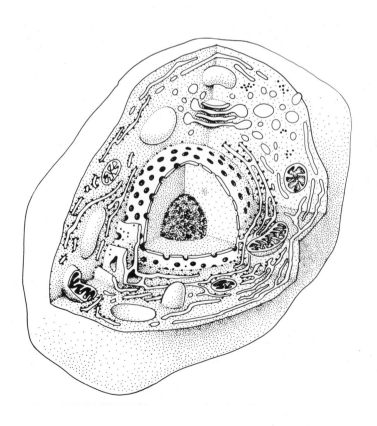

Cells are the basic functional units of living organisms. All the complex activities of animals and plants depend ultimately on the activities of their individual cells. A single cell removed from an organism, if kept under the proper conditions, may retain the activities of life indefinitely and may grow and reproduce. But if a cell is separated into its constituent parts the quality of life is lost. Thus the organization of matter that we recognize as life does not exist in units smaller than the cell.

The concept of the cell as the basic structural and functional unit of life developed gradually over a period of almost two centuries. This period of development extended from the initial description of cells in 1665 to the first statement of the cell theory in the 1840s and 1850s. During this time cell biology was largely a descriptive science. In the latter half of the nineteenth century, the study of cell chemistry began to have a greater influence on cell biology. In recent years the biochemical and molecular approach has dominated the research carried out in the study of cells.

4

AN INTRODUCTION TO CELLS

DEVELOPMENT OF THE CELL THEORY

Early developments in cell biology were tied closely to the invention and gradual improvement of the light microscope. Soon after the light microscope was invented in the seventeenth century, an Englishman, Robert Hooke, published the first description of cells. In his *Micrographia*, dated 1665, Hooke reported his observation of small compartments in the woody tissues of plants; he called these units "pores" or "cells." It is sometimes claimed that Hooke did not actually observe living cells because in some of the dead cork tissues he examined the cells were simply empty spaces outlined by residual cell walls. But in other tissues, such as the "inner pulp or pith of an Elder, or almost any other tree," Hooke "plainly enough discover'd these cells or Pores fill'd with juices" and thus saw living cells.

These first descriptions of cells were the beginnings of cell biology. Hooke's observations were extended by several investigators toward the end of the 1600s, most notably by Anton van Leeuwenhoek, a Dutchman and amateur microscopist who made remarkably accurate observations of the microscopic structure of protozoa, sperm cells, and a variety of other "animalcules," as he called them.

The imperfections of the light microscopes of this period were serious enough to retard further progress in the description of cells until the early nineteenth century, when practical methods to correct the lens defects of the early microscopes were finally discovered. These innovations made it possible to see the inner details of cell structure, and new discoveries in cell biology were quick to follow.

In the 1830s, the German scientists Theodor Schwann and M. J. Schleiden, working with the new microscopes, first discovered the cellular nature of animal tissues. Through this work and comparisons made with the cellular structure of plants, Schwann and Schleiden were able to conclude that all living organisms are composed of one or more cells and, moreover, that cells are the basic functional units of life. These conclusions, soon elaborated by a third hypothesis describing the origins of cells, formed the basic framework of the cell theory.

The third hypothesis was supplied by other scientists investigating cell origins. These workers observed that cells arise in both plants and animals by the division of a parent cell into two daughter cells. By 1855, this work had progressed far enough for the German scientist Rudolf Virchow to state that all cells arise only from preexisting cells by a process of division. This hypothesis completed the cell theory, which has remained virtually unchanged and has withstood all experimental tests to this day. The essence of the theory is that (1) all living organisms are

---

All living organisms are composed of nucleated cells.

Cells are the functional units of life.

All cells arise only from preexisting cells.

---

composed of cells, (2) cells are the functional units of life, and (3) cells arise only by division of preexisting cells.

THE ORGANIZATION OF CELLS

Cells are complex systems of molecules capable of carrying out all the interactions of life, including growth, reproduction, response to outside stimuli, and movement. Cells take highly varied forms in different plants, animals, and microorganisms. (Figures 4-1 and 4-2 give an idea of the wide range of cell types.) They may exist singly, as in the protistans and bacteria, or packed together by the billions as in the larger plants and animals. In size, cells range from the smallest bacteria, just barely visible in a light microscope, to units as large as a hen's egg. In fact, the yolky

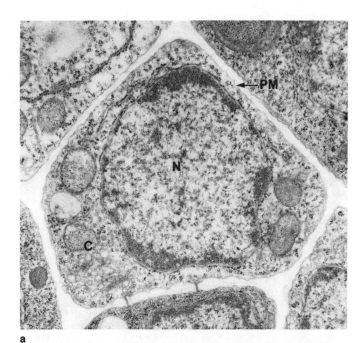

a

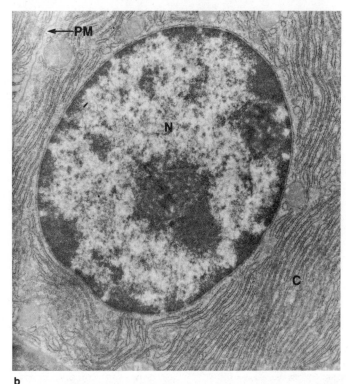

b

Figure 4-1 Electron micrographs of plant and animal cells. *N*, nucleus; *C*, cytoplasm; *PM*, plasma membrane. (**a**) An embryonic plant cell of *Sorghum bicolor*, a type of grass. ×12,800. Courtesy of Chin Ho Lin. (**b**) A cell from the pancreas of a rat. ×12,000. Photograph by the author.

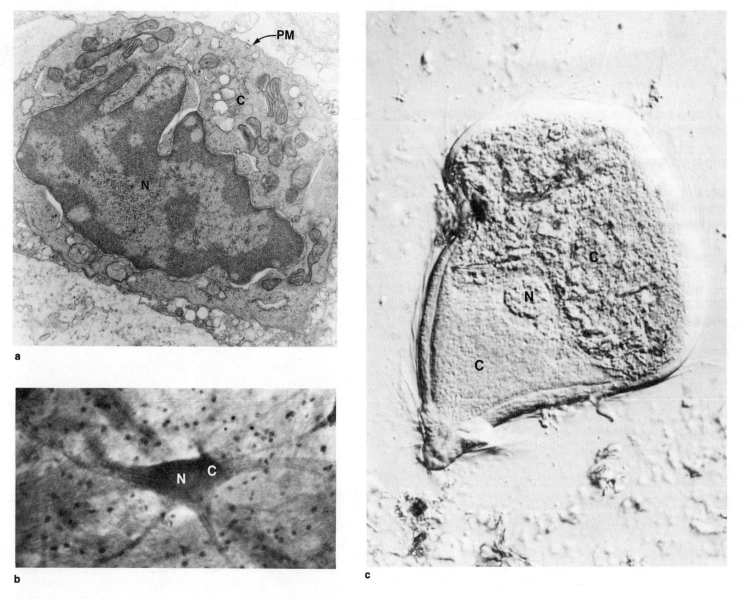

Figure 4-2 Additional cell types. *N*, nucleus; *C*, cytoplasm; *PM*, plasma membrane. (**a**) An electron micrograph of a human leukocyte (white blood cell). ×13,600. Courtesy of S. Brecher. (**b**) A light micrograph of a nerve cell from bovine spinal cord. ×1,750. (**c**) A light micrograph of a protozoan cell from the gut of a termite. ×400. Courtesy of T. K. Golder.

part of a hen's egg is a single cell, in this case several centimeters in diameter. Although most cells are microscopic and roughly spherical in shape, some, like the nerve cells of larger animals, may carry long extensions that are of microscopic diameter but more than a meter in length.

Membranes and Cell Structure

Cells are maintained as separate functional compartments by **membranes,** layers of lipids and proteins only a few molecules in thickness. The outermost boundary membrane of cells, called the **plasma membrane,** maintains

The special environments of life inside cells are maintained by membranes.

the cell interior as a distinct biochemical environment by controlling what molecules enter and leave the cell. Part of this control depends on the nonpolar solubility properties (see p. 14) of the lipid molecules forming the framework of membranes. The plasma membrane, because of these nonpolar lipid molecules, has an oillike character. By contrast the cell interior, and the fluids surrounding the cell, are polar, watery solutions. Because of these characteristics neither the molecules of the cell interior nor those of the surrounding fluids mix readily with the molecules of the plasma membrane. The result is an effective separation of the cell contents from the outside world. The cells of higher organisms, including all plants and animals, are subdivided internally into additional compartments by other systems of membranes.

Apart from acting as barriers to the free passage of ions and molecules, membranes also provide an organizing framework for enzymes, the protein molecules that speed the rate of biological reactions. Some of these enzymes are linked to the surfaces of cellular membranes; others are buried inside. Membranes also provide mechanical support to the cell surface and to internal cell structures.

The Fluid Mosaic Model   The importance of membranes in cells was recognized early in the development of cell biology. From the late 1800s onward, intensive research gradually revealed the chemical constituents of membranes and how they combine to form these thin surface coats of cells and internal cell structures. This early work established that cells are covered with a layer of lipid just two molecules in thickness called a **bilayer** (see p. 26). Proteins were also implicated in membrane structure, but the arrangement of the protein in or on the lipid bilayer was uncertain.

The uncertainties about membrane structure persisted until 1972, when S. J. Singer and his student Garth Nicolson combined information from a variety of sources into a hypothesis for membrane structure they called the **fluid mosaic model** (Figure 4-3). Singer and Nicolson agreed

Lipid bilayers form the basic framework of cellular membranes.

with earlier conclusions that the framework of membranes is supplied by a phospholipid bilayer. In the bilayer, phospholipid molecules (see p. 23) pack together with their nonpolar fatty-acid chains buried in the membrane interior

and their polar ends facing the surrounding watery medium (see Figures 2-16 and 4-3). Their model also proposed that the membrane lipids are in a mobile, fluid state and that individual lipid molecules are free to exchange places and move through the bilayer. This is the "fluid" part of the fluid mosaic model.

Singer and Nicolson proposed that the proteins of membranes float as globular or spherical units in the fluid bilayer like icebergs in the sea. Some of the membrane proteins, according to this model, penetrate entirely through the membrane and are exposed on both sides; other proteins penetrate only part way through. The distribution of proteins in membranes in dispersed units is the "mosaic" part of the fluid mosaic model.

In the fluid mosaic model, proteins float as individual units in or on the lipid bilayer.

According to the model, the membrane proteins are held in suspension in the fluid bilayer by their solubility properties. The proteins that pass entirely through the membrane have two hydrophilic ends and a hydrophobic middle region. The hydrophobic middle region is held in association with the hydrophobic membrane interior; the hydrophilic ends extend into the watery medium at the membrane surfaces. Proteins that extend only partially through the membrane have a hydrophobic end suspended in the membrane interior and a hydrophilic end facing the surrounding watery medium. The proteins remain in stable suspension in the bilayer because any change in orientation would expose their hydrophobic regions to the watery surroundings. Within these limitations, the proteins are free to displace phospholipid molecules and move laterally through the fluid bilayer. Most of the membrane proteins are enzymes that catalyze the specialized functions of membranes in different regions of the cell.

Most of the membrane proteins are enzymes that provide membranes with their functional activities.

Evidence Supporting the Model   The best evidence that proteins are embedded as individual units in membranes comes from the results of a method used to prepare cells for electron microscopy called the *freeze-fracture* technique. In this method a tissue sample is rapidly frozen by placing it in liquid nitrogen. The sample is then fractured by striking it with a sharp knife edge. Since cell membranes are

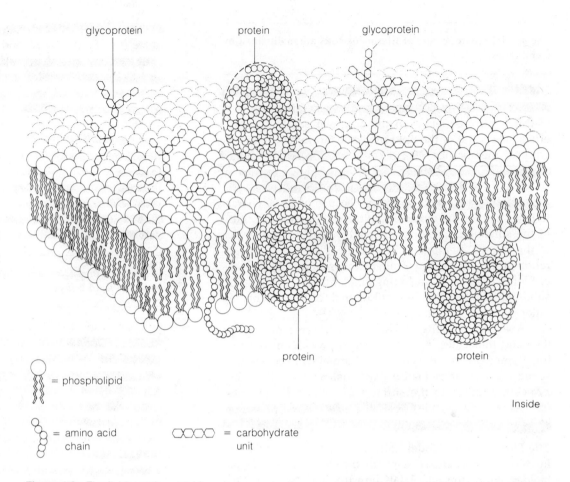

glycoprotein   protein   glycoprotein

protein   protein

Inside

= phospholipid

= amino acid
  chain

= carbohydrate
  unit

Figure 4-3   The fluid mosaic model for membrane structure proposed by S. J. Singer. Integral proteins forming a part of the membrane are deeply embedded in the lipid bilayer; more loosely bound proteins, called peripheral proteins, are attached to the membrane surface.

hydrophobic, nonpolar regions, they produce weakly frozen "faults" in the preparations that fracture more readily than the surrounding polar regions. As a consequence, the fracture tends to follow membranes, often splitting them into bilayer halves and revealing the hydrophobic membrane interior (Figure 4-4). (See Supplement 4-1 for further information on preparative techniques for electron microscopy.)

The freeze-fracture technique preserves membrane proteins and makes their distribution directly visible. It shows that membrane proteins are spaced as individual particles both on and within membranes (see Figure 4-4). Some of the proteins are attached at scattered points to the outside surfaces of membranes and some to the inner surfaces. Other proteins are embedded within membranes. Some of the embedded proteins extend from either surface and others extend all the way through from the inside to the outside membrane surfaces. All this evidence supports

the proposal that proteins are embedded as individual units in biological membranes as proposed in the fluid mosaic model.

Evidence that membrane bilayers are fluid and that proteins are free to move laterally in membranes comes from a graphic experiment carried out by L. David Frye and Michael A. Edidin at Johns Hopkins University. Frye and Edidin worked with mouse and human cells fused together by exposing them to the *Sendai* virus. When two cells are fused in this way, their plasma membranes join together into one continuous membrane covering the cytoplasm and nuclei of both cells.

Frye and Edidin used the cell fusion technique in combination with molecules that could bind specifically to membrane proteins of either cell type. The molecules were marked for identification with a dye that glows, or fluoresces, in ultraviolet light. The molecules specific for mouse and human membrane proteins were marked re-

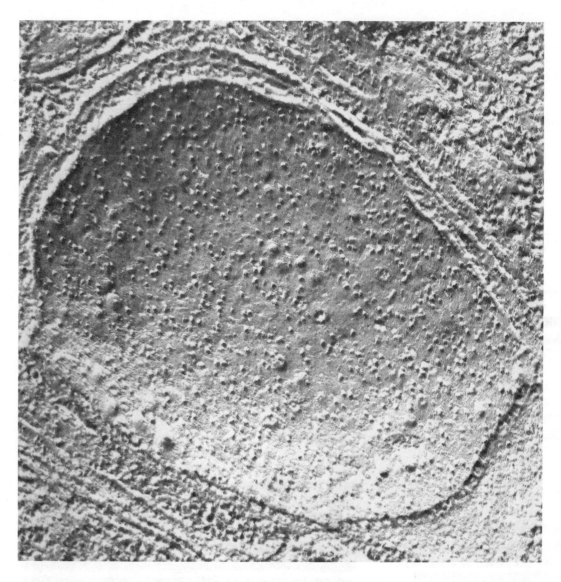

Figure 4-4 Segment of a cell containing several layers of membranes prepared for electron microscopy by the freeze-fracture technique. In the central region of the micrograph, the fracture has split the bilayer in half and exposed the membrane interior. The particles are protein molecules suspended as individual units in the membrane. ×110,000. Courtesy of D. W. Deamer.

spectively with dyes that fluoresce green and red. When mouse and human cells were fused together and reacted with the fluorescent molecules, the green and red colors were at first separated on the surface of the combined cells: half of the fused membranes glowed green and half glowed red. This pattern showed that membrane proteins derived from the mouse cells were initially restricted to one half of the fused plasma membranes and the human membrane proteins to the other. After about 40 minutes, most of the cells showed complete intermixing of the two colors, indicating that the membrane proteins could move readily and freely through the membrane bilayer. This movement would be unlikely unless the membrane bilayer is in a fluid state as proposed in the fluid mosaic model.

Membranes are held in stable structures in cells by the same forces that keep the hydrophobic portions of membrane proteins in suspension in the membrane interior. When surrounded on both sides by water molecules, as they normally are in cells, the phospholipid and protein molecules remain as stable membranes because any disturbance of the structure would expose their hydrophobic portions to the surrounding water molecules. This resistance to mixing, equivalent to the stable layering of oil on water when the two are mixed, keeps membranes intact and provides them with their unique structural properties as cell boundaries.

MEMBRANES AND CELL ORGANIZATION: PROKARYOTES AND EUKARYOTES

In spite of their varied sizes, shapes, and activities, all cells are organized internally into two major functional regions. The **nuclear region** contains the nucleic acid molecules that store and transmit the hereditary information required for cell growth and reproduction. This region is named differently according to its complexity and relationship to cellular membranes. The second region, the **cytoplasm,** uses the nuclear information to make most of the molecules required for growth and reproduction. The cytoplasm also provides the energy needed to carry out these activities.

There are two major subdivisions among living organisms according to the complexity of the nuclear region and the organization of cell membranes. The smallest, and most primitive, subdivision includes only two groups: the bacteria and blue-green algae. In these organisms, called **prokaryotes,** the cellular membrane systems are limited to the surface or plasma membrane and relatively simple internal membranes derived from it. No membranes separate the nuclear material from the surrounding cytoplasm

---

Prokaryotic cells have only a single membrane system based on the plasma membrane and its derivatives.

---

in the prokaryotes. (The name prokaryotes, from *pro* = before and *karyon* = nucleus, refers to the primitive organization of the nuclear material in these organisms.) Because of its primitive organization, the region containing the nuclear region in prokaryotes is termed the **nucleoid.** The second major division, the **eukaryotes** (*eu* = typical; *karyon* = nucleus), includes all the remaining plants, animals, and microorganisms on earth. Eukaryotic cells are divided by a large number of independent membrane systems into separate interior compartments. These separate

---

Eukaryotic cells contain systems of membranes forming internal organelles that are completely separate from the plasma membrane.

---

membrane-bound interior structures, called **organelles,** are specialized to carry out the various cellular functions of eukaryotes. The most conspicuous of the internal organelles is the **nucleus,** which contains the nuclear material. In contrast to prokaryotes, the nucleus in eukaryotes is completely separated from the surrounding cytoplasm by a continuous system of membranes.

Prokaryotic Cells

Prokaryotic cells (Figures 4-5 and 4-6) are usually not much more than a few micrometers long and a micrometer or slightly less in width. (Information Box 4-1 defines the units of measurement used in biology.) In almost all prokaryotes, the outer membrane of the cell, the plasma membrane, is surrounded by a rigid external layer of material: the **cell wall.** This exterior layer may range in thickness from 15 to 100 nanometers or more and may itself be coated with a thick, jellylike **capsule.** The cell wall provides rigidity and, along with the capsule, protects the cell within the wall.

The plasma membrane lining the inner surface of the cell wall may be smooth or may include folds that extend from the surface into the cell interior (as in Figure 4-6). In prokaryotes, the plasma membrane contains the molecules that oxidize fuel substances and release energy for cellular use. In the photosynthetic bacteria and blue-green algae, the enzymes and other molecules that absorb light and convert it to chemical energy are also associated with the plasma membrane and its interior derivatives. Moreover, the plasma membrane of bacteria is thought to play a part in replication and division of the nuclear material. Thus many functions that are located in separate membrane-bound organelles in eukaryotes are associated with the plasma membrane or its derivatives in prokaryotes.

The nucleoid containing the nuclear material in prokaryotes, as noted, is suspended directly in the cytoplasm without boundary membranes of any kind (see Figure 4-5). Within the nucleoid, which is irregular in shape, the electron microscope reveals a mass of fibers 3 to 5 nanometers in thickness. These fibers contain the DNA (deoxyribonucleic acid, the molecule containing encoded

---

The nucleoid of prokaryotic cells contains DNA fibers in combination with small amounts of proteins.

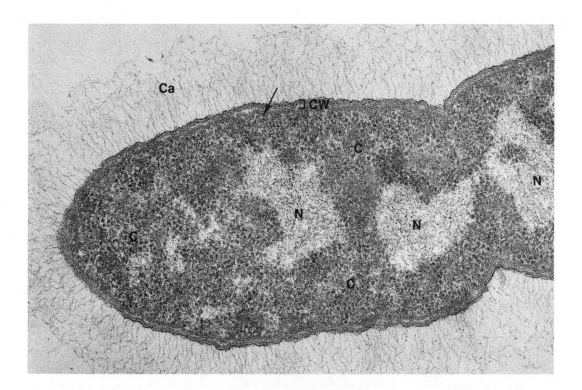

Figure 4-5 A dividing prokaryotic cell of the bacterium *Klebsiella pneumoniae*. The dividing nucleoid (*N*) appears in several pieces because the plane of section has passed near one edge of the irregularly shaped nuclear region. The cytoplasm (*C*) surrounding the nucleus is packed with ribosomes. The cell is surrounded by the cell wall (*CW*) and a capsule (*Ca*); the plasma membrane (arrow) lies just beneath the cell wall. ×62,000. Courtesy of E. N. Schmid, from *Journal of Ultrastructure Research* 75:41 (1981).

hereditary information) of the prokaryotic cell. In contrast to eukaryotic cells, prokaryotic DNA occurs without large quantities of associated proteins.

The cytoplasm surrounding the nucleoid usually appears darkly stained in electron micrographs. Most of this density is due to the presence of large numbers of small, roughly spherical particles 20 to 30 nanometers in diameter: the **ribosomes.** In bacteria, ribosomes are complex structures containing more than 50 different proteins in combination with several types of RNA (ribonucleic acid). These complex spherical bodies are the sites of protein synthesis in both prokaryotic and eukaryotic cells. (Eukaryotic ribosomes are slightly larger and contain more protein and RNA molecules.)

Other structures may be present in the cytoplasm of the more complex prokaryotes. In some, particularly the photosynthetic bacteria and the blue-green algae, the cytoplasm contains numerous closed sacs with walls formed by single, continuous membranes. In these prokaryotes, the molecules carrying out photosynthesis are associated with the internal sacs as well as the plasma membrane. No ribosomes are present inside the sacs; some may be filled with gas. Prokaryotes may also contain deposits of reserve polysaccharide or inorganic phosphates that appear as small, dense, spherical bodies scattered in the cytoplasm. Most of these structures, along with the ribosomes, internal photosynthetic membranes, and the nucleoid, can be identified in the bacterial and blue-green algal cells shown in Figures 4-5 and 4-6.

Many types of bacteria are capable of rapid movement generated by the action of threadlike structures covering the cell surface. These threads (Figure 4-7), called **flagella** (singular = *flagellum*), are made up from long chains of protein molecules that extend in bundles from the cell surface. The flagella, even though they may be as much as five times longer than the cell, are only 12–18 nanometers in diameter. Usually, the flagella of a single bacterial species contain only a single type of protein. Bacterial flagella are fundamentally different from the much larger and more complex flagella of eukaryotic cells.

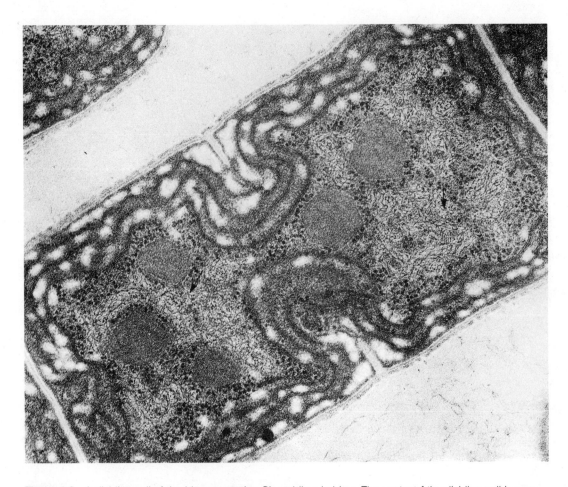

Figure 4-6 A dividing cell of the blue-green alga *Phormidium luridum*. The center of the dividing cell is occupied by the nucleoid, in which DNA fibers are clearly visible (arrows). Ribosomes are also clearly visible in the cytoplasm immediately surrounding the nucleoid. In contrast to the bacterial cell shown in Figure 4-5, the cytoplasm contains extensive collections of membranes. × 60,000. Courtesy of M. R. Edwards.

The apparent simplicity of prokaryotic cells is deceptive. Bacteria and blue-green algae contain all the biochemical mechanisms required to make the complex organic substances required for life. In most cases, they are able to make these substances from simple inorganic molecules. In many respects, in fact, prokaryotes are more versatile in their synthetic activities than eukaryotes.

Eukaryotic Cells

In contrast to prokaryotes, eukaryotic cells (Figures 4-8 and 4-9) are divided into distinct interior compartments by systems of internal membranes. The plasma membrane completely encloses the eukaryotic cell as in prokaryotes; all substances entering and leaving the cell must pass through this barrier. Although the enzymes that break down fuel substances or convert light to chemical energy are not associated with the eukaryotic plasma membrane as they are in prokaryotes, other enzymatic systems may be present in the surface membrane. Among the most important of these are the enzymatic systems that transport substances into or out of the cell. A second major membrane system, the **nuclear envelope,** separates the nucleus from the cytoplasm in eukaryotic cells.

Units of Measurement Used in Cell Biology

Much information about cell structure is given in terms of physical dimensions. The units most frequently used are the **micrometer**, μm (also called the **micron**, μ) the **nanometer**, nm (or millimicron, mμ), and the **angstrom** (Å or A). These units are compared in the table below.

The micrometer, equivalent to 1/1000 millimeter, is convenient for describing the dimensions of whole cells or larger cell structures such as the nucleus and is much used in this book. Most cells are between 5 and 200 micrometers in diameter, although some animal eggs may be much larger. Objects about 200 micrometers (0.2 millimeter) in diameter are just visible to the unaided eye.

Particles roughly the size of mitochondria and chloroplasts (objects in this range of dimensions are called the "ultrastructures" of the cell) are most often measured in nanometers or angstroms, units that are useful for descriptions from this level down to particles as small as molecules and atoms. The nanometer, equivalent to 1/1000 micrometer, is difficult to visualize, but with experience a relative appreciation can be made of the size of objects measured in this unit. Lipid molecules, for example, are about 2 nanometers long, and amino acids are about 1 nanometer long. Protein molecules may be 10 nanometers or so in diameter. On the level of cell organelles, membranes are 7.5 to 10 nanometers thick and ribosomes are about 25 to 30 nanometers in diameter. The electron microscope, incidentally, can "see" objects with diameters as small as 0.7 to 0.8 nanometer. The angstrom unit, equal to 0.1 nanometer or 1/10,000 micrometer, is employed for measurements in the same size range.

Unit	Equivalence in Millimeters	Equivalence in Micrometers	Equivalence in Nanometers	Equivalence in Angstroms
Millimeter (mm)	1	1,000	1,000,000	10,000,000
Micrometer (μm)	0.001	1	1,000	10,000
Nanometer (nm)	0.000001	0.001	1	10
Angstrom (A)	0.0000001	0.0001	0.1	1

The Eukaryotic Nucleus Most of the space inside the nucleus of eukaryotic cells is occupied by masses of very fine, irregularly coiled and folded fibers collectively called **chromatin.** The chromatin fibers, which average 10 to 20 nanometers in diameter, have been shown to contain the DNA of the nucleus along with two types of protein specifically associated with DNA in eukaryotes: the **histone** and **nonhistone** proteins. These proteins maintain the

---

The nucleus of eukaryotic cells contains DNA in combination with approximately equal quantities of histone and nonhistone proteins. A nucleolus is also present.

---

structure of the chromatin fibers and also regulate the biological activity of the DNA. Suspended within the chromatin of the nucleus are one or more irregularly shaped bodies called **nucleoli** (singular = *nucleolus;* see Figures 4-8 and 4-9). The nucleolar material is so densely packed that its boundaries are easily traced even though no membranes separate the nucleolus from the surrounding chromatin.

The nuclear envelope consists of two concentric membranes, one layered just inside the other (Figures 4-8 and

---

The eukaryotic nucleus is separated from the cytoplasm by the membranes of the nuclear envelope.

4-9). A narrow space called the **perinuclear compartment** separates the two membranes of the envelope. The nuclear envelope is crowded with large numbers of **pore complexes** that perforate the membranes and evidently serve as channels of communication between the nucleus

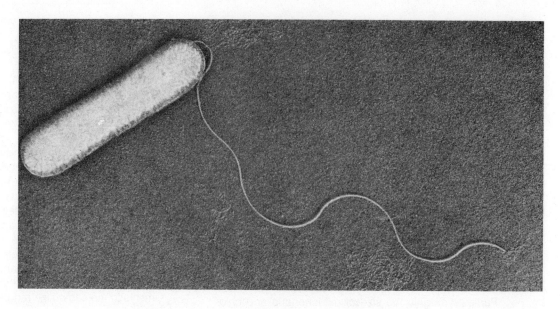

Figure 4-7 A bacterial cell with a single flagellum. Bacterial flagella are simple structures usually containing only a single kind of protein. ×29,600. Courtesy of J. Pangborn.

Figure 4-8 A eukaryotic animal cell from the pancreas of a rat. *N*, nucleus; *Nu*, nucleolus; *M*, mitochondrion; *ER*, endoplasmic reticulum; *Go*, Golgi complex; *PM*, plasma membrane; *NE*, nuclear envelope; arrows, nuclear pore complexes. ×24,000. Photograph by the author.

and cytoplasm (arrows in Figure 4-8). The pores in the envelope, which average about 70 nanometers in diameter, are filled in by a ring of dense material called the **annulus.** The molecular systems of the annulus, which have so far defied analysis, in some way control the movement of large molecules such as RNA and proteins between the nucleus and cytoplasm.

The nucleus is the ultimate control center for all cell activities.

The nucleus is the ultimate control center for all cell activities. Within the chromatin, the information required to synthesize proteins is coded into the DNA molecules. This information is copied into "messenger" RNA molecules that move to the cytoplasm through the nuclear envelope. In the cytoplasm, proteins are synthesized on ribosomes according to the directions in the messenger RNA. The nucleolus synthesizes the RNA of ribosomes and assembles this RNA with ribosomal proteins into ribosomal subunits. These subunits combine by twos in the cytoplasm to make finished ribosomes.

The nucleus also duplicates the chromatin as a part of cell reproduction. Just before a cell divides, all the chromatin components—including both DNA and chromosomal proteins—are copied exactly. During cell division, the two copies of the chromatin are separated and divided so that each of the two cells resulting from division receives a complete set of the directions for cell synthesis.

The Cytoplasm

Mitochondria Eukaryotic cytoplasm, the portion of the cell outside the nucleus, is packed with ribosomes and a variety of membrane-bound organelles. Among the most conspicuous of these organelles are the **mitochondria** (singular = *mitochondrion*; from *mitos* = thread, *chondros* = grain).

Reactions in the cytoplasm provide energy for cell activities, synthesize proteins and other molecules, and move cells or cell parts.

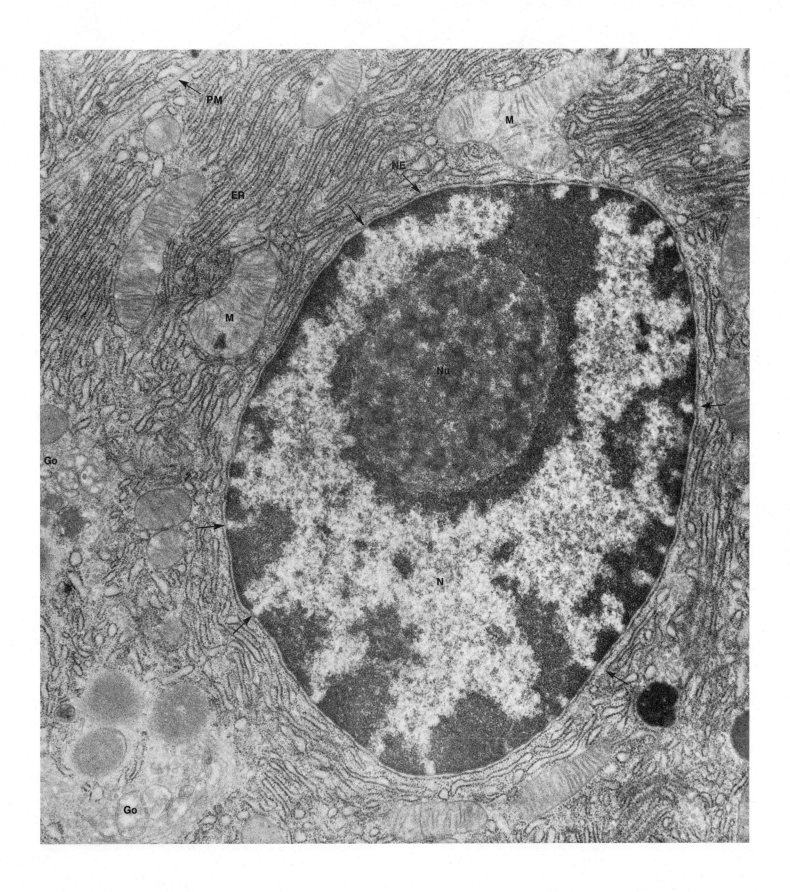

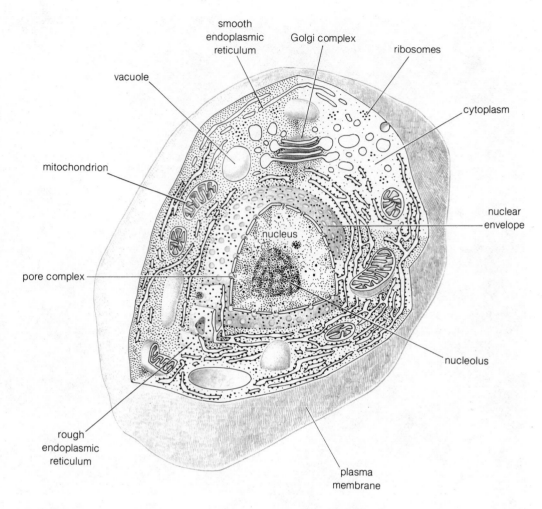

smooth endoplasmic reticulum

Golgi complex

ribosomes

vacuole

cytoplasm

mitochondrion

nuclear envelope

nucleus

pore complex

nucleolus

rough endoplasmic reticulum

plasma membrane

Figure 4-9 Structures typically seen in electron micrographs of eukaryotic cells.

The Greek words for "thread" and "grain" used to name this organelle refer to the fact that mitochondria are flexible structures that can change between threadlike and compact, granular forms. Mitochondria are constructed from two separate membrane systems, one enclosed within the other (Figure 4-10; see also Figures 7-2 and 7-3). The outer membrane is smooth and continuous; the inner

Mitochondria are the primary sources of chemical energy for cell activities and are therefore the "powerhouses" of the cell.

membrane is thrown into numerous folds or tubular projections called **cristae.** Mitochondria carry out most of the reactions that break down fuel substances such as carbohydrates and release energy for cellular activities. Because

of this function, mitochondria are frequently called the "powerhouses" of the cell.

Ribosomes, Endoplasmic Reticulum, and the Golgi Complex Eukaryotic ribosomes (see Figure 4-10), at about 25 to 35 nanometers, are somewhat larger than the ribosomes of prokaryotes. These ribosomes may either be freely suspended in clusters or attached to the surfaces of membranous sacs in the cytoplasm (see Figure 4-8 and Figures 9-6 and 9-7). The sacs with their attached ribosomes form extensive channels in the cytoplasm known as the **rough endoplasmic reticulum** or **rough ER.** (The word *rough* refers to the studded appearance of the ER membranes because of the attached ribosomes.) The ribosomes and membranes of the rough ER synthesize and transport pro-

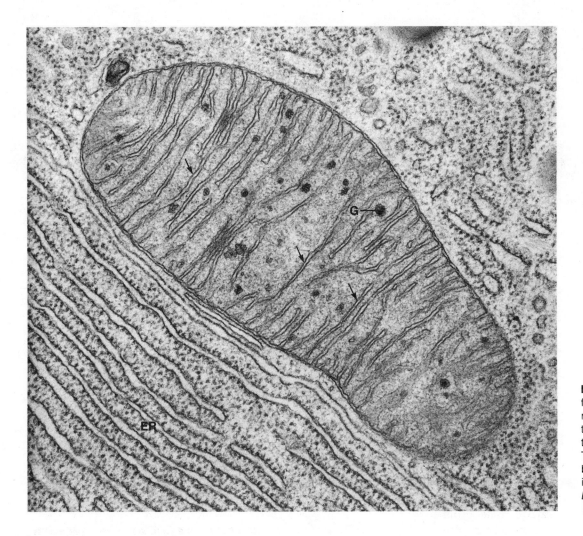

Figure 4-10 A mitochondrion from a bat pancreas. Cristae (arrows) extend into the interior of the mitochondrion as folds from the inner boundary membrane. The darkly stained granules are believed to be deposits of lipids in the mitochondrial interior. *ER*, endoplasmic reticulum. ×62,000. Courtesy of K. R. Porter.

teins. This system reaches its greatest development in animal cells that produce and secrete large quantities of

The enzymatic, structural, and motile proteins of cells are made on ribosomes.

proteins, such as the cells of the mammalian pancreas. Proteins synthesized on the ribosomes penetrate into the membranes of the ER, there to become part of the membrane structure, or into the enclosed ER channels. The channels enclosing the proteins eventually pinch off, forming spherical, membrane-bound bags of protein that have various fates in the cell. Some remain stored in the cytoplasm; others migrate to the plasma membrane to expel their contents to the cell exterior. Some merge with

Proteins that become part of cellular membranes, or are secreted to the outside of the cell, are made on ribosomes attached to the endoplasmic reticulum.

another membranous organelle, the **Golgi complex,** where the enclosed proteins undergo further chemical processing before storage or expulsion from the cell.

The Golgi complex or apparatus (named for its discoverer, Camillo Golgi) consists of a collection of membranous, ribosome-free sacs that often occur in the form of a flattened and layered stack (Figure 4-11). Within the Golgi complex, proteins are modified by the attachment of carbohydrates, lipids, or other groups. Following these modifications, sacs containing the proteins pinch off from the

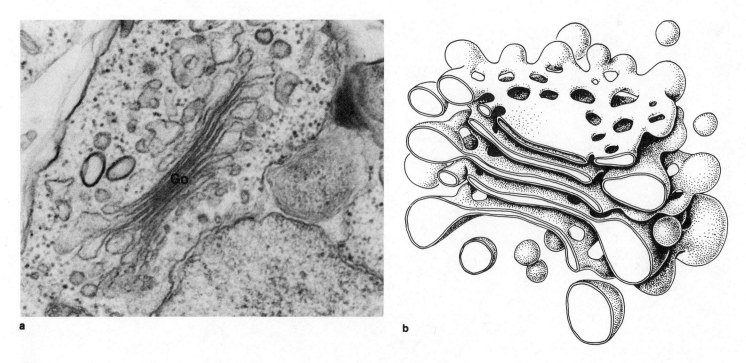

Figure 4-11 (a) Golgi complex *(Go)* in a plant cell. Courtesy of G. W. Grimes. (b) A three-dimensional reconstruction of a Golgi complex.

Golgi complex. These sacs may remain suspended in storage in the cytoplasm, or may be secreted to the exterior. Assembly of sugar units into carbohydrate groups or polysaccharides of various types has also been identified with the Golgi complex. The Golgi complex is particularly active in this role in plant cells, where sugar molecules are assembled into the subunits of cell walls in the sacs of the Golgi complex.

Proteins made in the endoplasmic reticulum are modified in the Golgi complex by attachment of lipid or carbohydrate groups.

Not all the interconnected membranous sacs collectively identified as endoplasmic reticulum are associated with ribosomes. The ribosome-free ER membranes, known as **smooth endoplasmic reticulum (smooth ER),** have various functions in the cytoplasm. One function is the transport of proteins synthesized in the rough ER. Other smooth ER membranes, probably unrelated to protein synthesis and transport, are involved in the initial breakdown of fats to release energy for cellular activities. Some segments of smooth ER have also been identified with the synthesis of steroids or with reactions that break down toxic substances absorbed by the cell.

Microtubules, Microfilaments, and Cell Movement Almost all cell movements in eukaryotes are generated by one or both of two cytoplasmic structures, **microtubules** and

---

Almost all cellular movements are generated by microtubules or microfilaments.

---

microfilaments. Microtubules (see Figure 4-12) are long, pipelike cylinders somewhat smaller in diameter than a ribosome and varying in length from a few nanometers to many micrometers. The cylindrical walls of microtubules enclose a central channel that is distinctly less dense than the walls. Microfilaments (see Figure 4-13) are extremely fine fibers no thicker than the wall of a microtubule. Each of the two motile structures is built up from proteins with a motile function—microtubules from a protein called *tubulin*, microfilaments from a different protein called *actin*.

Certain kinds of cellular movement, such as the movement of chromosomes during cell division, depend

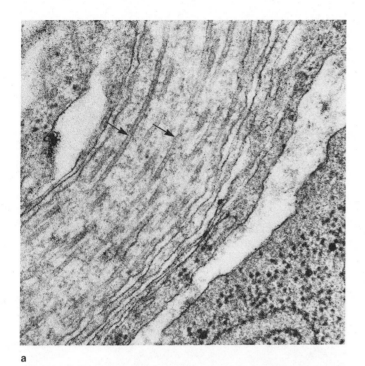

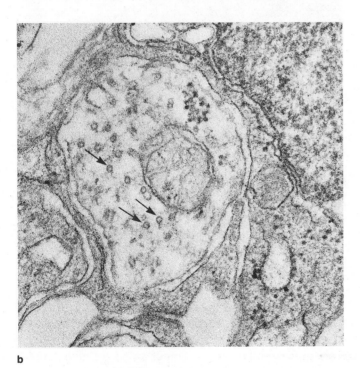

Figure 4-12 Microtubules (arrows) in longitudinal section (**a**) and cross section (**b**). The walls of microtubules, which are about 4 nanometers thick, enclose a central channel about 10 nanometers in diameter. The entire structure is about 25 nanometers in diameter. ×65,000. Courtesy of M. Daniels, reprinted from M. Daniels, *Annals of the New York Academy of Sciences,* 253:535 (1975).

Figure 4-13 Microfilaments (arrows) in the fingerlike surface extensions of a cell from the intestine of a chick. The surface extensions, called *microvilli,* are capable of movement; the microfilaments inside them are believed to provide the force for this movement. Other microfilaments (brackets) are present in the cytoplasm under the microvilli. ×80,000. Courtesy of C. Chambers.

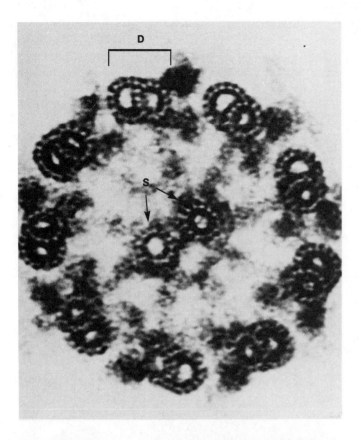

Figure 4-14 The 9 + 2 system of microtubules in a eukaryotic flagellum in cross section. *D*, peripheral doublet; *S*, central singlets. Subunits in the microtubule walls have been made visible by tannic acid staining. ×363,000. Courtesy of K. Fujiwara, reprinted from K. Fujiwara and L. G. Tilney, *Annals of the New York Academy of Sciences*, 253:27 (1975).

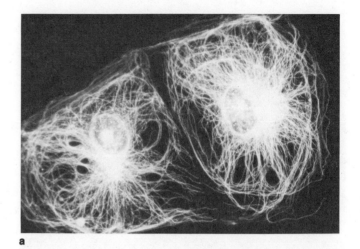

a

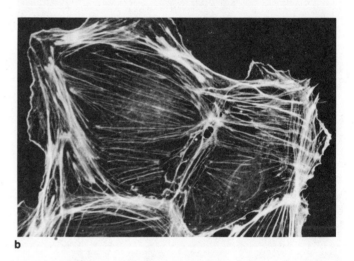

b

Figure 4-15 The cytoskeleton or cell matrix in cultured rat kangaroo cells. The staining method used for these preparations makes the cytoskeleton appear light against a dark background. (**a**) The microtubule network of the cytoskeleton. (**b**) The microfilament network. ×600. Courtesy of M. Osborn, reprinted from M. Osborn, W. W. Franke, and K. Weber, *Proc. Nat. Acad. Sci.* 74:2490 (1977).

on the activity of microtubules. Microfilaments are responsible for other types of movement, including the active flowing motion of cytoplasm called **cytoplasmic streaming.** More organized movements, such as the contraction of muscle cells, also depend on the activity of microfilaments.

Both microtubules and microfilaments produce the force for cell movement by an active sliding mechanism. The force for the sliding is believed to depend on chemical bonds that are alternately made and broken between adjacent microtubules or microfilaments. Although microtubules and microfilaments may coordinate to produce a complex cell movement, they apparently do not interact directly. That is, microtubules do not produce motion by sliding over microfilaments or vice versa.

Many types of plant and animal cells can move by means of whiplike surface structures called **flagella.** Within a flagellum is a remarkably complex system of microtubules consisting in cross section of a circle of nine peripheral double microtubules (the **doublets**) surrounding a central pair of single microtubules (the **central singlets;** see Figure 4-14). With rare exceptions, the same "9 + 2" arrangement of microtubules is found inside flagella of all types. These flagella include the tails of plant and animal

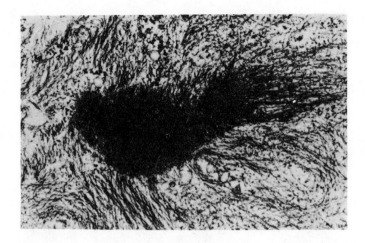

Figure 4-16 Intermediate filaments in the cytoplasm of a cultured mammalian cell. ×16,000. Courtesy of S. Brecher.

sperm cells, the flagella found on many animal cells such as the cells lining the respiratory tract of mammals, and the flagella of protistans and algae. In all these systems the microtubules of the flagellum are also believed to generate the force for movement by sliding actively over each other.

The Cytoskeleton or Cell Matrix Microtubules and microfilaments, in addition to their functions in cell motility, also form supportive networks in the cytoplasm called the **cytoskeleton** or **cell matrix** (Figure 4-15). Supportive structures are also formed by another class of nonmotile fibers somewhat larger than microfilaments called **intermediate filaments** or **fibers** (Figure 4-16). As the name suggests, the cytoskeletal networks of microtubules, microfilaments, and intermediate filaments form a framework providing structural rigidity to the cell. These networks are particularly highly developed in cells forming connective tissues in vertebrate animals.

Specialized Cytoplasmic Structures in Plants All the nuclear and cytoplasmic structures described up to this point, with the possible exception of intermediate filaments, occur in both plant and animal cells. Plant cells (Figures 4-17 and 4-18) have in addition a number of organelles and components not found in animals. The most conspicuous of these are **plastids,** large **vacuoles,** and the **cell walls.**

Plastids The plastids constitute a family of organelles with a variety of functions in plants. In green, photosynthesizing tissue the characteristic plastid is the **chloroplast** (Figure 4-19; see also Figures 6-1 and 6-2), a membranous organelle somewhat like mitochondria in structure. The chloroplast surface is formed by two concentric membranes. The outer membrane, as in mitochondria, is smooth; the inner one is highly folded and convoluted. Most of the

---

Animal cells contain a nucleus, mitochondria, ribosomes, endoplasmic reticulum, and Golgi complex. Plant cells contain in addition plastids, large vacuoles, and cell walls.

---

chloroplast interior is filled with a third system of saclike membranes called **thylakoids.** Within the thylakoid membranes, light is converted to chemical energy; this energy is then used inside the chloroplast to synthesize complex organic molecules such as sugars and starches from water, carbon dioxide, and other simple inorganic substances.

Chloroplasts convert light to chemical energy and use it to synthesize sugars and other organic molecules from simple inorganic substances.

The plastids of embryonic plant tissues are small and contain only a few inner membranes. These structures, called **proplastids,** develop into chloroplasts if the tissue is exposed to light. Other plastids, packed with stored lipid, protein, or starch rather than photosynthetic membranes, are called **leukoplasts** (*leukos* = colorless or white). If the primary storage material is starch, leukoplasts are frequently called **amyloplasts** (*amylon* = starch). In ripening fruit or in leaves changing to fall colors, leukoplasts or chloroplasts are transformed into **chromoplasts,** colored plastids in which red and yellow lipid pigments predominate.

Vacuoles Plant vacuoles are identified as distinct organelles of plant cells because they reach much larger dimensions than any of the membrane-bound sacs or vesicles of animal cells. Often most of the volume of a mature plant cell, up to nine-tenths or more, is occupied by one or more large vacuoles. These vacuoles, consisting of a single, continuous membrane enclosing an inner, fluid-filled compartment, are formed, coalesce, and increase in size during the maturation of plant cells (see Figure 4-17); much of the growth of the developing cells is due to the enlargement of their vacuoles. Plant vacuoles contain dilute solutions

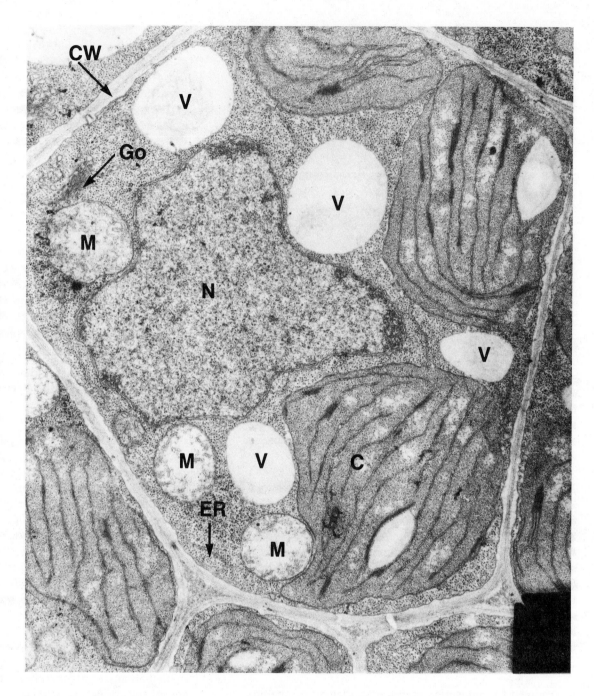

Figure 4-17 A maturing plant cell from a bean seedling *(Phaseolus vulgaris)*. A number of large chloroplasts *(C)* and a few mitochondria *(M)* are visible in the cytoplasm. A prominent nucleus *(N)* is present; the nucleolus has not been caught in the plane of section. Several vacuoles *(V)* are developing in the cytoplasm. Limited elements of the endoplasmic reticulum *(ER)* and Golgi complex *(Go)* are also present. The cell is surrounded by a cell wall *(CW)*. As the cell matures, the vacuoles will grow in size and coalesce into a single, large, central vacuole. ×17,000. Courtesy of Chin Ho Lin.

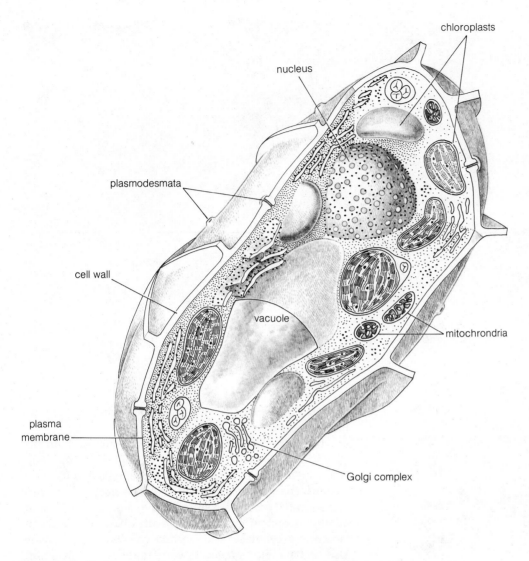

Figure 4-18 The structures typically visible in electron micrographs of plant cells.

Plant cells increase in size primarily by enlargement of the central vacuole.

---

of a variety of substances, including organic and inorganic salts, organic acids, sugars, and various pigments. Frequently the colors of flowers are due to pigments concentrated in cell vacuoles.

Plant vacuoles also support plant cells and tissues by exerting pressure against the rigid cell walls. In plants with small amounts of woody tissue, support of the stems and leaves depends primarily on pressure exerted by the vacuoles, which may be 20 or more times greater than at-

mospheric pressure. (The pressure is developed by a pattern of water movement called **osmosis;** see pp. 82–83; Chapter 5.) Significant loss of pressure or water from the cell vacuoles causes the softer tissues of plants to wilt.

Cell Walls Cell walls (see Figure 4-17) are layered structures that provide rigidity to plant cells. These structures are outside the plasma membrane and are considered to

---

Cell walls provide support to plant cells and to the roots, stems and leaves of plants.

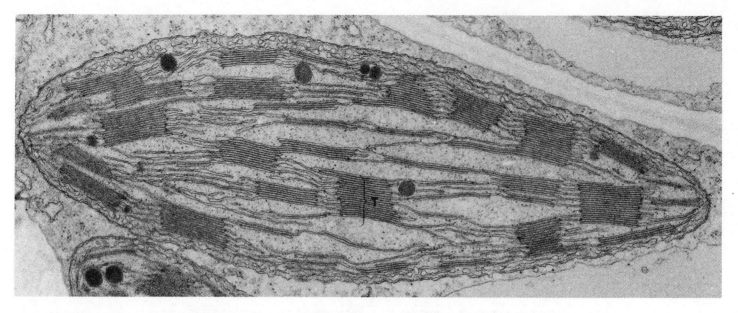

Figure 4-19 Chloroplast in a tobacco leaf. The reactions of photosynthesis converting light to chemical energy are concentrated in the thylakoid membranes *(T)* in the chloroplast interior. Courtesy of W. M. Laetsch.

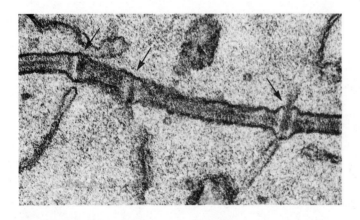

Figure 4-20 Plasmodesmata (arrows), which form channels or openings in cell walls through which the cytoplasm of adjacent plant cell walls remains in direct contact. × 70,000. Courtesy of A. Frey-Wyssling and Academic Press, Inc.

be external to the cell. The *primary* cell walls of embryonic plant cells are flexible and capable of extension during growth, but the walls of mature cells, called *secondary* walls, are usually rigid and inextensible. At intervals, cell walls in higher plants are perforated by minute channels, the **plasmodesmata** (see Figure 4-20). These openings, sometimes as narrow as a few nanometers, contain cyto-

plasmic bridges that extend between and directly connect the protoplasm of adjacent cells.

Both primary and secondary walls are built up primarily from fibers of **cellulose** (see Figure 2-11*a*)—long, chainlike molecules consisting of repeating subunits of the monosaccharide glucose. The cellulose fibers of primary and secondary walls are embedded in a network of **pectin,** another group of chainlike molecules basically similar in structure to cellulose. The secondary walls of higher plants also become impregnated to varying degrees with **lignin,** a hard, chemically resistant substance made up of chains of complex alcohols. The relative proportions of cellulose and lignin in the secondary cell walls of higher plants, along with smaller amounts of pectin and waxy substances, give the various woody tissues their characteristic texture, hardness, and strength.

Prokaryotes and Eukaryotes: A Summary

Prokaryotes are relatively small cells with a single membrane system based on the plasma membrane and its derivatives. The prokaryotic nuclear material, contained in the nucleoid, is suspended directly in the cytoplasm with no boundary membranes marking the division between the two regions. The nucleoid contains little more than DNA; no nucleolus is present, and no proteins equivalent

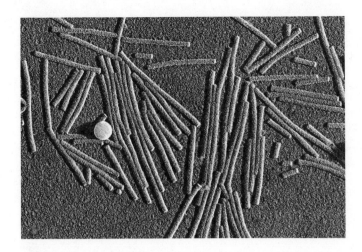

Figure 4-21 The tobacco mosaic virus, an example of a rod-shaped virus. ×52,000. Courtesy of R. B. Park.

Table 4-1   Major Cell Structures of Prokaryotes and Eukaryotes

Structure	Prokaryotes	Eukaryotes Plants	Eukaryotes Animals
Plasma membrane	+	+	+
Nucleus	–	+	+
Nucleoid	+	–	–
Nucleolus	–	+	+
Nuclear envelope	–	+	+
Mitochondria	–	+	+
Chloroplasts	–	+	–
Ribosomes	+	+	+
Endoplasmic reticulum	–	+	+
Golgi complex	–	+	+
Microtubules	–	+	+
Microfilaments	–	+	+
Intermediate filaments	–	?	+
Cell wall	+	+	–

to the histone and nonhistone proteins of eukaryotes occur in association with the DNA. Prokaryotic cytoplasm consists primarily of masses of ribosomes. Although small, membrane-bound sacs may occur in addition in the cytoplasm of complex prokaryotes, none of the discrete, membrane-bound organelles of eukaryotes are present.

Eukaryotic cells are divided into specialized compartments by systems of internal membranes. The nucleus is separated from the cytoplasm by a double layer of membranes that forms the nuclear envelope. Within the cytoplasm other membrane systems form mitochondria, chloroplasts, endoplasmic reticulum, and the Golgi complex. A plasma membrane, separate and distinct from the internal membranes, forms the outer boundary of the cell. The nucleus contains chromatin fibers, consisting of DNA in association with the histone and nonhistone proteins, and a prominent nucleolus. Motility is provided by microtubules and microfilaments, which produce a sliding motion through the activity of chemical bonds that are alternately formed and released at their surfaces. Microtubules and microfilaments, along with intermediate filaments, also form a cytoskeleton or cytoplasmic matrix that supports regions of the cell. The major differences between prokaryotic and eukaryotic cells are summarized in Table 4-1.

VIRUSES

Viruses are minute particles that are capable of infecting both prokaryotic and eukaryotic cells and converting them to the production of more particles like themselves. Virus particles outside their host cells consist of a *core* of nucleic acid, either DNA or RNA, surrounded by a *coat* of protein. These structures are concentrated in a particle smaller than a ribosome. All degrees of complexity exist in viruses.

Viruses are minute particles consisting of a core of DNA or RNA surrounded by a protein coat.

Some are simple particles consisting of a core with a single nucleic acid molecule and a coat of protein molecules of a single type. Others are complex particles with coats containing more than 50 different kinds of proteins. A few viruses infecting animal cells, such as the influenza and herpes viruses, are surrounded by an outer membrane derived from the plasma membranes of their host cells. (Information Box 4-2 surveys some important viral diseases of humans.)

The protein coat enclosing the nucleic acid core, depending on the virus, may be either rodlike, spherical, or lollipop-shaped. The coat of the much-studied tobacco mosaic virus (Figure 4-21), for example, is a rod-shaped structure built up from more than 2000 identical protein units. The RNA molecule of this virus winds into a spiral extending through the axis of the rod. Many of the viruses infecting bacteria, called **bacteriophages,** are lollipop-shaped and have a polyhedral *head* enclosing a DNA core and a *tail* containing several different proteins (Figure 4-22). The tail is complex in structure and consists of a *collar*, at the point of attachment to the head, a cylindrical *sheath*, and

a *baseplate*. The baseplate carries six long, hairlike extensions called the *tail fibers*.

Viruses reproduce by directing the synthetic machinery of infected cells to make virus particles.

The life cycle of a lollipop-shaped bacteriophage illustrates the general pattern by which virus particles infect their host cells (Figure 4-23). Free bacteriophage particles come into contact with bacterial cells by random collisions. If a virus particle collides with a bacterial cell, the tail fibers

---

Information Box 4-2

Viruses and Human Disease

Many human afflictions, ranging from mildly irritating to rapidly lethal, are caused by viruses. In some, the pattern of infection closely resembles the attacks on bacteria by bacteriophage particles. The cell is invaded by a virus particle, immediately begins to manufacture viral nucleic acid and protein, dies, and ruptures to release infective virus particles that infect healthy cells. Polio, influenza, and the common cold viruses follow this course of infection. Some diseases caused by such viruses are listed here.

Other viruses may remain in their human host cells for long periods of time without causing apparent damage. Among the best examples of this viral type are the herpes simplex viruses of humans. One herpes type, called herpes simplex I, is essentially universally distributed among humans. It causes cold sores around the mouth and nose. Normally, the virus inhabits nerve cells, where it is replicated in small quantity and passed from parent to daughter cells without ill effects. However, disturbances such as fever, sunburn, or exposure to extreme cold may cause the virus to enter a virulent, infective stage, in which it multiplies rapidly and destroys the host and surrounding cells, causing the familiar cold sores. A closely related virus, herpes simplex II, causes similar lesions on the genitals of both men and women. Herpes simplex II has also been implicated as a possible cause of cancers of the reproductive organs.

Other viruses that follow a slow pattern of infection similar to herpes are causes of human disease. Measles and chicken pox, caused by viruses that normally follow a fast, acute pattern of infection similar to polio and influ-

Disease	Severity
Chicken pox	Usually not serious
Common cold	Usually not serious
Encephalitis	10 to 50% of cases fatal
Hepatitis	Rarely fatal but debilitating, lasts several months
Influenza	Usually not serious unless other complications develop
Measles	Usually not serious
Mononucleosis	Usually not serious
Mumps	Usually not serious
Pneumonia, viral	Not usually fatal but recovery slow
Polio	1 to 4% fatal, can cause permanent disability
Rabies	Nearly always fatal
Smallpox	Often fatal
Yellow fever	Usually fatal

enza, may persist in some persons as apparently benign parasites after an apparent cure. In the infected cells, the virus particles are replicated in small numbers and passed bind to sites on the bacterial cell wall (Figure 4-23a). The head and tail sheath then contract and inject the DNA core into the cell (Figure 4-23b). The proteins of the virus remain outside.

Once inside the bacterial cell, the bacteriophage DNA immediately takes over the synthetic machinery of the infected cell and directs it to make from hundreds to thousands of new copies of the viral DNA and protein (Figure 4-23c and d). As the head and tail segments accumulate in the bacterial cytoplasm, the newly synthesized bacteriophage DNA condenses into the heads (Figure 4-23e). After synthesis of the virus particles is complete, a

final viral protein causes breakdown of the bacterial cell wall (Figure 4-23f). Rupture of the cell wall releases the newly completed bacteriophage particles to the surrounding medium, where chance collisions with additional bacterial cells may lead to another cycle of infection and release of virus particles.

Viruses are best classified as nonliving matter when they are outside their host cells. In this form, they carry out none of the activities of life and are inert except for the capacity to attach to their host cells. They can be purified and crystallized in this form and can be stored indefinitely without change or damage. The viral nucleic acid molecule

on in cell division without apparent damage to the host. Later the disease may break out again in a different form. For example, the chicken pox virus normally follows a cycle of acute infection, with almost immediate death and rupture of host cells. Within a few weeks, the infected person develops an immunity to the virus, which eliminates the virus particles and the disease. In some persons, however, the virus persists in small numbers in various cells of the body for the life of the individual. In these persons, the virus alternates between cycles of benign infection during which it is replicated in small numbers and passed on in cell division without cell rupture and death, and periods of acute outbreak in which it causes painful lesions called shingles along sensory nerves.

Slow viruses have been implicated recently as possible causes of progressive human diseases such as multiple sclerosis, rheumatoid arthritis, and some forms of diabetes. If current ideas about the viral origins of these diseases are correct, the disabilities caused are not so much an effect of the viruses as of the body's responses to their presence. Although slow viruses do not necessarily impair host cell function, they apparently always make their presence known by causing viral protein markers to be incorporated into the surface membranes of the infected cell. The body's immune system responds to the presence of these foreign proteins by producing antibodies that attack them and in the process impair or destroy the function of the cells carrying them. Over long periods of time, according to the slow virus hypothesis, enough cells are impaired to cause the disease to appear and become serious.

A number of viruses have been shown to cause cancerous growths in animals such as mice, chickens, frogs, and monkeys. Although no human cancers have ever been

definitely proved to be caused by a virus, the indications are good that some—such as breast cancer, leukemia, cervical cancer, and Burkitt's lymphoma (a cancerous growth of the lower jaw)—may have viral origins.

In cancer, a cell or group of cells enters a process of rapid, uncontrolled growth and division producing cell masses called **tumors**, which interfere with normal body functions. Proof that a virus causes cancer would be that a virus isolated from a cancerous growth consistently causes the same type of cancer if injected into other persons or test animals. Although this procedure has been successful with cancers of other animals, such as mammary cancer in mice and leukemia in mice, chickens, and cats, equivalent experiments with human cancers have not been possible.

There are several reasons why proving that viruses can cause cancer in humans has been difficult. One is that tumors supposedly caused by a virus frequently show no evidence of virus particles. In addition, if human cancers are actually caused by virus, they do not appear to be contagious as are the viral cancers of other animals or such viral diseases as polio and influenza. Finally, establishing a direct cause-and-effect relationship between a virus and cancer is complicated by the fact that so many environmental conditions can modify or prevent growth of a cancerous tumor. For example, small alterations in the diet of mice can prevent the appearance of leukemia even if the mice are directly infected with leukemia virus. In spite of these difficulties, most investigators believe that the human cancers mentioned above will eventually be shown to be caused by viruses.

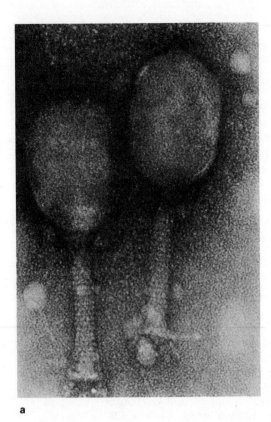

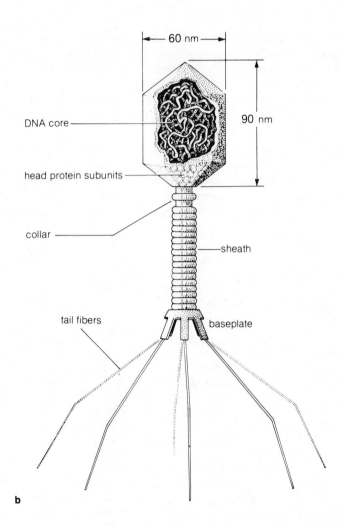

Figure 4-22 (a) "T-even" bacteriophages infecting *E. coli* cells. ×480,000. Courtesy of the Perkin-Elmer Corporation. (b) The structure of a T-even bacteriophage.

Labels in figure (b): 60 nm, 90 nm, DNA core, head protein subunits, collar, sheath, tail fibers, baseplate

carries only the information required to direct the host cell machinery to make more viral particles, and it is active in

---

Viruses are functional only when inside a host cell.

---

this function only when inside a host cell. Thus a virus particle probably represents nothing more or less than a fragment of a nucleoid or chromosome derived from a once-living cell, protected by a coat of protein, and reduced to a set of coded directions for making additional particles of the same kind.

Questions

1. Who was the first person to observe cells? Did he observe living cells?

2. Why did the study of cells progress so slowly from the time of their initial discovery until the mid-1800s?

3. The cell theory, as outlined by Schleiden and Schwann and Virchow, defines the structural and functional organization of living organisms and tells how they originate. What are the three parts of the cell theory?

4. What are the functions of plasma membranes?

5. What is a lipid bilayer?

6. Outline the structure of membranes according to the fluid mosaic model. What is the "fluid" part of the fluid mosaic model? What is the "mosaic" part?

7. What experiment supports the idea that membranes are fluid?

8. What experiment supports the idea that proteins are suspended in membranes as individual units?

9. What kinds of molecular forces hold membranes together?

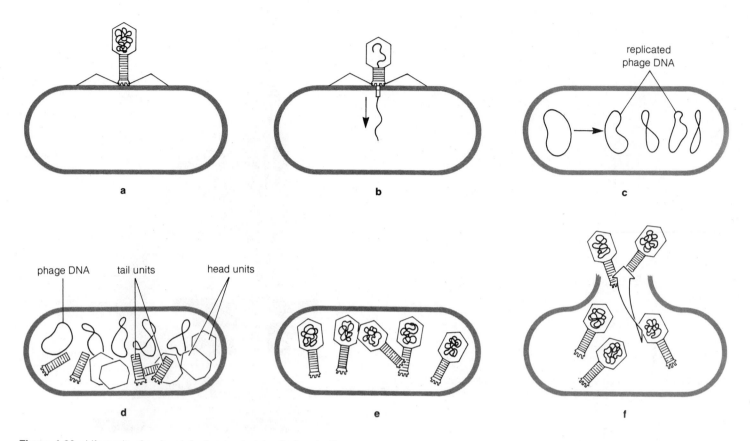

Figure 4-23 Life cycle of a virus infecting bacterial cells (see text).

10. How are membranes organized in prokaryotic cells? In eukaryotic cells? What do the terms *prokaryote* and *eukaryote* mean?

11. How is the nuclear material organized in prokaryotic cells? What structures are found in prokaryotic cytoplasm?

12. What are the differences between prokaryotic and eukaryotic flagella?

13. What structures occur in the nucleus of a eukaryotic cell?

14. How does the organization of the nuclear material differ in prokaryotic and eukaryotic cells?

15. Define chromatin, nucleolus, nuclear envelope, nuclear pores, ribosomes, mitochondria, chloroplasts, smooth and rough ER, microtubules, microfilaments, and intermediate filaments.

16. What is meant by the "9 + 2" system of microtubules?

17. What structures occur in plant cells but not in animal cells?

18. What types of plastids occur in plant cells? What are their functions?

19. How do vacuoles and cell walls function in plant cells?

20. What molecular elements occur in plant cell walls?

21. What are the basic structural elements of viruses?

22. How do viruses infect cells?

23. Are viruses living or nonliving when they are unassociated with cells?

Suggestions for Further Reading

Avers, C. J. 1978. *Basic Cell Biology.* Van Nostrand, New York.

Dyson, R. D. 1974. *Cell Biology.* Allyn & Bacon, Boston.

Karp, G. 1979. *Cell Biology.* McGraw-Hill, New York.

Wolfe, S. L. 1981. *Biology of the Cell.* 2nd ed. Wadsworth, Belmont, California.

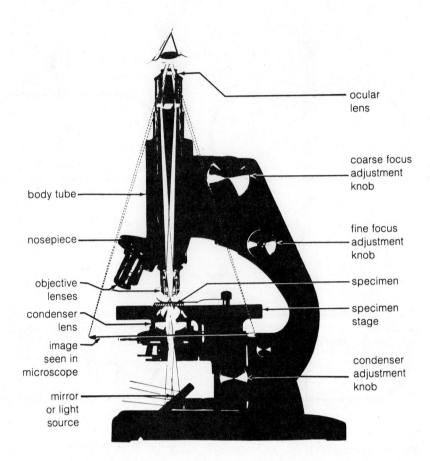

ocular
lens

coarse focus
adjustment
knob

body tube

fine focus
adjustment
knob

nosepiece

specimen

objective
lenses

specimen
stage

condenser
lens

image
seen in
microscope

condenser
adjustment
knob

mirror
or light
source

Figure 4-24 Construction of a light microscope. Objective lenses of different magnifying power can be selected by rotating the nosepiece. Light is focused on the specimen by the condenser lens, located just below the stage. Courtesy of the American Optical Company.

SUPPLEMENT 4-1: MICROSCOPY

Both light and electron microscopes are used extensively in biological research today. Prokaryotic cells and nucleoids and the larger organelles of eukaryotic cells are readily visible in the light microscope; objects as small as single protein molecules can be seen in the electron microscope. Many of the techniques used for preparing specimens for viewing in either instrument allow individual molecular types to be located within these cell structures.

The ability of microscopes to make small details visible depends on the wavelength of the illumination used to view the object. In general, the smaller the wavelength, the smaller the details that can be seen. The very best light microscopes, using the shortest light waves in the blue range (wavelength about 450 nanometers), can reveal objects in the specimen as small as 0.2 micrometer, or about 200 nanometers. At this level, cell organelles as small as mitochondria are just visible; smaller structures such as ribosomes or the chromatin fibers in the nucleus are invisible. Electron microscopes use electron beams focused by magnetic lenses as an illumination source. The wavelength of the electron beams routinely used in electron microscopes is as small as 0.0037 nanometer; at this very short wavelength, specimen details with dimensions as small as 0.5 nanometer can be seen. At this level, the smallest cell structures, such as ribosomes, membranes, and protein molecules, are readily visible. Under special operating conditions, large metallic atoms such as the uranium atom have actually been photographed in the electron microscope.

Light Microscopy

In the light microscope (Figure 4-24), light rays from the illumination source are focused on the specimen by a *condenser* lens. The rays leaving the specimen are focused into a magnified image of the specimen by two lenses placed

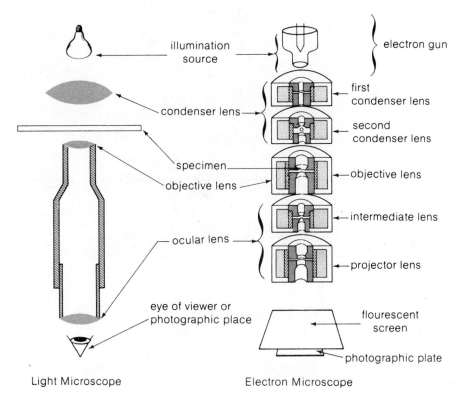

Figure 4-25 The arrangement of the illumination source and magnetic lenses in an electron microscope (right) resembles an inverted light microscope (left).

Light Microscope

Electron Microscope

at either end of a tube. The lens nearest the specimen is termed the *objective* lens. The lens at the opposite end of the tube is called the *ocular* lens. Each of these lenses, in order to correct for faults in the image, is actually constructed from a series of lenses placed close together— usually as many as eight to ten for the objective lens and two or three for the ocular lens. Since the individual elements in these lenses are placed close together, their net effect is to act as a single, highly corrected lens.

The image is observed by looking directly into the ocular lens. Coarse and fine controls are provided for movement of the specimen stage or lens tube to place the specimen in the correct position for focusing by the objective lens. The condenser lens can also be adjusted so that the light from this lens converges on the specimen and spreads after leaving the specimen into an inverted cone of light that completely fills the objective lens.

Although living specimens can be observed directly in the light microscope, the best viewing conditions are frequently obtained with material that has been stabilized by

chemical fixation, cut into sections, and stained by dyes that produce contrasting colors in cell components. The chemical fixatives used in light microscopy are substances that precipitate or coagulate cell structures or introduce chemical cross-links that anchor specimen molecules and structures in place. In either case, the fixatives give a more or less faithful preservation of the physical arrangement of structures within the cell.

Following fixation, specimens are embedded in a supportive material, usually following removal of water by exposing the tissue to successively higher concentrations of alcohol or acetone. For light microscopy, the embedding material of preference is often paraffin wax. The dehydrated tissues are impregnated with paraffin at temperatures above the melting point of the wax. After impregnation, the preparation is cooled, producing a block that is cut into sections between 5 and 10 micrometers thick. The sections are then placed on a glass slide and the paraffin replaced with a transparent material, usually a resin that dries to a hard, glassy film.

Cells may be stained at any point in the process from living material to finished sections. A wide variety of organic dyes are employed as stains in light microscopy. These can be used in procedures that differentially color structures that contain molecules such as RNA, DNA, and various proteins and polysaccharides.

Electron Microscopy

The Electron Microscope An electron microscope (Figures 4-25 and 4-26) is constructed much like an inverted light microscope. Electrons traveling at high velocities are released from a "gun" at the top of the microscope. These electrons are focused on the specimen by two magnetic lenses, called *condenser* lenses, just below the gun. The beam of electrons leaving the specimen is focused into an image of the specimen by a series of three magnetic lenses: the *objective, intermediate,* and *projector* lenses. These lenses are arranged so that the specimen image is projected on a fluorescent screen at the bottom of the microscope. The fluorescent screen, which resembles a television screen, converts the electron beam to a pattern of light that can be seen as a visible image by an observer. Permanent records of the image can be made by exposing photographic plates at the level of the screen. Ordinary films and plates can be used for this purpose, because the response of photographic emulsions to electrons and light is essentially the same.

Specimen Preparation Electrons have very low penetrating power and are easily scattered by molecules of gas in the microscope or absorbed by atoms in the specimen. Therefore the entire microscope column must be maintained at high vacuum, and the specimen must be dried and made as thin as possible—much thinner than even the smallest cells. These requirements for dehydration and thinness have frustrated attempts to view living material in the electron microscope. Instead, specimens for electron microscopy are prepared by one of several methods: sectioning, shadowing, negative staining, or freeze-fracture and etching.

Sectioning Producing sections for electron microscopy involves several steps. As in tissue preparation for sectioning in light microscopy, the tissues to be observed are stabilized by placing them in chemical fixatives. The tissue is then impregnated with plastic, hardened, and sliced into sections as thin as 40 to 60 nanometers with diamond or glass knives.

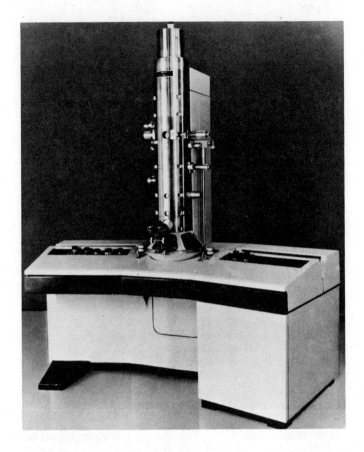

Figure 4-26 A high-performance electron microscope. The electron gun is at the top of the microscope; the lenses are stacked vertically and housed in the central column, which is maintained under a high vacuum. Controls for focusing the lenses are located on the panels to the right and left of the column. The image, formed on a fluorescent screen at the base of the column, can be viewed through the windows with the binocular microscope. Photographic plates for permanent records of the image can be exposed at the level of the screen. Courtesy of the Perkin-Elmer Corporation.

Sections of unstained cells have little contrast when viewed under the electron microscope because the atoms occurring most commonly in tissues—carbon, hydrogen, oxygen, and nitrogen—all scatter or deflect electrons to about the same extent. To improve the contrast, salts of various heavy metals are added to the tissue as stains. (Substances with high atomic numbers, such as the heavy metals, deflect electrons to the greatest extent.) In electron micrographs of stained sections (as in Figure 4-8) the dense areas are regions in which many atoms of the heavy-metal stain have accumulated. These regions appear dark in the micrograph because the electrons passing

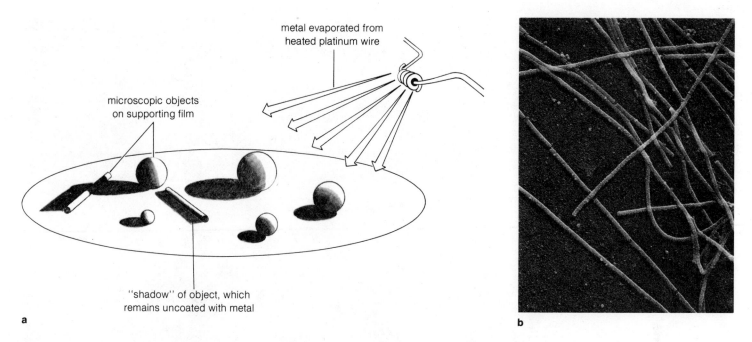

metal evaporated from
heated platinum wire

microscopic objects
on supporting film

"shadow" of object, which
remains uncoated with metal

a

b

Figure 4-27 (**a**) Shadowing of particles for electron microscopy by a metal evaporated from one side (see text). (**b**) The appearance of shadowed microtubules in an electron microscope. ×31,500. Photograph by the author.

through the heavy-metal deposits are either absorbed or deflected so strongly that they are lost from the electron beam and are not focused into the image by the electron lenses.

Shadowing and Negative Staining Cell parts and molecules are frequently isolated and examined in this form rather than in sections. The contrast of the isolated material, dried on thin plastic support films, is usually increased by one of two techniques, *shadowing* or *negative staining*, for viewing in the electron microscope.

In the shadowing technique, isolated cell organelles or molecules dried on the plastic supporting film are coated under vacuum by a heavy metal such as platinum evaporated from a source located on one side (Figure 4-27). Atoms of the metal, evaporated by electrically heating a small quantity of the metal to the boiling point, travel in straight lines and deposit on raised surfaces of the specimen facing the source (Figure 4-27a). The raised surfaces, because of their coating of heavy-metal atoms, have high contrast in the electron microscope. Depressions in the specimen, located in the shadow of higher points, are not coated by heavy-metal atoms and appear transparent in the electron microscope. Specimens prepared by shadowing therefore look as if a strong light is directed toward

the specimen from one side, placing surface depressions in deep shadow (Figure 4-27b).

In negative staining, a heavy-metal stain is allowed to dry around the surfaces of isolated cell particles or molecules. The stain molecules deposit into surface crevices in the specimen during the drying process, often outlining details with remarkable clarity. The technique typically produces a "ghost" image in which the specimen appears light against a dark background (as in Figure 4-22a).

Freeze-Fracture Preparations The freeze-fracture technique employs cells that have been frozen by plunging them into liquid nitrogen. At the temperature of liquid nitrogen (−196°C), the cells freeze so rapidly that the molecules of membranes and other cell structures are immobilized in the positions they occupy in the living state. The method induces so little change in cell structure that living cells frozen in liquid nitrogen after treatment with glycerol (to reduce the size of ice crystals formed inside the cells) can be stored for long periods and later thawed and brought back to life without apparent damage. Tissue culture cells and spermatozoa are routinely frozen and stored for later use in this way.

Once frozen, the specimen is placed under vacuum and fractured by a knife edge. The fracture travels through

Figure 4-28 A freeze-fracture preparation of a tobacco root-tip cell. The nuclear envelope surface *(NE)* has been exposed by the fracture; pores (arrows) are visible in the nuclear envelope; C, cytoplasm. ×34,700. Courtesy of J. H. Andrews.

the specimen and exposes membranes and other internal surfaces. The specimen is then frequently "etched" briefly by allowing water to dry from the fractured surface (which is still frozen and under vacuum). The surface is then shadowed by evaporating a heavy metal from a point source located to one side just as in the shadowing technique. Figure 4-28 shows a cell prepared for electron microscopy by this technique.

The Scanning Electron Microscope The scanning electron microscope (SEM) is useful for viewing the surfaces of cells and small organisms such as mites and insects. Although the SEM borrows some basic operating principles and lens systems from the standard electron microscope

shown in Figure 4-26, its theory of operation is radically different. In the SEM (Figure 4-29) an electron gun produces an electron beam. A magnetic lens system analogous to the condenser lens of a standard electron microscope focuses the beam into an intense spot on the specimen surface. Beam deflectors (charged plates that attract or repel the electron beam) placed below the condenser lenses *scan* the specimen by moving the focused spot back and forth rapidly over the specimen surface. The energy of the intense spot of electrons is absorbed by specimen molecules, which emit high-energy electrons of their own in response. These emitted electrons are collected by a detector located to one side of the specimen and converted electronically into an image of the specimen surface that is

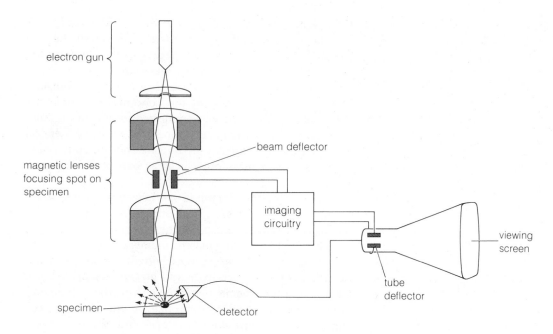

Figure 4-29 Construction of a scanning electron microscope (see text).

viewed on a television screen. (Figure 17-6*d* shows a protozoan photographed by this technique.)

Because the electrons emitted from the specimen surface in response to the scanning beam are easily scattered by gas molecules, the SEM column housing the lens system, specimen, and detector must be kept at high vacuum. This operating restriction requires that the specimen must be dry or, alternatively, must release gas or water molecules so slowly that the vacuum is not significantly disturbed. To meet this requirement, most biological objects, such as cells and tissues, must be fixed and dehydrated before viewing as in the standard electron microscope and cannot be examined in the living state. A few organisms have been placed in the SEM alive and have survived the vacuum and bombardment by the scanning beam without extensive damage.

The SEM produces excellent images of specimen surfaces ranging in size from whole cells up to small insects. Since its ability to make small objects visible is about 20 times better than the light microscope, the surfaces of objects in this size range are imaged with significantly greater fidelity. The SEM, however, is less useful for observing details inside cells than the standard electron microscope.

5

THE CELL SURFACE IN TRANSPORT, RECEPTION, AND RECOGNITION

Cells make contact with the outside world through the plasma membrane. Through this membrane all the fuel substances and raw materials necessary for life enter the cells from outside. Waste materials and cell secretions travel in the opposite direction, from inside to outside. These movements maintain the concentrations of molecules inside cells at the levels required for life and cellular functions. The plasma membrane also receives chemical and physical stimuli from the outside and, in many-celled organisms, recognizes other cells as part of the same organism or foreign.

The plasma membrane is specialized in various ways to carry out these tasks. Transport of substances in and out of cells reflects the properties of both the lipid and protein molecules of the plasma membrane. Cell reception and recognition are based on carbohydrate-containing lipids and proteins (glycolipids and glycoproteins) that also form part of the plasma membrane. The carbohydrate portions of these molecules extend like antennas from the cell surface, in effect providing the eyes and ears of the cell. The integrated functions of the plasma membrane supply cells with raw materials, excrete waste materials, and provide communication with the outside environment.

MOVEMENT OF SUBSTANCES THROUGH THE PLASMA MEMBRANE

Life depends on the organization of molecules inside cells. Any severe disturbance in the concentrations of substances inside the cell, or of the kinds present, will impair cell function or even cause cells to die. The internal concentrations of molecules and ions are maintained in their numbers and kinds by the plasma membrane, which regulates the passage of all substances in and out of cells.

The internal concentrations of molecules necessary for cellular life are maintained by the plasma membrane.

How does the plasma membrane accomplish this function?

There are two basic mechanisms by which materials are transported across plasma membranes. One, called

---

Movement of substances across membranes in response to concentration gradients is called passive transport.

---

passive transport, simply reflects the differences in concentration of substances inside and outside cells. If the molecules are more highly concentrated outside, the direction of movement will be from outside to inside. If the concentration is higher inside, movement will be from inside to outside. The rate of movement increases as the difference between concentrations outside and inside the cell increases. These differences in concentration inside and outside are called **concentration gradients.** Passive transport in response to simple concentration gradients requires no expenditure of cellular energy.

In the second mechanism transporting materials across cell membranes, **active transport,** substances are moved inside or outside *against* concentration gradients. Active

---

Active transport occurs against concentration gradients and requires an expenditure of cellular energy.

---

transport requires cells to expend energy since substances are moved against concentration gradients; it stops if the energy-producing mechanisms of cells are experimentally inhibited.

Not all substances enter or leave cells by penetrating directly across plasma membranes. Many eukaryotic cells, in addition, are able to take up molecules by enclosing them in pockets that form in the plasma membrane. Uptake by this route, called **endocytosis,** involves active infolding of segments of the plasma membrane. The pockets formed in this way subsequently pinch off as closed, membrane-bounded sacs that sink into the cytoplasm beneath the plasma membrane. Cells can also release molecules to the exterior by reversing the mechanisms of endocytosis. Release of materials in this manner, called **exocytosis,** forms the basis for secretion of proteins and other large molecules to the outside of the cell.

Passive Transport

The passive transport of substances along concentration gradients depends on the fact that the molecules in any space held at temperatures above absolute zero

---

Passive transport depends on the fact that molecules are in constant motion at temperatures above absolute zero.

($-273°C$) are in constant motion. The molecules travel in straight lines until they collide with other molecules in the space or with molecules forming the boundaries of the space. Each molecule has a specific energy that depends on the velocity of its movement and the direction and force of its collisions with other molecules. This energy of movement is the **kinetic energy** of the collection of molecules (see also p. 36).

An overall transfer of molecules from one region to another may result from this movement if the two regions have different concentrations of molecules—that is, different average numbers of molecules per unit volume. Imagine two collections of molecules in adjacent spaces of equal volume that are initially separated by a barrier that the molecules cannot pass. The absolute temperature of the spaces is the same, but one of them contains more molecules. As a result, there is a greater amount of kinetic energy in the space containing the greater concentration of molecules because there are more moving particles with mass and energy in this space.

Suppose now that the barrier between the two compartments is removed. In the region of the boundary between the two collections of molecules, the movement and collisions on the more concentrated side will propel molecules into the less concentrated side. Molecules will also move from the less concentrated side into the more concentrated region, but over any time interval there will be more collisions and movement from the more concentrated side. As a result, there will be a *net* movement of molecules from the side of greater to lesser concentration. This net movement in response to concentration differences, called **diffusion,** will continue until the molecules are evenly distributed throughout the available space.

Net movement of molecules from regions of greater to lesser concentration is called diffusion.

Diffusion shows how the laws of thermodynamics apply to all systems undergoing change, whether inside cells or in the world at large. According to the second law of thermodynamics (p. 37), systems run spontaneously toward a condition of greater disorder (or entropy). As they run to a less ordered state, they release energy to their surroundings. This energy is free energy, and it can accomplish work.

The system undergoing change in the example described above is the two collections of molecules, one more concentrated than the other. This arrangement is more ordered than one in which the molecules are equally distributed throughout all the available space (as they are after diffusion has taken place).[1] Once the barrier is removed,

---

[1]Why this is so can be understood by considering a somewhat improbable example. Suppose that you are assigned the task of stacking bricks in a pile during an earthquake. Stacking the bricks, and keeping them stacked, requires a considerable and continuous effort on your part. Any time you stop returning bricks to the pile, they cascade downward until they assume a layer of more or less even thickness on the ground. In falling from the pile to the ground, the bricks

the total system will run spontaneously toward the less ordered state in which the molecules are equally distributed.

Semipermeable Membranes and Diffusion A membrane placed between two regions containing collections of molecules at different concentrations will have no effect on the final outcome of diffusion if all the molecules can pass through it with equal ease. The net movement may be altered, however, sometimes in unexpected ways, if some molecules can pass through the membrane more readily than others or are excluded entirely. Membranes having this effect are said to be **semipermeable.** All biological membranes have this property and thus alter the net movement of molecules in response to concentration gradients.

Osmosis One of the most surprising effects of semipermeable membranes on molecular movement can be demonstrated by a simple apparatus—a deep dish separated into two halves by a thin layer of cellophane (Figure 5-1). On one side of the cellophane barrier the dish contains pure water. On the other side it contains a glucose solution. The water can pass freely in either direction, but the glucose molecules cannot penetrate through the cellophane and are retained on one side. Thus the cellophane acts as a semipermeable membrane. Within an hour enough water will move into the glucose solution to raise the level of the solution noticeably on the glucose side; that is, there will be a *net* movement of water from the pure water side into the glucose solution. The total system accomplishes *work* in raising the level of the glucose solution against the force of gravity. How is this work done?

The free energy accomplishing this work is released by the water molecules, which follow a concentration gradient in moving toward the side of the dish containing the glucose solution. Although it is clear that a concentration gradient exists for the glucose molecules, which are obviously more concentrated on one side of the dish than the other, the gradient for water molecules is not at first so apparent. On the side of the membrane containing the glucose solution, however, much of the space available for water molecules is occupied by the dissolved glucose molecules. As a result, there are fewer water molecules per unit volume on the side containing the glucose than on the pure water side. Therefore a concentration gradient between the two sides exists for the water. As a result, water will move across the membrane in response to the gradient.

Movement of water in response to a gradient of this type is so important in biology that it is given a special name: **osmosis.** Osmotic movement of water in living cells

---

A net flow of water molecules through a semipermeable membrane in response to gradients of water concentration is called osmosis.

---

involves the same conditions described in the apparatus described above: (1) a semipermeable membrane must completely separate two water solutions or a solution of molecules on one side and pure water on the other; (2) some of the nonwater molecules on one side must be unable to pass through the membrane; (3) the concentration of dissolved substances on the two sides of the membrane must be unequal, so that there are more molecules of water per unit volume on one side of the membrane than the other. Water will then flow osmotically toward the side of the membrane containing fewer water molecules per unit volume. As the water molecules run down the gradient toward a condition of greater disorder or entropy, the system releases free energy that can accomplish work.

It is easy to show that osmotic flow of water can accomplish work. In the apparatus shown in Figure 5-2, distilled water surrounds a tube with an opening at its base. A sheet of cellophane is stretched across the opening and sealed tightly to prevent leakage around its sides. The tube contains a solution of glucose in water. After a short time, the level of the solution in the tube will rise as water molecules move from the distilled water into the tube. The level of the solution will continue to rise until the pressure created by the weight of the raised solution in the tube exactly balances the tendency of water molecules to move from outside to inside. At this point, the system is in balance or equilibrium. Although water molecules will still move in both directions across the cellophane membrane, no further *net* movement of water will occur. The pressure

---

release the energy you expended in piling them. If harnessed in some way, the energy released as the bricks fall is free energy and could be used to do useful work. This situation is paralleled by molecules distributed unevenly on two sides of a barrier. The constant motion of molecules at all temperatures above absolute zero corresponds to the earthquake. The regions of greater concentration represent higher piles of molecular bricks on one side of the barrier. The piles tend

constantly to fall until the molecules are evenly distributed throughout the available space on both sides. The amount of energy available for this movement depends on the height of the pile of bricks—in other words, on the magnitude of the concentration difference on the two sides of the barrier. Whether the molecules can run to an even distribution on both sides in response to the concentration difference depends simply on whether they can get through the barrier.

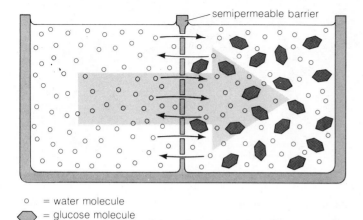

semipermeable barrier

○ = water molecule
⬡ = glucose molecule

Figure 5-1 Osmotic flow of water in a system in which a semipermeable barrier separates two compartments. The compartment on the left contains pure water. The compartment on the right contains a solution of glucose molecules in water. Although the water molecules can pass freely across the barrier, it is impermeable to the glucose molecules. A net movement of water molecules will occur from left to right in response to the gradient in concentration of water molecules, which is set up by the presence of the glucose molecules in the right compartment (see text).

required to counterbalance the tendency of water molecules to move into the tube is a measure of the **osmotic pressure** of the solution in the tube.

The force required to counterbalance the osmotic flow of water molecules is called osmotic pressure.

Osmosis in Living Cells Cells act as osmotic devices similar to the apparatus shown in Figure 5-2 because they contain solutions of proteins and other molecules that are retained inside by a membrane impermeable to these molecules but freely permeable to water. The resulting movement of water and the development of osmotic pressure is a force that operates constantly in living cells. This force may be utilized to accomplish part of the activities of life, or it may act as a disturbance that must be counteracted for survival of the cell. The root cells of most land plants, for example, contain proteins and other large molecules in solution but are surrounded by almost pure water. As a result, net movement of water due to osmosis occurs into the root cells; the pressure developed contributes part of the force required to raise water into the stems and leaves of plants. On the other hand, protozoa and other small cells or organisms living in fresh water, to keep from bursting, must

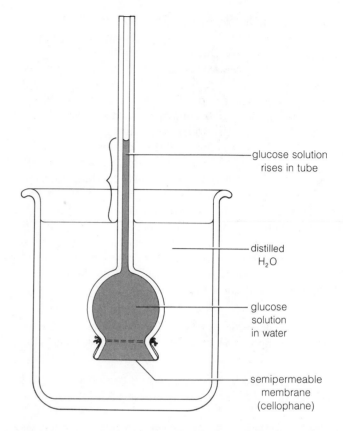

glucose solution rises in tube

distilled H_2O

glucose solution in water

semipermeable membrane (cellophane)

Figure 5-2 Apparatus demonstrating the ability of osmotic flow to accomplish work. Water will flow osmotically from the surroundings into the tube, raising the level of the solution in the tube. Flow continues until the weight of the column of water (bracket) develops sufficient pressure in the solution to counterbalance the tendency of water molecules to flow inward.

expend energy to excrete the water constantly entering by osmosis. In some organisms, among them bacteria, blue-green algae, and plants, thick cell walls (see Figures 4-5, 4-6, and 4-17) keep the cells from bursting from osmotic pressure. In such cells, the pressure developed by osmosis keeps the cell contents pressed tightly against the restraining wall. In plants, this pressure supports the softer tissues of stems and leaves against the force of gravity. Cells living in surroundings containing highly concentrated salt solutions, as in the ocean, have opposite problems and must constantly expend energy to replace the water lost to the outside through osmosis.

The Effects of Membrane Lipids and Proteins on Passive Transport The cellophane film used as a semipermeable membrane in an osmosis apparatus acts essentially as a uniform molecular sieve because it has no chemical or physical effects beyond restricting passage of molecules beyond a certain size. The lipids and proteins of biological membranes, however, arranged as they are in a mosaic of subregions with distinct hydrophobic, hydrophilic, and chemical properties (see p. 51), produce a semipermeable barrier that modifies the behavior of many molecules as they diffuse in response to concentration gradients.

The effects of membrane lipids and proteins on diffusion were first given detailed study at the turn of the century by a Swiss investigator, Ernst Overton, who followed the behavior of substances as they penetrated into plant and animal cells. Overton observed that the major difference between diffusion across biological membranes and artificial barriers such as cellophane is that lipid solubility modifies the rate of penetration of many substances into cells. Generally, the more soluble a substance in lipids, the more rapidly it penetrates into living cells, up to

---

Hydrophobic molecules penetrate through the phospholipid bilayers of cell membranes more readily than hydrophilic molecules.

---

a limit determined by molecular size. Overton's work provided the first clue that cells are surrounded by a lipidlike surface layer.

Most nonlipid substances, in contrast, do not pass through membranes very rapidly if at all. There are certain exceptions to this general rule, however, all of them important to the molecular economy of cells. Water passes through membranes rapidly even though it is a strongly polar substance. Other hydrophilic substances necessary for cellular life—including glucose, various amino acids, and ions such as Na^+ and Ca^{2+}—also pass through membranes rapidly even though they are polar or charged.

Recent research has demonstrated that the rapid penetration of these hydrophilic substances is due to the protein part of membranes. These transport proteins, suspended as individual particles or groups within the

---

Proteins enable hydrophilic substances to penetrate across cell membranes.

---

lipid bilayer framework of cellular membranes (see Figure 4-3), evidently form hydrophilic channels that extend through the membrane (Figure 5-3). The hydrophilic nature of the channels is determined by the amino acid chains of the membrane proteins, which fold so that the channels are lined by polar or charged groups. The channels may be formed by an opening that extends through a single protein (Figure 5-3a) or by several proteins that

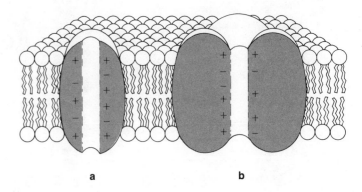

Figure 5-3 Formation of hydrophilic channels through the interior of single membrane proteins (**a**) or by alignment of several membrane proteins (**b**).

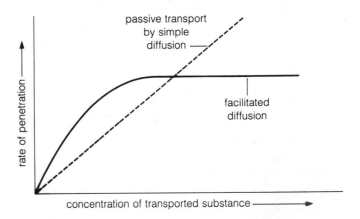

Figure 5-4 Passive transport by simple diffusion increases with concentration in an essentially linear fashion. Facilitated diffusion, in contrast, drops off at higher concentrations and reaches saturation at a maximum rate that does not respond to further increases in concentration.

become aligned to form a hydrophilic channel between them (Figure 5-3b).

Although the passive movement of certain polar substances through membranes is rapid, their passage still depends on diffusion: the energy required for this transport is provided by favorable concentration gradients and does not require an expenditure of cellular energy. Excep-

---

Passive movement of hydrophilic substances across membranes at higher rates than predicted by lipid solubility is called facilitated diffusion.

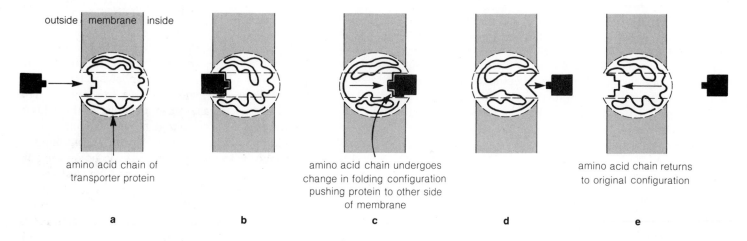

outside | membrane | inside

amino acid chain of transporter protein

amino acid chain undergoes change in folding configuration pushing protein to other side of membrane

amino acid chain returns to original configuration

a   b   c   d   e

Figure 5-5 Facilitated diffusion by a channel-forming carrier molecule (see text).

tional transport of hydrophilic substances of this type—passive transport that follows concentration gradients but proceeds at rates significantly higher than predictions based on lipid solubility—is called **facilitated diffusion.**

Measurement of the rate of penetration of substances transported by facilitated diffusion reveals a fundamental characteristic common to all. At successively higher concentrations of the transported molecule, the degree of enhancement drops off until, at some point, further increases cause no further rise in the rate of penetration (Figure 5-4). This pattern is in sharp contrast to the behavior of nonpolar molecules that penetrate according to their lipid solubility. For these molecules, permeability increases regularly with concentration; there is no dropoff at higher levels.

The dropoff noted for facilitated diffusion at high concentrations closely resembles the behavior of enzymes in catalyzing biochemical reactions (see p. 43 and Figure 3-7). As the concentration of reactant molecules increases, enzymes gradually become "saturated" and the rate of the reaction levels off. Another characteristic shared between the membrane proteins involved in facilitated diffusion and enzymes is **specificity:** only certain molecules, or groups of closely related molecules, are speeded in reaction rate by enzymes. Similarly, only certain molecules or molecule groups are speeded in transport by the membrane carrier molecules. These similarities suggest that the membrane proteins carrying out facilitated diffusion share many properties with enzymes.

These similarities to enzymes have suggested how the

membrane proteins facilitating the diffusion of polar substances might work (Figure 5-5). The membrane proteins are considered to be fixed in position, held in place by polar and nonpolar surface groups that anchor them in the lipid bilayer. The polar channel extending through the protein has an "active site" directed toward the side of the membrane facing the substance to be transported (Figure 5-5a). This site, like the active site of an enzyme, is tailored to fit the transported molecule. Molecular collisions result in binding of the transported substance to the active site (Figure 5-5b). This combination induces a change in the

---

Facilitated diffusion depends on channel-forming membrane proteins.

---

folding pattern of this limited portion of the protein molecule, resulting in movement of the active site through the polar channel to the other side of the membrane (Figure 5-5c). Change to this folding pattern alters the active site and reduces its affinity for the transported substance. As a consequence, the transported molecule is released on the other side (Figure 5-5d). This release changes the protein to its original folding pattern, with the active site exposed in a position to combine with a second molecule (Figure 5-5e). The energy required for movement of this type, involving limited internal regions of the carrier protein, is provided by the favorable concentration gradient.

Passive transport, involving simple and facilitated diffusion, accounts for the movement of a wide variety of substances to and from cells. By this mechanism cells are able to absorb many of the hydrophobic and hydrophilic molecules required for their biological reactions and release waste materials or secretions to the outside. These transport processes, driven by concentration gradients, take place without the requirement for the expenditure of cellular energy. They greatly enhance the transport of key substances that would otherwise penetrate cells too slowly to support cellular life.

Active Transport

Transport of substance in or out of cells also takes place *against* concentration gradients. Movement of this type is called **active transport.** The energy required to move

---

Active transport moves substances against concentration gradients.

---

substances against their gradients is supplied in cells by linking transport to oxidation of fuel substances, usually through the medium of the ATP molecule (see p. 37). This dependence on cellular energy is the distinguishing characteristic of active transport.

Other features of active transport resemble facilitated diffusion. The process depends on membrane proteins, and it stops if proteins are denatured or removed from membranes. Like facilitated diffusion, the mechanism of active transport has the property of specificity: only certain molecules or closely related groups of molecules are moved across membranes by a given active transport system. These characteristics provide the major criteria for identifying active transport: (1) it goes against concentration gradients; (2) it requires cellular energy; (3) it depends on the presence and activity of membrane proteins; and (4) it is specific for certain substances or closely related groups of substances. The work of active transport may make up a major part of the total energy expended by cells.

Active transport moves a variety of substances in and out of cells at the expense of energy, including ions, a number of different sugars and other fuel molecules, and several amino acids. Each ionic or molecular type in these categories is transported by a separate protein or group of proteins that "recognizes" and moves it across the membrane.

The active transport mechanisms moving these substances, like the facilitated diffusion mechanisms, are considered to depend on the ability of proteins to take up a variety of folding patterns. Figure 5-6 shows a model outlining how such conformational changes might carry out active transport. The end of the transport protein exposed on the inside surface of the membrane has a folded region

that exactly matches the ATP molecule (Figure 5-6a). ATP molecules are constantly colliding with the inside membrane surface; when one strikes the exposed binding site of the active transport protein, it attaches. The attachment site splits off one of the three phosphates (Figure 5-6b), converting ATP to ADP, and releases free energy. The binding and energy release cause a change in the folding pattern of the protein. The change moves a small segment of the protein to a new position in which (1) a binding site for the substance being transported is exposed on the outside membrane surface (Figure 5-6c) and (2) a strain is imposed on the protein chain. The effect is much like using the energy liberated by ATP breakdown to cock a spring within the protein molecule. The external site then binds a molecule of the transported substance colliding with the membrane surface (Figure 5-6d). This binding releases the transporter "spring," pushing the transported molecule through the central channel to the cell interior (Figure 5-6e). As the protein segment carrying the molecule reaches this position, changes in the folding pattern rearrange the binding site so that it no longer fits the transported molecule. As a result, the molecule is released into the cell interior (Figure 5-6f). This release triggers the final change in the folding pattern of the protein, returning it to the original state (Figure 5-6a) in which it is ready to bind a fresh ATP and restart the transport cycle.

The systems that actively transport ions are among the most important transport mechanisms of both prokaryotic and eukaryotic cells. By means of their active transport carriers, almost all animal cells maintain internal Na^+ and K^+ concentrations that differ from concentrations outside. Normally, in these animal cells Na^+ is excreted, so that internal Na^+ concentration becomes lower than in the surroundings. Moreover, K^+ is actively transported in the reverse direction so that it becomes more concentrated inside cells than outside.

Active transport of these ions is of special significance for animal cells because the movement of charged particles such as ions amounts to generation of an electrical current. The currents produced by active transport of Na^+ and K^+ ions across nerve cell membranes set up the conditions required for generation of electrical impulses by nerve cells. These impulses provide the basis for the complex activities of the brain and nervous system and for the sensory functions coordinating the rapid and complex behavior of animals. Excretion of Na^+ by the same pumping systems also helps control the internal osmotic pressure of animal cells. The active transport of Ca^{2+} is an important part of the mechanism regulating the activity of muscle cells in animals. Plant cells also concentrate some ions, including K^+, against a concentration gradient. Most plant

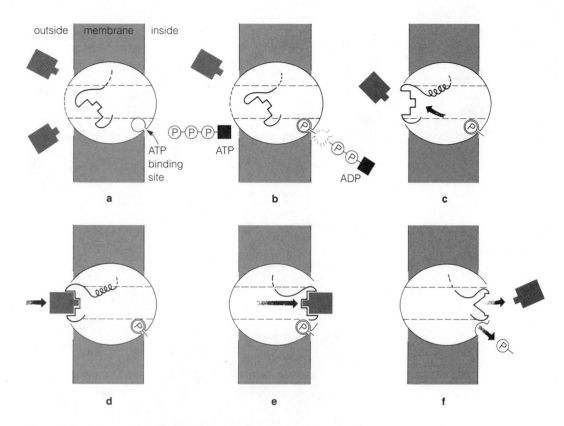

outside | membrane | inside

ATP binding site

ATP

ADP

a   b   c

d   e   f

Figure 5-6 Active transport by an ATP-dependent carrier molecule (see text).

cells do not regulate internal Na$^+$ concentration as do animal cells, however. The osmotic effects of high internal concentrations of this ion, which can cause animal cells to swell and burst if not eliminated, are counteracted in plants by the cell walls.

Endocytosis and Exocytosis

The processes of passive and active transport are limited to the movement of ions and relatively small molecules between a cell's surroundings and its interior. Cells also have mechanisms that allow them to absorb or release larger molecules such as nucleic acids or proteins or even cell parts or whole cells. These absorptive mechanisms, collectively called **endocytosis,** proceed by either of two mechanisms. One, called **pinocytosis** (meaning "cell drinking"), operates when the large molecules being taken in are in solution. The second mechanism, **phagocytosis** (meaning "cell eating"), involves the uptake of larger par-

Molecules or particles too large to penetrate directly across the plasma membrane are moved inward by pinocytosis or phagocytosis.

Pinocytosis is the uptake of large molecules in solution. Phagocytosis is the uptake of insoluble molecules, particles, or whole cells.

ticles. Both pinocytosis and phagocytosis, like active transport, require a cellular energy input to proceed. Release of materials from inside cells to the outside, called exocytosis, takes place by mechanisms that essentially reverse pinocytosis and phagocytosis. Apparently, exocytosis proceeds without an expenditure of cellular energy.

Pinocytosis and phagocytosis both require an input of cellular energy.

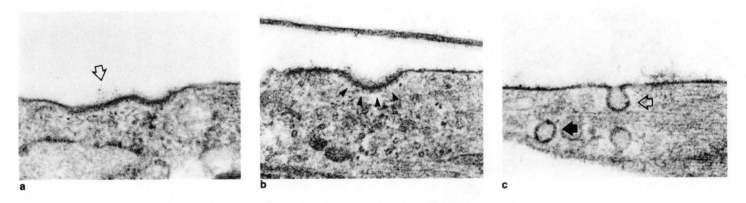

Figure 5-7 Formation of pinocytotic sacs in cultured human cells as seen under the electron microscope. The substance taken up, a lipoprotein, has been linked to a metallic stain to make it visible in the electron microscope. **(a)** Binding of the lipoprotein to the plasma membrane. **(b)** and **(c)** Infolding of membrane segments binding the lipoprotein. A free sac is visible in the cytoplasm beneath the plasma membrane in **(c)**. ×57,000. Courtesy of R.G.W. Anderson.

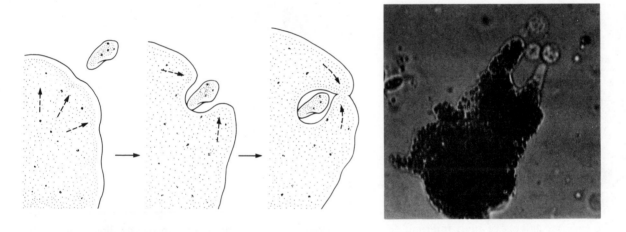

Figure 5-8 Phagocytosis. Part of the cell flows around a large particle to form a pocket in the cell membrane. The pocket pinches off as a vesicle, which sinks into the cytoplasm. The enclosed material may be broken down by enzymes secreted into the vesicle. The light micrograph shows an amoeba engulfing a food particle by phagocytosis. ×240.

Tracing the pattern of pinocytosis under the electron microscope shows that it occurs in three major steps (Figure 5-7). Initially, the substance being taken up is bound to the plasma membrane (as in Figure 5-7a). The membrane region binding the substance then folds inward (Figure 5-7b), producing a cup-shaped depression that deepens and pinches off, forming an unattached sac that sinks into the underlying cytoplasm (Figure 5-7c). The membrane of the sac may then break or disintegrate, introducing the

contained material directly into the cell interior. Alternatively, digestive enzymes may be secreted into the sac, causing the enclosed substances to break down into small molecules that can pass across the sac membranes.

Pinocytosis has been detected in many types of animal cells. In mammals, ingestion of large molecules by this route occurs regularly in intestinal, liver, blood, kidney, and tumor cells among others. Whether pinocytosis takes place in plants is controversial. Surface infolding and

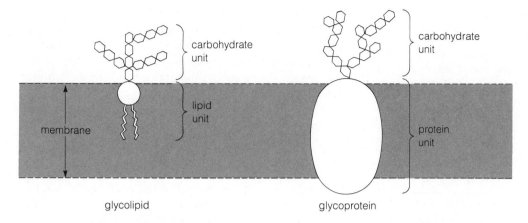

Figure 5-9 The structure of membrane glycolipids and glycoproteins.

channels have been observed in some plant cells, particularly in roots, but whether molecules actually enter by this route has not been established.

The particles taken in by phagocytosis, which may be as large as bacteria or even other eukaryotic cells, enter by a three-step mechanism basically similar to pinocytosis: (1) surface binding, (2) membrane infolding, and (3) sac formation (Figure 5-8). The sacs formed, however, are much larger, and they often persist visibly in the cytoplasm for extended periods of time. As in pinocytosis, the material in the sac may be released directly into the cell interior or may first be digested in the sac. Phagocytosis occurs commonly in protozoa such as *Amoeba* and in mammalian cells such as the white blood cells of the bloodstream, which remove foreign particles from the body by this mechanism.

In the reverse mechanisms of exocytosis, material enclosed in membrane-bound vesicles is carried through the cytoplasm to the inner surface of the plasma membrane. The vesicles then fuse with the plasma membrane and release their contents to the cell exterior. Many cell types in animals secrete proteins such as enzymes to the outside by this route. Residual material that resists digestion in vesicles taken in by pinocytosis or phagocytosis may also be expelled from the cell by the same pathway.

The basic transport mechanisms described in this section—passive transport, active transport, and endocytosis—account for the movement of substances between cells and their surroundings. As we have seen, passive transport, although it may operate with some specificity as in facilitated diffusion of sugars and amino acids, requires no direct energy input by the cell. The remaining transport mechanisms require energy in amounts some-

times representing a significant part of the cell's total output. All the transport mechanisms depend on intact membrane structure and the activity of both membrane lipids and proteins.

THE CELL SURFACE IN RECEPTION AND RECOGNITION

Carbohydrate-bearing lipids and proteins of the plasma membrane receive signals of various kinds from outside the cell and recognize other cells as either part of the same individual or foreign. The molecules active in these functions, the membrane glycolipids and glycoproteins, consist of a phospholipid or protein structure embedded within the membrane from which a branched carbohydrate chain extends outward from the cell surface (Figure 5-9). The varieties in which sugar units may combine to form surface carbohydrates are almost endless. One sugar unit commonly found in surface carbohydrates, glucose, can combine with itself through different linkages to form

---

Cell reception, recognition, and adhesion depend on membrane glycolipids and glycoproteins.

---

11 different types of disaccharides and 176 distinct trisaccharides. Other sugars combine with glucose to form combinations containing mixtures of sugar units in both straight and branched chains.

Surface Carbohydrates and Cell Reception

One of the most exciting developments in recent years is the discovery that cells contain external receptors that can recognize and bind a wide variety of substances. Binding of many of the molecules recognized by the receptors results in transfer of a "signal" across the plasma membrane to the cell interior. This signal in turn triggers a specific internal response, such as an increase in oxidation of

Cell reception involves detection of signals from the outside by the glycoproteins of the plasma membrane.

cellular fuels, protein synthesis, or cell division. The molecules active in surface reception in most cases have been identified as glycoproteins—that is, membrane proteins carrying a carbohydrate surface "antenna."

Binding and cellular response to insulin provide an excellent example of the role of carbohydrate groups in surface reception and the cellular response that follows. Insulin is a protein-based hormone secreted by the pancreas in mammals. Cells exposed to insulin immediately increase their rate of glucose transport and oxidation. The hormone is recognized and bound by glycoprotein receptor molecules at the cell surface.

The rate of glucose breakdown increases almost immediately after cells bind the hormone. The initial internal response occurs without penetration of the insulin molecule through the plasma membrane. In some way, the receptor glycoprotein transmits a signal across the membrane indicating to the cell interior that the hormone has been recognized and bound at the cell surface.

Persons with the disease *diabetes mellitus* are unable to produce sufficient quantities of the insulin hormone. Body cells in affected individuals, particularly in muscles, fail to take up and oxidize glucose because of the low levels of insulin in the bloodstream. Uptake by liver cells, which store glucose by converting it to glycogen, is also reduced. The condition leads to muscular weakness and an accumulation of glucose in the bloodstream; the high blood content of glucose leads directly and indirectly to a wide variety of other effects, including alterations in nerve and brain function. If untreated, the effects of the disease can produce coma and death. Fortunately for individuals with diabetes, injection or oral administration of the insulin hormone restores glucose uptake by body cells and glucose metabolism returns to normal levels.

Insulin and several other peptide hormones eventually enter their target cells through endocytosis. After binding to the receptors and initiating the cellular response, insulin is taken in through pockets in the cell surface that pinch off as sacs and sink into the underlying cytoplasm. After entrapment in the sacs, the hormone is broken down by enzymes introduced from the cytoplasm. This process, since it takes place after initiation of the cellular response through hormone binding to surface receptors, is probably a mechanism "clearing" the cell surface for reception of further messages.

Depending on the cell type, the cell surface may contain from hundreds to thousands of individual receptor molecules. For the different protein-based hormones, receptors may number from 500 to as many as 250,000 per cell. Only a fraction of these receptors need be bound to the hormone for development of full response by the target cell. Only 2 or 3 percent of the insulin receptors at the surface of target cells, for example, are required to interact with insulin molecules for a maximum internal increase in glucose metabolism to occur in response.

Cell Recognition by Surface Carbohydrates

Carbohydrate-containing lipid and protein molecules are also important in cell-to-cell recognition. Individual cells of mammals and other animals contain surface "markers"

Cell recognition depends on the surface carbohydrate groups of membrane glycoproteins and glycolipids.

Cells recognize each other as self or foreign by their carbohydrate surface markers.

that identify them as belonging to a given individual as well as recognition sites that allow identification and detection of these markers. The plasma membranes of most human body cells, for example, contain a group of molecules called the HL-A markers. Each of us has our own combination of HL-A markers; this combination is present on almost all the cells of the body. Through the HL-A combination, body cells can recognize each other as belonging to the same individual. If cells from another person, containing different HL-A markers, are transplanted into an individual, the introduced cells are recognized as foreign and destroyed by the same mechanisms that protect the body against invasion by disease organisms. Both the HL-A markers, which form the main barriers to successful tissue transplants between unrelated persons, as well as the surface receptors binding them have been identified as membrane glycoproteins. The part of the foreign HL-A glycoproteins that stimulates the defensive response is the carbohydrate portion of the molecule. In some cell recognition systems, such as the blood groups in humans, the surface carbohydrate chains marking the cells are attached to membrane lipids as well as proteins.

The ability to recognize cells as part of self or foreign is also important in the cell-to-cell identification and assembly that occurs as a part of embryonic development in animals (described in Chapters 14 and 16). As a part of embryonic development, individual cells or cell groups frequently move from one part of an embryo to another, releasing old cell-to-cell associations and forming new ones. The capacity of these cells to identify their final positions in the embryo has been shown to depend on recognition glycoproteins carried in the plasma membranes of both the migrating cells and the cells at the final destination.

The receptor and recognition glycoproteins embedded in the plasma membrane thus act as a sensory system at the cell surface. The glycoprotein and glycolipid molecules of this system allow cells to recognize stimuli in the form of hormones and other molecules and to respond to them. It also provides the basis for recognition of the cells of an individual as self or foreign. Finally, it supplies a mechanism for adhesion and combination of cells into higher levels of organization.

Questions

1. What is the difference between passive and active transport?

2. What provides the energy required for passive transport?

3. What is a concentration gradient?

4. What is diffusion? What causes diffusion?

5. What is a semipermeable membrane?

6. What is osmosis? What conditions are necessary for osmosis to occur? What provides the energy required for the osmotic movement of water?

7. What is osmotic pressure? How is osmotic pressure related to cellular life?

8. Compare a cell with the osmosis apparatus shown in Figure 5-2.

9. How is passive transport of substances through the plasma membrane related to lipid solubility?

10. In what ways do proteins modify the passive transport of substances through the plasma membrane?

11. What is facilitated diffusion? How does facilitated diffusion resemble the action of enzymes?

12. Compare passive transport through a cellophane film, an artificial phospholipid bilayer, and a natural membrane. What are the primary differences in diffusion through the three types of barriers?

13. Compare facilitated diffusion and active transport.

14. What are the characteristics of active transport mechanisms?

15. What supplies the energy for active transport?

16. What is the difference between pinocytosis and phagocytosis? What is the primary difference between these mechanisms and active and passive transport?

17. How do facilitated diffusion, active transport, and cell reception depend on the ability of proteins to take up different folding conformations?

18. How is the capacity for cell-to-cell recognition related to organ transplantation?

Suggestions for Further Reading

Weissman, G., and R. Claiborne, eds. 1975. *Cell Membranes*. HP Publishing Company, New York.

Wolfe, S. L. 1981. *Biology of the Cell*. 2nd ed. Wadsworth, Belmont, California.

6

THE FLOW OF ENERGY IN LIVING ORGANISMS I: ENERGY CAPTURE IN PHOTOSYNTHESIS

The activities of life require a continuous input of energy. At the cellular level, the required energy is derived either directly from sunlight or from the breakdown of complex organic molecules such as carbohydrates and fats. If traced to its ultimate source, the energy bound into the complex molecules used as fuels also has its origin in the sun. Consequently, the sun is the primary energy source for life on earth.

Part of the radiant energy of sunlight falling on earth is absorbed by plants and transformed into chemical energy. The captured energy is then used to synthesize all the biological molecules needed by plants. Usually this synthesis proceeds from raw materials no more complex than water, carbon dioxide, and inorganic minerals. Some of the potential energy of these biological molecules is then used by plants themselves as an energy source, particularly during periods of reduced light or darkness. The complex molecules synthesized by plants also form the primary energy source for animals that live by eating plants: the **herbivores.** Herbivores are eaten in turn by **carnivores,** animals that live by ingesting other animals. Carnivores are eaten by other carnivores, and so on down the line, until the last bits of chemical complexity are broken down by bacteria and fungi as the uneaten carnivores and any leftover herbivores and plants die. Thus the energy required for life flows from the sun through plants and animals and finally to organisms of decay. Without energy from the sun, all of these life forms would die.

The energy required to support life on the earth flows from the sun.

Only a few organisms lie outside this pathway of energy flow. These organisms, including a few species of bacteria, live by using relatively high-energy inorganic molecules as an energy source. The inorganic molecules used by these *chemosynthetic* bacteria, as they are called, exist in mineral deposits in the earth and do not contain energy originating from the sun. Even these organisms would probably cease to exist without sunlight, however, because the conditions necessary for their survival, such as environmental temperatures within the ranges necessary for life, could not be maintained.

The basic processes in the capture and use of solar energy take place at the cellular level. In the initial phase,

The basic processes in the flow of energy through living organisms take place at the cellular level.

photosynthesis, light is absorbed and converted to chemical energy. This phase takes place in the chloroplasts of eukaryotic plants and in the cytoplasm of blue-green algae and photosynthetic bacteria. The chemical energy captured in this phase is used to assemble complex organic molecules from simple inorganic substances in the photosynthetic cells. In the second major phase, **respiration,** the organic molecules built up by photosynthesis are broken down to release energy for cellular activities. Respiration occurs largely in mitochondria in both plant and animal eukaryotes. In prokaryotes, the reactions of respiration, like photosynthesis, are distributed throughout the cytoplasm.

This chapter describes the first phase of energy flow in living organisms: the capture and utilization of light energy in photosynthesis. In this discussion, primary emphasis is given to the way photosynthesis takes place in the eukaryotic plants. (Photosynthesis in prokaryotes is described in Supplement 6-2.)

CHLOROPLAST STRUCTURE IN THE EUKARYOTIC PLANTS

The chloroplasts of the eukaryotic algae and higher plant cells are easily seen under the light microscope as lens-shaped bodies about 5–10 micrometers in diameter. Usually, chloroplasts are considerably larger than the mitochondria in the same plant cell and occur in smaller numbers. The number may vary from a single chloroplast, as found in *Micromonas* and several other green algae, to several hundred in the cells of most higher plants.

The green pigments of chloroplasts can be seen under the light microscope to be concentrated in disklike substructures, the **grana,** in most eukaryotic plants. Grana are suspended in a colorless background substance inside chloroplasts called the **stroma.** This pigment distribution is maintained by a system of membranes and compartments visible in chloroplasts under the electron microscope (Figures 6-1 and 6-2). Two continuous boundary membranes separate the chloroplast interior from the surrounding cytoplasm. The region between these membranes is usually difficult to trace in electron micrographs (as in Figure 6-1) because of extensive folding of the inner boundary membrane. The two boundary membranes completely enclose the stroma. Within the stroma is a complex system of membranous sacs that forms the grana and their interconnections.

The individual unit of a granum is a flattened sac or vesicle, the **thylakoid** (*thylakos* = pouch), consisting of a single, continuous membrane that completely encloses an interior *thylakoid compartment*. Grana are formed from a closely fused pile of these individual sacs, set one on top

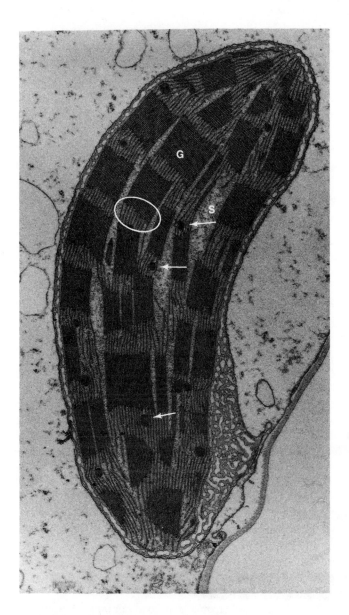

Figure 6-1 Chloroplast from a corn leaf. The chloroplast is surrounded by a smooth outer membrane and a much-folded inner membrane. Thylakoid membranes stacked into grana (G) are clearly visible inside the chloroplast. Membranous connections (circled) run between the thylakoids of adjacent grana. The thylakoid membranes are suspended in the inner substance of the chloroplast—a solution of proteins and other molecules called the stroma (S). ×20,000. Courtesy of L. K. Shumway and T. E. Weier, from *Amer. J. Bot. 54:* 773 (1967).

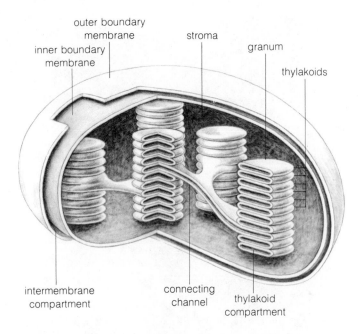

outer boundary
membrane

inner boundary
membrane

stroma

granum

thylakoids

intermembrane
compartment

connecting
channel

thylakoid
compartment

Figure 6-2 The arrangement of membranes and compartments inside a chloroplast.

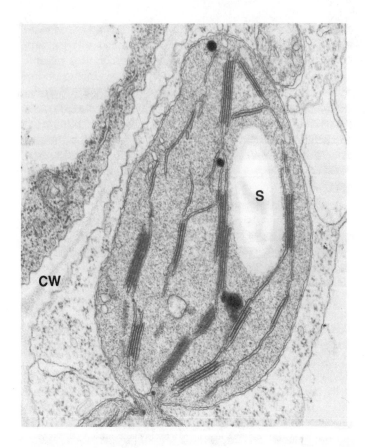

Figure 6-3 A starch granule(s) in a tobacco chloroplast. CW, cell wall. Courtesy of D. Stetler.

of another much like a stack of coins. Chloroplasts may contain from 40 to 60 grana, each formed from stacks of two or three to as many as a hundred individual thylakoid sacs.

Frequent membrane connections can be seen between the thylakoids of adjacent grana (circle in Figure 6-1). These connections form a channel that is continuous with the thylakoid compartments of adjacent grana (diagrammed in Figure 6-2). The channels may connect the thylakoid compartments into a single, continuous internal region inside chloroplasts.

The chloroplast stroma may also contain a variety of inclusions suspended in the regions surrounding the grana. Most conspicuous of these are the *starch granules* (Figure 6-3) found in chloroplasts after a period of active photosynthesis. The chloroplast stroma also contains DNA, ribosomes, and all the biochemical factors required for making DNA and RNA and synthesizing proteins. Of the visible elements of this system, the DNA appears at scattered locations in the stroma as faintly visible threads (see Figure 9-22). The ribosomes, conspicuously smaller than the ribosomes of the surrounding cytoplasm, are distributed throughout the stroma. The DNA and ribosomes of chloroplasts, which resemble the equivalent structures of prokaryotes much more closely than eukaryotic DNA and ribosomes, are evidently necessary for the production of

some of the proteins required for photosynthesis. The similarities between the DNA and ribosomes of chloroplasts and prokaryotes suggest that chloroplasts may have evolved from prokaryotes that became established as permanent, beneficial residents in the ancestors of present-day eukaryotic plant cells. (See Supplement 9-2 and Chapter 21 for details.)

PHOTOSYNTHESIS IN CHLOROPLASTS

The Overall Reactions of Photosynthesis

In 1772 an Englishman, Joseph Priestley, showed that air in an enclosed vessel, "injured" by the presence of a burning candle, could no longer support life. Priestley found that the injured air could be "restored" if a green plant was placed in the vessel and left in the light for a time. It was soon established that air is "injured" by a burning candle, a breathing animal, or even a plant in darkness because all produce carbon dioxide and use up oxygen. Green plants "restore" the injured air in light by removing CO_2 and releasing O_2. In the 1800s water was found to be necessary for this reaction to take place. Not long after this discovery, once the basic raw materials and products were known, the mechanism of photosynthesis was first written in skeletal form:

$$6CO_2 + 6H_2O \xrightarrow[\text{energy}]{\text{light}} C_6H_{12}O_6 + 6O_2 \qquad (6\text{-}1)$$

In plain words, this reaction states that in photosynthesis light energy is used to drive the uphill synthesis of car-

---

In photosynthesis light energy drives the synthesis of carbohydrates and other substances from CO_2 and H_2O.

---

bohydrates from carbon dioxide and water; oxygen is given off as a by-product of the reaction.

In the 1930s C. B. van Niel of Stanford University developed a more general statement of the basic reactions of photosynthesis through his study of photosynthetic bacteria. These bacteria do not use water as a raw material. Instead, they use other hydrogen-containing substances such as hydrogen sulfide (H_2S), alcohols, or even hydrogen itself (see Supplement 6-2). Significantly, oxygen is not given off as a by-product of photosynthesis by these bacteria. Bacteria using H_2S as a raw material, for example, release molecular sulfur instead of oxygen. Noting the similarity between H_2S and H_2O as raw materials, and the evolution of either sulfur or oxygen as by-products, van Niel proposed that the oxygen in eukaryotic photosynthesis is derived from the H_2O entering the process and not the CO_2 as previously assumed. Van Niel rewrote the overall reaction for photosynthesis in a more general form to take his observations and hypothesis into account:

$$CO_2 + 2H_2D \rightarrow (CH_2O) + H_2O + 2D \qquad (6\text{-}2)$$

In this reaction the substance H_2D (D = donor) represents any raw material that can donate electrons and hydrogen to the photosynthetic mechanism. The D in H_2D represents the atom (if any) attached to the hydrogen of this raw material. The D atom is eventually released in free form as a by-product of photosynthesis. In higher plants, the H_2D in van Niel's reaction is H_2O; the O is released as molecular oxygen. The (CH_2O) in Reaction 6-2 represents a unit of carbohydrate. Six of these units are combined in plants to produce a carbohydrate such as glucose. Van

Niel also found, as shown in Reaction 6-2, that water molecules are assembled as by-products of photosynthesis.

Van Niel's general equation for photosynthesis, and his proposal that the oxygen evolved by higher plants originates from water, were confirmed when researchers were able to isolate and track the various isotopes of the elements. The heavy oxygen isotope ^{18}O was used to label oxygen atoms in CO_2 or H_2O, and then it was followed through photosynthesis.[1] If an organism carrying out photosynthesis was supplied with water containing ^{18}O, the label showed up in the oxygen given off as a by-product. If carbon dioxide containing ^{18}O was supplied, the label showed up instead in the carbohydrates produced in photosynthesis and in the water molecules assembled as a by-product of the reaction. These results confirmed van Niel's hypothesis that the oxygen evolved in photosynthesis is derived from the water molecules entering the process, not the CO_2.

The Light and Dark Reactions of Photosynthesis

During the early part of this century biochemical investigations established that the overall reactions of photosynthesis occur in two major, interdependent parts that can be experimentally separated. One part, the **light reactions,** is directly dependent on light and stops if light energy is unavailable. The second part, the **dark reactions,** is light-independent and may continue in darkness if all the necessary reactants are present.

Later work concentrated on identifying the individual steps that occur in the light and dark reactions of photosynthesis and working out the chemical mechanisms that link the two reaction series together. This work showed that the rapid, light-dependent reactions synthesize two high-energy chemical products that are required by the dark reactions (Figure 6-4). One of these products, as you might have guessed, is ATP. The second is **NADP** (see Figure 6-16), a molecule that carries high-energy electrons and hydrogen. The light reactions also use H_2O as a raw material and release oxygen as a by-product. The dark reactions use the high-energy products of the light reactions as an energy source to synthesize carbohydrates, lipids, amino acids, and a host of other substances from CO_2 and other simple inorganic molecules. As part of the dark reactions, ATP is broken down to ADP and the hydrogen

---

[1]Oxygen in its most common form has eight neutrons and eight protons in its atomic nucleus, giving ^{16}O; "heavy" oxygen has two neutrons, giving the isotope ^{18}O; see p. 8.

and high-energy electrons are removed from NADP. These substances, ADP and NADP depleted of its electrons and hydrogen, then cycle back to the light reactions to be re-charged with chemical energy derived from sunlight.

The dark reactions do not necessarily occur in darkness as their name suggests. In fact, they generally occur in the daytime, when the light reactions are active in producing the high-energy products used as the energy source for the dark reactions. The name simply means that the dark reactions do not use light energy directly.

THE LIGHT REACTIONS OF PHOTOSYNTHESIS

Light and Light Absorption

Visible light is a form of radiant energy that travels in waves through space, with wavelengths varying from about 390 nanometers (seen as blue light) to 760 nanometers (seen as red light). The energy of light depends on its wavelength. The shorter the wavelength, the greater the energy delivered by the light.

Certain molecules can absorb the energy of light at one or more wavelengths. These molecules appear colored or pigmented because of the unabsorbed wavelengths, which are transmitted or reflected. Absorption of light at specific wavelengths is a property of electrons occupying orbitals in the absorbing molecules.

Electrons in these orbitals exist at characteristic energy levels. If one of the electrons absorbs the energy of a wavelength of light, it moves to a new orbital that lies at a greater distance from its atomic nucleus. In the new orbital, the electron is said to be in an **excited state.** The difference in energy level between the unexcited state (called the **ground state**) and the excited state is exactly equivalent to the amount of energy contained in the light absorbed by the electron.

The excited state is so unstable that an electron can remain in an excited orbital for only a billionth of a second or less. Unless the excited electron is "trapped" by another molecule, it quickly returns to the unexcited or ground state and the absorbed energy is released from the pigment molecule as heat or a mixture of heat and light. (The release of some of the absorbed energy as light in this way is called **fluorescence.**)

The alternative to release of the absorbed light energy, trapping of the excited electron by another molecule, is the essential step in conversion of light to chemical energy in photosynthesis. If the pigment molecule containing an

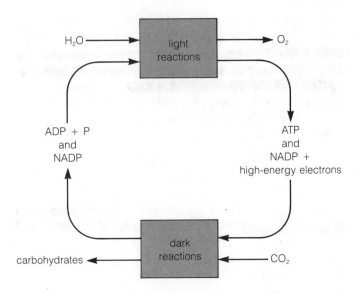

Figure 6-4 Distribution of photosynthesis between the light and dark reactions. The two reaction sequences are linked by ATP and NADP.

excited electron lies close enough to another molecule that can accept electrons, the excited orbital may extend outward far enough to overlap vacant orbitals in the acceptor. Transfer of the excited electron to a stable orbital in the acceptor molecule may then occur. Once transferred to a suitable acceptor in this way, the energy of the excited

---

Once an excited electron is transferred to a stable orbital in an acceptor molecule, its energy has been trapped in chemical form.

---

electron has been trapped as chemical energy. The acceptor molecule is said to be *reduced* (see Information Box 6-1) by accepting the high-energy electron. The energy absorbed by the pigmented molecules active in photosynthesis is converted to chemical energy in this way: by transfer of excited electrons to stable orbitals in acceptor molecules.

The Molecules Absorbing Light in Photosynthesis

The molecules that absorb light energy in the eukaryotic plants are all concentrated inside chloroplasts. These molecules are lipidlike and can be easily extracted from plant tissue by lipid solvents such as ether or acetone. Analysis of the extracted molecules reveals that chloroplasts contain two major classes of pigments: the green **chlorophylls** and yellow **carotenoids.**

The chlorophylls (Figure 6-5) absorb light most strongly at red and blue wavelengths and transmit green light. The different types of chlorophyll molecules are built up from a complex central ring structure to which is attached a long, nonpolar tail that gives the chlorophylls their lipid-like solubility. A single magnesium ion is bound into the center of the ring structure. The two major chlorophylls found in all higher plants, called *chlorophyll a* and *chlorophyll b*, differ only in a single substitution in one side group bound to the central ring (see Figure 6-5).

Both chlorophyll *a* and *b* have an extensive series of electrons that can absorb light energy and jump to excited orbitals. These electrons each absorb light at a different

---

Information Box 6-1

Oxidation and Reduction

Many substances can accept or donate electrons. Much of the energy passed from one substance to another in photosynthesis is transferred in the form of electrons removed from a donor molecule and accepted in the same instant by an acceptor molecule. Removal of electrons is called **oxidation** and acceptance is termed **reduction.** A substance from which electrons are removed is said to be *oxidized,* and the accepting substance is *reduced.* In general, a substance contains more energy in its reduced state than in its oxidized state.

For any reactions involving transfer of electrons from a donor to an acceptor there is always a simultaneous reduction and oxidation; the two substances acting in the joint reduction-oxidation are often termed a **redox couple.** In biological reductions, a hydrogen ion (H^+) is often attached to the acceptor molecule at the instant of reduction. This is the case with NADP, the important reduced substance produced by the light reactions of photosynthesis (see Reaction 6-4).

The electrons removed in oxidation have a characteristic energy that depends on the orbitals they occupied in the oxidized substance. The energy associated with the removed electrons can be measured and expressed as voltage (see the figure in this box). The standard used for comparisons is the energy associated with electrons removed from hydrogen in the reaction $H_2 \rightarrow 2H^+ + 2$ electrons. The voltage or potential of these electrons is arbitrarily designated as 0.00 millivolt; and all other voltages (often called **redox potentials**) are compared to this standard. Although all electrons carry a negative charge, their *relative* voltage may be greater or less than the 0.00 standard and can therefore be called positive or negative. In general, the more negative a substance is on the redox scale, the greater is the amount of energy contained in electrons removed from the substance. Electrons are only transferred downward on the scale; that is, the donor must be higher on the scale than the acceptor. As electrons are transferred from a donor to an acceptor, an amount of free energy equivalent to the distance between the donor and acceptor on the redox scale is released.

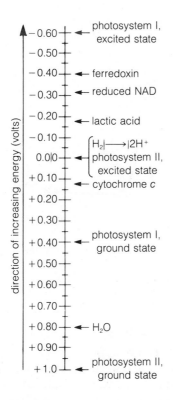

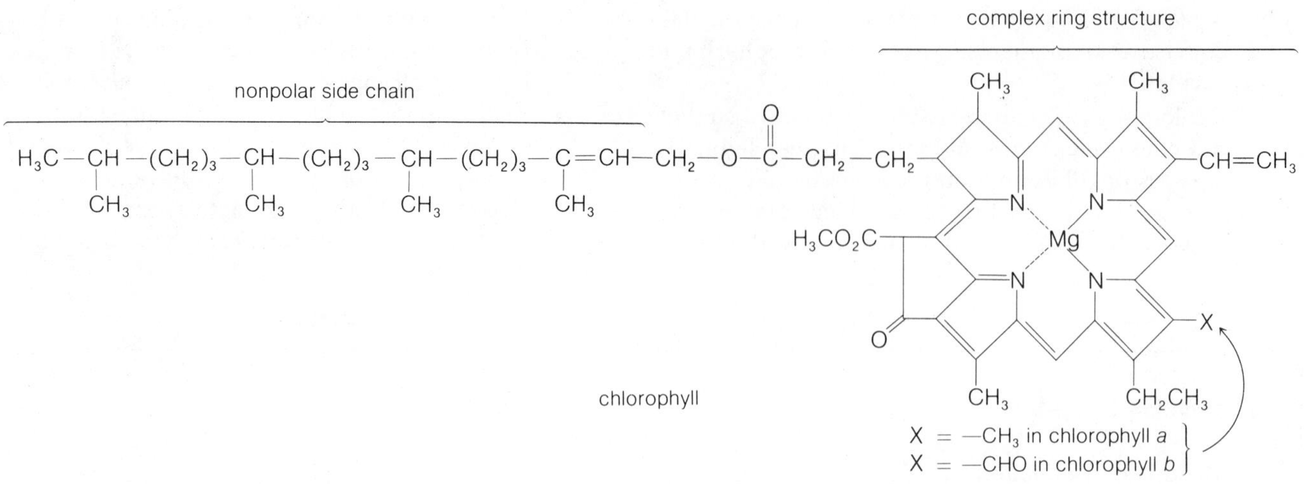

nonpolar side chain

complex ring structure

$H_3C-CH-(CH_2)_3-CH-(CH_2)_3-CH-(CH_2)_3-C=CH-CH_2-O-\overset{O}{\overset{\|}{C}}-CH_2-CH_2-$

chlorophyll

$X = -CH_3$ in chlorophyll *a*
$X = -CHO$ in chlorophyll *b*

Figure 6-5 Structures of chlorophyll *a* and *b*. Chlorophyll *a* has a methyl (—CH₃) group at the position marked X (arrow) in the figure; chlorophyll *b* has an aldehyde (—CHO) group in this position. Additional chlorophylls, with minor substitutions at other points in the ring structure, are found in some algae and plants.

All photosynthetic organisms use chlorophylls and carotenoids as major light-absorbing pigments in photosynthesis.

wavelength, which has the effect of greatly broadening the distribution of wavelengths absorbed also from a single peak to a broad, smooth curve with several peaks (Figure 6-6). The particular wavelengths absorbed depend on the proteins and other chemical groups associated with chlorophyll molecules inside chloroplasts.

The carotenoids (Figure 6-7), which form a separate family of pigmented lipid molecules, are all built up from a single, long carbon chain containing 40 carbon atoms. Various substitutions in the side groups attached to the 40-carbon backbone give rise to the different carotenoid pigments. The carotenoids absorb light at blue wavelengths (Figure 6-8) and transmit yellow, green, orange, and red light. The transmitted combination appears predominantly yellow to the eye.

The net effect of the entire combination of photosynthetic pigments is to broaden the spectrum of wavelengths usable as energy sources for photosynthesis so that light energy ranging over most of the visible wavelengths can be absorbed (Figure 6-9). Of the chlorophyll and carotenoid pigments, however, only chlorophyll *a* appears to be directly involved in the electron transfer that converts light to chemical energy.

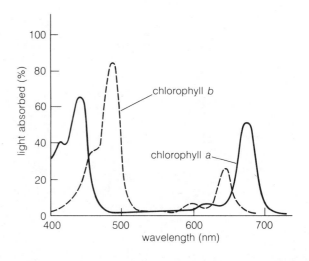

Figure 6-6 Amount of light absorbed by chlorophyll *a* and *b* at different visible wavelengths.

Recent research has revealed that the pigments of chloroplasts are organized into groups containing about 200 chlorophylls and about 50 carotenoid molecules, linked together with proteins into light-absorbing assemblies. The energy of light absorbed anywhere in one of these assemblies is passed along until it reaches a chlorophyll *a*

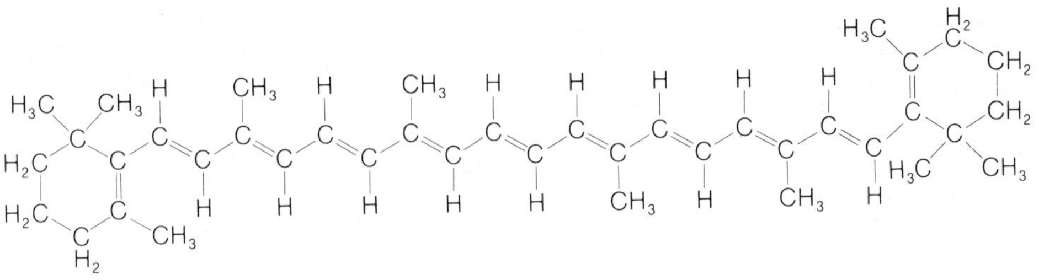

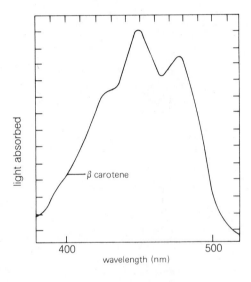

β-carotene

Figure 6-7   The carotenoid pigment β-carotene.

Figure 6-8   Amount of light absorbed by the carotenoid β-carotene at different visible wavelengths.

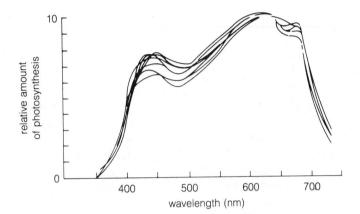

Figure 6-9   The amount of photosynthesis in various crop plants driven by different light wavelengths. The curves represent the combined activity of the chlorophyll and carotenoid pigments. Redrawn courtesy of K. J. McCree and Elsevier Scientific Publishing Company.

molecule that acts as the *reaction center* for the entire assembly (Figure 6-10). At the reaction center, the energy of

---

Absorbed light energy is transferred from chlorophyll to acceptor molecules in the form of high-energy electrons. This transfer converts the light energy to chemical energy.

---

the absorbed light is converted to chemical energy by passage of an excited electron from chlorophyll *a* to a stable orbital in an acceptor molecule.

Although it is clear that light energy is actually passed along from the absorbing molecules to the chlorophyll *a* molecules at the reaction center of a light-absorbing assembly, it is not yet certain how this transfer takes place. One idea is that the absorbing molecules vibrate as a result

of absorbing the light energy. Any adjacent molecule is induced to vibrate at the same frequency, and, in so doing, absorbs the excitation energy of the first molecule. In the pigment assemblies the energy of absorbed light is thought to travel in this way from one pigment molecule to another until it reaches the reaction center. Once at the reaction center, the energy is used to raise electrons in the chlorophyll *a* molecules at the center to excited orbitals.

Much research effort has gone into tracing the pathways followed by the electrons after their transfer from chlorophyll *a* to the electron acceptor substances. This

---

The light reactions of photosynthesis make ATP and reduced NADP.

---

work has revealed the identity of some of the acceptor substances and has shown that in the light reactions the energy carried by the electrons is used to synthesize ATP and reduce NADP.

Generation of ATP and Reduced NADP

Several kinds of acceptor molecules are involved in the generation of ATP and reduced NADP in the light reactions. These molecules have been found to act in series, passing electrons from one to another after the first member of the series accepts excited electrons from chlorophyll *a*. As the electrons are passed from one carrier to the next, some of their energy is released as free energy; during several of the transfers enough energy is liberated to drive the synthesis of ATP. Eventually the electrons, still containing much of the energy originally absorbed from sunlight, are delivered to NADP. As high-energy electrons flow from one carrier to the next in the carrier chains, the carriers are alternately reduced and oxidized (see Information Box 6-1). As each carrier accepts one or two electrons, it is reduced; as it releases electrons to the next carrier in the chain, it is oxidized.

The electron carriers are linked with two separate light-absorbing assemblies of chlorophyll and carotenoid molecules into what has been called the **Z-pathway** of the light reactions. (Note that the "Z" of the pathway is turned on its side in Figure 6-11.) In the pathway, which was first pieced together by R. Hill and Fay Bendall at Cambridge University in England, low-energy electrons derived from water are delivered through a short series of

---

Electrons flow through the Z-pathway from water to NADP.

---

carriers to the first light-absorbing assembly of the pathway: **photosystem II**.[2] The electrons are pushed to higher energy levels by light absorbed in photosystem II and are then transferred to a longer chain of electron carriers that links photosystem II to **photosystem I**, the second light-absorbing assembly of the pathway. As the electrons flow down the series of carriers linking the two photosystems, some of their energy is tapped off and used to drive the synthesis of ATP. After delivery to photosystem I, a second absorption of light pushes them to the highest energy

---

[2]The photosystems were identified and named "I" and "II" some time before they were placed in sequence in the light reactions of photosynthesis; as luck would have it, photosystem II was found to precede I in the light reactions. For better or worse, the original designations have been retained.

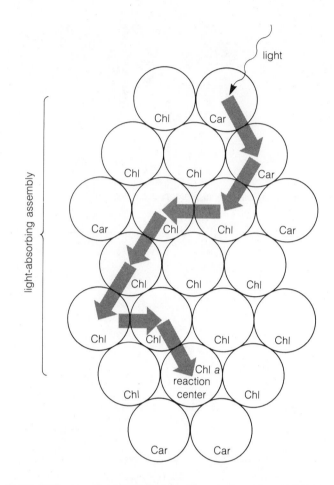

Figure 6-10 How light absorbed anywhere in the chlorophyll (Chl) and carotenoid (Car) molecules of a light-absorbing assembly is passed from molecule to molecule until it reaches the reaction center (see text).

level of the pathway. From this point the electrons enter a short carrier chain for delivery to NADP.

All but one of the carrier molecules consist of a protein molecule in combination with a nonprotein group that serves as the actual electron carrier. Four major kinds of these protein-based electron carriers have been identified. One, the *cytochromes*, consists of a protein molecule linked to a complex ring structure similar to that of chlorophyll, except that the metal atom bound into the center of the ring is iron instead of magnesium (Figure 6-12). The several cytochromes of the Z-pathway differ only in minor substitutions of side groups on the central ring structure. The second major carrier type, the *iron–sulfur proteins*, is as yet poorly characterized and is known only to contain "centers" containing iron atoms in close association with

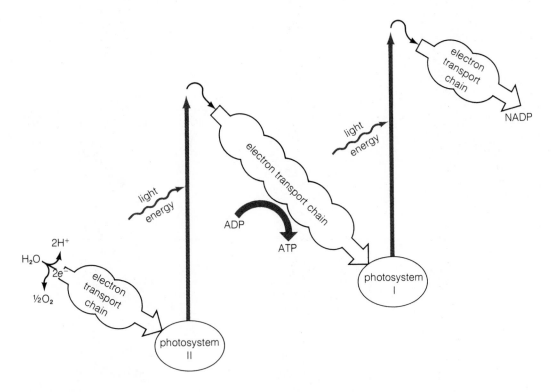

Figure 6-11 Flow of electrons through the Z-pathway (see text).

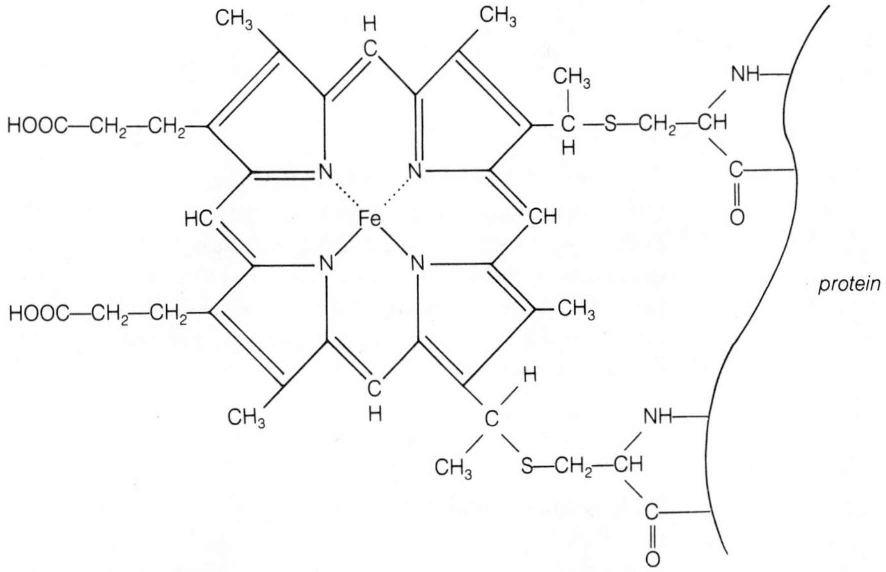

Figure 6-12 The electron-carrying group of a cytochrome. The iron in the center of the ring structure, by changing alternately from the ferrous (Fe^{+2}) to the ferric (Fe^{+3}) form, acts as the electron carrier. Similar rings occur as central structures in hemoglobin, chlorophyll, and other molecules of biological importance (compare with Figure 6-5).

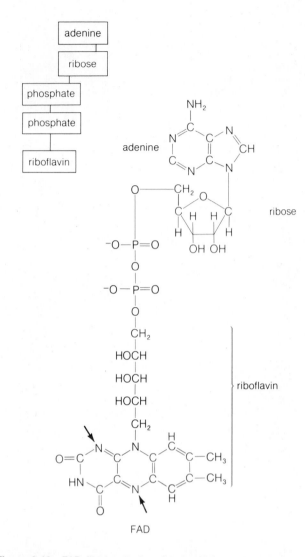

Figure 6-13 FAD (flavin adenine dinucleotide), an electron carrier linked to a protein in a flavoprotein. A hydrogen and electron are added at each of the two positions marked by arrows when FAD is reduced.

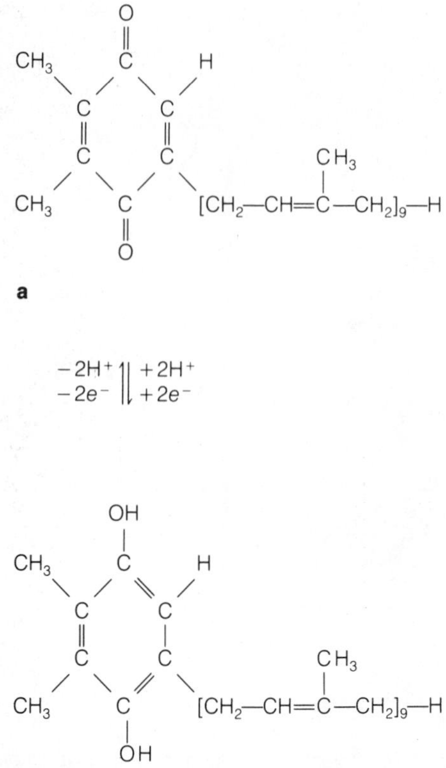

a

$$-2H^+ \uparrow\downarrow +2H^+$$
$$-2e^- \;\;\; +2e^-$$

b

Figure 6-14 Plastoquinone, a nonprotein electron carrier of the Z-pathway. (**a**) Oxidized form; (**b**) reduced form.

sulfur-containing groups. The third carrier type, a molecule known as a *flavoprotein*, contains a carrier group based on a nucleotide. In the single carrier of this type occurring in chloroplasts, *FAD* (*flavin adenine dinucleotide;* see Figure 6-13), the nucleotide contains a structure derived from *riboflavin*, a vitamin of the B group. The final protein-based carrier, *plastocyanin*, has a carrier group containing copper. The single nonprotein electron carrier, *plastoquinone*, is a lipidlike substance built up from a relatively simple ring-shaped structure (Figure 6-14).

Figure 6-15 shows the tentative positions of the individual carriers of the Z-pathway as far as they are known. Following the course of electrons through the pathway indicates how the two photosystems and the electron transport chains are considered to work in detail. The electrons are removed from water at the beginning of the sequence in the reaction:

$$H_2O \rightarrow 2H^+ + 2 \text{ electrons} + \tfrac{1}{2}O_2 \qquad (6\text{-}3)$$

Little is known about the reaction splitting water or of the enzymes involved in speeding this step; the carriers conducting electrons from water to the reaction center of photosystem II also remain to be identified.

After passing along the initial short sequence of carriers, the electrons enter ground-state orbitals in chlorophyll *a* molecules at the reaction center of photosystem I.

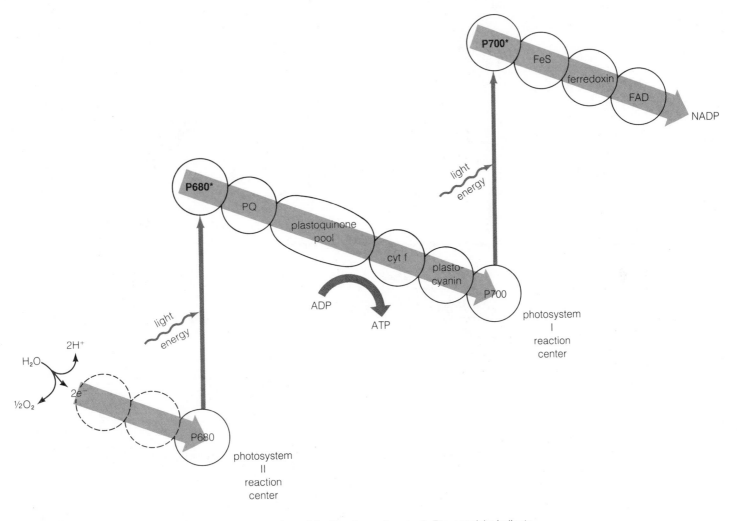

Figure 6-15 The tentative positions of the electron carriers of the Z-pathway (see text). The asterisks indicate the excited forms of chlorophyll P680 and P700.

These chlorophyll *a* molecules, since they absorb light most strongly at a wavelength of 680 nanometers, are called *chlorophyll P680*. The light-absorbing assembly of photosystem II then absorbs light and transfers energy to the P680 molecules, raising the electrons derived from water to excited orbitals. The excited electrons are transferred immediately to the first acceptor of photosystem II, known as the *primary acceptor* of this system. As the electrons enter stable orbitals in the primary acceptor, their energy is trapped in chemical form in the acceptor molecule. The acceptor molecule is reduced, and the P680 chlorophyll at the reaction center is oxidized as this transfer takes place. Of the various candidates proposed for the primary acceptor of this photosystem, it now appears that

the first substance to accept excited electrons from chlorophyll P680 is a specialized form of plastoquinone (identified as PQ in Figure 6-15).

From plastoquinone PQ, electrons flow to other plastoquinones and then through the remainder of the chain, consisting of cytochrome *f* and plastocyanin. From plastocyanin, electrons are transferred to the chlorophyll *a* molecules forming the reaction center of photosystem I. These chlorophyll *a* molecules, since they absorb light most strongly at 700 nanometers, are identified as *chlorophyll P700*. (The differences in light absorption between the various forms of chlorophyll *a* are produced by the manner in which the chlorophyll molecules link to proteins in the light-absorbing assemblies of the two photo-

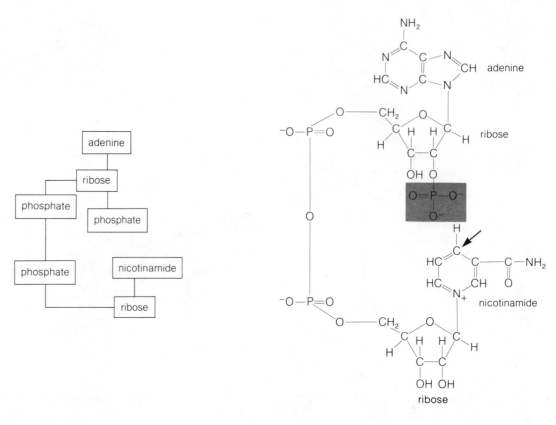

Figure 6-16 NADP (nicotinamide adenine dinucleotide phosphate), a carrier of hydrogen and high-energy electrons in cells. Hydrogen and electrons are added at the position marked by the arrow as NADP is reduced.

systems.) Through the absorption of light energy in photosystem I, the electrons are excited to high energy levels and are immediately transferred to the primary acceptor of photosystem I, tentatively identified as an iron–sulfur protein. From the primary acceptor the electrons then flow along the short final chain from ferredoxin (another iron–sulfur protein) through FAD to NADP. NADP is reduced at the final step in the Z-pathway.

As NADP accepts a pair[3] of high-energy electrons at the end of the sequence, it also binds a hydrogen ion from the surrounding water solution at the same time:

$$NADP + 2 \text{ electrons} + H^+ \rightarrow NADPH \text{ (or reduced NADP)} \quad (6\text{-}4)$$

The electrons accepted by reduced NADP are still at such high energy levels that they are capable of reducing almost all other cellular electron acceptors. These electrons are therefore a source of "reducing power" in the cell—an energy source as vital to cellular activities as ATP. Many of the reactions synthesizing complex molecules in the cell involve a reduction in which high-energy electrons are transferred to the substance being synthesized (as, for instance, in the dark reactions of photosynthesis building up

---

[3]Excited electrons are released one at a time by chlorophyll P700 and P680. Many of the electron carriers in photosynthesis, such as ferredoxin and the cytochromes, also carry electrons singly. Other carriers, such as FAD, plastoquinone, and the final electron acceptor, NADP, carry electrons in pairs. To balance the flow, the chlorophyll P700 and

P680 molecules and the single electron carriers may be considered to cycle twice for every NADP molecule reduced at the end of the pathway or, alternatively, to be present in the pathway in twice the amount of NADP.

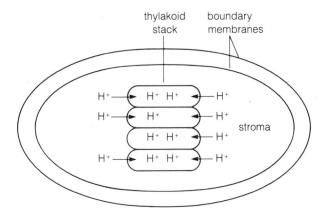

thylakoid stack   boundary membranes

stroma

Figure 6-17 According to Mitchell's chemiosmotic hypothesis, movement of electrons through the Z-pathway leads to a buildup of H^+ ions inside the thylakoid compartments. The resulting concentration gradient, with H^+ ions at higher concentrations in the thylakoid compartments than in the stroma, is used as a source of energy to drive ATP synthesis (see text).

carbohydrates from CO_2). Where reductions of this type occur, the required high-energy electrons are usually delivered to the system by reduced NADP. The NADP molecule (*nicotinamide adenine dinucleotide phosphate;* Figure

---

Reduced NADP carries high-energy electrons to cellular reactions that require a reduction.

6-16) thus acts as a carrier of high-energy electrons, just as ATP carries high-energy phosphates. The hydrogen carried by NADP in the reduced state (see Reaction 6-4) is also important in the production of carbohydrates, a reaction in which CO_2 is reduced to (CH_2O).

ATP Synthesis in the Z-Pathway: Photophosphorylation

The electrons flowing through the transport chains of the Z-pathway are able to pass from one carrier to the next because the orbitals in each successive carrier exist at lower energy levels than the orbitals of the carrier preceding it in the series. Thus the energy level of the electrons is lower in each successive member of the chains. As the electrons pass from one carrier to the next, the difference in energy level is released as free energy. Some of this released energy is used to drive the synthesis of ATP.

Scientists have worked intensively since the 1950s to determine how the energy released from electrons flowing

through the electron transport systems of the Z-pathway is used to drive ATP synthesis. The mechanism was finally explained by the **chemiosmotic hypothesis,** first proposed in 1961 by Peter Mitchell of the Glynn Research Laboratories in England. Briefly, Mitchell's hypothesis states that in chloroplasts the flow of electrons through the Z-pathway leads to a buildup of H^+ ions inside the thylakoid sacs (Figure 6-17). The resulting concentration gradient, with H^+ ions in high concentration inside the thylakoids and low in the stroma, is then used as a source of free energy to synthesize ATP from ADP and phosphate.

An H^+ gradient created by electron flow through the Z-pathway provides the energy required to drive ATP synthesis.

To understand how this is possible, remember that in active transport (see p. 86 and Figure 5-6), the energy of ATP is used to push ions *against* a concentration gradient, so that they become increasingly more concentrated on one side of a membrane. The mechanism synthesizing ATP in chloroplasts simply uses the same mechanism running *in reverse.* The entire system of chemical springs and levers that would normally break down ATP and use the energy released from this breakdown to push an H^+ ion across a membrane is forced to run backward by the very high H^+ ion concentration built up inside the thylakoid membranes. As a result, the H^+ ions run back through the membrane channels, and the membrane carrier molecules *add phosphate groups to ADP to form ATP.*

The chemiosmotic mechanism depends on two facts observed about the Z-pathway in chloroplasts. One is that all the photosystems and electron carriers of the pathway are embedded within the thylakoid membranes of the chloroplast. The second is that some electron carriers of the Z-pathway carry both hydrogen ions and electrons (plastoquinone and FAD) and some carry only electrons (the cytochromes, plastocyanin, and the iron–sulfur proteins).

According to Mitchell, each time electrons pass from a nonhydrogen carrier to a hydrogen carrier in the chain, hydrogen is picked up from the H^+ ions in the stroma solution (Figure 6-18a). As the electrons pass from an electron–hydrogen carrier to a nonhydrogen carrier, the H^+ ions are expelled into the thylakoid compartment (Figure 6-18b). This movement of H^+ ions builds the H^+ ion concentration inside the thylakoid compartments and reduces it in the stroma so that a steep concentration gradient is formed between the thylakoid compartments and the stroma.

The thylakoid membranes housing the Z-pathway carriers contain a membrane carrier protein that would

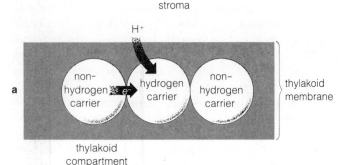

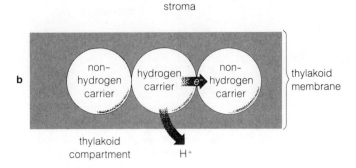

Figure 6-18 Movement of H⁺ from the stroma to the thylakoid compartment as a result of transfer of electrons between hydrogen and nonhydrogen carriers in the Z-pathway. (**a**) As electrons pass from a nonhydrogen to a hydrogen carrier, H⁺ ions are removed from the stroma. (**b**) As electrons pass from a hydrogen carrier to a nonhydrogen carrier, electrons are expelled into the thylakoid compartment.

normally break down ATP to transport H⁺ ions from the stroma into the thylakoids. The H⁺ concentration inside the thylakoid compartments becomes so high as a result of electron transport through the Z-pathway that the H⁺ pump is forced to reverse: instead of breaking down ATP, the carrier protein synthesizes ATP from ADP and phosphate. Synthesis of ATP in this way, through the energy released by electron transport in the light reactions of photosynthesis, is termed *photophosphorylation.*

The total amount of ATP synthesized as each pair of electrons follows the Z-pathway from H_2O to NADP remains somewhat in doubt. The most recent experiments indicate that movement of an electron pair through the entire pathway drives the synthesis of two molecules of ATP. The light reactions of photosynthesis thus use light as an energy source to reduce NADP and convert ADP to ATP. These products provide the energy (in the form of ATP) and the reducing power (in the form of reduced NADP, or NADPH) required for the production of complex organic molecules from CO_2 in the dark reactions.

THE DARK REACTIONS OF PHOTOSYNTHESIS

In the dark reactions, chemical energy produced in the light reactions is used to "fix" carbon dioxide into carbohydrates and a variety of other organic products. Although termed the dark reactions because they do not depend directly on light, the various interactions synthesizing carbohydrates, as noted, actually take place primarily in the daytime, when ATP and reduced NADP are readily available from the light reactions.

Tracing the Dark Reactions

Little progress was made in unraveling the dark reactions until the 1940s, when radioactive compounds first became available to biochemists. One substance in particular, CO² labeled with the radioactive carbon isotope ¹⁴C, made possible the first real breakthroughs in research into the reactions that assemble carbohydrates inside chloroplasts.

Melvin Calvin, Andrew A. Benson, and their colleagues at the University of California at Berkeley used this radioactive form of CO_2 to trace out the biochemical pathways of the dark reactions. In their experiments, Calvin and his colleagues allowed photosynthesis to proceed in *Chlorella,* a eukaryotic green alga, in the presence of radioactive CO_2. Extracts of carbohydrates and other substances were then made from the *Chlorella* cells.

If the carbohydrate extracts were made within a few seconds after exposing the cells to radioactive CO_2, most of the radioactive carbon in the extract could be found in a three-carbon sugar. Thus the CO_2 taken in by *Chlorella* was incorporated very rapidly into this three-carbon substance, evidently one of the earliest products of photosynthesis. If the cells were allowed to photosynthesize for longer periods before extracts were made, radioactive label showed up in more complex substances, including glucose, a six-carbon sugar. In other experiments, Calvin and his colleagues reduced the amount of CO_2 to levels so low that photosynthesis could proceed only very slowly in the *Chlorella* cells, even though adequate light was supplied. Under these conditions, a nonradioactive five-carbon sugar accumulated in quantity in the cells. Accumulation of this five-carbon sugar at low CO_2 levels suggested that it is the first substance to combine chemically with CO_2 and, moreover, that it "piles up" in chloroplasts when the CO_2 supply is abnormally low. By similar methods, an extensive series of intermediate compounds between CO_2 and glucose was identified.

Using this information, Calvin, Benson, and James A. Bassham were able to piece together the dark reactions of photosynthesis. For his brilliant work with this system and his successful model for the photosynthesis of carbohydrates, now called the **Calvin cycle,** Calvin was awarded the Nobel Prize in 1961.

The Calvin Cycle

The sequence of dark reactions described by Calvin and his coworkers occurs in a repeating cycle. In this cycle CO_2, ATP, and reduced NADP are the raw materials, and fuel (hydrogen is carried to the cycle by the reduced NADP) and carbohydrate units (CH_2O) are the primary product. (See Information Box 6-2 for an overview of the

The dark reactions produce carbohydrates from CO_2, using ATP as an energy source and reduced NADP as a source of hydrogen and high-energy electrons.

cycle.) Each step of the Calvin cycle is catalyzed by a specific enzyme.

The intermediate compounds of the Calvin cycle are complex. The overall progress of the cycle can easily be understood, however, if the carbon chains and phosphate groups are isolated and followed through the sequence (Figure 6-19). (A detailed outline of the cycle is presented in Supplement 6-1.) The first steps of the cycle (reactions 1 through 3 in Figure 6-19) take up CO_2 and combine it with a five-carbon sugar, producing two molecules of a three-carbon, one-phosphate sugar at the expense of ATP and reduced NADP. These reactions proceed in the following way. In the first reaction, CO_2 combines with the five-carbon sugar, which contains two phosphate groups; this five-carbon sugar is designated 5C, 2P in Figure 6-19. The reaction produces two molecules of a three-carbon sugar, each containing one phosphate group (3C, 1P in Figure 6-19). In the next step (reaction 2) another phosphate group is added to each of these sugars at the expense of 2ATP derived from the light reactions, yielding two molecules of a three-carbon, two-phosphate sugar (3C, 2P in Figure 6-19). These highly reactive products are reduced in reaction 3 by accepting electrons and hydrogen from two molecules of reduced NADP, also derived from the

Information Box 6-2

An Overview of the Calvin Cycle

The Calvin-Benson cycle is a series of reactions that uses CO_2, ATP, and reduced NADP as raw materials and generates units of carbohydrate (CH_2O) as a product.
The ADP and oxidized NADP released from the cycle return to the light reactions to be converted into ATP and reduced NADP again. All of the intermediate molecules in the cycle are continuously regenerated as the cycle turns.

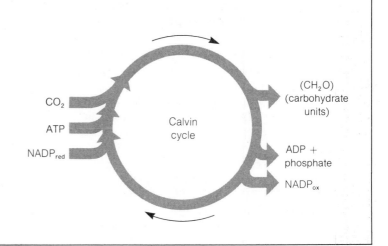

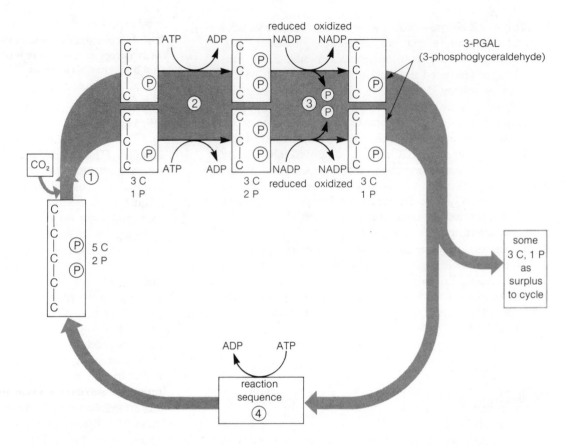

Figure 6-19 The Calvin cycle, showing only the carbon chains and phosphate groups of the intermediates of the cycle (see text).

light reactions. As this reduction takes place, one phosphate group is removed from each of the sugars. The final product of these reactions is two high-energy molecules of *3-phosphoglyceraldehyde (3PGAL)*. This three-carbon, one-phosphate sugar is the central product of the dark reactions.

Much of the 3PGAL produced is used to replace the five-carbon sugar used in reaction 1 of the cycle. This sequence, which breaks down one additional molecule of ATP for each turn of the cycle, is designated reaction sequence 4 in Figure 6-19. Some 3PGAL is also released as a product of the cycle. The ATP and reduced NADP used in the cycle, converted in the process to ADP + phosphate and oxidized NADP, return to the light reactions to be converted back to ATP and reduced NADP again.

The primary product of the Calvin cycle—the three-carbon, one-phosphate sugar 3PGAL—is the starting point

for synthesis of a variety of more complex carbohydrates and polysaccharides. Two molecules of 3PGAL combine through a stepwise series of reactions to produce glucose, which contains six carbons but no phosphates, and other six-carbon sugars.[4] The surplus 3PGAL may also be oxidized directly by the cell as a fuel substance.

Free glucose is actually formed in only limited quantities in the chloroplasts of most plants. Instead, glucose and other six-carbon sugars are used as the starting points for making sucrose (common table sugar), starch, cellu-

---

[4]The reactions that convert 3PGAL to glucose reverse part of the sequence that breaks down glucose in cellular oxidations (see Chapter 7).

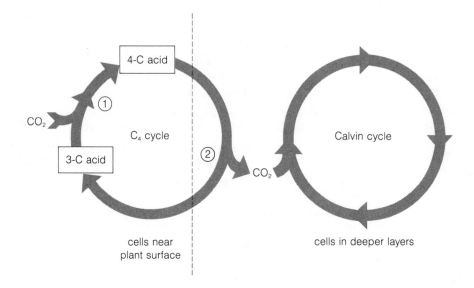

Figure 6-20 The C_4 cycle, a pathway that increases the efficiency of CO_2 utilization in some plants (see text).

lose, and a wide variety of other organic molecules. In addition to carbohydrates, amino acids, lipids, and proteins also become labeled rapidly in illuminated chloroplasts supplied with radioactive CO_2. All the amino acids required for protein synthesis can be synthesized by most plants, either inside chloroplasts or in the surrounding cytoplasm, by pathways starting from products of the dark reactions. Protein synthesis, occurring on ribosomes suspended in the chloroplast interior, can also be detected inside chloroplasts, along with linkage of nucleotides into DNA and RNA (see Supplement 9-2).

The balance among carbohydrates, fats, and amino acids synthesized in chloroplasts varies widely among different species of plants. In some, nearly all the CO_2 absorbed is incorporated into carbohydrates such as sucrose, formed from glucose and another six-carbon sugar, fructose. In other species, such as the alga *Chlorella*, the synthesis of fats and amino acids greatly exceeds sucrose synthesis, which may account for 5 percent or less of the

---

Chloroplasts have a synthetic capacity essentially equivalent to entire cells.

CO_2 absorbed. In any event, the chloroplasts of most species contain the enzymes and chemicals required to make a wide variety of substances in addition to carbohydrates—in fact, their capacity to synthesize molecules is practically equivalent to that of entire cells.

The C_4 Cycle

The Calvin cycle is supplemented in some plants by an important secondary cycle that carries out a biochemical trick to get around a deficiency in one of the Calvin cycle enzymes. The deficient enzyme is the one that catalyzes the first step in the Calvin cycle in which CO_2 combines with the 5C, 2P sugars to produce two molecules of a 3C, 1P product (reaction 1 in Figure 6-19). The defect in the enzyme results from the fact that oxygen can substitute for CO_2 in this reaction, leading to the formation of products that cannot be utilized in the Calvin cycle and are lost to the dark reactions.

The supplementary C_4 cycle gets around this deficiency by combining CO_2 with a three-carbon acid to produce the four-carbon acid for which the cycle is named (step 1 in Figure 6-20). This step occurs instead of the Calvin cycle in cells at the surface of the plant, where oxygen is abundant enough to interfere with the deficient Calvin cycle enzyme. The four-carbon acid then diffuses to deeper layers of the plant, where it is broken down to release CO_2 to the Calvin cycle in regions where oxygen is present in concentrations that are too low to interfere with the enzyme (step 2 in Figure 6-20). The three-carbon acid produced at this step diffuses back to the surface of the plant to pick up another molecule of CO_2. The C_4 cycle, which greatly increases the efficiency of photosynthesis, occurs in important crop grasses such as corn, sugarcane, and sorghum. (The C_4 cycle is discussed in greater detail in Supplement 6-1.)

WHERE PHOTOSYNTHESIS OCCURS
INSIDE CHLOROPLASTS

Chloroplasts are relatively easy to isolate in quantity from the leaves of plants such as spinach. Once isolated and cleaned of cellular debris, the chloroplasts can be further broken down by grinding or homogenization. Two fractions are produced by this treatment: (1) a membrane fraction, including thylakoid membranes and fragments of the boundary membranes, and (2) a soluble fraction without membranes, originating from the chloroplast stroma. These membrane and stroma fractions retain much of their biochemical activity when isolated. Using preparations of this type, it has been possible to determine where the various light and dark reactions occur in the inner compartments of the chloroplast.

Preparations of thylakoid membranes contain all the parts and pieces of the light reactions—including the chlorophylls, carotenoids, photosystems I and II, and all the

---

The light reactions are associated with thylakoid membranes in chloroplasts.

---

molecules associated with electron transport, reduction of NADP, and ATP synthesis. The soluble fraction derived from the stroma contains all the enzymes and chemicals of the dark reactions. These evidently occur in solution

---

The dark reactions are in solution in the chloroplast stroma.

---

inside chloroplasts, without attachment to the internal membrane systems.

The light and dark reactions of chloroplasts provide the raw materials needed for synthesis of the proteins, lipids, carbohydrates, and nucleic acids of the eukaryotic plants. All that most plants require in addition is a supply of inorganic minerals. The energy needed for synthesis of these molecules is derived from the light reactions of chloroplasts. The molecules of the eukaryotic plants are used in turn as energy sources for most of the remaining organisms of the world. (The pathways of photosynthesis in the prokaryotes are described in Supplement 6-2.)

Questions

1. Trace the flow of energy from sunlight through the living organisms on earth.

2. What structures are visible inside chloroplasts in the electron microscope? What is a thylakoid? What is the relationship between thylakoids and grana?

3. Write the equation for photosynthesis of glucose from CO_2 and H_2O. How is this equation related to van Niel's general equation for photosynthesis?

4. What experiment proved that the oxygen evolved in photosynthesis is derived from water and not CO_2?

5. What substances link the light and dark reactions together?

6. What makes some molecules appear pigmented or colored? What happens when light is absorbed by a pigmented molecule?

7. What molecules absorb light in chloroplasts? What colors do they absorb? Why do chloroplasts appear green?

8. What is the central event in the conversion of light to chemical energy?

9. What kinds of molecules are present in photosystems I and II?

10. What are chlorophyll P700 and P680?

11. What kinds of electron carriers appear in the Z-pathway?

12. Trace the flow of electrons through the Z-pathway. Where does hydrogen enter and leave the pathway? What significance does this have for the mechanism synthesizing ATP?

13. How is ATP synthesis in chloroplasts related to active transport?

14. How is ATP synthesis in chloroplasts related to thylakoid membranes?

15. How were the dark reactions worked out?

16. How are ATP and reduced NADP used in the dark reactions? How does CO_2 enter the dark reactions?

17. What major kinds of biological molecules are synthesized in chloroplasts?

18. Where are the light and dark reactions located in chloroplast ultrastructure?

19. What is the C_4 cycle? What is the significance of this cycle to plants that possess it?

Suggestions for Further Reading

Clayton, R. K. 1970. *Light and Living Matter.* Vols. 1 and 2. McGraw-Hill, New York.

Govindjee, and R. Govindjee. 1974. "The Primary Events of Photosynthesis." *Scientific American* 231:68–82.

Hinkle, P. C., and R. E. McCarty. 1978. "How Cells Make ATP." *Scientific American* 238:104–123.

Kirk, J.T.O., and R.A.E. Tilney-Bassett. 1978. *The Plastids.* 2nd ed. Elsevier/North-Holland, New York.

Lehninger, A. L. 1971. *Bioenergetics.* 2nd ed. Benjamin, Menlo Park, California.

Miller, G. T., Jr. 1971. *Energetics, Kinetics and Life.* Wadsworth, Belmont, California.

Wolfe, S. L. 1981. *Biology of the Cell.* 2nd ed. Wadsworth, Belmont, California.

SUPPLEMENT 6-1:
FURTHER LIGHT ON THE DARK REACTIONS

The Calvin Cycle

The Calvin cycle uses CO_2, ATP, and reduced NADP as raw materials. As a primary product the cycle releases a three-carbon sugar, 3-phosphoglyceraldehyde (3PGAL), which is used to synthesize more complex carbohydrates. Through various reactions, 3PGAL also enters the synthesis of lipids, amino acids, proteins, and nucleic acids. The cycle releases ADP, inorganic phosphate, and oxidized NADP. As the cycle turns, all the intermediate compounds in the sequence are continuously regenerated.

The important intermediate reactions are shown in Figure 6-21. In the first reaction of the cycle, CO_2 combines directly with the five-carbon, two-phosphate substance *ribulose 1,5-diphosphate (RuDP)*. The reaction, catalyzed by the enzyme *ribulose 1,5-diphosphate carboxylase (RuDP carboxylase)*, produces two molecules of *3-phosphoglyceric acid*, a three-carbon substance. One of these contains the newly incorporated CO_2 in the position marked by an asterisk in Figure 6-21. The overall reaction requires no input of energy because the two three-carbon products exist at a much lower energy level than RuDP, which can be considered a high-energy substance. The RuDP carboxylase enzyme catalyzing this reaction makes up as much as 25 to 50 percent of the total protein of chloroplasts. This high RuDP carboxylase content is believed to be related to the fact that the enzyme reacts fairly slowly with CO_2. The high concentration of RuDP carboxylase in chloroplasts evidently compensates effectively for the low reactivity of the enzyme.

In the next step in the cycle (reaction 2) another phosphate group is added to 3-phosphoglyceric acid to produce *1,3-diphosphoglyceric acid*. This conversion, an "uphill" reaction, is made possible by the simultaneous breakdown of ATP to ADP, from which the added phosphate is derived:

3-phosphoglyceric acid + ATP →

$$1,3\text{-diphosphoglyceric acid} + ADP \quad (6\text{-}5)$$

This highly reactive three-carbon, two-phosphate sugar is reduced in the next step (reaction 3) by accepting two electrons and hydrogen from NADP to produce 3PGAL. At the same time, one of the phosphate groups added in reaction 2 is removed and released to the medium as inor-

ganic phosphate. The products contain less energy and order than the reactants, and the total reaction proceeds spontaneously with the release of free energy:

1,3-diphosphoglyceric acid + reduced NADP →

$$3PGAL + \text{oxidized NADP} + HPO_4^{2-} \quad (6\text{-}6)$$

Reactions 2 and 3 in Figure 6-21 are shown multiplied by a factor of 2 because two molecules of 3PGAL are produced for each molecule of CO_2 interacting with RuDP in reaction 1.

The remainder of the cycle replaces the RuDP used in the first reaction in the sequence and leaves one molecule of 3PGAL as a surplus to the cycle. These reactions operate in the following way. Three turns of the cycle as far as reaction 3 produce six molecules of 3PGAL. These three molecules collectively contain 18 atoms of carbon. Five of the 3PGAL molecules, containing a total of 15 carbons, enter the complex series designated as reaction system 4 in Figure 6-21, yielding three molecules of ribulose 5-phosphate (15 total carbons). These are converted to RuDP (reaction 5 in Figure 6-21) at the expense of one additional molecule of ATP for each molecule of RuDP produced. This replaces the three molecules of RuDP used in the three turns of the cycle. The surplus molecule of 3PGAL is the starting point of a wide variety of reactions yielding sugars, starch, and other complex molecules in the chloroplast. Because three turns of the cycle are required to produce one of these three-carbon molecules as a surplus, each single turn may be considered to yield a one-carbon carbohydrate unit (CH_2O). Six turns would be required to generate enough units to manufacture one six-carbon molecule of glucose.

Tracing one complete turn of the cycle reveals that for each molecule of CO_2 fixed and each unit of carbohydrate generated, three molecules of ATP and two molecules of reduced NADP are converted to 3ADP and 2 oxidized NADP:

$$CO_2 + 3ATP + 2 \text{ reduced NADP} \rightarrow$$

$$(CH_2O) + 2 \text{ oxidized NADP} + 3ADP + 3HPO_4^{2-} \quad (6\text{-}7)$$

The oxidized NADP, ATP, and inorganic phosphate cycle back to the light reactions of photosynthesis to be converted back to ATP and reduced NADP in the Z-pathway.

More on the C_4 Pathway

The C_4 cycle was discovered when P. Kortschak and his colleagues in Hawaii and M. D. Hatch and C. R. Slack in

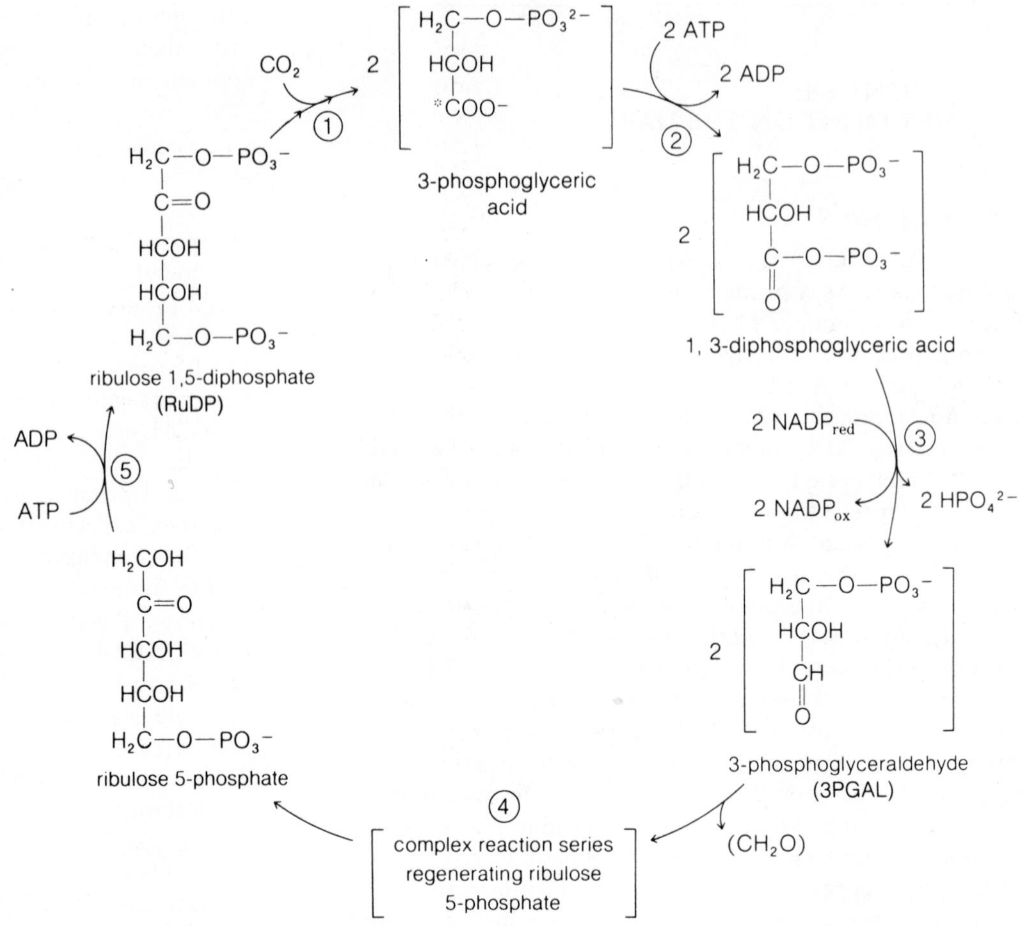

Figure 6-21 Major reactions and intermediate compounds of the Calvin cycle (see text).

Australia looked for the earliest labeled intermediates produced in corn, sugarcane, and other grasses of tropical origin after exposure of the plants to labeled CO_2. Surprisingly, the earliest label appeared in a mixture of four-carbon acids instead of the three-carbon 3PGAL of the Calvin cycle. (Hence the name "C_4 cycle"; the Calvin cycle, which produces a three-carbon molecule as its end product, is sometimes called the "C_3 cycle.") Intermediates of the Calvin cycle were also found to be labeled after a delay of some seconds following appearance of label in the four-carbon acids.

The appearance of label in the four-carbon acids was proposed by Hatch and Slack to be part of a side cycle linked to the main Calvin cycle (Figure 6-22). As a preparatory step in the side cycle, *pyruvic acid*, a three-carbon

substance, is pushed to a higher-energy form by linkage to a phosphate group derived from ATP (step 1 in Figure 6-22). The product, a three-carbon, one-phosphate acid called *phosphoenolpyruvic acid*, then reacts with CO_2 to produce *oxaloacetic acid*, a four-carbon substance (step 2 in Figure 6-22). The phosphate group is removed and released to the medium in this step. Oxaloacetic acid is then reduced to *malic acid* in the next step in the cycle (step 3 in Figure 6-22). The electrons added to form malic acid are donated by NADP, which is converted to the oxidized form as a result. The CO_2 carried by malic acid is released in the next step (step 4 in Figure 6-22), an oxidation in which electrons as well as CO_2 are removed from malic acid. NADP accepts the electrons removed in this step and is converted back to reduced NADP, replacing the mole-

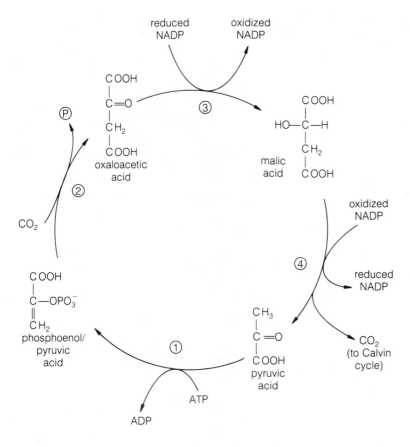

Figure 6-22 Major reactions of the C_4 cycle (see text).

cule of reduced NADP used in step 3 of the cycle. NADP is thus converted between its oxidized and reduced forms within the cycle so that there is no net gain or loss of this electron carrier. The CO_2 released enters the Calvin cycle by the regular route—combination with RuDP—and the Calvin cycle turns as usual. The other product of reaction 4, pyruvic acid, replaces the pyruvic acid used in the first step of the C_4 pathway.

The C_4 pathway appeared at first to be a "futile" cycle that results only in net breakdown of ATP. Investigators soon realized, however, that it increases availability of CO_2 to the Calvin cycle by compensating for the imperfection in the activity of RuDP carboxylase, the enzyme catalyzing the first uptake of CO_2 in the Calvin cycle. As noted in the text, oxygen can compete effectively with CO_2 for the active site on the RuDP carboxylase enzyme, diverting it from its central role in the Calvin cycle to one in which it adds oxygen to RuDP. The products of this reaction are lost to the Calvin cycle, and the cycle fails to turn. Plants

with the C_4 pathway, such as grasses, are able to get around this deficiency in RuDP carboxylase through the C_4 cycle.

In grasses, the C_4 cycle occurs in the chloroplasts of cells that lie close to the external surface of the plant. Because these cells are near the plant's surface, oxygen is present inside them in relatively high concentration. These cells lack RuDP carboxylase, however, effectively preventing loss of RuDP to carbohydrate production through the faulty activity of the enzyme. Carbon dioxide is taken up instead in the C_4 pathway, leading to production of an extensive pool of malic acid. Malic acid from these cells diffuses to deeper tissues of the plant, where it is broken down to release CO_2 in quantity. The chloroplasts of these internal cells contain RuDP carboxylase in normal amounts. Since oxygen is present in reduced concentration in the deeper layers of the plant, the RuDP carboxylase carries out its Calvin cycle function of CO_2 fixation normally, without diversion to the use of oxygen. The end result is much greater efficiency in the activity of RuDP carboxylase in CO_2 fixation—and, consequently, much greater efficiency in photosynthesis. Under optimum conditions, plants using the C_4 pathway carry out photosynthesis at about twice the rate of plants lacking the pathway.

Plants with the C_4 pathway are found primarily in tropical and subtropical regions, particularly in arid habitats. Water loss is reduced in these plants through a side effect of the C_4 pathway on the minute openings in leaves that admit CO_2 and allow water vapor to escape. The greater efficiency in CO_2 uptake leads to reduction in the size of these openings and in the amounts of water lost through them. As a result, C_4 plants are about twice as economical in water use as plants lacking the pathway.

SUPPLEMENT 6-2:
PHOTOSYNTHESIS IN THE PROKARYOTES

The two groups of prokaryotes, the bacteria and blue-green algae, differ fundamentally in their photosynthetic mechanisms. The photosynthetic bacteria possess a comparatively primitive system limited essentially to the activities carried out by photosystem I in the eukaryotic plants. Because photosystem II is absent, the photosynthetic bacteria cannot use water as an electron and hydrogen donor and do not release oxygen in photosynthesis. In contrast, the blue-green algae, although typically prokaryotic in cellular organization, carry out photosynthesis by essentially the same mechanisms as higher plants.

The differences between the photosynthetic bacteria and the remaining photosynthetic organisms of the world are primarily in the photosystems and light reactions. All photosynthetic groups, prokaryotic and eukaryotic, use the Calvin cycle to fix CO_2 and carry out the dark reactions in essentially the same way.

Bacterial photosynthesis is limited to two groups named by the color of their cells: the *purple* and *green* photosynthetic bacteria. These bacteria contain a distinct family of chlorophylls, called **bacteriochlorophylls,** that differs slightly from the chlorophylls of eukaryotic plants (Figure 6-23). These bacterial chlorophylls absorb light most strongly in the far red wavelengths, which are almost invisible to humans, and transmit almost all wavelengths of visible light. As a result, the bacteriochlorophylls contribute no distinctive color to either bacterial group. Their colors come instead from carotenoid pigments also present in the bacteria. In one group, the bacterial carotenoids absorb

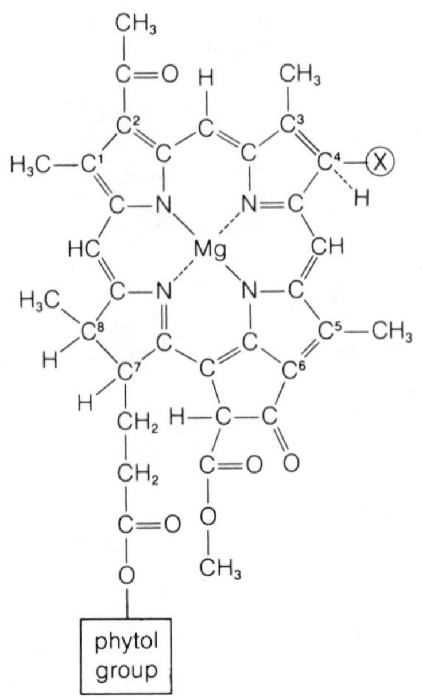

Figure 6-23 Bacteriochlorophyll. In bacteriochlorophyll *a*, X = C_2H_5; an additional —H occurs at C_4 in bacteriochlorophyll *a*. In bacteriochlorophyll *b*, X = CH—CH_3.

most visible wavelengths but transmit green; in the other, purple light is transmitted.

Since the purple and green photosynthetic bacteria do not possess photosystem II, they cannot split water to provide electrons and hydrogen for photosynthesis. Thus in the overall reaction

$$CO_2 + 2H_2D \rightarrow (CH_2O) + H_2O + 2D \qquad (6\text{-}8)$$

the substance H_2D may be hydrogen (H_2)

$$CO_2 + 2H_2 \rightarrow (CH_2O) + H_2O \qquad (6\text{-}9)$$

or hydrogen sulfide (H_2S)

$$CO_2 + 2H_2S \rightarrow (CH_2O) + H_2O + 2S \qquad (6\text{-}10)$$

Other donors of electrons and hydrogen are also used as raw materials by different photosynthetic bacteria. All these usable donors release electrons at relatively high en-

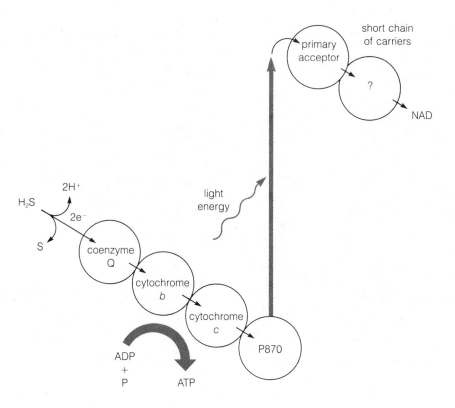

Figure 6-24 The tentative arrangement of carriers and the photosystem in a bacterial system using H_2S as electron donor.

ergy levels when oxidized; the electrons released from water exist at energy levels too low to enter the system.

From the donor the electrons flow through a chain of carriers believed to be set up as shown in Figure 6-24. The chain includes coenzyme Q (see Figure 7-10), a molecule closely similar to plastoquinone in structure, and two cytochromes, *b* and *c*. From the chain, the electrons pass to the reaction center of the bacterial photosystem, which consists of a specialized bacteriochlorophyll called P870. From P870 the electrons, after absorbing light energy, flow to the primary acceptor and finally to the final electron acceptor, which in bacteria is *NAD* (*nicotinamide adenine dinucleotide;* see Figure 7-5). NAD is identical to NADP except for an extra phosphate group (shaded in Figure 6-16) that is present in NADP but absent in NAD.

The electron carriers and the single photosystem in bacteria are tightly bound to membranes, as the equivalent systems are in eukaryotes. In bacteria, these photosynthetic membranes form part of the plasma membrane or are suspended as closed sacs in the bacterial cytoplasm (Figure 6-25). No structures equivalent to chloroplasts are present in bacteria.

Movement of electrons through the carriers produces an H^+ gradient across the photosynthetic membranes of bacteria, as it does in eukaryotic plants. This gradient leads to ATP synthesis through the activity of a membrane transport protein also bound to the bacterial membranes.

The blue-green algae use chlorophyll *a* in photosynthesis in combination with essentially the same carotenoid pigments and electron carriers as the eukaryotic plants. As in the eukaryotes, the pigments are organized into two photosystems interconnected by electron carriers into a Z-pathway. Consequently, the blue-green algae are able to use water as a source of electrons and hydrogen for photosynthesis; oxygen is evolved as a by-product. As far as can be determined, photosynthesis in these prokaryotes

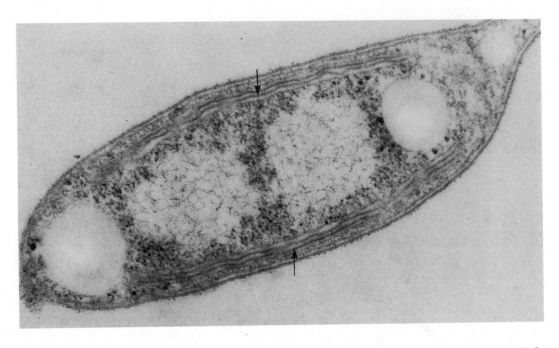

Figure 6-25 Photosynthetic membranes (arrows) in the cytoplasm of a bacterium. ×52,500. Courtesy of W. C. Trentini and the American Society for Microbiology.

resembles the systems of eukaryotic plants rather than bacteria, and all or most of the pathway shown in Figure 6-16 occurs in the blue-green algae. However, the molecules carrying out the light reactions in the blue-green algae are bound to membranous sacs suspended directly in the cytoplasm as they are in bacteria (Figure 6-26); no chloroplasts are present in these prokaryotic organisms.

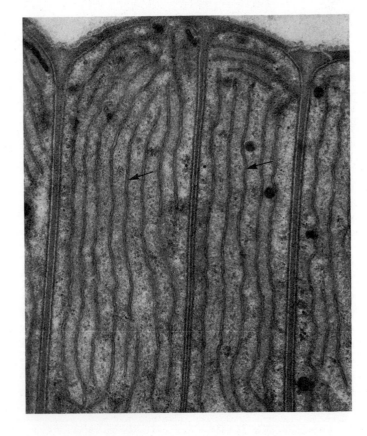

Figure 6-26 Thylakoid membranes (arrows) suspended directly in the cytoplasm of a row of blue-green algae cells. ×44,500. Courtesy of N. J. Lang.

The products of photosynthesis are used as an energy source by both plants and animals. Plants break down the photosynthetic products directly as an energy source; animals use the same products directly or indirectly by eating plants or each other. Of the food molecules used as an energy source, carbohydrates and lipids of various kinds are most important to both plant and animal cells. Almost all biological molecules, however, including proteins and nucleic acids, can be broken down by cells to release energy.

The products of photosynthesis are used as fuel substances for cellular oxidation.

The reactions breaking down these substances are oxidations in which high-energy electrons are removed from the fuel molecules and transferred to acceptors of various kinds. Much of the energy released in these transfers is used to synthesize ATP—the energy "dollar" of the cell.

The reactions producing ATP take place in two major stages in the cytoplasm of eukaryotic cells (Figure 7-1). In the first stage, the food molecules—whether carbohydrates, lipids, proteins, or nucleic acids—are partly oxidized and converted into short two- or three-carbon segments. Only limited quantities of ATP are produced in this first stage. The products of the first stage are the immediate fuels of the second major stage, in which the short carbon chains are oxidized completely to carbon dioxide and water. Almost all the ATP produced in cells arises from the reactions of the second stage.

The reactions in the first stage of cellular oxidations are distributed primarily in the cytoplasm outside mitochondria. All the reactions of the second stage, and thus the major production of ATP, occur inside mitochondria.

Mitochondria are the powerhouses of the cell.

THE STRUCTURE OF MITOCHONDRIA

In most cells, mitochondria are spherical or slightly elongated bodies that approximate the dimensions of a bacterial cell. Apparently mitochondria are not rigid; they may take on a variety of shapes in the cytoplasm. They can be observed under the light microscope to change slowly between spherical and longer filamentous forms and to fuse together and divide.

The electron microscope shows that mitochondria contain two separate membrane systems (Figures 7-2 and

7

THE FLOW OF ENERGY IN LIVING ORGANISMS II: ENERGY RELEASE IN GLYCOLYSIS AND RESPIRATION

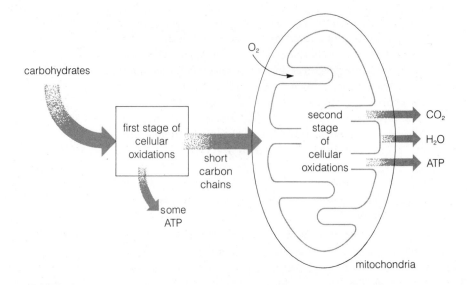

Figure 7-1 The major stages of cellular oxidation (see text).

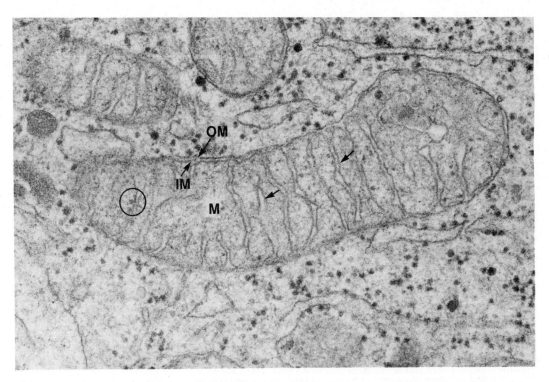

Figure 7-2 A mitochondrion from an intestinal cell of the chick. The outer boundary membrane (*OM*) is smooth and covers the entire mitochondrion; the inner boundary membrane (*IM*) is thrown into folds called cristae (arrows) that extend into the mitochondrial interior. The matrix (*M*), the inner substance enclosed by the two membranes, contains proteins and other molecules in solution. Mitochondrial ribosomes (circle) are also present in the matrix. ×83,000. Courtesy of J. J. Mais.

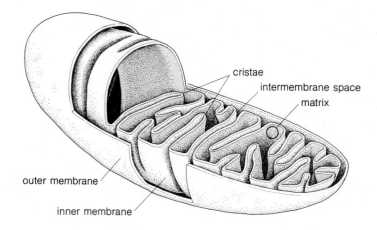

cristae
intermembrane space
matrix
outer membrane
inner membrane

Figure 7-3 The membranes and compartments of a mitochondrion.

Almost all eukaryotic cells contain mitochondria. The exceptions are a few metabolically inert types such as the red blood cells of higher animals, which contain no cytoplasmic organelles of any kind when fully mature. A few algae and some protozoa contain only a single mitochondrion. Most cells, however, contain from several hundred to a thousand mitochondria. Some may contain many more. Liver cells in higher animals may contain from slightly less than a thousand to more than 2500 mitochondria. A few very large cells, such as the egg cells of animals, may have hundreds of thousands of mitochondria.

THE FIRST STAGE OF CELLULAR OXIDATIONS: GLYCOLYSIS

The initial reactions breaking cellular fuels into three-carbon segments take place in the cytoplasm outside mitochondria. These reactions all follow a similar pattern, which is best illustrated by the oxidation of glucose, a central fuel molecule in all eukaryotic plants and animals. The series of reactions breaking this molecule into shorter carbon chains is called **glycolysis.**

The Reactions of Glycolysis

The overall reactions of glycolysis break the six-carbon glucose molecule into two three-carbon fragments. (For an overview of the reactions of glycolysis, see Information Box 7-1.) During glycolysis a single oxidation occurs and a small amount of ATP is generated. The three-carbon fragments produced become the fuel for the oxidations taking place inside mitochondria, which produce much larger quantities of ATP.

The reactions of glycolysis are easiest to follow if only the carbon chains and phosphate groups are considered (Figure 7-4). (A more thorough description of glycolysis is given in Supplement 7-1.) The first series of reactions (reactions 1 through 3 in Figure 7-4) converts glucose (six carbons) into a more reactive substance by adding two phosphate groups at the expense of two molecules of ATP. At the end of this sequence, a six-carbon, two-phosphate sugar is produced (6C, 2P in Figure 7-4). These initial reactions of glycolysis go "uphill" in terms of the energy content of the products. They proceed to completion only because they are coupled to the breakdown of 2ATP, converted in the process to 2ADP.

The ATP used in these initial steps is recovered with a net gain in the remaining reactions of glycolysis (reactions 4 through 9 in Figure 7-4). These reactions break the

7-3). One membrane forms a single, continuous, and relatively smooth outer layer around the mitochondrion, completely separating the interior from the rest of the cytoplasm. Closely lining this membrane, inside the mitochondrion, is a second membrane. The two boundary membranes, visible in Figure 7-2, are roughly analogous to the outer and inner boundary membranes of chloroplasts. The inner boundary membrane is thrown into folds or tubular extensions called **cristae** (singular = *crista*) that reach into the inner cavity of the mitochondrion. These folds take many forms in the mitochondria of different tissues and species. Among the most common is the arrangement shown in Figures 7-2 and 7-3, in which the cristae consist of flattened, saclike folds extending across the interior of the mitochondrion, more or less at right angles to the boundary membranes.

The outer and inner boundary membranes separate the mitochondrial interior into two distinct regions (see Figure 7-3): the compartment between the inner and outer membranes, and the **matrix,** the innermost compartment enclosed by the inner mitochondrial membrane. The space inside the cristae is continuous with the compartment between the inner and outer boundary membranes.

The mitochondrial matrix, which is analogous in location to the chloroplast stroma, may contain a variety of structures, including large dense granules, fibrils, and crystals. Other smaller granules, the *mitochondrial ribosomes,* occur in the matrix in large numbers (circle, Figure 7-2). Also embedded within the matrix are scattered deposits of DNA, closely similar in structure and properties to bacterial DNA (for details, see Supplement 9-2).

Glycolysis breaks glucose into three-carbon segments and yields ATP and reduced NAD.

six-carbon, two-phosphate product of the first series into two three-carbon fragments—yielding reduced NAD (nicotinamide adenine dinucleotide; see Figure 7-5) in addition to a net gain in ATP. In the first of these reactions (reaction 4 in Figure 7-4), the six-carbon, two-phosphate sugar is broken into two three-carbon, one-phosphate fragments. The two three-carbon sugars produced are different; only

one of them, 3PGAL (3-phosphoglyceraldehyde, already familiar as a product of the dark reactions of photosynthesis), directly enters the next step in glycolysis. As this three-carbon sugar is used, however, a rearranging enzyme converts the second three-carbon sugar into 3PGAL. Thus both sugars produced by reaction 4 are used in the remainder of the sequence.

The 3PGAL produced in reaction 4 is a high-energy substance. Some of its energy is tapped off in the next step (reaction 5 in Figure 7-4), an oxidation. Remember that an oxidation involves removal of electrons and does not require reaction with an atom of oxygen (see Information

---

Information Box 7-1   Glycolysis: An Overview

The reactions of glycolysis break glucose (with six carbons) into two molecules of pyruvic acid (with three carbons each). In the process, ATP and reduced NAD are produced. Glycolysis takes place in two parts. The first part raises six-carbon molecules derived from glucose to higher energy levels, using up ATP in the process:

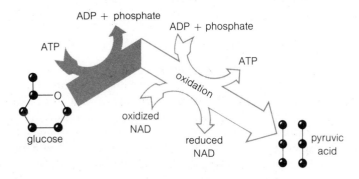

In the second part the high-energy derivatives of glucose are oxidized and split into three-carbon units of pyruvic acid. The second part produces reduced NAD and yields more ATP than the first part uses:

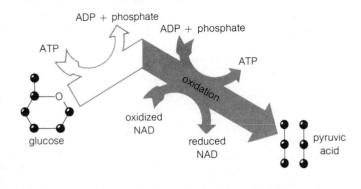

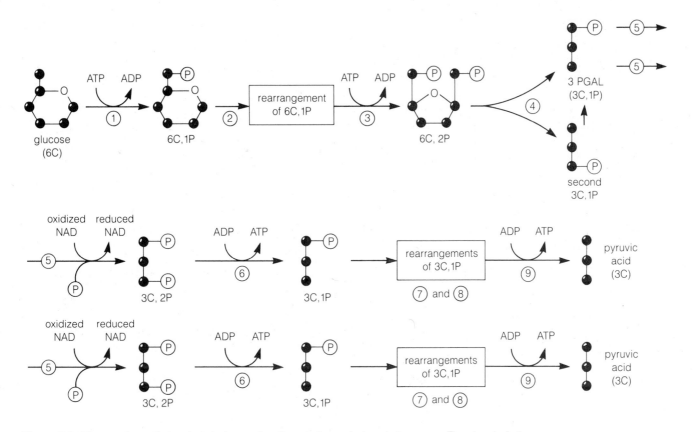

Figure 7-4 The reactions of glycolysis in terms of carbon chains and phosphate groups. The chemical structures of the intermediate molecules and the enzymes in the sequence are shown in Figure 7-16.

Box 6-1). In this step, two high-energy electrons are removed from 3PGAL and transferred to the acceptor molecule NAD. Except for the absence of one phosphate group, NAD (Figure 7-5) is identical to the primary electron acceptor of photosynthesis, NADP. (Compare Figures 7-6 and 6-16.) Two hydrogens ($2H^+$) are removed from 3PGAL at the same time. One of the hydrogens attaches to NAD during oxidation of the three-carbon sugar to form NADH. The second is released to the medium as an H^+ ion. The reduced NAD (or NADH) formed at this step is another high-energy substance. Most of the energy it carries is eventually used to convert ADP to ATP.

Transferring electrons from 3PGAL to NAD releases a large amount of free energy because, even though reduced NAD is a high-energy substance, the energy level of the transferred electrons is much lower in NAD than in 3PGAL. Some of the energy lost by the electrons as they are transferred to NAD is used to attach a second phosphate group to 3PGAL, yielding a highly reactive three-carbon, two-phosphate sugar (the 3C, 2P substance following reaction 5 in Figure 7-4). This second phosphate is

derived from inorganic phosphate ions in the surrounding cytoplasm and not from ATP.

The final reactions of glycolysis (reactions 6 through 9 in Figure 7-4) remove the two phosphate groups one at a time and transfer them to 2ADP, yielding 2ATP. In the first of these transfers (reaction 6), one phosphate is removed from the three-carbon, two-phosphate sugar to produce a three-carbon, one-phosphate sugar. Part of the free energy released at this step is captured in the transfer of the phosphate group to ADP. The reaction yields one molecule of ATP for each three-carbon, two-phosphate molecule entering the reaction. Because two molecules of the three-carbon, two-phosphate sugar enter this reaction for every molecule of glucose originally entering glycolysis, there is a gain of 2ATP at this step, replacing the two molecules used in reactions 1 through 3.

The reaction removing the first of the two phosphates is worth discussing in some detail, because it illustrates one method by which the energy of oxidation is converted into a usable chemical form in the cell. If 1 mole of the 3C, 2P molecule entering the reaction, a substance called

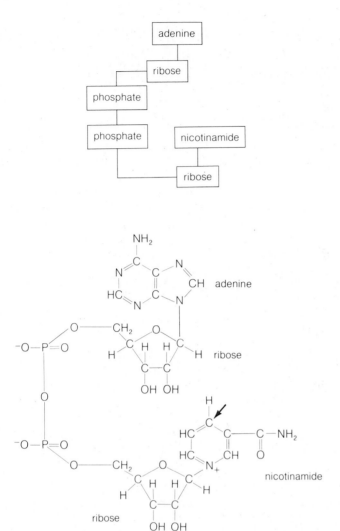

Figure 7-5 NAD, nicotinamide adenine dinucleotide, a carrier of high-energy electrons, has properties closely similar to those of NADP (compare with Figure 6-16). In the reduced form, electrons and a hydrogen (H⁺) are added at the position marked by the arrow.

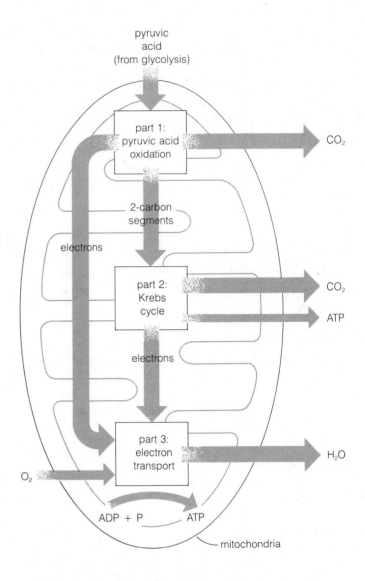

Figure 7-6 The three major parts of the oxidative reactions of mitochondria (see text).

1,3-diphosphoglyceric acid, breaks down directly to the 3C, 1P product, called *3-phosphoglyceric acid*, the reaction releases between 10,000 and 15,000 calories. In glycolysis, the reaction is coupled to ATP synthesis. The combined reaction still releases energy, but at the reduced level of about 4500 cal/mole:

1,3-diphosphoglyceric acid + ADP →

$$\text{3-phosphoglyceric acid} + \text{ATP} \qquad (7\text{-}1)$$

The difference in calories released between the two reactions represents energy captured in the formation of ATP.

The final phosphate group is removed from the 3C, 1P product of reaction 6 after two reactions (reactions 7 and 8 in Figure 7-4) that simply rearrange the atoms of the molecule. In the final reaction of glycolysis (reaction 9) the phosphate group of the rearranged 3C, 1P sugar is transferred to ADP, yielding one molecule of ATP for each molecule of 3C, 1P entering reactions 7 to 9. As in reaction 6, two molecules of 3C, 1P enter this final series for each molecule of glucose starting down the glycolytic pathway; hence a total of two molecules of ATP is produced at this step. This reaction thus provides a net gain of 2ATP for

---

Glucose oxidation produces two molecules of pyruvic acid, two molecules of reduced NAD, and two molecules of ATP.

---

the entire glycolytic sequence. The final product of glycolysis is *pyruvic acid*, a three-carbon substance containing no phosphate groups. It is still a relatively high-energy substance and provides the major fuel for the subsequent cellular oxidations inside mitochondria.

The final product of glycolysis, pyruvic acid, is the immediate fuel for the next series of cellular oxidations.

Each of the reactions in glycolysis is increased in rate by a different enzyme. (The glycolytic enzymes are listed individually in Figure 7-15.) All of the enzymes, and the

---

Each reaction of glycolysis is increased in rate by a specific enzyme.

---

reactants, intermediate substances, and products of glycolysis are suspended in solution in the background substance of the cytoplasm.

Net Products of Glycolysis

Subtracting the consumption of ATP in glycolysis from the amount produced shows that the overall reaction sequence provides a net gain of 2ATP for each molecule of glucose entering the sequence. For every molecule of glucose entering glycolysis, a molecule of ATP is broken down to ADP in reactions 1 and 3. Thus a total of 2ATP is lost in the initial part of the sequence. Two molecules of ADP are converted to ATP at reaction 6, however, and two more at reaction 9 for each molecule of glucose entering the pathway. The 4ATP gained in the second, oxidative part of glycolysis provide the net gain of 2ATP for each molecule of glucose oxidized to pyruvic acid. Another significant product is the reduced NAD (two molecules of reduced NAD for each molecule of glucose), which is reduced by accepting electrons at reaction 5. The electrons accepted by NAD, which exist at a high energy level, can do chemical work if passed from reduced NAD to a suitable electron acceptor elsewhere in the cell. The total reactants and products of glycolysis are therefore

$$\text{Glucose} + 2\text{ADP} + 2\text{HPO}_4^{2-} + \text{oxidized NAD} \rightarrow$$

$$2 \text{ pyruvic acid} + 2\text{ATP} + 2 \text{ reduced NAD} \qquad (7\text{-}2)$$

Fermentations and Other Variations of Glycolysis

Glycolysis is the central pathway for the initial breakdown of carbohydrates in both plant and animal cells. Starch in plants and glycogen in animals are both long, chainlike molecules made up from repeating glucose links (see Figure 2-11). These molecules enter the glycolytic pathway after being broken into individual glucose molecules. Other sugars, including a wide variety of monosaccharides and disaccharides, enter glycolysis after being converted by enzymes into one of the initial molecules of the pathway.

At the opposite end of the sequence, pyruvic acid may be modified to yield other final products. In one of the most important of these modifications, pyruvic acid is converted into *lactic acid* after accepting electrons from the NAD reduced earlier in the sequence. In this form of glycolysis, lactic instead of pyruvic acid accumulates as the final product of the pathway:

$$\text{Pyruvic acid} + \text{reduced NAD} \rightarrow \text{lactic acid} + \text{oxidized NAD} \qquad (7\text{-}3)$$

This modification is significant because it regenerates oxidized NAD, which is then free to cycle back to accept electrons in the oxidations of glycolysis. Because oxidized NAD is continually regenerated by this alternate pathway, glycolysis can continue to run with the net production of ATP. This pathway is vital to cells living temporarily or permanently in the absence of oxygen. (Reduced NAD transfers its electrons through a series of carriers to oxygen when the gas is present.) Lactic acid production occurs in the muscle cells of animals, including humans, if intensive, sustained physical activity is carried out before increases in breathing and heart rate have a chance to meet the demand for oxygen in muscle tissues. The lactic acid accumulating as a by-product is oxidized later, when the oxygen content of the muscle cells returns to normal levels (see Information Box 7-2).

Another glycolytic variation of importance to humans occurs in organisms such as yeast. In this modification, pyruvic acid accepts electrons from reduced NAD and is converted by additional reactions into *ethyl alcohol* (a two-carbon substance) and CO_2:

$$\text{Pyruvic acid} + \text{reduced NAD} \rightarrow$$

$$\text{ethyl alcohol} + CO_2 + \text{oxidized NAD} \qquad (7\text{-}4)$$

This variation is of central importance in human economy. The CO_2 released by yeast cells carrying out this reaction raises the dough used in baking bread, a staple of the human diet over much of the world. Yeast cells carrying out the same reaction also produce the alcohol and CO_2 that provide the basis of the brewing industry.

These variations in the glycolytic pathway, in which the electrons carried by reduced NAD are traded off to an *organic* substance such as pyruvic acid, are collectively

---

Fermentations are variations of glycolysis in which the electrons carried by reduced NAD are delivered to an organic substance.

---

called **fermentations.** In the alternate pathway, the electrons carried from glycolysis by reduced NAD eventually reach an *inorganic* substance: oxygen. Fermentations of various kinds, producing a wide variety of end products, are used as an ATP source by many species of bacteria. Some of these species, called **strict anaerobes,** are limited to glycolytic fermentations for ATP production and cannot use oxygen at any time as a final electron acceptor. Others can use either fermentations alone or both glycolysis and reactions equivalent to mitochondrial oxidations if oxygen is available. Bacteria in this category are called **facultative anaerobes.** A number of species, termed **strict aerobes,** are unable to live by fermentations alone. Many cells of higher organisms, including the muscle cells of vertebrates described above, are facultative and can switch between fermentation and complete oxidation depending on their oxygen supply. Others, such as brain cells, are strict aerobes.

The glycolytic pathway is also important in photosynthesis. The three-carbon sugar 3PGAL, which appears as an intermediate compound in glycolysis (reaction 4 in Figure 7-4), is also the primary product of the Calvin cycle in the dark reactions of photosynthesis (see Chapter 6). This molecule is converted into glucose in chloroplasts essentially by reversing the first four reactions of glycolysis. The glucose molecules produced in these steps may then be linked end to end into starch. Alternatively, 3PGAL may serve as a fuel substance by entering the glycolytic pathway at reaction 4 and proceeding in the other direction toward pyruvic acid or other products.

Appreciation of the central role of glycolysis in cell biology and biochemistry came as a gradual development of research first taken up in the late 1800s. In Germany, Eduard and Hans Buchner discovered that alcoholic fermentation could be carried out by nonliving extracts of yeast cells. Through this work, the first enzymes were discovered and described. Research with alcoholic fermentation, involving a great many investigators in various parts of the world, continued until well into this century. By the 1930s the emphasis in these investigations had shifted from alcoholic fermentation to the more general pathway of glycolysis. Our present understanding of the glycolytic pathway is based primarily on research carried out in the 1930s by two German scientists, Gustav Embden and Otto Meyerhoff. These researchers played so central a part in tracing the reactions and enzymes of glycolysis that the sequence is frequently called the Embden–Meyerhoff pathway in their honor. By 1940, glycolysis was known essentially as it is today.

THE SECOND STAGE OF CELLULAR OXIDATIONS: OXIDATION AND ATP SYNTHESIS IN MITOCHONDRIA

In the second major stage of cellular oxidations, pyruvic acid, the three-carbon molecule produced in the final step of glycolysis, is completely oxidized to CO_2 and water. A very large quantity of free energy is released in this

---

The second stage of cellular oxidation oxidizes pyruvic acid to CO_2 and H_2O.

---

stage, and the ATP produced in these reactions far exceeds that obtained through glycolysis. The reactions occur in three parts, all located inside mitochondria. In the first part, called **pyruvic acid oxidation,** pyruvic acid is shortened into a two-carbon segment and CO_2 is released. In the second part, the **Krebs cycle,** this two-carbon segment is completely oxidized to two molecules of CO_2. In the final part, **electron transport,** the electrons remoed in these oxidations are delivered through a series of electron carriers to oxygen. Much of the free energy released by this electron transport is used to drive the synthesis of ATP. Because oxygen is the final electron acceptor for these reactions, the oxidative activities of mitochondria are frequently termed **respiration.** When we breathe, we take in the oxygen needed to support these mitochondrial reactions.

Part 1 of Mitochondrial Respiration: Oxidation of Pyruvic Acid to Two-Carbon Segments

After entering mitochondria, pyruvic acid is shortened to a two-carbon segment in a series of reactions that remove two electrons (an oxidation), two hydrogens (as H^+), and one carbon from the pyruvic acid chain (as CO_2; see Information Box 7-3). The two-carbon segment produced is an *acetyl* (—CH$_3$CO$^-$) group:

$$\text{Pyruvic acid} \rightarrow \text{acetyl group} + 2 \text{ electrons} + 2\text{H}^+ + \text{CO}_2 \qquad (7\text{-}5)$$

The two electrons and one of the hydrogens removed from pyruvic acid in this oxidation are transferred to NAD, which is reduced in the process. The CO$_2$ and the remaining hydrogen ion are released to enter the surrounding medium. The two-carbon acetyl group is transferred to a carrier molecule called *coenzyme A* (Figure 7-7) to produce the high-energy substance *acetyl coenzyme A*. Most of the free energy released by the oxidation of pyruvic acid is captured as chemical energy in this substance.

Coenzyme A is another carrier molecule based on nucleotide structure that closely resembles ATP, NAD, and NADP (compare Figures 3-1, 6-16, 7-5, and 7-7). Coenzyme A accepts and carries two-carbon acetyl units, just as ATP carries phosphates and the NAD transports electrons.

Information Box 7-2

The 100-Yard Dash

The muscles of a runner poised at the start of a 100-yard dash are loaded with a high-energy substance capable of releasing the ATP required for the run. This substance, *creatine phosphate,* forms a storage reservoir for high-energy phosphate groups. Comparatively little ATP is stored in muscle; the reserves of creatine phosphate are about five times that of the ATP phosphate. However, phosphate groups are readily transferred from creatine phosphate to ATP in the reaction

$$\text{Creatine phosphate} + \text{ADP} \rightarrow \text{creatine} + \text{ATP}$$

The runner's muscles at the starting line also contain reserves of glycogen (animal starch), made up from long chains of glucose units linked end to end.

At the starting gun the runner springs from rest into sudden physical exertion, straining every muscle to the limit. The immediate energy for this contraction comes from the breakdown of the small reserves of ATP. Replenishment of these reserves by transfer of phosphate groups from creatine phosphate to ATP then begins, probably within the first second after the starting gun is fired.

Besides initiating the transfer of phosphate from creatine phosphate, the ADP produced by ATP breakdown after the gun also stimulates oxidation of fuel substances within the runner's muscles to provide energy for converting ADP to ATP. Glucose units, broken off one at a time from the glycogen molecules stored in muscle, are the fuel for this oxidation. Because the runner's breathing rate has not yet increased to its maximum level, the oxygen required for complete oxidation of glucose to water and CO$_2$ in his mitochondria is quickly depleted. As a result, the main weight of ATP formation falls on glycolysis.

Continued glycolysis forms reduced NAD and depletes oxidized NAD. The reduced NAD accumulating passes its electrons to the pyruvic acid produced by glycolysis, forming lactic acid that gradually increases in concentration in the runner's muscles. By the 50-yard mark, significant quantities of lactic acid have formed; by the 75-yard mark, lactic acid has begun to reach high concentrations in the runner's muscles and is released into his bloodstream. This contributes to a feeling of fatigue that the runner now begins to notice.

As the runner crosses the finish line his breathing and heart rate, in response to the demand for oxygen in his tissues, have reached high levels. Breathing and heart rate remain high, even though the runner has now finished the race. His body begins to oxidize the lactic acid, converting it to pyruvic acid and eventually back to glycogen. (These reactions take place in the liver.) Breathing and heart rate continue above resting levels until most of the lactic acid is oxidized. The extra oxygen required by the runner during this period pays off what is called the *oxygen debt*. This debt was accumulated during the maximal exertion of the 100-yard dash, when oxygen in the runner's muscles was depleted below the level required for oxidation of glucose to CO$_2$ and water. While the runner is resting, the creatine phosphate reserves return to their normal level through ATP produced by mitochondrial oxidations:

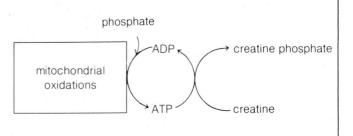

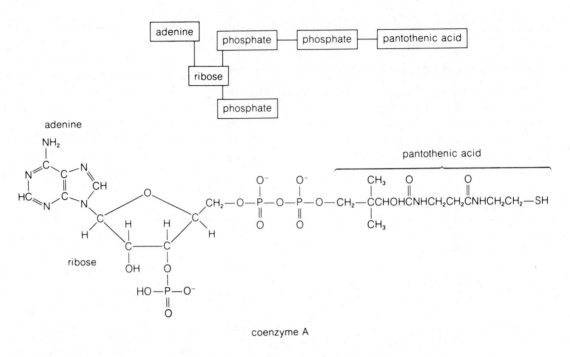

Figure 7-7 Coenzyme A, a carrier of acetyl (—CH_3CO^-) groups.

Information Box 7-3

Pyruvic Acid Oxidation: An Overview

The oxidation of pyruvic acid produces two-carbon acetyl groups, which are the main fuel for the second part of mitochondrial respiration. The reaction sequence uses pyruvic acid (with three carbons), oxidized NAD, and coenzyme A as raw materials. It produces carbon dioxide, reduced NAD, and two-carbon acetyl groups attached to coenzyme A:

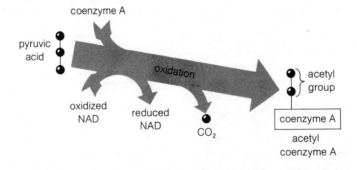

Pyruvic acid oxidation produces acetyl coenzyme A, reduced NAD, and CO$_2$.

The overall reaction sequence in pyruvic acid oxidation thus yields as net products acetyl coenzyme A, reduced NAD, and CO$_2$:

2 pyruvic acid + 2 oxidized NAD + 2 coenzyme A →

$\qquad$ 2 acetyl coenzyme A + 2 reduced NAD + 2CO$_2$ $\quad$ (7-6)

All these reactants and products are multiplied by 2 in Reaction 7-6 because pyruvic acid oxidation is considered as a continuation of glycolysis, in which two molecules of

The product of pyruvic acid oxidation, acetyl coenzyme A, is the fuel for the final reactions of cellular oxidations.

pyruvic acid are produced for each molecule of glucose entering the pathway.

Part 2 of Mitochondrial Respiration: Oxidation of Acetyl Units to CO$_2$ in Mitochondria

The Krebs Cycle The two-carbon acetyl groups carried by coenzyme A are oxidized to CO$_2$ in a cycle of reactions first described in 1937 by a British investigator, Hans Krebs, who received the Nobel Prize for his brilliant work with cellular oxidation. In the Krebs cycle there is a continuous input of two-carbon acetyl groups as reactants. Carbon dioxide, ATP, reduced NAD, and another electron carrier in reduced form, FAD (see Figure 6-13), are released

The Krebs cycle uses two-carbon acetyl groups as fuel and produces CO$_2$, ATP, and reduced NAD and FAD.

as products. As in the Calvin cycle of photosynthesis, the molecules forming intermediate parts of the Krebs cycle are continuously regenerated as the cycle turns. (For an overview of the Krebs cycle, see Information Box 7-4.)

Only one molecule of ATP is formed as a direct product of each turn of the Krebs cycle. Most of the energy released by the several oxidations of the cycle is trapped in the electrons carried from the cycle by reduced NAD and FAD. The energy carried by reduced NAD and FAD is used in the final reactions of mitochondrial oxidation to generate most of the ATP produced in cellular respiration.

The Krebs cycle (Figure 7-8) works in the following way. In the first reaction of the cycle (reaction 1 in Figure 7-8) the two-carbon acetyl unit carried by coenzyme A is transferred to a four-carbon molecule, *oxaloacetic acid*, to produce *citric acid*, a six-carbon molecule. The reaction is catalyzed by a specific enzyme, as are all the reactions of the Krebs cycle. (The enzymes are listed individually in

All the reactions of the Krebs cycle are catalyzed by specific enzymes.

Figure 7-17.) Coenzyme A, relieved of its acetyl unit by this reaction, is free to enter another cycle of pyruvic acid oxidation:

Acetyl coenzyme A + oxaloacetic acid →

$\qquad$ citric acid + coenzyme A $\quad$ (7-7)

The citric acid formed in the initial reaction is converted by a rearranging enzyme into a closely similar six-carbon acid, which enters the next reaction of the cycle. In this reaction (reaction 2 in Figure 7-8), two electrons and two hydrogens are removed from the six-carbon acid. The carbon chain is also shortened at this step, with the release of one molecule of carbon dioxide, yielding a five-carbon product.

Although NAD is the usual acceptor for the electrons removed in reaction 2, either NAD or NADP may be reduced, depending on the relative concentrations of ATP, ADP, and reduced NAD in the cell. Which of the acceptors is actually used illustrates one of the many control systems regulating cellular oxidation. The oxidation is catalyzed by either of two enzymes that are identical except that one uses NAD and the other NADP as acceptor for the electrons removed. When ATP concentrations are high, the enzyme using NAD is inhibited and NADP becomes the favored acceptor. When ATP concentrations are low, inhibition of the enzyme using NAD is released and NAD replaces NADP as the electron acceptor.

These differences are significant for ATP production because reduced NAD normally transfers its electrons to the electron transport system of mitochondria that generates ATP. Reduced NADP instead acts as electron donor for reactions of cell synthesis in which a reduction is required. As a result of this regulatory system, reaction 2 is finely tuned to the needs of the cell for ATP. Under the usual conditions, in which cellular activity demands a more or less continuous supply of ATP, the enzyme using NAD as electron acceptor predominates in Krebs cycle oxidation at this step, leading to ATP synthesis in response.

For this discussion we will consider NAD as the primary electron acceptor in the reaction.

The electrons removed in the next step in the Krebs cycle (reaction 3) are accepted by NAD, and the chain is shortened to four carbons with release of one molecule of CO_2. Enough free energy is released in the step to synthesize one molecule of ATP. Therefore reaction 3 yields a four-carbon product and a molecule each of CO_2, ATP, and reduced NAD. The ATP produced in this reaction is the only ATP originating directly from the Krebs cycle.

The electrons removed in the next step in the cycle (reaction 4) are accepted by *FAD* (*flavin adenine dinucleotide;* see Figure 6-13), a molecule that forms part of the electron transport system of mitochondria. In reaction 4, FAD is reduced by accepting two electrons and both of the two hydrogens removed, forming $FADH_2$ as the reduced product.

The four-carbon product of reaction 4 is rearranged, with the addition of a molecule of water, to form another four-carbon acid. This acid is then oxidized to oxaloacetic acid in the final reaction in the cycle (reaction 6). NAD is the electron acceptor for this final oxidation of the Krebs cycle. The molecule of oxaloacetic acid produced at this step replaces the oxaloacetic acid used in reaction 1, and the cycle is ready to turn again.

Overall Products of the Krebs Cycle We can now total up the overall products of the Krebs cycle and summarize the cellular oxidation of glucose to this point. As the Krebs cycle proceeds through one complete turn, one two-carbon acetyl group is consumed and two molecules of CO_2 are released. High-energy electrons are removed at each of four reactions in the cycle. At three of these steps, NAD is the acceptor (assuming that NAD is used in reaction 2),

---

Information Box 7-4

An Overview of the Krebs Cycle

The Krebs cycle oxidizes two-carbon acetyl groups into carbon dioxide. Acetyl coenzyme A, oxidized NAD, oxidized FAD, and ADP enter the cycle; coenzyme A, reduced NAD and FAD, ATP, and CO_2 are released as products.

Most of the energy released by the oxidations of the cycle is captured in the reduced NAD produced. All the intermediate molecules of the cycle are continuously regenerated as the cycle turns.

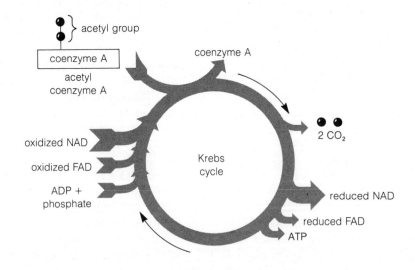

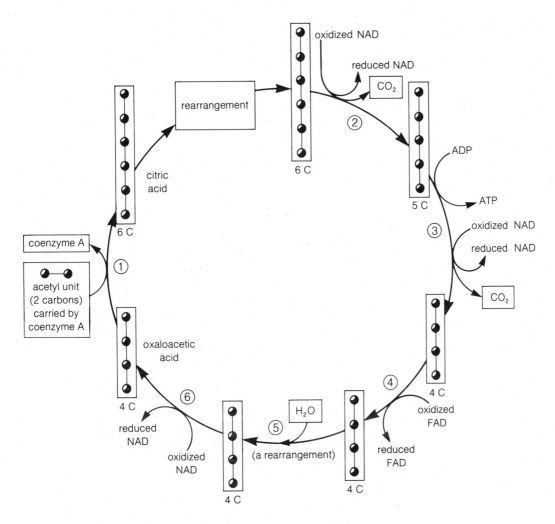

Figure 7-8 The Krebs cycle (see text). Only the carbon chains of intermediate molecules are shown in this diagram; the structures of these intermediates are shown in Figure 7-17.

producing three molecules of reduced NAD; one step reduces a molecule of FAD instead. In reaction 3 of the cycle, one molecule of ATP is generated. Oxaloacetic acid, used in the initial reaction of the cycle, is regenerated in the final reaction. Thus, as overall reactants and products, the Krebs cycle includes

An acetyl group (2 carbons) + 3 oxidized NAD +

oxidized FAD + ADP + HPO_4^{2-} →

$\qquad$ $2CO_2$ + 3 reduced NAD + reduced FAD + ATP $\quad$ (7-8)

With this information we can sum up the total products of the oxidation of glucose to carbon dioxide from glycolysis through the Krebs cycle. For each molecule of

glucose entering the series, the Krebs cycle will turn twice. Glycolysis and acetyl coenzyme A formation together yield 4 reduced NAD, 2ATP, and $2CO_2$ (from Reactions 7-2 and 7-6). Adding to this the products of two turns of the Krebs cycle for every glucose entering oxidation we have

Glucose + 4ADP + $4HPO_4^{2-}$ +

10 oxidized NAD + 2 oxidized FAD →

$\qquad$ 4ATP + 10 reduced NAD + 2 reduced FAD + $6CO_2$ $\quad$ (7-9)

Note at this point that little ATP has been produced. However, the electrons carried by the 10 reduced NAD and 2 reduced FAD molecules occupy orbitals at high energy levels and contain most of the energy obtained from

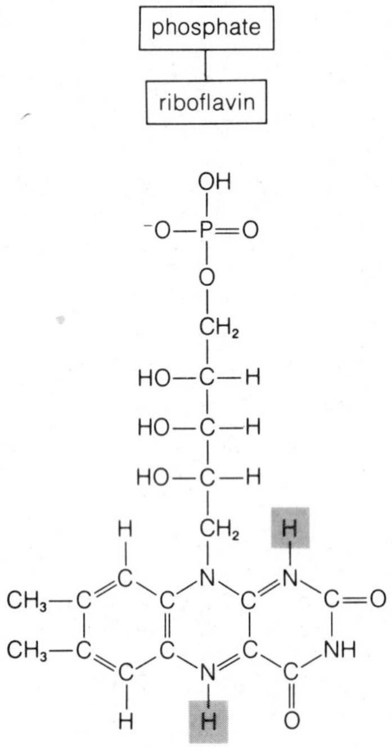

Figure 7-9 FMN (flavin mononucleotide). FMN is another nonprotein group of the flavoprotein carriers of the electron transport system. The riboflavin group in FMN binds two hydrogens in the shaded positions to form $FMNH_2$.

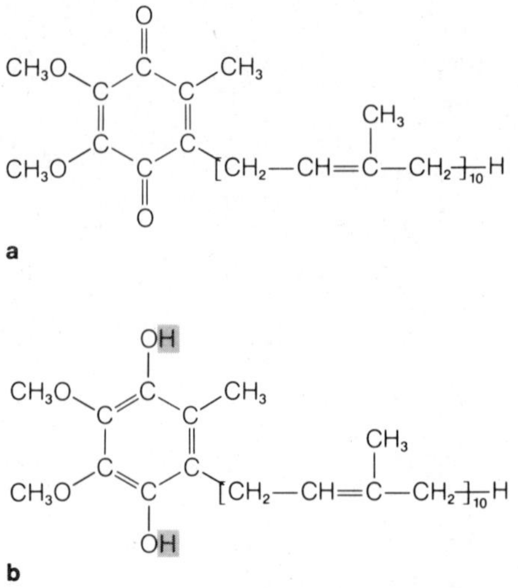

Figure 7-10 Coenzyme Q, the only carrier of the electron transport system not linked to a protein. (**a**) Oxidized form; in going to the reduced form (**b**), electrons and two hydrogens (H^+) are added at the shaded positions.

One turn of the Krebs cycle produces three molecules of reduced NAD, one reduced FAD, one molecule of ATP, and 2CO$_2$.

---

oxidation of glucose to $6CO_2$. The free energy released in the transfer of these electrons to oxygen is used as the major source of energy for cellular ATP synthesis in part 3 of mitochondrial oxidations. (For additional details of the Krebs cycle, see Supplement 7-1.)

Part 3 of Mitochondrial Respiration: ATP Synthesis in the Electron Transport System

The mitochondrial system transporting electrons from the Krebs cycle to oxygen is basically similar to the electron transport system of the light reactions of photosynthesis (see pp. 100–106). As electrons flow through the system, they lose part of their energy at each step in the chain of

carriers. At some of these steps, enough free energy is released to drive the synthesis of ATP from ADP. By the time the electrons reach oxygen, most of their energy has been tapped and conserved in the form of ATP.

Most of the ATP generated in cellular oxidation is produced during transport of electrons from reduced NAD to oxygen.

All but one of the known carriers in the mitochondrial electron transport system are proteins combined with a nonprotein subunit. The nonprotein subunit is the part of the total carrier molecule that actually accepts and releases electrons, and it is alternately oxidized and reduced as electrons flow along the chain. The carriers of the electron chain in mitochondria include FAD and another flavoprotein, FMN (*flavin mononucleotide*; see Figure 7-9), several iron–sulfur proteins, and four cytochromes. (Details of these carrier types are given in Chapter 6.) The single non-

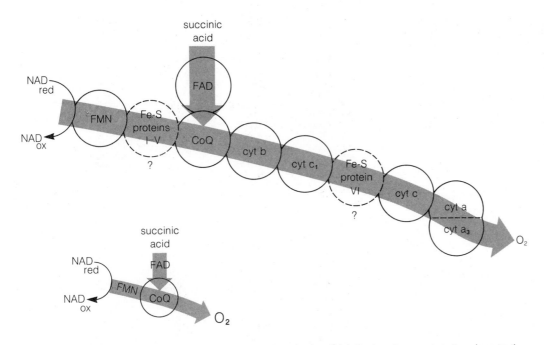

Figure 7-11 The probable sequence of carriers in the mitochondrial electron transport system (see text).

protein carrier of the mitochondrial electron transport

The carriers of the mitochondrial electron transport chain include flavoproteins, cytochromes, iron–sulfur proteins, and coenzyme Q (a quinone).

---

chain, *coenzyme* Q (Figure 7-10), closely resembles the plastoquinone carrier of the chloroplast carrier chain. All these carriers are alternately reduced and oxidized as electrons flow through the chain to oxygen.

The Mitochondrial Carrier Chain The mitochondrial electron carriers are believed to accept and release electrons in the sequence shown in Figure 7-11. Note in the figure that electrons have two routes of entry into the chain: one at FMN and one at FAD. These routes join at coenzyme Q, which thus forms a collection point for electrons entering from the two sources. All the electron carriers of the mitochondrial system have been shown to be embedded in the inner mitochondrial membranes, including the folds of these membranes that form the mitochondrial cristae. Reduced NAD, originating from cellular oxidations, transfers its electrons and hydrogen to FMN, the first carrier in this branch of the chain. In the transfer, NAD is oxidized and FMN reduced. The oxidized NAD is then free to cycle back

to the reactions of pyruvic acid oxidation and the Krebs cycle. The electrons probably then flow through one or more iron–sulfur proteins (the position of these carriers in the chain is still uncertain) and then to the coenzyme Q collection point. From coenzyme Q, electrons flow through a series of cytochromes—including cytochromes b, c_1, c, and the final carrier, a two-part cytochrome known as a–a_3. The a–a_3 carrier consists of two different cytochrome groups (see Figure 6-12), a and a_3, linked to the same protein molecule. At least one iron–sulfur protein is believed to be located between the cytochrome carriers in the position shown in Figure 7-11. The final carrier in the chain, cytochrome a–a_3, reduces oxygen to water in the reaction:

$$\text{Reduced cytochrome } a\text{–}a_3 + \tfrac{1}{2}O_2 + 2H^+ \rightarrow$$
$$\text{oxidized cytochrome } a\text{–}a_3 + H_2O \quad (7\text{-}10)$$

The hydrogens required for this reduction are derived from the H^+ ions in the water solution surrounding the electron transport chain.

Electrons entering the chain at the alternate point, through FAD, originate from the single reaction of the Krebs cycle that uses FAD directly as the electron acceptor (reaction 4 in Figure 7-8). While FAD acts as the electron acceptor for this reaction, it actually forms a fixed part of the mitochondrial electron chain and does not cycle freely between the mitochondrial electron transport chain and the Krebs cycle reactions as does NAD. Therefore passage of electrons from the four-carbon substance to FAD in reaction 4 occurs only when the substance collides with the inner mitochondrial membranes housing FAD in the electron transport system. From FAD, the electrons flow to the coenzyme Q collection point and then, through the cytochrome chain, to oxygen.

ATP Synthesis in Mitochondria   The mechanism coupling electron transport to ATP synthesis in mitochondria, according to Peter Mitchell's chemiosmotic hypothesis (see p. 105), is considered to depend on the same two-step process that occurs in chloroplasts. (1) The free energy released as electrons flow through the carriers of the electron transport system is used to move H^+ ions across the

---

Transport of electrons to oxygen builds an H^+ gradient.

---

mitochondrial membranes housing the electron transport carriers, creating an H^+ gradient. (2) The gradient is then used as a source of free energy to synthesize ATP from ADP and phosphate. As in chloroplasts, operation of the mechanism depends on the fact that some of the electron carriers in the mitochondrial chain are nonhydrogen carriers (the cytochromes and iron–sulfur proteins) whereas others carry hydrogens as well as electrons (FMN, FAD, and coenzyme Q). As electrons move from a nonhydrogen carrier to a dual hydrogen–electron carrier, hydrogens are picked up from the solution in the mitochondrial matrix (Figure 7-12a); as electrons move from dual hydrogen–electron carriers to nonhydrogen carriers, the hydrogens are released and expelled into the intermembrane compartment (Figure 7-12b; see also Figure 6-18). Electron transport thus produces an H^+ ion gradient: H^+ ions are at high concentration in the intermembrane compartment and low in the matrix (Figure 7-13). As in chloroplasts, this gradient forces an H^+ transport protein in

---

The H^+ gradient produced by electron transport drives ATP synthesis.

---

the inner membranes to run backward, driving the synthesis of ATP from ADP and phosphate.

The H^+ gradient created as an electron pair flows from NAD to oxygen is sufficient to drive the synthesis of 3ATP. Electrons entering the pathway from FAD (see Figure 7-11) flow through only part of the pathway, with the

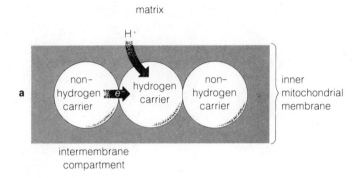

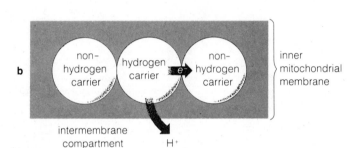

Figure 7-12   How electron transport through the mitochondrial chain sets up an H^+ gradient (see text).

result that fewer H^+ ions are added to the gradient. The smaller gradient reduces the quantity of ATP synthesized

---

Electron pairs carried by NAD release enough energy to synthesize 3ATP; electrons carried by FAD drive the synthesis of 2ATP.

---

to 2ATP for each pair of electrons traveling the portion of the transport chain between FAD and oxygen.

Total ATP Production from Glucose Oxidation

The ATP production driven by reduced NAD and FAD provides the information needed to calculate the total ATP production of mitochondrial electron transport and to derive a grand total for the entire sequence of oxidations from glucose to CO_2 and H_2O. Oxidation of a glucose molecule from glycolysis through the Krebs cycle yields a total

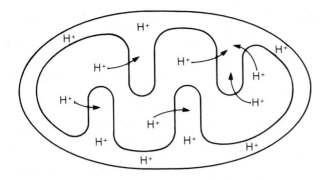

Figure 7-13 Electron transport in mitochondria increases the concentration of H^+ ions in the intermembrane compartment.

of 4ATP, 10 reduced NAD, and 2 reduced FAD (from Reaction 7-9). The electrons carried by the ten molecules of reduced NAD will drive the synthesis of $10 \times 3 = 30$ molecules of ATP as they pass through the electron transport system. The electrons carried by the two molecules of reduced FAD drive the synthesis of only four molecules of ATP because they enter the sequence of carriers farther along in the chain. This gives a total of 34ATP from electron transport. This 34ATP, when added to the ATP produced in glycolysis and the Krebs cycle (Reaction 7-9), gives a grand total of 38ATP for each molecule of glucose[1] completely oxidized to CO_2 and H_2O:

$$Glucose + 38ADP + 38HPO_4^{2-} + 6O_2 \rightarrow$$
$$6CO_2 + 44H_2O + 38ATP \quad (7\text{-}11)$$

Under standard conditions (pH 7, reactants and products at 25°C and one molar concentration), the hydrolysis of ATP to ADP yields 7000 cal/mole. Using this value as the energy required to synthesize ATP from ADP and phosphate, the total energy trapped during the oxidation of glucose amounts to $38 \times 7000 = 266,000$ cal/mole if

38ATP is produced. Combustion of glucose in air yields 686,000 cal/mole. On this basis the efficiency of glucose metabolism in cells approximates $266/686 \times 100 = 39\%$.

Cellular oxidations capture about 40 percent of the energy of glucose in the form of ATP.

At this level, the efficiency of mitochondrial energy conversion is considerably higher than most of the energy conversion systems designed by human engineers, which rarely perform above 5 to 10 percent efficiency.

The electron transport system of mitochondria was discovered in investigations carried out during the first three decades of this century. In 1913 the German scientist Otto Warburg proposed that an iron-containing enzyme must be directly involved in the use of oxygen by respiring cells. Warburg's enzyme, later identified as a cytochrome, was linked to the enzymes that remove electrons from substances oxidized in respiration by the work of Albert Szent-Gyorgyi, a Hungarian scientist who later became an American citizen. Szent-Gyorgyi proposed the basic idea that electrons flow from oxidized substances through electron carriers to oxygen. Research still continues in the effort to identify all the carriers of the system and to establish their exact sequence in electron transport in both mitochondria and chloroplasts.

Where Oxidative Reactions Occur Inside Mitochondria

Several methods are available for breaking mitochondria into pure fractions. In the most simple of these techniques, mitochondria are exposed briefly to a detergent, which causes breaks in the outer membrane. The outer membrane can then be easily separated from the inner membranes and purified in a centrifuge. The inner membranes and the enclosed matrix remain intact and can be separately concentrated and purified. The inner membranes can then be broken by prolonged treatment with

---

[1]The total of 38ATP given in this equation assumes that the two molecules of NAD reduced in glycolysis will each drive the synthesis of 3ATP inside mitochondria. Reduced NAD from glycolysis cannot directly enter mitochondria to pass electrons to the transport system, however, because the mitochondrial membranes are impermeable to NAD. Instead, the electrons carried by reduced NAD are transferred to other substances that can shuttle back and forth across the mitochondrial membranes. Two of these shuttle mechanisms are known. The more efficient one transfers electrons from NAD outside to NAD inside mitochondria. The less efficient mechanism results instead in

the reduction of FAD inside mitochondria. If the more efficient shuttle is operating, no loss occurs and 38ATP results from each molecule of glucose completely oxidized, as shown in Reaction 7-11. If the less efficient shuttle is operating, the electrons from reduced NAD are transferred to FAD inside mitochondria. These enter the electron transport system farther along in the chain and result in the synthesis of 2ATP for each molecule of reduced FAD. If this less efficient shuttle operates, complete oxidation of glucose will result in a grand total of 36ATP. The predominating shuttle seems to depend on the particular species involved.

detergent and isolated from the matrix by centrifugation. The three fractions—outer membrane, inner membrane, and matrix—retain much of their separate biochemical activity when separately purified in this way.

The outer membrane in such preparations proves to contain several enzymes associated with the initial breakdown of fatty acids and amino acids. The inner membrane fraction contains the enzymes and carriers engaged in electron transport and ATP formation. The matrix contains in solution the various enzymes and intermediates of the Krebs cycle and pyruvic acid oxidation in solution. Only

The enzymes of the Krebs cycle are in solution in the mitochondrial matrix.

one enzyme of the Krebs cycle—the enzyme catalyzing the removal of electrons in reaction 4 of Figure 7-8 (succinic

The carriers transporting electrons from reduced NAD to oxygen and the system generating ATP are tightly bound to the inner mitochondrial membranes.

acid dehydrogenase)—is tightly bound to the inner mitochondrial membranes. The electron acceptor for this reaction, FAD, is also tightly bound to the cristae membranes, where it forms part of the chain of electron transport carriers.

As might be expected, the presence of enzymatic activity of this magnitude is reflected in a high percentage of proteins in mitochondria. The mitochondrial matrix is gel-like and about 50 percent protein; an even greater percentage of the inner membrane, about 70 to 80 percent, is protein. The outer boundary membrane, which is less active enzymatically than the inner membrane, is about 50 percent protein. The remainder of the mitochondrial membranes is lipid.

CELLULAR OXIDATION OF FATS AND PROTEINS

When used as an energy source, both fats and proteins are split by preliminary reactions into shorter segments that enter cellular oxidation at various levels. In many cases, the shorter segments derived from these molecules are acetyl groups that are carried to the Krebs cycle by coenzyme A.

The breakdown of fats and proteins is linked to mitochondrial respiration by coenzyme A.

Fats

Initially fats are split into glycerol and fatty acids in the cytoplasm outside mitochondria. A phosphate group is then added to glycerol, at the expense of ATP, to produce a three-carbon, one-phosphate sugar—the same one produced by reaction 4 in glycolysis (see Figure 7-4). The one-phosphate sugar is converted to 3PGAL and enters the glycolytic pathway directly at this point. After preliminary reactions the fatty-acid chains enter mitochondria, where they are broken into two-carbon acetyl groups linked to coenzyme A. From this point, oxidation is the same as for acetyl groups derived from glucose. Because the fatty-acid chains are long, containing from 14 to 22 carbons, a large number of acetyl groups are supplied to the Krebs cycle as the chains split into two-carbon segments. In the overall conversion of fats to CO_2 and H_2O, approximately 17ATP is produced for each two-carbon segment in the fatty-acid residues. This ATP yield, about double that of carbohydrates by weight, explains why fats are such an excellent energy source in the diet.

Proteins

Individual amino acids removed from proteins are also oxidized through the Krebs cycle. Proteins are first split into amino acids in the cytoplasm outside the mitochondria. Some amino acids, after removal of their amino groups, are further broken into two-carbon acetyl groups linked to coenzyme A, which then enter the Krebs cycle. Other amino acids are converted into intermediates of the Krebs cycle, such as oxaloacetic acid, and enter the cycle directly in this form.

The cellular oxidation of carbohydrates, fats, and proteins is summarized in Figure 7-14. Inspection of this figure reveals the central position of coenzyme A in linking together the oxidation of many different substances in the cell.

MITOCHONDRIAL OXIDATION AND CELLULAR ACTIVITY

The rate at which acetyl groups are oxidized to CO_2 and H_2O in mitochondria is regulated by several control systems in cells. Of these controls, the primary one is based on the concentration of ADP in the cytoplasm surrounding mitochondria. As ADP concentration increases, synthesis of ATP from ADP and phosphate begins in mitochondria. ATP synthesis in mitochondria is tightly coupled to electron transport; as ATP is synthesized, electrons

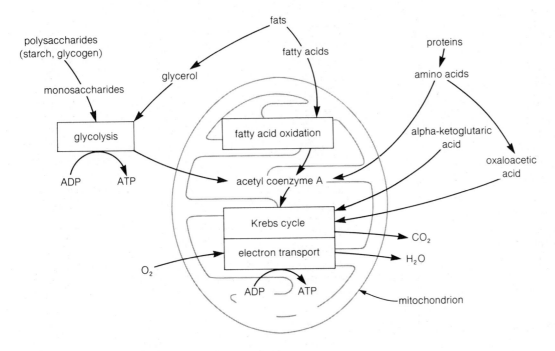

Figure 7-14 Coordination of the oxidation of carbohydrates, fats, and proteins in cells. Coenzyme A plays a central role in linking together the oxidation of these molecules.

flow without restriction from NAD to oxygen, and the NAD end of the transport chain becomes relatively oxidized. NAD, once oxidized, is free to recycle as an electron acceptor for the oxidations of the Krebs cycle. The Krebs cycle is then free to run. This system closely links the rate of oxidation in the Krebs cycle to the concentration of ADP in the cell.

The rate of mitochondrial oxidation of fuels is regulated primarily by ADP concentration in the cell.

If cellular activity is limited, ATP concentration remains high and ADP becomes unavailable to the mechanism synthesizing ATP in the cristae membranes. As a result, the reactions synthesizing ATP, and electron transport, stop. According to the Mitchell hypothesis, stoppage of electron transport is due to a buildup in the H^+ gradient that is unrelieved by backflow through the H^+ membrane carrier. The H^+ gradient gradually builds until it reaches levels high enough to oppose further expulsion of H^+ across the cristae membranes by the electron transport system. As a consequence, electron transport—and, in turn, oxidation in the Krebs cycle—stops. The rate of glycolysis is also linked to some extent to mitochondrial activ-

ity, since the NAD reduced in glycolysis may transfer its electrons to the mitochondrial carrier systems.

The sensitivity of mitochondrial activity to ADP concentration coordinates oxidation with cellular activities requiring an energy input. If such cellular activities as growth, reproduction, and motility proceed at very low levels, little ATP is converted to ADP. In this case, the oxidative activity of mitochondria remains low, allowing oxidizable fuel substances to be conserved. If cellular activity increases, the demand for energy results in conversion of ATP to ADP somewhere in the cell, eventually increasing the concentration of ADP in the mitochondrial environment. Increase of ADP concentration stimulates mitochondrial oxidation, which continues until ATP concentration is restored and ADP levels are low. Mitochondrial activity is thus finely tuned to cellular activity, ready to run on demand if ATP is converted to ADP anywhere in the cell.

In plant cells with both mitochondria and chloroplasts, the combined activity of the two organelles provides all the ATP required for cell activities (Figure 7-15). Carbohydrates and fats are synthesized in the chloroplasts (to the left of the dotted line in Figure 7-15) by reactions using energy derived from light; these reactions use CO_2, H_2O, and various minerals as raw materials and liberate

CO₂, H₂O, and O₂ cycle between photosynthesis and cellular oxidations.

oxygen as a by-product. These carbohydrates and fats are oxidized in mitochondria as an energy source (to the right of the dotted line in Figure 7-15), yielding ATP, CO₂, and H₂O as products. The oxygen liberated in photosynthesis is used as the final electron acceptor in mitochondria; the H₂O and CO₂ produced in mitochondria complete the cycle by serving as raw materials for the synthetic activities of chloroplasts. The entire cycle of activity is ultimately driven by the light energy absorbed and converted to chemical energy in the chloroplasts of green plants.

No mitochondria occur in bacteria and blue-green algae. In these prokaryotes, energy-yielding mechanisms are distributed between the cytoplasm and the plasma membrane. (The oxidative systems of bacteria and blue-green algae, which enable these organisms to obtain ATP through the breakdown of carbohydrates and other fuel molecules, are described in Supplement 7-2.)

Questions

1. Why is ATP called the energy "dollar" of the cell?

2. Define oxidation and reduction.

3. What kinds of molecules can be used as energy sources in cells?

4. What are the major steps in the cellular oxidation of carbohydrates?

5. Outline the structure and function of the membranes and compartments of mitochondria.

6. Outline the overall sequence of events in glycolysis. What are the primary chemical inputs and outputs of glycolysis?

7. What is a fermentation? What advantages does the ability to carry out fermentations have for cells? Can cells live by fermentations alone? Do fermentations occur in human cells?

8. What are aerobes, anaerobes, facultative anaerobes, and strict aerobes?

9. How are glycolysis and mitochondrial oxidations integrated in the cell?

10. Outline the major chemical inputs and outputs of pyruvic acid oxidation. What overall processes occur in pyruvic acid oxidation?

11. Outline the major chemical inputs of the Krebs cycle. What overall processes occur in the Krebs cycle?

12. Is the Krebs cycle a major direct source of ATP?

13. Compare the functions and structure of ATP, NAD, NADP, FAD, and coenzyme A.

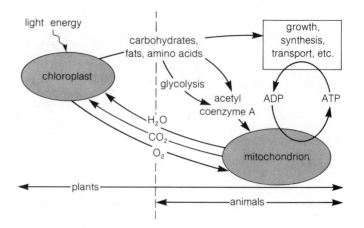

Figure 7-15 How chloroplasts and mitochondria interact in the flow of energy through cells. Plant cells contain the functions shown in the entire diagram; animal cells contain only the functions shown to the right of the dotted line. Energy flows from left to right in the diagram.

14. What is a cytochrome? A flavoprotein? A quinone? How many of these molecular groups are linked to proteins?

15. How is the mitochondrial electron transport chain believed to set up an H⁺ gradient?

16. Compare the flow of electrons and hydrogens (H⁺ ions) through the electron transport system of chloroplasts and mitochondria.

17. How is the H⁺ gradient believed to drive ATP synthesis?

18. How many molecules of ATP are synthesized through the reactions of glycolysis? Pyruvic acid oxidation? The Krebs cycle? (Consider that all electrons carried by NAD and FAD are transferred through the mitochondrial electron transport system to oxygen.)

19. The total efficiency of glucose oxidation to CO₂ and H₂O can be calculated as 39 percent. What part of this efficiency is contributed by pyruvic acid oxidation? By the Krebs cycle? By glycolysis? (Consider that all electrons carried by NAD and FAD flow through the mitochondrial electron transport system to oxygen.)

20. Why do electron pairs carried by NAD result in the synthesis of 3ATP and FAD 2ATP, as they flow along the mitochondrial electron chain to oxygen?

21. Where are fats oxidized in cells? Proteins?

22. Trace the cycles of chemical raw materials and products between chloroplasts and mitochondria in plant cells.

23. What effects do the relative concentrations of ATP and ADP have on the rate of cellular oxidations? What significance does this have for the efficiency of energy utilization in cells?

24. How many ATPs could be made from a molecule of 3PGAL? From a molecule of pyruvic acid? From a two-carbon segment removed from a fatty acid and attached to acetyl coenzyme A? Consider that all electrons are delivered to oxygen.

Suggestions for Further Reading

Hinkle, P. C., and R. E. McCarty. 1978. "How Cells Make ATP." *Scientific American* 238:104–123.

Lehninger, A. L. 1971. *Bioenergetics*. 2nd ed. Benjamin, Menlo Park, California.

Lehninger, A. L. 1973. *Biochemistry*. 2nd ed. Worth, New York.

Miller, G. T., Jr. 1971. *Energetics, Kinetics and Life*. Wadsworth, Belmont, California.

Quinn, P. J. 1976. *The Molecular Biology of Cell Membranes*. University Park Press, Baltimore.

Wolfe, S. L. 1981. *Biology of the Cell*. 2nd ed. Wadsworth, Belmont, California.

SUPPLEMENT 7-1: FURTHER INFORMATION ON GLYCOLYSIS AND THE KREBS CYCLE

Glycolysis

The individual reactions in glycolysis, and the enzymes catalyzing them, are shown in Figure 7-16. In the first three reactions of the pathway, glucose is converted into a more reactive substance containing two phosphate groups. The energy required for this part of the pathway is provided by the breakdown of two molecules of ATP. In the first reaction of the pathway (reaction 1 in Figure 7-16), a phosphate group derived from ATP is added to glucose, producing *glucose 6-phosphate*. This substance has a higher energy content as a result of the added phosphate, and the additional energy enables it to enter the next reactions in glycolysis. The activity of the enzyme catalyzing the first reaction of the pathway, *hexokinase*, illustrates one of the many controls regulating the rate of oxidation in cells. The product of the first reaction, glucose 6-phosphate, is an inhibitor of hexokinase. If glucose 6-phosphate accumulates because the remainder of the sequence is running slowly, the hexokinase enzyme is inhibited. This inhibition blocks further entry of glucose into the pathway.

Reaction 2 of the pathway simply rearranges glucose 6-phosphate into the closely related sugar *fructose 6-phosphate*. A second phosphate is added in reaction 3 to produce the highly reactive substance *fructose 1,6-diphosphate*. The second phosphate comes from another molecule of ATP, which is converted to ADP in this step. The overall sequence of reactions 1 through 3, yielding the high-energy substance fructose 1,6-phosphate, requires energy and proceeds to completion because the reactions are coupled to the simultaneous breakdown of ATP to ADP:

$$\text{Glucose} + 2\text{ATP} \rightarrow \text{fructose 1,6-diphosphate} + 2\text{ADP} \quad (7\text{-}12)$$

The rest of the glycolytic sequence oxidizes segments derived from this highly reactive sugar and proceeds without further input of ATP. In the first of these reactions (reaction 4 in Figure 7-16), fructose 1,6-diphosphate is split into two three-carbon sugars. One of these, 3-phosphoglyceraldehyde (3PGAL), enters the remaining steps in glycolysis. As 3PGAL is depleted from the pool, it is replaced by conversion of the second three-carbon product.

The next series of five reactions (5 to 9 in Figure 7-16) oxidizes 3PGAL and generates ATP and reduced NAD. The oxidation occurs in reaction 5. Two electrons and two hydrogens are removed from 3PGAL; a part of the free energy released is used to attach another phosphate group to the three-carbon sugar, yielding *1,3-diphosphoglyceric acid*. The phosphate group added in this reaction is derived from inorganic phosphate in the surrounding solution and not from ATP. The electrons removed are accepted by NAD along with one of the two hydrogens; the second hydrogen enters the pool of H^+ ions in the surroundings. The overall reaction releases free energy and proceeds spontaneously to completion:

$$\text{3-phosphoglyceraldehyde} + \text{oxidized NAD} + HPO_4^{2-} \rightarrow$$
$$\text{1,3-diphosphoglyceric acid} + \text{reduced NAD} \quad (7\text{-}13)$$

Removing one of the two phosphate groups from 1,3-diphosphoglyceric acid at the next step (reaction 6) produces *3-phosphoglyceric acid* and a second large increment of free energy. Some of this free energy is captured when the phosphate group removed is attached to ADP, yielding a molecule of ATP:

$$\text{1,3-diphosphoglyceric acid} + \text{ADP} \rightarrow$$
$$\text{3-phosphoglyceric acid} + \text{ATP} \quad (7\text{-}14)$$

The 3-diphosphoglyceric acid formed in reaction 6 is rearranged in two steps (reactions 7 and 8) into *phosphoenolpyruvic acid*. The remaining phosphate group is removed in the last step in glycolysis (reaction 9), yielding pyruvic acid. Part of the free energy released in this reaction is used to attach the phosphate to ADP, forming another molecule of ATP.

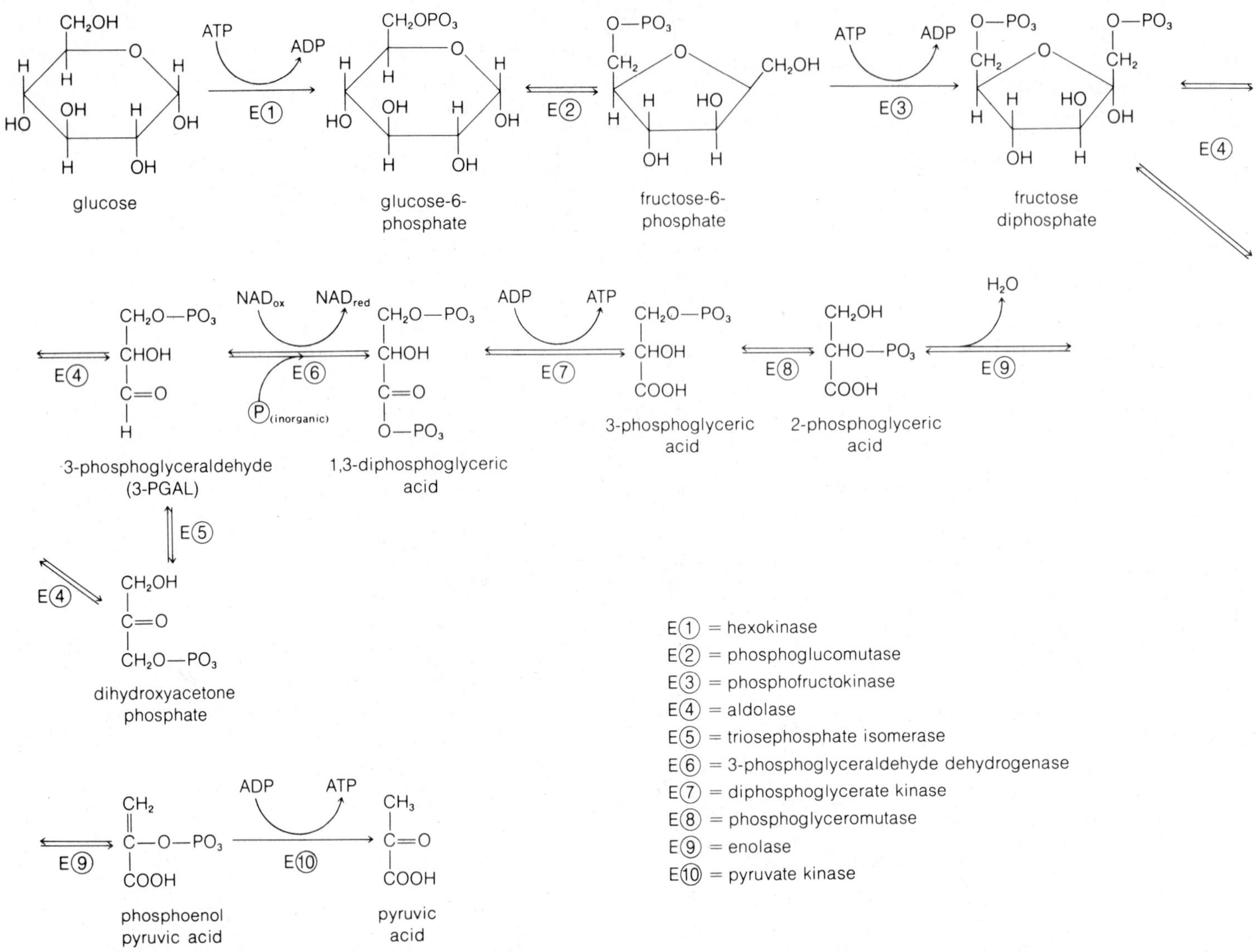

Figure 7-16 The reactions and enzymes of glycolysis (see text).

Many of the intermediate reactions of glycolysis are reversible and may go in either direction in response to high concentrations of either reactants or products. (The reversible steps are indicated in Figure 7-16 by ⇌ arrows.) Alternate enzymatic pathways are available in the cytoplasm for the irreversible steps, with the result that glycolysis can in effect be made to run in reverse. The first reaction in glycolysis, for example, the conversion of glucose to glucose 6-phosphate by hexokinase, is essentially irreversible. However, an alternate enzyme, *glucose 6-phosphatase*, can catalyze the reverse reaction.

Note that many of the intermediates and products of the final series of reactions in glycolysis are also part of the Calvin and C_4 cycles of the dark reactions of photosynthesis. (Compare Figure 7-16 with Figures 6-21 and 6-22.) Any of these intermediates of the dark reactions in plants may enter glycolysis at their respective places. The first half of the glycolytic sequence is also important in photosynthesis; glucose is synthesized from 3PGAL, the primary product of the Calvin cycle. This synthesis is achieved essentially by the reversal of reactions 1 through 4 of glycolysis.

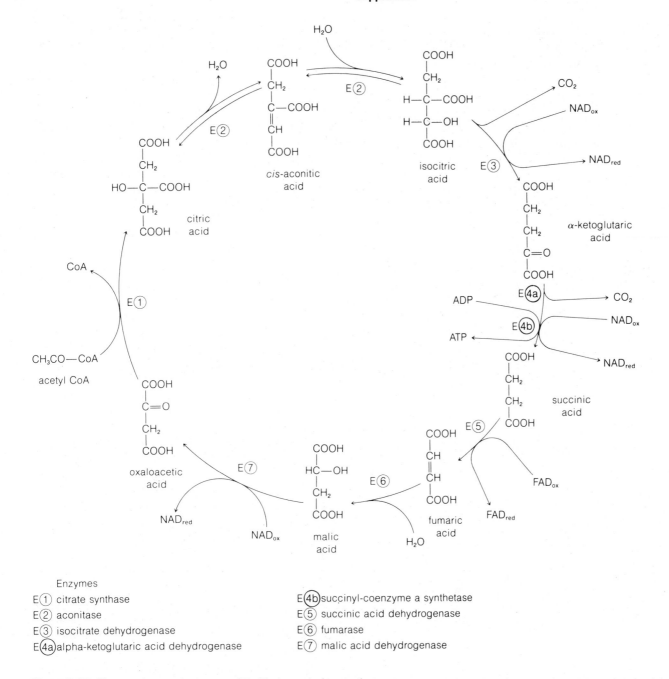

Enzymes

E① citrate synthase
E② aconitase
E③ isocitrate dehydrogenase
E④a alpha-ketoglutaric acid dehydrogenase

E④b succinyl-coenzyme a synthetase
E⑤ succinic acid dehydrogenase
E⑥ fumarase
E⑦ malic acid dehydrogenase

Figure 7-17 The reactions and enzymes of the Krebs cycle (see text).

The Krebs Cycle

The pyruvic acid (with three carbons) produced by glycolysis is subsequently broken into acetyl (two-carbon) groups in mitochondria. In this reaction, an oxidation, one molecule of CO_2 is released for every molecule of pyruvic acid entering the sequence. The resulting acetyl groups, at-

tached to coenzyme A in a high-energy complex, then enter the Krebs cycle for oxidation to CO_2.

In the first reaction of the Krebs cycle (Figure 7-17), an acetyl group is transferred from coenzyme A to oxaloacetic acid (four carbons), yielding citric acid (six carbons). The overall reaction is driven by the free energy released by the removal of the acetyl unit from coenzyme

A. The citric acid formed in reaction 1 enters a pool of acids that are readily converted from one to the other by the enzyme *aconitase*. As *isocitric acid*, the reactant entering the next step in the cycle, is depleted, the enzyme replenishes this substance by converting the citric acid in the pool.

In the second step in the cycle (reaction 2), two electrons and two hydrogens are removed from isocitric acid. At the same time, the carbon chain is reduced in length from six to five carbons, yielding *alpha-ketoglutaric acid*. The carbon removed is released as CO_2. The electron acceptor for this reaction, which also combines with one of the two hydrogens removed, may be either NAD or NADP (see p. 127).

The next reaction (reaction 7) is complex and is shown only in skeleton form in Figure 7-17. Alpha-ketoglutaric acid is oxidized to *succinic acid* in this step, which removes two electrons and two hydrogens. (NAD acts as electron acceptor in this oxidation.) The carbon chain is again shortened and CO_2 is released. As part of the complex side cycle accomplishing this step, one molecule of ATP is synthesized. As overall products, therefore, reaction 3 yields

$$\alpha\text{-Ketoglutaric acid} + \text{oxidized NAD} + \text{ADP} + \text{HPO}_4^{2-} \rightarrow$$

$$\text{succinic acid} + \text{reduced NAD} + \text{ATP} + CO_2 \quad (7\text{-}15)$$

The next oxidation of the Krebs cycle (reaction 4) is catalyzed by the enzyme *succinic acid dehydrogenase*. This enzyme and its electron acceptor, FAD, are the only molecules of the Krebs cycle attached to the internal membranes of mitochondria. The product of reaction 4, *fumaric acid*, is then rearranged at the next step (reaction 5), with the addition of a molecule of water, to form *malic acid*. In the final reaction of the Krebs cycle (reaction 6) malic acid is oxidized to oxaloacetic acid, completing the cycle. NAD is the electron acceptor for the final oxidation.

SUPPLEMENT 7-2: CELLULAR OXIDATIONS IN PROKARYOTES

Bacteria

Almost all bacteria are able to use carbohydrates in some form as an energy source. Organic acids and amino acids are also commonly oxidized by these organisms. Some bacteria can oxidize a wide variety of organic molecules; others are so restricted that their preferences in fuel substances can be used to identify them. In most bacteria, however, carbohydrates are the primary energy source as in eukaryotes.

Most bacteria are capable of oxidizing carbohydrates by sequences closely resembling the glycolytic pathway. Because oxygen itself is not used as the final electron acceptor in glycolysis, this type of metabolism, if it is the only type used by the bacteria, occurs as a fermentation and oxidation is anaerobic. Often pyruvic acid is the final acceptor in a reaction similar to lactic acid generation in muscle tissue:

$$\text{Pyruvic acid} + \text{reduced NAD} \rightarrow$$

$$\text{lactic acid} + \text{oxidized NAD} \quad (7\text{-}16)$$

By different pathways, bacterial fermentations may yield a wide variety of final products, including most commonly acids such as lactic and acetic acid or alcohols such as ethyl and butyl alcohol. Many bacterial fermentations yield products useful to humans, such as sauerkraut, vinegar, and some types of cheese. Other bacteria are aerobic and carry out oxidations through the complete sequences of glycolysis, pyruvic acid oxidation, the Krebs cycle, and electron transport to oxygen.

The enzymes of glycolysis are suspended in solution in the cytoplasm of bacterial cells just as they are in eukaryotes. The components of the Krebs cycle, if present, are also suspended in the cytoplasm; no membranes separate these reaction sequences from their cytoplasmic surroundings as in eukaryotic cells. The enzymes and carriers of the electron transport chain are bound tightly to the plasma membrane in bacteria and remain with the membrane fraction if bacterial cells are disrupted for biochemical analysis. Most of the enzymes of glycolysis and the Krebs cycle in bacteria closely resemble their counterparts in eukaryotes. The bacterial electron transport chain also consists of flavoproteins, quinones, and cytochromes, as in eukaryotes. The electron carriers show much variation in bacteria, however, and include flavoproteins, quinones, and cytochromes that are different from the electron transport molecules in higher organisms.

Bacterial electron transport systems are typically more "branched." That is, they usually have supplemental points for entry and exit of electrons in addition to the routes typical of eukaryotes. The example shown in Figure 7-18, which diagrams the electron transport system in *Escherichia coli*, contains one such extra input, which accepts electrons removed from lactic acid. (*Escherichia coli* is the primary bacterium inhabiting the intestinal tract in hu-

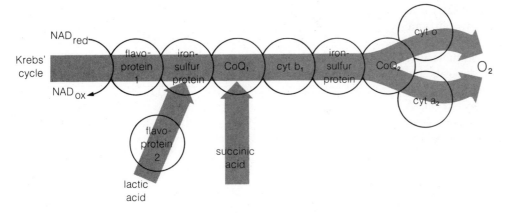

Figure 7-18 The electron transport system of the bacterium *E. coli*. The electron carriers are embedded in the plasma membrane in bacteria.

mans.) The final segment of the pathway in *E. coli* is also branched, offering two routes to oxygen: one through cytochrome *o* and one through cytochrome *a₂*. Neither of these cytochromes occurs in eukaryotes. Although the electron carrier systems of aerobic bacteria use oxygen as the final electron acceptor, some anaerobic species are adapted to different acceptors and use substances such as sulfate or nitrate in this role instead.

ATP synthesis is linked to electron transport in bacteria by the chemiosmotic mechanism as it is in chloroplasts and mitochondria. Peter Mitchell and his coworkers have shown that electron transport in bacteria results in expulsion of H^+ ions across the plasma membrane, from the inside of the bacterial cell to the outside. The H^+ gradient that is established, as in eukaryotes, provides the driving force for the synthesis of ATP.

Blue-Green Algae

Comparatively little is known of the oxidative mechanisms of the blue-green algae. These prokaryotes appear to live primarily by photosynthesis, and evidently they obtain most of the ATP they require through electron transport in the light reactions of photosynthesis. If grown in the dark, some blue-green algae are able to take up a carbohydrate, such as glucose, and slowly oxidize it to CO_2. The pathways of this oxidation are not yet clear, and it is not certain whether blue-green algae contain all the components of glycolysis and the Krebs cycle. Although certain enzymes of each of these oxidative sequences have been isolated from the blue-green algae, others have never been detected or identified. For example, the enzymes ox-

idizing an intermediate compound of the Krebs cycle—α-ketoglutaric acid (produced by reaction 2 in Figures 7-8 and 7-17)—have never been identified in blue-green algae. Thus it is not known whether these prokaryotes have a complete Krebs cycle or are able to oxidize carbohydrates only through part of the cycle. If all the components are present, the various enzymes and electron transport carriers are probably located in solution in the cytoplasm as in bacteria, since blue-green algae do not have mitochondria.

SUPPLEMENT 7-3:
HUMAN DIET AND THE MITOCHONDRION

Many of the reactions of cellular oxidation are intimately involved in human diet. Gaining or losing weight, for example, depends on the balance between the potential energy contained in the food eaten, tapped off in the reactions of cellular oxidation, and the energy expended in activities such as physical exercise. The first law of thermodynamics applies to dieters as well as to biochemical reactions: Energy can neither be created nor destroyed. If the caloric content of the food eaten exceeds the amounts expended, we gain weight; if the calories expended exceed the calories taken in, we lose weight.

Much of the effect of overeating or undereating is expressed through reactions linked by acetyl-coenzyme A. If the intake of fuel substances is high, particularly in sweet or starchy foods that are rich in carbohydrates, the acetyl-coenzyme A produced in pyruvic acid oxidation is shunted into a reaction series in the liver that reverses fatty-acid oxidation: the two-carbon acetyl units carried by coenzyme A are linked end to end to form fatty acids. After diffusing from the liver to storage tissues, these fatty-acid molecules are linked with glycerol and laid down in fat deposits that increase in proportion to the amount of excess carbohydrate eaten. Fatty foods also contribute acetyl-coenzyme A units that, if produced in excess of energy requirements, enter the fat storage deposits by the same pathway. The only way to prevent the deposition of fat is to reduce the caloric intake or increase the amount of physical exercise so that the intake of calories in food is balanced by energy used in physical activity.

When we exercise, ATP is converted to ADP in our muscle tissues. The increased ADP concentration, in the vicinity of mitochondria, promotes mitochondrial electron transport. As a result, the Krebs cycle turns and acetyl-coenzyme A units are oxidized to CO_2 and H_2O instead of entering the pathways leading to fat deposition.

Coenzyme A is also involved in the effects of extreme starvation. In this case energy is supplied almost entirely through rapid breakdown of the body's fat reserves. In starving individuals carbohydrate intake in the diet is reduced to the point that the intermediate compounds of the Krebs cycle, such as oxaloacetic acid, are seriously depleted. As a result, oxidation of the acetyl-coenzyme A units derived from fatty-acid breakdown accumulate faster than they can be oxidized in the Krebs cycle. In response, the liver links acetyl-coenzyme A units together by a series of reactions to form molecules called **ketone bodies** (Figure 7-19). (Ketone bodies are so called because of the carbonyl or C=O groups located in the interior or ketone positions in the carbon chains of the molecules—see Table 2-2.) The ketone bodies produced accumulate in the blood, where they may lead to a condition known as *ketoacidosis*, in which the blood becomes acid because of the presence of the ketone body acids. Since the ketone bodies register the same as alcohol in the breath tests used to determine sobriety, someone on a starvation diet could conceivably be arrested for drunken driving! In extreme cases, the presence of ketone bodies in the bloodstream can impair brain function and produce coma and death.

The mitochondrial reactions and coenzyme A also figure in the attempt to control blood cholesterol through reduced intake of cholesterol in the diet—often recommended as a means to prevent cholesterol deposits from accumulating in the arteries. Many of the possible beneficial effects of such diets are thwarted by the liver, which readily synthesizes cholesterol in reactions that use acetyl-coenzyme A as the starting point. When cholesterol is reduced

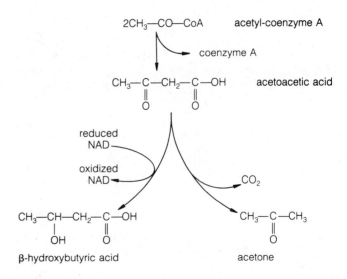

Figure 7-19 The reactions producing ketone bodies from acetyl coenzyme A in the liver.

in the diet, the liver compensates by producing cholesterol from coenzyme A derived from carbohydrate or fatty-acid breakdown, and the amount in the bloodstream remains constant.

Even Mitchell's chemiosmotic hypothesis, which may seem more academic than practical in relation to human affairs, has entered the dieting picture. For a time, the drug DNP (2,4-dinitrophenol) was given to overweight persons to control obesity. DNP is an agent that causes membranes to become "leaky" to H^+ ions. The leaky mitochondrial membranes destroy the H^+ gradient created by electron transport in the treated persons and completely uncouple electron transport from ATP synthesis. The effect is much like running an automobile engine at high speed with the transmission in neutral, since fuel substances can be oxidized with no relationship to the amount of ATP consumed in exercise. The treatment, however, proved to be too drastic an alteration in the body chemistry of the treated persons, and, after a few deaths, it was abandoned for other methods of weight control. Reduced food intake and increased exercise remain the method of choice in healthy persons.

The primary use of the cellular energy captured in mitochondria and chloroplasts is in the synthesis of the molecules required for cell growth. This growth occurs by the assembly of all the major classes of cellular molecules— carbohydrates, lipids, proteins, and nucleic acids—powered ultimately by the ATP energy captured in chloroplasts and mitochondria.

The synthetic activity of cells follows a closely similar pattern in all living things from the simplest bacteria to the most complex plants and animals. In all these organisms, the directions for protein synthesis are coded into DNA molecules in the cell nucleus or nucleoid. (In a few viruses, the information required for synthesis is coded in RNA molecules instead.) The coded information is transcribed in the nuclear region into RNA "messages" that are copied from the DNA. The RNA messages are then transferred into the cytoplasm, where they direct the assembly of proteins—among them enzymes, motile proteins, and structural proteins. The enzymes, in turn, speed and direct the synthesis of the remaining molecules of the cell, including carbohydrates, lipids, a wide variety of smaller molecules, and even the nucleic acid information molecules themselves.

8

CELL SYNTHESIS I: THE ROLE OF THE NUCLEUS

HOW CELL SYNTHESIS WORKS: A BRIEF OVERVIEW

Figure 8-1 outlines the mechanism of cellular synthesis. The information required for protein synthesis in eukaryotes is coded into different sequences of the four nucleotides making up the DNA (deoxyribonucleic acid)

The information required for synthesis is coded into different sequences of the four nucleotides of DNA.

---

molecules of the nucleus. Two primary kinds of information are stored in the DNA sequences: directions for making proteins and directions for making accessory RNA molecules. The directions for making proteins are spelled out by a code that uses the four DNA nucleotides, three at a time, in all possible combinations. Each three-nucleotide code word stands for an amino acid (see Figure 9-3). Reading the code words in sequence along the DNA spells out the sequences of amino acids in a protein. These protein-encoding regions are duplicated into RNA copies called **messenger RNAs (mRNAs)** that carry the directions for making proteins to the cytoplasm.

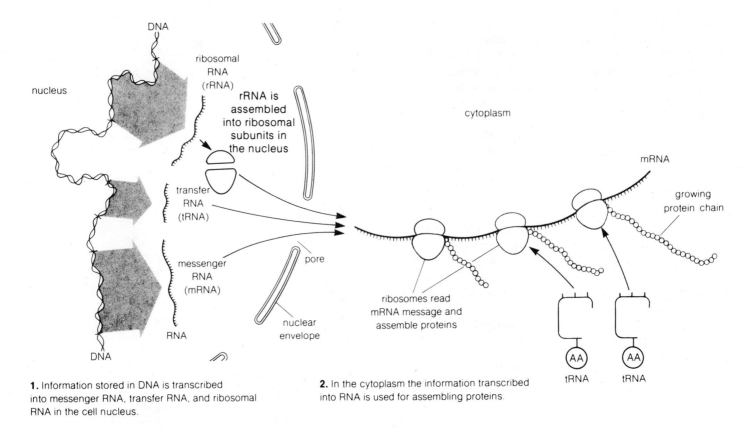

1. Information stored in DNA is transcribed into messenger RNA, transfer RNA, and ribosomal RNA in the cell nucleus.

2. In the cytoplasm the information transcribed into RNA is used for assembling proteins.

Figure 8-1   The major steps in cellular synthesis (see text).

Messenger RNA carries the coded information for making proteins to the site of protein synthesis in the cytoplasm.

Other DNA regions store the directions for making two types of accessory RNAs that act in parts of the protein synthesis mechanism. One, **ribosomal RNA (rRNA),** forms part of the ribosomes—the RNA–protein structures that assemble amino acids into proteins in the cytoplasm. The second, **transfer RNA (tRNA),** binds directly to amino acids during protein synthesis and provides the necessary link between the nucleic acid code and the amino acid sequence of proteins.

Synthesis of the mRNA, rRNA, and tRNA copies of DNA, called **transcription,** occurs within the cell nucleus. Following transcription, the RNA copies pass through the nuclear envelope and enter the cytoplasm. Messenger RNA, once in the cytoplasm, attaches to one or more ribosomes. The ribosomes then assemble amino acids into

proteins, using the information carried in the mRNA as a guide. In this synthesis, called **translation,** a ribosome starts at one end of an mRNA molecule and moves along the sequence of nucleotides in the mRNA until it reaches the other end. As it moves along the mRNA, it assembles amino acids into a gradually lengthening amino acid chain according to the directions coded into the mRNA. At the

---

Ribosomal RNA forms part of ribosomes. Transfer RNA provides the link between the nucleic acid code and the amino acid code of proteins.

---

end of the message, both the ribosome and the completed protein molecule detach from the mRNA. At any instant, several ribosomes may be at different places on a single mRNA molecule, engaged in reading the message and assembling protein chains. Ribosomes and mRNA may recycle through this mechanism many times. In this way,

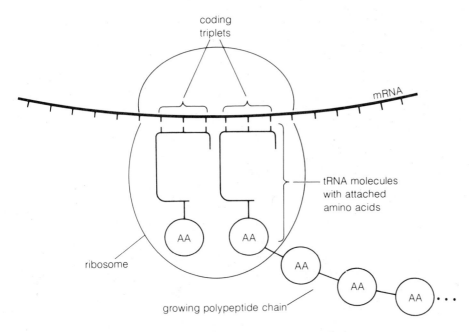

coding
triplets

mRNA

tRNA molecules
with attached
amino acids

ribosome

AA

AA

AA

AA

AA

. . .

growing polypeptide chain

Figure 8-2 The tRNA–mRNA pairing mechanism at the ribosome that enters successive amino acids into a growing polypeptide chain (see text).

each mRNA molecule may serve as a template for hundreds of identical protein molecules.

Transfer RNA molecules function as the "dictionary" in the translation mechanism. Each kind of tRNA corresponds to one of the 20 amino acids used in protein synthesis. The different tRNAs are attached to their amino acids by enzymes that can recognize both a particular amino acid and the tRNA (or tRNAs) corresponding to it. As a result of the activity of these enzymes, each amino acid is linked to a specific kind of tRNA. The tRNAs, in turn, are capable of recognizing and binding to the coding triplets in mRNA specifying their attached amino acid.

Transcription is the synthesis of RNA in the nucleus. Translation is the synthesis of proteins in the cytoplasm.

Binding of tRNAs and their attached amino acids to the mRNA coding triplets takes place on ribosomes. As a ribosome encounters an mRNA coding triplet specifying a given amino acid, the tRNA carrying that amino acid binds to the ribosome. This binding places the amino acid in its correct location in the protein chain growing from the ribosome (Figure 8-2). The ribosome then moves to the next mRNA coding triplet, causing the next tRNA–amino acid complex specified by the code to bind. As each successive amino acid arrives at the ribosome it is split from

its tRNA carrier and linked into the gradually lengthening protein chain. The process repeats until the ribosome reaches the end of the message and completes assembly of the protein.

The cell structures and mechanisms functioning in transcription and translation are taken up in this chapter and the next. This chapter concentrates on RNA transcription and the structure of the cell nucleus and chromosomes. Protein synthesis and the structure of the cytoplasmic organelles engaged in it are covered in the following chapter.

DNA, RNA, AND TRANSCRIPTION

DNA and RNA Structure

The two nucleic acids, DNA and RNA, are built up from nucleotides (see Figure 2-24 and Information Box 8-1) linked together into long chains. DNA (Figure 8-3a) is constructed from nucleotides containing the sugar *deoxyribose* and one of the four nitrogenous bases *adenine (A), thymine (T), guanine (G),* and *cytosine (C).* RNA (Figure 8-3b) contains nucleotides built up from the sugar *ribose* and one of the four nitrogenous bases *adenine (A), uracil (U), guanine (G),* and *cytosine (C).* The nucleotide chains of

Information Box 8-1   The Nucleotides

The nucleotides of DNA and RNA consist of a nitrogenous base, a five-carbon sugar, and from one to three phosphate groups:

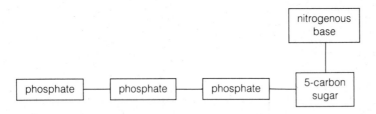

The nitrogenous bases of DNA and RNA are either purines or pyrimidines. The nucleotides of DNA contain any one of the bases adenine (A), thymine (T), guanine (G), and cytosine (C); the nucleotides of RNA contain any one of the bases adenine (A), uracil (U), guanine (G), and cytosine (C):

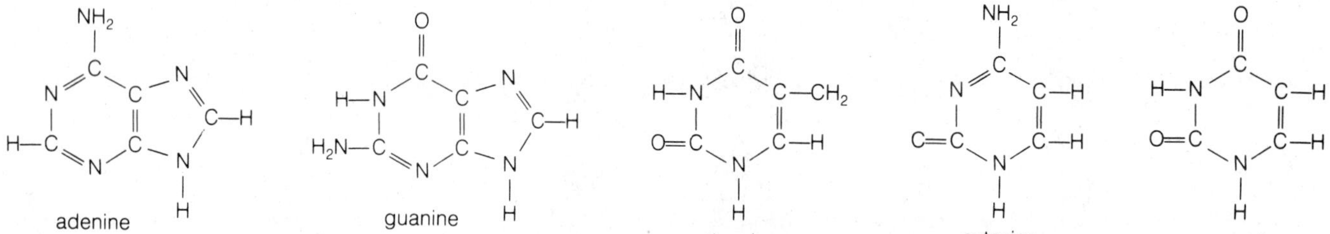

adenine

guanine

thymine

cytosine

uracil

purines

pyrimidines

These nitrogenous bases combine with the five-carbon sugar deoxyribose in DNA and with ribose in RNA:

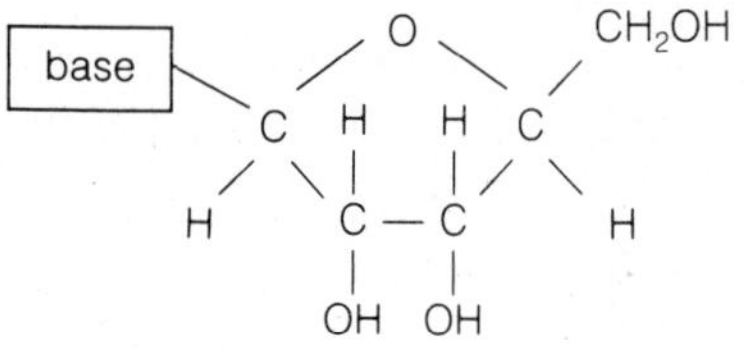

base

CH₂OH

ribose

Adding one, two, or three phosphate groups to the base–sugar units forms the mono-, di-, and triphosphates of the nucleotides.

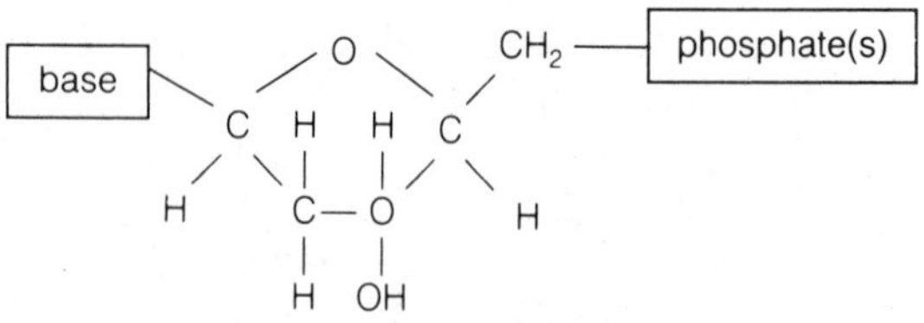

base

CH₂ — phosphate(s)

deoxyribose

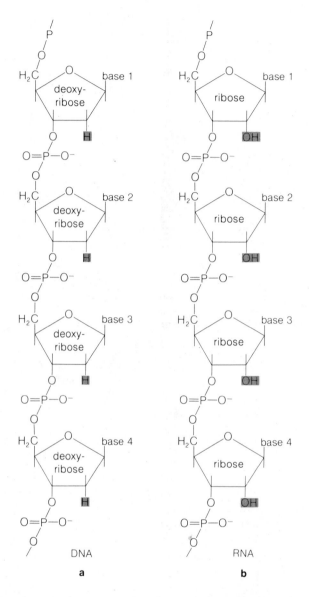

Figure 8-3 Linkage of nucleotides into the nucleotide chains of DNA (**a**) and RNA (**b**). In DNA, any one of the four nitrogenous bases adenine (A), thymine (T), cytosine (C), or guanine (G) may be bound at the positions marked as bases. In RNA, any one of the bases A, G, C, and uracil (U) occurs at these sites. The deoxyribose and ribose sugars of DNA and RNA differ only in the presence or absence of a single oxygen atom at the shaded positions; this oxygen is absent in DNA.

DNA contains the sugar deoxyribose and the four bases A, T, G, and C. RNA contains the sugar ribose and the four bases A, U, G, and C.

DNA and RNA thus differ in two ways: (1) DNA contains the base thymine but not uracil whereas RNA contains uracil but not thymine; and (2) DNA contains the sugar deoxyribose whereas RNA contains the sugar ribose. (The ribose and deoxyribose sugars differ only in the presence or absence of a single oxygen atom—see Figure 8-3.) Otherwise the nucleotide chains of the two nucleic acid molecules are identical.

The nucleotides are linked into nucleic acid chains by covalent bonds that extend between the phosphate group of one nucleotide and the sugar of the next nucleotide in the chain, as shown in Figure 8-3. These alternating sugar–phosphate bonds form the "backbone" of a nucleic acid molecule. The nitrogenous bases extend outward from the sugar–phosphate backbone.

DNA normally occurs in cells as a **double helix:** a double molecule containing two nucleotide chains twisted together into a double spiral (Figure 8-4). RNA, in contrast, is usually found as a single nucleotide chain. In some regions, however, the single RNA chain may fold back and twist upon itself to form a double helix. Hybrid double helices, containing one RNA and one DNA chain twisted together, may also form as temporary structures during transcription in the cell nucleus.

The arrangement of the two sugar–phosphate backbone chains in a spiral around the outside of the DNA molecule leaves a cylindrical space that extends through the central axis of the molecule. The nitrogenous bases attached to the sugar–phosphate backbones extend into this space, forming pairs consisting of one base from each chain. The space is just wide enough to allow a purine to pair with a pyrimidine base. (Purine bases, with two carbon rings, are about twice as wide as pyrimidine bases;

The space available in the interior of a DNA molecule restricts the bases to purine–pyrimidine pairs.

---

see Information Box 8-1.) Purine–purine pairs are too wide, and pyrimidine–pyrimidine pairs too narrow, to fit this space.

Further pairing restrictions come from the shape of the bases and the possibilities for hydrogen bonding between purine–pyrimidine base pairs. The shape of the bases allows adenine and thymine to pair together like pieces of a jigsaw puzzle and form two hydrogen bonds that stabilize the arrangement (Figure 8-5). Similarly, cytosine and guanine fit together perfectly, forming a pair that is stabilized by three hydrogen bonds. The other purine–pyrimidine pairing possibilities, such as adenine with cytosine, do not work: the pieces of the puzzle do not fit together, and stabilizing hydrogen bonds cannot form.

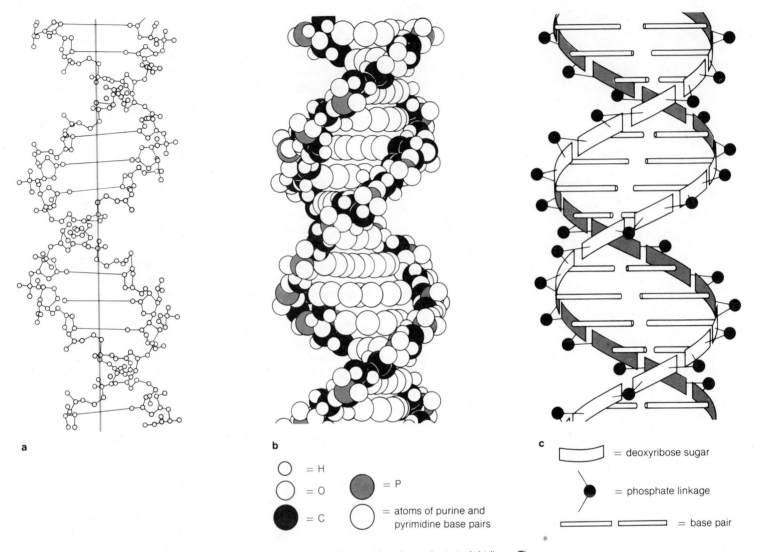

Figure 8-4 The DNA double helix. In (**a**) atoms are shown as circles and bonds as short straight lines. The bases, which lie in a flat plane, are seen on edge from this viewpoint and are shown as straight horizontal lines running between the backbone chains. Redrawn from an original courtesy of M.H.F. Wilkins. Copyright 1963 by the American Association for the Advancement of Science. In (**b**) the spaces occupied by the atoms of the double helix are shown as spheres. (**c**) Diagram showing the arrangement of sugars, phosphate groups, and base pairs in the DNA double helix.

The shape and hydrogen-bonding capabilities of the nitrogenous bases thus limit the purine–pyrimidine pairing in the center of the DNA molecule to adenine–thymine and guanine–cytosine pairs. These pairs are seen from the

The pattern of hydrogen bonding restricts the purines and pyrimidines to adenine–thymine and guanine–cytosine pairs.

edge in Figure 8-4a as horizontal lines; when viewed from one end of the molecule, the pairs appear as shown in Figure 8-5. One complete turn of the double helix includes ten of the base pairs.

Two kinds of forces hold the DNA double helix together. One force is supplied by the hydrogen bonds formed between the base pairs in the interior of the molecule. Although the hydrogen bonds are individually rel-

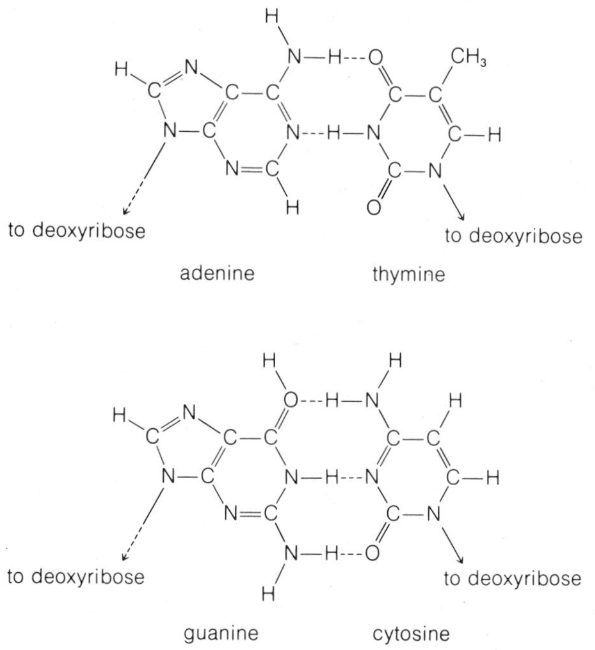

Figure 8-5 The two kinds of base pairs formed in DNA. **Top**: An adenine–thymine (A–T) base pair. **Bottom**: A guanine–cytosine (G–C) base pair. The pattern of hydrogen bonds in the two base pairs is shown by dotted lines. The total width of either base pair just fills the space between two sugar–phosphate backbone chains wound into a double helix.

The DNA double helix is held together by hydrogen bonds and hydrophobic associations between the paired bases.

---

atively weak, their combined effect along the double helix forms a stable structure if the DNA molecule is more than about ten base pairs in length. The second stabilizing force is provided by hydrophobic associations (see p. 14) between the paired bases in the interior of the molecule. In this region, the bases, which are primarily nonpolar in character, pack tightly enough to exclude water and form a stable, nonpolar environment.

The base-pairing rules are of great significance for the activities of DNA in the storage and transfer of coded information. Because of the requirement that adenine must pair with thymine, and cytosine with guanine, a sequence

---

The sequence of nucleotides in one chain of a DNA double helix fixes the sequence in the opposite chain.

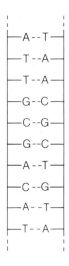

Figure 8-6 Complementarity of base sequences in the two nucleotide chains of a DNA double helix. Because A always pairs with T, and G with C, the sequence in one chain fixes the sequence of the other.

in one chain, once fixed, can pair with only one sequence in the opposite chain. If the sequence in one chain is A–T–T–G–C–G–A–C–A–T (A = adenine, T = thymine, G = guanine, C = cytosine), for example, the opposite chain is restricted to the sequence T–A–A–C–G–C–T–G–T–A (Figure 8-6). In the parlance of molecular biologists, the two sequences of the two chains are said to be *complementary*. Complementary base pairing provides the basis for information transfer from DNA to RNA in transcription (see below) and also for the exact duplication of DNA

Complementary base pairing forms the basis for both RNA transcription and DNA replication.

---

molecules as part of cell division. (DNA duplication, called **replication,** is discussed in detail in Chapter 10.)

The structure of DNA was discovered in 1953 by James D. Watson, an American, and Francis H. C. Crick, an Englishman, both working at Cambridge University, and Maurice H. F. Wilkins, another Englishman working at Kings College in London. Watson, Crick, and Wilkins were awarded the Nobel Prize for their work in 1962. Few discoveries in the history of biology have provided as much insight into the molecular nature of life or have had as great an impact on research in all biological fields. (The research leading to the discovery of DNA structure and establishing DNA as the informational and hereditary molecule of cells is described in Supplement 8-1.)

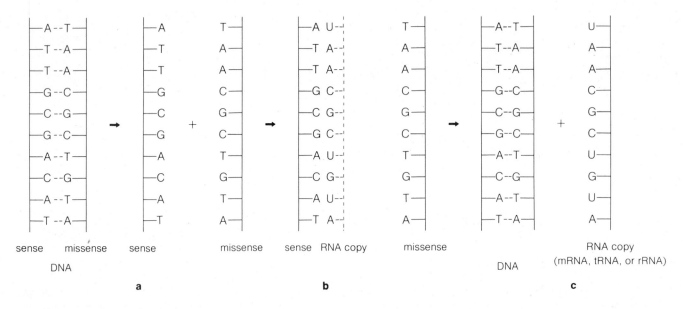

Figure 8-7 Transcription of RNA from the sense chain of the DNA molecule shown in Figure 8-6 (see text).

Complementarity and Transcription

The fact that the two nucleotide chains of a DNA molecule are complementary—that is, that the sequence in one chain fixes the sequence in the opposite chain—provides the basis for transcription because RNA copies made from the nucleotide chains of DNA molecules must also obey the rules of complementarity. The mechanism works like this. Of the two nucleotide chains in a DNA molecule, only one carries the coded directions for synthesizing a protein; this code, as noted, is determined by the sequence of bases in the chain. This nucleotide chain is called the **sense chain** of the two; the opposite chain of the DNA double helix, which has a complementary sequence but

---

Only the sense chain of a DNA molecule is copied in RNA transcription.

---

does not carry coded information, is called the **missense chain.** The missense chain is important in the duplication of DNA molecules in cell reproduction but does not enter directly into transcription.

During transcription, the two nucleotide chains of a DNA molecule unwind, and the sense chain becomes the pattern or **template** for RNA synthesis (Figure 8-7a). The new RNA chain is assembled on the DNA template (Fig-

ure 8-7b); as the RNA chain is built, bases are added according to the rules for complementary pairing. Wherever cytosine (C) occurs in the DNA template chain, a guanine (G) is placed in the growing RNA chain; wherever thymine (T) occurs in the DNA template an adenine (A) is placed in the RNA chain; and wherever guanine (G) appears in the template a cytosine (C) is placed in the RNA. Although thymine does not occur in RNA, uracil (U) forms the same pattern of hydrogen bonds as thymine and substitutes for it wherever adenine appears in the template. Thus the new RNA molecule contains the bases G, C, A, U in a sequence that is exactly complementary to the C, G, T, A sequence of the template DNA nucleotide chain. When transcription is complete (Figure 8-7c), the RNA copy unwinds from the DNA template and the sense chain of the DNA molecule rewinds with its complementary missense chain into the double helix. RNA transcription is catalyzed by a group of enzymes called *RNA*

Genes are segments of a DNA molecule coding for mRNA, tRNA, or rRNA molecules.

---

polymerases. (Information Box 8-2 outlines the mechanism of assembly of RNA by the RNA polymerase enzymes.)

Each DNA molecule in a eukaryotic nucleus contains the codes for hundreds or even thousands of different mRNA molecules. Codes for transfer and ribosomal RNA molecules are also concentrated by the hundreds or thousands at different points on the DNA. These individual segments coding for either mRNA, tRNA, or rRNA molecules are the **genes** of the nucleus. Each of the long DNA molecules containing multiple mRNA, tRNA, and rRNA sequences is a **chromosome**. In eukaryotes, the segments coding for mRNA molecules—that is, the segments coding for proteins—are copied individually, producing molecules that code for a single protein or polypeptide. In order to make these copies of individual segments within the long DNA molecules of the cell nucleus, the RNA polymerase enzymes must begin and complete transcription at points just preceding and following the sequences to be copied. Presumably, the RNA polymerase enzymes recognize specific control sequences in the DNA, located at the begin-

---

Information Box 8-2

The Mechanism of RNA Transcription

The mechanism of RNA transcription was worked out in experiments using **cell-free systems** in which the various enzymes and molecules involved are isolated, purified, and placed together in a test tube. In these systems, independently pioneered in the late 1950s by the American investigators Jerard Hurwitz, Samuel B. Weiss, and Audrey Stevens, RNA transcription can be detected and followed in a medium containing only a DNA template, Mg^{2+} or Mn^{2+} ion, the enzyme RNA polymerase, and the four bases adenine, guanine, cytosine, and uracil. The bases must be present as the nucleoside triphosphates ATP, GTP, CTP, and UTP (ATP = adenosine triphosphate, GTP = guanosine triphosphate, CTP = cytidine triphosphate, and UTP = uridine triphosphate). GTP, CTP, and UTP are all high-energy compounds with properties similar to ATP.

Research with cell-free systems has revealed that the mechanism of RNA transcription takes place as shown in the figures below:

In the first step, the RNA polymerase enzyme binds to the DNA template. If the first DNA base in the sequence to be transcribed is adenine, as shown in **a,** the enzyme binds uridine triphosphate (UTP) from the surrounding medium. With cytosine as the next base in the DNA template, the enzyme binds GTP from the medium (as shown in **b**). The total complex at this point consists of the DNA template, the enzyme molecule, and the two triphosphates UTP and GTP. The enzyme now catalyzes a reaction in which two of the three phosphates of GTP are split off (**c**), leaving one phosphate attached to the ribose sugar of guanine. This phosphate is simultaneously linked to the ribose sugar of the first nucleotide attracted, forming the first sugar–phosphate backbone linkage in the growing RNA chain. The energy required to form this linkage comes from the removal of the terminal phosphates from GTP.

The enzyme then moves to the next exposed base in the DNA template (guanine in **d**), and the entire process is repeated, producing the short RNA chain U–G–C. This process repeats sequentially until the end of the template is reached.

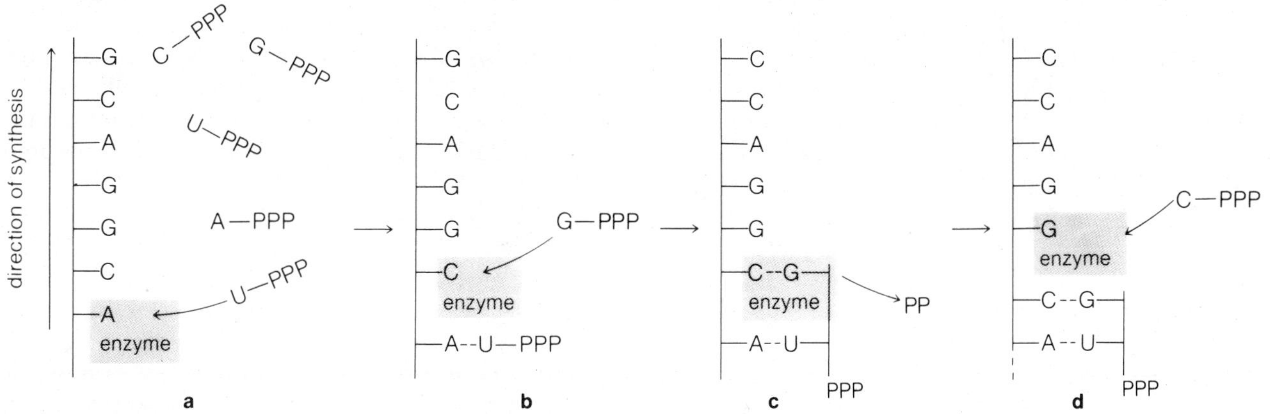

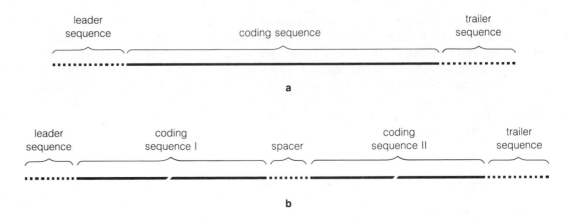

Figure 8-8 The structure of eukaryotic (**a**) and prokaryotic (**b**) messenger RNAs (see text).

ning and end regions of the genes, that tell the enzymes where to start and stop the transcription process.

Most of the RNA molecules transcribed in both prokaryotic and eukaryotic cells are made in the form of **precursor** molecules that are longer than the finished RNA products active in protein synthesis in the cytoplasm. Once transcribed, these precursors are then *processed*, as it is called, to produce the finished RNA molecules. In processing, the precursor is shortened by enzymes that clip surplus segments from the ends or middle of the molecule. If any segments are clipped from the interior of the molecule, the free ends produced are rejoined to form a single, continuous RNA. Other enzymes may add additional RNA nucleotides to either end of the precursor while the clipping and any necessary rejoining are in progress. These additional nucleotides, since they are added after transcription is complete, are not complementary to any sequences in the DNA template. As a final step in processing, individual G, C, A, and U bases in the precursor may be chemically modified by enzymes to other forms. Thus the finished mRNA, tRNA, and rRNA products may contain other purine and pyrimidine bases in addition to the original G, C, A, and U types inserted during transcription. These **modified bases,** as they are called, occur in particularly high proportions in tRNA molecules.

Characteristics of the Major RNA Classes

Messenger RNA The completely processed mRNA molecules entering the cytoplasm to direct protein synthesis in eukaryotes have several unusual features. All contain, as expected, a continuous sequence of nucleotides coding for the sequence of amino acids in a protein or a polypeptide subunit of a protein. The coding sequence lies in the middle of an mRNA molecule, separated from the ends by stretches of nucleotides that are never translated into amino acid sequences (Figure 8-8a). These extra nucleotides form "leader" and "trailer" sequences that may function to thread mRNAs through ribosomes—much like the blank leader and trailer segments used to thread a movie film through a projector.

Bacterial mRNA molecules (Figure 8-8b) are similar in structure. However, several coding sequences, each spelling out the sequence of amino acids in distinct and different proteins or polypeptides, may be present in a single bacterial mRNA molecule. Usually, the several proteins coded together in bacterial mRNA molecules are enzymes that carry out sequential steps in the same biochemical pathway. When several coding sequences are present in a bacterial mRNA, each is separated from the next by a short spacer sequence that is not translated. As in eukaryotes, the coding sequences in bacterial mRNAs are preceded by blank leader and trailer sequences.

Eukaryotic mRNA molecules are transcribed in the nucleus in the form of precursor molecules that may be as much as five to ten or more times as long as their finished mRNA products. Once transcribed, the surplus segments are clipped from the precursor molecules by the processing enzymes. Why so much extra RNA is transcribed into the mRNA precursors and then degraded almost immediately in mRNA processing is anybody's guess at the present time.

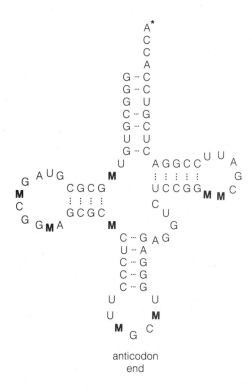

anticodon
end

Figure 8-9 The nucleotide sequence of a eukaryotic tRNA from yeast cells that carries the amino acid alanine in protein synthesis. Modified bases occur at the positions marked "M"; the amino acid binds at the position marked by the asterisk.

Transfer RNA Finished tRNAs are small molecules containing from 73 to 93 nucleotides. They are distinguished from the other RNA molecules of the cell by their high content of modified bases, which may make up as much as 15 percent of the total. Cells contain a variety of different tRNAs corresponding to the triplets specifying different amino acids in the nucleic acid code. There are 20 families of these tRNAs, one family for each of the 20 amino acids.

The first nucleic acid molecule of any kind to be completely sequenced was a tRNA from yeast that binds to the amino acid *alanine* in protein synthesis. This pioneering work was accomplished by Robert W. Holley and his coworkers at Cornell University. Their research occupied seven years of effort and used a full gram of alanine tRNA, purified from more than 300 pounds of yeast cells! Holley received the Nobel Prize in 1968 for his work in nucleic acid sequencing. Since this first success, more than 350 additional tRNA molecules have been isolated, purified, and fully sequenced. The alanine tRNA is shown in Figure 8-9.

As part of their original investigations, Holley and his colleagues noted that the sequences in many regions in the alanine tRNA molecule occur as *reverse repeats*. In these regions the sequence to the left of a point is repeated in reverse order to the right of the point in the form of complementary bases (Figure 8-10*a*). The reverse repeat allows the single nucleotide chain to fold back on itself in these regions, pairing into a short double helix called a "hairpin" (Figure 8-10*b*). Holley found enough of the reverse repeats to fold the tRNA he studied into the cloverleaf

All tRNA molecules can fold into the cloverleaf pattern.

structure shown in Figure 8-9. Holley's cloverleaf was subsequently shown to be compatible with the internal reverse-repeat sequences of all tRNA molecules, and all can evidently fold in this way.

Transfer RNA molecules are processed from somewhat longer precursor molecules in both eukaryotes and prokaryotes. Processing includes clipping out surplus segments, modifying bases at points in the tRNA sequences, and adding a few bases at the end of the molecule that binds the amino acid.

Ribosomal RNA Ribosomal RNA is defined as the RNA that can be extracted from ribosomes and the rRNA precursors that occur in the cell nucleus. At least four different rRNA molecules can be detected in eukaryotic ribosomes; three are found in prokaryotes. Sequencing

Eukaryotic ribosomes contain four different rRNA molecules in combination with ribosomal proteins.

studies have revealed that the rRNAs of all eukaryotes are closely similar. In the same way, bacterial rRNAs are closely related in sequence among different bacteria. The rRNAs of bacteria, however, appear to be unrelated to eukaryotic rRNAs.

Three of the four rRNA molecules found in eukaryotic ribosomes are transcribed in the form of a single large precursor molecule that is subsequently split into smaller pieces to release the three rRNAs (Figure 8-11). The fourth rRNA of eukaryotes is a relatively small RNA, not much larger than a tRNA molecule, that is transcribed separately as a second rRNA precursor molecule. Processing includes removal of a small surplus segment from the precursor to this rRNA.

The rRNAs of bacterial cells are also transcribed in the form of a large precursor that contains one sequence of each of the three types. These are released by enzymes that clip the precursor into segments, as in eukaryotes.

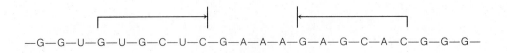

—G—G—U—G—U—G—C—U—C—G—A—A—A—G—A—G—C—A—C—G—G—G—

a

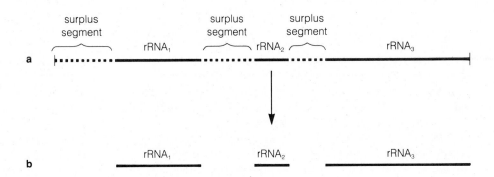

b

Figure 8-10 (**a**) A reverse repeat in an RNA molecule. The sequence to the left of the point marked by the arrow is repeated in reverse order in complementary bases to the right of the arrow. (**b**) Formation of foldback pairs in the reverse-repeat region to form a double helix.

Figure 8-11 Ribosomal RNAs are also transcribed in the form of large precursors (**a**) that are subsequently clipped during processing to release the finished rRNA molecules (**b**).

As rRNA processing takes place in both eukaryotes and prokaryotes, the proteins that also form part of ribosomes are added to the various rRNA types to form completed ribosomal subunits. In this form, the ribosomal subunits enter the cytoplasm, where they join by twos to form ribosomes and interact with mRNA and tRNA molecules in protein synthesis.

THE CELL NUCLEUS AND RNA TRANSCRIPTION

Nuclear Structure

The most intense and prolonged RNA transcription occurs in cells during the period of the cell cycle between divisions. During this nondividing stage, called **interphase,**

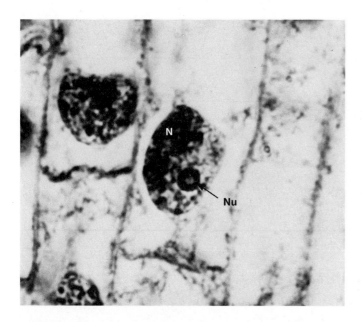

Figure 8-12 Light micrograph of a cell from an onion root tip. The nucleolus (*Nu*) is clearly visible within the nucleus (*N*).

the organization of the nucleus is similar in all eukaryotes (Figures 8-12 and 8-13). Two major structures can be seen in the nucleus at this time: (1) the **chromatin,** consisting of very fine fibers distributed throughout most of the nucleus, and (2) one or more roughly spherical bodies, the

---

A eukaryotic nucleus contains masses of chromatin fibers and one or more nucleoli.

---

nucleoli (singular = *nucleolus*). The chromatin is usually densely packed in some regions and more loosely packed elsewhere in the nucleus. The densely packed chromatin is frequently located in a layer just inside the borders of the nucleus. The chromatin fills the entire nucleus except for spaces taken up by the nucleoli.

The nucleus is most frequently viewed under the electron microscope in the form of very thin sections of chemically fixed and embedded tissue. (See Supplement 4-1 for details.) In such preparations the fibers of nuclear chromatin appear to be about 10 nanometers in diameter, or about one-third the diameter of a ribosome. The chromatin fibers always appear to be highly folded in both the densely packed and the more diffuse regions of the nucleus.

Nucleoli are easily recognized in thin sections as dense, irregularly shaped masses of fibers and granules suspended in the chromatin (Figures 8-13 and 8-14). The granules, which are the most conspicuous and characteristic feature of the nucleolus, are simply called the *nucleolar granules*. The *nucleolar fibrils* are indistinct and thinner than chromosome fibers. Chromatin fibers also extend into spaces within the nucleolus. Investigations into nucleolar structure and function have established that the fibrils and granules are the forms taken by successive steps in rRNA processing and the assembly of ribosomal subunits, both of which take place in the nucleolus.

The entire nucleus is surrounded by a system of two membranes (Figure 8-13). The outermost membrane covers the surface of the nucleus and faces the surrounding cytoplasm. The inner membrane lies just under the outer membrane and faces the nucleoplasm. A narrow compartment about 20 to 30 nanometers wide, the **perinuclear compartment,** separates the two membranes. The two membranes together are referred to as the **nuclear envelope.**

At frequent intervals the membranes of the nuclear envelope are perforated by *pores* that form channels of communication between the nucleus and cytoplasm (see arrows, Figures 8-13 and 8-15). At the margins of the pores, the outer nuclear membrane folds inward and becomes continuous with the inner membrane, forming a circular opening in the nuclear envelope about 70 nanometers in diameter. At this diameter, the pores are much wider than even the largest molecules or molecular complexes in the cell; they are more than twice the diameter of a ribosome.

The pores are not completely open but appear instead to be filled in with a "plug" of electron-dense material known as the **annulus.** The annulus appears to be perforated by a narrow central channel. The relationships of the

---

The nuclear envelope is formed from two concentric membranes perforated by pore complexes.

---

pore and annulus, which together form the **pore complex,** are diagrammed in Figure 8-15.

Pores may take up as much as one-third of the total area of the nuclear envelope. If the pores were completely open, ions and molecules would be expected to diffuse freely between nucleus and cytoplasm. However, several experiments have shown that although ions and small molecules of the size of monosaccharides, disaccharides, or amino acids pass freely and rapidly between the nucleus and the cytoplasm, the pores do control the passage of larger molecules on the order of RNA and proteins.

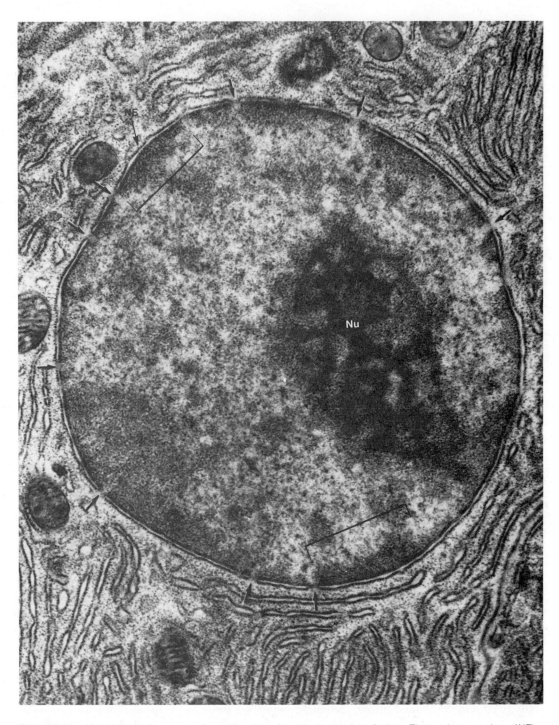

Figure 8-13 An electron micrograph of a cell nucleus from the pancreas of a bat. The nuclear envelope (*NE*) and pores (arrows) are clearly visible. Within the nucleus, the nucleolus (*Nu*) and chromatin fibers can be seen. The chromatin fibers are typically packed tightly together in some regions and less densely packed in others. The tightly packed chromatin is frequently concentrated in a layer just inside the nuclear envelope (brackets). Courtesy of D. W. Fawcett, from *The Cell* © 1966 by W. B. Saunders Company.

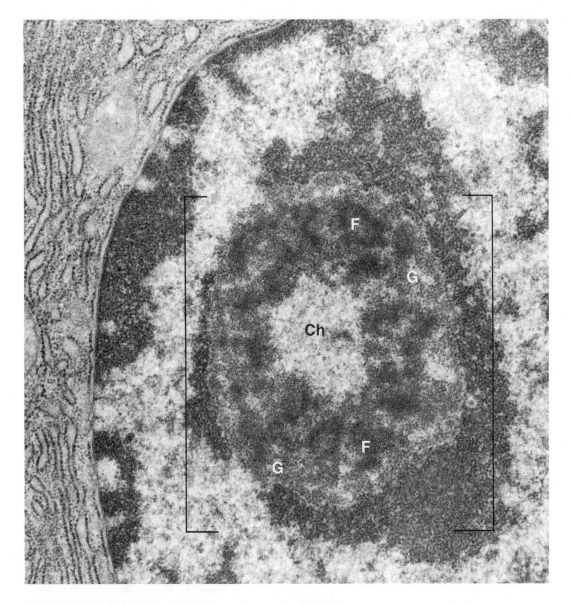

Figure 8-14 A nucleolus (brackets) from a rat pancreas cell at higher magnification. The fibrillar zones (*F*) and granular zones (*G*) are clearly visible within the nucleolus. A part of the chromatin (*Ch*) extends into the center of the nucleolus. Photograph by the author.

The nuclear pore complexes control the movement of RNA and protein molecules between the nucleus and cytoplasm.

The pore complexes probably act also as a complete barrier to some molecules, such as the DNA of the chromosomes. The membranes and pore complexes of the envelope ob-

viously function too to separate the major organelles of the nucleus and cytoplasm—or as one investigator put it, "to keep the chromosomes in and the mitochondria out."

Prokaryotic nucleoids, the equivalent of eukaryotic nuclei, also contain masses of dense fibers (Figure 8-16). In both bacteria and blue-green algae, the fibers are noticeably thinner than eukaryotic chromatin fibers. Neither

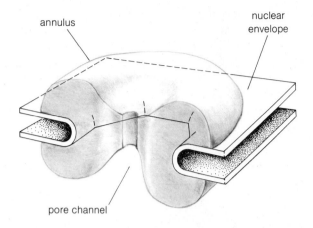

Figure 8-15 The relationship between the pore and annulus in the nuclear envelope. The channel formed by the annulus can probably accommodate molecules as large as proteins and nucleic acids.

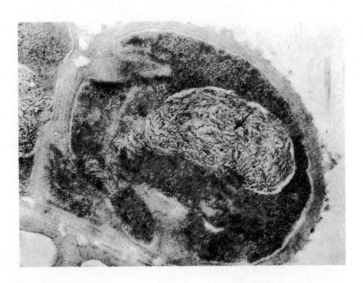

Figure 8-16 The nucleoid of a bacterium *Staphylococcus*. DNA fibers (arrow) are visible inside the nucleoid. ×54,500. Courtesy of A. K. Kleinschmidt.

bacteria nor blue-green algae have nucleoli, and no membranes equivalent to the nuclear envelope separate prokaryotic nucleoids from the surrounding cytoplasm.

The Molecular Structure of Chromatin

Chromatin fibers can be isolated from eukaryotic cells and purified in large quantities without difficulty. Analysis of such preparations has revealed that chromatin fibers

---

Chromatin fibers consist of DNA in combination with the histone and nonhistone chromosomal proteins.

---

consist of DNA in association with two major classes of nuclear proteins: the **histone** and **nonhistone** proteins. Recent research with isolated chromatin has begun to reveal how these proteins, particularly the histones, are arranged with DNA in chromatin.

The Histone Proteins In eukaryotic nuclei, DNA interacts with the histone proteins to form the basic structural units of chromatin. The histones are easily recognized against

---

The histones are primarily structural molecules in chromatin.

---

the background of proteins isolated with chromatin because they are relatively small proteins that carry a strongly positive charge at the pH ranges characteristic of living cells—properties shared by few other cellular proteins. They are easily released from chromatin that has been isolated from cells by increasing the salt concentration of the

isolating solution. Release by salt in this manner indicates that the bonds linking the histones to DNA in chromatin are electrostatic (see p. 10) rather than covalent bonds. Five major kinds of histones, called H1, H2A, H2B, H3, and H4, can be released from chromatin in this way.

All eukaryotic nuclei contain five major kinds of histones.

The Histones and the Nucleosome The five histones are believed to combine with DNA to form a spherical, bead-like structure called the **nucleosome.** This structure, first proposed in detail by Roger D. Kornberg of Harvard University, is built around a **core particle** consisting of two molecules each of histones H2A, H2B, H3, and H4 (Figure 8-17). Wrapped around this core particle is a length of DNA long enough to coil through about 1½ to 1¾ turns. (A piece of DNA this long includes 140 base pairs.) The entire structure, including the histone core particle and the

---

Nucleosomes consist of a length of DNA wrapped around a histone core.

---

length of DNA wrapped around it, forms the nucleosome. Each nucleosome is connected to the next by a short length of DNA running between them called a **linker.** The remaining histone, H1, binds to the linking DNA segment

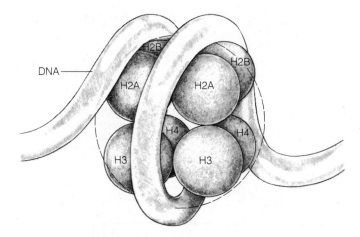

Figure 8-17 The structure of nucleosomes as proposed by R. D. Kornberg. Histones form the protein core of a nucleosome. DNA is wrapped around the histone core in a coil of 1½ to 1¾ turns. Histones H2A, H2B, H3, and H4 combine by twos to form the core; histone H1 is associated with the linker.

in the ratio of one H1 molecule per linker. Thus chromatin resembles a series of pulleys (the histone core particles) with the DNA forming a rope that winds around each pulley and connects it to the next one in line. The nucleosome particles, which are about half the diameter of ribosomes, are easily seen in electron microscope pictures of isolated chromatin (Figure 8-18).

Nucleosomes are one of the few cell structures that can be induced to self-assemble. This assembly can be accomplished simply by mixing DNA together with the five histones in salt solutions approximating the concentrations in living nuclei. The nonhistone proteins are unnecessary for this self-assembly; evidently they do not occur as major constituents of the nucleosome structures of chromatin.

The Nonhistone Proteins The nonhistone proteins are defined as the proteins, excluding the histones, that occur with DNA in chromatin. Most of the proteins in this group are negatively charged or neutral at cellular ranges of pH. A few are basic, positively charged proteins, but none of these is as basic as the histones. Typically, the nonhistone proteins occur in many more types in cell nuclei than the histones and show much greater variability between cells and organisms.

Since nucleosomes can be reassembled from DNA and the five histones alone, the nonhistone chromosomal proteins are thought to play a functional rather than structural role in the nucleus. Included in this functional category are the enzymes catalyzing transcription and replication as

---

The nonhistone proteins are primarily enzymes of transcription and replication and the proteins that control these activities.

---

well as the proteins that regulate these activities. Of the nonhistone proteins, only the enzymes carrying out transcription and replication, such as the RNA polymerase enzymes, have been isolated and identified. (The possible roles of nonhistone proteins in the control of transcription are discussed further in Chapter 9.)

The Structure of Prokaryotic DNA

Proteins cannot be detected in combination with prokaryotic DNA. Instead, the DNA in bacteria and blue-green algae is believed to combine with metallic ions, such as calcium (Ca^{2+}), or with relatively small, positively charged, organic molecules. The small diameter of the nuclear fibers in these organisms probably reflects the absence of chromosomal proteins. Since no proteins resembling the eukaryotic histones are present in any quantity in prokaryotes, there are evidently no nucleosomes in these organisms.

The DNA of a bacterial nucleoid occurs in the form of a single, large closed circle with no free ends. In *Escherichia coli*, the best-studied bacterium, the nucleoid circle con-

---

Bacterial nucleoids contain a single DNA molecule in the form of a closed circle with no associated histone or nonhistone proteins.

---

tains 1360 micrometers of DNA, equivalent to 4 million base pairs. All of this DNA is packed into cells only 1 to 2 micrometers long. Comparatively little is known about the structure of the DNA in blue-green algae. Considerably more DNA, up to several times the amount in bacteria per cell, is present in these prokaryotes. Whether the DNA of blue-green algae exists in the form of closed circles is as yet unknown.

RNA Transcription in Chromatin Fibers

Several experiments have shown that RNA transcription is related to the distribution of chromatin fibers in the nucleus. One experiment demonstrating this fact uses a radioactive form of uracil, the nucleotide base that occurs in RNA but not DNA. As cells transcribe RNA in the presence of radioactive uracil, the regions of the nucleus carrying out the transcription become labeled by radioactivity and can be identified. Exposing cells to radioactive uracil in this way reveals that RNA transcription occurs almost entirely in the loosely packed chromatin fibers of the nucleus. By different techniques, the RNA transcribed in

Messenger RNA and transfer RNA are transcribed in regions of the nucleus outside the nucleolus.

these regions has been identified as messenger and transfer RNA. Ribosomal RNA transcription is confined almost entirely to the nucleolus.

The nucleolus is the primary site of ribosomal RNA synthesis.

One of the most interesting discoveries resulting from research into ribosomal RNA transcription is that the genes for rRNA are repeated many times in chromatin. The genes for the larger rRNA types, repeated from a few hundred to thousands of times in different eukaryotes, are arranged end to end, usually at one place in the DNA of one chromosome or chromosome pair. All these multiple rRNA genes are active simultaneously in transcribing rRNA precursors, which are subsequently processed and combined with the ribosomal proteins in the same location. This extensive biochemical activity becomes visible as the fibrils and granules of the nucleolus. The genes for the smallest RNA type, which may exist in as many as 20,000 copies, are distributed in blocks elsewhere in the chromatin. The many repeats of rRNA genes are reflected in the fact that cells make more rRNA than any other type; in most cells, rRNA accounts for 85 to 95 percent or more of the total RNA transcribed. The large number of rRNA genes thus increases the cell's "horsepower" for transcribing rRNA and making ribosomes.

Figure 8-19, one of the most striking series of electron micrographs ever taken, shows rRNA genes caught in the act of transcription by Oscar L. Miller of the University of Virginia. The long, filamentous strand (arrow) in the picture is a part of the DNA molecule containing the repeated rRNA coding sequences. The "hairy" regions are collections of rRNA molecules being transcribed on the DNA; each hair is a single rRNA precursor molecule. The granule at the base of each rRNA precursor is a molecule of RNA polymerase engaged in transcribing the nucleolar organizer DNA; many RNA polymerase molecules are copying the DNA template at one time. At the end of the hairy region with short rRNA precursors, the RNA polymerase molecules have just begun transcription. At the opposite end, the rRNA precursor molecules are longer and transcription is nearly complete. A nontranscribed DNA spacer lies on either side of the rRNA precursor gene; no RNA

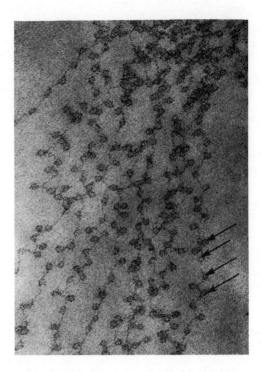

Figure 8-18 The beads-on-a-string image produced by nucleosomes (arrows) in chromatin fibers isolated in dilute salt concentrations. ×172,000. Courtesy of A. L. Olins. Reproduced from *The Molecular Biology of the Mammalian Genetic Apparatus*, originally published by Elsevier Biomedical Press B. V.

polymerase enzymes or rRNA molecules occupy this region. The nucleolus consists of many such genes separated by nontranscribed spacers.

THE STRUCTURE AND FUNCTION OF THE CELL NUCLEUS: A SUMMARY

The information required to synthesize mRNA, tRNA, and rRNA is coded into the DNA of the cell nucleus. Instead of existing as one large molecule, the nuclear DNA coding this information in eukaryotes is broken into shorter lengths. Each of these DNA lengths is a *chromosome* of the cell nucleus. The nuclei of human cells contain 46 of these DNA lengths. The total amount of DNA per nucleus is immense, considering the size of a cell. The 46 molecules of one human nucleus, for example, would measure about 1 meter in length if laid end to end! (Each DNA molecule is 2 to 3 centimeters long.)

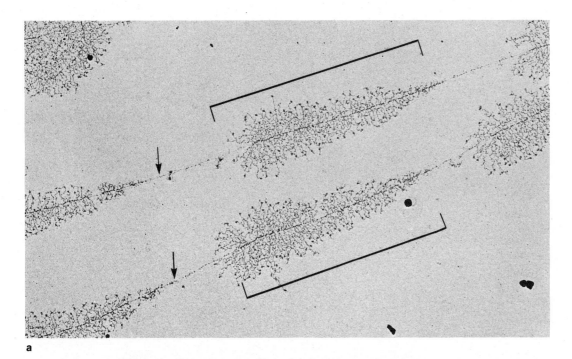

a

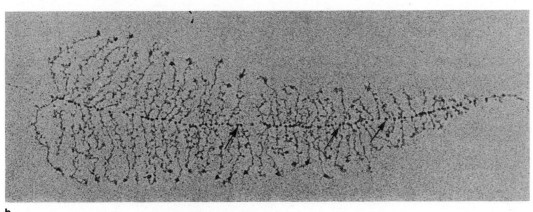

b

Figure 8-19 (**a**) DNA molecules (arrows) in nucleoli isolated from developing egg cells of a salamander caught in the process of RNA transcription. The RNA molecules being transcribed appear as shorter fibers running at right angles to the DNA strands (brackets). (**b**) One of the transcribing regions at higher magnification. The spherical granules (arrows) at the base of each RNA molecule are molecules of RNA polymerase engaged in transcribing the DNA. Courtesy of O. L. Miller, Jr., and B. R. Beatty, Biology Division, Oak Ridge National Laboratory.

The coded information in one of these chromosomes might be arranged as shown in Figure 8-20. Long stretches are taken up by mRNA coding sequences. Spacer sequences occur both in and between the regions coding for mRNA in eukaryotes. Although most of the mRNA coding sequences occur only in single copies, a few are repeated. (One known example of repeated mRNA genes is the gene group coding for the histones, which may occur in hundreds or thousands of copies.) Clustered in one region are from hundreds to thousands of repeats of the sequence coding for the larger rRNA molecules, each separated by a non-transcribed spacer sequence. This clustered region, when surrounded by rRNA precursors in various stages of processing and assembly with the ribosomal proteins, forms

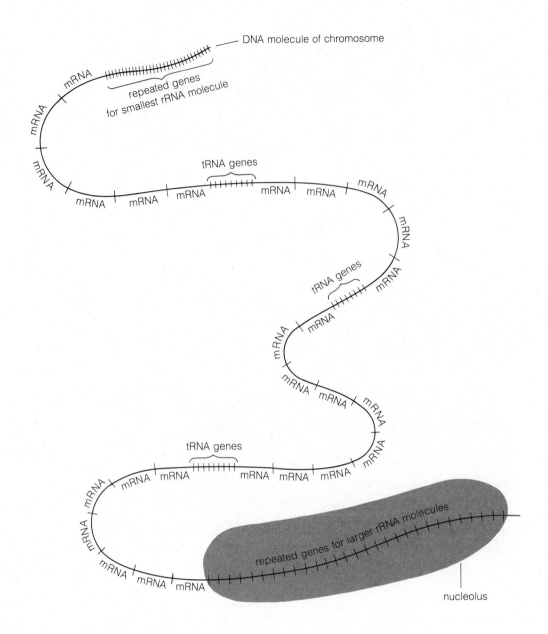

Figure 8-20 The arrangement of mRNA, tRNA, and rRNA genes in a eukaryotic chromosome (see text).

Labels in figure: DNA molecule of chromosome; repeated genes for smallest rRNA molecule; tRNA genes; repeated genes for larger rRNA molecules; nucleolus; mRNA

the *nucleolus* of the nucleus. In most organisms, only one chromosome or a pair of identical chromosomes carries the cluster of rRNA sequences. Located at some distance from the nucleolar cluster, on the same or a different chromosome, are the repeated sequences coding for the smallest rRNA type. These regions also consist of clustered repeats of the small rRNA sequence separated by nontranscribed spacers. Clustered at other points are repeats of the tRNA precursor sequence, each also separated by spacer sequences. Each sequence coding for an mRNA, rRNA, or tRNA is called a *gene*.

The DNA molecules of the chromosomes combine with the histone and nonhistone proteins to form the *chromatin fibers* of the cell nucleus. Within the chromatin fibers, the DNA winds around the histone core particles to form the *nucleosomes* of the chromatin.

The RNA polymerase molecules copy segments of the DNA molecules in the nucleus, producing the precursors of mRNA, tRNA, and rRNA. These precursors are subsequently clipped into shorter pieces and modified chemically to produce the finished RNA types. Only a few of the thousands of mRNA coding segments are transcribed at any one time; these transcribed segments form the *active* chromatin of the cell nucleus, which is typically more loosely packed in the nucleus than inactive segments. Usually, the dense clumps of inactive chromatin are concentrated in a layer just inside the nuclear envelope.

In bacteria, the various RNA coding sequences are distributed along a single DNA molecule that forms a closed circle. Located at points around the circle are a few repeats of the genes coding for ribosomal RNA. Transfer RNA genes are spotted at other locations around the circle. Other segments are filled in with mRNA coding sequences. These sequences may be continuous, with no intervening spacers, or they may be separated by small amounts of nontranscribed spacer DNA. The distribution of sequences in blue-green algae is probably similar to the bacterial pattern. Since histones cannot be detected in bacteria, the DNA circle probably exists in "naked" form without winding into nucleosomes. The DNA molecules of blue-green algae are likely to take a similar form.

Questions

1. Outline the overall pattern of protein synthesis.

2. What is mRNA? tRNA? rRNA?

3. Define transcription and translation.

4. Outline the structures of nucleotides, DNA, and RNA. How do DNA and RNA differ? How is information coded in DNA and RNA?

5. What factors determine base pairing in DNA? What holds a DNA molecule together?

6. What is complementarity? What significance does complementarity have for the process of transcription?

7. What are the sense and missense chains of a DNA molecule? What is the function of the sense chain? How do you think the missense chain might function in DNA replication?

8. Draw a sense chain containing 25 A, T, C, and G nucleotides in a random sequence. What is the sequence of the missense chain that is complementary to this sequence? What RNA sequence would be copied from the DNA sense chain?

9. What is a chromosome? What is the difference between the chromatin and the chromosomes of a nucleus?

10. Outline the mechanism of RNA transcription.

11. What is RNA processing? What steps may take place in RNA processing?

12. How do bacterial and eukaryotic mRNAs differ?

13. Outline the structure of a tRNA molecule. What are reverse repeats? What are foldback double helices?

14. How is rRNA defined?

15. What major structures occur in eukaryotic nuclei? In prokaryotic nuclei?

16. What is the relationship between the nucleolus and the rRNA genes?

17. Outline the structure of the nuclear envelope.

18. What are the histone and nonhistone proteins? How many major histones occur in eukaryotic nuclei?

19. What is a nucleosome? How do DNA and the histones combine to form nucleosomes?

20. Why are the nonhistone proteins not considered to be essential parts of nucleosome structure?

21. What are the possible functions of the nonhistone proteins?

22. Outline an experiment demonstrating that RNA is transcribed in the nucleus.

23. Each nucleus in an amphibian such as the leopard frog *(Rana pipiens)* contains about 15×10^{-6} microgram of DNA. Using the fact that DNA weighs approximately 1.5×10^{-12} microgram per micrometer, what is the total length of the DNA in a frog nucleus?

Suggestions for Further Reading

Brown, D. D. 1973. "The Isolation of Genes." *Scientific American* 229:21–29.

Crick, F.H.C. 1970. "Split Genes and RNA Splicing." *Science* 204:264–271.

Rich, A., and S. H. Kim. 1978. "The Three-Dimensional Structure of tRNA." *Scientific American* 238:52–62.

Watson, J. D. 1968. *The Double Helix.* Atheneum, New York.

Watson, J. D. 1976. *Molecular Biology of the Gene.* 3rd ed. Benjamin, Menlo Park, California.

Wolfe, S. L. 1981. *Biology of the Cell.* 2nd ed. Wadsworth, Belmont, California.

SUPPLEMENT 8-1: DISCOVERY OF THE STRUCTURE AND FUNCTION OF DNA

Discovery of DNA's Informational and Hereditary Function

The first definitive evidence that DNA contains the directions for cell synthesis and heredity came with two sets of experiments conducted during the 1940s and 1950s. The first of these, by Oswald Avery and his coworkers at the Rockefeller Institute, was carried out with different types of *Pneumococcus*, the bacterium causing pneumonia. The

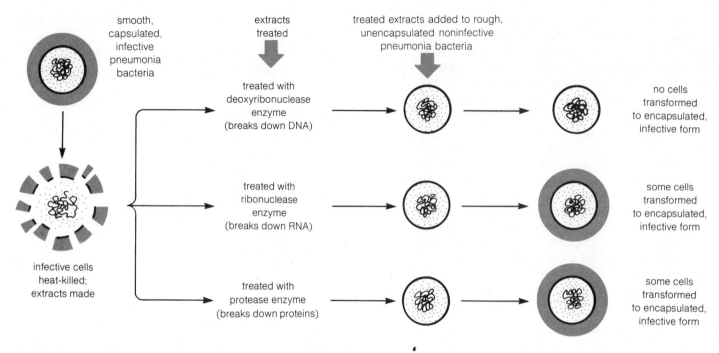

Figure 8-21 The bacterial transformation experiment carried out by Avery and his colleagues (see text).

Avery experiments took advantage of two known forms of this bacterium. An infective form that can cause pneumonia is surrounded by a polysaccharide capsule and produces smooth, gelatinous colonies when grown in pure cultures. A noninfective variety of the same bacterium has no capsule and forms colonies that appear lusterless or rough in texture when the bacteria are grown in culture dishes.

Genetic experiments conducted earlier had revealed that the two kinds of *Pneumococcus* bacteria are different hereditary types. When kept in pure colonies, the hereditary types bred true—that is, the cells of either type gave rise only to cells of the same type when dividing. Exposing rough, noninfective cells to smooth, infective cells, however, frequently caused transformation of the noninfective cells to the infective type. Transformation of this kind could be detected even if the infective bacteria were heat-killed before exposing the noninfective cells to them. Once transformed into the infective type, the bacteria bred true, indicating that the change was hereditary.

Avery and his colleagues were interested in identifying the substance in killed cells that could permanently transform rough, noninfective cells into smooth, infective cells. To identify the substance, they treated an extract of killed cells with enzymes that can catalyze the breakdown

of DNA, RNA, or proteins (Figure 8-21). Only the enzyme capable of breaking down DNA destroyed the capacity of the extract to transform noninfective cells into the infective type. Enzymes breaking down RNA or proteins had no effect on the transforming ability of the extract. From these results, Avery proposed in 1944 that DNA is the substance responsible for transforming noninfective pneumonia bacteria to the infective kind. Since DNA can carry hereditary information in this way, it also seemed a likely candidate for the normal carrier of genetic information in the cell nucleus.

Avery's conclusion was directly supported and extended by a second series of experiments carried out in 1952 at the Carnegie Laboratory of Genetics by Alfred D. Hershey and Martha Chase, who studied the infection of bacteria by bacterial viruses. These viruses, called *bacteriophages* (see Figure 4-22), cause bacterial cells to cease producing their own molecules and to make instead the DNA and protein of new virus particles. (Viruses consist only of a core of DNA or RNA surrounded by a surface coat of proteins; see p. 69.) When Hershey and Chase began their experiments, biologists assumed that during infection of a bacterium by a virus, either the DNA or the protein of the bacteriophage must enter the bacterial cell to alter its genetic and synthetic capacity.

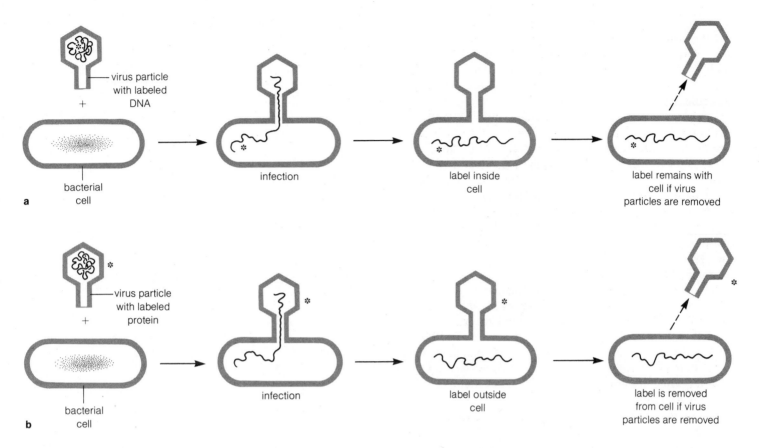

Figure 8-22 The Hershey and Chase experiment demonstrating that the DNA of viruses enters bacterial cells to alter their genetic machinery (see text).

To decide between these alternatives Hershey and Chase used radioactive isotopes of phosphorus and sulfur, newly available at that time. Using the isotopes, they were able to label the proteins and DNA specifically because DNA contains phosphorus but no sulfur whereas the viral protein being studied contained sulfur but no phosphorus. To label the virus, bacterial cells were infected and allowed to make virus particles in a culture medium containing either radioactive phosphorus or sulfur. By this method, mature virus particles were obtained that contained either radioactive DNA or protein.

Hershey and Chase then mixed the labeled virus particles with fresh bacterial cells and allowed infection to take place (Figure 8-22). Radioactivity was found inside the infected bacterial cells only if the virus particles contained radioactive DNA (Figure 8-22a). When the experiment was conducted with virus particles containing labeled protein, all radioactivity remained outside the bacterial cells (Figure 8-22b). This showed that only the DNA of the infecting virus entered the bacterial cells. Since only the

DNA entered, this molecule must have carried the information required to convert the bacterial cell machinery to the synthesis of the new bacteriophage particles.

The experiments of Avery and his colleagues and Hershey and Chase established that DNA, and not protein, carries genetic information and is likely to form the basis of information storage in the cell. These experiments touched off intensive research that eventually revealed not only the molecular structure of DNA but also how information is coded into the nucleic acids.

Discovery of DNA's Molecular Structure

When Hershey and Chase's results were announced, the complete molecular structure of DNA was unknown. Some information was available, however. It was known that DNA is formed from long chains of nucleotide subunits and that these subunits are connected into chains by sugar–phosphate linkages in the molecule. The molecular structure of the individual pyrimidine and purine bases and the structure of the deoxyribose sugar had long since been worked out by organic chemists. The number of nucleotide chains in a DNA molecule, however, and the manner in which the chains fold or twist to form the intact molecule, were unknown.

This was the state of affairs when James D. Watson and Francis H. C. Crick took up their study of DNA structure at Cambridge University in the early 1950s. In their work, Watson and Crick relied heavily on data gathered from the x-ray diffraction of DNA molecules in the laboratory of Maurice H. F. Wilkins at Kings College, London.[1] Wilkins' evidence, developed in collaboration with his co-worker Rosalind Franklin, indicated that the molecule is cylindrical with an outside diameter of about 2 nanometers. Within the molecule, the x-ray data suggested that the nucleotide chains forming the backbone of the DNA molecule twist into a helix of regular diameter and pitch. Other evidence, from a chemical analysis of DNA carried out by Erwin Chargaff at the Rockefeller University, indicated that the amounts of adenine and guanine in DNA, while usually not equal, are always exactly paralleled by the amounts of thymine and cytosine respectively. That is, regarding their amounts in DNA, adenine = thymine and guanine = cytosine.

From this information, Watson and Crick developed the model for DNA structure outlined in this chapter. They deduced first that the molecule consists of two nucleotide chains wound into a double helix—not a triple helix containing three chains as others had proposed. Later they constructed scale models of the nucleotides and fit the pieces together in various ways until they arrived at the correct relationship for base pairing and the three-dimensional structure of the DNA molecule. Their discovery was announced in a brief paper published in the British journal *Nature* in 1953. In the years following the publication of their work, all tests of the model have supported their hypothesis and confirmed that DNA is arranged in the double helical, internally paired structure they proposed.

---

[1]In x-ray diffraction of DNA, a beam of x rays is directed at a purified, partially dried DNA sample. (X rays are a form of radiation with much shorter wavelengths than visible light.) Within the DNA molecules, any patterns of atoms that occur at regular, periodic distances cause the beam to bend (diffract) at these points. If a pattern is repeated often enough in the specimen, the diffracted x rays separate into smaller beams that can be imaged as spots and recorded on photographic film. Analysis of the angles and distances between the spots provides an estimate of the likely arrangement of atoms in the specimen.

Suggestions for Further Reading

Watson, J. D. 1968. *The Double Helix*. Atheneum, New York.

Watson, J. D., and F.H.C. Crick. 1953. "Molecular Structure of Nucleic Acids: A Structure for Deoxyribose Nucleic Acid." *Nature* 171:737–738.

The function of the cell nucleus, as outlined in the previous chapter, is in storing and transmitting the information required for protein synthesis. This information is coded into the sequence of nucleotide bases in DNA. It is transcribed and transferred to the cytoplasm as the directions for protein synthesis in the form of messenger RNA (mRNA) molecules. Other RNA molecules also important in protein synthesis, including ribosomal RNA (rRNA) and transfer RNA (tRNA), are transcribed at the same time. These RNAs also enter the cytoplasm after transcription and interact with mRNA in the synthesis of proteins.

Cytoplasmic protein synthesis takes place in two major steps. In a preliminary reaction sequence, the different transfer RNA molecules link to their specific amino acids. This step takes place in the cytoplasmic solution. In the second step, these tRNA–amino acid complexes interact with messenger RNA in protein synthesis on ribosomes. All the reactions of the second step take place on ribosomes.

THE REACTIONS OF PROTEIN SYNTHESIS

Step 1: Attachment of tRNAs to Amino Acids

A large number of different tRNAs occur in the cytoplasm of eukaryotes—many more than the minimum of 20 required to provide one tRNA for each of the 20 amino acids used in protein synthesis. (The structure of tRNA molecules is outlined on p. 153.) This means that most of the 20 amino acids link to more than one type of tRNA.

Most of the amino acids link to more than one type of tRNA molecule.

Because tRNAs are the molecules that match amino acids with "words" in the messenger RNA code, most of the amino acids are therefore coded for by more than one word in the message. The amino acid serine, for example, is designated by any one of six different words in the nucleic acid code, and it may link to any one of six different tRNA molecules. How this system of multiple code words and tRNAs actually operates in protein synthesis is explained later in this chapter.

The amino acids attach to their respective tRNA molecules in a two-part reaction sequence that takes place in the cytoplasm surrounding ribosomes but not directly connected to them. The final product of the sequence, an amino acid linked to tRNA, is a high-energy complex that contains energy directly derived from ATP. Because the amino acids are joined with their tRNAs in a high-energy

9

CELLULAR SYNTHESIS II: PROTEIN SYNTHESIS IN THE CYTOPLASM

complex, the entire two-part sequence is usually called **amino acid activation.** The reaction sequence is really more extensive than this name suggests, however, because, in addition to increasing in energy level, the amino acids are also matched with their specific tRNA molecules.

In amino acid activation, amino acids are increased in energy content and attached to their correct RNAs.

In the first part of the series, an amino acid interacts directly with ATP. This enzyme-catalyzed reaction removes two phosphate groups from ATP and attaches the AMP (adenosine monophosphate) product to the amino acid:

$$\text{AA} + \text{ATP} \rightarrow \text{AA—AMP} + 2 \text{ phosphate groups} \qquad (9\text{-}1)$$
amino
acid

The phosphate groups are released to the surrounding cytoplasmic solution. Much of the energy released as the phosphate groups are split from the ATP entering the reaction is trapped in the AA—AMP complex.

At this point in the process the AA—AMP complex remains attached to the same enzyme molecule, which also speeds the next part of the reaction sequence. In this step, the amino acid is transferred from AMP to one of its specific tRNA molecules:

$$\text{AA—AMP} + \text{tRNA} \rightarrow \text{AA—tRNA} + \text{AMP} \qquad (9\text{-}2)$$
aminoacid–tRNA

The AA—tRNA product and AMP are released from the enzyme when the reaction is complete. As in part 1 of the sequence, much of the energy released by converting ATP to AMP in Reaction 9-1 is retained in the AA—tRNA complex produced as a final product of Reaction 9-2. As a result, this complex is a high-energy substance. This energy is eventually used to drive formation of a peptide bond when amino acids link together on ribosomes.

The synthetase enzymes match amino acids to their correct tRNAs.

The same enzyme, called a *synthetase,* catalyzes both the steps shown in Reactions 9-1 and 9-2. There are 20 different synthetase enzymes, one for each of the 20 different amino acids. In the reaction, the enzymes recognize both a specific amino acid and one of the correct transfer RNA molecules for that amino acid and match the two together. Thus, although tRNA molecules provide the necessary link between the nucleic acid code and the amino

acid sequence of proteins, the accuracy of the system ultimately depends on the ability of the synthetase enzymes to match amino acids with their respective tRNAs.

In summary, amino acid activation satisfies two requirements for protein synthesis. One of these is provision of the energy required for formation of peptide bonds. The second is matching tRNAs with their correct amino acids, which provides the basis for translation of the sequence of nucleotides in an mRNA into the sequence of amino acids in a protein. The aminoacid–tRNA complexes formed as the final product of amino acid activation are the raw material as well as the primary energy source for the next phase of protein synthesis: the assembly of amino acids into polypeptide chains on ribosomes.

Step 2: Polypeptide Assembly on Ribosomes

To simplify protein synthesis we will consider the assembly of a hypothetical protein containing only two different amino acids, arginine and serine, alternating in the sequence arg–ser–arg–ser . . . and so on. The information required for synthesis of this protein is coded in DNA in the sequence of the four nitrogenous bases A, T, G, and C. These are taken by threes to form the DNA code words or **codons.** One of the DNA code words for arginine is the sequence guanine–cytosine–guanine (GCG); one codon for serine is adenine–guanine–adenine (AGA). Therefore the sense chain for our hypothetical protein would carry the code GCG/AGA/GCG/AGA . . . and so on.

In the first step of the overall process, the DNA code is transcribed into an mRNA for the protein. In our example, the DNA coding sequence GCG/AGA/GCG/AGA . . . will be transcribed into the complementary mRNA sequence CGC/UCU/CGC/UCU . . . and so on.

After transcription in the nucleus, the mRNA enters the cytoplasm and attaches to a ribosome. In the cytoplasm surrounding this mRNA–ribosome complex are all the various kinds of RNA, each attached to its specific amino acid. Among the tRNAs are several types that carry the amino acids arginine and serine. Of these, the ones of interest are the tRNAs able to recognize and bind to the CGC and UCU codons in our hypothetical messenger RNA. This recognition capacity depends on the **anticodon** of the tRNA molecule (see Figures 9-1 and 8-9), a region able to form complementary base pairs with the codons in the message. The tRNA chain folds in such a way that the anticodon is exposed at one end of the molecule and the amino acid at the other end. The anticodon pairs with a codon on RNA by forming the appropriate base pairs. We will consider that the tRNA carrying arginine has the anticodon GCG (Figure 9-1a), which is able to recognize and

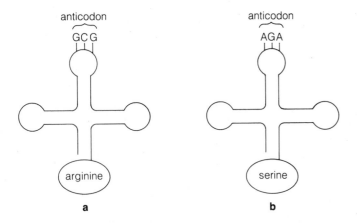

anticodon
GCG

anticodon
AGA

arginine

serine

a

b

Figure 9-1 Transfer RNA molecules for arginine and serine, showing the anticodons assumed in the text.

bind to the CGC codon for arginine in the messenger RNA. The tRNA carrying serine has the anticodon AGA (Figure 9-1b), which is able to recognize and bind to the UCU codon for serine in the message.

These aminoacyl–tRNAs interact with the messenger at the ribosome in the following way. If the mRNA sequence CGC is the first codon exposed at the ribosome (Figure 9-2a), a tRNA carrying arginine will attach to the ribosome and form complementary base pairs between the CGC codon on mRNA and the GCG anticodon of the tRNA (Figure 9-2b). As the ribosome moves to the next codon on the messenger RNA, it encounters the mRNA triplet UCU. A tRNA with the anticodon AGA, carrying the amino acid serine, pairs with the messenger at this point (Figure 9-2c).

This interaction places the two amino acids now at the ribosome in a position favorable for the formation of the first peptide bond. Formation of the bond, catalyzed by an enzyme that is part of the ribosome itself, involves separating the amino acid from the first tRNA binding to the ribosome and transferring it to the amino acid carried by the most recently bound tRNA. In our example, arginine is removed from its tRNA and linked by a peptide bond to the serine carried by the second tRNA (Figure 9-2d). The energy required to form the peptide bond comes from the free energy released when arginine separates from its tRNA. This tRNA, now free of its amino acid, is released from the ribosome.

The complex now consists of the ribosome, the mRNA, and a tRNA carrying the short polypeptide chain ser–arg. The ribosome now moves to the next codon on the message. In our example, this is the triplet CGC. A tRNA car-

rying arginine attaches to the ribosome at this point (Figure 9-2e) in a position that favors formation of the second peptide bond. In formation of this bond, the short ser–arg polypeptide is transferred to the arginine tRNA most recently attached to the ribosome (Figure 9-2f), producing the three-amino-acid chain arg–ser–arg. The entire process repeats as the ribosome moves to the next codon and continues until the end of the coding portion of the mRNA molecule is reached. With some differences in detail, protein synthesis occurs in this general pattern in both prokaryotes and eukaryotes.

Once synthesis is complete, the amino acids in new polypeptide chains may be variously modified to other forms by the addition of acetyl, methyl, phosphate, hydroxyl, or other chemical groups. As a result of these modifications, as many as 140 different modified amino acids may occur in various natural proteins. Some proteins are also altered by addition of metallic ions or by combination with complex lipid or carbohydrate-containing groups.

Once protein synthesis is complete, individual amino acids may be chemically modified to other forms.

Completed proteins probably fold automatically into their final three-dimensional shape. Once the final folding pattern is taken up, it may be stabilized by formation of disulfide (S—S) linkages (see p. 28) located at different points in the polypeptide chain. The finished proteins, depending on their amino acid sequences, may become enzymes, antibodies, peptide hormones, or the structural, motile, transporter, or receptor molecules of the cell.

The Genetic Code for Protein Synthesis

Working out the nucleic acid code words used in protein synthesis was one of the most exciting endeavors in the history of biology. The project attracted the interest and attention of scientists of all types: biologists, mathematicians, physicists, and even an astronomer. It stands as an unusual example of a biological conclusion contributed to by workers from practically all the major branches of science.

The initial problem in solving the genetic code was to determine how four different nucleotides code for 20 different amino acids.

When the structure of DNA was discovered, it became apparent that the directions for assembling amino acids into proteins are stored in the sequence of nucleotides in DNA. The problem to be solved was how different

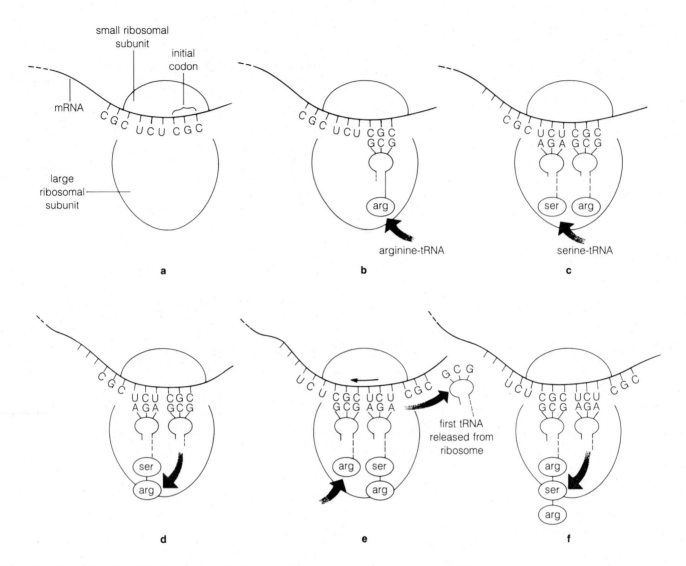

Figure 9-2 Interaction of mRNA, ribosomes, and aminoacyl–tRNAs in protein synthesis (see text). Only the anticodon arms of the tRNA molecules are shown in this diagram.

combinations of only four different bases could designate 20 amino acids and, later, the identity of the individual code words for the 20 amino acids.

Four different nucleotides can be used to spell 64 different three-letter code words.

By the early 1950s investigators realized that code words would have to be at least three nucleotides long to specify 20 amino acids. Only four code words could be made if the bases were taken just one at a time ($4^1 = 4$). Although 16 code words could be made if all possible two-

letter combinations were put together ($4^2 = 16$), 16 would still be insufficient to code for all 20 amino acids. Three-letter code words, however, provide 64 combinations ($4^3 = 64$)—more than enough for the total requirements of the code vocabulary. Therefore three-letter code words seemed reasonable as a starting point for breaking the code.

Several different experiments established that the genetic code does in fact use three-letter code words. One, carried out by H. Gobind Khorana and his colleagues at the Massachusetts Institute of Technology, used artificial messenger RNA molecules with different repeating sequences. These mRNAs were added to a cell-free system containing ribosomes, amino acid–tRNA complexes, and all the other factors required for protein synthesis. Artificial mRNAs with a repeating sequence of the form ABABABAB . . . , where A and B are any two RNA nucleotides, coded for the synthesis of a polypeptide containing only *two* amino acids when added to the cell-free protein synthesis system. In the polypeptide produced by this artificial mRNA, the two amino acids alternated regularly. Assuming three-letter code words, this is the expected result, because the code in the artificial mRNA would then be read $_{ABA-BAB-ABA-BAB}^{1\ \ \ 2\ \ \ 1\ \ \ 2}$ . . . ; only two different "words" are contained in the message. Repeating artificial mRNAs of the form ABCABCABCABC . . . coded for the synthesis of polypeptides containing only *one* type of amino acid. This is also expected because in a triplet code this message would be read $_{ABC-ABC-ABC-ABC}^{1\ \ \ \ 1\ \ \ \ 1\ \ \ \ 1}$ . . . . Finally, repeating artificial mRNAs of the form ABCDABCDABCDABCD . . . coded for chains of polypeptides containing alternating sequences of *four* different amino acids. This is expected only if the message ABCDABCDABCDABCDABCDABCD . . . . is read in triplets as $_{ABC-DAB-CDA-BCD-ABC-DAB-CDA-BCD}^{1\ \ \ \ 2\ \ \ \ 3\ \ \ \ 4\ \ \ \ 1\ \ \ \ 2\ \ \ \ 3\ \ \ \ 4}$ . . . Khorana's results are therefore possible only if the code words in the nucleic acid code are triplets.

The next task was to identify the code words. Although many investigators solved parts of the puzzle, Marshall W. Nirenberg of the National Institutes of Health carried out the most definitive work. Nirenberg synthesized very short mRNA molecules only three nucleotides in length. These short mRNAs thus contained only a single code word. All 64 possibilities for these short mRNAs were made and then added one at a time to cell-free systems containing ribosomes and all the factors required for protein synthesis. The three-letter mRNAs caused single tRNA molecules with the amino acid corresponding to the mRNA code word to bind to the ribosomes. By identifying the amino acid binding to the ribosomes for each three-letter code word, the coding assignments of the triplet codons were worked out. Nirenberg and Khorana received the Nobel Prize in 1968 for their work in solving the genetic code.

This approach allowed all the code words to be identified (Figure 9-3). By convention, the code words are designated in RNA rather than DNA equivalents. The RNA codon UUU for phenylalanine, for example, corresponds to the DNA codon AAA. Three codons, UAG, UAA, and UGA, were found to have no coding assignments. These codons, which were at first termed "nonsense codons" because they do not code for an amino acid, were later found to be a signal for the ribosome to terminate protein synthesis. Thus the codons UAG, UAA, and UGA, now called **terminator codons,** represent the word *stop* in the coded message for protein synthesis. When a ribosome encounters one of these three codons, protein synthesis ceases and the mRNA and finished polypeptide are released.

The three-letter code words were identified and given their amino acid coding assignments by testing the binding of short nucleotides to RNA molecules. Three of the 64 codons say "stop" instead of specifying amino acids.

Inspection of the code reveals that all but two of the amino acids are specified by more than one code word. This feature of the code, called **degeneracy,** means simply that many of the code words are synonyms. The amino acid proline, for example, was found to have the four synonymous codons CCU, CCC, CCA, and CCG in the code.

There are many synonyms in the genetic code.

A second feature of the genetic code is its almost completely universal usage in living organisms. With very few exceptions, the same codons stand for the same amino acids in all organisms from bacteria to the higher eukaryotes and also in viruses. The exceptions, recently discovered, occur in the DNA–protein synthesis systems of mitochondria, where UGA evidently codes for tryptophan instead of termination, AUA codes for methionine instead of isoleucine, and CUA codes for threonine instead of leucine in some polypeptides. These coding substitutions apparently depend on the unusual properties of the tRNA molecules of mitochondria (see Supplement 9-1 for details). The otherwise universal nature of the code indicates that it has very ancient origins and became established as one of the earliest events in the evolution of cellular life. If life based on the same nucleic acid–protein system is ever discovered on another planet, it will be interesting to see if the code words are the same as our own—that is, if the code is really universal in the cosmic sense.

THE CYTOPLASMIC STRUCTURES ACTIVE IN PROTEIN SYNTHESIS

Several cytoplasmic structures function in the synthesis and modification of proteins. The simplest of these are the ribosomes. Ribosomes occur in large numbers in the cytoplasm, either freely suspended or attached to a system of cytoplasmic membranes: the **endoplasmic reticulum (ER).** Ribosomes assemble proteins in large quantities in either location. Apparently the existence of ribosomes in

first base of codon	U	C	A	G	third base of codon
U	UUU UUC } phe / UUA UUG } leu	UCU UCC UCA UCG } ser	UAU UAC } tyr / UAA * / UAG *	UGU UGC } cys / UGA * / UGG try	U C A G
C	CUU CUC CUA CUG } leu	CCU CCC CCA CCG } pro	CAU CAC } his / CAA CAG } gln	CGU CGC CGA CGG } arg	U C A G
A	AUU AUC } ileu / AUA / AUG met	ACU ACC ACA ACG } thr	AAU AAC } asn / AAA AAG } lys	AGU AGC } ser / AGA AGG } arg	U C A G
G	GUU GUC GUA GUG } val	GCU GCC GCA GCG } ala	GAU GAC } asp / GAA GAG } glu	GGU GGC GGA GGG } gly	U C A G

ala = alanine
arg = arginine
asn = asparagine
asp = aspartic acid
cys = cysteine
gln = glutamine
glu = glutamic acid
gly = glycine
his = histidine
ileu = isoleucine
leu = leucine
lys = lysine
met = methionine
phe = phenylalanine
pro = proline
ser = serine
thr = threonine
try = tryptophan
tyr = tyrosine
val = valine

Figure 9-3 The genetic code in RNA code words. To find the DNA code-word equivalents, let A = T, U = A, G = C, and C = G. For example, the RNA code word UCA is equivalent and complementary to the DNA code word AGT. The triplets marked with an asterisk are terminator codons that signal the end of a message and cause the ribosome to release a finished protein.

either freely suspended or attached form depends on the ultimate destinations of the proteins. Freely suspended ribosomes make proteins that become part of the soluble substance of the cytoplasm or part of structures such as microtubules that become suspended directly in the cytoplasm. The ribosomes attached to ER membranes assemble proteins that may either become part of cellular membranes or may become enclosed in membranous vesicles. These vesicles are stored in the cytoplasm or released to the outside of the cell in cellular secretions. Another cytoplasmic structure, the **Golgi complex,** modifies some of the proteins assembled in the endoplasmic reticulum by attaching lipid or carbohydrate groups to them to form lipoproteins or glycoproteins (see p. 30). Finished proteins may also be packed into storage or secretion vesicles in the Golgi complex.

Freely suspended ribosomes assemble proteins that enter the watery solution in the cytoplasm.

Ribosomes attached to the endoplasmic reticulum assemble proteins that are secreted or become parts of membranes.

Ribosome Structure

Ribosomes can be easily isolated from cells and purified. Analysis of these preparations reveals that ribosomes are about half RNA and half protein by weight. Prokaryotic ribosomes contain 54 different proteins in combination with the prokaryotic rRNAs; eukaryotic ribosomes,

Ribosomes are combinations between rRNAs and ribosomal proteins.

which are somewhat larger and more complex, are built up from 70 to 80 different proteins. Among these, in addition to any proteins that play a purely structural role, are enzymes involved in protein synthesis and proteins that recognize and bind mRNAs and tRNAs to ribosomes. Although the exact function of the rRNA molecules of ribosomes is unknown, they are certain to be directly involved in the assembly of amino acids into proteins. They probably bind the various parts of the ribosome together, as well, and coordinate with ribosomal proteins in binding mRNA and tRNA–amino acid complexes to the ribosome.

The subparts of eukaryotic ribosomes are assembled in the cell nucleus. The four kinds of ribosomal RNA that

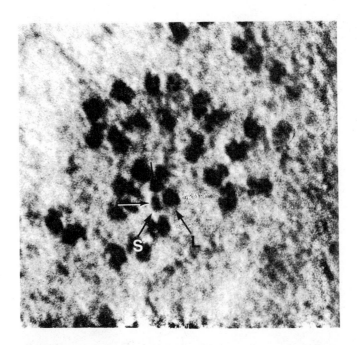

Figure 9-4 Ribosomes in a thin-sectioned rat liver cell. The large *(L)* and small *(S)* subunits and the cleft (arrows) marking the division between the subunits are visible in many of the ribosomes. Ribosomes occurring in rows such as these are probably reading different segments of the same messenger RNA molecule. Courtesy of N. T. Florendo.

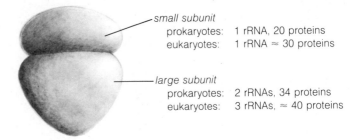

small subunit
prokaryotes: 1 rRNA, 20 proteins
eukaryotes: 1 rRNA ≈ 30 proteins

large subunit
prokaryotes: 2 rRNAs, 34 proteins
eukaryotes: 3 rRNAs, ≈ 40 proteins

Figure 9-5 Combination of a small and large ribosomal subunit to form a complete ribosome. The figures show the number of rRNAs and proteins in prokaryotic and eukaryotic ribosomes.

occur in eukaryotic ribosomes are transcribed in the nucleus, three of them within the nucleolus. The ribosomal proteins are made in the cytoplasm and then enter the nucleus to be assembled with the rRNAs into ribosomal subunits in the nucleolus. The completed subunits are released from the nucleolus as separate particles. From the nucleus they travel through the nuclear pores to reach the cytoplasm. The process is similar in prokaryotes except that there is no nucleolus; the assembly of ribosomal subunits in prokaryotes probably takes place around the edges of the nucleoid. The two subunits of the complete ribosomes of both prokaryotes and eukaryotes are often clearly visible in electron micrographs, in which a cleft or line marks the junction between them (Figures 9-4 and 9-5).

Since complete ribosomes do not occur in the nucleus, protein synthesis probably occurs only in the cytoplasm.

Although separate ribosomal subunits are easily isolated from the nucleus, complete ribosomes have never been detected anywhere in cells except in the cytoplasm. Since functional, complete ribosomes do not occur in the nucleus, all protein synthesis probably takes place in the cytoplasm. Experiments using radioactive amino acids as labels support this conclusion; when the labeled amino acids are added, radioactive proteins can be detected in the cytoplasm but not in the nucleus.

The Endoplasmic Reticulum

The membranes of the endoplasmic reticulum take two forms in eukaryotic cytoplasm: smooth and rough. Membranes with ribosomes attached are termed the **rough endoplasmic reticulum** (or **rough ER**) because the ribosomes covering their surfaces give them a serrated appearance in the electron microscope (see Figures 9-6 to 9-8). Similar cytoplasmic membranes without attached ribosomes are called **smooth endoplasmic reticulum,** or **smooth ER** (see Figures 9-6 and 9-9).

Ribosomes are attached to the surfaces of rough ER membranes; smooth ER membranes have no ribosomes.

The membranes of the rough ER take the form of tubules, vesicles, and large flattened sacs—all with ribosomes attached to the surfaces facing the surrounding cytoplasm. The largest ER sacs (Figures 9-6 and 9-7) are called **cisternae.** The sacs and tubules form a branched network that extends through much of the cytoplasm in

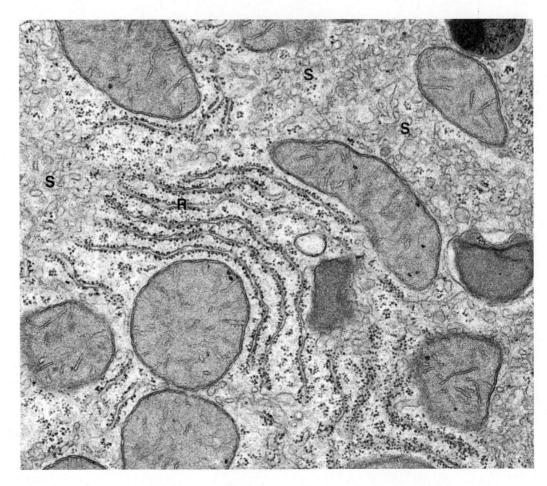

Figure 9-6 A liver cell from a hamster, showing both rough (R) and smooth (S) endoplasmic reticulum. ×34,000. Courtesy of D. W. Fawcett, from *The Cell*, copyright 1966 by W. B. Saunders Company.

many cell types. In these networks the boundary membrane is continuous and unbroken and completely separates the channel enclosed in the endoplasmic reticulum from the surrounding cytoplasm.

In most cells ribosomes are also attached to the outermost membrane of the nuclear envelope, on the side facing the surrounding cytoplasm (see Figure 9-8). Connections can sometimes be seen between the outer nuclear membrane and the rough ER membranes (arrow, Figure 9-8), making the compartment between the two membranes of the nuclear envelope (the perinuclear compartment) continuous with the ER channels at these points. These connections indicate that the outer nuclear membrane is probably closely related in structure and function to the rough ER.

The relative amounts of free and membrane-bound ribosomes vary considerably according to cell type. At one extreme are cells such as the pancreatic or salivary gland cells of mammals. In these cells, which actively secrete large quantities of protein, rough ER membranes almost completely fill the cytoplasm (see Figure 4-8). At the other extreme are cells such as muscle and kidney cells, in which rough ER is almost completely absent and most ribosomes are unattached. Other animal cell types fall at all possible points between these two extremes. Plant cells typically contain more sparsely distributed deposits of rough ER.

Smooth ER membranes, which carry out a transport function in protein synthesis, lack ribosomes and form primarily tubular sacs that are generally less distinct and of smaller dimensions than the rough ER (Figures 9-6 and

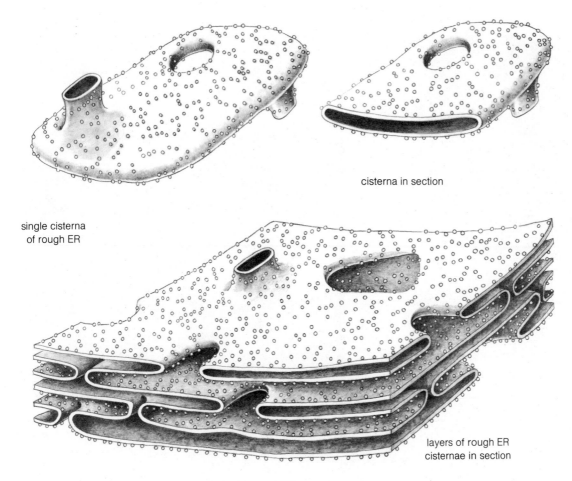

cisterna in section

single cisterna
of rough ER

layers of rough ER
cisternae in section

Figure 9-7 The arrangement of cisternae in the rough ER.

9-9). At many points, the smooth and rough ER membranes may connect together to form a continuous channel enclosed by the two systems.

The Golgi Complex

The Golgi complex, named for the cell biologist who first described it in the 1800s, is formed from flattened, saclike vesicles assembled into a stack (see Figures 9-10 and 4-11a and b). Each flattened vesicle in the stack consists of a single, continuous membrane enclosing an inner space. The spacing in and between the stacks is very regular except at the edges, which appear swollen or dilated. The edges of the sacs fragment or bud off, producing large numbers of vesicles at the margins of the Golgi complex (see Figure 4-11b). These vesicles may fuse together to form larger storage vesicles that remain in the cytoplasm or to form secretion vesicles (see Figure 9-10) that are released to the

outside of the cell. No ribosomes occur between the sacs or in the region immediately surrounding the Golgi complex.

Golgi complexes are usually closely associated with the rough ER, separated from it only by a layer of small, smooth-walled vesicles. Some of the vesicles in this layer can be seen to arise as buds from the rough ER; others join with the nearest Golgi sac (Figure 9-11). From this and other evidence the vesicles of this layer, called *transition vesicles,* are believed to link the two membrane systems together—probably by budding off from the endoplasmic reticulum and fusing with the Golgi complex.

Golgi complexes are found in multiple numbers in cells. In plants, they are more or less evenly scattered through the cytoplasm. In animal cells, Golgi complexes frequently occur in masses near the endoplasmic reticulum or just outside the nucleus. The number of Golgi complexes is extremely variable in cells from different tissues

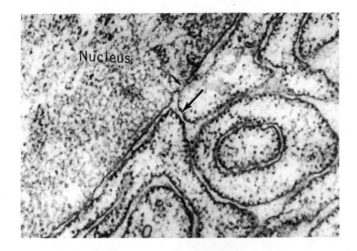

Figure 9-8 A connection between the outer membrane of the nuclear envelope and the rough ER (large arrow) in a nucleus of mouse salivary gland. The outer membrane of the nuclear envelope is typically covered with ribosomes. The small arrow points to a pore complex. ×62,000. From *The Cell* by D. W. Fawcett, 1966. Courtesy of D. W. Fawcett, H. Parks, and the W. B. Saunders Company.

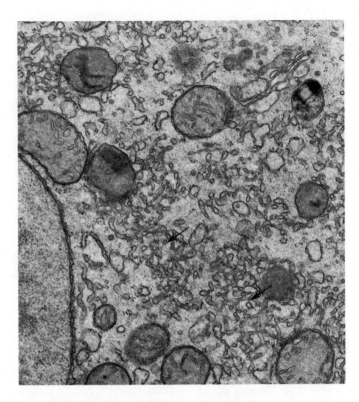

Figure 9-9 Smooth ER membranes (arrows) in a hamster liver cell. ×20,000. Courtesy of D. W. Fawcett, from *The Cell,* copyright 1966 by the W. B. Saunders Company.

Figure 9-10 A group of Golgi complexes *(Go)* and secretion vesicles *(SV)* in a cell lining the digestive tract of an animal. ×22,500. Courtesy of H. W. Beams and R. G. Kessel and Academic Press, Inc., from *International Review of Cytology* 23:209 (1968).

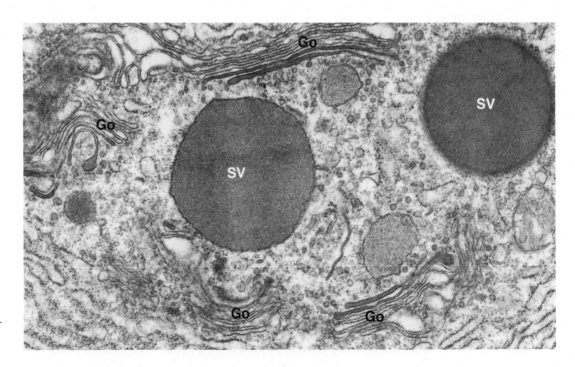

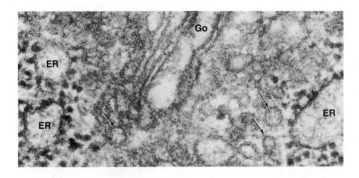

Figure 9-11 Small vesicles called *transition vesicles* (arrows) are believed to conduct proteins from the rough ER (ER) to the Golgi complex (Go). Cat pancreas, × 108,000. From *Cytology and Cell Physiology*, ed. G. H. Bourne, 1964. Courtesy of F. S. Sjöstrand and Academic Press, Inc.

or species. While the average is about 20 per cell, there may be many more. Corn root tip cells, for example, contain several hundred complexes; more than 25,000 have been counted in cells of *Chara,* a green alga. Some animal tissues, such as cells of the salivary glands in insects, may also contain thousands of Golgi complexes.

The ER and Golgi Complex in Protein Synthesis

The coordination of the endoplasmic reticulum and Golgi complex in the synthesis and transport of proteins has been followed by tracing the uptake and movement of amino acids labeled with a radioactive isotope. In studies of this type, pioneered by George E. Palade and his colleagues at the Rockefeller University, cells actively engaged in protein synthesis and secretion are exposed to the radioactive amino acids. If the cells are fixed and sectioned within a few minutes after exposure, radioactivity can be detected only over the rough ER, indicating that the amino acids are first assembled into proteins in this region (Figure 9-12*a*). If the cells are prepared about 10 to 15 minutes after exposure, radioactivity can be detected within vesicles of the smooth ER (Figure 9-12*b*). This indicates that the proteins made on the ribosomes of the rough ER are transported across the ER membranes and concentrated inside the ER cisternae. The proteins then move within the enclosed channels to the smooth ER. If cells are fixed after 20 to 30 minutes, the radioactive label also appears over the transition vesicles between the smooth ER and the Golgi complex and over the Golgi complex itself. Subsequently, label appears over the larger se-

cretion vesicles lying between the Golgi complex and the plasma membrane (Figure 9-12*c*). Finally, within 1 to 4 hours after initial exposure to the labeled amino acids, radioactivity appears in the spaces just outside the secretory cells, showing that the large vesicles eventually discharge their contents to the cell exterior.

The smooth ER transports newly synthesized proteins from the rough ER to other cellular locations.

The Golgi complex modifies proteins and packs them into secretion vesicles.

These studies link the various membranous elements of the cytoplasm into a coordinated system for synthesis, transport, and secretion of proteins. They show, moreover, that proteins follow the route rough ER → smooth ER → transition vesicles → Golgi complex → secretion vesicles → cell exterior through the system. Since ribosomes also occur on the outside of the nuclear envelope, and connections can be found between the nuclear envelope and the ER, the nuclear envelope and the compartment enclosed between its two membranes are probably also connected directly to the cytoplasmic system for synthesizing, transporting, and secreting proteins.

Several variations of this basic route are observed in different cell types. In some cells, newly synthesized proteins may bypass the Golgi complex. In others, secretion vesicles are stored in the cytoplasm—either temporarily as a prelude to later release or more or less permanently as a specialized structure of the cytoplasm. (The synthesis and functions of **lysosomes,** specialized vesicles of this type, are described in Supplement 9-2.) Another much-used pathway for proteins assembled in the rough ER inserts newly synthesized proteins into cellular membranes. In this pathway proteins made on ER ribosomes penetrate into the ER membrane bilayers and are retained as membrane proteins as the membranes successively form parts of the smooth ER, Golgi complex, secretion vesicles, and plasma membrane. Figure 9-13 summarizes the activity of ribosomes and cytoplasmic membrane systems in protein synthesis and secretion.

THE REGULATION OF PROTEIN SYNTHESIS

The cells of all organisms, both prokaryotic and eukaryotic, have the capacity to make thousands of different proteins. At any given time, however, only a fraction of the total is actually synthesized. Probably no cells of any

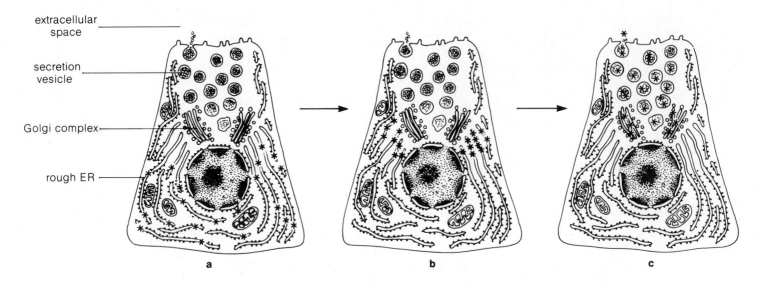

extracellular space

secretion vesicle

Golgi complex

rough ER

a

b

c

Figure 9-12 Migration of labeled proteins (asterisks) after injection of radioactive amino acids into secretory cells. (**a**) Three minutes after injection; only the rough ER is labeled. (**b**) After 10–15 minutes; label is distributed in the smooth ER near the Golgi complex. (**c**) After 30 minutes; label is located in the Golgi complex and the secretory vesicles. Adapted; original courtesy of P. Favard.

type assemble all their encoded proteins simultaneously. For example, red blood cells in vertebrate animals, as they mature, synthesize only hemoglobin and a few enzymatic proteins. This limitation to a few proteins occurs even though these cells retain all the chromosomes and the entire complement of genes found in any other cell of the same organism. These observations point to precise control of protein synthesis at the cellular level.

Control of protein synthesis to increase the levels of some proteins and inhibit or stop the production of others is termed *regulation*. Regulation may take place at several points in the entire mechanism. It may be imposed by inhibiting messenger RNA synthesis; control at this level is termed *transcriptional regulation*. Control at the level of protein synthesis in the cytoplasm is termed *translational regulation*. Translational regulation may include any mechanism controlling the availability of mRNA, ribosomes, or other factors necessary for protein synthesis, or adjusting the rate of amino acid activation or polypeptide assembly. There is good evidence that both transcriptional and translational regulation control the number and types of proteins made in prokaryotic and eukaryotic cells.

Transcriptional Regulation in Prokaryotes

Control of RNA transcription in bacteria has been studied extensively and is better understood than transcriptional regulation in eukaryotes. Bacteria can quickly regulate the synthesis of enzymes to suit the biochemical conditions of their surroundings. Bacteria living in a growth medium containing only inorganic salts and a food substance such as glucose, for example, produce the enzymes required to make all 20 amino acids. If an amino acid is added to the medium, the enzymes synthesizing that amino acid quickly drop in quantity in the bacteria and soon reach undetectably low levels. If the amino acid is then removed from the medium, the enzymes required to synthesize it reappear and quickly reach their former levels. Responses of this type in bacteria often take place within minutes after changes are made in the culture medium.

Enzymes that can be made to appear or disappear within bacterial cells by changes in the culture medium are called *inducible enzymes*. Bacteria have a variety of inducible enzymes and can respond to hundreds of organic compounds by synthesizing specific enzymes for their utilization. There are two classes of these enzymes. One is normally absent from the cell but is induced to appear if a substance acted upon by that enzyme is added to the medium. The other class is reduced in quantity in the cell if the substance normally made by that enzyme is supplied in the medium (as in the amino acid system described above).

The mechanism controlling inducible enzymes was explained in the 1950s by two investigators at the Pasteur

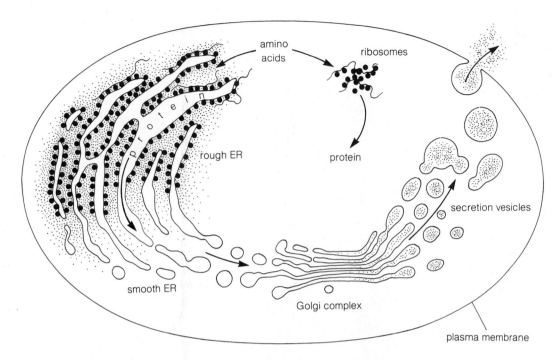

Figure 9-13 The integration of ribosomes, rough and smooth ER, Golgi complex, and secretion vesicles in protein synthesis and secretion. Proteins are assembled on ribosomes that are either freely suspended in the cytoplasm or bound to membranes of the rough ER. The proteins made on freely suspended ribosomes enter the solution in the surrounding cytoplasm. The proteins synthesized on rough ER ribosomes penetrate into the membranes or cisternae of the endoplasmic reticulum. From here, the proteins are transported through the smooth endoplasmic reticulum to the vicinity of the Golgi complex. The proteins are then transferred to the complex by transition vesicles that bud off from the smooth ER and fuse with the Golgi complex. After modification within the Golgi complex, which may include attachment of lipid, carbohydrate, or other groups, the proteins are enclosed in vesicles that bud off from the margin of the complex. These vesicles gradually fuse together to form large secretion vesicles, which are stored in the cytoplasm or fuse with the plasma membrane. After fusion with the plasma membrane, the proteins inside the vesicles are released to the cell exterior.

Institute in Paris, Francois Jacob and Jacques Monod. Jacob and Monod studied a group of enzymes that allows bacteria to use sugars called *galactosides*. If no galactosides are present in the medium surrounding the bacterial cells, few or no molecules of the enzymes catalyzing the breakdown of these sugars are made in the bacterial cytoplasm. If galactosides are added to the medium, synthesis of the required enzymes begins very quickly and the bacteria are soon able to utilize the added sugar. All the genes coding for the enzymes breaking down the galactoside sugars were found to be clustered together at one point in the bacterial DNA. These genes are transcribed as a unit into a single mRNA that directs the synthesis of the galactoside enzymes in the cytoplasm. Other elements were also found to be involved in the control mechanism. One, which Jacob and Monod termed the **operator,** consists of

a DNA sequence that lies just in advance of the galactoside genes—in the region where an RNA polymerase molecule would bind to the DNA to begin transcribing the genes. The operator was found to act as an "on/off" switch controlling the transcription of the galactoside genes. The operator, in turn, was found to be controlled by another gene lying at some distance from the galactoside genes. Jacob and Monod called this gene the **regulator gene.**

Jacob and Monod proposed that these elements are integrated into a control system that regulates the synthesis of the galactoside enzymes and other inducible enzymes in bacteria. According to their model, which they termed the *operon hypothesis* (Figure 9-14), the genes coding for the enzymes breaking down galactosides are controlled as a unit by the operator; this unit is the **operon.** Without the normal activity of the regulator gene and the operator

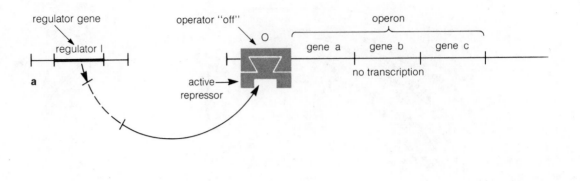

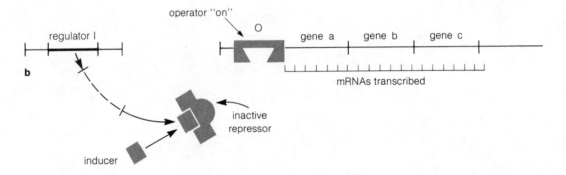

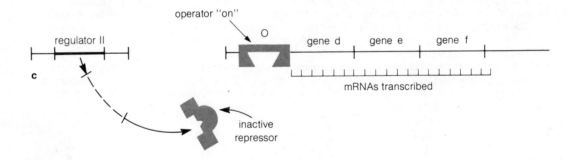

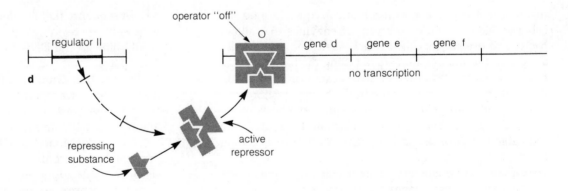

Figure 9-14 Jacob and Monod's operon mechanism for transcriptional regulation in prokaryotes (see text).

the operon would be continuously transcribed. As a result, the galactoside enzymes would be present at all times in high concentrations in the cell. According to the operon hypothesis, the regulator gene codes for synthesis of a **repressor** protein. The repressor has two sites: one that can bind to the inducing substance and one that can bind to the operator. If no galactoside sugars are present in the medium, the repressor site that binds to the galactoside operator is active and the repressor binds to the operator next to the galactoside genes. Since this binding blocks access to the genes by RNA polymerase enzymes, the genes are not transcribed. As a result, no galactoside enzymes are made in the cytoplasm (Figure 9-14*a*).

Transcription is regulated by the operon mechanism in prokaryotes.

If galactoside sugars are added to the medium, some of them enter the cell and bind to the inducer site on the repressor. This combination changes the folding conformation of the repressor, causing it to lose its affinity for the operator site. The operator is then vacated, allowing access by the RNA polymerase enzyme to the galactose genes. The genes controlled by the operator are then turned on and transcribed continuously (Figure 9-14*b*), producing a continuous flow of galactoside enzymes.

The same operon mechanism also explains how synthesis of enzymes can be turned off by substances added to the culture medium—as in the amino acid systems that are turned off when the amino acid they synthesize becomes available in the medium. To explain this effect, Jacob and Monod proposed that the regulator gene for the amino acid systems also codes for a repressor with two binding sites. One of these sites can bind the amino acid, and one can bind to the operator region of the genes that code for the enzymes catalyzing synthesis of the amino acid. In this case, the site on the repressor that binds to the operator is inactive, and the repressor has no affinity for the operator. As a result, transcription of the genes coding for the enzymes goes on continuously (Figure 9-14*c*). If the amino acid is added to the surrounding medium, some of it diffuses into the cell and combines with the repressor. The combination causes a conformational change in the repressor that activates its operator binding site. The repressor then binds to the operator, turning off transcription of the mRNAs for the enzymes that synthesize the amino acid (Figure 9-14*d*).

The differences in the alternate forms of the operon hypothesis thus depend on whether the repressor, coded for by the regulator gene, is synthesized in an *active* or an *inactive* state. The alternate forms can be summarized as follows:

Induction of enzyme synthesis:
Active repressor + inducing substance → inactive repressor

In this case the inactive repressor does not combine with the operator, and the genes are continuously transcribed.

Repression of enzyme synthesis:
Inactive repressor + inducing substance → active repressor

In this case the active repressor combines with the operator, and the genes are turned off.

Since Jacob and Monod first proposed the operon hypothesis, experimental support for their model has come from many sources. Some of the strongest evidence for the model, for which Jacob and Monod received the Nobel Prize in 1965, has come through isolation and identification of the repressors of various operons. W. Gilbert and B. Müller-Hill of Harvard University, for example, isolated and purified a protein coded by the galactoside regulator gene. This protein has strong affinity for the operator site controlling the galactoside genes. The protein also has strong binding affinity for galactoside sugars; combination with a galactoside sugar destroys its affinity for the galactoside operator region. Thus the protein isolated by Gilbert and Müller-Hill has all the characteristics predicted for the galactoside repressor by the operon hypothesis.

Transcriptional Regulation in Eukaryotes

It is doubtful whether an automated system equivalent to the operon mechanism controls transcription in eukaryotes. The elements necessary for the operon mechanism, such as regulator and operator genes, have never been detected in eukaryotes. Further, clusters of genes with related function that are transcribed as a unit, such as the galactoside operon, are rare or nonexistent in eukaryotes.

The operon mechanism allows prokaryotes to adjust rapidly to environmental changes.

The apparent absence of an operon mechanism in eukaryotes is probably a reflection of the fundamental differences in the way transcription is regulated in prokaryotes and eukaryotes. The operon mechanism provides sensitivity to changes in the environment and allows prokaryotic cells to adjust rapidly to them. The resulting changes in enzyme concentrations persist only as long as the environment remains the same. Most of the transcriptional regulation observed in eukaryotes, however, is more or less permanent. These long-term changes are part of the cell specialization that is a characteristic feature of embryonic development in many-celled organisms (outlined in Unit Five).

Transcriptional regulation in eukaryotes produces long-term or permanent changes in the enzymes and other proteins synthesized.

Although many questions remain about the mechanisms producing long-term transcriptional changes in eukaryotes, these regulatory processes are now thought to depend on the histone and nonhistone proteins that combine with the DNA in eukaryotic chromatin (see Chapter 8).

Histones and Transcriptional Regulation A variety of experiments have shown that histones can control the transcription of genes. "Naked" DNA, without associated histone or nonhistone proteins, serves as an excellent template for RNA transcription in cell-free systems. All of the DNA, including mRNA, tRNA, and rRNA genes, is transcribed in such systems—including "filler" sequences that are normally never transcribed in cells. Addition of histones from any source to DNA in such cell-free systems inhibits transcription by as much as 90 percent or more.

This inhibition is apparently random. The amount of mRNA, tRNA, and rRNA synthesized is simply reduced in overall quantity without significant changes in the particular genes transcribed. The greater the amount of histones added, the greater the reduction in transcription;

The histones probably control the accessibility of large blocks of genes to RNA polymerase enzymes.

---

histones from one source are as effective as another in producing this effect. Thus inhibition of transcription by the histones, as far as anyone has been able to determine, is a general effect that does not distinguish between individual genes.

This conclusion is not surprising, since there is relatively little variability in the histones from cell to cell within the same organism, or even between different species. The number of different genes in eukaryotic nuclei probably amounts to thousands or even tens of thousands. To control each gene separately, the proteins con-

trolling them would be expected to occur in many different kinds. However, there are only five major kinds of histones in eukaryotic nuclei. Thus the histones do not seem to vary enough to control specific genes in a eukaryotic nucleus.

Nonhistones and Transcriptional Regulation In contrast to the apparently general, nonspecific control of RNA transcription by the histones, there is good evidence that the nonhistone proteins can regulate the activity of individual genes. The histones, as noted, inhibit transcription generally when added to DNA in a cell-free system. Addition of the nonhistones to such DNA–histone complexes partially reverses the inhibition: when the nonhistones are added, some DNA sequences are turned on again.

The nonhistones control the transcription of specific genes.

Recent experiments show that the reversal of inhibition in DNA–histone preparations by the nonhistones is specific and, moreover, that particular genes are turned on when nonhistones are added. In one of these experiments (Figure 9-15), R. S. Gilmour and J. Paul, at the Beatson Institute for Cancer Research in Scotland, tested the ability of histone and nonhistone proteins to control the transcription of an mRNA coding for a liver cell protein called a *globin*. An mRNA coding for this protein is transcribed in liver cells of a mouse, but not in mouse brain cells. Gilmour and Paul isolated the chromatin from these two mouse cell types and then split the chromatin into separate DNA, histone, and nonhistone fractions. The fractions were subsequently recombined in all possible combinations. The recombined chromatin was then added to a cell-free system containing all the factors required for RNA transcription.

Adding the histones back to the DNA in any combination (brain cell DNA + brain or liver cell histones; liver cell DNA + brain or liver cell histones) had no effect other than a general inhibition of RNA transcription. Adding the nonhistone proteins to the DNA–histone combinations, however, reversed the inhibition by the histones in a specific way. If nonhistone proteins from liver cells were added to any of the DNA–histone combinations, the globin mRNA was transcribed and appeared in the cell-free systems (see Figure 9-15). The globin mRNA did not appear if brain cell nonhistones were added to any of the DNA–histone combinations. These effects were observed even if the histones from both cell types were pooled together and combined with either type of DNA before adding the nonhistones. Thus, in this experiment, one or more proteins in the nonhistone fraction from the liver

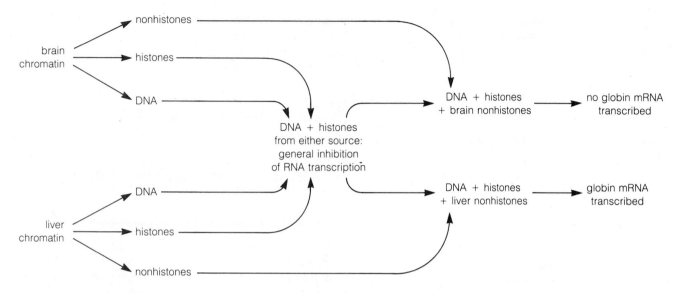

Figure 9-15 The Gilmour and Paul experiment demonstrating specific transcriptional regulation by the nonhistone chromosomal proteins (see text).

cells were able specifically to turn on a gene coding for a typical liver protein.

These experimental findings suggest that the histone and nonhistone chromosomal proteins cooperate in regulating transcription. The histones act as general inhibitors that turn off all the genes in the nucleus. This general inhibition is reversed at specific points by nonhistone proteins that can recognize and turn on single genes within the large blocks of DNA inhibited by the histones. The advantage of this system is that eukaryotic cells do not have to make specific nonhistone proteins to control each gene in the nucleus. (This would require as many different nonhistone control proteins as there are genes—perhaps as many as 30,000 to 40,000 per cell.) Instead, the genes are inhibited generally by the histones, which occur in only five major kinds. Turning on specific genes by the nonhistones then requires only that a relatively few specific nonhistone proteins need be made by each cell, one or a few for each gene that is to be turned on.

Translational Regulation

The reactions of amino acid activation and polypeptide assembly on ribosomes include many steps that can be controlled by regulatory mechanisms. However, the translational controls detected most frequently in prokaryotes and eukaryotes operate at the initial steps of polypeptide assembly on ribosomes. These controls include

regulation of both the activity or availability of mRNAs in the cytoplasm or other factors required for the first steps in protein synthesis.

Several mechanisms are known to regulate the activity or availability of mRNAs in both eukaryotes and prokaryotes. One of these mechanisms works by controlling the rate at which messengers are degraded by RNAase enzymes. Translational control by this pathway is best known in bacteria, where differences can clearly be detected in the rate at which different mRNAs are broken down. For example, in the mRNA synthesized by the galactoside operon, which includes codes for three separate enzymes, the coding segment for the middle enzyme is attacked and degraded by enzymes most frequently.

Translational controls are coordinated with translational regulation to fine-tune synthesis to suit conditions in and around the cell. Transcriptional controls determine *which* proteins are to be made in a cell at a given time. Translational controls, since they affect the rate of protein synthesis, determine *how much* of the proteins are to be made. The result of the two regulatory mechanisms working together is the highly specific and controlled synthetic activity characteristic of living cells.

Questions

1. Outline the steps in amino acid activation. Why is the reaction called an *activation*? Trace the flow of energy derived from ATP through the reaction sequence.

2. Compare the roles of tRNA and the synthetase enzymes in protein synthesis. How does the accuracy of protein synthesis depend on the function of tRNAs? On the functions of the synthetase enzymes?

3. Outline the mechanism of polypeptide assembly on ribosomes. What is the source of energy for formation of peptide linkages as each successive amino acid is added to a growing polypeptide chain?

4. How did Khorana's experiment establish that the "words" in the genetic code are triplets?

5. How were codons identified and assigned to amino acids? Are all 64 possibilities used? What are "nonsense" codons?

6. Define degeneracy, synonym, codon, terminator, universality, and anticodon.

7. Refer to Figure 9-3 and make a chart listing all 20 amino acids and the codons assigned to them. Can you detect any common patterns or similarities in the multiple codons for the amino acids? Which amino acids have only one codon?

8. Write out a random DNA sequence 30 nucleotides long and see what polypeptide sequence it would code for. Then introduce "mutations" by changing the sequence at five random places. Does a mutation in the DNA code necessarily cause a change in the sequence of the polypeptide coded for? Why?

9. Outline the structure of ribosomes, rough and smooth ER, and the Golgi complex. What is the function of each structure in protein synthesis?

10. Trace the pathway of a secreted protein from the rough ER to the outside of the cell. Are variations possible in this pathway? How was the pathway determined experimentally?

11. Trace the pathway followed by a membrane protein from the ER to the plasma membrane.

12. How is the nuclear envelope related to protein synthesis and secretion?

13. What is transcriptional regulation? How does transcriptional regulation differ in prokaryotes and eukaryotes?

14. Outline the operon mechanism. How does the mechanism induce synthesis of an enzyme? How does it repress synthesis of an enzyme? Define operon, operator, regulator, and repressor.

15. How are the histones believed to function in transcriptional regulation? The nonhistones?

16. Outline an experiment demonstrating that the nonhistone proteins can control the transcription of individual genes.

Suggestions for Further Reading

Palade, G. 1975. "Intracellular Aspects of the Process of Protein Synthesis." *Science* 189:347–358.

Satir, B. 1975. "The Final Steps in Secretion." *Scientific American* 233:28–37.

Stein, G. S., and L. J. Keinsmith. 1975. *Chromosomal Proteins and Their Role in Regulation of Gene Expression.* Academic Press, New York.

Watson, J. D. 1976. *Molecular Biology of the Gene.* 3rd ed. Benjamin, Menlo Park, California.

Wolfe, S. L. 1981. *Biology of the Cell.* 2nd ed. Wadsworth, Belmont, California.

SUPPLEMENT 9-1: TRANSCRIPTION AND TRANSLATION IN MITOCHONDRIA AND CHLOROPLASTS

The DNA of Mitochondria and Chloroplasts

Both mitochondria and chloroplasts contain DNA molecules that are active in transcription. All the major RNAs, including mRNA, tRNA, and rRNA, are transcribed from the organelle DNA. These RNA species interact in protein synthesis in the organelle interior. In most characteristics, transcription and translation in mitochondria and chloroplasts resemble the mechanisms of prokaryotes more closely than those of eukaryotes.

The DNA molecules of both organelles are visible as fibrous deposits suspended in the innermost cavity: the matrix in mitochondria and the stroma in chloroplasts (Figures 9-16 and 9-17). Typically, the DNA fibers visible in the organelles are much thinner than nuclear chromatin fibers. They approach the dimensions of the fibers in bacterial nucleoids—that is, the dimensions expected for a protein-free DNA double helix.

The analysis of DNA isolated from the two organelles shows that the similarities to prokaryotic DNA are more than superficial. Both mitochondrial and chloroplast DNA take the form of closed circles with no associated histone or nonhistone proteins. The circles in animal mitochondria are typically very small and contain only about 5 micrometers of DNA. The mitochondrial circles of fungi and plants are generally larger, ranging in size from about 20 to 30 micrometers of included DNA. Chloroplast circles, with about 40 micrometers of included DNA, are larger still. None of the organelle circles approach the dimensions of bacterial DNA circles, however, which contain about 1000 to 1500 micrometers of DNA.

The amount of DNA associated with mitochondria varies from as little as 0.2 percent of the total cellular DNA in mouse liver cells, to 20 percent in haploid yeast, and more than 99 percent in amphibian oocytes. Dividing the total mitochondrial DNA fraction by the number of mitochondria gives enough DNA in most cells for five to ten circles per mitochondrion. Chloroplast DNA may make up as much as 6 percent of the total DNA of a plant cell. At

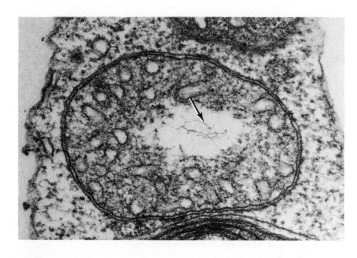

Figure 9-16   DNA deposits (arrow) within the matrix of a mitochondrion of the brown alga *Egregia*. ×58,000. Courtesy of T. Bisalputra, A. A. Bisalputra, and The Rockefeller University Press, from *Journal of Cell Biology 33*:511 (1967).

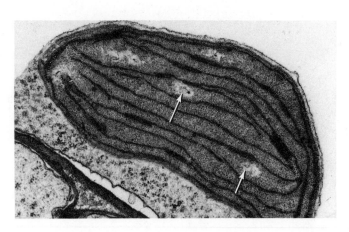

Figure 9-17   DNA deposits (arrows) in the stroma of a corn chloroplast. ×20,000. Courtesy of L. K. Shumway and T. E. Weier.

these levels, there is enough DNA to form 60 to 80 circles per chloroplast.

The ribosomal RNAs of mitochondria and chloroplasts are associated with ribosomes that, in dimensions, resemble the ribosomes of prokaryotes. As such, they are significantly smaller than the ribosomes in the cytoplasm surrounding the organelles. Mitochondrial ribosomes occur in the matrix, where they may be freely suspended or attached to the surfaces of cristae membranes (Figure 9-18). Chloroplast ribosomes occupy the equivalent inner compartment, the stroma, where they may be either freely suspended or connected to thylakoid membranes (Figure 9-19).

Protein Synthesis in Mitochondria and Chloroplasts

The ribosomes of mitochondria and chloroplasts carry out protein synthesis by the same overall mechanism as the cytoplasmic ribosomes of prokaryotes and eukaryotes. However, the details of the mechanism in the two organelles most clearly resemble polypeptide assembly in prokaryotes. One of the similarities between protein synthesis in the two cytoplasmic organelles and bacteria is the reaction to two antibiotics that interfere with protein synthesis: *cycloheximide* and *chloramphenicol.* Cycloheximide, which interferes with protein synthesis on eukaryotic but not bacterial ribosomes, has no effect on polypeptide assembly in either chloroplasts or mitochondria. Conversely, chlor-

amphenicol, which inhibits prokaryotic but not eukaryotic protein synthesis, stops polypeptide assembly on chloroplast or mitochondrial ribosomes.

The responses of protein synthesis on cytoplasmic, mitochondrial, and chloroplast ribosomes to antibiotics has provided one of the key approaches for determining which polypeptides are synthesized inside the two organelles. Cycloheximide, for example, which interrupts cytoplasmic protein synthesis in eukaryotes, inhibits the production of almost all the mitochondrial or chloroplast ribosomal proteins. The opposite inhibitor, chloramphenicol, which stops organelle protein synthesis, has no effect on the synthesis of these proteins. These results indicate that the proteins of the organelle ribosomes, although typically prokaryotic in structural and functional characteristics, are actually made on the ribosomes in the cytoplasm outside mitochondria and chloroplasts.

Other polypeptides prove to be encoded in mitochondrial or chloroplast DNA and synthesized inside the organelles. The list is necessarily short, since the coding capacity of the organelle DNA molecules is small. In both organelles, the polypeptides assembled inside include parts of the enzyme system synthesizing ATP in the inner organelle membranes. Other polypeptides assembled inside the organelles include parts of the carrier molecules transporting electrons in mitochondria and a subunit of the RuDP–carboxylase enzyme of chloroplasts—the enzyme that fixes CO_2 in the dark reactions of photosynthesis (see p. 111). In no case, however, is a complete

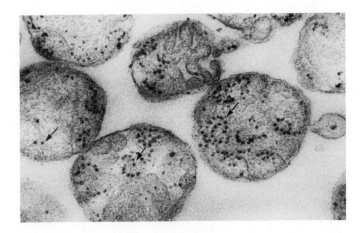

Figure 9-18 Mitochondrial ribosomes (arrows) in a yeast cell, in a section that just grazes the surfaces of the cristae membranes. ×46,000. Courtesy of J. André.

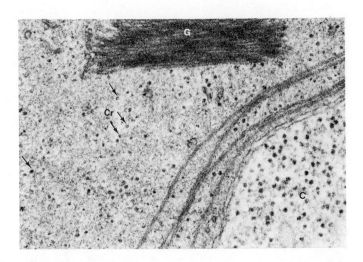

Figure 9-19 Ribosomes inside a chloroplast of *Chlamydomonas*, a green alga. The chloroplast ribosomes *(Cr)* are distinctly smaller than the ribosomes in the cytoplasm outside the chloroplast *(C)*. *G*, granum. ×62,000. Courtesy of U. W. Goodenough, R. P. Levine, and The Rockefeller University Press, from *Journal of Cell Biology* 33:511 (1967).

protein synthesized inside either organelle. All the polypeptides made inside mitochondria and chloroplasts are subunits of complex proteins that also require additional polypeptides made in the surrounding cytoplasm for their completion.

This fact raises one of the most interesting questions concerning the organelle transcription/translation systems. Since no complete proteins are made in either mitochondria or chloroplasts, why are any polypeptides encoded and synthesized inside the organelles at all? This question is especially pertinent if you consider that 100 enzymatic and ribosomal proteins, almost all synthesized in the cytoplasm *outside* mitochondria and chloroplasts, must be supplied to the organelles just to run the transcription and translation systems required to synthesize the relatively few and incomplete proteins made inside.

Some of the polypeptides synthesized in mitochondria have been completely sequenced. Comparisons between the amino acid sequences of these polypeptides and the nucleic acid sequences coding for them in the mitochondrial DNA of yeast and human cells have revealed the unexpected fact that several substitutions have been made in the genetic code in mitochondria. Where the DNA specifies the mRNA triplet CUA, which codes for leucine elsewhere in living organisms, one of the mitochondrial polypeptides carries a threonine at the corresponding point in its amino acid sequence. Comparisons between another polypeptide and its mitochondrial DNA

gene reveal two additional coding substitutions. This polypeptide (part of the electron carrier chain) carries the amino acid tryptophan at points corresponding to UGA, normally a terminator codon, in its coding sequence. At other points, methionine is inserted at points coded for by the AUA triplet, which normally specifies isoleucine. The coding changes discovered in mitochondria are the first known differences in the coding assignments of the genetic code, which were heretofore thought to be completely universal.

The many similarities between the transcription/translation systems of mitochondria, chloroplasts, and prokaryotes have led to the hypothesis that the two organelles evolved from prokaryotic cells that were taken up as food particles by cells destined to become eukaryotes. Instead of breaking down, the prokaryotes persisted and gradually evolved into mitochondria and chloroplasts in the cytoplasm of their host cells. By this route, mitochondria may have originated from ancient bacteria; chloroplasts may have arisen from ancient blue-green algae. The transcription and translation mechanisms of the organelles, according to this hypothesis, are the last remnants of the once-independent existence of their prokaryotic ancestors. (For details, see Chapter 21.)

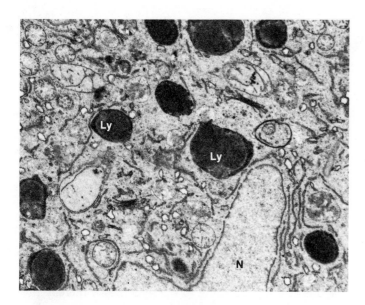

Figure 9-20 Lysosomes in the cytoplasm of a corn root cell. *Lys*, lysosomes; *N*, nucleus. ×8,000. Courtesy of P. Berjak and Academic Press, Inc., from *Journal of Ultrastructure Research* 23:233 (1968).

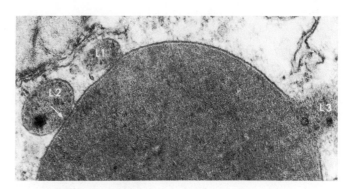

Figure 9-21 Lysosomes (L1, L2, L3) of a rat liver cell fusing with a large membrane-bound droplet of absorbed protein. The arrow marks a point where one of the lysosomes is fusing with the protein droplet. ×45,000. Courtesy of S. Goldfischer, A. B. Novikoff, A. Albala, L. Biempica, and The Rockefeller University Press, from *Journal of Cell Biology* 44:513 (1970).

SUPPLEMENT 9-2: LYSOSOMES

One of the most important products of the secretory pathway in cells is a specialized class of secretion vesicles known as **lysosomes** (Figures 9-20, 9-21, and 9-22). The proteins of lysosomes are synthesized in the rough ER and subsequently packed into vesicles that remain in storage for various lengths of time in the cytoplasm. The proteins in the lysosomes include a variety of enzymes that, collectively, are capable of breaking down all the major biological molecules of the cell. The reactions catalyzed by the lysosome enzymes are all *hydrolytic* reactions that split biological molecules into their building-block subunits by adding the elements of water:

Nucleic acids + H_2O → nucleotides and nucleosides

Proteins + H_2O → amino acids

Polysaccharides and carbohydrate groups + H_2O →
disaccharides and monosaccharides

Lipids + H_2O → fatty acids, glycerol

One of the many interesting and as yet unanswered questions about lysosomes is how, considering their impressive battery of enzymes, the lysosome boundary membranes enclosing them remain intact.

Lysosomes have been identified in all animal cells examined and in protistans, several types of fungi, and lower and higher plants. Lysosomes are especially abundant in animal cells—such as the white blood cells of vertebrates—that ingest large quantities of extracellular material. They are also prominent in the cells of animal tissues undergoing degenerative changes or metabolic stress, as in starvation, aging, or hormonal stimulation.

Lysosomes break down substances in cells by several different processes. One involves digestion of material brought into the cytoplasm by endocytosis (see p. 87). In endocytosis the plasma membrane contacts the extracellular material and folds inward, creating a pocket in which the material is trapped. This pocket then pinches off from the plasma membrane as a vesicle and sinks into the cytoplasm, where it fuses with one or more lysosomes. Fusion activates the enzymes contained in the lysosomes, which proceed to break down the biological molecules inside the vesicle. (Figure 9-17 shows several lysosomes in a liver cell carrying out this process.) After breakdown by the lysosomal enzymes, the products enter the cytoplasm of the cell by diffusing through the vesicle membrane. Any remaining undigested debris may be retained within the lysosome as a membrane-bound deposit in the cytoplasm or may be expelled to the cell exterior.

In this function, lysosomes act as a digestive system for the cell. This activity forms a regular part of the digestive process in cells that ingest food particles, as in the

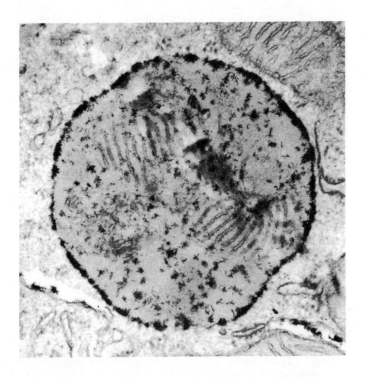

Figure 9-22 A lysosome in the cytoplasm of a mouse kidney cell. Remnants of several mitochondria are visible inside the lysosome. ×50,000. Courtesy of F. Miller.

normal developmental changes in both plants and animals (as, for example, in breakdown and absorption of the tail in developing frog embryos) as well as in cell degeneration.

The effects of lysosomes in some cases extend from their microscopic and molecular world to touch the conditions and affairs of humans. The degenerative changes in bones and joints associated with arthritis are suspected by some medical scientists to be caused in part by the abnormal release of lysosome enzymes from bone or lymph cells into the extracellular fluids. On a somewhat different plane, lysosomal activity has also been offered as an explanation for the contraceptive action of IUDs (intrauterine devices). According to this idea, an IUD is recognized as a foreign body by cells of the uterine lining, which release lysosomal enzymes into the uterus in response. The resultant high concentration of lysosomal enzymes in the uterine fluids interferes with attachment of the embryo to the uterus and, according to the hypothesis, thereby prevents pregnancy.

protistans or lower invertebrates such as the sponges and coelenterates. In higher animals, the lysosomal digestive process figures as an important part of the defense mechanisms against bacteria, virus particles, and toxic molecules. The white blood cells of vertebrates, for example, eliminate foreign cells and particles by ingesting them in vesicles that subsequently fuse with lysosomes.

In the second process, cell organelles rather than extracellular material are digested by lysosomes. In some manner, as yet unknown, a cell organelle penetrates a lysosome membrane and is digested inside. (Figure 9-18 shows the remains of two mitochondria undergoing breakdown by this process.) Although the significance of self-digestion of this type, called **autolysis**, is obscure, it probably plays an important role in the rapid breakdown of organelles that sometimes occurs in cells undergoing physiological stress.

The third lysosomal mechanism is a similar but more spectacular process in which the lysosomes rupture and release their enzymes into the surrounding cytoplasm. This process, if large numbers of lysosomes are involved, either causes or accompanies death of the cell. Operation of lysosomes as "suicide bags" in this way forms part of

Dividing Cells

10

MITOTIC CELL DIVISION

All living organisms, from single-celled forms to the most complex animals and plants, grow and reproduce by processes taking place at the cellular level. Growth, or the increase in the mass of cells, is accomplished by the formation of new cellular molecules, including proteins, nucleic acids, lipids, and carbohydrates. Following a period of growth, cells reproduce by dividing. In the type of cell growth and division that increases the body size of higher plants and animals, the division mechanism produces daughter cells that contain exactly the same hereditary information as the parent cells. That is, the DNA sequences are copied exactly and are passed on to daughter cells without change. All other cell parts are divided approximately equally between daughter cells, although not with the same precision as the DNA.

The cellular processes underlying cell division take place in three clearly defined phases. In the first phase, called **interphase**, a cell increases in mass by synthesizing biological molecules, among them two exact copies of its DNA. Other cellular molecules necessary for division are stockpiled. In the second phase, termed **mitosis**, the replicated DNA molecules, with their histone and nonhistone proteins, are divided and placed in two separate daughter nuclei. Following mitosis, in the final phase of cell reproduction, the cytoplasm divides and two completely separate daughter cells are produced. The cytoplasmic division taking place in the last phase, called **cytokinesis**, usually divides the cytoplasmic organelles and molecules approximately equally between the daughter cells. The entire interphase-mitosis-cytokinesis sequence makes up the **cell cycle**.

Cell division takes place in three phases: (1) chromosome duplication during interphase, (2) nuclear division (mitosis), and (3) cytoplasmic division (cytokinesis).

AN OVERVIEW OF THE CELL CYCLE

Cells from almost all eukaryotic organisms follow a closely similar pathway through the cell cycle (Figure 10-1). A newly formed daughter cell enters a period of interphase synthesis and growth during which large amounts of protein, lipid, and carbohydrate molecules are made. This stage, called G_1 of interphase, is of variable length in different cell types. In some cells, as in rapidly dividing embryonic tissue, G_1 may last for only a few minutes. Other cells, such as the muscle and nerve cells of adult mammals, remain in G_1 for the life of the animal and never divide again. In cells destined to divide again, DNA rep-

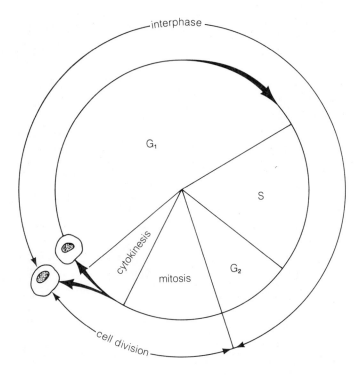

Figure 10-1 The cell cycle. The G_1 period of interphase may be of variable length, but for a given cell type the remaining S, G_2, and division segments of the cycle are usually of uniform duration.

The cellular mechanisms regulating the cell cycle are presently the subject of one of the most intensive research efforts in cell biology. The outcome of this research is of fundamental importance to our understanding of the ways in which cells work—and, perhaps of greatest significance to human well-being, why cells grow uncontrollably in cancer to produce malignant tumors.

Many conditions modify the progress of cells through the cell cycle.

The fundamental control of the cell cycle probably involves a series of genes that are activated in sequence to initiate replication, mitosis, and cytoplasmic division.

Because the beginning of the S phase is the first significant event in the processes leading to cell division, most of the current research effort is directed toward identifying the mechanisms that trigger DNA replication. Although many factors that modify the entry into S have been discovered, the ultimate control mechanisms have yet to be established. It is likely, however, that the basic controls involve genes that are turned on and off in sequence to initiate DNA replication and the remaining stages of cell division.

CHROMOSOME DUPLICATION DURING S

The primary activity of the S period of interphase is duplication of the chromosomes. Replication of chromosomal DNA produces two copies that, with very rare exceptions, are exact, sequence-by-sequence duplicates of their parent molecule. The duplication of chromosomal proteins produces essentially the same result: the histone and nonhistone chromosomal proteins are duplicated and combined with the replicated DNA molecules to produce two daughter chromosomes that are exact duplicates of each other and their G_1 chromosomal parent. Of the mechanisms accomplishing chromosome duplication, only DNA replication is understood in any detail. Many uncertainties still surround the synthesis of chromosomal proteins and the pattern in which they combine with replicated DNA to form daughter chromosomes.

DNA Replication

When Watson and Crick discovered the molecular structure of DNA (see Supplement 8-1), they pointed out that the complementarity of the two nucleotide chains of the DNA molecule provides a mechanism for DNA duplication (Figure 10-2). Complementarity refers to the fact that

lication eventually begins, terminating G_1 and initiating the **S** period of interphase. (S stands for synthesis, meaning DNA synthesis.) This stage, in most eukaryotes, lasts about 6 to 8 hours. After the S phase is completed, another interval, called G_2, passes before mitosis begins. (The G in G_1 and G_2 stands for "gap" and indicates the periods of interphase in which there is a pause in DNA synthesis.) The G_2 phase in most cells lasts for 2 to 5 hours.

Interphase thus consists of the G_1, S, and G_2 phases. Although G_1 may last from hours to months or years, S and G_2 are of uniform length in cells of the same type. As a result, cells usually enter mitosis and cytokinesis within 8 to 12 hours after the beginning of the S phase. When mitosis and cytokinesis are complete, the resulting daughter cells enter the G_1 period of the next interphase. This sequence makes it obvious that cells are fixed into the pathway leading to division at the onset of S. Once S begins, G_2, mitosis, and cytokinesis usually follow without delay.

Cells are fixed into the pathway leading to division by the initiation of DNA replication.

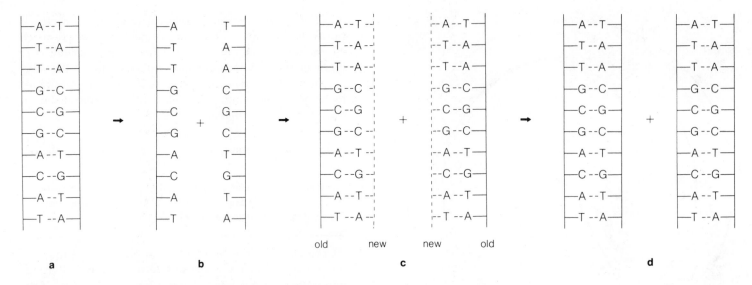

Figure 10-2 DNA replication. (**a**) An intact DNA double helix before replication. (**b**) The two nucleotide chains unwind. (**c**) Each half of the "old" molecule acts as a template for the synthesis of a complementary, "new" nucleotide chain. (**d**) The two molecules produced are exact duplicates of the molecule entering replication; each consists of one old and one new chain.

a given base in one of the two nucleotide chains of a DNA molecule will pair only with one kind of base in the opposite chain: thymine will pair only with adenine and guanine only with cytosine in the opposite chain (Figure 10-2*a*). As a result of complementarity, the sequence of bases in one chain fixes the sequence of bases in the opposite chain. From this, as Watson and Crick pointed out, the two nucleotide chains of a DNA molecule, if unwound and separated (Figure 10-2*b*), can act as patterns or **templates** for the synthesis of their missing halves (Figure 10-2*c* and *d*). The two molecules of DNA produced from these templates are each identical in sequence to the original molecule.

The precision of DNA replication depends on complementarity.

This model of replication, as first outlined by Watson and Crick, set the stage for subsequent work. It was soon established that in replication the two original nucleotide chains, after serving as templates, remain paired with their newly synthesized copies (Figure 10-3). As a result, each of the two molecules formed contains one old and one new nucleotide chain. This pattern of DNA replication is called **semiconservative** replication because one-half of the original parent DNA molecule is conserved in each of the daughter molecules. (One of the key experiments

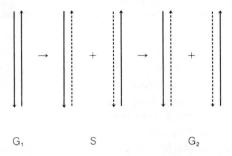

Figure 10-3 The pattern of semiconservative replication (see text). All living organisms replicate their DNA by the semiconservative pathway.

demonstrating that replication is semiconservative is described in Supplement 10-1.)

DNA replication follows the semiconservative pathway.

Semiconservative DNA replication proceeds by a mechanism that closely resembles RNA transcription (see p. 150 and Information Box 8-2), at least as far as the addition of nucleotides to the growing nucleotide chains is concerned. The reaction is catalyzed by *DNA polymerase*, an enzyme similar in its activity to the RNA polymerase enzymes catalyzing transcription. In the reaction (see Information Box 10-1), the DNA polymerase enzyme moves along the template chain, adding nucleotides to the growing new chain by matching up complementary base pairs. The energy required for the DNA replication is carried to the reaction by the nucleoside triphosphates dATP, dGTP, dCTP, and dTTP; see p. 32) that are linked into the new chain—each of them is a high-energy molecule with properties similar to ATP. The new nucleotide chain, once synthesized, remains wound into a double helix with its template chain.

Information Box 10-1

The Mechanism of DNA Replication

DNA replication requires, among other factors, a DNA template chain, the DNA polymerase enzyme, and the four nucleotides occurring in DNA: *deoxyadenosine triphosphate (dATP), deoxyguanosine triphosphate (dGTP), deoxycytidine triphosphate (dDTP),* and *deoxythymidine triphosphate (dTTP;* see Information Box 8-1). Each of these nucleotides differs from its counterpart in RNA synthesis by the presence of deoxyribose, rather than ribose, as the five-carbon sugar of the molecule. (The small *d* in front of dATP indicates this fact.) Since the DNA polymerase enzyme can add nucleotides only to the end of an existing nucleotide chain, a short segment called the **primer** must already be in place opposite the template chain. Although both short DNA and RNA chains can serve as primers for DNA synthesis, almost all DNA molecules in nature are assembled on RNA primers. (The short RNA primers are assembled by specialized RNA polymerase molecules that act in this role in replication; see Supplement 10-2.)

Replication proceeds as shown in the accompanying diagram. In the first step in the reaction sequence (part *a*), DNA polymerase binds to the template chain at the end of the short length of primer. Here the first base exposed at the end of the primer is cytosine. Base pairing and the activity of the enzyme match the complementary nucleoside triphosphate dGTP to the template at this point (part *b*). The dGTP triphosphate binds to the enzyme in a position close to the end of the primer. The total complex now consists of the enzyme, the template DNA with primer, and the molecule of dGTP. The enzyme then catalyzes removal of the two end phosphates from dGTP; at the same time, the remaining phosphate is linked to the terminal primer nucleotide (part *c*). Energy for the formation of this linkage comes from the removal of the phosphates from dGTP. The enzyme now moves one step along the template and binds at the end of the nucleotide just added (part *d*). Adenine is exposed on the template DNA at this point; the complementary nucleoside triphosphate dTTP is bound from the medium and the cycle repeats. The enzyme moves along the template in this way, catalyzing the assembly of the complementary strand in stepwise fashion until it reaches the end of the template.

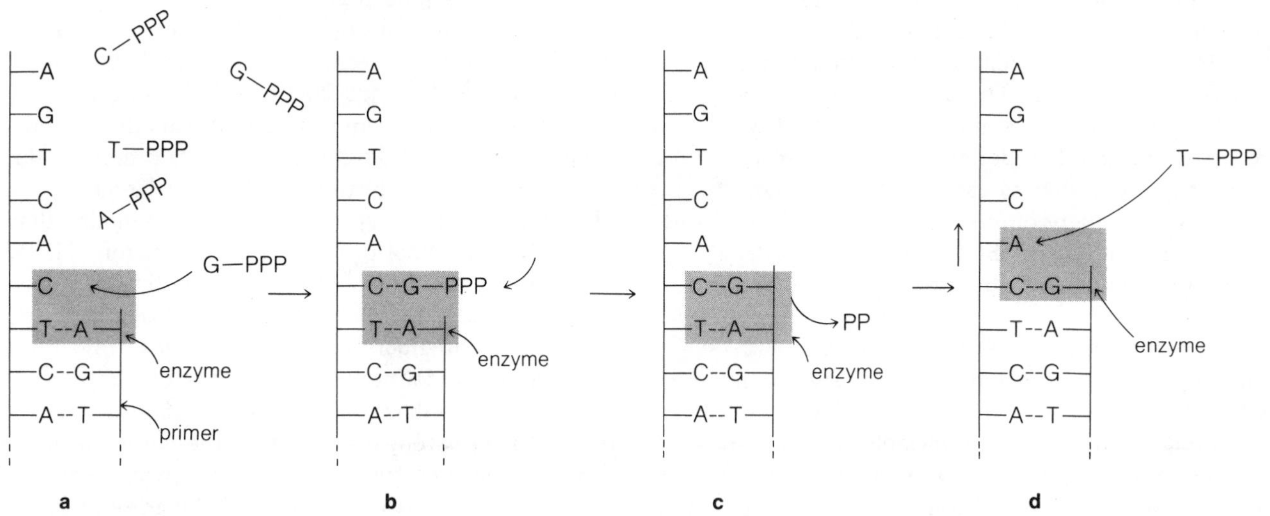

Although the addition of nucleotides to a growing DNA chain resembles RNA transcription, the process of replication differs fundamentally in the fact that DNA is a double helix. Because it is double helical in structure, DNA must unwind for replication to occur; in fact, DNA unwinds at a rate estimated to approach 13,000 revolutions per minute in bacteria and about 1000 to 2000 revolutions per minute in eukaryotes. Unwinding DNA involves an additional group of enzymes that have no parallels in RNA transcription. (How these enzymes unwind the template DNA chains and the details of other steps in DNA replication are outlined in Supplement 10-2.)

Duplication of Chromosomal Proteins

The histone and nonhistone chromosomal proteins of eukaryotes are also duplicated during interphase. Histones are synthesized during the S stage of interphase at the same time as DNA replication. The total quantity of histones doubles during S, as would be expected if duplication of these chromosomal proteins is closely coupled with DNA replication. The duplicated chromosomes at the completion of S, as a result, contain old and new histones in approximately equal quantities as well old and new DNA nucleotide chains. The nonhistone chromosomal proteins are also duplicated during interphase, some of them during G_1 and some during S, so that by the time S is complete the duplicated chromosomes have a full complement of these proteins.

MITOSIS

Once chromosome duplication is complete and the brief G_2 stage of interphase is over, the cell is ready to enter mitosis. In contrast to the molecular synthesis that occurs during S and G_2, which produce no visible changes, mitosis involves extensive alterations in cell organelles. These changes are easily observed under the light microscope and were correctly interpreted as early as 1880 by many investigators. Most prominent of these early workers were the European scientists Eduard Strasberger and Walter Flemming, who are credited with the first description and interpretation of mitosis.

Mitosis is a continuous process and takes place with no significant pauses or interruptions. For convenience in study, however, the process is usually broken down into four stages: **prophase** (*pro* = before), **metaphase** (*meta* = between), **anaphase** (*ana* = back), and **telophase** (*telo* = end). These stages are shown in light micrographs in Figure 10-4 and in diagrammatic form in Figure 10-5.

Prophase

Eukaryotic nuclei contain a collection of long DNA molecules combined with histone and nonhistone proteins. The individual DNA molecules, with their associated proteins, are the **chromosomes** of the nucleus (see Information Box 10-2). During interphase, the chromosomes, taken together, form the **chromatin** of the nucleus, which is so generally distributed at this stage that individual chromosomes cannot be distinguished (as in Figure 8-13). Other than the chromatin fibers, the interphase nucleus shows little internal differentiation except for the nucleolus.

During prophase, the chromosomes condense and the spindle takes form. The nucleoli disappear, and the nuclear envelope breaks down.

The Chromosomes at Prophase The beginning of prophase (Figures 10-4*a* and 10-5*a*) is marked by the first appearance of chromosomes as recognizable threads in the nucleus. (The word *mitosis*, from *mitos* = thread, is derived from the threadlike appearance of the chromosomes as they begin to pack into thicker structures as prophase begins.) This progressive folding and packing of the chromatin fibers into thicker structures is called **condensation**.

As soon as the chromosomes appear as distinct threads, it is clear that each is split lengthwise into two subunits (double arrows, Figure 10-5*a* and *b*). The two subunits, called **chromatids**, are the result of the duplication of DNA and chromosomal proteins during interphase. As such, the two chromatids of each chromosome are exact duplicates of each other and contain exactly the same genetic information. The longitudinal split marking the division between chromatids becomes more distinct as the chromosomes thicken during prophase. By the end of prophase, the chromosomes have condensed down to relatively short, thick double rods (Figures 10-4*b* and 10-5*b*).

Just how the chromosomes fold from the extended interphase state to the tightly packed condition of late prophase remains one of the unsolved problems of mitosis. The most widely accepted hypothesis maintains that one of the chromosomal proteins, possibly histone H1 (see p. 158), forms cross-links between chromatin fibers that fold and hold them together. As more cross-links form, the long chromatin fibers gradually condense into the compact, tightly folded form characteristic of late prophase.

By late prophase, when condensation is nearly complete, it is apparent that many of the chromosomes have distinctive shapes and sizes. This distinctive structure depends on the length and width of the chromosomes and on the locations of narrow regions called **constrictions**.

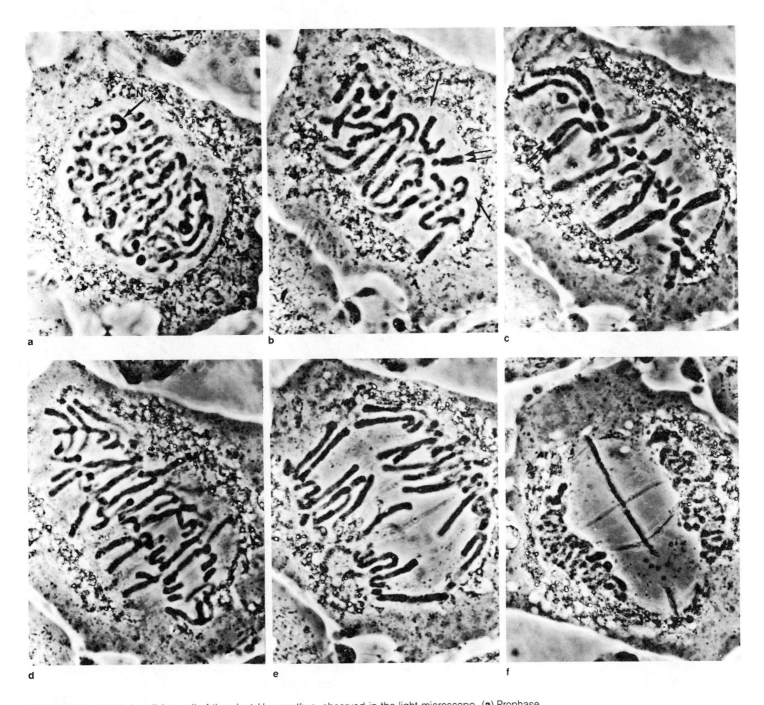

Figure 10-4 Mitosis in a living cell of the plant *Haemanthus*, observed in the light microscope. (**a**) Prophase. The chromosomes have condensed into distinct structures. The nucleolus (single arrow) is still visible at these early stages. The clear zone around the nucleus is the first indication of spindle formation in plants. Electron micrographs at this stage show the clear zone to be occupied by large numbers of spindle microtubules. (**b**) Beginning of metaphase. The nuclear envelope has broken down but fragments (single arrows) are still visible. (**c**) Metaphase. Although the kinetochores lie in a plane at the midpoint of the spindle, the chromosome arms extend in random fashion toward either pole. (**d**) and (**e**) Anaphase. (**f**) Telophase. The chromosomes at the poles are no longer individually distinguishable. The double arrows in (**b**) and (**c**) mark points where both chromatids of a chromosome are visible. ×700. From *Cinematography in Cell Biology*, 1963. Courtesy of M. Bajer and Academic Press, Inc.

Figure 10-5 Diagram of mitosis; compare with Figure 10-4.

Each chromosome usually has at least one prominent narrow region, called the *primary constriction*, that marks the attachment point for the microtubules that pull the chromatids apart during division. Usually one or more chromosomes also have other narrow regions known as *secondary constrictions*. The pattern and location of primary and secondary constrictions—along with the chromosome number and variations in the length and thickness of the chromosomes—provide a collective form known as the **karyotype** of an organism. Frequently, the karyotype is so distinctive that a species can be identified from this information alone. Figure 10-6 shows the distinctive structure of the human karyotype.

Other Nuclear Changes During Prophase Nucleoli usually disappear after the early stages of prophase. As condensation proceeds in early prophase, it becomes clear that the nucleolar material is attached to one or more of the chromosomes at prominent secondary constrictions. (Other chromosomes may have constrictions with no obvious

connections to the nucleolus.) Toward the close of prophase the nucleoli grow smaller and usually disintegrate by the onset of metaphase. This happens because rRNA transcription, as well as synthesis of other RNA types, slows considerably or stops entirely during prophase and metaphase of mitosis.

As prophase draws to a close, the nuclear envelope breaks down, releasing the chromosomes and other contents of the nucleus into the surrounding cytoplasm. By this time the chromosomes have condensed into short, double rodlets that are distributed more or less at random in the region formerly occupied by the nucleus. The nucleolar material, if still visible at this time, disperses and becomes indistinguishable from the surrounding cytoplasm.

Spindle Formation During Prophase As these events in the nucleus follow their course, changes also take place in the cytoplasm, some beginning as early as the S stage of interphase. Almost all animal cells, and the cells of many lower plants, contain a pair of small, barrel-shaped struc-

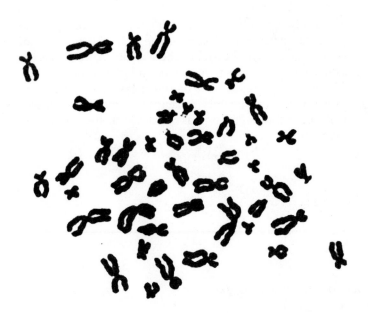

Figure 10-6 Human chromosomes isolated from a metaphase cell. ×1,500. Courtesy of S. Brecher.

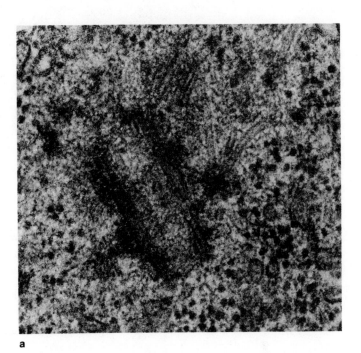

a

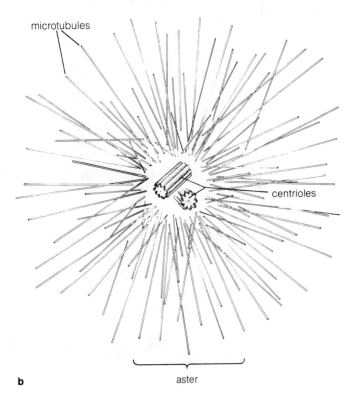

microtubules

centrioles

b   aster

Figure 10-7 Centrioles. (**a**) Lengthwise section of a centriole from a mouse. ×110,000. From *The Nucleus*, ed. A. J. Dalton and F. Hagenau, 1968. Courtesy of E. de Harven and Academic Press, Inc. (**b**) The relationship of the centriole and aster. Note that none of the astral microtubules touch the centrioles.

tures called **centrioles** (see Figure 10-7) in the cytoplasm just outside the nuclear envelope. (Supplement 10-3 describes centriole structure in detail.) The centrioles at G_1 are surrounded by short lengths of microtubules that radiate outward in all directions from the centriole pair. As the S period begins, the centrioles separate slightly and duplicate; this duplication produces two pairs of centrioles still contained within the radiating microtubules. During centriole duplication, these surrounding microtubules increase in length and number, shaping the surrounding granules and vesicles of the cytoplasm into a starlike array called the **aster** (shown in Figure 10-7b).

At the initiation of prophase, the two pairs of centrioles begin to separate; by late prophase the pairs have moved to opposite ends of the nucleus (Figure 10-8). As this shift takes place, bundles of microtubules lengthen between the separating centrioles, stretching in the direction of movement. By late prophase, when the centrioles have reached the opposite ends of the nucleus, the microtubules form a mass extending completely around one side of the nucleus. This mass is the **primary spindle** (Figure 10-8c).

The cells of some higher plant species, particularly in the flowering plants, do not contain centrioles. (In plants, centrioles are limited to species that produce motile reproductive cells.) Some animal cells, such as certain protozoa and the developing eggs of many higher animals, also lack centrioles. In these cells, spindle formation follows a different pathway. No conspicuous changes take place until prophase, when under the light microscope a clear space becomes visible in a narrow zone surrounding the nucleus (Figure 10-9a). This zone increases in size until a spindle-shaped formation develops (Figure 10-9b). Electron microscopy reveals that the "clear" zone is packed with microbutules extending around the nucleus from pole to pole. No asters are present at the poles in this type of spindle (see Figure 10-10b). The absence of centrioles or asters has no apparent effect on spindle formation and function. (The replication of centrioles, and the significance of the movements of centrioles and asters in animals and other organisms, is discussed further in Supplement 10-3.)

Information Box 10-2   An Overview of Mitosis

If all the DNA in a eukaryotic cell, containing all the genetic information of the cell nucleus, existed in a single piece, it could be represented as a single line of definite length:

No eukaryotic cells contain DNA in a single piece like this, however. Instead, the DNA is broken into shorter subunits:

These DNA subunits, with their attached histone and nonhistone proteins, are the **chromosomes** of the cell nucleus. During interphase, the chromosomes are duplicated exactly:

In mitosis, the duplicated chromosomes first condense into shorter, thicker units:

They then attach to spindle microtubules in such a way that the two parts of each chromosome connect to microtubules leading to opposite ends of the cell:

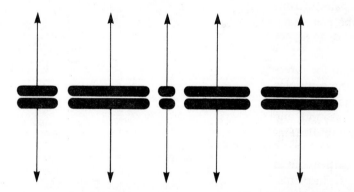

Metaphase

The transition from prophase to metaphase (Figures 10-4c and 10-5c) is gradual but can be conveniently marked by the fragmentation and breakdown of the nuclear envelope. Three major rearrangements take place after the nuclear envelope breaks down. First, the spindle moves into the region formerly occupied by the nucleus. Second, and at the same time, the chromosomes move to the midpoint of the spindle. Third, during movement to the midpoint of the spindle each chromosome attaches to bundles of spindle microtubules. These three events establish metaphase: the spindle is completely formed and the chromosomes, each attached to spindle microtubules, are aligned at the midpoint of the spindle.

The spindle microtubules then develop tension, separating the duplicate parts of each chromosome and moving them to opposite ends of the cell:

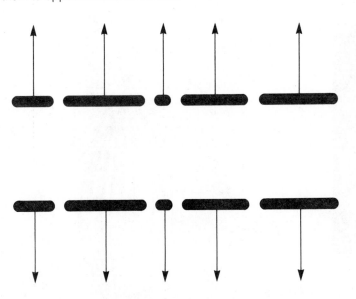

The chromosomes then unfold and become enclosed in separate cell nuclei:

When the cytoplasm divides, the two nuclei are separated into two daughter cells, the products of mitotic division.

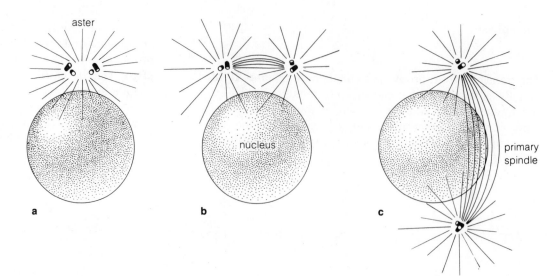

Figure 10-8 Formation of the primary spindle in animals (see text).

aster

nucleus

primary spindle

a   b   c

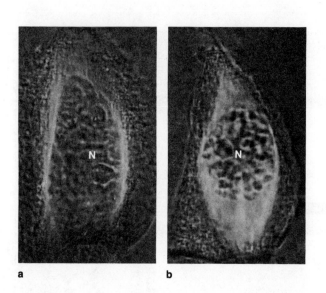

N

N

a   b

Figure 10-9 Spindle formation in plants, as seen in the light microscope. (**a**) A clear zone forms around the nucleus *(N)*. Under an electron microscope, this clear zone proves to be packed with microtubules. (**b**) The clear zone develops into the spindle by continued formation of microtubules. ×700. Courtesy of S. Inoué and A. Bajer.

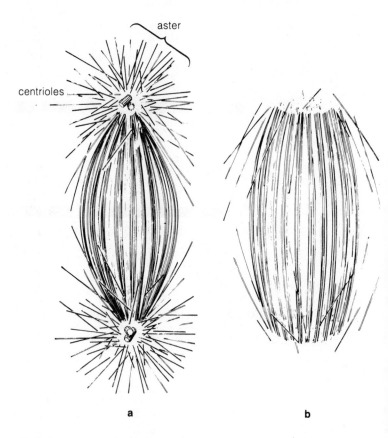

aster

centrioles

a   b

Figure 10-10 The two types of spindles at maturity: (**a**) with centrioles and asters; (**b**) with no centrioles or asters. Either spindle type functions normally in mitosis.

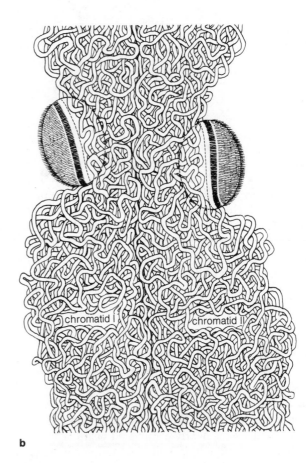

a

b

Figure 10-11 (a) Kinetochores *(K)* of a Chinese hamster. Both kinetochores of this chromosome have been caught in the plane of section. ×45,000. Courtesy of B. R. Brinkley and Springer-Verlag. (b) Diagram outlining the kinetochores visible in the electron micrograph.

By this time, the spindle in higher eukaryotes contains from 500 to 1000 or more microtubules stretching from one end of the cell to the other (Figure 10-10). The microtubules are more closely spaced near the ends or **poles** of the spindle, producing a pronounced narrowing of the structure at the tips.

One of the most significant events of mitosis is the way in which the chromosomes attach to the spindle at metaphase. Much of the precision of mitosis depends on this attachment. Remember that each chromosome at this stage consists of two longitudinal subparts, the chromatids, which are exact duplicates of each other. At the primary constriction of the chromosome, each of the two chromatids of the metaphase chromosomes has a **kinetochore** (brackets, Figure 10-11), a disklike structure that forms the point of attachment for spindle microtubules. The two kinetochores of each chromosome attach to the spindle in such a way that they face and connect to microtubules leading to *opposite* poles of the spindle (Figure 10-12). This pattern ensures that the two chromatids of each chromosome separate and move to opposite poles when the spindle microtubules develop their pulling force at anaphase. As a result, the chromatids are divided equally between the two poles when separation and movement of all the chromatids is complete.

During metaphase, the chromosomes attach to spindle microtubules. The two chromatids of each chromosome connect to microtubules leading to opposite ends of the spindle.

By the time the kinetochore connections are made, growth of the spindle has progressed to such an extent that most of the cytoplasm is filled by spindle microtubules. This development, and the connection of chromosomes to spindle microtubules at the spindle midpoint, sets the stage for the next part of mitosis: anaphase.

Anaphase

Tension is developed by the spindle microtubules as the chromosomes align at the midpoint of the spindle. This tension is sufficient to pull the two chromatids apart slightly at the primary constriction, so that the attachment points, the kinetochores, stretch toward the opposite spindle poles. Tension on the kinetochores continues to develop, and after a brief pause in this condition, the two chromatids of each chromosome separate completely and begin moving to opposite poles of the spindle. The separation of the chromatids and the initiation of movement to the spindle poles marks the transition from metaphase to anaphase (Figures 10-4d and 10-5d).

During anaphase, the two chromatids of each chromosome separate and move to opposite poles of the spindle.

The movement of the chromatids to opposite ends of the spindle, the most spectacular feature of mitosis, has intrigued scientists for nearly a hundred years. The basis for this rapid movement is still not completely understood. Most investigators now believe that the spindle microtubules generate the force for poleward movement by a combination of controlled growth and active sliding movements.

Anaphase movement continues until the separated chromatids are collected at the opposite spindle poles (Figures 10-4e and 10-5e). Since each end of the spindle receives one chromatid from every metaphase chromosome, the two collections of chromatids have exactly the same hereditary information. Once the chromatids have completed their movement to the spindle poles, the cell enters telophase, the final stage of mitosis.

Telophase

During telophase (Figures 10-4f and 10-5f) the chromatids at the poles unfold and become indistinct. Segments of new nuclear envelope appear at the borders of the decondensing chromatids. These segments gradually extend around the mass until the chromatids, by now closely resembling interphase chromatin, are completely separated from the surrounding cytoplasm by a new, continuous nuclear envelope.

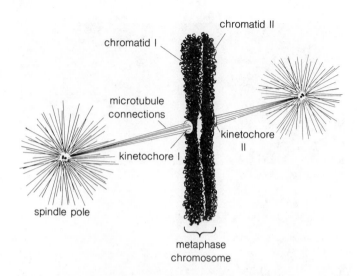

Figure 10-12 Kinetochore connections at metaphase. Two kinetochores are present on each metaphase chromosome, one for each of the two chromatids. The kinetochores are directed toward opposite poles of the spindle and make microtubule connections leading to them.

During telophase, the separated chromatids unfold and return to the interphase state. The nucleoli reappear, and the nuclear envelope reforms.

Soon after the beginning of telophase, nucleolar material begins to appear at one or more secondary constrictions in the chromatids. If nucleoli form at more than one secondary constriction, the several nucleoli produced may fuse together or remain separate, depending on the species. By the time nucleoli are completely formed, the polar masses of chromatin are again surrounded by nuclear envelopes and are indistinguishable from G_1 chromosomes. This transition is accompanied by a gradual return to full transcription of all RNA types.

The distinction between chromatids and chromosomes at this stage is somewhat arbitrary. For most purposes, the transition from chromatids to chromosomes at telophase can be considered complete when the chromatids of each chromosome entering mitosis have been separated and enclosed in separate daughter nuclei.

The spindle starts to disappear as soon as the change from anaphase to telophase begins. The asters, if present, become smaller until the centriole pair near each daughter nucleus is surrounded by a limited number of relatively short microtubules. The centrioles then take up their characteristic interphase position at one side of the nucleus, just outside the nuclear envelope. All that remains of the spindle after telophase is a layer of short microtubules at the former spindle midpoint. This persistent layer marks the region where the cytoplasm will divide to separate the daughter cells at the close of mitosis.

The entire mitotic sequence, from the onset of prophase to the close of telophase, may require as little as 5 to 10 minutes in rapidly developing animal embryos or as much as 3 hours in various plant and animal tissues. Of the different stages, prophase is usually the most extended. Metaphase and telophase generally require less time. Anaphase proceeds most rapidly, rarely taking more than a few minutes in most species.

The Significance of Replication and Mitosis

The result of replication and mitosis is the production of two daughter nuclei, each with genetic capacity equivalent to the original parent nucleus. The accuracy of mitotic division depends on two basic features of the mechanism: (1) the parallel arrangement of spindle microtubules into two distinct poles in the cell and (2) the opposite spindle pole connections made by the two chromatids of each chromosome. These connections result in the separation and delivery of the two chromatids of each chromosome to the opposite poles.

CYTOKINESIS

Cytokinesis, the division of the cytoplasm, completes the process of cell division by enclosing the daughter nuclei produced by mitosis in separate cells. Although the details of cytokinesis differ in plants and animals, spindle remnants function similarly in both processes.

Cytokinesis in Animals: Furrowing

In animals, short lengths of microtubules persist at the spindle midpoint after mitosis is complete. The rest of the spindle breaks down and disappears after anaphase, except for the short lengths of microtubules left around the centrioles. The microtubules at the spindle midpoint become surrounded with patches of dense, apparently structureless material. This material forms a layer called the **midbody**, which soon extends completely across the cell (Figure 10-13).

After the midbody develops, a depression or **furrow** appears in the plasma membrane around the outside of the cell at the level of the midbody (Figure 10-14). This furrow gradually deepens, following the plane of the midbody, until the two daughter cells are completely separated by continuous plasma membranes. As the deepening furrow penetrates into the cell, the midbody is compressed and becomes smaller, usually disappearing as the furrow cuts off the daughter cells. Mitochondria, endoplasmic reticulum, Golgi complex, vesicles, and other cytoplasmic organelles are roughly divided between the daughter cells as the furrow deepens.

Cytoplasmic division occurs in animals by furrowing, which results from the activity of microfilaments.

The manner in which the furrow develops makes it obvious that the original position of the spindle determines the plane of cytoplasmic division. Ordinarily, the spindle is situated with its midpoint at the cell equator, so that the daughter cells formed by cytokinesis are of equal size. In some cases, as in the unequal division of cytoplasm during egg development in animals (see Figure 14-5), the spindle is positioned at one side of the dividing cell. The furrow forms opposite the spindle midpoint as usual, cutting the cytoplasm into two unequal parts. The factors governing the alignment and position of the spindle in animals, which determine the later plane of furrowing, remain unknown.

In cells that are unattached to their neighbors, such as certain developing eggs or cells in tissue culture, the cell looks as if a drawstring is being tightened around it during furrowing. This impression is directly supported by electron micrographs of the advancing furrow in dividing cells, which reveal large numbers of microfilaments at the furrow edge (see Figure 10-15). These microfilaments, which are also found in many motile systems in cells, are believed to produce contractile force for furrowing by actively sliding over each other, progressively tightening the furrow until the dividing cell is separated into two parts.

Cytokinesis in Plants: Formation of the Cell Plate

The initial stages in plant cytokinesis resemble midbody formation in animals. During telophase, portions of the spindle microtubules persist at the spindle midpoint. These microtubules become surrounded by a layer of dense material (Figure 10-16), as in midbody formation in animals. In plants, however, much of this dense material is enclosed in membrane-bound vesicles that originate

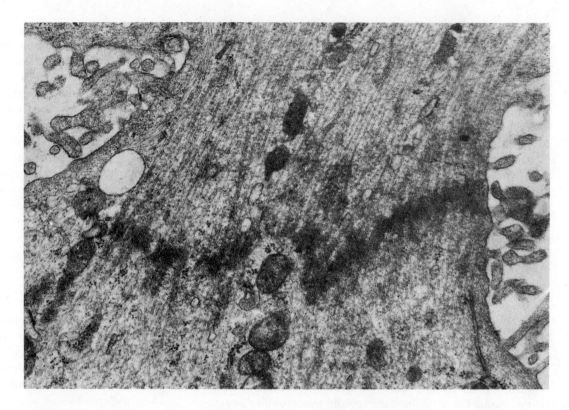

Figure 10-13 Midbody of a human cell at late telophase. ×35,000. Courtesy of A. Krishan, R. C. Buck, and The Rockefeller University Press, from *Journal of Cell Biology* 24:433 (1965).

from the Golgi complex. These vesicles gradually increase in number until a continuous layer extends across the equator of the cell at the former spindle midpoint. This layer of microtubules and vesicles determines the position of the new wall that will separate the daughter cells.

Cell wall formation begins in the central area of the layer of microtubules and vesicles. Gradually the new wall extends outward toward the plasma membrane, in a direction opposite to furrowing in animals. This process takes place as the vesicles fuse together, forming two layers of continuous plasma membranes (Figure 10-17a and b). As the vesicles fuse, their contents are released into the developing extracellular space between the daughter cells. When the fusion process reaches the original cell walls, the daughter cells are completely separated by two continuous plasma membranes with an enclosed space between them (Figure 10-17c). This space is filled with the dense material from the fused vesicles. On analysis, this material proves to include cellulose and other components of a new cell wall. When fully formed, the layer of wall material separating the daughter cells is called the **cell plate**.

Once formed, the cell plate is progressively thickened and strengthened by the deposition of new cell wall material between the membranes separating the daughter cells. At points this new wall is perforated by cytoplasmic connections that remain intact between the daughter cells (see Figure 4-20). These narrow connections, termed **plasmodesmata** (singular = *plasmodesma*), evidently serve as channels of chemical communication between the cells of plant tissue.

The plane of division in plant cells, as in animal cells, is determined by the position of the spindle at metaphase. Usually the spindle occupies the center of the cell, so that the subsequent cytokinesis separates the dividing cell into two equal parts. In some plant tissues, however, the spindle takes a position at one side, so that the following cytoplasmic division is unequal. As in animals, the factors determining the position of the spindle and the resultant plane of cytoplasmic division are unknown.

Cytoplasmic division in plants occurs by cell plate formation.

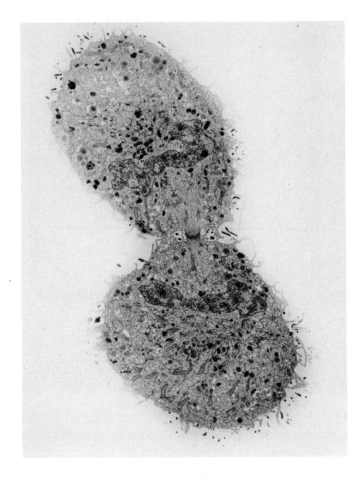

Figure 10-14 Furrow formation in a dividing human cell in culture. The furrow gradually deepens until the cytoplasm is divided into two parts. ×4,000. Courtesy of G. G. Maul.

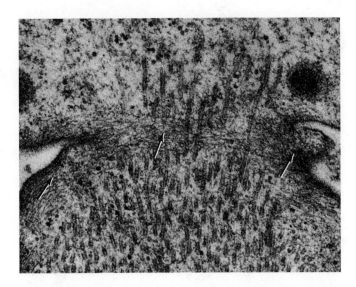

Figure 10-15 Microfilaments (arrows) at the edge of the advancing furrow in a dividing egg cell of a rat. ×35,000. Courtesy of D. Szollosi and The Rockefeller University Press, from *Journal of Cell Biology 44*:192 (1970).

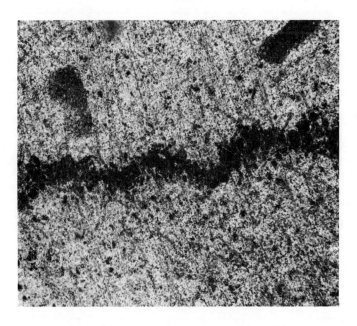

Figure 10-16 The equivalent of midbody formation in a plant cell of *Haemanthus*. Microtubules are embedded in the midbody-like layer. ×17,000. Courtesy of A. Bajer and Springer-Verlag.

THE OVERALL EFFECTS OF REPLICATION, MITOSIS, AND CYTOKINESIS: A REVIEW

To review the overall effects of mitotic cell division we will consider a diploid organism (see Information Box 10-3) with only two pairs of chromosomes: one long pair and one short pair. At interphase (Figure 10-18*a*), the chromosomes extend throughout the nucleus. As a result of replication during the S period (Figure 10-18*b*), each chromosome becomes double at all points and now consists of two chromatids. During prophase of mitosis, the chromosomes condense into short rodlets (Figure 10-18*c*). At metaphase (Figure 10-18*d*), they align at the midpoint of the spindle. Spindle attachments are made so that the two kinetochores of each chromosome connect to microtubules

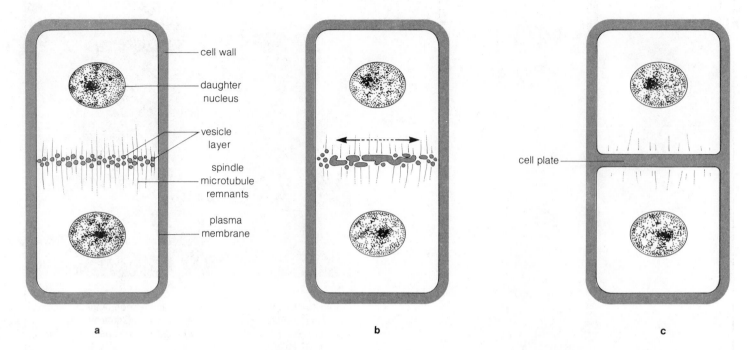

cell wall

daughter nucleus

vesicle layer

spindle microtubule remnants

plasma membrane

cell plate

a

b

c

Figure 10-17 Cell plate formation in plants (see text).

Information Box 10-3

Haploidy and Diploidy

Almost all body cells of a species have the same number of chromosomes. Depending on the life history of the organism, this number is usually the **haploid** or **diploid** number for the species, although it may be some higher multiple of these numbers. In a haploid nucleus, only one copy of each chromosome is present. (Haploid nuclei are also sometimes called **monoploid** nuclei.) Diploid nuclei contain two copies of each chromosome, so that all the genetic information is represented twice.

The two copies of each chromosome in diploids make up a pair. The pairs are described as **homologs** because they contain the same genes in the same sequence. (Slightly different forms of the genes, called **alleles,** may be present on either chromosome of a pair.) Because higher plants and animals are diploid throughout most of their life cycles, mitosis is described in this chapter as it would occur in a diploid organism. The mechanism and outcome of mitosis are independent of the chromosome number, however, and follow the same pattern in both haploids and diploids. During mitotic cell division, the two members of the homologous pairs of diploids remain separate and proceed independently through all the stages of mitosis.

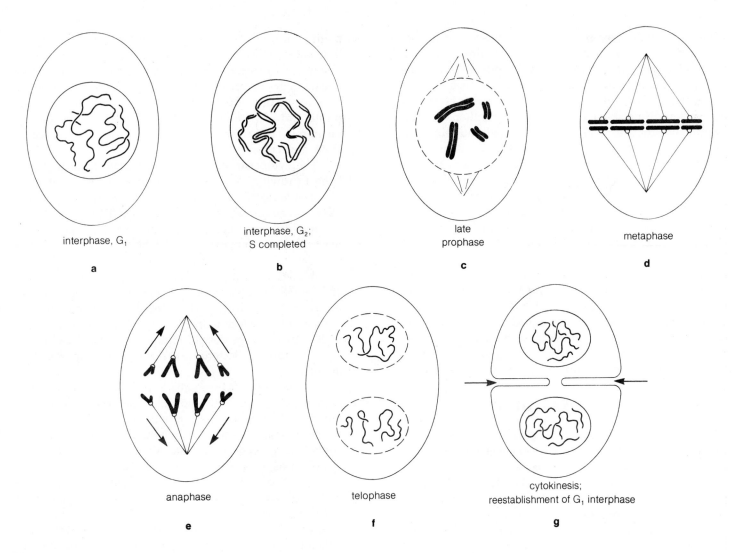

interphase, G$_1$

a

interphase, G$_2$;
S completed

b

late
prophase

c

metaphase

d

anaphase

e

telophase

f

cytokinesis;
reestablishment of G$_1$ interphase

g

Figure 10-18 A review of mitosis (see text).

leading to opposite poles of the spindle. At anaphase (Figure 10-18e), the two chromatids of each chromosome separate and move to the opposite spindle poles. At the poles, the chromatids unfold and return to the interphase state during telophase (Figure 10-18f). The formation of new nuclear envelopes during telophase, and the subsequent cytoplasmic division (Figure 10-18g), completes the separation of the two daughter cells. As a result of replication and mitosis, the nucleus in each of these daughter cells contains exactly the same number and types of chromosomes as the original parent cell.

Mitosis works equally well in haploid and diploid organisms or in organisms with any multiple of the basic haploid number of chromosomes.

The processes of cell division usually occur in sequence as outlined in this chapter. Replication, mitosis, and cytokinesis are potentially separable, however, and in some organisms may proceed independently. In the salivary glands of *Drosophila* (the fruit fly) larvae, for example,

DNA replication takes place without mitosis or cytokinesis, producing large nuclei containing hundreds or thousands of copies of the basic DNA complement. In some groups of fungi, replication and mitosis proceed without cytokinesis. This sequence produces cells with many nuclei enclosed in a common cytoplasm. Another variation eventually produces cells containing single nuclei, but it does so in a pattern in which mitosis and cytokinesis are widely separated in time. In developing insect eggs, for example, replication and mitosis proceed rapidly for a time without cytokinesis, producing an early embryo with several hundred nuclei suspended in a common cytoplasm. Eventually, rapid cytokinesis occurs in these embryos, enclosing the many daughter nuclei in separate cells.

Mitotic cell division may serve as a method of reproduction for whole organisms if one or more daughter cells are released from the parent and grow separately into complete individuals. Reproduction of this type, which occurs in many kinds of plants, animals, and single-celled organisms, is called **vegetative** or **asexual reproduction.** Because the cell products in asexual reproduction result from mitosis, all the offspring are genetically identical.

CELL DIVISION IN PROKARYOTES

Cell division in the bacteria and blue-green algae, although evidently as precise as division in eukaryotes, does not proceed by mitosis. In bacteria, where replication and cell division have been most intensively studied, replicated DNA is probably separated not by mitosis but by the activity of the plasma membrane. Current hypotheses of bacterial division are based on the observation that bacterial DNA, when isolated from the cell, frequently has one or more attachments to the plasma membrane. These attachments, which are sometimes visible in sectioned bacteria, are believed to divide replicated bacterial DNA molecules equally between daughter cells according to the following mechanism.

In bacteria the hereditary information of the nucleoid is coded into a single DNA molecule. This DNA molecule, the single "chromosome" of a bacterial cell, exists as a closed circle with no free ends. At one point, and possibly more, this circle is considered to be attached to the plasma membrane. Replication and division of the circle then take place as shown in Figure 10-19. Replication starts at one point on the molecule and then proceeds in both directions from this point, producing two replication "forks" that gradually advance around the circle (Figure 10-19a and b). As the forks complete their circuit around the DNA molecule, the duplicated circles completely separate (Figure 10-19c). Presumably the membrane attachment point is also duplicated at the same time, with the result that both circles are now attached to the plasma membrane at two closely spaced points. The membrane then begins to grow between the attachment points (Figure 10-19d), separating and pushing the two DNA circles to opposite ends of the cell (Figure 10-19e). Thus the plasma membrane rather than a spindle divides the replicated DNA molecules.

The cytoplasm then divides by a mechanism resembling furrowing in animal eukaryotes. An indentation in the plasma membrane appears around the midpoint of the cell between the separated DNA circles (Figure 10-19f). As this indentation or furrow deepens, it eventually pinches the cytoplasm into two completely separate halves. The cell wall follows the inward path of the furrow; as the membranes separate, the growing wall becomes complete and extends as a partition between the daughter cells (Figure 10-19g). Inward growth of the plasma membrane and cell wall of a bacterium can be seen in progress in Figure 10-20. The entire process of replication and cell division in rapidly growing bacteria takes no more than about 20 minutes.

The prokaryotic mechanism works effectively because there is only one "chromosome" per cell. The greater number of chromosomes in eukaryotic cells would no doubt cause many mistakes in distribution if division proceeded as it does in bacteria. Mitosis probably arose in evolution as an adaptation in response to the increasing length and complexity of the genetic message as eukaryotes first appeared. As the length of the chromosomal DNA increased, restrictions on the total time required for replication, the danger of breakage, and the mechanical difficulties of dividing long chromosomes favored subdivision of the genetic information into the shorter subunits that become the chromosomes of eukaryotes. As the chromosomes appeared, a mechanism was required to divide their replicated products precisely and equally between daughter cells.

The precision of mitotic cell division depends on (1) the arrangement of the spindle, which establishes two "ends" in the dividing cell, and (2) the pattern of attachment of chromatids to spindle microtubules leading to opposite ends of the cell.

The required precision, as noted in this chapter, is supplied by the mitotic spindle and the attachments made by the chromosomes to the spindle at metaphase. The parallel arrangement of microtubules in the spindle, running from end to end of the cell, sets up two distinct poles to receive the divided chromosomes. The chromatids are separated into two equal groups and delivered to these poles

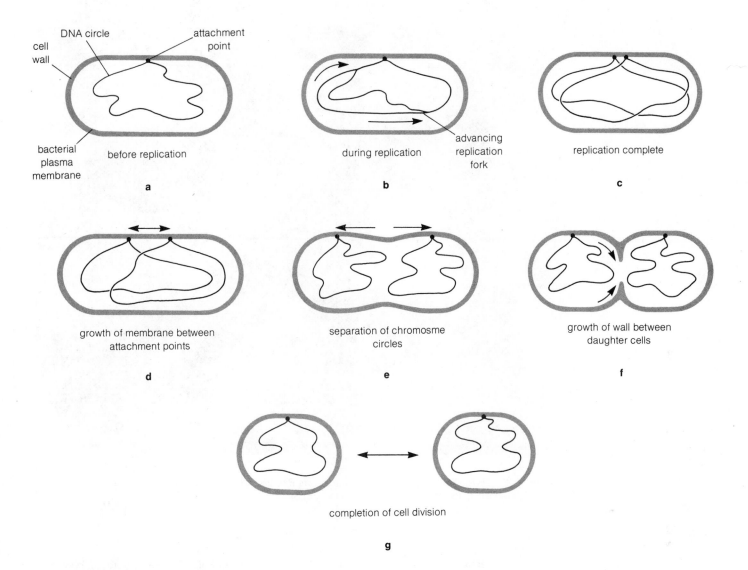

Figure 10-19 Division of the nucleoid and cytoplasm in a bacterial cell (see text).

as a result of the connections made by the kinetochores of the chromosomes at metaphase, connections that always lead to opposite spindle poles.

The entire cycle of replication, mitosis, and cytokinesis, in all its elegant complexity, occurs countless billions of times in the growth, development, and maintenance of structure in many-celled eukaryotes. In the maintenance of red blood cells alone in humans, mitotic divisions occur in each individual at the rate of more than 2 million per second. The perfection of the mechanism is such that these repeated cycles of division occur almost without error throughout the lifetime of the organism.

Questions

1. List the major stages in the cell cycle. What happens in each stage?

2. What happens in G_1, S, and G_2 of interphase?

3. What does the complementarity of the two nucleotide chains of a DNA molecule have to do with replication?

4. Define semiconservative replication. What happens to the "old" and "new" nucleotide chains in semiconservative replication?

5. Outline the steps that occur in the addition of nucleotides to a growing DNA chain during replication.

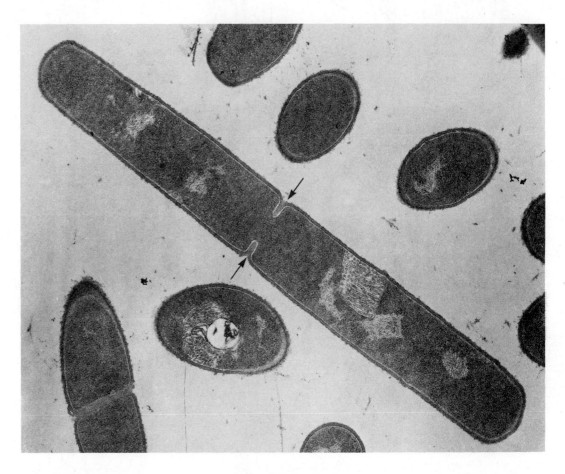

Figure 10-20 Division of the cytoplasm in a bacterium, *Bacillus cereus*. Inward growth of a new wall (arrows) is separating the daughter cells. ×23,000. Courtesy of L. Santo.

6. In what ways are DNA replication and RNA transcription alike? In what ways do they differ?

7. What happens during mitotic prophase? What events mark the beginning and end of this stage?

8. What structures are visible on the chromosomes during prophase and metaphase? Define chromosome and chromatid. At what times in the life cycles of cells are chromatids present? What is a karyotype? A kinetochore?

9. Trace the development of the spindle in organisms with and without centrioles.

10. What happens during metaphase? What events mark the beginning and end of this stage?

11. What happens during anaphase? What events mark the beginning and end of this stage? How are the spindle microtubules believed to move the chromosomes during anaphase?

12. What happens during telophase? What events mark the beginning and end of this stage? What events are considered to convert the separated chromatids to chromosomes?

13. What is the outcome and significance of mitosis? What features of mitosis ensure that the chromatids are equally divided between daughter nuclei?

14. Trace the patterns of cytoplasmic division in plants and animals. In what ways is cytoplasmic division similar in plants and animals? In what ways is it different?

15. How is the plane of cytoplasmic division related to the position taken by the spindle in mitosis?

16. How are the activities of microtubules and microfilaments coordinated in animal cell division?

17. Define midbody, furrow, cell plate, and plasmodesma.

18. Trace mitotic cell division in an organism with three pairs of chromosomes (a diploid). Do the same for an organism that has three chromosomes that exist singly rather than in pairs (a haploid or monoploid; see Information Box 10-3).

19. Does the existence of chromosomes singly or in pairs make any fundamental difference in the way mitosis proceeds?

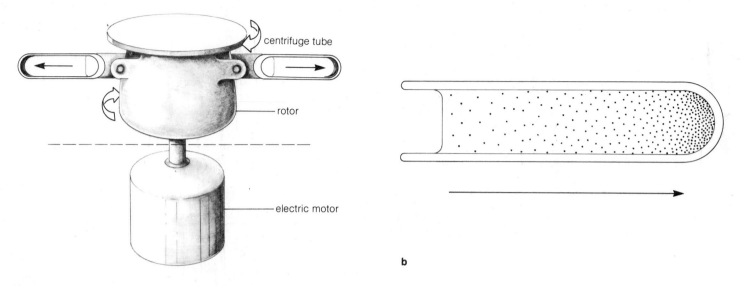

Figure 10-21 Buoyant density centrifugation. (**a**) The arrangement of tubes and rotor. The tubes are hinged so that they spin at right angles to the axis of rotation. (**b**) Creation of a density gradient in a centrifuge tube by centrifugal force (see text). The dots represent cesium chloride molecules.

20. How is cell division believed to occur in prokaryotes? Compare prokaryotic and eukaryotic cell division.

21. What happens if replication occurs without mitosis and cytokinesis? What happens if replication and mitosis occur without cytokinesis?

22. What is vegetative or asexual reproduction?

Suggestions for Further Reading

Alberts, B., and R. Steinglanz. 1977. "Recent Excitement in the DNA Replication Problem." *Nature* 269:655–661.

Baserga, R. 1976. *Multiplication and Division in Mammalian Cells.* Marcel Dekker, New York.

Kornberg, A. 1980. *DNA Replication.* Freeman, San Francisco.

Mazia, D. 1974. "The Cell Cycle." *Scientific American* 230:54–64.

Prescott, D. M. 1976. *Reproduction of Eukaryotic Cells.* Academic Press, New York.

Rost, T. L., and E. M. Gifford, Jr., eds. 1977. *Mechanisms and Control of Cell Division.* Dowden, Hutchinson, and Ross, Stroudsburg, Pennsylvania.

Wolfe, S. L. 1981. *Biology of the Cell.* 2nd ed. Wadsworth, Belmont, California.

Yeoman, M. M. 1976. *Cell Division in Higher Plants.* Academic Press, New York.

SUPPLEMENT 10-1:
THE EVIDENCE FOR SEMICONSERVATIVE REPLICATION

When DNA structure was first discovered, semiconservative replication (see Figure 10-3) was only one of several pathways considered as possibilities for DNA duplication. Semiconservative replication was established as the pathway followed in nature by Matthew Meselson and Franklin W. Stahl at the California Institute of Technology, who investigated replication in a prokaryote, and J. Herbert Taylor at Columbia University, who worked with a eukaryote.

Meselson and Stahl's experiment used the bacterium *Escherichia coli.* First they grew *E. coli* for several generations in a medium containing the heavy nitrogen isotope ^{15}N. (The most common form of nitrogen is the less dense isotope ^{14}N.) This period of growth was continued long enough to ensure that all the *E. coli* DNA contained the ^{15}N isotope. Meselson and Stahl then removed the bacteria from the ^{15}N source and placed them in a medium containing only ^{14}N nitrogen. After the transfer to unlabeled medium, cells were removed at intervals and the DNA was isolated and purified.

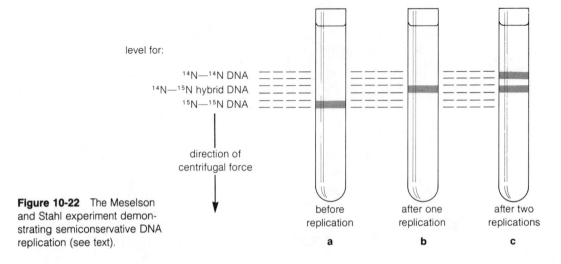

level for:

^{14}N—^{14}N DNA
^{14}N—^{15}N hybrid DNA
^{15}N—^{15}N DNA

direction of
centrifugal force

before
replication

after one
replication

after two
replications

a

b

c

Figure 10-22 The Meselson and Stahl experiment demonstrating semiconservative DNA replication (see text).

The DNA was then compared to pure ^{15}N and ^{14}N DNA standards. To carry out these comparisons, Meselson and Stahl used a technique known as *buoyant density centrifugation* (Figure 10-21, preceding page). In this technique, a centrifuge tube is filled with a solution of cesium chloride (CsCl) made up to a concentration approximating the density of DNA. As the tube spins at high speed in the centrifuge, the cesium chloride molecules become more concentrated toward the bottom of the tube, producing a gradient of density from top to bottom. At the top of the tube, the CsCl is less concentrated and thus lower in density than the DNA sample. At the bottom of the tube, centrifugal force packs the CsCl molecules more closely together, producing a region higher in density than the DNA sample. In response, the DNA descends or ascends in the tube until it reaches the level at which it matches the density of the surrounding CsCl solution. If the DNA is of uniform density, it will form a sharply defined band at this point. Because it contains the heavier nitrogen isotope, ^{15}N DNA is denser and forms a distinct band in the centrifuge tube at a lower level than ^{14}N DNA. This process enables the two kinds of DNA to be separated and identified.

The results of Meselson and Stahl's experiment are summarized in Figure 10-22. DNA extracted from bacterial cells at the instant of transfer from ^{15}N to ^{14}N growth medium, before any DNA replication in ^{14}N medium could occur, formed a single band in the centrifuge at the level expected for ^{15}N DNA (Figure 10-22*a*). After about 20 minutes following transfer from ^{15}N to ^{14}N medium (the time required for one complete bacterial cell cycle in ^{14}N medium, including DNA replication and cell division), all the DNA isolated from the cells formed a band at a level in-

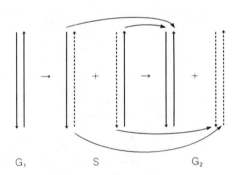

G_1 S G_2

Figure 10-23 Conservative DNA replication (see text).

termediate in density between ^{15}N and ^{14}N DNA (Figure 10-22*b*). DNA removed from the bacteria after two generations of growth in the unlabeled medium formed two bands when centrifuged, one at the intermediate level and one at a level characteristic of pure ^{14}N DNA (Figure 10-22*c*).

This finding effectively eliminated a possible pathway for DNA replication called **conservative replication.** In conservative replication (Figure 10-23) the nucleotide chains of a parent DNA molecule, after serving as templates, would unwind from their copies and rewind together again. The two newly synthesized chains would likewise wind together into a double helix. Thus one of the DNA molecules produced by conservative replication would consist entirely of "old" DNA and one entirely of "new" DNA. If replication were conservative, two bands would therefore be expected to appear in the centrifuge after one

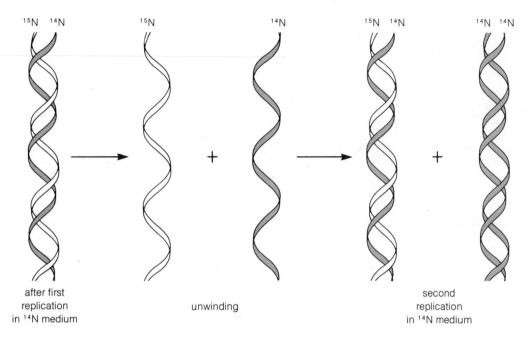

^{15}N ^{14}N ^{15}N ^{14}N ^{15}N ^{14}N ^{14}N ^{14}N

after first
replication
in ^{14}N medium   unwinding   second
replication
in ^{14}N medium

Figure 10-24 Explanation of Meselson and Stahl's results (see text).

replication in the ^{14}N medium: one band of pure ^{15}N DNA and one band of pure ^{14}N DNA. This result would be expected because the two "old" ^{15}N nucleotide chains of the parent DNA molecule, after serving as templates for replication, would reassociate into an all "old" ^{15}N–^{15}N DNA molecule. The two newly synthesized chains, assembled entirely from nucleotides containing ^{14}N nitrogen (because the bacteria were placed in medium containing only ^{14}N nitrogen as a growth source), would wind together into an all-new ^{14}N–^{14}N DNA molecule. The appearance of a *single* band of intermediate density at this point in the experiment, rather than two distinct bands at the ^{14}N and ^{15}N levels, therefore ruled out the possibility of conservative replication.

The single intermediate band observed at this time is fully explained if replication is semiconservative. In semiconservative DNA replication, the parent DNA molecule, consisting of two ^{15}N nucleotide chains, unwinds and serves as template. After the synthesis of its complementary chain from nucleotides containing ^{14}N nitrogen, each template remains wound into a double helix with the new ^{14}N copy, producing a hybrid molecule consisting of one old ^{15}N nucleotide chain wound into a double helix with a new ^{14}N nucleotide chain. This hybrid DNA molecule is of intermediate density, between pure ^{15}N and ^{14}N DNA,

and will form a band at an intermediate level in the centrifuge. After one semiconservative replication, all the bacterial DNA would consist of ^{15}N–^{14}N hybrids of this type and would produce only one intermediate band in the centrifuge. As we have seen, this was in fact the result obtained by Meselson and Stahl.

DNA removed from bacteria allowed to grow for two full generations after transfer from ^{15}N to ^{14}N medium produced two bands after centrifugation—one at the ^{15}N–^{14}N intermediate density and one at the level of pure ^{14}N DNA. This observation is also exactly as expected if replication is semiconservative. We can explain these results by following one of the ^{15}N–^{14}N hybrid DNA molecules produced in the first generation through another cycle of semiconservative replication in ^{14}N medium (Figure 10-24). During the second replication, the two nucleotide chains separate and serve as templates for DNA synthesis. The ^{15}N chain remains with its newly synthesized copy, producing another ^{15}N–^{14}N hybrid of intermediate density. The ^{14}N template chain also remains with its ^{14}N complementary copy, producing a "pure" ^{14}N–^{14}N DNA molecule, with both chains synthesized from nucleotides containing only ^{14}N nitrogen. This DNA centrifuges to a position characteristic of ^{14}N DNA. Thus, after two generations, the DNA isolated from large numbers of bacteria

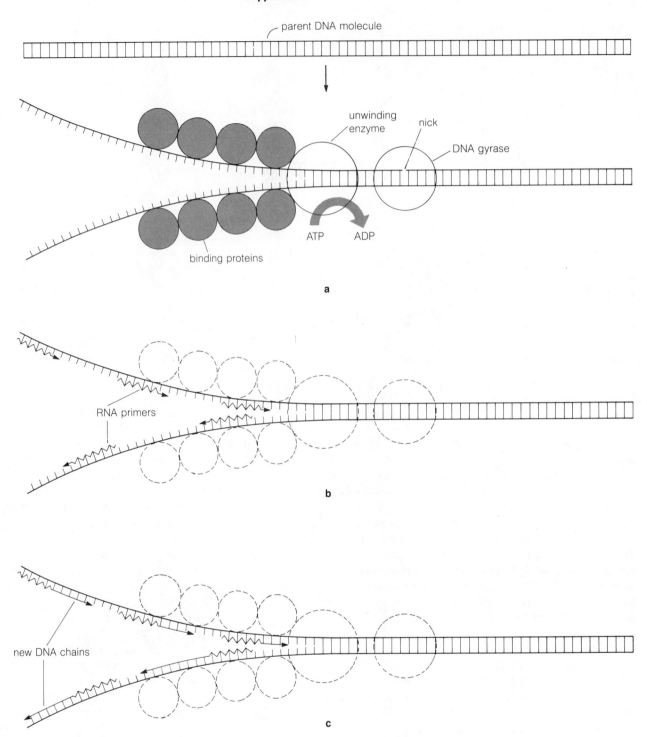

Figure 10-25 Details of the mechanism of DNA replication (see text). (**a**) Unwinding by the unwinding enzyme, DNA gyrase, and the binding proteins. (**b**) Primer synthesis. (**c**) Synthesis of new DNA chains. (**d**) Removal of primers. (**e**) Completion of DNA chains to remove gaps left by primer removal. Nicks still separate the short DNA chains. (**f**) Closure of nicks to join the short DNA lengths into continuous nucleotide chains.

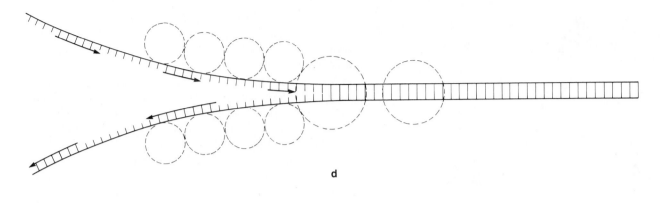

d

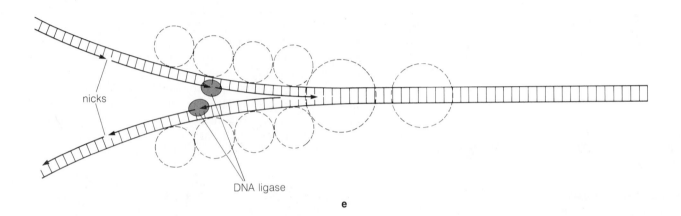

nicks

DNA ligase

e

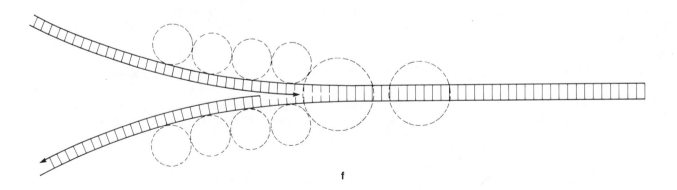

f

would be expected to centrifuge into two bands: one characteristic of hybrid ^{15}N–^{14}N DNA and one equivalent to pure ^{14}N DNA. Since these were in fact the results obtained by Meselson and Stahl, the outcome of the experiment is completely explained by semiconservative replication.

Meselson and Stahl's results with *E. coli* were reported in 1958. At about the same time Taylor, working independently with cells of a higher plant, *Vicia faba* (the broad bean), showed by equivalent experiments that replication in eukaryotes also proceeds by the semiconservative pathway.

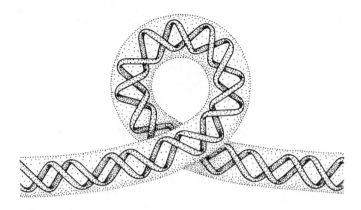

Figure 10-26 A supercoil thrown into DNA by the activity of the unwinding enzyme (see text).

SUPPLEMENT 10-2:
UNWINDING AND PRIMER SYNTHESIS IN DNA REPLICATION

The basic mechanism assembling nucleotides into nucleic acid chains is similar in DNA and RNA synthesis (see Information Boxes 8-2 and 10-1). However, the double helical structure of DNA, and the fact that the DNA polymerase enzymes need a primer to begin synthesis, introduce complexities in DNA replication that require a battery of enzymes with no counterparts in RNA transcription.

Unwinding the Double Helix

Two major enzymes cooperate to unwind the nucleotide chains of a DNA double helix during replication. The first of these, called the *unwinding enzyme,* uses the energy of ATP to open the DNA helix and unwind it into single nucleotide chains. The enzyme works in such a way that one molecule of ATP is broken down to ADP for each turn of the helix unwound (Figure 10-25a). Unwinding is promoted by a group of nonenzymatic *binding proteins* (see Figure 10-25b). These proteins bind strongly to each single nucleotide chain exposed by the unwinding enzymes. By stabilizing the chains in single-stranded form, the binding proteins greatly reduce the energy required to unwind the DNA from its double helical state. Unwinding creates a "fork" in the DNA: two single nucleotide chains trail be-

hind the fork and the intact double helix lies in front of it.

DNA molecules are so long, and the cellular medium containing them so viscous, that neither the single nucleotide chains behind the replication fork nor the intact helix in front of it is free to rotate in response to the unwinding. As a result, *supercoils* are created in advance of the fork at the rate of one supercoil for each turn of the helix unwound. To understand why this happens, take two lengths of string and twist them together into a double helix. Now place a weight on one end of the twisted string to keep it from rotating. Take one of the two strings at the opposite end of the helix in each hand, and create a replication fork by pulling them apart. Notice that pulling the ends apart throws the helix in front of the fork into extra turns or supercoils at the rate of one supercoil for each turn of the helix pulled apart. Eventually, you will reach a point at which the strain imposed by the supercoils prevents the strings from unwinding further. The strain created by unwinding at the fork is even more intense in DNA than in the string example because the DNA double helix cannot wind into a tighter helix (more turns per unit length) as the string can to compensate for the turns pulled apart. Instead, the helix in front of the fork is thrown into loops (see Figure 10-26) at the rate of one loop for each turn unwound.

These loops, and the strain imposed on the helix in front of the fork, are relaxed by *DNA gyrase,* the second major enzyme involved in unwinding DNA (see Figure 10-25a). This enzyme opens a "nick" in one of the two template chains in front of the replication fork. The free end created rotates around the other chain, which acts as a swivel. A single rotation relaxes the supercoil. Subsequently, the same gyrase enzyme reseals the nick by a covalent bond, leaving the template strand intact in advance of the fork. The energy required to drive the nicking-rotation-sealing reaction catalyzed by the gyrase enzyme is apparently derived from free energy released as the supercoils unwind.

Primer Synthesis and DNA Replication at the Fork

The DNA polymerase enzymes can add nucleotides only to the end of an existing nucleotide chain. Therefore, they cannot attach to the free ends generated by the unwinding enzymes to begin synthesis until short lengths of nucleotides are laid in place as primers by other enzymes.

The necessary primers are synthesized by special RNA polymerase molecules called *primases.* These enzymes, in contrast to the DNA polymerases, can begin synthesis of a complementary RNA chain whether other nucleotides are already in place or not. Instead of laying down a single primer at one point on each of the two template chains, the primase enzymes synthesize a series of primers. Each primer in the series is separated from the next by an open space on the template (Figure 10-25b). The short RNA segments laid down by the primases are then used as primers by the DNA polymerases, which fill in the open spaces between the primers with DNA nucleotide chains (Figure 10-25c). This pattern of synthesis produces a series of short DNA lengths opposite the template chain, each with an RNA primer still attached at one end.

The final steps in replication remove the RNA primers, fill in the gaps created by their removal, and seal the new DNA chains into continuous molecules. The primers are removed by a *ribonuclease* enzyme (Figure 10-25d) that removes the RNA nucleotides of the primers one at a time until the DNA part of the new chains is reached. The gaps left by primer removal are filled in by DNA polymerase enzymes. These enzymes add DNA nucleotides in the usual pattern until the remaining bases in the gaps are paired and filled in (Figure 10-25e). Since the DNA polymerase enzymes cannot link the last nucleotide added in each segment to the end of the next segment, a nick is left open in the new DNA chain at each of these points. These remaining nicks are closed by the final enzyme active in replication: *DNA ligase.* This enzyme forms the final phosphate linkages required to seal the short DNA segments into a single, continuous DNA molecule, at the expense of one ATP molecule for each gap closed (Figure 10-25f). Table 10-1 summarizes the roles of the various enzymes active in DNA replication.

Replication proceeds at an overall rate of about 1000 nucleotides per second in prokaryotes and 100 per second in eukaryotes. The entire process is so rapid that the RNA primers and the gaps left by primer removal persist for only a few seconds. As a consequence, the enzymes and factors of replication operate only in the region of the fork. At a distance of only a few micrometers behind the fork, the new DNA chains are already continuous and fully wound around their template chains into complete DNA molecules.

Many scientists contributed bits and pieces to the picture of DNA replication outlined in this supplement, particularly Arthur Kornberg and his associates at Stanford University. Kornberg was the first to isolate and characterize a DNA polymerase enzyme, and later he filled in many of the steps in the replication process. Kornberg received the Nobel Prize in 1959 for his work with DNA polymerase and the mechanism of DNA replication.

Table 10-1 Major Enzymes Active in DNA Replication

Enzyme	Activity
Unwinding enzyme	Unwinds template DNA helix
DNA gyrase	Relaxes supercoils induced by unwinding
Primase	Synthesizes RNA primers
DNA polymerase	Synthesizes new DNA nucleotide chains
Ribonuclease	Removes primer
DNA polymerase	Fills gaps left by primer removal
DNA ligase	Seals nicks left by filling gaps

SUPPLEMENT 10-3: THE ROLE OF ASTERS AND CENTRIOLES IN MITOSIS

Centrioles and Centriole Replication

Centrioles are small, barrel-shaped structures about 200 nanometers in diameter that occur in the cytoplasm just outside the nucleus in eukaryotic cells. They are made up of a system of nine microtubule *triplets* arranged in a circle (Figure 10-27). Each triplet consists of a row of three microtubules fused together. There are no other microtubules in centrioles; no central microtubules equivalent to the central singlets of flagella occur in these structures (compare Figures 10-27 and 4-14).

The pattern of centriole replication was first worked out in detail by Joseph G. Gall, then at the University of Minnesota. Gall noted that at the time of DNA replication, during S, the two "parent" centrioles move apart slightly, and each produces a small, budlike **procentriole** at one end (Figure 10-28). The budding procentrioles, which are separated from the parent centrioles by a narrow space, extend outward at right angles from the parents.

As the procentrioles form, the typical pattern of nine triplets of microtubules becomes visible inside them, although somewhat less clearly than in mature centrioles. The procentrioles, about 70 nanometers long when they

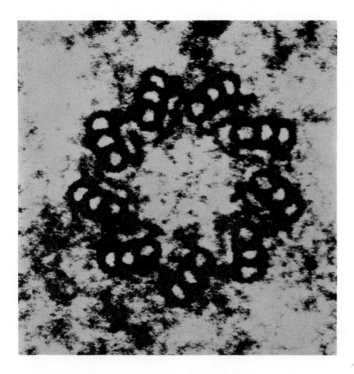

Figure 10-27 A centriole from a mouse cell in cross section. ×270,000. From *The Nucleus*, ed. A. J. Dalton and F. Hagenau, 1968. Courtesy of E. de Harven and Academic Press, Inc.

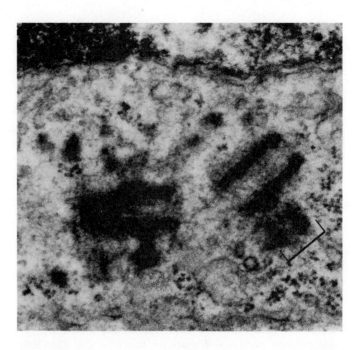

Figure 10-28 Procentriole (bracket) budding from a parent centriole in a cell of the rat. ×61,000. Courtesy of R. G. Murray and The Rockefeller University Press, from *Journal of Cell Biology 26*:601 (1965).

first appear, gradually lengthen through S, G₂, and prophase of mitosis until the mature length of about 200 nanometers is reached. The replicating centrioles, surrounded by the microtubules of the asters, then separate gradually and move to opposite ends of the nucleus. As they do, the developing spindle forms between them, gradually lengthening until the primary spindle is formed.

Centrioles and Asters in Mitosis

To investigators at the turn of the century, these movements of the centrioles and asters seemed to be related to spindle formation. The spindle was later discovered, however, to form and function equally well in plant and other cells without centrioles. Thus it became obvious that the pattern of centriole replication and separation observed in animal cells cannot be directly required for spindle formation.

That this is actually the case was demonstrated by the experiments of Roland Dietz at the Max Planck Institute in Germany. Dietz isolated living cells from a line that normally contains asters and centrioles. By flattening the cells

under a microscope slide during division of the asters, Dietz succeeded in preventing migration of the asters and centrioles around the nucleus. The spindle formed nonetheless, and the subsequent mitosis and cytokinesis produced two lines of cells, one containing twice the normal number of centrioles and asters and one lacking these structures. The cells without centrioles or asters formed a spindle at the next division and divided normally. This result confirmed that centrioles are not necessary for spindle formation, even in animal cells that normally possess them.

These experiments, and the absence of centrioles and asters in many cell types under normal conditions, suggest that the complex division cycle of centrioles and asters probably serves a functional role other than generation of the spindle. Some cell biologists, Dietz among them, have suggested that the asters separate the replicated centrioles and place them at opposite poles of the spindle. In these locations, the centrioles are in a position to be incorporated into separate daughter cells at the following mitosis and cytokinesis. Thus, instead of giving rise to the spindle, the division of the asters and centrioles is simply a mechanism ensuring that both daughter cells receive a

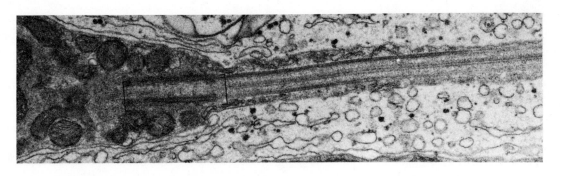

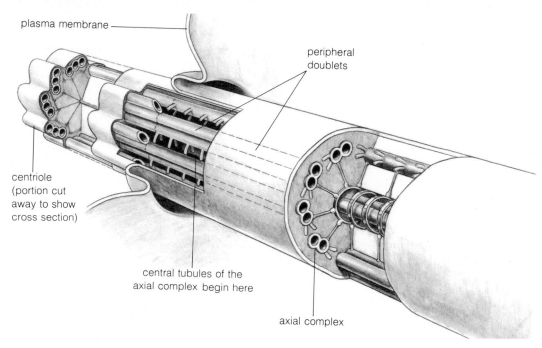

plasma membrane

peripheral
doublets

centriole
(portion cut
away to show
cross section)

central tubules of the
axial complex begin here

axial complex

Figure 10-29 The centriole persisting as a basal body (brackets) of the flagellum in a sperm cell. ×19,000. Courtesy of B. R. Zirkin.

pair of centrioles during cell division. According to this hypothesis, therefore, the spindle microtubules divide centrioles just as they do chromosomes.

If centrioles do not give rise to the spindle, what is their function in eukaryotic cells? The presence of centrioles seems to be correlated with the existence of cells that possess cilia or flagella at some time in the life cycle of the organism. In plants, for example, the species containing centrioles reproduce by means of flagellated, motile gametes. During development of flagella, a centriole gives rise to the system of microtubules that forms the axis of the flagellar shaft. Two of the three microtubules of

each triplet lengthen at the end of the centriole, giving rise to the doublets of the 9 + 2 system of the flagellum. The central singlets of the axial complex grow from the region where the flagellum joins the centriole; there are no connections to any microtubules of the centriole. In most flagella, no structures arise from the third or outermost microtubules of the centriole triplets.

As the 9 + 2 system forms, the centriole usually moves to a position close to the plasma membrane and the growing microtubules are directed toward the cell surface. As the 9 + 2 microtubules push outward, they remain covered by a gradually lengthening extension of the plasma membrane. Eventually the flagellum forms a long, whip-like tail extending from the cell surface. When development is complete, the centriole giving rise to a flagellum remains attached to the microtubules of the 9 + 2 system at the base of the flagellum (Figure 10-29, preceding page). In this position, the centriole is frequently termed the **basal body** of the flagellum.

Because of the importance of flagella to reproduction and other functions in plants and animals, it is not difficult to imagine that the activity of asters evolved as an adaptation ensuring that centrioles, as well as chromosomes, are duplicated and equally distributed to daughter cells during cell division.

The sequence of events in replication and mitosis ensures that the cellular products of division receive identical copies of the genetic information. During the late 1800s it became apparent that another, modified form of cell division must occur at some time in the life cycle of organisms that reproduce sexually. At this time, studies of sexual reproduction revealed that the nuclei of egg and sperm fuse together during fertilization to form a composite nucleus. This observation made it obvious that, at some point before or after fertilization, a special process must reduce by half the number of chromosomes in the egg and sperm nuclei. Otherwise, the fusion of egg and sperm nuclei at fertilization would double the number of chromosomes each generation.

Meiosis halves the chromosome number and mixes genetic information into new combinations.

The anticipated division sequence was soon discovered. Investigators found that in animals the number of chromosomes is reduced by one-half in a series of divisions that occur just before the maturation of eggs and sperm. Within a few years, an equivalent series of reduction divisions was described in plants. The highly specialized process of division was called **meiosis** (from *meioun* = diminish).

In addition to reducing the chromosome number of eggs and sperm, meiosis involves two processes that have great significance for the development and survival of sexually reproducing organisms. One is **recombination,** a mechanism by which segments of chromosomes are exchanged to mix gene sequences into new combinations. This process provides an almost infinite variety of new genetic types to meet the demands of a changing environment. The second additional process of meiosis, of greatest importance in animals, is synthesis of the RNA and protein molecules required for the development of eggs and sperm and at least part of the development of the embryo after fertilization.

MEIOTIC CELL DIVISION

Haploids, Diploids, Genes, and Alleles

Meiosis occurs only in eukaryotic organisms that are **diploids** or have some higher multiple of the diploid number of chromosomes. In diploids, the nuclei of body cells contain two copies of each chromosome (see Information Box 10-3). Thus the chromosomes occur in pairs. In humans, for example, there are 23 different chromosomes; since

11

MEIOSIS: THE CELLULAR BASIS OF SEXUAL REPRODUCTION

these occur in pairs, human body cells contain a total of 46 chromosomes. Meiosis reduces the chromosome number by *separating the chromosome pairs*. The sperm or egg nuclei, as a result, receive one member of each pair present in body cell nuclei. Such nuclei, since they contain only half the number of chromosomes in body cells, are called **haploids** or **monoploids**. Human egg and sperm cells, for example, contain 23 chromosomes, the haploid number for our species.

Meiosis occurs only in diploid eukaryotes or in eukaryotic organisms with some multiple of the diploid number.

Each chromosome in a diploid body cell nucleus consists of a single DNA molecule containing sequences coding for mRNA, rRNA, and tRNA molecules. Each individual coding sequence within a chromosome is a **gene.** The two members of each chromosome pair in a diploid contain the same genes in the same order. For this reason they are called **homologous** chromosomes. Although the two members of a homologous pair contain the same genes in the same order, there may be differences in the sequence of nucleotides for a given gene in either chromosome of a pair. These different nucleotide sequences are called the **alleles** of a gene. If mRNAs coding for a protein are transcribed from two different alleles of a gene (differences in rRNA and tRNA genes rarely occur), the mRNAs will differ in sequence. This difference, in turn, may produce different amino acid sequences in the proteins synthesized under the direction of the transcribed mRNAs.[1] Depending on the position of the changes, the biochemical properties of the two versions of the protein synthesized may differ only slightly or they may be drastically different. The effect of recombination is to mix the alleles of the two chromosomes of a homologous pair into new combinations. This is the primary source of the genetic variability produced by meiosis.

Meiosis: How it Works

The overall mechanism of meiosis is simple in broad outline. In mitosis, a single replication is followed by a single division of the chromosomes. In meiosis, a single replication is followed by *two* sequential divisions of the chromosomes. As a result, in meiosis the chromosome number

---

[1]Certain differences in mRNA sequence do not change the amino acid sequence of the protein coded for. This happens if the sequences spell out different triplet code words for the same amino acid.

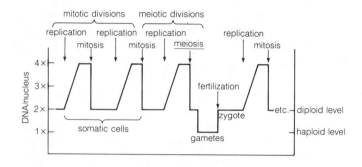

Figure 11-1 The relative amounts of DNA per nucleus during mitotic and meiotic cell cycles. The base level, $1X$, represents the amount of DNA in a sperm or egg nucleus.

is halved in the products of the division. This reduction in chromosome number is paralleled by changes in the quantity of DNA per nucleus (Figure 11-1). If the DNA content of the eggs and sperm of a sexually reproducing organism is considered as the base or $1X$ level, then diploid body cells of the organism at the G_1 stage before replication contain the $2X$ amount of DNA (see Figure 10-1). After replication, the resultant G_2 cells contain the $4X$ amount of DNA. If replication is followed by mitosis, the daughter nuclei contain the $2X$ amount at the completion of division. If the division sequence is meiotic, the two sequential divisions of meiosis reduce the DNA quantity to the $1X$ or haploid value. This quantity is restored to the $2X$ diploid value after fusion of the haploid egg and sperm nuclei during fertilization.

The two chromosomes of each homologous pair have the same genes in the same order, but different forms (alleles) of these genes may be present in either member of the pair.

In meiosis, DNA replication is followed by two divisions of the chromosomes.

In the pattern most familiar to us, that of animals, meiosis directly produces eggs and sperm. In animals, the cells undergoing meiosis are located in the testes in males and in the ovaries in females. Union of the reproductive cells, or **gametes,** in fertilization restores the chromosome number to the diploid level and initiates development of the new individual. In the flowering plants, meiosis producing the female gamete takes place in the ovaries of the flower; male gametes are produced in the anthers.

Meiosis follows the same basic pattern in all diploid organisms. (For an overview of the entire process, see Information Box 11-1.) After a premeiotic interphase, during

Information Box 11-1 An Overview of Meiosis

Meiosis occurs only in organisms that have at least the diploid number of chromosomes. Let a single line represent all the DNA coding for one copy of the genetic information of a cell:

Then break the line into shorter subunits representing the chromosomes:

This set of lines represents the chromosome complement of a haploid cell. Diploid cells have *two* copies of the genetic information:

Each vertical pair of lines represents a homologous pair of chromosomes. The two chromosomes of a homologous pair have the same genes in the same sequence. Different forms of a gene, however, called alleles, may be present in either member of a pair. During the interphase before meiosis, the chromosomes duplicate:

Each chromosome now contains two chromatids.
   In meiosis, the chromosomes condense into shorter, thicker units:

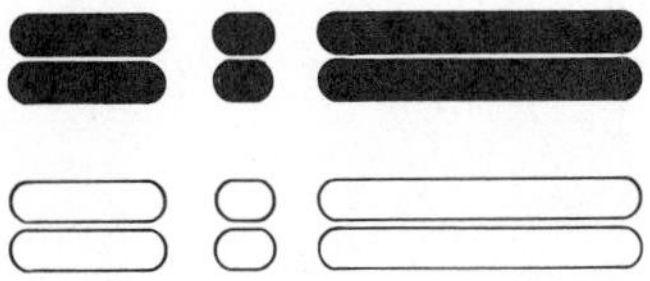

Homologous pairs then come together and pair:

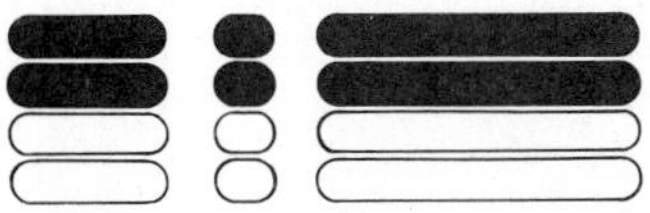

While they are paired, some of the homologous chromosomes may exchange segments through a process of breakage and reunion of reciprocal parts:

(continued on next page)

This exchange creates new combinations of alleles in the chromosomes undergoing the exchange. The paired chromosomes then line up on the spindle for the first of the two meiotic divisions. In this division, the two *chromosomes* of each homologous pair make microtubule connections leading to opposite ends of the cell:

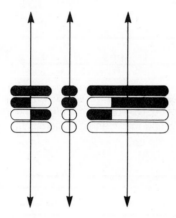

Note that both *chromatids* of a chromosome connect to microtubules leading to the same end of the cell. This division separates the chromosomes of homologous pairs:

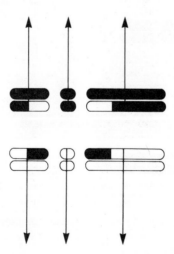

The two chromatids of each chromosome, however, are still together.

The chromosomes then line up on spindles for the second meiotic division. There is no DNA replication preceding this division. In the division, the two chromatids of each chromosome now make connections to microtubules leading to opposite ends of the cell:

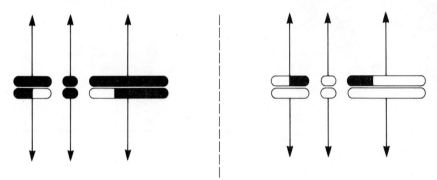

This division separates the chromatids of each chromosome:

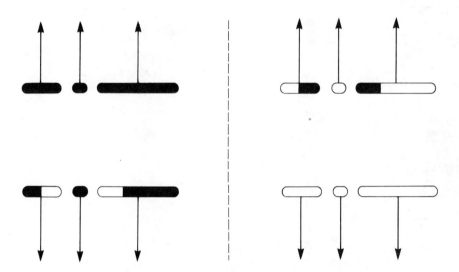

These chromosomes are then enclosed in separate nuclei or cells as the four haploid products of meiosis:

nucleus 1

nucleus 2

nucleus 3

nucleus 4

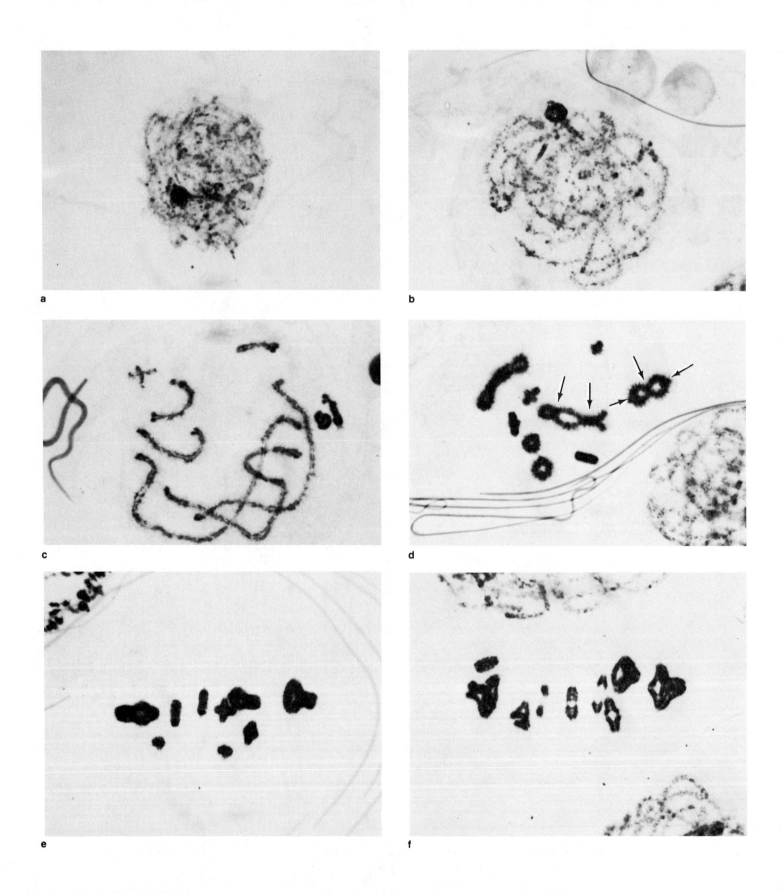

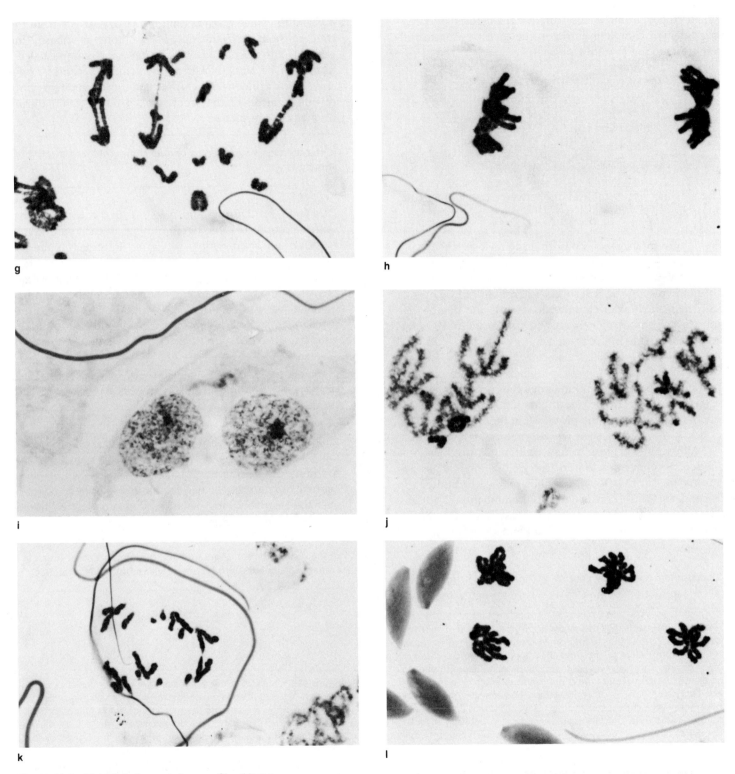

Figure 11-2 Meiosis in the grasshopper *Chorthippus*.
(**a**) Condensation stage. (**b**) Pairing stage. (**c**) Recombination. The narrow separation between the paired homologs is visible at many points. (**d**) Transcription. In the grasshopper, uncoiling during the transcription stage is not pronounced. The chromosomes are held to-gether by crossovers at several points (arrows). (**e**) Recondensation. (**f**) Metaphase I. (**g**) Anaphase I. (**h**) Telophase I. (**i**) The stage be-tween divisions resembling interphase. (**j**) Prophase II. (**k**) Anaphase II. (**l**) Telophase II. ×1000. Courtesy of J. F. Walters.

which the chromosomes duplicate, meiosis proceeds through two complete chromosome divisions, each basically resembling a mitotic division:

Division I	Division II
Prophase I	Prophase II
Metaphase I	Metaphase II
Anaphase I	Anaphase II
Telophase I	Telophase II

There is no DNA replication between the two meiotic divisions.

Premeiotic Interphase Cells entering meiosis are usually the immediate products of a series of mitotic divisions. These cells enter the interphase before meiosis, which is not markedly different from an interphase preceding mitotic division. Premeiotic interphase ends and meiosis begins as the chromosomes start to condense and become visible as threads in the nucleus.

Meiotic Prophase I Meiotic prophase I is more complex than mitotic prophase and includes the chromosome rearrangements that cause recombination. This stage lasts much longer than mitotic prophase in the same organism; in fact, it may extend over weeks, months, or even years.

Although it is more or less continuous, prophase I is divided for convenience into five well-defined stages according to the functional activities of the chromosomes: (1) **condensation**, (2) **pairing**, (3) **recombination**, (4) **transcription**, and (5) **recondensation**. These stages are shown in photographs in Figure 11-2 and in diagrams in Figure 11-3.

Condensation Stage The condensation stage (traditionally called **leptotene**)[2] begins as the chromosomes first condense into visible threads (Figures 11-2a and 11-3a). In the condensation mechanism, as in mitotic prophase, the chromosome fibers fold into shorter, more compact structures that become thicker as the folding becomes extensive. Each condensing chromosome contains two chromatids, which are the result of DNA replication during premeiotic interphase.

In the condensation stage the chromosomes fold into short, thick structures.

Pairing Stage During interphase and the condensation stage of prophase I, the two homologs of each pair occupy random and often widely separated locations in the nucleus. The condensation stage changes to the pairing stage (traditionally termed **zygotene**) as the two chromosomes of each homologous pair come together and begin to line up side by side (Figures 11-2b and 11-3b). The pairing mechanism, called **synapsis**, proceeds until the homologous pairs are lined up in exact side-by-side register.

In the pairing stage the two members of each homologous pair come together and line up side by side.

As with many mechanisms of mitosis and meiosis, the molecular forces that bring the homologous chromosomes together in synapsis are unknown. The questions surrounding the pairing mechanism are especially interesting because no intermolecular forces are known to operate over the relatively long distances that separate the homologs as synapsis begins.

In the recombination stage homologous chromosomes break and exchange segments, producing new combinations of alleles.

---

[2]The traditional names for the stages of prophase I are derived from the structural form of the chromosomes at each stage. *Leptotene* (*leptos* = fine or thin) refers to the threadlike appearance of the chromosomes during this stage. *Zygotene* (*zygon* = yoke-shaped) refers to the Y-shaped appearance of the chromosomes during the pairing process. *Pachytene* (*pachus* = thick) describes the shorter, more compact appearance of the chromosomes at this stage. *Diplotene* (*diplos* = double) is derived from the fact that the separation between the two chromatids of each chromosome cannot be seen distinctly until this stage. The final traditional name, *diakinesis* (*dia* = across, *kinesis* = movement), refers to a change in position of the crossovers between chromatids, which are pushed toward the tips of the chromosomes by the tight condensation occurring at this stage. Frequently, the chromosomes become so compact and rounded at the recondensation or diakinesis stage that they are held together only at their tips.

Figure 11-3 Meiosis in diagrammatic form. (**a**) Condensation stage. Although both chromatids are shown for diagrammatic purposes in the magnified circles of parts *(a)*, *(b)*, and *(c)* of this figure, the split between the two chromatids of a chromosome is not actually visible during these stages. (**b**) Pairing stage. Pairing is in progress at several points (arrows). (**c**) Recombination. (**d**) Transcription. The chromosomes are held in pairs only by crossover points (arrows) remaining from the recombination stage. In most organisms, all four chromatids usually become visible in the tetrads at this time. (**e**) Recondensation. (**f**) Metaphase I. (**g**) Anaphase I. (**h**) Telophase I. (**i**) Prophase II. (**j**) Metaphase II. (**k**) Anaphase II. (**l**) Telophase II.

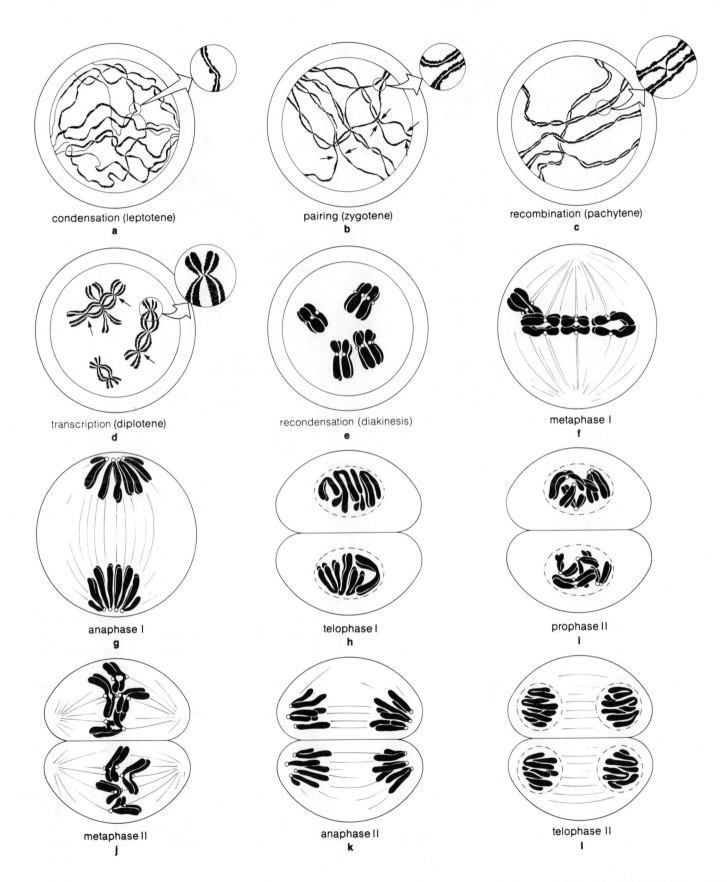

condensation (leptotene)
a

pairing (zygotene)
b

recombination (pachytene)
c

transcription (diplotene)
d

recondensation (diakinesis)
e

metaphase I
f

anaphase I
g

telophase I
h

prophase II
i

metaphase II
j

anaphase II
k

telophase II
l

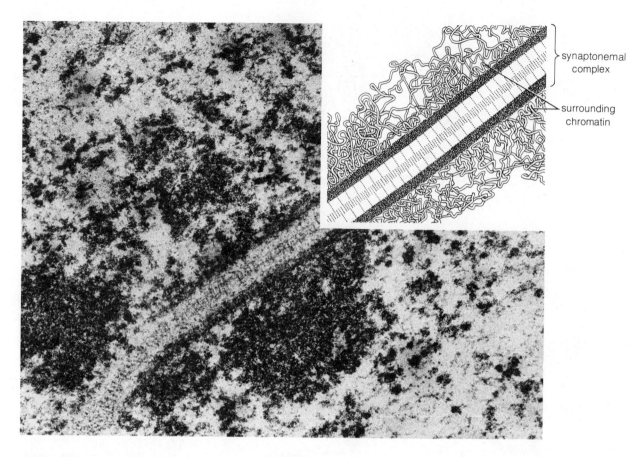

synaptonemal
complex

surrounding
chromatin

Figure 11-4 The paired chromosomes of a grasshopper. The narrow space between the chromosomes contains the synaptonemal complex. The inset shows the relationship of the synaptonemal complex to the chromatin of the paired chromosomes. Micrograph courtesy of P. B. Moens and The Rockefeller University Press, from *Journal of Cell Biology* 40:542 (1969).

Recombination Stage   The recombination stage (or **pachytene**) begins as the pairing of homologous chromosomes becomes complete (Figures 11-2c and 11-3c). Each paired chromosome, as noted, consists of two chromatids, the products of replication during premeiotic interphase. Thus a total of four closely associated chromatids are present in each paired structure during the recombination stage. Two different terms are commonly used for the paired chromosomes. When considered at the level of chromatids, the paired homologs are referred to as **tetrads** (*tetra* = four) because four chromatids are present in each paired structure. When considered at the level of chromosomes, the same structure is called a **bivalent** because two homologous chromosomes are synapsed in each pair.

Although the synapsis of homologs is very close at this stage, the two chromosomes of each pair remain separated by a regular space about 0.15 to 0.2 micrometer wide. This space, which is just visible in the light microscope, proves under the electron microscope to contain a highly specialized structure called the **synaptonemal complex** (Figure 11-4). Genetic recombination has been shown to take place between the four closely associated chromatids at this stage by a process of physical breakage and exchange of segments between the chromatids of homologous chromosomes (see Figure 11-9). The synaptonemal complex is thought to be directly involved in these exchanges, possibly by holding the chromatids in the closely paired configuration and also by containing and aligning the enzymes required for the breakage and exchange mechanism.

In contrast to the condensation and pairing stages, which usually run their course within a matter of hours, the recombination stage may take weeks, months, or even years. The length of this stage probably reflects the com-

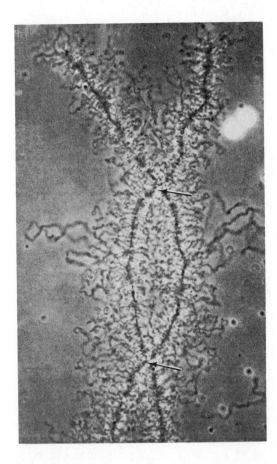

Figure 11-5 A lampbrush chromosome from a salamander nucleus. Loops active in RNA transcription extend outward from all parts of the tetrad. The tetrad is held together by crossovers at the arrows. ×470. Courtesy of J. G. Gall.

plex events associated with the breakage and exchange of chromatid segments during genetic recombination.

Transcription Stage As the recombination stage comes to a close, the homologs separate at many points (Figures 11-2*d* and 11-3*d*). This separation signals the beginning of the transcription stage (or **diplotene**). As the homologs separate, the synaptonemal complex disappears from between the chromosomes.

The homologous chromosomes become so widely separated at this stage that they almost seem to repel each other except at scattered attachment points (arrows, Figure 11-3*d*). On close inspection, these attachment points prove to be regions in which two of the four chromatids cross over between homologous chromosomes (see also Figure 11-5). These crossing places, called **crossovers** or **chiasmata** (singular *chiasma* = crosspiece), are remnants of the

chromatid exchange that took place during the recombination stage.

During the transcription stage, the chromosomes unfold to a greater or lesser extent and become active in RNA synthesis. In organisms in which RNA transcription is relatively limited at this time, the unfolding produces only a slight fuzziness in the outline of the chromosomes. In other species, such as many insects, the chromosomes unfold almost completely and revert to a state apparently like interphase. In some organisms, particularly in female amphibians, birds, and reptiles, loops extend outward from all parts of the chromatids during this stage (Figure 11-5). The tetrads are called *lampbrush chromosomes* when in this configuration. The loops in these chromosomes are sites of intensive RNA synthesis throughout the transcription stage.

In the transcription stage the chromosomes unfold and synthesize RNA.

Both ribosomal and messenger RNA are synthesized in quantity during the transcription stage. Structurally the ribosomal RNA synthesis is marked by extensive growth of the nucleolus or production of extra nucleoli numbering in the hundreds or thousands. The entire meiotic cell also grows at this time through the synthesis of large quantities of proteins, fats, carbohydrates, and other molecules in the cytoplasm. This period of growth is most pronounced in the developing eggs (oocytes) of amphibians, birds, and reptiles, which may grow from microscopic size to diameters of several centimeters during the transcription stage. Most of this growth is in the cytoplasm. The nucleus in these cells remains microscopic, or it may become just large enough to be visible to the naked eye.

The transcription stage may last for a long time. In developing amphibian egg cells, for example, this part of meiotic prophase may last nearly a year. In humans, oocytes reach this stage in unborn females at about the fifth month of fetal life and remain arrested at this point during the remainder of prenatal life, through birth and childhood, until the female reaches sexual maturity. Then, just before ovulation, one oocyte each month becomes active again and continues the meiotic sequence. The time between the onset and completion of the transcription stage in human females may thus range from 12 or so to more than 50 years. Intensive RNA and protein synthesis, however, occur only during the part of the transcription stage preceding meiotic arrest in the developing fetus.

Recondensation Stage During the recondensation stage (or **diakinesis**) the chromosomes condense again into short, tightly packed structures (Figures 11-2*e* and 11-3*e*). If little unfolding has occurred during the transcription stage, the transition to the recondensation stage may be difficult to identify. By the end of the recondensation stage the chromosomes are packed so tightly that they may be almost spherical.

Completion of the recondensation stage marks the close of prophase I of meiosis. The various events of prophase I produce three results of great significance for the outcome of meiosis: (1) recombination, (2) synthesis of most or all of the RNA, protein, lipid, and carbohydrate molecules required for the growth of gametes and the early stages of embryonic development, and (3) condensation of the chromosomes into short rodlets. The remaining stages of meiosis are concerned primarily with division of the chromosomes to the haploid number.

In the recondensation stage the chromosomes fold into short rodlets again.

The Meiotic Divisions Near the end of prophase I the spindle for the first meiotic division forms by a pattern similar to mitosis. The centrioles, if present, have replicated during the previous interphase. The two pairs of centrioles migrate with the asters to opposite sides of the nucleus and take up positions at the spindle poles just as in mitosis (see p. 197). This pattern is most typical of the meiotic divisions in male animals leading to sperm formation. In many animal eggs, the centrioles disappear and cannot be detected in developing eggs at any stage of meiosis. In these eggs, the spindle forms without the involvement of centrioles and asters, essentially as it does in the higher plants. In either case, the spindle, built up from hundreds to thousands of microtubules stretching from end to end of the cell, soon fills most of the cytoplasm around the nucleus.

Metaphase I of Meiosis Just as in mitosis, the breakdown of the nuclear envelope provides a convenient reference point for the transition from prophase I to metaphase I of meiosis. The spindle moves to the position formerly occupied by the nucleus, and the tetrads, scattered by the breakdown of the nuclear envelope, make their way to the spindle midpoint (Figures 11-2*f* and 11-3*f*). Except for the pairing of homologous chromosomes, meiosis at this point closely matches the transition from prophase to metaphase in mitosis.

The first major divergence from the pattern observed in mitosis, and in fact the distinctive event of the meiotic divisions following prophase I, occurs as the paired chro-

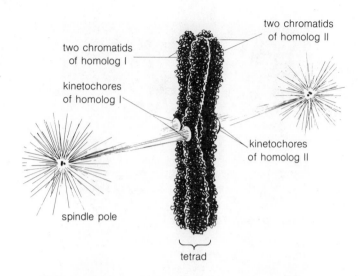

Figure 11-6 Kinetochore connections at metaphase I of meiosis. The two kinetochores of a homologous chromosome make microtubule connections leading to the same pole of the spindle (compare with Figure 10-12).

mosomes attach to the spindle microtubules. Each homologous chromosome of a pair has two kinetochores, one for each of its two chromatids (see Figure 11-6). Although two kinetochores are present in each homologous chromosome, both connect to the *same* pole of the spindle at metaphase I of meiosis (Figure 11-6; compare with Figure 10-12). The two kinetochores of the other chromosome of a homologous pair connect to the opposite pole. Thus the two kinetochores of each homologous chromosome act as a functional unit during metaphase I of meiosis.

Anaphase I Separation of the homologous pairs by the spindle initiates anaphase I (Figures 11-2*g* and 11-3*g*). Because of the opposite connections made by the kinetochores, the anaphase movement separates the two chromosomes of each homologous pair and delivers them to opposite spindle poles. At the completion of anaphase I, as a consequence, the poles receive the haploid number of chromosomes. Each chromosome is still double, however, and contains two chromatids (from the replication during the previous interphase). As a result, the two groups of chromosomes at the spindle poles still contain twice the haploid amount of DNA.

Anaphase I separates the homologous pairs and halves the chromosome number.

The pattern of separation of the homologs during anaphase I of meiosis is as important for variability as recombination during prophase I. This source of variability depends on the maternal and paternal origins of the two members of each homologous pair. The haploid egg and sperm nuclei giving rise to an individual each contain only one member of each homologous pair. Fusion of the egg and sperm nuclei in fertilization brings the homologous pairs together, creating the diploid condition. In each chromosome pair of the diploid, the chromosome originating from the sperm nucleus is called the **paternal** chromosome; the chromosome originating from the egg is called the **maternal** chromosome. As an individual grows by cell division from the fertilized egg, replication and mitosis ensure that the same pairs are equally distributed. Thus all the homologous pairs in the organism contain one maternal and one paternal chromosome. Although the maternal and paternal chromosomes of a homologous pair contain the same genes in the same order, different alleles of these genes may be present in either member of the pair.

The spindle poles may receive any combination of maternal and paternal chromosomes as a result of the first meiotic division.

Since the different homologous pairs separate independently from each other at anaphase I of meiosis, any combination of chromosomes of maternal and paternal origin may be delivered to a spindle pole. The random combinations of maternal and paternal chromosomes delivered to the poles contribute to the total genetic variability of the products of meiosis. In humans, for example, with 23 pairs of chromosomes, 2^{23} possible combinations of maternal and paternal chromosomes can be delivered to the spindle poles at anaphase I. Even without recombination, therefore, the probability that two children of the same parents will receive the same combination of maternal and paternal chromosomes would be one chance out of $(2^{23})^2$— or 1 out of 70,000,000,000,000! This variability arises from random segregation alone. The further variability introduced by recombination, which mixes chromatid segments randomly between maternal and paternal chromatids, makes it practically impossible for an individual to produce genetically identical offspring (except for identical twins, which arise not from random combinations but by division of the fertilized egg—see p. 296).

Telophase I A well-defined telophase I (Figures 11-2h and 11-3h) does not always occur between anaphase I and the second meiotic division. All possible gradations are found in nature—from organisms with no detectable unfolding of the chromosomes at telophase I, as in some insects, to species with almost complete reversion to a state much like interphase. In the latter organisms, such as corn and certain other plants, the chromosomes partly unfold and a nuclear envelope temporarily surrounds the polar masses of chromatin. In most species, however, telophase I is a transitory stage in which meiotic cells pause only briefly before entering prophase II of meiosis. No DNA replication occurs at this time in any known organism.

During telophase I the single metaphase spindle reorganizes into two spindles that form in the regions of the telophase I spindle poles. If centrioles and asters are present, they too divide at this time, placing a single centriole at each pole of the two spindles. These events complete the cellular rearrangements for the second meiotic division.

The Second Meiotic Division The second meiotic division follows essentially the same pattern as ordinary mitotic division. After a brief or even nonexistent interphase II and prophase II (Figures 11-2i and j and 11-3i), the chromosomes left at the two poles by the first meiotic division move to the midpoints of the two newly formed spindles. If unfolding has occurred, the chromosomes condense back into tight rodlets as movement to the metaphase II spindles takes place. The kinetochores of the chromosomes attach to the spindle microtubules at metaphase II of meiosis (Figure 11-3j) exactly as they would at mitotic metaphase. Each chromosome at this stage contains two chromatids; the kinetochores of these chromatids make attachments to microtubules leading to *opposite* poles of the spindle as in mitosis.

At anaphase II (Figures 11-2k and 11-3k) the two chromatids of each chromosome, because of their connections to spindle microtubules, separate and move to opposite poles of the spindle. This separation and movement delivers the haploid number of chromatids to each pole of the spindle. As a result, each pole now contains the haploid quantity of DNA. This quantity is one-fourth of the DNA present at G_2 in the original cell entering the two sequential meiotic divisions.

At telophase II (Figures 11-2l and 11-3l), the chromatids unfold and nuclear envelopes form around the four division products. When unfolding is complete and the nuclear envelopes have completely formed, the chromatids in these nuclei are considered to be chromosomes. The four nuclei formed at this point have widely divergent fates in various species of plants and animals. In the males of animal species, all four nuclei are enclosed in separate cells by cytoplasmic divisions and differentiate into functional male gametes, the sperm. In females, only one of the nuclei resulting from meiosis becomes functional as

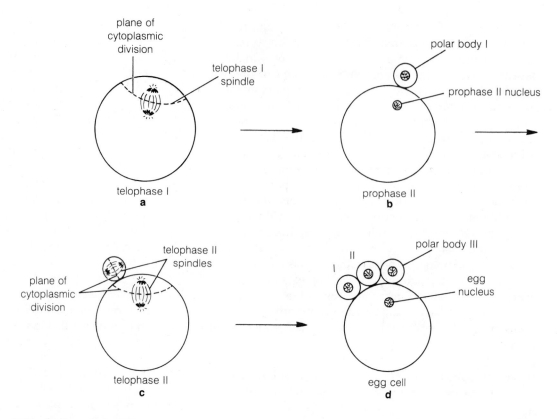

Figure 11-7 Unequal division of the cytoplasm during the two meiotic divisions in female animals. The unequal division produces one large egg cell and three small, nonfunctional cells called polar bodies. In some species the first polar body does not divide during the second meiotic division. This variation in the sequence produces an egg cell with two polar bodies instead of three.

the egg nucleus. The other three are compartmented by unequal division of the egg cytoplasm into small, nonfunctional cells at one side of the oocyte called **polar bodies** (Figure 11-7). This unequal division concentrates most of the cytoplasm into the single large egg cell resulting from meiosis. Higher plants develop similarly: in males all four cells give rise to sperm nuclei; in females only one of the meiotic products gives rise to an egg nucleus. (Gamete development in higher plants and animals is discussed further in Chapters 14 and 15.)

The Overall Effects of Meiosis: A Review

For simplicity we will consider meiosis in a hypothetical cell containing only two pairs of chromosomes—one long pair and one short pair, as in the review of mitosis presented in Chapter 10. At premeiotic G_1, these uncondensed pairs are unassociated and randomly distributed in the nucleus (Figure 11-8a). During premeiotic S, each chromosome replicates and becomes double at all points. As a result, each contains two chromatids (Figure 11-8b). After G_2, as the condensation stage begins, the chromosomes fold into thicker threads that become visible in the light microscope (Figure 11-8c). Synapsis at the pairing stage (Figure 11-8d) brings homologous chromosomes together, producing two tetrads in the nucleus. Each tetrad contains two chromosomes and four chromatids. During the recombination stage, segments of the homologous chromosomes exchange (Figure 11-8e), leading to new combinations of alleles. At the transcription stage, at least some unfolding and RNA synthesis take place. The chromosomes condense again at the final, recondensation stage of prophase I and enter into metaphase I (Figure 11-8f).

Figure 11-8 A review of meiosis (see text).

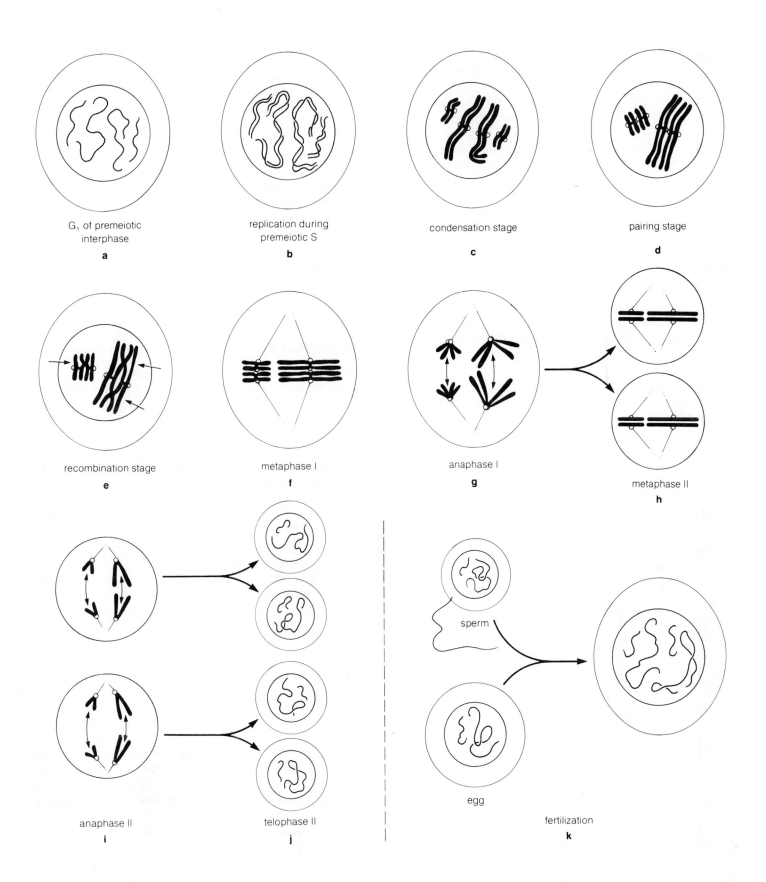

G₁ of premeiotic
interphase

a

replication during
premeiotic S

b

condensation stage

c

pairing stage

d

recombination stage

e

metaphase I

f

anaphase I

g

metaphase II

h

anaphase II

i

telophase II

j

sperm

egg

fertilization

k

At metaphase I, the two homologous chromosomes of each pair connect to microtubules leading to opposite poles of the spindle. Anaphase I (Figure 11-8g) therefore separates the homologs and moves the haploid number of chromosomes to each spindle pole. Thus only one long chromosome and one short chromosome are present at each pole, and pairs no longer exist. Note that at this point the chromosomes at the poles still contain two chromatids.

During telophase I, any intervening period, and prophase II two new spindles form at the metaphase I division poles. At metaphase II (Figure 11-8h), the chromosomes move to the spindle midpoints and make microtubule attachments as in mitosis. They attach in such a way that the two chromatids of each chromosome connect to microtubules leading to opposite spindle poles. Separation of the chromatids at anaphase II (Figure 11-8i) therefore delivers the haploid number of chromatids to the poles. Each of the four division products at telophase II (Figure 11-8j) now contains the haploid quantity of DNA and possesses only two chromatids (now technically called chromosomes), one long and one short.

If meiosis and gamete formation produce eggs and sperm in this hypothetical organism, the end products will each contain two chromosomes, one long and one short. Fusion of an egg and sperm nucleus in fertilization (Figure 11-8k) rejoins the pairs and returns the chromosome complement to the G_1 level of premitotic cells.

Thus the cycle is completed and sexual reproduction, through meiosis and fertilization, is accomplished without changing the basic chromosome number. Meiosis supplies three important parts to the overall scheme: (1) it reduces the chromosomes to the haploid number in gametes, (2) it produces genetic variability among the gametes produced by the same individual, and (3) it supplies RNA molecules, synthesized during the transcription stage, needed for fertilization and the early stages of development.

The meiotic divisions also reduce the chromosome number in species such as our own, in which one pair of chromosomes is different in males and females. Supplement 11-1 describes these **sex chromosomes,** as they are called, and shows how they are distributed to gamete cells in meiosis.

The factors controlling the progress of cells through meiosis have proved as elusive as the controls of the mitotic cycle (see Chapter 10). In fungi and protistans, essentially any cell can be induced to enter a meiotic rather than mitotic cell division by simple changes in the environment, usually in the direction of less favorable growth conditions. In higher plants and animals, the induction of meiosis probably involves an interaction between hormones and the cell surface. Various internal responses

then trigger meiosis. The observations made to date suggest, as in mitotic cycles, that the ultimate controls rest on a series of genes activated in sequence to trigger meiosis.

THE MECHANISM OF RECOMBINATION

The genetic variability arising from meiosis results from two processes: (1) recombination during meiotic prophase I and (2) the independent separation of maternal and paternal chromosomes during anaphase I. We have already considered how the kinetochore connections made by the homologs during metaphase I result in new combinations of maternal and paternal chromosomes. Recombination generates variability through a different mechanism: by exchanging segments between homologous chromosomes.

The Genetic Consequences of Recombination

The exchanges of chromosome segments in recombination take the form shown in Figure 11-9. Consider a homologous pair in which both chromosomes contain the genes A and B. In this example, one of the two chromosomes carries the A form of gene A and the other chromosome has the slightly different a allele of the same gene. The B gene is similarly present as the two different alleles B and b. At the beginning of meiosis the chromosomes carry the alleles in the combinations shown in Figure 11-9a. One contains the A and B alleles and the other the a and b alleles of the two genes.

At G_1 of premeiotic interphase the two chromosomes of the pair are unreplicated, as shown in Figure 11-9a. After replication (Figure 11-9b), both chromatids of the A–B chromosome will have the A and B alleles at these sites. Both chromatids of the opposite member of the homologous pair will have the alleles a and b at the corresponding locations. At this stage, these chromosomes are identical in sequence to chromosomes of the same homologous pair in any G_2 cell of the same organism. During prophase I of meiosis, the homologs pair closely (Figure 11-9c). If recombination occurs between these genes, two of the four paired chromatids break in equivalent locations (Figure 11-9d). The broken ends then cross over and rejoin with the respective broken ends in the opposite chromatid. As a result, the two chromatids exchange precisely matched segments (Figure 11-9e). The end products after recombination consist of two chromatids of unchanged sequence and two that have been changed. The unchanged chromatids, called **parentals,** are identical to the chromosomes of the original parent cell; that is, they contain al-

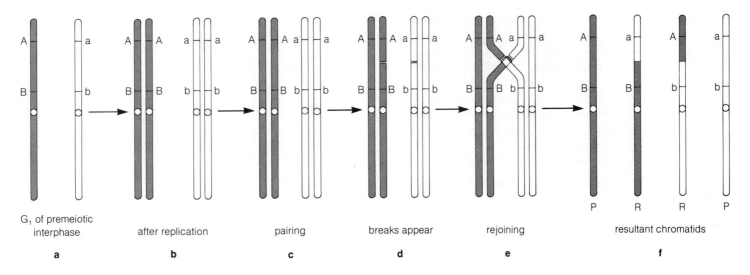

G₁ of premeiotic interphase	after replication	pairing	breaks appear	rejoining	resultant chromatids
a	**b**	**c**	**d**	**e**	**f**

Figure 11-9 The pattern of the exchanges between chromatids during recombination (see text). P = parental; R = recombinant.

leles *A* and *B* together or *a* and *b* together. Two are **recombinants:** they have the new combinations *a–B* or *A–b*. During the subsequent meiotic divisions, the four chromatids separate and are enclosed in four different nuclei. Thus each of the four nuclei produced contains one of the four chromosomes shown in Figure 11-9*f*.

Each recombination event involves only two of the four chromatids in a tetrad.

As this example shows, a single recombination event involves only two of the four chromatids and produces two parental chromosomes and two recombinants. The location of a recombination event is random; recombinations may take place at essentially any position along the chromosomes and between any two of the four chromatids of a tetrad. Recombinations usually occur at a frequency of about one per chromosome "arm" (the segment of a chromosome between the kinetochore and the tip).

How Recombination Takes Place

Physical breakage and exchange between homologous chromosomes was suspected to be the mechanism underlying recombination as long ago as the first decade of this century. However, the experimental demonstration that recombination actually takes place this way was not ac-

complished until the mid-1960s. The critical experiment was carried out at Columbia University by J. H. Taylor, the scientist who first established that replication is semiconservative in eukaryotes (see Supplement 10-1).

Recombination occurs by physical breakage and exchange between homologous chromosomes.

Taylor developed a way to label one of the two chromatids of each homologous chromosome with radioactive DNA. (His methods are described in Supplement 11-2.) As a result, two of the four chromatids in each tetrad entering meiotic prophase were labeled and two were not (Figure 11-10*a*). At the close of meiosis, Taylor was able to recover some chromatids that contained both labeled and unlabeled segments (Figure 11-10*b*; see also Figure 11-15). Chromatids of this type could be produced in meiosis only by breakage and exchange of parts between labeled and unlabeled chromatids of the paired homologs. Thus Taylor's work established that recombination occurs by physical breakage and exchange.

The molecular mechanisms underlying breakage and exchange in recombination, which are still not completely understood, are presently the subject of intensive investigation. From results obtained by genetic crosses, it is clear that the breaks and exchanges are normally so precise that neither of the recombined chromatids receives any extra DNA or is missing any. As one molecular biologist put it, this is equivalent to breaking two lengthy books at the same letter of the same word, exchanging them, and splicing the pieces back together so perfectly that not one letter is missing from either copy! The enzymes active in recombination are probably organized on the synaptonemal complex, the framework visible between the paired homologs during the recombination stage (the use of a mechanism resembling recombination to produce genetic clones is described in Supplement 11-3).

The enzymes active in recombination are probably organized in the synaptonemal complex.

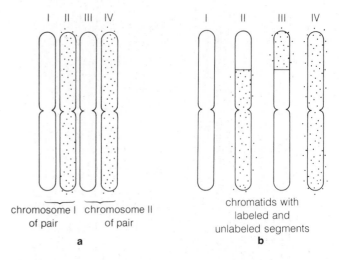

chromosome I   chromosome II
of pair   of pair

a

chromatids with labeled and unlabeled segments

b

Figure 11-10   The Taylor experiment demonstrating that recombination between homologous chromosomes takes place by physical breakage and exchange between chromatids (see text). The dotted regions in the diagram represent radioactive segments of the chromatids.

THE TIME AND PLACE OF MEIOSIS IN DIFFERENT ORGANISMS

There are three major variations in the time and place of meiosis in the life cycle of eukaryotic organisms (Figure 11-11). The most familiar pattern occurs in animals, many protistans, and a few lower plants. In this case, meiosis occurs immediately before the gametes form (Figure 11-11a). The four haploid products of meiosis become sperm in males; in females, usually only one of the four nuclei becomes the egg nucleus.

Many plants, including all the higher plants (the angiosperms and gymnosperms), undergo meiosis at an intermediate stage in the life cycle (Figure 11-11b). Organisms possessing this meiotic pattern alternate in each generation between haploid and diploid individuals (see Figure 15-1). Fertilization produces a diploid generation, in which the individuals are called **sporophytes.** After growing to maturity by mitosis, the sporophytes produce by meiosis asexual reproductive cells called **spores.** These are diversified in genetic makeup as a result of recombination and the meiotic divisions. The spores grow directly by mitosis into haploid individuals called **gametophytes.** At some point, cells in the gametophytes differentiate into eggs and sperm following ordinary mitotic divisions. All the eggs or sperm produced from an individual gametophyte, since they arise through mitosis, are genetically identical. Fusion of the haploid gametes returns the cycle to the diploid, sporophyte generation.

A third major variation in the time and place of meiosis in life cycles is observed in fungi, some algae, and a few protistans (Figure 11-11c). This variation has proved to be of great value to researchers in cell biology. In these organisms, meiosis takes place immediately after fertilization. The two haploid gamete nuclei designated as egg and

sperm fuse to produce a diploid nucleus. The diploid nucleus immediately enters meiosis, producing four haploid nuclei. The nuclei eventually develop into haploid spores, which grow by mitotic divisions into the haploid generation. The haploid generation thus makes up the dominant phase of the life cycle, and the diploid generation is reduced to a single nucleus. This nucleus remains diploid only during meiotic prophase I and metaphase I.

In a number of the organisms with the third variation, DNA replication takes place in the haploid gamete nuclei before the gametes fuse to produce the diploid nucleus. In these organisms, therefore, replication is clearly separate in time and place from genetic recombination, which can occur only in the brief diploid phase of the life cycle during prophase I. This fact has been most useful in establishing that recombination occurs in meiotic prophase instead of during replication, as many biologists had previously assumed. The organisms of this group include various yeasts and the fungus *Neurospora*, which are much used in biochemical and genetic studies.

Whenever it occurs in the life cycle of sexually reproducing organisms, meiosis has the three primary results outlined in this chapter. (1) It reduces the chromosome number to the haploid level, so that the chromosome number in the species does not double at each fertilization. (2) Through the genetic rearrangements occurring in recombination and the independent separation of maternal and paternal chromatids at anaphase II, meiosis produces

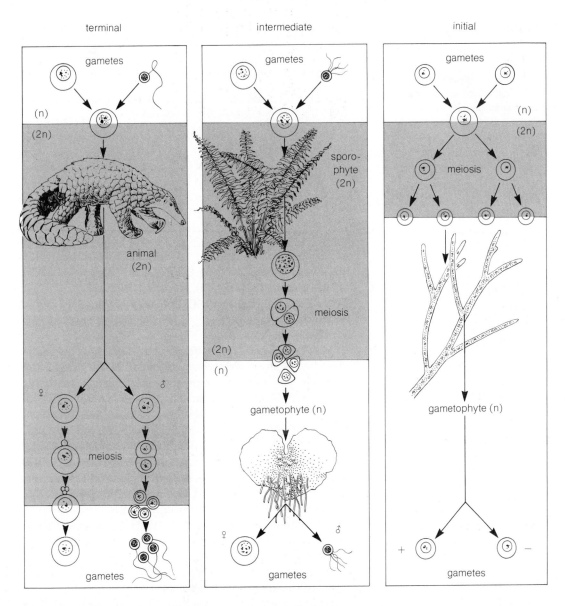

Figure 11-11 The time and place of meiosis in the life cycles of various organisms (see text). The shaded portion marks the diploid phase of the life cycle. Redrawn with permission of the Macmillan Company, from *The Cell in Development and Heredity* by E. B. Wilson. Copyright 1925, The Macmillan Company, renewed in 1953 by Anne M. K. Wilson.

genetic variability in the haploid products of the division sequence. (3) Finally, through the RNA and protein synthesis that occurs during the transcription stage, meiosis provides the ribosomes, enzymes, structural proteins, and raw materials needed for gamete production, fertilization, and the early stages of embryonic development.

The variations in the meiotic products resulting from recombination and independent separation of maternal and paternal chromosomes are responsible for most of the variability noted in the offspring of sexually reproducing organisms. The production of both blue-eyed and brown-eyed offspring from brown-eyed parents in human families, for example, results from chromosome rearrangements during meiosis. Since these rearrangements rest on ordered processes in meiosis, their genetic outcome can be predicted. The mathematical analysis of the products of meiosis provides the basis of the science of genetics, the subject of Unit Four.

The genetic variability arising from meiosis results from (1) recombination in prophase I and (2) independent separation of maternal and paternal homologs during anaphase I.

Questions

1. How do the results of mitosis and meiosis differ?

2. Why was meiosis suspected to occur before it was actually discovered?

3. What mechanisms produce genetic variability in meiosis?

4. Define haploid and diploid.

5. What is the difference between genes and alleles?

6. What happens during each stage of meiotic prophase I? What events mark the beginning and end of each stage?

7. What is the synaptonemal complex? When and where does it appear and disappear in meiosis?

8. Define tetrad, bivalent, homolog, synapsis, and crossover.

9. Compare the pattern of kinetochore connections in the two meiotic divisions. What significance do these connections have for the outcome of meiosis?

10. What are maternal and paternal chromosomes? How are maternal and paternal chromosomes distributed in meiosis? At what point in the meiotic divisions do maternal and paternal chromosomes separate?

11. A chromosome pair has the alleles shown below at G_1 in a cell destined to enter meiosis. If a crossover occurs at the point shown by the arrow, what kinds of chromatids will appear in the products of meiosis?

12. What are parental chromatids? Recombinant chromatids?

13. How did Taylor's experiment demonstrate that recombination takes place by an exchange of segments between homologous chromosomes?

14. How do generations alternate in the higher plants? What is the relationship of meiosis to this alternation? Is there an equivalent alternation of generations in animals?

15. The text says that meiosis in fungi and protozoa is frequently induced by adverse environmental conditions. Does this fact provide any advantage for the survival of a population? Why?

16. How does the outcome of meiosis differ in male and female animals? What are polar bodies? What is the significance of polar body formation in egg development?

17. What are the three major results of meiotic cell division?

Suggestions for Further Reading

Bostock, C. J., and A. T. Sumner. 1978. *The Eukaryotic Chromosome.* Elsevier/North Holland, Amsterdam.

Schultz-Schaeffer, J. 1980. *Cytogenetics: Plants, Animals, Humans.* Springer-Verlag, New York.

Wolfe, S. L. 1981. *Biology of the Cell.* 2nd ed. Wadsworth, Belmont, California.

SUPPLEMENT 11-1: DIVISION OF THE SEX CHROMOSOMES IN MEIOSIS

In some plants and most animals, the males and females of a species show differences in chromosome structure or number. The chromosomes that differ regularly in the males and females of a species are called **sex chromosomes.** In most of these species, including humans, cells in females contain a homologous pair of the sex chromosomes whereas cells in males contain only one of this pair. By convention, the chromosomes present in a pair are called *X chromosomes.* In the male, the single representative of this pair, the X, may exist alone ("XO" males) or may be associated with another sex chromosome, the Y, which is not found in females (Figure 11-12). In most animals and the relatively few plants with sex chromosomes, XX females and XY or XO males are the rule. In a few groups, such as the birds, butterflies, and moths, the situation is reversed: males are XX and females are XY. All other chromosomes in the nucleus are found in homologous pairs in both males and females. These chromosomes, which show no differences in number or morphology in either sex, are called **autosomes.**

The sex chromosomes, as the name suggests, frequently have a direct developmental role in determining the sex of an individual. Further, in individuals with the XY combination (or XO) unusual genetic effects may occur because the X and Y chromosomes often carry entirely different genes. These unusual patterns of inheritance related to the sex chromosomes, called **sex linkage** or **sex-linked inheritance,** are important in several genetically determined diseases of humans, such as color blindness and hemophilia. (Sex-linked inheritance is discussed in Chapter 13.)

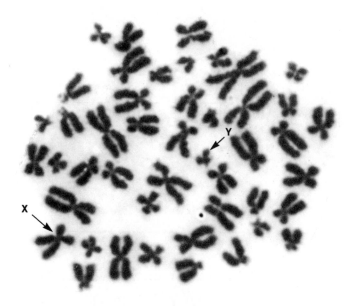

Figure 11-12 The X and Y sex chromosomes in a human male, isolated from a mitotic cell at metaphase. ×2,000. Courtesy of S. Brecher.

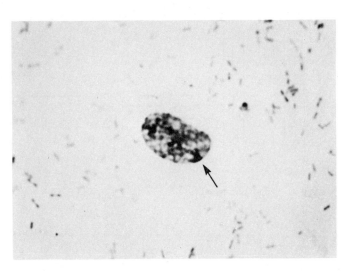

Figure 11-13 The Barr body (arrow) in the nucleus of a cell from a human female. The Barr body, an X chromosome that remains tightly coiled and genetically inactive in body cells, often lies next to the nuclear envelope, as it does in this micrograph. ×1,000.

The presence of sex chromosomes introduces no mechanical difficulties during the meiotic divisions. In the sex with the XX pair (the female in humans), the sex chromosomes go through meiosis exactly like autosomes. Pairing and recombination occur between the two X chromosomes, and each product of meiosis receives a single X at the close of meiosis. All the eggs of species with XX females therefore contain a single X chromosome; the eggs of normal females never carry a Y.

In XY individuals the X and Y chromosomes may or may not contain homologous genes and pair during prophase. Whether or not pairing occurs, the X and Y line up on the spindle along with the autosomes at metaphase I of meiosis. Both the X and Y contain two chromatids at this stage as a result of replication at premeiotic interphase. The two subsequent meiotic divisions separate the four chromatids of these two chromosomes and distribute one to each of the four products of meiosis: two receive an X chromatid, and two receive a Y chromatid. In XY individuals (such as the human male), therefore, sperm nuclei may contain either an X or a Y chromosome.

In individuals without a Y chromosome (XO individuals, common in insects) both chromatids of the single X usually go to the same pole of the spindle at anaphase I. The other pole receives no X. At anaphase II, the nucleus receiving the X chromosome divides again and distributes one X chromatid to each of the two division products. The other nucleus, which receives no X during anaphase I, divides normally, yielding two nuclei with no X chromosome. Of the four nuclei resulting from meiosis in these XO individuals, two contain an X chromosome and two have no X chromosome (O nuclei).

At fertilization the X gametes produced by an XX individual may fuse with either an X or a Y gamete produced by an XY individual (or an X or O gamete in XO individuals). The resultant diploid nuclei are therefore either XX or XY (or XX or XO), and the chromosome complements of the two sexes are reconstituted.

The sex chromosomes often behave differently from the autosomes during interphase. In human females, for example, one of the two X chromosomes remains tightly coiled and is inactive throughout interphase in all cells of the body. The coiled X is visible in the interphase nuclei of human females as a triangular or hourglass-shaped block of tightly packed chromatin called the **Barr body** (Figure 11-13). Isolated human cells can often be identified as female or male by the presence or absence of the Barr body. The Barr body test is used to determine whether Olympic athletes are genetically male or female. The test is relatively easy and painless because it can be carried out using cells removed by lightly scraping the cheeks inside the mouth.

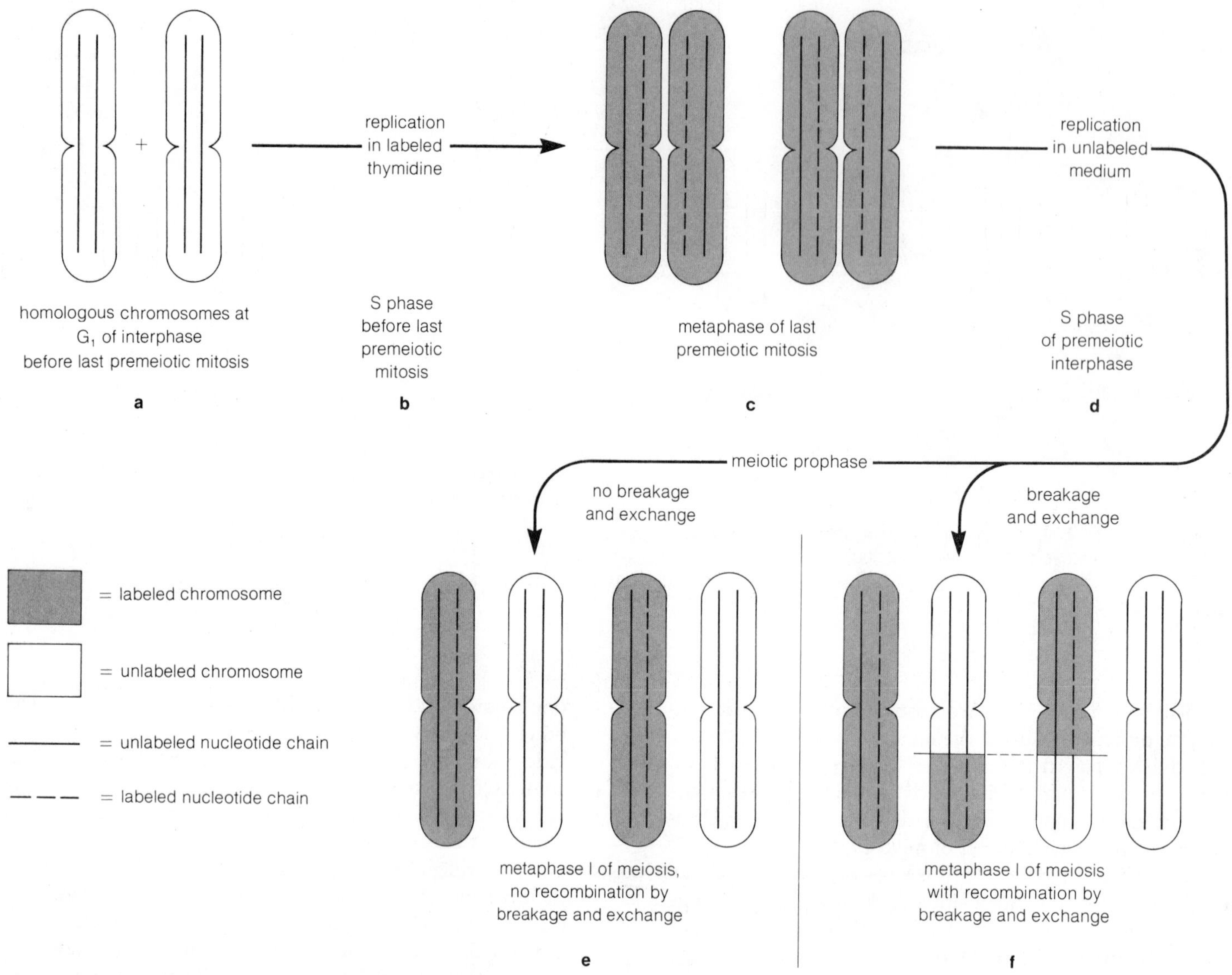

Figure 11-14 The Taylor experiment demonstrating that recombination occurs by breakage and exchange (see text).

SUPPLEMENT 11-2:
EVIDENCE THAT RECOMBINATION TAKES PLACE BY BREAKAGE AND EXCHANGE

J. H. Taylor's experiment showing that recombination takes place by physical breakage and exchange between homologous chromatids (see p. 230) was carried out with meiotic cells of a grasshopper, *Romalea*. To label the chromosomes of the grasshopper, Taylor injected them with radioactive thymidine, a chemical that is incorporated in quantity only into DNA. The thymidine was injected only once, under conditions that assured that only one DNA replication would take place in the presence of the radioactive label. Taylor then followed the distribution of label in meiotic cells by removing them at different stages and testing them for radioactivity.

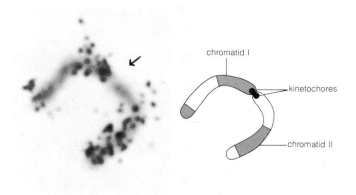

Figure 11-15 A chromosome recovered at metaphase II of meiosis in Taylor's *Romalea* experiment. The shaded regions in the tracing show the labeled segments. Reciprocal exchange of labeled segments in this manner could take place only by physical breakage and exchange. Micrograph ×4,000. Courtesy of J. H. Taylor and The Rockefeller University Press, from *Journal of Cell Biology* 25:57 (1965).

Taylor was interested in obtaining cells with only one chromatid of each homologous chromosome labeled. This was possible, according to the experimental design, in cells that picked up label during the last premitotic interphase before meiosis. To understand why only one chromosome of each homologous pair is labeled during the subsequent meiosis, we must follow these cells through two cycles of semiconservative replication (Figure 11-14*a*). As a result of semiconservative replication in the presence of labeled thymidine during the last premitotic interphase (Figure 11-14*b*), each replicated DNA molecule contains one "old" unlabeled nucleotide chain and one "new" labeled chain. At metaphase of the mitotic division following this replication, both chromatids of each chromosome are labeled (Figure 11-14*c*).

These cells now enter the interphase before meiosis. First, the cells undergo another DNA replication. Labeled thymidine has been washed from the tissues, and this replication occurs in unlabeled medium. The DNA molecules, each containing one labeled and one unlabeled nucleotide chain, unwind for replication to occur (Figure 11-14*d*). The labeled nucleotide chain serves as a template for replication; because no label is present in the medium, the newly synthesized nucleotide chain copied from it is unlabeled. These nucleotide chains, the old labeled chain and the new unlabeled chain, according to semiconservative replication, remain together to form one of the two chromatids of the replicated chromosome. This chromatid, at the subsequent metaphase, shows the presence of label. The

other old nucleotide chain, the unlabeled one, also serves as a template at this replication. This nucleotide chain, and the new chain copied from it during replication, are both unlabeled and remain wound together to form the other chromatid of the chromosome. This chromatid shows no label at the subsequent metaphase. Therefore, all the chromosomes at the subsequent meiotic metaphase contain one labeled and one unlabeled chromatid (Figure 11-14*e*).

At metaphase I of meiosis, these chromosomes reveal whether recombination has occurred by breakage and exchange. If no breakage and exchange occurs, all chromatids are either completely labeled or completely unlabeled at metaphase I (as in Figure 11-14*e*). If any recombination occurs by physical breakage and exchange between labeled and unlabeled chromatids, some chromosomes should appear at metaphase with single chromatids segmented into labeled and unlabeled regions (Figure 11-14*f*).

Taylor was able to find segmented label in many of the chromatids in meiotic cells of his experimental animals (Figure 11-15). This finding demonstrated conclusively that physical breakage and exchange take place between homologous chromatids during meiosis.

SUPPLEMENT 11-3:
RECOMBINANT DNA AND CLONING

Through a series of recently developed techniques essentially any DNA sequence, including whole genes, can be introduced into living cells. The DNA is introduced into recipient cells by linking it to DNA molecules normally forming part of the host cell DNA or to the DNA of an infecting virus. The linkage process is called *recombination* since the end result, in which DNA from two different sources is covalently linked into continuous, intact molecules, resembles a recombinant chromosome. Once introduced into a host cell, the DNA is replicated along with the host cell DNA and passed on in cell division. This replication creates a line of descendants called a **clone,** all containing the DNA sequence introduced into the original cell.

The recombinant DNA technique depends on a group of enzymes called *restriction endonucleases* that can be extracted from bacteria. These enzymes are capable of recognizing short, specific DNA sequences and breaking the DNA at or near the recognition sequence. Most useful of these are the enzymes that attack reverse-repeat sequences (see p. 154) in such a way that free ends are produced

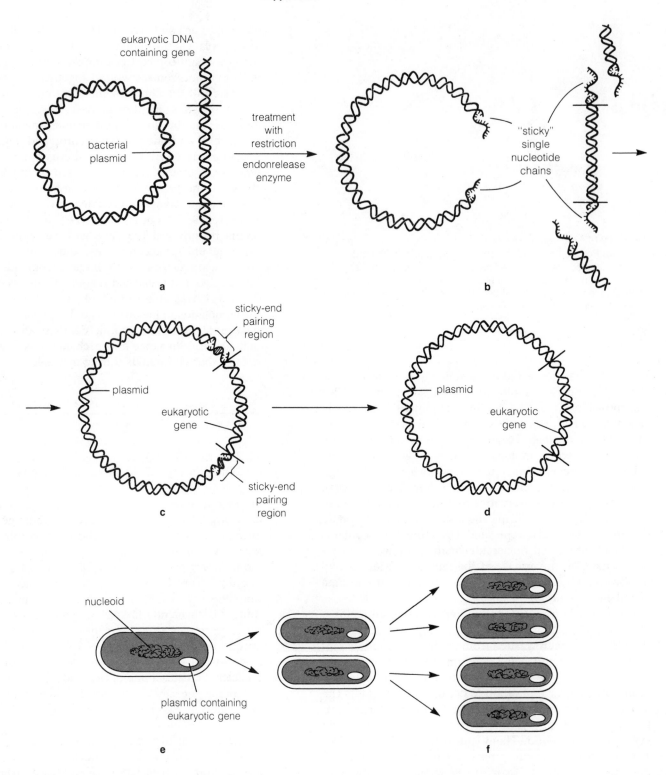

Figure 11-16 Cloning recombinant DNA produced by linking a gene to a bacterial plasmid (see text).

containing complementary single nucleotide chains. For example, one restriction endonuclease attacks the reverse-repeat sequence

$$
\begin{array}{ccccccc}
 & & \downarrow & & & & \\
--- & G & A & T & T & C & --- \\
 & | & | & | & | & | & \\
--- & C & T & T & A & G & --- \\
 & & & & & \uparrow &
\end{array}
$$

at the arrows, producing free ends with exposed, complementary single nucleotide chains:

$$
\begin{array}{ccccccccc}
--- & G & & & & A & A & T & C & --- \\
 & | & | & | & | & + & | & | & | & | \\
--- & C & T & T & A & & & & G & ---
\end{array}
$$

Since the enzyme attacks only this sequence, all the products of the attack have complementary ends and any fragments produced can rewind at the ends and join together.

The sequence attacked by the endonuclease appears randomly spaced at intervals separated by several thousand base pairs in DNA from almost any source. As a result, essentially any DNA molecule can be broken into a number of fragments, all with "sticky" ends that can pair and join with any other fragment produced by the same enzyme. To produce recombinant DNA, sticky fragments obtained from different DNA molecules by treatment with the enzyme are mixed and then exposed to DNA ligase (see p. 217), which seals any paired ends that wind together into covalently linked, hybrid DNA molecules.

The "recombinant" DNA molecules produced in this way are cloned in bacteria by taking advantage of **plasmids.** These are small, circular DNA molecules that exist in the cytoplasm of bacterial cells without direct connection to the major DNA circle of the bacterial nucleoid. Plasmids are replicated in bacterial cells and passed on in division as is the primary DNA circle of the bacterial nucleoid (see p. 208). To clone a sequence, both the DNA of interest and plasmids isolated from a bacterium (Figure 11-16a) are digested with a restriction endonuclease (Figure 11-16b). Digestion by the endonuclease produces sticky ends from both molecules that can pair in different combinations (Figure 11-16c). These combinations are subsequently sealed into closed circles by the DNA ligase enzyme (Figure 11-16d). The recombinant plasmids containing the DNA of interest are then introduced into a bacterium such as *E. coli* (Figure 11-16e). Exposing the cells briefly to elevated temperature—42°C—for a few minutes in the presence of Ca^{2+} ions promotes uptake of the recombinant plasmids.

The recombinant plasmids are then replicated in the bacterial cytoplasm and passed on to daughter cells during division. After many rounds of replication and division, a large population of bacterial cells, all containing the recombinant DNA plasmid with the DNA sequences of interest, is produced (Figure 11-16f).

The same techniques have also been used to create clones in cells isolated from eukaryotes and grown in cultures, most frequently from the African green monkey. In this case, the restriction endonuclease enzymes are used to recombine the DNA molecules of interest with the DNA of a virus infecting the monkey cells. Green monkey cell cultures are then infected with the recombined viral DNA. In the infected cells, the inserted DNA sequences are replicated along with the viral DNA.

Cloning is not without risk. There is danger that highly infective types may be created by the inclusion of recombinant DNA sequences in bacterial cells. The risk is reduced, however, by growing the cloned cells in carefully monitored environments from which escape is unlikely. The danger of escape is further minimized by the use of mutant bacteria with nutritional requirements so stringent that their chance of survival outside the laboratory is essentially zero.

The cloning technique has revolutionized DNA sequencing studies because it provides a way to increase the number of copies of a DNA sample to levels permitting biochemical analysis. Many genes have now been completely sequenced by this approach. Cloning also provides a method for introducing genes coding for various proteins into bacterial cells, where the encoded proteins may be produced in quantity. The protein insulin, required for the survival of persons suffering from diabetes, has already been produced in bacterial cells in this way, and clones producing other proteins of human interest and benefit are presently being developed.

It is also possible that recombinant DNA techniques may eventually be developed to allow introduction of DNA sequences containing the normal forms of genes into persons suffering from hereditary diseases. This potential recombinant DNA technique, called **genetic engineering,** shows promise of revolutionizing the treatment of genetically based human diseases if workable techniques can be developed for its application.

Suggestions for Further Reading

Gilbert, W., and L. Villa-Komaroff. 1950. "Useful Proteins from Recombinant Bacteria." *Scientific American* 242:74–94.

Guarente, L., T. M. Roberts, and M. Ptashne. 1980. "A Technique for Expressing Eukaryotic Genes in Bacteria." *Science* 209:1428–1430.

Inheritance

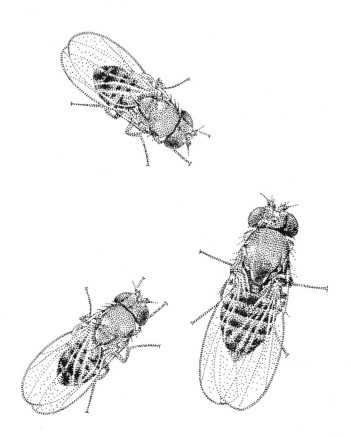

12

MENDELIAN GENETICS: DEVELOPMENT OF THE GENE THEORY OF INHERITANCE

Until about 1900, the most popular view of inheritance among both scientists and the general public was that hereditary traits "blend" in offspring through the mixing of parental blood. These ideas persist even today in popular usage, since we still speak of persons of "mixed blood" and "half-breeds." Most people find it reasonable to suppose that characteristics of the parents, such as skin color, body size, and facial features, will be evenly blended in their offspring.

The first events leading to a change in these attitudes came with the work of Gregor Mendel in the 1860s. Mendel revolutionized the study of inheritance by establishing that characters do not blend at all but are passed on to offspring in the form of discrete hereditary units that we now know as genes. Mendel's research and conclusions were so advanced for his time that many years were to pass before his findings were fully appreciated by other scientists. Although Mendel's work has been completely supported by all subsequent experimentation, it has still not entirely converted our everyday attitudes about the transmission of hereditary traits from parents to offspring.

Mendel revolutionized the study of inheritance by showing that hereditary characteristics are passed from parents to offspring in units: the genes.

Mendel's findings lay unnoticed until 1900. At this time, several independent investigators rediscovered the principles of inheritance found by Mendel some 35 years earlier. Because work in other fields had progressed far enough to allow full appreciation of the findings, a burst of interest and work in genetics immediately followed. This research, which continues unabated to the present day, has provided some of the most significant and far-reaching conclusions of biology.

THE BEGINNING OF GENETICS: MENDEL'S GARDEN PEAS

It is rare in the history of science that the development of a major area of study can be traced largely to the work of one man. This, however, is the case with Gregor Mendel (Figure 12-1) and the relationship of his research to the science of genetics. Mendel's method of research illustrates, perhaps as well as any series of experiments in the entire history of science, how scientific work at its best is carried out—through observation, hypothesis, and experimental tests.

Figure 12-1 Gregor Mendel.

Gregor Mendel was an Austrian priest who lived and worked in a monastery in Brünn (now part of Czechoslovakia). Although a priest, Mendel had an unusual education in science for his time. He had studied mathematics, chemistry, zoology, and botany at the University of Vienna under some of the foremost scientists of his day.

Mendel got started in his work through his curiosity about the manner in which hereditary traits are transmitted from parents to offspring. He chose the ordinary garden pea for his research because peas could be grown easily at his monastery without elaborate equipment. Moreover, many varieties of peas were known, including some in which hereditary characters bred true—that is, were passed on without change from one generation to the next. Mendel was able to order more than 30 true-breeding varieties of peas from commercial suppliers, and in his experiments he eventually used a total of seven.

Mendel's success in analyzing the transmission of hereditary traits in peas was based partly on a good choice of experimental organism. But more than this, Mendel, because of his training and natural inclination, analyzed his results in quantitative terms, a radically new departure for his time. Using this method, he was able to describe the characteristics of the unseen hereditary determiners passed on in the peas and, incidentally, to establish the basic methods for future work in genetics.

Mendel's Experiments

The Single-Trait Crosses Among the inherited traits that Mendel selected for study were two that affected the shape of seeds. One variety of peas produced pods with round, smooth seeds; another produced pods with wrinkled seeds. If these lines were self-pollinated, the wrinkled or round seed traits would breed true. If crossed between themselves, for example, plants with wrinkled seeds always produced offspring with wrinkled seeds. Would these traits blend evenly if the two types of pea plants were crossed? That is, would crossing them produce an intermediate plant with slightly wrinkled seeds?

To test the inheritance of the traits affecting seed shape, Mendel took pollen from a plant that normally produced round seeds and placed it on the flowers of a variety that produced wrinkled seeds. The reverse cross was also carried out: pollen from wrinkled-seed plants was placed on the flowers of plants normally producing round seeds. Mendel noted that all the seeds produced in the pollinated plants were round and smooth, as if the trait for wrinkled seeds had disappeared or was masked. These seeds, which represent the first generation of offspring of the cross, are called the F_1 generation. (F stands for filial, from the Latin *filius* = son.) The seeds were then planted and grown to maturity. Some of the F_1 plants were then crossed among themselves. In the seeds resulting from this cross, which formed the F_2 generation, Mendel counted 5474 of the round variety and 1850 of the wrinkled variety, in an approximate ratio of three round to one wrinkled or about 75 percent round and 25 percent wrinkled seeds. Other pairs of characters, seven in all, were tested in the same way (Table 12-1). In all cases, a uniform F_1 generation was obtained in which only one of the two characters of the pair appeared. In the second generation (the F_2 generation) both of the two characters appeared again. Moreover, the character present in the F_1 generation always represented about 75 percent of the offspring in the F_2 generation. When analyzed quantitatively, the pattern in which these traits were transmitted from parents to offspring proved to be entirely unlike an even blending of

Table 12-1 Results of Mendel's Crosses Using Seven Different Characters in Peas

Character	F_1	F_2
Round × wrinkled seeds	All round	5474 round; 1850 wrinkled
Yellow × green seeds	All yellow	6022 yellow; 2001 green
Red × white flowers	All red	705 red; 224 white
Green × yellow pods	All green	428 green; 152 yellow
Large × small pods	All large	882 large; 299 small
Tall × short plants	All tall	787 tall; 277 short
Flowers along stem × flowers at ends of stems	All flowers along stems	651 along stems; 207 at ends of stems

characters. Instead, one trait of each pair was completely absent in the F_1 generation but reappeared in the F_2 generation in a definite proportion among the offspring.

Mendel's Hypotheses Mendel's interpretation of these results was nothing short of brilliant. He saw that the results could be explained if he assumed that the adult plants carry a *pair* of factors governing the inheritance of each trait. He proposed that these characters separate as gametes are formed, so that each gamete cell receives only one of the pair. As the maternal and paternal gametes fuse, the resulting diploid nucleus (today called the zygote nucleus) receives one factor for the trait from the male gamete and one factor for the same trait from the female gamete, reuniting the pair. Because a gamete receives only one member of each pair of factors governing a trait, Mendel termed this phenomenon the *segregation* of the factor pairs.

Mendel hypothesized that heredity is controlled by *pairs* of factors in each individual.

Mendel proposed that the factor pairs separate when gametes are formed, so that only one member of any pair is carried by a gamete.

The observation that one of the traits, such as wrinkled seeds, disappeared in the F_1 generation and reappeared in the F_2 was interpreted by Mendel to mean that although it was undetected in the F_1, the factor was still present and merely masked in some way by the "stronger" factor. Mendel called this effect **dominance**. He assumed

that when the two different factors occurred together in an organism, only the factor for roundness in seeds had effect. The factor for wrinkled seeds was thought to be **recessive**, expressed only when both members of the pair in an individual represented the factor for wrinkling.

Mendel proposed that when different factors (alleles) of a pair are present in the same individual, one is dominant over the other and masks its expression.

Using these assumptions, Mendel was able to explain the results of his crosses along the following lines (Figure 12-2). Both factors in the parent plant with round seeds (the plant used in the original cross) are assumed to be the same and to govern the production of round seeds. This factor is given the symbol *R*; in this parent, therefore, the genetic constitution of the pair is *RR*. An individual that carries a pair of factors of the same type is today called a **homozygote**. In the production of gametes the two members of the pair separate and, because only *R* factors are

An individual carrying a pair of factors (alleles) of the same type is a homozygote. In heterozygotes the two members (alleles) of a factor pair are different.

present in the pair, all gametes receive one *R* factor (the left-hand heading in Figure 12-2). In the original parent with wrinkled seeds, both factors are also assumed to be the same. This factor is given the symbol *r*, and the wrinkled parent, also a homozygote, has the genetic constitu-

Figure 12-2 The result of Mendel's cross between pea plants with round seeds and wrinkled seeds in the F₁ generation (see text).

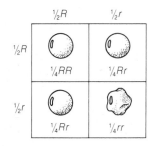

Figure 12-3 The results of Mendel's cross between plants with round and wrinkled seeds in the F₂ generation (see text).

tion *rr*. These factors are also separated and distributed singly to each gamete. All gametes from this parent thus receive one *r* factor (the top heading in Figure 12-2). When these gametes cross-fertilize to produce the F₁ generation, they combine at random. However, the results of the combination are always the same in the F₁ generation. All zygotes produced by fertilization between gametes from the round and wrinkled plants will produce the combination *Rr* (shown in the block in Figure 12-2). An individual of this type, in which the two members of the factor pair are different, is today called a **heterozygote**. Because *R* is dominant over *r*, all the offspring will have round seeds even though the factor for wrinkled seeds is present in all individuals of the F₁ generation.

These assumptions also explain the results of Mendel's F₁ crosses. According to his explanation, all F₁ plants produce two kinds of gametes. Because the *Rr* pairs separate to produce gametes, one-half will receive the *R* factor and one-half the *r* factor. (These gametes are entered in both the horizontal and vertical headings in Figure 12-3; the blocks show the possible combinations.) Combining two gametes that carry the *R* factor produces an *RR* F₂ plant; combining *R* from one parent and *r* from the other produces an *Rr* F₂ plant. Finally, combining *r* from both F₁ parents produces an *rr* F₂ plant. The *RR* and *Rr* plants in the F₂ generation will have round seeds; however, the *rr* offspring will have wrinkled seeds. Thus Mendel's assumptions explain the reappearance of the wrinkled trait in the F₂ generation. These assumptions are that (1) the factors governing a trait occur in pairs, (2) one factor is dominant over the other if different forms of the factors are present in a pair, and (3) the two members of a pair are separated and delivered singly to gametes during their formation.

Mendel's three hypotheses were: (1) the hereditary factors governing a trait occur in pairs, (2) one factor is dominant over another if different forms of the factor are present in an individual's pair, and (3) factor pairs separate during formation of gametes.

Mendel's assumptions also explain the proportions of offspring obtained in the F₂ generation. In the crosses between round and wrinkled-seed plants, according to Mendel's principle of segregation of factors in gamete formation, one-half of the gametes of the F₁ generation will contain the *R* factor and one-half will contain the *r* factor (see Figure 12-3). To produce an *RR* zygote, two *R* gametes must be selected from among the gametes and combined. To produce an *rr* zygote, two *r* gametes must be selected from the gametes and combined.

The chance of two independent choices occurring at the same time is determined by multiplying together their individual probabilities.

The chance of accomplishing these combinations is predicted by a simple law of probability: The chance of two independent choices occurring together is determined by multiplying their individual probabilities. Since *R* and *r* gametes occur in equal numbers, the chance of selecting either an *R* or *r* gamete from the gamete pool is 1 out of 2 or ½. The chance of selecting two *R* gametes in a row to produce an *RR* zygote is then equal to the chance of selecting one *R* gamete (½) times the chance of selecting a second *R* gamete, or ½ × ½ = ¼. Therefore one-fourth of the F₂ offspring of the F₁ cross *Rr* × *Rr* is expected to be *RR* and to have round, smooth seeds. By the same line of reasoning, one-fourth of the F₂ offspring will be *rr* with wrinkled seeds. If one-fourth of the F₂ generation is *RR* and one-fourth is *rr*, it follows that the remaining one-half of the F₂ generation must be *Rr*. Therefore three-fourths of the F₂ offspring will be expected to have round, smooth seeds (¼*RR* + ½*Rr*) and one-fourth wrinkled seeds, in the ratio 3:1. This ratio, as shown in Table 12-1, is the approximate ratio Mendel actually obtained in his cross. The same ratios hold for all other factor pairs used in his experiments. Thus Mendel's three assumptions about the hereditary factors successfully explain both the types of offspring and their proportions in the F_1 and F_2 generations.

Figure 12-3 illustrates a simplified method for determining the expected proportions of *RR*, *Rr*, and *rr* plants in the F_2 generation. The chance of obtaining each type of gamete from the "male" parent is entered with the gametes on the horizontal heading to the diagram, and the gametes from the "female" parent are entered on the vertical heading. The various combinations possible in the offspring can be obtained by combining the factors carried by the gametes in the squares opposite the headings. The frequency of the combination in each square is calculated by multiplying the frequencies of the gametes used to produce it. In Figure 12-3, two squares, each with a frequency of one-fourth, contain the combination *Rr*. Added together, these two squares provide the total chance of obtaining the *Rr* class in the F_2 generation: ¼ + ¼ = ½. This device, called a *Punnett square* after R. C. Punnett, an early geneticist, is quite useful for determining the expected outcome of a genetic cross. (For an algebraic method, see Supplement 12-1.)

Mendel's Experimental Proofs Mendel noted that the validity of his assumptions could be proved by observing how closely his hypotheses predict the outcome of a cross different from any tried so far. To carry out this test, Mendel crossed an F_1 plant with round seeds (assumed to have the factor pair *Rr*) with a wrinkled-seed plant of the original parental type (assumed to have the pair *rr*). The predicted outcome of this cross (*Rr* × *rr*) is most easily understood by means of a Punnett square. According to Mendel's hypothesis, all the gametes of the *rr* plant will contain a single *r* factor. Therefore the probability of selecting an *r* gamete from this parent is 1. The gamete and its probability are entered in the left-hand heading of the Punnett square in Figure 12-4. The *Rr* parent will produce two types of gametes, half containing the *R* factor and half containing the *r* factor. These values are entered along the top heading of the square. Filling in the possible combinations in Figure 12-4 gives the two classes *Rr* and *rr*, both with the expected frequency of one-half. Therefore this cross is expected to produce one-half round-seed plants and one-half wrinkled-seed plants in a 1:1 ratio. Mendel's actual results in one experiment testing this cross were 49 round-seed plants and 47 wrinkled-seed plants, closely approximating the expected 1:1 ratio. (A contemporary statistical technique for evaluating how closely actual results match expectations in experimental tests of this type, called the *chi square* method, is described in Supplement 12-1.) Other tests by Mendel and countless experiments by

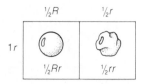

Figure 12-4 The results of Mendel's testcross proving the hypothesis (see text).

investigators since Mendel's time have supported his hypotheses about hereditary traits and their patterns of inheritance.

The cross used by Mendel to test his hypotheses, in which the F_1 generation is crossed with the homozygous recessive parent, is still used in genetics to determine whether individuals carrying a dominant trait are homozygotes or heterozygotes for that trait. This question often arises because the homozygote (*RR* in the case of round versus wrinkled seeds) is externally indistinguishable from the heterozygote (*Rr* in the case of round versus wrinkled seeds). In the cross used to determine whether an individual is a homozygous or heterozygous dominant, called a **testcross,** the individual in question is always crossed with a homozygous recessive. If the offspring of the cross are of two types, with one-half displaying the recessive trait and one-half the dominant trait, as in Mendel's testcross, then the individual in question must be a heterozygote (*Rr* in the case of round versus wrinkled seeds). If all the offspring display the dominant trait, the individual in question must be a homozygote (*RR* in the case of round versus wrinkled seeds). The cross producing these results is *RR* × *rr*.

In a testcross the experimental individual is mated with a homozygous recessive.

Dihybrid Crosses All of Mendel's initial crosses, outlined in Table 12-1, involved single pairs of factors determining the same characteristic such as seed form. Crosses of this type are called **monohybrid** crosses. Mendel was also curious about the effects of crossing parental stocks with differences in two sets of hereditary characteristics. For a cross of this type, called a **dihybrid** cross, Mendel chose seed shape and a second characteristic: seed color (green versus yellow).

In Mendel's experiment, plants with round, yellow seeds were crossed with plants with wrinkled, green seeds. The F_1 generation produced from this cross con-

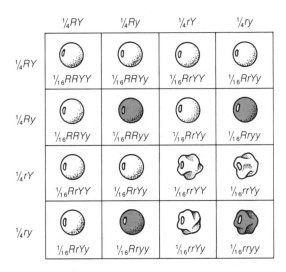

Figure 12-5 The results of crossing F₁ *RrYy* plants (see text).

	¼*RY*	¼*Ry*	¼*rY*	¼*ry*
¼*RY*	¹⁄₁₆*RRYY*	¹⁄₁₆*RRYy*	¹⁄₁₆*RrYY*	¹⁄₁₆*RrYy*
¼*Ry*	¹⁄₁₆*RRYy*	¹⁄₁₆*RRyy*	¹⁄₁₆*RrYy*	¹⁄₁₆*Rryy*
¼*rY*	¹⁄₁₆*RrYY*	¹⁄₁₆*RrYy*	¹⁄₁₆*rrYY*	¹⁄₁₆*rrYy*
¼*ry*	¹⁄₁₆*RrYy*	¹⁄₁₆*Rryy*	¹⁄₁₆*rrYy*	¹⁄₁₆*rryy*

sisted entirely of round, yellow-seed offspring. Mendel then crossed individuals of the F₁ generation among themselves and obtained, in the F₂ offspring of these crosses, 315 round, yellow-seed plants, 101 wrinkled, yellow-seed plants, 108 round, green-seed plants, and 32 wrinkled, green-seed plants. Mendel noted that in numbers these offspring approximate a 9:3:3:1 ratio.

Mendel's principle of independent assortment states that different factors (genes) controlling different traits separate independently in the formation of gametes.

The ratios that Mendel obtained were consistent with his previous findings if one additional assumption was made—that the two sets of factors separate independently during formation of gametes. That is, separation of the pair of factors for seed shape in gamete production has no effect on the separation of the pair of factors for seed color. Mendel termed this assumption **independent assortment.**

To understand the effect of independent assortment in the cross, assume that the parental types are *RRYY* (round, yellow seeds) and *rryy* (wrinkled, green seeds). The round, yellow parent will produce only *RY* gametes; the wrinkled, green parent will produce only *ry* gametes. In the F₁ generation all possible combinations of these gametes will produce only one class of offspring: *RrYy*. Assuming that the factor for yellow is dominant over green, all the F₁ will be, as observed, round, yellow-seed plants.

If these factor pairs for shape and color separate independently in gamete formation, each F₁ pea plant will produce four types of gametes (indicated on the horizontal and vertical headings in the Punnett square in Figure 12-5). The *R* factor for seed shape can combine with either the *Y* or *y* factor of the pair for color. Similarly, the *r* factor can be delivered to a gamete with either *Y* or *y*. This independent assortment of *RrYy* factors would thus be expected to produce four types of gametes appearing with equal frequency: ¼*RY*, ¼*Ry*, ¼*rY*, and ¼*ry*. These gametes, and their expected frequencies, are entered for both parents in the headings to Figure 12-5.

The results of all possible combinations of these gametes can be obtained by filling in the squares of Figure 12-5. Sixteen classes are obtained, all with an equal frequency of 1 out of every 16 offspring. Of these, the classes *RRYY*, *RRYy*, *RrYY*, and *RrYy* will have the same appearance: round, yellow seeds. These combinations occur in 9 out of the 16 squares in the diagram and therefore will be expected in a total frequency of ⁹⁄₁₆. The combinations producing wrinkled, yellow seeds, *rrYY* and *rrYy*, are found in three squares for a total expected frequency of ³⁄₁₆. Similarly, the combinations *RRyy* and *Rryy*, yielding round, green-seed plants, occur in three squares for a total expected frequency of ³⁄₁₆ in the offspring. Finally, the combination *rryy*, producing wrinkled, green-seed plants, occurs in only one square and will thus be expected in ¹⁄₁₆ of the offspring. These expected frequencies of round yellow:wrinkled yellow:round green:wrinkled green, in a ratio of 9:3:3:1, closely approximate the actual results of 315:101:108:32 obtained by Mendel. (A chi square test of this experiment is given in Supplement 12-1.) Thus Mendel's hypotheses, with the added assumption that multiple pairs of factors separate independently, explain the observed results. Mendel's tests of the added hypothesis completely confirmed it: the testcross *RrYy* × *rryy*, for example, gave the expected 1:1:1:1 ratio in the offspring.

Mendel's Results: A Summary

Mendel's hypotheses concerning the inheritance of traits in sexually reproducing organisms were thus confirmed by his experiments: (1) the hereditary factors controlling a single characteristic occur in pairs in an individual; (2) the pairs of factors are separated and occur singly in gametes (now called the *principle of segregation*); (3) one member of the pair is dominant in its effects over the other if different forms are present in a single individual (now called the *principle of dominance*); and (4) every pair of factors separates independently of every other pair of factors (now called the *principle of independent assortment*). In modern terminology the factors controlling a character or trait are called *genes*, and the alternate forms of a gene, such as *R* and *r*, are termed the *alleles* of the gene. In his discovery of genes and their patterns of transmission Mendel showed that traits are passed from parents to offspring as hereditary *units* in predicable ratios and combinations, completely disproving the notion of genetic blending.

Mendel also realized that the genetic makeup of an individual cannot be determined from its outward appearance alone. That is, Mendel showed that individuals may look the same but differ genetically—as in his experiments demonstrating that *RR* and *Rr* plants, although different genetically, still have the same outward appearance. In modern terminology, the genetic makeup of an organism is termed the **genotype** to distinguish it from the outward appearance, called the **phenotype** (*phainein* = to show). According to this terminology, the two genotypes *RR* and *Rr* produce an identical phenotype with round seeds.

An individual's genetic makeup is its genotype. Its outward appearance is its phenotype.

Mendel's hypotheses and his methods for deriving and testing them are all the more remarkable if you consider that at the time of his discovery the structure and function of chromosomes and the processes of meiosis were completely unknown. His results amounted to a completely abstract view of the hereditary material, a view based on mathematical reasoning rather than morphological description, in direct opposition to the usual techniques of the day. For these reasons, Mendel's paper describing his experiments, published in 1866, was not understood by the scientists who had an opportunity to read it. Moreover, Mendel, an obscure priest working with garden peas in an out-of-the-way monastery, was probably not even taken seriously by his fellow investigators. His paper lay unnoticed until the early 1900s, when three investigators—Hugo de Vries, Carl Correns, and Erich von Tschermak—independently repeated the same line of reasoning and experimentation used earlier by Mendel and came up with the same results. These men, in researching the published literature as they prepared their results for publication, discovered to their surprise that they had been scooped 34 years earlier by an Austrian monk. Each man gave credit to Mendel's earlier discoveries, and the quality and far-reaching implications of Mendel's work were at last realized. Mendel had died in 1884, however, sixteen years before the rediscovery of his factors and their manner of inheritance, and never in his lifetime received the recognition he deserved.

HOW MENDEL'S FACTORS RELATE TO CELL STRUCTURES

The Chromosome Theory of Inheritance

By the time Mendel's results were rediscovered in 1900, most of the details of meiotic cell division had been worked out and it was not long until the similarities between chromosomes and Mendel's factors were noted. In a historic paper published in 1903, Walter Sutton, then a graduate student at Columbia University, drew all the necessary parallels. Sutton drew attention to the fact that chromosomes occur in pairs in sexually reproducing organisms, as do Mendel's factors. The two chromosomes of each pair are separated and delivered singly to gametes, as are Mendel's factors. Further, the separation of every pair of chromosomes in gamete formation is independent of the separation of every other pair, as in the behavior of different pairs of factors in Mendel's dihybrid crosses. Finally, he noted, one member of each chromosome pair is derived in fertilization from the male parent and one from the female parent, in an exact parallel to Mendel's factors. On the basis of this total coincidence in behavior, Sutton concluded that Mendel's factors, the genes, are carried on the chromosomes.

Through the coincidence in their behavior, Sutton concluded that genes are carried on the chromosomes in the cell nucleus.

Sutton's paper was written in the form of a hypothesis without experimental tests. Proof of the *chromosome theory of inheritance,* as his hypothesis came to be known, was established in 1916 by Calvin B. Bridges in his work with a phenomenon known as **nondisjunction.** (Supplement 13-1 describes Bridges' experiments.)

Mendel, DNA, RNA, and Protein Synthesis

How do the factors or genes discovered by Mendel relate to interactions of DNA and RNA in transcription and protein synthesis as described in Chapters 9 and 10? We now know that genes consist of DNA sequences coding for mRNA, tRNA, and rRNA molecules. The alleles studied by Mendel are slightly different sequences of the sequence of a gene coding for an mRNA molecule. (Differences in tRNA or rRNA sequences rarely occur.) Thus the dominant form *B* of a gene coding for blue flower color might have the sequence

. . . ATAGAGATTGCATTAGACATAGGC . . .

in one region within its boundaries on the DNA. (The brackets indicate the coding triplets.) The recessive *b* allele of this gene might have the same sequence except for a base substitution at one point:

$$...\overline{ATA}\overline{GAG}\overline{ATT}\overline{GCA}\overline{TTA}\overline{GTC}\overline{ATA}\overline{GGC}...$$

These alternate forms of the gene are copied into mRNA molecules by the transcription mechanism:

$$...\overline{UAU}\overline{CUC}\overline{UAA}\overline{CGU}\overline{AAU}\overline{CUG}\overline{UAU}\overline{CCG}...\;(B\ \text{form})$$

or

$$...\overline{UAU}\overline{CUC}\overline{UAA}\overline{CGU}\overline{AAU}\overline{CAG}\overline{UAU}\overline{CCG}...\;(b\ \text{form})$$

The mRNA molecules then attach to ribosomes and direct the synthesis of protein molecules. The proteins produced will be identical except for the substitution of a single amino acid. The CUG codon of the mRNA transcribed from the *B* allele will direct the insertion of leucine at the corresponding point in the amino acid sequence of its protein; the CAG codon of the *b* allele mRNA will substitute glutamine at this point.

For this example we will consider that the protein coded by the *B* gene is an enzyme catalyzing a step in the pathway leading to the production of blue pigment in the cells of flower petals:

Substance A $\xrightarrow{\text{enzyme 1}}$ substance B $\xrightarrow{\text{enzyme 2}}$

substance C $\xrightarrow{\text{enzyme 3}}$ blue pigment

The protein coded by the *B* gene is enzyme 2 in the pathway. In the form coded by the *B* allele, the presence of leucine allows the amino acid chain of the enzyme to fold into a form in which it is fully active. All the steps in the pathway then run in sequence, and the blue flower pigment is produced. The presence of glutamine at the same position alters the folding conformation of the enzyme at the active site, however, completely destroying its ability to catalyze the conversion of substance B to C. As a result, the pathway is blocked and blue pigment is not produced.

Note that *BB* individuals, with the *B* allele on both chromosomes of the pair carrying this gene, produce blue pigment and therefore have blue flowers. Although *Bb* individuals carry one of the recessive alleles, the *B* allele on the other chromosome of the pair codes for the active form of the enzyme; this form is produced in sufficient quantity to keep the pathway going. Therefore *Bb* individuals will

also have blue flowers, and the *B* allele is said to be dominant to the *b* allele. Individuals with the *b* allele on both chromosomes of the pair (*bb*) will produce no blue pigment; this condition is seen as white flowers in the plant.

Mendel's hypotheses have withstood the test of repeated experiments and are now accepted as laws of genetics: the principles of segregation, independent assortment, and dominance. Continued work in genetics has revealed, however, that there are important exceptions to two of these principles, independent assortment and dominance. It is now known that independent assortment applies without qualification only to genes that are carried on *separate* chromosome pairs within an individual; the existence of different genes on the *same* chromosome modifies their pattern of transmission from parents to offspring. The research following the rediscovery of Mendel's results also revealed that some alleles are inherited without dominance, so that the effects of both can be detected in heterozygotes. These and other modifications, surveyed in the next chapter, extend rather than contradict Mendel's hypotheses. They still rest upon his fundamental discovery that heredity is controlled by individual factors inherited as units: the genes.

Questions

1. How were hereditary traits believed to be transmitted to offspring before Mendel did his research?

2. What is the basic difference between Mendel's hypotheses of inheritance and the earlier notion of blending?

3. Why did Mendel choose garden peas for his research?

4. Explain what Mendel meant when he said that factors (genes) segregate in the formation of gametes.

5. What are dominant and recessive traits?

6. What is a homozygote? A heterozygote?

7. What were Mendel's basic hypotheses about the hereditary factors (genes) and their pattern of transmission of offspring?

8. What is a monohybrid cross? A dihybrid cross?

9. What is a testcross? How did Mendel use a testcross to test his hypotheses concerning monohybrid crosses?

10. Explain what *independent assortment* means. How are independent assortment and segregation related to meiosis?

11. What is a genotype? A phenotype?

12. Explain alleles in terms of DNA sequences.

13. Outline the steps Mendel followed in applying the scientific method to his studies of garden peas.

Problems*

1. The *C* allele of a gene controlling color in corn produces kernels with color; plants homozygous for a recessive *c* allele of this gene have colorless or white kernals. What kinds of gametes, in what proportions, would be produced by the parent plants in the following crosses? What seed color would be expected in the offspring of the crosses? In what proportions?

$$CC \times Cc \qquad Cc \times Cc \qquad Cc \times cc$$

2. In peas the allele *T* produces tall plants and the allele *t* produces dwarfs. The *T* allele is dominant to *t*. If a tall plant is crossed with a dwarf, the offspring are distributed about equally between tall and dwarf plants. What are the genotypes of the parents?

3. Assume that eye color in humans is controlled by a pair of alleles of a single gene, with brown dominant to blue. Can two brown-eyed parents have a blue-eyed child? Can blue-eyed parents have a brown-eyed child? A brown-eyed couple, both of whom had one blue-eyed parent and one brown-eyed parent, are expecting their first child. What is the chance that the child will have blue eyes? Suppose the first child actually has blue eyes. What is the chance that the second child of the same couple will have blue eyes?

4. Two brown-eyed parents have four blue-eyed children. What is the chance that their next child will have brown eyes? Blue eyes?

5. In guinea pigs, an allele for rough fur (*R*) is dominant over an allele for smooth fur (*r*); an allele for black coat (*B*) is dominant over that for white (*b*). You have an animal with rough, black fur. What cross would you use to determine whether the animal is homozygous for these characteristics? What results would you expect if your animal is homozygous?

6. You cross a lima bean plant breeding true for green pods with another lima bean that breeds true for yellow pods. You note that all the F$_1$ plants have green pods. These green-pod F$_1$ plants, when crossed, produce 675 plants with green pods and 217 with yellow pods. How many genes probably control pod color? What are the alleles? Which one is dominant?

7. Some people can roll their tongue into a complete circle, and some cannot. This ability is inherited as if it is controlled by two alleles of a single gene; the ability to roll the tongue is the dominant trait. Two tongue-rolling parents have a child who cannot roll his tongue. Can you explain this? Write the genotypes and phenotypes for the three individuals.

8. Some recessive alleles have such a detrimental effect that they are lethal when present in both chromosomes of a pair. That is, the homozygous recessive cannot survive and dies at some point during embryonic development or early life. Suppose that the allele *r* is lethal in the homozygous *rr* condition. What ratios of offspring would you expect from the following crosses?

$$RR \times Rr \qquad Rr \times Rr$$

9. In garden peas, green pods (*GG* or *Gg*) are dominant over yellow (*gg*); tall plants (*TT* or *Tt*) are dominant over dwarfs (*tt*); and round seeds (*RR* or *Rr*) are dominant over wrinkled (*rr*). What offspring are expected in the F$_1$ generation if a true-breeding tall plant with green pods and round seeds is crossed with a true-breeding dwarf plant with yellow pods and wrinkled seeds? What offspring are expected if F$_1$ individuals are crossed?

Suggestions for Further Reading

Goodenough, U. 1978. *Genetics*. Holt, Rienhart & Winston, New York.

Herskowitz, I. H. 1979. *Elements of Genetics*. Macmillan, New York.

Strickberger, M. W. 1976. *Genetics*. Macmillan, New York.

Sturtevant, A. H. 1965. *A History of Genetics*. Harper & Row, New York.

Watson, J. D. 1976. *Molecular Biology of the Gene*. 3rd ed. Benjamin, Menlo Park, California.

Woeller, B. R. 1968. *The Chromosome Theory of Inheritance*. Appleton-Century-Crofts, New York.

SUPPLEMENT 12-1: PREDICTING RATIOS AND TESTING RESULTS IN GENETIC CROSSES

Determining Expected Ratios in Offspring

Constructing a Punnett square to determine expected ratios of alleles in offspring is convenient for monohybrid or dihybrid crosses. If used for more than two pairs of genes, however, the technique is cumbersome and time-consuming. The outcome of such crosses can be easily calculated by a simple algebraic method instead of a Punnett square.

The method works this way in monohybrid crosses. In the first of Mendel's experiments described in this chapter, two heterozygous (*Rr*) parents carrying both the *R* allele for smooth seeds and the *r* allele for wrinkled seeds were crossed. In algebraic terms, the gametes expected are ($\frac{1}{2}R + \frac{1}{2}r$), produced by the "male" parent, and ($\frac{1}{2}R + \frac{1}{2}r$), produced by the "female" parent. Multiplying these individual alleles and probabilities algebraically to obtain the combined outcome and probabilities gives

$$(\tfrac{1}{2}R + \tfrac{1}{2}r)(\tfrac{1}{2}R + \tfrac{1}{2}r) = \tfrac{1}{4}RR + \tfrac{1}{2}Rr + \tfrac{1}{4}rr \qquad (12\text{-}1)$$

These are the expected genotypes and their ratios in the offspring. Because the RR and Rr classes are phenotypically identical in the offspring of this cross, the outcome can also be designated

$$(\tfrac{1}{2}R + \tfrac{1}{2}r)(\tfrac{1}{2}R + \tfrac{1}{2}r) = \tfrac{3}{4}R\_ + \tfrac{1}{4}rr \qquad (12\text{-}2)$$

where $R\_$ indicates either RR or Rr.

The algebraic method is most useful when the cross under study involves more than a single pair of genes. In Mendel's experiment testing the inheritance of seed shape and color, the two parents in the F_1 cross had the genotype $RrYy$ with round, yellow seeds. The Rr and Yy pairs in the cross $RrYy \times RrYy$ are assumed to be completely independent of each other. Since they are independent, the outcome of the cross with respect to the Rr pair is $\tfrac{3}{4}R\_$ + $\tfrac{1}{4}rr$, where $R\_$ can be either RR or Rr (from Equation 12-2). Similarly, the outcome of the cross with respect to the Yy pair is $\tfrac{3}{4}Y\_$ + $\tfrac{1}{4}yy$. The combined probability of these events occurring independently and simultaneously is the product of their individual probabilities:

$$(\tfrac{3}{4}R\_ + \tfrac{1}{4}rr)(\tfrac{3}{4}Y\_ + \tfrac{1}{4}yy)$$
$$= \tfrac{9}{16}R\_Y\_ + \tfrac{3}{16}rrY\_ + \tfrac{3}{16}R\_yy + \tfrac{1}{16}rryy \qquad (12\text{-}3)$$

The product gives the expected 9:3:3:1 ratio in the offspring. (See if you can redo Problems 5, 6, and 9 by using the algebraic method.)

Testing the Fit of Observed to Expected Results

In experimental genetic crosses it is rare that the ratios of offspring exactly match the expected frequencies. In Mendel's dihybrid cross outlined in this chapter, for example, the expected 9:3:3:1 ratio is approximated but not precisely matched by the observed distribution of different classes in the offspring. The numbers actually obtained, 315:101:103:32, are obviously close to the expected ratios but do not duplicate them exactly. How far can observed frequencies depart from the expected results and still be considered to support a hypothesis? To answer this question one must determine whether the departure from expected results is simply due to chance rather than to mistakes in the hypothesis.

To make this determination geneticists often use a statistical technique called the *chi square* (χ^2) *method* to decide whether the outcome is close enough to expectations to support the hypothesis. To illustrate its use we will apply the method to Mendel's experiment described on p. 253.

In the experiment, 556 offspring were obtained with differences in seed form and color distributed among four classes. The offspring from the cross ($RrYy \times RrYY$) were expected in a 9:3:3:1 ratio—or approximately 313 round, yellow seeds ($\tfrac{9}{16} \times 445$); 104 wrinkled, yellow seeds ($\tfrac{3}{16} \times 556$); 104 round, green seeds ($\tfrac{3}{16} \times 556$); and 35 wrinkled, green seeds ($\tfrac{1}{16} \times 556$). As noted, the numbers actually obtained were 315:101:108:32.

Table 12-2 Calculation of χ^2

Calculation	Column 1 Round, Yellow	Column 2 Wrinkled, Yellow	Column 3 Round, Green	Column 4 Wrinkled, Green
1. Observed results	315	101	108	32
2. Expected results (e)	313	104	104	35
3. d (observed − expected)	+2	−3	+4	−3
4. d^2	4	9	16	9
5. d^2/e	0.01	0.08	0.15	0.2
6. $\chi^2 = \Sigma\,(d^2/e) = (0.01 + 0.08 + 0.15 + 0.2) = 0.44$				

Table 12-3 Chi Square Values

Degrees of Freedom	P = 0.99	P = 0.98	P = 0.95	P = .90	P = 0.80	P = 0.70	P = 0.50	P = 0.30	P = 0.20	P = 0.10	P = 0.05	P = 0.02	P = 0.01
1	0.00006	0.00063	0.0039	0.016	0.064	0.148	0.455	1.074	1.642	2.706	3.841	5.412	6.635
2	0.0201	0.0404	0.103	0.211	0.446	0.713	1.386	2.408	3.219	4.605	5.991	7.824	9.210
3	0.115	0.185	0.352	0.584	1.004	1.424	2.366	3.665	4.642	6.251	7.815	9.837	11.341
4	0.297	0.429	0.711	1.064	1.649	2.195	3.357	4.878	5.989	7.779	9.488	11.668	13.277
5	0.554	0.752	1.154	1.610	2.343	3.00	4.351	6.064	7.289	9.236	11.070	13.388	15.086

These results are evaluated by the chi square method according to this equation:

$$\chi^2 = \Sigma(d^2/e) \tag{12-4}$$

where χ^2 is chi square, d is the difference between observed and expected numbers for each class, and e is the expected number for each class. The symbol Σ means "the sum of." Calculation of chi square from this equation using the data from Mendel's experiment is shown in Table 12-2. The observed results for each class of offspring are entered in line 1 of the table and the expected results are listed in line 2. The difference between observed and expected numbers of offspring is found (line 3) and then squared (line 4). Then, for each column, line 4 is divided by line 2 (d^2/e) and the result is entered in line 5. Finally, the chi square value for the experiment is calculated by adding the results from each column together ($\Sigma[d^2/e]$) in line 6.

The figure obtained is then located on a table of chi square values (Table 12-3). To use the table, the *degrees of freedom* for the experiment must first be determined. Usually, the degrees of freedom are equivalent to the total number of classes in the experiment minus 1:

$$\text{Degrees of freedom} = n - 1 \tag{12-5}$$

where n is the number of classes. In the experiment entered in Table 12-2, there are four expected classes, so the degrees of freedom are $4 - 1 = 3$. Locating 3 degrees of freedom in Table 12-3 and following across that line to the figures closest to the value for chi square obtained in our calculation shows that our number falls between columns 3 and 4 of Table 12-3—that is, between $P = 0.95$ and

$P = 0.90$, closer to 0.95. By definition, values for P greater than 0.05 indicate that chance variations are responsible for the observed deviations from expected results and the hypothesis is therefore acceptable. Our P value, falling between 0.95 and 0.90, is well above this arbitrary limit. If lower values for P, between 0.05 and 0.01, are obtained, the differences between expected and observed ratios are considered significant and indicate that further experiments should be done before the hypothesis is accepted or rejected. Values for P less than 0.01 indicate that the hypothesis is probably wrong.

The rediscovery of Mendel's findings and conclusions in the early 1900s produced an immediate burst of interest and research in genetics. This work led to a series of additions and modifications to Mendel's basic principles of segregation, dominance, and independent assortment. The first of these modifications arose from the analysis of genes linked together in inheritance by their location together on the same chromosome pair. This analysis, in turn, led to the discovery of recombination, the result of exchanges between segments of homologous chromosomes. Other modifications developed from a study of the inheritance of genes carried on the sex chromosomes, variations in the degrees of dominance between the alleles of a gene, and genes that interact to produce combined rather than distinct and separate effects. Some genes were also discovered to be carried in the cytoplasm rather than in the cell nucleus. These new findings, rather than supplanting Mendel's basic principles, added to them and greatly expanded our understanding of genes and how they are transmitted and interact to produce the characteristics we detect as the phenotype of an individual.

13

ADDITIONS AND MODIFICATIONS TO MENDEL'S GENETIC PRINCIPLES

LINKAGE AND RECOMBINATION

Mendel studied seven pairs of factors, all of which separated independently in the formation of gametes. By chance the total number of pairs chosen for Mendel's experiments matches the number of chromosome pairs in his experimental organism: there are seven pairs of chromosomes in peas. If Mendel had extended his study to eight or more pairs, his hypothesis that the pairs always separate independently might have required some modification.

This fact was first pointed out shortly after Sutton published his hypothesis of the equivalence of behavior between genes and chromosomes. Several people, including Sutton, drew attention to the fact that completely independent separation of gene pairs is not always expected if the number of gene pairs exceeds the number of chromosome pairs in an organism. Because it was obvious that there are at least hundreds or thousands of genes in an organism (there are probably tens of thousands or more), it seemed likely that many sets of genes would be found to be linked together on the same chromosomes and transmitted together to offspring.

The Discovery of Linkage and Recombination

The first of these linked patterns was discovered in peas by the English scientists W. Bateson and R. C. Punnett (for whom the Punnett square is named) in 1906. These investigators discovered that two pairs of genes, one controlling flower color (purple versus red) and one the length of pollen grains (long versus short), were transmitted to offspring in the same combinations and rarely separated independently. Instead, if one parent had purple flowers and long pollen grains, these tended to be transmitted to offspring together as if flower color and pollen length were controlled by a single gene. It was obvious that the two characteristics could not be controlled by a single gene, however, because other plants could be found in which purple flower color was combined with short pollen grains. In this case, the purple color and short pollen length were usually transmitted together as a unit. Even more disturbing was the observation that, on occasion, offspring of a parent with the linked characters sometimes showed the opposite arrangement. A parent with purple flowers and long pollen grains, for example, might sometimes produce offspring with purple flowers and short pollen grains.

A satisfactory explanation for observations of this type was not advanced until 1910, when Thomas H. Morgan and his coworkers at Columbia University correctly interpreted their equivalent discoveries with linked genes in the fruit fly *Drosophila*. Morgan noted that linkage could be explained if the genes usually inherited together are carried on the same pair of chromosomes. At the same time, he also offered the correct explanation for the occasional changes in the combination of genes carried on the same chromosome, a mechanism later called **recombination**.

Different genes carried on the same chromosome are linked together in their inheritance.

An Experiment Demonstrating Linkage and Recombination Since Morgan's experiments demonstrating linkage and recombination were complex, we will review instead a later experiment with corn carried out by C. B. Hutchinson in which the results are easier to follow. Two pairs of genes were studied in Hutchinson's experiment. One pair, designated *S*, controlled the shape of grains in the ears; these were either full and round (*S*) or shrunken (*s*). The second pair of genes, designated *C*, affected the color of grains in the ears, which were either colored (*C*) or colorless (*c*). In initial crosses between plants that had round, colored grains with plants that had shrunken, colorless grains, all the offspring had round, colored grains. This result is similar to Mendel's dihybrid crosses and can be explained by assuming that the parent with round, col-

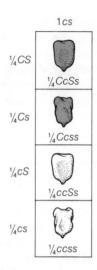

Figure 13-1 Expected results of the testcross *CcSs* × *ccss* if the *C* and *S* genes separate independently (see text).

ored seeds possessed the gene pairs *CCSS* whereas the parent with shrunken, colorless grains had the genetic constitution *ccss*. The F_1 offspring, with round, colored seeds, would then have the pairs *CcSs* and would all show the round, colored traits because these are dominant over the shrunken, colorless condition.

In the next experiment, Hutchinson backcrossed a number of *CcSs* F_1 plants with one of the parent varieties: the *ccss* type with shrunken, colorless seeds. Figure 13-1 shows what would happen if these gene pairs separated independently as in Mendel's dihybrid crosses. The gametes expected in the *CcSs* parent are shown in the left-hand headings in the Punnett square. The *ccss* parent can produce only one type of gamete (entered in the top heading of the Punnett square): *cs*.

Filling in the squares of the diagram reveals that four classes of offspring would be expected from this cross if the two pairs of genes separated independently. Each of these classes, *CcSs*, *Ccss*, *ccSs* and *ccss*, occurs with an expected frequency of one-fourth. If separation of the *C* and *S* gene pairs is independent, the cross would therefore yield offspring with colored, round:colored, shrunken:colorless, round:colorless, shrunken grains in a 1:1:1:1 ratio.

In Hutchinson's experiment these expected ratios were not even remotely approximated. Of the 8368 offspring, almost all were like the parents: 4032 with col-

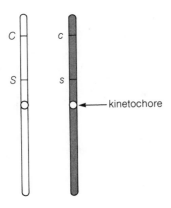

Figure 13-2 Arrangement of the alleles *Cc* and *Ss* on the two chromosomes of a homologous pair (see text).

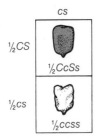

Figure 13-3 Results of a cross between *CcSs* and *ccss* parents if the *C* and *S* genes are located on the same chromosome (see text).

ored, round kernels (*CcSs*) and 4035 with colorless, shrunken kernels (*ccss*). These parental types amounted to 8067/8368 × 100 or about 96.4 percent of the total offspring. The few remaining offspring fit into the other two classes. Only 149 colored, shrunken (*Ccss*) and 152 colorless, round (*ccSs*) individuals were counted, amounting to only 301/8368 × 100 or about 3.6 percent of the total.

The preponderance of combinations resembling the parental types can be explained by assuming that the genes are linked on the same chromosome. In Hutchinson's recombination experiment, suppose that the two gene pairs *Cc* and *Ss* in the colored, round parent are located at different points on the same pair of chromosomes (Figure 13-2). One chromosome of the pair carries the two alleles *C* and *S*; the other has the two alleles *c* and *s* at corresponding locations. As the chromosomes separate in meiosis, gametes will receive either the chromosome bear-

ing the *C* and *S* alleles together or the chromosome carrying *c* and *s*. Thus only two types of gametes, *CS* or *cs*, are expected with an equal frequency of one-half. The other parent used in the cross, *ccss*, possesses two identical chromosomes with respect to the gene pair studied: both would contain the *c* and *s* alleles. These gametes are entered in the Punnett square in Figure 13-3. Filling in the squares in the diagram reveals that only two classes of offspring are expected, *CcSs* and *ccss*, in a 1:1 ratio. Thus the assumption of linkage can explain the predominance of parental combinations of alleles in the offspring. But what about the small percentage of new combinations, *Ccss* and *ccSs*, observed with a frequency of about 3 in 100 in the crosses?

These offspring are called **recombinants** (see also p. 237) because they do not contain the parental combination of alleles. Combinations that resemble the *CcSs* parent (called **parentals**) in the offspring of Hutchinson's first cross contain either the alleles *CS* or *cs* linked together on the same chromosomes. These, combined with the *cs* chromosome from the other parent, produce the two dominant classes making up 97 percent of the offspring. The remaining 3 percent contain chromosomes obviously originating from the *CcSs* parent but with the new combination *Cs* or *cS* (the recombinants).

Morgan's Explanation of Recombination Morgan was the first to see that linkage could be explained by assuming that linked genes are inherited together because they are carried on the same chromosome. He also realized, from his work with *Drosophila*, that the low frequency of recombinants was probably related in some way to the **chiasmata** or **crossovers** appearing between chromatids of the tetrads in prophase 1 of meiosis (see p. 231). He suggested that the recombinants arose through these crossovers by the mechanism shown in Figure 13-4. In most of the cells of the *CcSs* parent entering meiosis, no chiasmata were formed between the points on the tetrad bearing the alleles for seed shape and color (Figure 13-4*a*). These four chromatids separated and entered the gametes singly, producing the two gamete types shown in Figure 13-4*b*. But in a small percentage of the cells entering meiosis in the *CcSs* parent, a crossover or chiasma formed between the two genes (Figure 13-4*c*). This crossover, by breakage and exchange of parts in two of the four chromatids, brought together the new combinations *C* and *s* in one recombinant chromatid and *c* and *S* in the other. (Note that two of the four chromatids of the tetrad remained unrecombined.) These four chromatids separated during the two meiotic divisions and entered separate gametes, giving

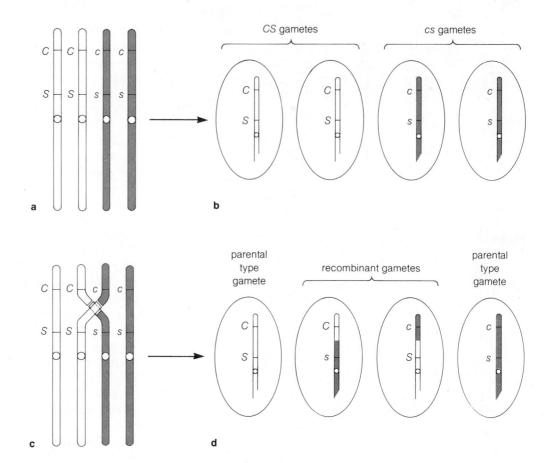

Figure 13-4 Morgan's explanation of linkage and recombination (see text).

rise to four different types: two parentals with the *CS* or *cs* combination (from the two chromatids not entering into the crossover) and two recombinants with the new *cS* and *Cs* combinations (Figure 13-4*d*; see also Figure 11-9). When combined in fertilization with the *cs* gametes from the *ccss* parent, the recombinants produced the two additional *ccSs* and *Ccss* classes noted as 3 percent of the offspring.

Morgan's hypothesis of linkage and recombination was able to reconcile the new observations with Mendel's hypothesis that different pairs of genes separate independently when gametes are formed. It was now realized that completely independent assortment, as proposed by Mendel, applies only to genes carried on separate chromosomes. Linked genes, carried on the same chromosome, are inherited according to Mendel's pattern for single genes except for rearrangements of the linkage due to recombination.

Linked genes are inherited according to Mendel's pattern for single genes except for rearrangements of the linkage due to recombination.

Recombination of linked genes depends on the breakage and exchange (crossing over) of segments between homologous chromosomes.

Chromosome Mapping by Recombination Studies

The observation that 3 percent of the offspring in Hutchinson's experiment were recombinants means that 3 percent of the gametes originating from the *CcSs* parent contained recombined chromosomes. In other words: for every 100 gametes formed, 3 contain recombinants. To produce an average of 3 percent recombinants among the gametes, 6 percent of the cells in meiosis must undergo a recombination between the *C* and *S* genes. (Any cell undergoing a single recombination or crossover between the *C* and *S* genes to produce two recombined chromosomes also yields two unrecombined chromosomes; see Figure 13-4*d*.) In general, then, the percentage of cells undergoing recombination is always twice the percentage of gametes containing recombinant chromosomes.

Morgan noted that the percentage of cells undergoing recombination depends on the genes being studied. Some linked genes are almost always found in the parental combinations, with recombinants appearing in very low frequency, sometimes in less than 1 percent of the total gametes. Other genes show a much higher rate of recombination, so high that it is difficult to detect that the genes are linked together. In other words, the alleles, even though known by other crosses to be carried on the same chromosome, segregate in an approximate 1:1:1:1 ratio if the basic *AaBb* × *aabb* testcross used in Hutchinson's experiment is carried out. From observations of this kind, Morgan and one of his students, Alfred M. Sturtevant, proposed that the amount of recombination observed between any two genes located on the same chromosome pair is a reflection of the distance between them on the chromosome. The greater this distance, the greater the chance that a crossover can form between the genes and the greater the recombination frequency. In more formal terms, Morgan and Sturtevant's hypothesis stated that the frequency of recombination between two genes is proportional to the distance between them on the chromosome.

The frequency of recombination between different genes carried on the same chromosome is proportional to the distance between them.

It was quickly realized that if the recombination frequency is proportional to the distance between linked genes, these frequencies could be used to *map* the chromosome and assign the genes to relative locations. Assume, for example, that three genes, *A*, *B*, and *C*, are known to be linked together on the same chromosome. In genetic crosses, 10 percent recombination is detected between genes *A* and *B*. Crosses between *A* and *C* reveal 8 percent recombination, and crosses between *B* and *C* show 2 percent recombination. These frequencies are compatible with only one possible arrangement of genes on the chromosome. Note that both the order of the genes and their

relative degree of separation can be estimated from their recombination frequencies:

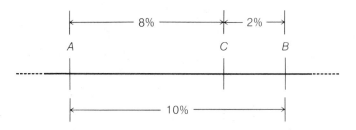

Using this method, most of the known genes of the best-studied organisms, *Drosophila*, *Neurospora*, *Escherichia coli*, corn, and various other species have been mapped and assigned positions on the chromosomes of these species. (Figure 13-5 shows the linkage map of one of the chromosomes of a mouse.) Note that these positions are only relative; the positions of the genes are indicated only by recombination percentage distances (called **crossover** or **map units**) and not by physical distances in micrometers or nanometers.

Recombination maps show the relative positions of genes on a chromosome but not their actual location.

Sex Linkage

In 1910 Morgan noted a curious pattern of inheritance in *Drosophila* in which eye color seemed to vary according to the sex of the fly. All the genetic traits studied up to that time had been transmitted equally between male and female offspring, and the sex of the individuals could be ignored in analyzing the frequency of various classes. Morgan's discovery of an exceptional pattern of inheritance dependent on the sex of the offspring, called **sex linkage**, was to have great significance for the later understanding of several hereditary diseases in humans that are transmitted as sex-linked traits.

Morgan first detected sex linkage in a line of flies that had white eyes instead of the usual red. If white-eyed males of this line were crossed with red-eyed females, all of the F_1 generation, both males and females, had red eyes

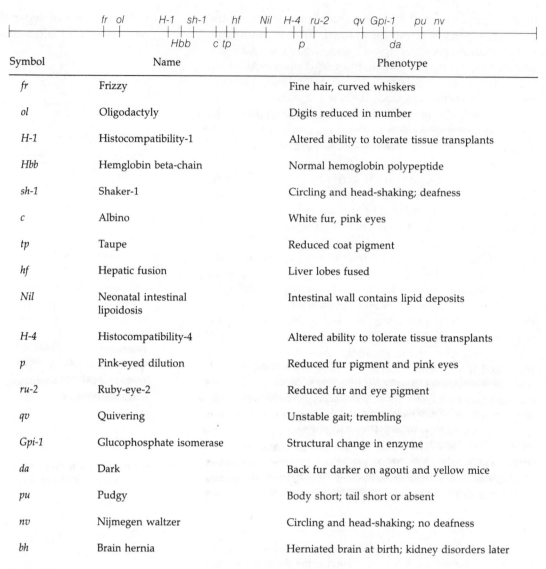

Symbol	Name	Phenotype
fr	Frizzy	Fine hair, curved whiskers
ol	Oligodactyly	Digits reduced in number
H-1	Histocompatibility-1	Altered ability to tolerate tissue transplants
Hbb	Hemglobin beta-chain	Normal hemoglobin polypeptide
sh-1	Shaker-1	Circling and head-shaking; deafness
c	Albino	White fur, pink eyes
tp	Taupe	Reduced coat pigment
hf	Hepatic fusion	Liver lobes fused
Nil	Neonatal intestinal lipoidosis	Intestinal wall contains lipid deposits
H-4	Histocompatibility-4	Altered ability to tolerate tissue transplants
p	Pink-eyed dilution	Reduced fur pigment and pink eyes
ru-2	Ruby-eye-2	Reduced fur and eye pigment
qv	Quivering	Unstable gait; trembling
Gpi-1	Glucophosphate isomerase	Structural change in enzyme
da	Dark	Back fur darker on agouti and yellow mice
pu	Pudgy	Body short; tail short or absent
nv	Nijmegen waltzer	Circling and head-shaking; no deafness
bh	Brain hernia	Herniated brain at birth; kidney disorders later

Figure 13-5 Linkage map of chromosome I of the mouse. From data compiled by M. C. Green, in *Handbook of Biochemistry* (H. A. Sober, ed.), Chemical Rubber Company, Cleveland, 1970.

(Figure 13-6a). This result suggested that red and white eye color is controlled by two alleles of a single gene; red is dominant over white if both alleles are present in the same individual. In the offspring of a cross between individuals of the F_1 generation, these alleles would then be expected to segregate without respect to sex according to a 3:1 ratio. Red eyes would be expected as the most frequent class if the gene for eye color followed the usual pattern of Mendelian inheritance. Surprisingly, Morgan

found that all the females in the F_2 generation were red-eyed and, among the males, half were red-eyed and half white-eyed (Figure 13-6b).

In the reciprocal cross, in which white-eyed females were crossed with red-eyed males, different results were obtained. In the F_1 offspring of this cross, all the females were red-eyed and all the males white-eyed. Crosses between F_1 individuals produced two classes of eye color in each sex of the F_2 generation: one-half of all males and

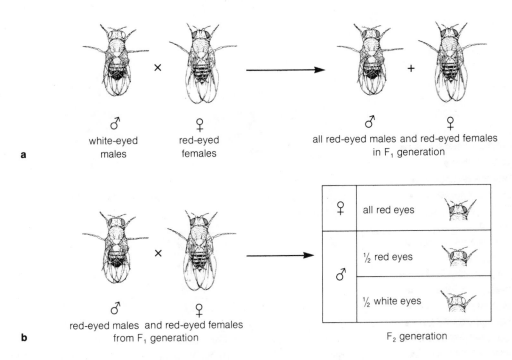

Figure 13-6 Sex linkage in a cross with *Drosophila* (see text). (**a**) A cross between white-eyed males and red-eyed females. (**b**) The F₁ cross between the red-eyed males and red-eyed females resulting from *(a)*.

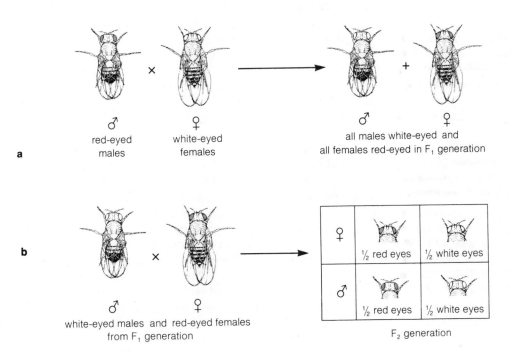

Figure 13-7 Sex linkage in the reciprocal cross of Figure 13-6 (see text). (**a**) Cross between red-eyed males and white-eyed females. (**b**) F₁ cross between the red-eyed females and white-eyed males resulting from *(a)*.

females were red-eyed, and one-half were white-eyed (Figure 13-7).

Morgan realized that this unexpected pattern of inheritance followed the transmission of sex chromosomes in *Drosophila.* (See Supplement 11-1 for a discussion of sex chromosomes and how they are distributed to gametes in meiosis.) These chromosomes are transmitted from parents to offspring according to the following patterns in *Drosophila.* Both males and females have three pairs of chromosomes (the **autosomes**) found in both sexes (Figure 13-8). Females, however, have one more pair: the X pair. Males have only a single X chromosome, which pairs during meiosis with another single chromosome: the Y chromosome. Females have no Y chromosome. Thus, with respect to the sex chromosomes, females are XX and males are XY. When gametes are formed in females, the X pair separates and each egg receives one X chromosome. In males, the X and Y separate and sperm cells of two types are produced, each containing the usual autosomes and either an X or a Y sex chromosome. An egg fertilized by a Y sperm results in an XY zygote, which develops into a male; eggs fertilized by an X sperm produce an XX zygote, which develops into a female.

Traits are sex-linked because the genes controlling them are located on the sex chromosomes.

Morgan reasoned that the inheritance of eye color in *Drosophila* could be explained if the gene for this trait is carried on the X chromosome but not on the Y. (The Y chromosome in *Drosophila* is almost inert genetically and carries few active genes.) Inheritance of eye color would then follow the pattern shown in Figure 13-9. In the first cross carried out by Morgan (from Figure 13-6), red-eyed females were crossed with white-eyed males. Suppose that the two X chromosomes of the females both carry the dominant allele W for red color and that the single X in the male carries the recessive allele w for white. These females produce one kind of egg with respect to eye color, all carrying an X chromosome with the W allele for red eyes. Males produce two kinds of sperm cells—one carrying an X with the recessive w allele for white eyes and one with the Y, which does not carry a gene or allele for eye color. These two types of sperm cells occur with an equal frequency of one-half. The egg and sperm cells are entered in the Punnett square in Figure 13-9a; note in the results of the cross that all F_1 females are Ww with red eyes. All the males receive an X carrying the W form of the allele from their mothers and are also red-eyed.

Crossing these F_1 individuals gives the results shown in Figure 13-9b. The Ww females of the F_1 generation pro-

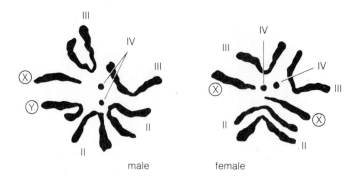

Figure 13-8 The chromosomes of *Drosophila melanogaster* at metaphase of mitosis.

duce two types of eggs with an equal frequency of one-half—one with an X chromosome carrying the W allele for red eyes and one with an X carrying the w allele for white eyes. The F_1 males produce two types of sperm cells in equal numbers. One type carries the X chromosome with the W allele for red eyes, and the second type the inert Y. These egg and sperm cells and their frequencies are entered in the headings to the Punnett square. The filled-in squares of the diagram predict that one-half of the females (one-fourth of the total offspring) will receive an X chromosome with the W allele from both the egg and sperm and thus will have red eyes. The other half of the females receive an X with the recessive w allele for white eyes from the egg and the dominant W-bearing X from the sperm and will also be red-eyed. Among the males, one-half receive an X through the egg with the dominant W allele for red eyes and one-half receive an X with the recessive allele w for white eyes. Thus one-half of the males will be red-eyed and one-half white-eyed.

These predictions match the results obtained by Morgan. Moreover, all other crosses and tests involving the W gene for eye color support the hypothesis that these alleles are carried on the X chromosome. (See if you can set up the Punnett square for Morgan's second experiment and predict the frequency of red and white eyes among the F_1 and F_2 males and females.) Since Morgan's first discovery of sex-linked inheritance, more than 100 genes with a similar pattern of inheritance have been described in *Drosophila.* Among the many important results arising from this work was the first proof of the chromosome theory of inheritance, which followed soon after Morgan's discovery of sex linkage (see Supplement 13-1).

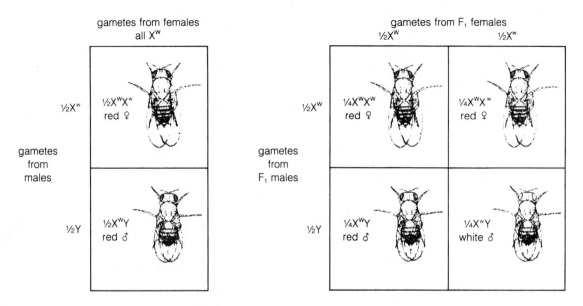

Figure 13-9 The explanation of Morgan's results (see text). **(a)** Explanation of the parental cross in Figure 13-6. **(b)** The F₁ cross of Figure 13-6.

Sex Linkage and Human Disease

Several human diseases are transmitted by the same patterns noted for sex-linked traits in *Drosophila*. Humans have XX–XY sex chromosomes as in *Drosophila*, with XX females and XY males. Two of the best-studied examples of inheritance linked to the sex chromosomes in humans are certain types of *color blindness* and *hemophilia*.

Color Blindness Human red-green color blindness is a recessive trait that follows a pattern of inheritance similar to the red-white eye color pattern studied by Morgan in *Drosophila*. Although testcrosses cannot be carried out with humans, a similar analysis can often be made by looking into family records and constructing a chart called a **pedigree**. A pedigree shows all marriages and offspring for as many generations as possible, the sex of individuals in the different generations, and the presence or absence of the trait of interest.

A pedigree of a family with a history of color blindness is shown in Figure 13-10. Females are designated by the symbol ♀ and males by ♂. Solid black circles in the pedigree indicate the presence of the trait, and white circles signify its absence. At the very top of the pedigree

(generation 1) the earliest recorded marriage in the family involved a color-blind male and a normal female. The single child of this marriage (generation 2), a daughter, married a normal male. This female, who received the X chromosome bearing the recessive allele for color blindness from her father, passed the trait to five of her seven sons (generation 3). The two remaining sons received the other X from their mother, carrying the dominant allele for normal color vision.

Three of the color-blind males of this generation married normal females. From the pattern of inheritance of the offspring (generation 4) in the marriages of the color-blind sons, it seems probable that none of their wives carried a recessive allele for color blindness because none of the sons or daughters in the fourth generation showed the trait. All the daughters of color-blind fathers were **carriers** of the recessive gene for the trait, however, because each received the X chromosome bearing the recessive allele from her father. As a result, the trait showed up again in sons in the next generation (generation 5) who received the X with the recessive allele from their mothers. Note that appearance of the trait alternates from generation to generation in the males. It does so because a father does not pass on his X chromosome, with either recessive or dominant traits, directly to his sons; the X chromosome

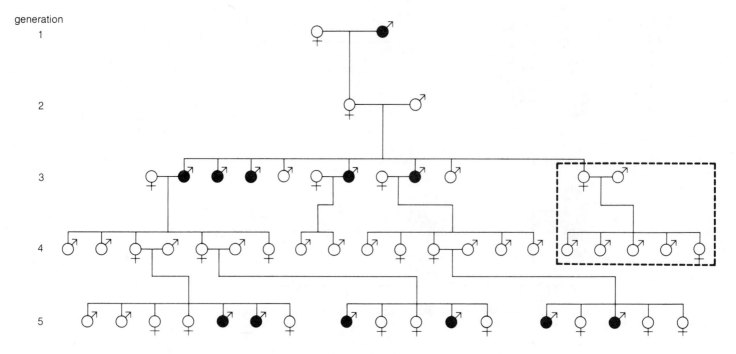

Figure 13-10 Pedigree of a family with a history of color blindness. Females are designated by the symbol ♀ and males by ♂. Solid black circles indicate the presence of color blindness; white circles signify its absence. Marriages are indicated by a horizontal bar between two circles.

received by a male always comes from his mother. As a result, skipping of a trait in a human pedigree suggests that the allele under study is recessive and carried on the X chromosome.

None of the females in the pedigree in Figure 13-10 developed color blindness. A color-blind daughter could be produced, however, if a female with normal vision who carried a recessive allele for color blindness on one chromosome married a color-blind male. A possible outcome of this type is shown in Figure 13-11, which shows what might have happened if the marriage enclosed in dotted lines in Figure 13-10 was to a color-blind male. Some of the sons would be likely to receive the X carrying the dominant allele for normal vision from the mother, and some would receive the X with the recessive allele for color blindness. Thus both color-blind and normal sons could arise from this marriage. The same possibilities hold true for the daughters. The father's X chromosome brings one recessive allele to each daughter. This allele, if combined with the mother's X bearing the recessive allele, would produce a color-blind daughter. If combined with the mother's X carrying the dominant allele for normal vision, the daughter would have normal color vision but would carry the trait.

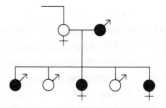

Figure 13-11 A possible outcome of the marriage enclosed in dotted lines in Figure 13-10 if the daughter carrying the trait married a color-blind male. Both color-blind and normal sons and daughters could arise from this marriage.

The Inheritance of Hemophilia Hemophilia is another human disability that is recessive and carried on the X chromosome. Hemophilic individuals are "bleeders"; they do not possess the normal mechanism for clotting blood if injured. Although affected persons, with luck and good care, can reach maturity, their lives are tightly circumscribed by the necessity to avoid injury of any kind. Even slight bruises can cause internal bleeding and prove fatal.

One of the best-known cases of hemophilia occurred in the royal families of Europe descended from Queen Vic-

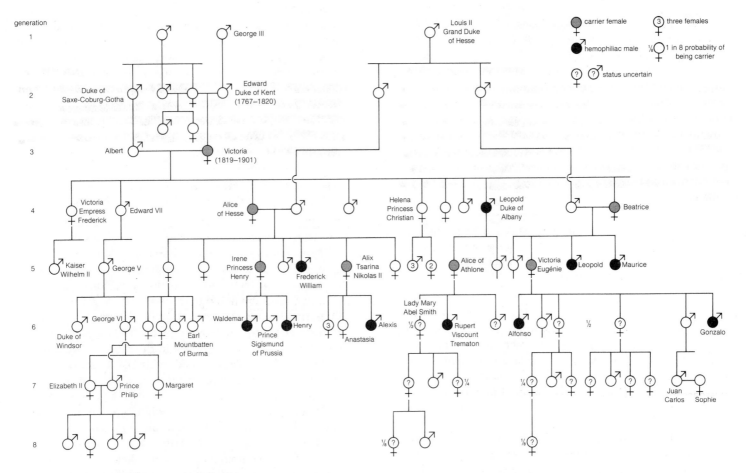

Figure 13-12 The inheritance of hemophilia in descendants of Queen Victoria. From V. A. McKusick, *Human Genetics*, 2nd ed., © 1969. Reprinted by permission of V. A. McKusick and Prentice-Hall, Inc.

toria (Figure 13-12). Hemophilia had not been recorded in the ancestors of Victoria in the royal line, so the recessive trait probably arose in the queen or possibly her father as a mutation.[1] Because so many sons of European royalty were affected, the trait had an important and sometimes tragic influence on the course of history. In Russia, for example, the preoccupation of the royal family with their hemophilic son has been considered to be a major factor in the events leading to the revolution and the eventual installation of a communist government in that country.

The disease affected only sons in the royal lines but presumably could have affected daughters if a hemophilic son married a female carrying the recessive allele on one

of her X chromosomes. Because the disease is very rare, the chance of a hemophilic son marrying a female unaffected but carrying the trait is exceedingly low, and only a few marriages of this type have been recorded. The daughters of the marriages in the Queen Victoria pedigree, although carriers of the trait, were normal and obviously received the X chromosome carrying the normal, dominant allele from their mothers rather than the X carrying the recessive allele for hemophilia. (Other important inherited human diseases are described in Supplement 13-2.)

Sex-Influenced Traits

Not all characteristics that differ in expression between the sexes are controlled by genes located on sex chromosomes. Many traits, such as balding in men, growth and

---

[1]The origins of mutations, which are caused by spontaneous changes in the DNA sequence of a gene arising from a variety of chemical and physical mechanisms, are discussed in detail on pp. 449–452.

distribution of body hair in both sexes, enlargement of the breasts in women, voice pitch, and even the form of the external sex organs, are controlled by genes and alleles present in both males and females. These genes produce different external effects in men and women because of the influence of other factors, primarily the sex hormones, and not because certain genes or alleles are present in one sex and absent in the other. Imbalance in the hormones affecting the genes carried on autosomes can cause the appearance of opposite sexual characteristics, such as enlargement of the breasts in men or growth of facial hair in women.

Sex-influenced traits are controlled by genes and alleles that are present in both males and females. These genes and alleles produce different effects in males and females through the influence of other factors such as hormones.

Pattern baldness, for example, is inherited in a manner suggesting that the trait is controlled by two alleles, H and h, of a gene present in both men and women. In males, the H form of the gene acts as a dominant allele. Males with the HH or Hh combination of alleles begin to lose hair from the scalp with the onset of puberty, as soon as the male hormone *testosterone* is secreted in quantity in the body. (The same hormone, however, at the same time, increases the growth of facial hair.) Homozygous recessive (hh) males, who occur as a small percentage of the population, are relatively unaffected and do not develop pattern baldness to nearly so great a degree in response to the hormone. While many "cures" for pattern baldness are sold to anxious HH and Hh males, the only effective treatment is castration. Most men would consider that too high a price to save the hair.

Women also carry the H and h alleles of the same gene for baldness. The H allele, however, which acts as a dominant in males, acts as a recessive under the influence of the group of sex hormones secreted in females. As a result, only HH females suffer hair loss (which, in any event, is less severe than in men); Hh and hh females are unaffected. Thus the same alleles produce different external effects in men and women through the influence of the sex hormones.

LACK OF DOMINANCE

Mendel's principle of dominance assumes that if two different alleles of a gene are present in the same individual, one masks the effects of the other and the individual displays only the trait of the dominant allele. Although this assumption was true for the traits Mendel studied in his garden peas, further work has revealed that the effects of recessive alleles are not always completely masked by dominant alleles. This phenomenon, termed **lack of dominance** or **incomplete dominance**, means simply that the effects of both alleles can be detected in a diploid, heterozygous individual with two different alleles of a gene. In other words, each genotype has a distinct and distinguishable phenotype.

In lack of dominance the effects of both alleles carried by a heterozygous individual can be detected in the phenotype.

The inheritance of feather color in blue Andalusian chickens follows a pattern showing lack of dominance. Among breeders this blue Andalusian is much admired for its color. In trying to maintain this color in offspring, however, breeders noted that these birds never breed true. Blue Andalusians, if crossed among themselves, always produce offspring with two new colors—black and speckled white in addition to blue—in the ratio 1 black:2 blue:1 speckled white. Black mated with black always breeds true, giving only black offspring. Speckled white chickens, if mated among themselves, also maintain the speckled white color. Black and speckled white birds, however, give rise to all blue Andalusian offspring if mated.

These results suggested that the three colors come from different combinations of two alleles of the gene for feather color. These alleles are C for black and c for speckled white. Black does not completely dominate white; as a result, in heterozygous Cc individuals the mixture of effects produces the Andalusian blue color. A cross between black (CC) and speckled white (cc) yields all Cc (blue Andalusian chickens) in the F_1 generation, as observed in fact. Crossing these individuals produces the combinations CC, Cc, and cc in the F_2 offspring in the ratio 1:2:1. These combinations, which produce black (CC), blue Andalusian (Cc), and speckled white birds (cc), thus explain the results obtained by breeders in their attempts to produce true-breeding strains from blue Andalusian birds. The hypothesis may be tested by backcrossing individuals of the F_2 generation with the original black and speckled white parents. (See if you can set up the Punnett squares predicting these results and ratios.) The predictions are

Black parent × blue F_2 ($CC \times Cc$) = ½ black + ½ blue offspring

Speckled white parent × blue F_2 ($cc \times Cc$)

= ½ blue + ½ speckled white offspring

These results, in the predicted ratios, are in fact obtained.

In lack of dominance, the various genotypes for feather color (*CC*, *Cc*, and *cc*) result in different phenotypes—that is, a different outward appearance in the individuals obtained in crosses. Completely dominant alleles such as that for round seed shape produce homozygous and heterozygous individuals that are identical in phenotype even though their genotypes—*SS* and *Ss*—are different.

A allele	...ATGCAGATACCGATTACAGACCATAGG...
a_1 allele	...ATGCAGA<u>G</u>ACCGATTACAGACCATAGG...
a_2 allele	...ATGCAGAT<u>G</u>CCGATTACAGACCATAGG...
a_3 allele	...ATGCAGATACCGATTACAG<u>G</u>CCATAGG...

Figure 13-13 The multiple alleles of a gene consist of small differences in the nucleotide sequence at one or more points (underlined).

MULTIPLE ALLELES

One of Mendel's basic assumptions was that each hereditary trait is controlled by pairs of factors, the alternate alleles of each gene. Mendel's assumption has withstood the tests of time and experimentation: the alleles for each gene in diploid individuals do occur in pairs. It soon became apparent, however, that more than two alleles of a gene may exist if all the individuals in a population are taken into account. Imagine a population in which some diploid individuals have the *A* and a_1 alleles of the *A* gene, others the *A* and a_2 alleles, and still others the a_1 and a_2 alleles. (Note that *AA*, A_1a_1, and a_2a_2 individuals are also possible.) Thus although there are only two alleles of the *A* gene in any one individual, there are three alleles of the gene—*A*, a_1, and a_2—in the entire population. These several forms of a gene are called **multiple alleles.**

We now know that most genes probably occur in a large number of allelic forms. There are at most only two alleles in any one individual, of course, because only two copies of each chromosome are present in diploids. In the cells of haploid organisms, moreover, only one copy of each chromosome and one allele of each gene are present in any one individual. Nevertheless, in haploids as well as diploids, many alleles may be present in a population as a whole. In contemporary terms, these different alleles represent small differences at one or more points in the sequence of nucleotides making up the gene (Figure 13-13).

Although an individual may carry only two alleles of a gene, more than two alleles may be distributed among the individuals of a population.

One of the best examples of multiple alleles is the inheritance of coat colors in rabbits. Wild rabbits have brown fur that is fairly uniform in color over the whole animal. At the other end of the spectrum is the all-white fur of the **albino** domestic rabbit. Crosses between **wild-type** and albino rabbits reveal that the wild-type and white colors are controlled by two alleles of a single gene. A dominant *C* allele of this gene produces the brown fur color of wild-type rabbits; individuals homozygous for the recessive *c* allele are albinos. Thus *CC* or *Cc* rabbits have brown fur and *cc* rabbits are albinos; crosses between *Cc* parents produce brown and albino offspring in a 3:1 ratio. (Albinism in higher animals is frequently the result of a homozygous recessive gene.)

A number of other fur colors have been discovered that also give a 3:1 segregation in the F_2 generation in crosses with wild-type rabbits. *Chinchilla* rabbits are silver-colored instead of brown. If chinchillas are crossed with wild-type rabbits, the offspring in the F_1 generation are brown; crosses among the F_2 offspring give three-fourths brown and one-fourth chinchilla, indicating that chinchilla is also recessive to the wild type. Crosses between chinchilla and albino, however, reveal that the chinchilla allele is dominant to albino. One other allele of interest, called *Himalayan*, produces white rabbits with pigmented nose, ears, tail, and legs. This coat is also produced by a recessive form of the gene for fur color; Himalayan rabbits crossed with the wild type also give a 3:1 ratio of wild type to Himalayan in the F_2 generation.

Multiple alleles may occur with or without dominance.

Crosses between different combinations of the Himalayan, chinchilla, and albino alleles reveal that Himalayan is recessive to chinchilla but dominant to albino. These alleles are identified in genetic shorthand as C = wild type, c^{ch} = chinchilla, c^h = Himalayan, and c^w = albino. Their dominance relationships are summarized as follows:

C with C, c^{ch}, c^h, or c^w = wild-type brown fur (CC, Cc^{ch}, Cc^h, Cc^w)

c^{ch} with c^{ch}, c^h, or c^w = chinchilla ($c^{ch}c^{ch}$, $c^{ch}c^h$, $c^{ch}c^w$)

c^h with c^h or c^w = Himalayan (c^hc^h, c^hc^w)

c^w with c^w = albino (c^wc^w)

Most genes exist as a series of multiple alleles rather than in only two alternate forms. This additional complexity presents no real difficulty in genetic analysis, however, because a given diploid individual has only two alleles.

GENE INTERACTIONS

Polygenic Inheritance

Some traits follow a pattern of inheritance that seems to mimic the blending of genetic characters. Certainly human characteristics such as skin color, body size, and intelligence appear to take on intermediate levels in the offspring of parents that differ markedly in these traits. However, careful analysis indicates that the apparent blending results from interactions between the alleles of two or more different genes controlling the same inherited characteristics. This pattern is termed **polygenic inheritance**. The inheritance of skin color in humans, first investigated by C. B. Davenport in studies carried out with mulattoes in Bermuda and Jamaica, follows the polygenic pattern.

In polygenic inheritance the same phenotypic trait is controlled by two or more different genes.

Davenport discovered that "blending" of parental skin color to produce offspring with intermediate color was the most frequent outcome of matings between mulattoes, but not the only outcome. In his investigations of family pedigrees, Davenport found that such matings can also produce children that are either darker or lighter than the parents with a low but regular frequency. In fact, matings between mulattoes can produce a spectrum of skin pigmentation ranging from apparently pure black to colors indistinguishable from Caucasian white. Intermediate colors resembling the parents are most common, however.

Davenport saw that this distribution of skin pigmentation could be explained if pigmentation in humans is controlled by more than one gene. In what was probably an oversimplification, he proposed that two different genes, P_1 and P_2, control human skin color. Although there are at least two alleles of each gene, P and p, no dominance exists between any of the alleles. Pure African Negroes, with no white ancestors, were assumed to carry the four alleles $P_1P_1P_2P_2$. Persons of pure Caucasian ancestry were assumed to carry the four alleles $p_1p_1p_2p_2$. Matings between Negro and white individuals would produce the genotype $P_1p_1P_2p_2$, which is assumed to be a mulatto offspring of an intermediate brown color. Because no dominance is exerted between the P and p alleles of either

gene, $P_1p_1P_2p_2$ individuals can be distinguished in external appearance, or phenotype, from either $P_1P_1P_2P_2$ or the $p_1p_1p_2p_2$ parent.

A mating between two $P_1p_1P_2p_2$ mulattoes with the same degree of pigmentation would produce very different results, some of them unexpected if blending were to occur. The possible combinations of gametes from the two individuals are shown in the Punnett square in Figure 13-14. Both persons will produce the four gametes shown with equal frequencies of one-fourth. These gametes, when combined in the squares of the figure, produce a range of pigment combinations varying more or less evenly from pure black, $P_1P_1P_2P_2$, through a range of intermediate values (shaded in Figure 13-14) to pure white, $p_1p_1p_2p_2$. Either of the two extremes, white or black, would be expected in 1 out of 16 births if pigmentation is controlled by two genes as Davenport suggested. This range of pigments would not be expected if the inheritance of skin color depended simply on the blending of parental pigmentation. If blending occurred, two mulatto parents of intermediate color would always produce children of the same intermediate color.

Analysis of pigmentation in the offspring of mulatto parents does in fact indicate a distribution of this type. The children of such parents show the expected range of color from black to white, and intermediate values arise more frequently than either extreme. The extremes, in fact, are much rarer than would be expected if skin pigmentation in humans is controlled by only two genes. The low frequency with which pure black or white offspring are encountered in matings between persons of intermediate color suggests that skin pigmentation may be controlled by as many as five genes, each with at least two alleles P and p. In that case, crosses between $P_1p_1P_2p_2P_3p_3P_4p_4P_5p_5$ mulattoes would be expected to produce either pure black or white skin color in approximately 1 out of every 1000 children—an expectation that is close to the results observed.

Plotting the distribution among individuals of a trait that is controlled by polygenic inheritance, such as skin color, typically produces a bell-shaped curve. Figure 13-15 plots the frequencies of various body heights in men; the bell-shaped curve indicates that this human trait too is controlled by polygenic inheritance. Other organisms show equivalent patterns for certain characteristics, such as ear length in corn, seed color in wheat, and color spotting in mice.

Another Type of Gene Interaction: Epistasis

Interactions also occur in a pattern indicating that the alleles of one gene may override the effects of a different

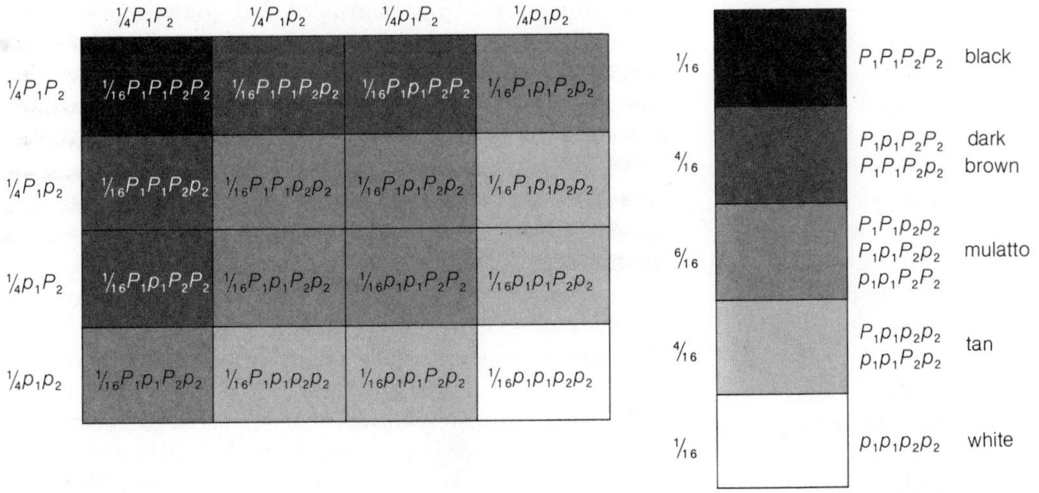

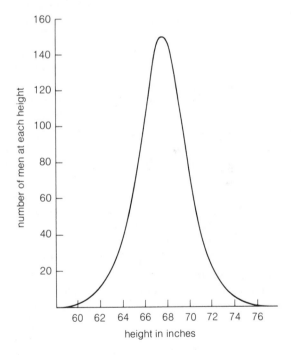

	¼P_1P_2	¼P_1p_2	¼p_1P_2	¼p_1p_2
¼P_1P_2	$\frac{1}{16}P_1P_1P_2P_2$	$\frac{1}{16}P_1P_1P_2p_2$	$\frac{1}{16}P_1p_1P_2P_2$	$\frac{1}{16}P_1p_1P_2p_2$
¼P_1p_2	$\frac{1}{16}P_1P_1P_2p_2$	$\frac{1}{16}P_1P_1p_2p_2$	$\frac{1}{16}P_1p_1P_2p_2$	$\frac{1}{16}P_1p_1p_2p_2$
¼p_1P_2	$\frac{1}{16}P_1p_1P_2P_2$	$\frac{1}{16}P_1p_1P_2p_2$	$\frac{1}{16}p_1p_1P_2P_2$	$\frac{1}{16}p_1p_1P_2p_2$
¼p_1p_2	$\frac{1}{16}P_1p_1P_2p_2$	$\frac{1}{16}P_1p_1p_2p_2$	$\frac{1}{16}p_1p_1P_2p_2$	$\frac{1}{16}p_1p_1p_2p_2$

$\frac{1}{16}$	$P_1P_1P_2P_2$	black
$\frac{4}{16}$	$P_1p_1P_2P_2$ $P_1P_1P_2p_2$	dark brown
$\frac{6}{16}$	$P_1P_1p_2p_2$ $P_1p_1P_2p_2$ $p_1p_1P_2P_2$	mulatto
$\frac{4}{16}$	$P_1p_1p_2p_2$ $p_1p_1P_2p_2$	tan
$\frac{1}{16}$	$p_1p_1p_2p_2$	white

Figure 13-14 The proposed inheritance of skin color in the offspring of parents with the $P_1p_1P_2p_2$ genotype (see text).

gene. Inheritance of coat color in guinea pigs follows this pattern. In guinea pigs, the gene *B* determines whether coat color is black or brown. The dominant allele *B* produces black color in *BB* and *Bb* individuals; *bb* individuals are brown. However, the expression of the *B* gene is controlled by a different gene *C* for color. In this gene, the *C* allele is dominant and both *CC* and *Cc* guinea pigs develop the black or brown fur color determined by the *B* allele. Homozygous recessive *cc* individuals develop no fur color, however, and are albinos no matter what alleles are present at the *B* gene. This type of interaction, in which one gene overrides the effects of another, is called **epistasis** (*epi* = over; *stasis* = standing or stopping).

In epistasis one gene overrides the expression of one or more different genes.

In spite of its rather uncommunicative name, epistasis has an easily understood molecular basis. The genes involved usually code for different enzymes that catalyze steps in a common pathway leading to production of a substance such as a skin, hair, or eye pigment:

Substance A $\xrightarrow{\text{enzyme 1}}$ substance B $\xrightarrow{\text{enzyme 2}}$

substance C $\xrightarrow{\text{enzyme 3}}$ pigment

Suppose that in an animal the *B* gene codes for enzyme 1 in the pathway. In its dominant *B* allele the gene codes for

Figure 13-15 A graph plotting men's height in inches against the frequency of each height. The distribution of frequencies produces a typical bell-shaped curve, indicating that height is controlled by more than one gene pair.

a highly efficient form of enzyme 1, leading to intensive activity of the pathway and production of large quantities of a fur pigment. The recessive *b* allele codes for a modified form of the enzyme that catalyzes the first step in the pathway very slowly, leading to greatly reduced pigment production by the remainder of the pathway when the *b* allele is present on both chromosomes carrying the gene. This condition could lead to black (*BB* or *Bb*) or brown (*bb*) fur or skin colors, since the two colors are produced by different concentrations of the same brown-black pigment, *melanin*.

Production of the pigment, however, depends on a different gene that codes for another enzyme catalyzing a subsequent step in the pathway. Assume that this is the enzyme catalyzing step 3. In the active form coded by the dominant *C* allele of this gene, step 3 proceeds at a rate sufficient to convert all available molecules of substance C into the pigment, whether the individual has the genotype *CC* or *Cc*. The recessive *c* allele, however, codes for a completely inactive form of the enzyme. Individuals homozygous for this allele (*cc* individuals) produce no active molecules of enzyme 3, and the pathway is blocked at step 3. As a consequence, these individuals are albinos and have white fur no matter which alleles of the *B* gene are present.

The two genes involved in an epistatic interaction may be on the same or different chromosome pairs. When they are on different chromosome pairs, as is often the case, the ratios resemble the products of a dihybrid cross (see p. 252) with some of the expected categories lumped together. In guinea pigs, for example, epistasis by the *C* gene combines the *ccBB*, *ccBb*, and *ccbb* genotypes together as albinos, producing the following distribution of phenotypes:

$$\frac{9}{16}\text{ black } + \frac{3}{16}\text{ brown } + \frac{4}{16}\text{ albino}$$

instead of the 9:3:3:1 ratio expected in a dihybrid cross. (See if you can set up the Punnett square producing these ratios.) There are many additional examples of epistatic interaction between different genes that lead to the same 9:3:4 or other ratios.

CYTOPLASMIC INHERITANCE

Not long after the rediscovery of Mendel's work in the early 1900s, a number of factors were observed in plants that did not appear to follow any of the usual Mendelian ratios in their pattern of inheritance. This unusual pattern was first noted in 1909 by the German scientist C. Correns, who studied the inheritance of leaf color in the four-o'clock (*Mirabilis*; Figure 13-16). In this flowering plant, individuals sometimes show patterns of white and green segments (called **variegation**) on leaves and stems. Chloroplasts within the cells of the bleached segments are colorless and do not turn green. If enough normal, green tissue is also present, plants containing the white segments can survive and may produce flowers in both the white and green segments.

By choosing flowers from different regions of the variegated plants, Correns was able to obtain pollen or eggs originating from entirely white or green portions of the plants or from regions in which the white and green segments were intimately mixed. Crosses between pollen and eggs from these segments revealed that the color of the offspring, instead of segregating in typical Mendelian ratios, always reflected the color of the tissue from which the egg was derived. The color of the parts of the plants contributing the pollen had no effect on the outcome of the cross. For example, eggs in flowers on white segments fertilized by pollen from flowers on green segments produced completely white seedlings; this condition was lethal and the seedlings died. Fertilizing eggs in flowers on green segments by pollen from flowers on white segments produced normal, green seedlings with no variegation into white and green sections. Eggs in flowers on segments of the plant with closely mixed white and green segments produced white, green, or variegated seedlings, no matter what segment the fertilizing pollen was derived from.

Correns noted that this pattern of inheritance follows the inheritance of cytoplasm in flowering plants. In fertilization in these plants (see p. 335), only the nucleus of the pollen normally enters the female gametophyte and fertilizes the egg. All the cytoplasm of the fertilizing pollen is left outside the female gametophyte in the supporting tissues of the flower (the ovary). As a result, all the chloroplasts and other organelles of higher plants are derived from the cytoplasm of the fertilized egg and not from the pollen. Noting this pattern, Correns reasoned that the gene affecting color in the chloroplasts was carried somewhere in the cytoplasm and not in the nucleus.

This pattern of transmission of heredity, called **cytoplasmic inheritance**, has since been observed in hundreds of eukaryotic species, including other plants, algae, animals, fungi, and protistans. As linkage was discovered and chromosome maps were developed, geneticists noted that the cytoplasmically inherited factors, as expected, showed no linkage to any genes carried on the chromosomes of the cell nucleus.

Usually, cytoplasmic genes affect the biochemistry and function of mitochondria or chloroplasts. Inheritance

Figure 13-16 The four-o'clock, *Mirabilis*.

often follows the pattern noted in four-o'clocks, in which most or all of the cytoplasm of the offspring is derived from one of the two parents. Establishment of DNA as the molecular basis of heredity in the 1950s prompted a search for DNA in mitochondria and chloroplasts. If cytoplasmic genes are indeed present in these structures, and if DNA is the genetic material, close inspection of these organelles ought to reveal the presence of DNA.

Some genes and alleles are carried on the DNA molecules in mitochondria and chloroplasts.

By the 1960s, the techniques for detecting nucleic acids had been sufficiently refined to show that DNA is in fact present in both mitochondria and chloroplasts. The genes in this DNA have been shown to code for polypeptide subunits of a variety of proteins in mitochondria and chloroplasts. Segments coding for ribosomal and transfer RNA are also included in the DNA of both organelles. (For details, see Supplement 9-1.)

THE CHANGING GENE CONCEPT

The results of experimentation in heredity described in this and the previous chapter—monohybrid and dihybrid crosses, linkage and recombination, sex linkage, incomplete dominance, polygenic inheritance, epistasis, and cytoplasmic inheritance—are usually grouped together as the findings of classical genetics. These patterns of inheritance were discovered and fully described by 1920.

From these studies certain physical characteristics of genes could be determined. Nuclear genes were known to be located on the chromosomes; this was demonstrated finally by Bridges' study of nondisjunction (see Supplement 13-1). Through recombination studies and calculations of the frequency of crossing over, it was determined that the genes are aligned on chromosomes in unbranched, linear order.

The idea that genes are units of function was also developed at this time. Although the mechanism was unknown, genes were understood to control more or less clearly defined characteristics of organisms. This functional definition was first translated into molecular terms by the biochemical research of the 1950s and 1960s. This work led to the discovery of DNA structure, the genetic code, RNA transcription, and the mechanism of protein synthesis. From this work, genes were initially proposed to be coding units that spell out the directions for making enzymes. As a result, a "one gene, one enzyme" concept arose and held favor for a time. Gradually it became apparent that not all proteins are enzymes; some, such as the protein subunits of microtubules, serve motile or structural roles in cells. Others are hormones, membrane receptor proteins, or antibodies. Moreover, some genes were found to code for polypeptide subunits of complex, multichain proteins rather than complete proteins. As a consequence of these findings, the concept was modified to state that each gene codes for a polypeptide.

By the mid to late 1960s, it became obvious that even this concept was too narrow to define genes adequately. At this time, research into the kinds of RNA transcribed in the nucleus revealed that genes code for ribosomal RNA and tRNA as well as polypeptides. Regulatory genes or regions, such as the operators of prokaryotic DNA (see p. 179), were also discovered; these DNA regions are never transcribed into RNA of any kind.

These findings have led to a return to the concept that a gene is a segment of a DNA molecule that codes for a unit of function. The function may be to code for the entire polypeptide chain of a simple protein or for a polypeptide subunit of a complex protein with several polypeptide subunits. These proteins may be enzymes, structural units, peptide hormones, recognition and receptor proteins, antibodies, or other protein-based cellular molecules. Alternatively, the unit of function may code for ribosomal or transfer RNA, or it may be involved in regulating the activity of other functional units within the DNA. Genes with these functions (with the possible exception of regulatory genes) are found in chloroplasts and mitochondria as well as in the cell nucleus. (The contributions to genetics of studies with bacteria and viruses are outlined in Supplement 13-3.)

Genes are segments of DNA that act as functional units in coding for mRNA, tRNA, or rRNA molecules or in regulating transcription.

Questions

1. What is linkage? How was linkage discovered?

2. How does recombination affect linkage?

3. Why are the recombinant chromosomes produced by a single recombination event found in only one-half of the gametes?

4. Explain recombination in terms of chromosomes and chromatids.

5. What is the relationship between the frequency of recombination between genes and their positions on chromosomes? Why are the maps developed from recombination frequencies laid out in crossover units instead of nanometers?

6. What is sex linkage? How does it differ from ordinary linkage? What are sex chromosomes? Autosomes?

7. Explain sex-linked inheritance in terms of the meiotic division sequence.

8. What observations led to the discovery of sex linkage? How does the inheritance of sex-linked genes differ from genes carried on the autosomes?

9. What is a pedigree? How are pedigrees used in studies of human inheritance?

10. What is a carrier? Explain why a human male cannot be a carrier of color blindness or hemophilia.

11. Why was hemophilia so prominent in European royal families?

12. How do sex-influenced traits differ from sex-linked traits?

13. What is incomplete dominance?

14. Explain the difference between the number of possible alleles of a gene in individuals and in populations of individuals.

15. What is polygenic inheritance? How does it differ from the inheritance of traits governed by a single gene? In what way is the fact that mulattoes can have children with unpigmented skin related to the reappearance of recessive traits in the F_2 generation of Mendel's crosses?

16. What is epistasis?

17. What is cytoplasmic inheritance? How was it discovered? How could you use a testcross to determine whether a trait is inherited through genes contained in the cytoplasm?

18. Why is cytoplasmic inheritance frequently linked to female rather than male gametes?

19. What is a gene? Why were the "one gene, one enzyme" and "one gene, one polypeptide" hypotheses inadequate?

20. Explain the difference between the alleles of a gene in terms of DNA sequences.

Problems*

1. Suppose that a man has the following combination of alleles for the genes A, B, C, D, E, F, and G:

$$AaBBCcddEeFfGg$$

How many different kinds of gametes can be produced with respect to these genes? (Assume no linkage.)

2. Persons with the dominant allele A can taste the chemical phenylthiocarbamide (PTC); those who are homozygous for the alternate a allele of this gene cannot taste PTC. Suppose a brown-eyed taster marries a blue-eyed nontaster. What kinds of children, in what proportions, could this couple expect if one of the brown-eyed taster's parents was a blue-eyed nontaster? What kinds of children, in what proportions, could a pair of brown-eyed taster parents expect if each had a blue-eyed nontaster parent? (Assume no linkage, and assume further that eye color in humans is controlled by a pair of alleles of a single gene, with brown dominant to blue.)

3. In four-o'clock flowers, red plants are homozygous for the dominant allele R of a gene for flower color. White plants are homozygous for the r allele of the same gene. Heterozygotes, with one R and one r allele, have pink flowers. What offspring and in what proportions would be expected from the following crosses?

$$RR \times Rr \qquad RR \times rr \qquad Rr \times Rr \qquad Rr \times rr$$

4. In humans, red-green color blindness is a sex-linked recessive trait. If a man with normal vision and a color-blind woman have a son, what is the chance that the son will be color-blind? What is the chance that a daughter will be color-blind?

5. The following pedigree shows the pattern of inheritance of color blindness in a family (persons with the trait are indicated by black circles):

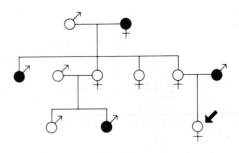

What is the chance that a son of the third-generation female indicated by the arrow will be color-blind if she marries a normal male? If she marries a color-blind male?

6. Persons affected by a condition known as polydactyly have extra fingers or toes. The trait was present (black circles) in the following members of one family:

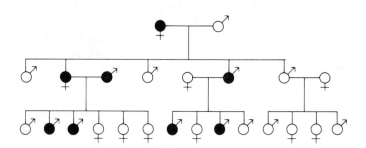

From the pedigree, can you tell if polydactyly comes from a dominant or a recessive allele? Is the trait sex-linked? As far as you can determine, what is the genotype of each person in the pedigree with respect to the trait?

7. In chickens, feathered legs are produced by a dominant allele *F*. Another allele *f* of the same gene produces featherless legs. The dominant allele *P* of another gene produces pea combs; a recessive allele *p* of this gene causes single combs. A breeder makes the following crosses with birds 1, 2, 3, and 4; all parents have feathered legs and pea combs (assume no linkage):

1 × 2: all feathered, all pea comb

1 × 3: ¾ feathered, ¼ featherless, all pea comb

1 × 4: ⁹⁄₁₆ feathered, pea comb; ³⁄₁₆ featherless, pea comb;

³⁄₁₆ feathered, single comb; ¹⁄₁₆ featherless, single comb

What are the genotypes of the four birds?

8. A brown rabbit that is crossed with a chinchilla produces three brown rabbits and one white. What are the genotypes of the parents?

9. In humans, the blood groups M and N are determined by two alleles of a single gene without dominance: M individuals are homozygous *MM*; MN individuals are heterozygous *Mm*; and N individuals are homozygous *mm*. A mixup in a hospital ward caused a mother with type O and MN blood to think that a baby given to her really belonged to someone else. Tests in the hospital revealed that the doubting mother was able to taste PTC. The baby given to her had type O and MN blood and could not taste PTC. The mother had four other children with the following blood types and tasting abilities (the ABO blood types and their inheritance are described in Supplement 13-2):

Type A and MN blood, taster

Type B and N blood, nontaster

Type A and M blood, taster

Type A and N blood, taster

Without knowing the father's blood types and tasting ability, can you determine whether the child is really hers? (Assume that all her children have the same father.)

10. A number of genes carried on the same chromosome are tested and show the following crossover frequencies. What is their sequence in the map of the gene?

Genes	Crossover frequency between them
C and A	7%
B and D	3%
B and A	4%
C and D	6%
C and B	3%

11. In *Drosophila*, the recessive genes for black body color and purple eyes are carried on the same chromosome. You make a cross between a fly with normal eye and body color and a black-bodied fly with purple eyes. Among the offspring, about half have normal eye and body color and half have purple eyes and black bodies. A small percentage have (1) normal eye color and black bodies and (2) purple eyes with normal body color. What alleles are carried together on the chromosomes in each of the flies used to make the cross? What alleles are carried together on the chromosomes of the F_1 flies with normal eye color and black bodies and those with purple eyes and normal body color?

12. Using the chi square method, determine how closely the following proportions of offspring fit a 9:3:3:1 ratio (see Supplement 12-1):

a.	74	33	38	1
b.	880	310	330	100
c.	1810	597	603	198
d.	807	410	400	205

13. You carry out a cross in *Drosophila* that produces only half as many males as females in the offspring. What might you suspect as a cause? (*Hint:* Review Problem 8 in the set following Chapter 12.) Show the genetic basis for your answer.

14. In cats the genotype *AA* produces the tabby fur color; *Aa* is also a tabby, and *aa* is black. Another gene pair is epistatic to the gene for fur color. When present in its dominant *W* form (*WW* or *Ww*), this gene blocks the formation of fur color and all the offspring are white; *ww* individuals develop normal fur color. What fur colors, in what proportions, would you expect from this cross:

AaWw × *AaWw*

(Assume no linkage.)

---

*For answers to these problems, see pp. 563–564.

Suggestions for Further Reading

Corwin, H. O., and J. B. Jenkins. 1975. *Foundations of Modern Genetics*. Houghton Mifflin, Boston.

Goodenough, U. 1978. *Genetics*. Holt, Rinehart & Winston, New York.

Herskowitz, I. H. 1979. *Elements of Genetics*. Macmillan, New York.

Strickberger, M. W. 1976. *Genetics*. 2nd ed. Macmillan, New York.

Watson, J. D. 1976. *Molecular Biology of the Gene*. 3rd ed. Benjamin, Menlo Park, California.

Wolfe, S. L. 1981. *Biology of the Cell*. 2nd ed. Wadsworth, Belmont, California.

SUPPLEMENT 13-1:
BRIDGES' PROOF OF THE CHROMOSOME THEORY OF HEREDITY

When Morgan first described sex-linked inheritance, there were still doubters who argued that the apparent similarity of behavior between genes and chromosomes was only coincidence. Their arguments were considerably weakened by Morgan's discovery of sex-linked inheritance, which for the first time allowed provisional assignment of a gene to a particular chromosome: the X chromosome of *Drosophila*. Another of Morgan's experiments soon eliminated any chance that the parallel between genes and chromosomes was coincidental.

In one of Morgan's crosses, white-eyed females with red-eyed males (Figure 13-7a), all females in the F_1 generation were expected to have red eyes and all males white eyes. The majority of the offspring followed this expected distribution, but there were exceptions. Occasionally, in about one individual out of 2000 to 3000 flies in the F_1 generation, a single red-eyed male or white-eyed female appeared. These exceptional individuals, rather than disproving Morgan's hypothesis, gave additional support when their presence was correctly interpreted. They also suggested a method for proving the chromosome theory of heredity.

C. B. Bridges, one of Morgan's graduate students at Columbia University, reasoned that the exceptional flies could be explained if it was assumed that, very rarely, the two X chromosomes of a female failed to separate during meiosis and were delivered to the same egg. (Bridges termed this failure to separate **nondisjunction.**) For every egg produced with two X chromosomes, another egg would be formed with no X at all. If this occurred among the white-eyed females used in Morgan's cross, some of these exceptional eggs would contain two X chromosomes, both carrying the recessive allele for white eyes. Other eggs, deprived of an X chromosome as a result of

nondisjunction of the X, would carry no allele for eye color. All other chromosomes of the egg, the autosomes, were assumed to be normally distributed to the egg.

Figure 13-17 illustrates the results expected when these rarely produced eggs are fertilized by sperm from red-eyed males. These sperm (entered along the top of the diagram) are normal; they carry either an X chromosome with the dominant *W* allele for red eyes or a Y chromosome with no allele for eye color. If fertilized by a sperm carrying an X, the egg carrying two X chromosomes, with the recessive allele *w* for white eyes, forms a zygote with three X chromosomes. These, as it happens, usually die before reaching adulthood and do not enter into the results. The eggs with no X chromosome, if fertilized by an X-bearing sperm, produce an XO zygote. (The O indicates that no Y is present.) In *Drosophila*, these XO zygotes develop into sterile males that are externally indistinguishable from normal males. Because the single X donated by the father carries the allele for red eyes, this combination would account for the rare appearance of red-eyed males in Morgan's cross. If fertilized by a sperm carrying a Y chromosome, the XX eggs, produced as a result of nondisjunction, form an XXY zygote. These, in *Drosophila*, develop into functional females. These females would have white eyes because both the X chromosomes present, originating from the mother, carry the recessive allele. These XXY females would account for the rarely occurring white-eyed females in the F_1 generation of Morgan's cross.

In *Drosophila*, the sex of a fly is determined by a balance between autosomes and X chromosomes. The Y, although necessary for development of normal fertility in males, has little effect on whether a fly develops into a female. If the three pairs of autosomes are combined with only one X chromosome, the fly develops into a male; if two or more Xs are present, the fly develops into a female.

Bridges tested his hypothesis by making other crosses with the functional XXY females; these crosses and the ratios obtained supported his hypothesis without exception. But, of greatest importance, Bridges examined the cells of the various exceptional flies under a light microscope and counted the numbers of X and Y chromosomes present at metaphase. The exceptional white-eyed females, assumed to be XXY in his hypothesis, proved indeed to have two X chromosomes and a Y when directly examined. Similarly, the exceptional red-eyed males, expected to have the XO combination according to the nondisjunction hypothesis, had an X but no Y visible in their cells. This degree of correspondence between genetic and cellular results could hardly be explained by coincidence and finally established that Mendel's factors, the genes, are carried on the chromosomes.

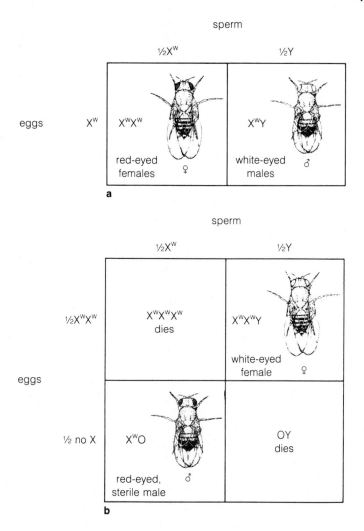

sperm

½Xʷ   ½Y

eggs   Xʷ

XʷXʷ
red-eyed
females ♀

XʷY
white-eyed
males ♂

a

sperm

½Xʷ   ½Y

eggs

½XʷXʷ

XʷXʷXʷ
dies

XʷXʷY

white-eyed
female ♀

½ no X

XʷO

red-eyed,
sterile male ♂

OY
dies

b

Figure 13-17 The explanation for the rare appearance of red-eyed males or white-eyed females in the cross shown in Figure 13-7a. (**a**) The usual outcome of the cross; (**b**) the rare appearance of red-eyed males or white-eyed females due to nondisjunction (see text).

SUPPLEMENT 13-2: HUMAN GENETICS

Inherited Traits and Diseases

A large number of inherited traits in humans cause disease or are otherwise of interest because of their effect on human life. Several of these traits and their patterns of in-

heritance have already been discussed, including color blindness and hemophilia (sex-linked traits), and skin pigmentation (controlled by multiple genes). This supplement surveys a number of additional human traits and diseases of importance.

Eye Color in Humans Eye color is evidently controlled by alleles of a single gene in a pattern that is basically simple but complicated by a large number of modifying factors. Human eye color is based on only two colors, blue or brown, with brown dominant. Modifying factors vary these basic colors from black (considered a shade of deepest brown) through browns from dark to amber and blues from deep blue to very light blue or gray. If both parents have blue eyes, in the simplest case, the children too will have blue eyes. Brown-eyed parents may carry either two dominant alleles or a recessive allele masked by the dominant allele for brown. Although the situation is more complicated than the following ratios suggest, these parents are likely to have children in the colors and chances listed below. Consider that in the gene for eye color B is the blue allele for brown and b the recessive allele for blue:

Parents: $BB \times BB$ = all children brown-eyed

$BB \times Bb$ = all children brown-eyed

$BB \times bb$ = all children brown-eyed

$Bb \times Bb$ = any child has a 75 percent chance of being brown-eyed

$Bb \times bb$ = any child has a 50 percent chance of being brown-eyed

$bb \times bb$ = all children blue-eyed

Table 13-1 lists other human traits that are inherited as if they are controlled by single genes.

The ABO Blood Groups in Humans The ABO blood groups of humans were discovered in 1900 by Karl Landsteiner, who was investigating the sometimes fatal outcome of attempts to transfer blood from one person to another. Landsteiner found that only certain combinations of four blood types, designated A, B, AB, and O, can be successfully mixed in blood transfusions.

The incompatibility between some blood groups was found to depend on differences between the red blood cells and the colorless fluid or **serum** in which they are suspended. Landsteiner determined that, in the wrong combinations, red blood cells from one blood group are agglutinated or clumped by an agent in the serum of another group. Agglutination depends on naturally occurring **antibodies** in the serum of different groups. These

Table 13-1 Human Traits Known or Suspected to Be Caused by Simple Dominant-Recessive Alleles of Single Genes

Dominant	Recessive
Brown eyes	Blue eyes
Mongolian eye fold	No fold
Nearsightedness	Normal vision
Farsightedness	Normal vision
Astigmatism	Normal vision
Dark hair	Blond hair
Curly hair	Straight hair
Early balding (in males)	Normal rate of hair loss
Normal body pigment	Albino
Free ear lobes	Attached ear lobes
Normal hearing	Congenital deafness
Ability to roll tongue	No ability to roll tongue
Polydactyly (more than five fingers or toes)	Normal
Syndactyly (webbing between fingers or toes)	Normal
Short fingers or toes	Normal

Table 13-2 Human ABO Blood Types

Blood Type	Antigens on Red Blood Cells	Has Antibodies Against
A	A	B, AB
B	B	A, AB
AB	A and B	None
O	None	A, B, AB

Although the I^A and I^B alleles are both dominant over the I^O allele, there is a lack of dominance between the I^A and I^B alleles. Persons with both these alleles produce blood cells with both the A and B surface antigens.

The series of multiple alleles in the ABO blood groups are also of value in determining the legal paternity of children. The ABO groups and other genes known to affect blood types can be used to eliminate some persons as the possible father of a child. A man with type O blood, for example, could not possibly be the father of an AB child. Although some males can be ruled out as possible fathers, coincidence of blood types does not of course prove paternity.

Rh Factor Human blood antigens are controlled by a large number of genes in addition to the ABO system. One of these is the Rh gene, which has six or more allelic forms. This gene was discovered through the study of a disease of newborn infants known as *erythroblastosis fetalis*, a form of anemia. Infants with the disease have an antigen called the *Rh factor* associated with their blood cells. Although the gene is not sex-linked, the Rh antigen is usually found on the red blood cells of the father but not the mother when the infants are affected by the disease. Tests of the blood of mothers carrying the afflicted infants reveal large quantities of antibody against the Rh antigen of the infant. Supposedly, the Rh antigen in the infant, designated Rh$^+$, stimulates the production of anti-Rh$^+$ antibodies in the bloodstream of the Rh$^-$ mother during fetal development. These antibodies, which are capable of crossing the barriers between the circulation of mother and fetus, enter the bloodstream of the developing fetus. In the fetus, they cause breakdown of the Rh$^+$ blood cells of the developing child, leading to anemia or death. Usually, the effects are most pronounced and dangerous in second and later pregnancies involving an Rh$^+$ fetus in an Rh$^-$ mother, after

antibodies are protein molecules that interact with specific substances called **antigens** on the surfaces of the red blood cells. Persons with type A blood have natural antibodies against antigens on the red blood cells of type B blood; persons with type B blood have antibodies against antigens on the red blood cells of type A blood; and type O individuals have antibodies against the antigens of both A and B blood cells (see Table 13-2). Type AB individuals carry no antibodies, but their red blood cells are coagulated by antibodies in the blood of all the other groups.

Analysis of human pedigrees revealed that the four blood groups are produced by three alleles of a single gene I that affect the antigens of red blood cells. The three alleles, designated I^A, I^B, and I^O, produce the following blood types:

$$I^A I^A = \text{type A blood} \qquad I^A I^B = \text{type AB blood}$$
$$I^B I^B = \text{type B blood} \qquad I^O I^O = \text{type O blood}$$

the initial pregnancy has developed a significant antibody response in the mother's bloodstream.

In the infant attacked by anti-Rh antibodies from the mother, red blood cells break down and release hemoglobin into the bloodstream. The hemoglobin breaks down into a related pigment called *bilirubin*, which colors the skin a deep yellow. If present in large amounts, the pigment may damage the brain, leading to mental retardation and even death.

The inheritance of the Rh factor is complex because the Rh gene exists in so many different allelic forms. The disease generally follows a pattern resembling that of a single gene, however, with Rh^+ dominant to Rh^-. Determining whether an individual is Rh^+ or Rh^- can usually be accomplished simply by mixing a blood sample with purified anti-Rh antibodies. Many doctors suggest a test of Rh blood types before marriage to determine whether difficulties relating to the Rh factor should be expected. Various steps can then be taken before or during pregnancy to reduce the infant's chance of death or impairment from the effects of anti-Rh antibody. One of the most effective techniques involves ridding the mother's bloodstream of anti-Rh^+ antibodies after each pregnancy. This step eliminates the chance that subsequent pregnancy will produce an affected child.

Diabetes   Some forms of diabetes, a disease caused by an inability of the body to use glucose properly, are genetically determined. Glucose may be concentrated in the bloodstream of diabetics, but it is unable to enter most types of body cells for breakdown as an energy source. The disease is caused by a deficiency in the production of insulin, a protein that facilitates penetration of glucose into cells from the bloodstream. The disturbance in sugar breakdown results in progressive muscular weakness and a gradual loss of protein and fatty tissues, excessive thirst, and the appearance of toxic products such as ketones (see p. 142) in the blood. Muscular weakness extends to the walls of vessels in the circulatory system, including the small arteries that supply the retina of the eye. Rupture of these arteries frequently leads to blindness, a common symptom in advanced cases of diabetes. Some cells, such as those in the lens of the eye, do not require insulin for glucose uptake, and the concentration of glucose in these cells rises to match the high concentration in the bloodstream. The excess glucose in these cells is then metabolized into other substances with deleterious effects. One of these substances, sorbitol, causes cataracts if it accumulates in the lens of the eye. The deficiencies in sugar metabolism in diabetes, if untreated, affect the entire body and can eventually lead to coma and death.

At least two major forms of genetically determined diabetes occur: juvenile and adult. The most serious form develops in childhood and is inherited in a pattern suggesting control by the recessive allele of a single gene. (A number of additional genes modify this basic pattern of inheritance.) An adult form of diabetes, which appears later in life, is evidently controlled by the dominant allele of another gene unrelated to the juvenile form. The time at which the dominant, adult form of diabetes appears is modified by many factors, including diet, excessive consumption of alcohol, and pregnancy. Juvenile diabetes requires continued injection of insulin for survival. Adult diabetes, once brought under control by injected insulin after its initial appearance, can frequently be regulated further by careful diet and oral administration of drugs that stimulate natural insulin production.

Diabetes is one of the major medical problems of our age. In addition to the genetically determined forms of the disease, there are indications that viral infections of the pancreas, the organ that normally produces insulin, can also bring on diabetes. The number of diabetics in the United States increased from 1.2 million in 1950 to about 5 million in 1975—a rate of increase many times greater than the rate of population growth. Part of this increase comes from the fact that juvenile diabetics, once unlikely to reach adulthood, now grow to maturity and have children that also carry the trait. Diabetes causes one death in every five in this country and is the second most frequent cause of blindness.

Sickle-Cell Anemia   Sickle-cell anemia, which affects red blood cells and impairs their ability to carry oxygen, is found almost exclusively in this country in persons descended from African Negroes. The trait, caused by the recessive allele of a gene coding for part of the hemoglobin molecule, results from the change of a single amino acid in hemoglobin. Persons homozygous for the allele are seriously impaired and rarely survive past late adolescence. The red blood cells in these individuals are exceptionally fragile and take on a peculiarly curved "sickle" shape when blood samples are deprived of oxygen (Figure 13-18).

The sickle shape assumed by red blood cells under low oxygen conditions is caused by a property of the hemoglobin molecules coded for by the recessive allele. This hemoglobin, called *hemoglobin S* or HbS, forms long, fibrous crystals under conditions of low oxygen concentration. These crystals give the red blood cells their elongated, sickle shape. Crystallization of the HbS hemoglobin also makes red blood cells rigid. Normal red cells, in contrast, are highly flexible, a property that allows them to

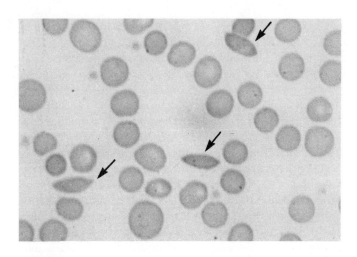

Figure 13-18 Sickle-shaped red blood cells (arrows) caused by crystallization of hemoglobin S under low oxygen conditions.

change shape readily and slip through narrow channels in the capillaries, the smallest blood vessels of the circulatory system. In regions where oxygen supply is low, as in the capillaries in many regions of the body, crystallization of HbS causes red blood cells to take on the rigid, sickle shape and block the small vessels. Crystallization also leads to breakdown and elimination of red blood cells in great numbers in the spleen and other locations where damaged cells are normally removed from the circulation. The combination of circulatory blockage and rapid elimination of sickled red blood cells leads to disability and eventually death in afflicted persons.

Carriers of the trait, who number between 10 and 15 percent of the black population in the United States, have one normal and one recessive allele of the gene. Although carriers produce enough normal hemoglobin through the activity of the dominant allele to be essentially unaffected, the HbS hemoglobin produced by the recessive allele is also present in their red blood cells. The presence of the HbS hemoglobin can be detected through a relatively simple blood test.

Each child of matings between carriers has about one chance in four of having sickle-cell anemia. In the American black population, matings of this type result in the fatal anemia in about one out of every 500 children born. If only one parent is a carrier, none of the children is expected to have the disease. However, each child of a marriage between a carrier and a noncarrier has a 50 percent chance of being another carrier.

The recessive allele has the unusual side effect of protecting heterozygous carriers from malaria. The blood parasites causing malaria cannot tolerate the HbS hemoglobin present in the red blood cells of carriers; as a result, malarial infections of carriers are mild and nonfatal. Infections of noncarriers, in contrast, kill about one out of every ten persons in malarial areas, usually in childhood. Thus the protection conferred by the recessive allele provides a definite survival advantage in highly malarial areas. As a result, even though the homozygous recessive combination is usually fatal, the gene persists in relatively high frequencies in individuals living in malarial regions in Africa.

Nondisjunction and Human Disabilities

Several human conditions are caused by a failure of chromosomes to separate normally during meiosis. This error of meiosis, called **nondisjunction** (see Supplement 13-1), results in the omission of a chromosome or the inclusion of an extra chromosome in an egg or sperm cell. The various abnormal combinations of sex chromosomes observed among humans, most commonly XO females and XXY males, probably arise in this way. In XO females (called *Turner's syndrome*, occurring in about one out of 5000 births) the ovaries are underdeveloped and the affected individuals, although females with normal external genitalia, are sterile. The XO females are typically short in stature, with underdeveloped breasts. The XXY males (*Klinefelter's syndrome*, occurring in about one out of every 400 to 600 births) have male external genitalia but very small and underdeveloped testes. Body hair is sparse, and some development of the breasts is usually noted. These XXY males are also sterile. Other abnormal combinations of X and Y chromosomes probably arising through nondisjunction are found in humans (see Table 13-3); in most cases these unusual combinations produce sterile individuals who are both physically and mentally abnormal. Sex in humans is determined primarily by the presence or absence of the Y chromosome. Individuals with a Y chromosome are externally male-like, no matter how many X chromosomes are present. If no Y is present, X chromosomes in various numbers give rise to female-like individuals.

Down's syndrome (Mongoloid idiocy), in its most common form, arises through nondisjunction of one of the 22 pairs of autosomal chromosomes in humans. These individuals develop from eggs receiving the extra chromosome; evidently eggs deprived of the chromosome by nondisjunction fail to develop if fertilized. As a result, those afflicted by Down's syndrome have one chromosome of the set present in three copies. These individuals are short in stature and mentally retarded. More than half do not live past the age of ten. The disease is relatively common, occurring once in every 500 to 600 births among women under 30 and increasing tenfold among mothers at ages 40 to 45. The age of the father apparently has no effect on the frequency of the disease. Down's syndrome is the major cause of severe mental retardation in humans. The disruptions caused by additions or deletions of whole chromosomes through disjunction, such as Turner's, Klinefelter's, and Down's syndromes, are incurable and likely to remain so.

Control of Hereditary Diseases: Eugenics

Many genetically determined human diseases, such as diabetes, were once fatal or so seriously debilitating that affected individuals were unlikely to reach reproductive age. Thus the frequency of the alleles in the human population as a whole remained constant at low levels. In more recent times, however, medical science has provided successful treatments for the symptoms of many of these diseases, with the result that affected people can lead a normal life, marry, and have children. Unfortunately, these medical successes also mean that the deleterious alleles are passed on to offspring in greater numbers than before and thus are steadily increasing in the population. Consequently, in contrast to diseases caused by infective agents such as bacteria and viruses, the development of treatments for hereditary diseases is likely to result in an *increase* of the malady in the human population.

What can be done about the situation? Selective breeding, called **eugenics**, is possible among domestic animals but difficult to accomplish in human society. Involuntary sterilization, although proposed now and then as a solution, constitutes a serious violation of human rights that most people are unable to accept even as a punitive measure. Therefore eugenic controls can generally be practiced only if carriers of a deleterious trait are identified and agree voluntarily not to have children.

Two methods for voluntary eugenics are gradually becoming established in our society. The first is *genetic counseling*, in which a geneticist analyzes family medical histories of people contemplating marriage or childbearing and attempts to construct pedigrees for both families. From the pedigrees, it is frequently possible to establish whether the people are carriers of a genetically determined disease and to indicate their chances of having either affected or normal children. On the basis of this information, they may decide not to marry or not to have children.

Table 13-3 Sex Chromosome Abnormalities in Humans

Sex Chromosomes	Sex	Fertility
XY	Normal male	+
XX	Normal female	+
XO	Turner's syndrome, female	−
XXY	Klinefelter's syndrome, male	−
XYY	Male	+
XXX	Triple X syndrome, female	±
XXXY	Triple X-Y syndrome, male	−
XXXX	Tetra X female	?
XXXXY	Tetra X male	−
XXXXX	Penta X female	?

The second method, called **amniocentesis**, provides a means to determine whether a developing fetus is afflicted by a genetic disorder. In this procedure, a needle is inserted directly through the abdominal wall of the pregnant mother and into the fluid-filled sac surrounding the fetus, the *amnion*. Some of the fluid in the amnion, which contains living cells released from the fetus as a normal part of embryonic development, is drawn through the needle and placed under conditions favoring cell growth. The cells are then tested for chromosomal abnormalities or biochemical malfunctions. The method has been successfully employed to detect Down's syndrome, abnormalities in the sex chromosomes, and more than a hundred other human traits. While there is some risk in the technique, it can provide the information parents need to decide whether to terminate a pregnancy when an incurable genetic disease is shown to be present.

Eugenics has sometimes been proposed as a method to increase the frequency of favorable traits such as high intelligence, health, and physical beauty in the human population. Individuals favored with these qualities would be deliberately selected as breeders or as suppliers of stock for sperm banks. Farfetched as it might sound, this plan is regularly proposed. In practice, it is doubtful that this technique could improve the population very much as long as the control of deleterious traits remains voluntary. And, as pointed out, involuntary control of carriers of harmful genes by sterilization is not likely to be accepted as an alternative. As a result, human difficulties caused by genetically determined diseases are likely to increase gradually unless some controlled way of altering the genotype of affected individuals can be found.

The latter approach, called **genetic engineering**, offers some hope for the future. Genetic engineering involves the addition of chromosome segments to fertilized eggs or embryos missing them or the replacement of deleterious genes by DNA containing the normal form of the alleles (see Supplement 11-3). While techniques for accomplishing this feat with humans have not yet been worked out, some success has been obtained with microorganisms and animals such as insects. In time, it may be possible to perfect human genetic engineering.

Many of these techniques and approaches for solving the problems of genetic disorders in humans raise serious moral questions for which there are no ready answers. Does society have the right to regulate childbearing by those carrying deleterious hereditary traits? Should a pregnancy be terminated if the fetus is shown by amniocentesis to have a seriously debilitating genetic disorder such as Down's syndrome? Do scientists have the right to alter the hereditary makeup of individuals or even to experiment toward this end? Who should decide in these matters—individuals, governments, scientists and physicians, religious authorities, or society at large? As one biologist has put it, "Who should play God?"

SUPPLEMENT 13-3:
THE GENETICS OF BACTERIA AND VIRUSES

During the 1940s and 1950s, geneticists working in several laboratories established that genetic recombination takes place in the DNA of bacteria and viruses as well as in eukaryotes. This was a most significant discovery, because bacteria and viruses can be grown with ease by the millions or even billions in generations that pass in a matter of minutes. In comparison, a generation of *Drosophila* takes two weeks to grow from zygote to sexual maturity, and geneticists working with corn are lucky to get two generations a year.

The rapid generation times of bacteria and viruses allowed crosses and their outcomes to be traced much more quickly. In addition, the millions or billions of offspring obtained from a cross permitted rare events to be detected, such as recombination between very closely spaced points along DNA sequences. This, in turn, allowed mapping to be extended to points within genes, sometimes to dis-

tances as short as individual nucleotides. Even more important to the study of genes and their activity was the fact that the effects of different alleles could be defined in terms of biochemical effects rather than morphological differences in offspring.

Recombination in Bacteria

Recombination in eukaryotes occurs in diploids and involves the exchange of segments between the chromatids of a homologous chromosome pair. Although bacteria are normally haploids, recombination may occur in individual cells temporarily converted to the diploid condition by the entry of DNA originating from another bacterial cell. The extra DNA may include only a segment of the DNA circle of another bacterium; such cells are called **partial diploids**. In this case, which is the most frequent condition observed in bacteria, recombination occurs only in the segment of the host cell DNA that is homologous to the piece of DNA entering from outside. More rarely, a complete DNA circle may enter from another cell, creating a complete diploid. Three routes have been recognized by which DNA from one bacterial cell may enter another to create partial or complete diploids: **conjugation, transformation,** and **transduction.**

In conjugation, which resembles sexual reproduction in diploid eukaryotes, bacterial cells of the same species make contact and fuse together at one point, forming a cytoplasmic bridge that directly connects the two cells (Figure 13-19). Once the bridge forms, a copy of part or all of the donor cell's DNA may cross the bridge and enter the recipient cell.

In transformation, pieces of DNA released from fragmenting bacterial cells are absorbed from the surrounding solution by healthy cells. For the DNA pieces to be absorbed, they must originate from the same species as the recipient cell. Although transformation is best known from laboratory studies in which recipient cells are exposed to DNA artificially extracted and purified from donor cells, the process probably also occurs under natural conditions. The molecular basis for penetration of the absorbed DNA through the wall and plasma membrane of the recipient cell is unknown.

Transduction occurs through a transfer of DNA fragments from one cell to another in virus particles. The process depends on the cycle of infection and multiplication of viral particles inside bacteria (see p. 70). Occasionally a fragment of the DNA of a bacterial host cell is packed into a virus particle along with the viral DNA. These viral particles, which are released along with the normally consti-

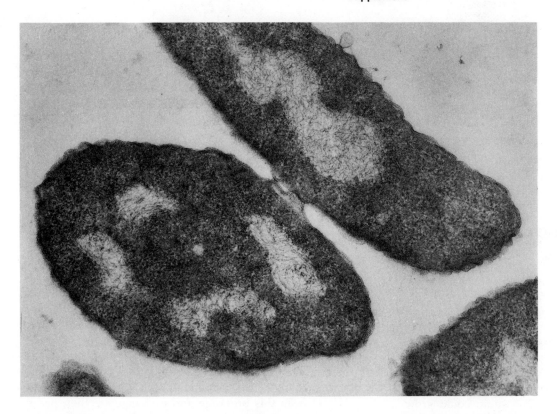

Figure 13-19 Conjugating cells of *E. coli*. A cytoplasmic bridge has formed between the cells in the region of contact. ×68,000. Courtesy of L. G. Caro and Academic Press, Inc. (London) Ltd., from *Journal of Molecular Biology 16*:269 (1966).

tuted particles, are infective and can attach to another cell of the same species. When attachment occurs, the fragment of bacterial DNA contained inside the viral particle is injected into the recipient cell just as the viral DNA would be.

Recombination in Partial Diploids

All three processes, conjugation, transformation, and transduction, result in the insertion of an extra piece of DNA of the same species inside a bacterial cell, creating a partial or complete diploid. Once the extra piece of DNA is inside, it can pair with the homologous sequences of the DNA circle of the bacterial cell (Figure 13-20a) and enter into recombination. During recombination, which occurs by essentially the same molecular mechanisms as in eukaryotes, homologous segments break and exchange between the bacterial DNA circle and the extra piece of DNA. If different alleles of the genes in the diploid region are present in the two molecules entering into the exchange, the result is a new combination of alleles in the bacterial DNA circle (Figure 13-20b). Following recombination, the bacterial DNA replicates and the cell divides normally, producing a cell line with the new combination of alleles.

Bacterial recombinants are detected through changes in phenotype, just as they are in eukaryotes. However, most of the bacterial phenotypes studied in recombination experiments involve biochemical characteristics. These characteristics may include the chemical requirements of the growth medium, differences in cell processes such as protein synthesis, transcription, or replication, or changes in products secreted by the cells.

One of the experiments that first detected recombination in bacteria, conducted by J. Lederberg and E. L. Tatum at Yale University, illustrates how bacterial recombination studies are carried out. Lederberg and Tatum

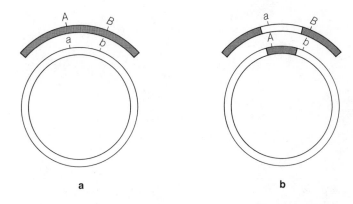

Figure 13-20 Recombination in a region in which a bacterial cell has been converted to a partial diploid. (**a**) Pairing between the extra DNA segment and the DNA circle of the recipient cell; (**b**) the result of recombination in the paired region.

worked with several genetically controlled strains of *Escherichia coli* that had lost their ability to grow on the minimal medium required by normal, "wild-type" cells. (Wild-type *E. coli* requires only a solution of inorganic salts and an energy source such as glucose.) One strain could grow only if the substance *biotin* and the amino acid *methionine* were added to the medium but did not require either of the amino acids *threonine* or *leucine*. The second strain had no requirement for biotin or methionine, but it could grow only if threonine and leucine were added. These strains can be represented as:

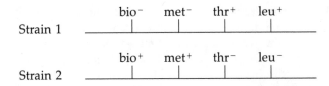

Here bio⁺ indicates the wild-type allele providing cells with the ability to synthesize biotin from inorganic precursors whereas bio⁻ represents the allele producing cells that cannot synthesize biotin. Similarly, met⁺, met⁻, thr⁺, thr⁻, leu⁺, and leu⁻ are the respective wild-type and deficient alleles for methionine, threonine, and leucine synthesis.

Lederberg and Tatum mixed about 100 million cells of the two deficient strains together and placed them on a minimal medium. Most of the cells were unable to survive, but a few hundred viable colonies of bacteria were formed. These contained cells descended from single individuals that were able to grow on the minimal medium.

Since these cells could grow, they must have had the following genetic constitution:

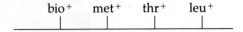

These cells were therefore completely wild-type bacteria.

It was obvious from this result that recombination had taken place between DNA containing the different alleles to produce the wild-type strain (Figure 13-21). In this case, the recombination probably occurred in partial diploids created by conjugation.

Recombination in Viruses

Recombination can also occur between DNA molecules originating from different virus particles if two or more viruses simultaneously infect the same cell. At some point during the growth of the new bacteriophage particles within an infected cell, viral DNA molecules from the different parental types may pair and cross over by breakage and exchange. Any recombinant viral chromosomes produced from this breakage and exchange are then packed into protein coats and, upon rupture of the host cell, are released to the medium to infect further hosts.

Viral recombinants are detected primarily through changes in the biochemistry of cell infection or the structure of virus particles produced by the infection. In the viruses infecting *E. coli*, for example, various alleles cause differences in enzymes replicating the viral DNA or differences in the proteins of the viral coats. These phenotypes can be detected by biochemical tests or electron microscopy.

The Results of Genetic Studies

Recombination studies with bacteria and viruses offer several advantages of fundamental importance to the science of genetics and the field of biology in general. One advantage is that allelic differences can be followed biochemically. Bacteria are particularly suited to biochemical research because they can be grown in large quantities in culture solutions that are defined chemically. Thus what goes into the bacterial cells, what comes out, and what changes occur inside can in many cases be detected and followed. These studies have revealed how replication, transcription, translation, and other biochemical processes take place in cells. Moreover, they have inevitably led to a more thorough and accurate definition of genes and alleles and their activities at the biochemical and molecular level.

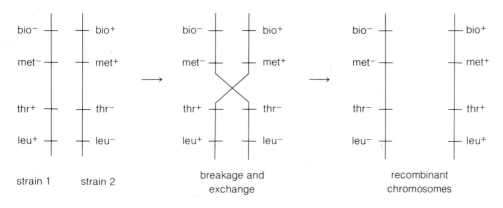

Figure 13-21 Recombination between the bacterial strains used in the Lederberg and Tatum experiment (see text).

Work with DNA ligase, one of the enzymes active in DNA replication (see Supplement 10-2), illustrates how different alleles have been used to trace the steps in biochemical pathways. Bacteria with the wild-type allele for DNA ligase are able to carry out DNA replication normally. In bacteria with an allele coding for a defective form of the DNA ligase enzyme, however, newly synthesized DNA is left in short, disconnected pieces. This finding indicates that DNA ligase probably catalyzes the last step in DNA replication, in which the short pieces assembled in earlier steps are joined into long, continuous molecules (see Figure 10-25e and f). Other steps in the process were traced by a similar approach.

Genetic experiments with bacteria and viruses have also been of fundamental importance in research unraveling the biochemical steps and characteristics of recombination. Some of this research has been concerned with the enzymes and other factors active in recombination and also the sequence of molecular changes that produce breaks and exchanges between homologous DNA molecules. This investigation has been approached by the same techniques used in studies of replication, transcription, and protein synthesis. In this case the steps in the mechanism are traced by noting the effects of alleles coding for deficient forms of enzymes and other factors suspected to be active in recombination.

Suggestions for Further Reading

Goodenough, U. 1978. *Genetics.* Holt, Rinehart & Winston, New York.

Watson, J. D. 1976. *Molecular Biology of the Gene.* 3rd ed. Benjamin, Menlo Park, California.

Wolfe, S. L. 1981. *Biology of the Cell.* 2nd ed. Wadsworth, Belmont, California.

Development

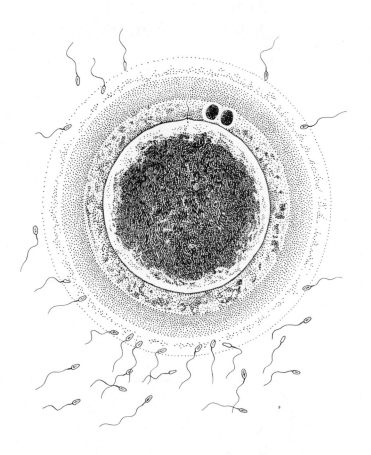

14

BASIC PROCESSES OF DEVELOPMENT IN ANIMALS

All higher plants and animals are multicellular—that is, they are made up of many cells, numbering in the billions or trillions in the larger animals and plants. Within multicellular organisms, individual cells are specialized in structure and function to carry out the different activities of life. These cells originate from a single cell, the fertilized egg, by the processes of growth and division. As they grow and divide during development, cells become different in structure and function. These differences are normally retained throughout the life of the individual.

A developing plant or animal is known as an **embryo** during the period in which its basic cell structures are laid down. With few exceptions, the body cells of the developing embryo are produced by mitotic divisions without genetic modifications of any kind. As a result, no matter how different in structure and function they become, almost all the body cells of an individual remain genetically identical.

Almost all the body cells of an individual remain genetically identical no matter how different they become through development.

Cells specialize during development through a series of changes that are highly coordinated and take place in an ordered sequence. The program for these developmental changes is built into the fertilized egg; no information from outside the embryo is required. How these changes are programmed, and how cells that are genetically identical become different in structure and function during development, are among the major questions of developmental biology.

Of equal interest is the question of how cells migrate and find their final places in developing animal embryos. During embryonic development, individual cells or groups of cells leave the sites in which they are laid down by cell division and move to other locations in the embryo. As they arrive at their final locations, they organize into the complex structures of the adult animal. Thus these final structures frequently include cells originating from several different initial sites in the embryo.

The changes in cell structure and function during embryonic development that produce the many cell types of the adult organism are known as cellular **differentiation**. The cell growth and division, and the cell movements that organize the complex structures of the adult, are called **morphogenesis.** The two processes of cellular differentiation and morphogenesis, acting together, transform the fertilized egg into a functioning, complex, and fully developed animal.

THE BEGINNINGS OF DEVELOPMENT IN ANIMALS: GAMETOGENESIS AND FERTILIZATION

The first significant events of animal development take place before fertilization—during the formation of the **gametes**, the eggs and sperm that meet in fertilization. In the formation of gametes, termed **gametogenesis**, the master program of development for the embryo is laid down. Meiosis is an integral part of gamete formation in animals; the meiotic divisions rearrange alleles and reduce the chromosomes to the haploid number (see Chapter 11). Cytoplasmic changes in the developing egg and sperm cells also prepare the gametes for fertilization and subsequent embryonic development.

Once an egg is fully mature, the egg cytoplasm contains the information required to direct its development as far as an embryo containing hundreds or thousands of cells with some degree of differentiation. In many animal eggs this information is even compartmented into regions that will give rise to different major parts of the embryo. The raw materials and energy sources for part or all of embryonic development are also stored in the egg cytoplasm in the form of proteins, fats, carbohydrates, and the building blocks of the nucleic acids. Although the sperm contributes little more than the paternal set of chromosomes to the embryo (few structures of the sperm besides the nucleus persist), the developmental changes giving rise to sperm are necessary for fertilization to occur. These changes produce a cell that is highly specialized to travel to the egg, carry out fertilization, and trigger embryonic development.

The Formation of Mature Egg Cells in Animals: Oogenesis

A mature egg cell originates from a cell line that is set apart early in the embryonic life of an animal. These cells, usually originating elsewhere in the embryo, migrate into the developing female sex organs (the ovaries), where they divide rapidly by mitosis until hundreds or thousands of potential egg cells are produced. These cells develop into mature eggs by a process called **oogenesis**.

Oogenesis The development of egg cells is closely linked to meiotic cell division. The future egg cells, termed **oocytes** during the maturation process, proceed through the last premeiotic duplication of the chromosomes and enter meiosis. Usually they pass through the stages in which their chromosomes condense, pair, and recombine (see Figure 11-3) within a matter of weeks. During these early

stages, oocytes grow very little. As they reach the transcription stage of meiotic prophase (see p. 231), however, further rearrangements of the chromosomes stop and the oocytes enter an extended period of cytoplasmic growth that may last from a few weeks to as long as a year.

Synthesis of RNA and Other Substances During the transcription stage of meiotic prophase, the chromosomes in the oocyte nucleus synthesize messenger, transfer, and ribosomal RNA. Ribosomal RNA is produced most intensively at this time, and the egg cytoplasm becomes packed with ribosomes.

Transfer and messenger RNA also enter the cytoplasm in quantity after transcription in the egg nucleus. Some of the mRNA produced is used immediately to direct the synthesis of proteins in the oocyte, but much of it is stored in the cytoplasm in inactive form and remains unused until the egg is fertilized.

The mRNA that is immediately active in oogenesis directs the synthesis of massive amounts of enzymes and ribosomal proteins. Other proteins are made and stored in the oocyte cytoplasm. Some of the enzymes direct the synthesis of carbohydrates, lipids, and nucleic acid building blocks, which are also stored in quantity in the cytoplasm of the maturing egg.

Much of the protein synthesized at this time is stored in the cytoplasm as **yolk**. Depending on the species, the yolk may contain lipids and carbohydrates as well as proteins. The stored yolk may be suspended freely in the cytoplasm, or it may be enclosed in membranes in structures called **yolk bodies** or **yolk platelets** (Figure 14-1). By the time the egg is mature (Figure 14-2), much of the volume of the oocyte may be occupied by yolk; deposits of lipids may also be present in the cytoplasm (as in Figure 14-2). These yolk and lipid deposits furnish the raw materials and energy required for the developing embryo.

In many animals, the cytoplasmic materials stored as yolk may also originate in cells located outside the ovaries. In birds and amphibians, for example, yolk components are synthesized in the liver and released into the circulation. From there they are absorbed by the developing oocytes and stored in yolk bodies.

Because of the intensive synthesis of yolk and other components during meiotic prophase, the developing oocytes in many animals grow from microscopic size to dimensions large enough to be easily visible to the naked eye. In fish and amphibians, eggs regularly become several millimeters in diameter; bird eggs may reach several centimeters in width. The "yolk" of a hen's egg, for example, is a mature oocyte. Almost all of this growth occurs

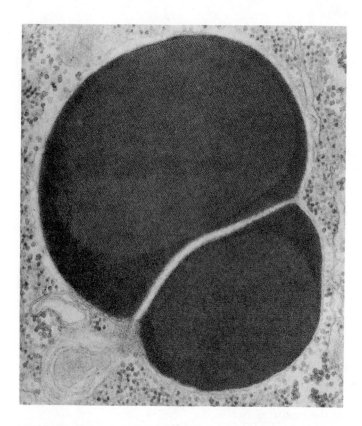

Figure 14-1 A yolk body from the amphibian *Xenopus* at high magnification. The regular striations or periodicity visible in the central portion of the yolk body show that the enclosed material is packed in a crystalline form. ×49,600. Courtesy of P. B. Armstrong.

Figure 14-2 The cytoplasm of a surf clam *(Spisula)* oocyte containing yolk bodies *(YB)* and lipid droplets *(LD)*. Mitochondria *(M)*, pigment granules *(PG)*, and the egg coat *(EC)* are also visible. ×18,500. Courtesy of F. J. Longo and Academic Press, Inc.

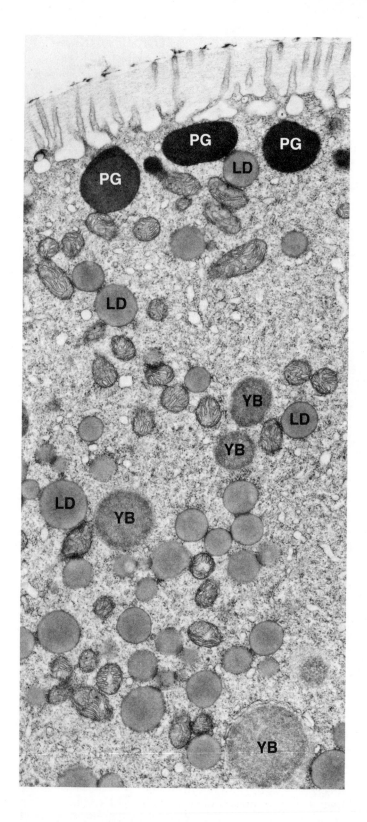

in the cytoplasm. The nucleus usually remains microscopic, often not much bigger than the nuclei in body cells of the same organism. In some animals, such as mammals (including humans), there is little or no yolk formation and the eggs remain relatively small (Figure 14-3).

The amount of yolk and the degree of cytoplasmic growth in a maturing egg are related to the animal's mode of embryonic development. If the period of development is short, or if nutrients are supplied to the embryo from the tissues of the mother (as in mammals), the oocyte cytoplasm grows very little during oogenesis and the mature eggs are small. Species in which the embryos begin feeding before development is complete, such as insects and

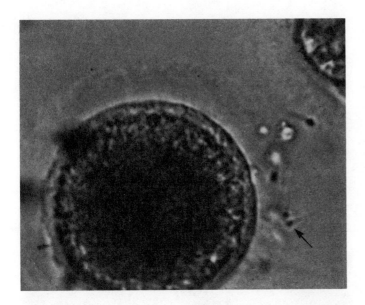

Figure 14-3 A mature human egg. Several sperm cells (arrow) have attached to the egg coat, which is faintly visible in the micrograph. ×580. From B. G. Brackett et al., "The Mammalian Fertilization Process," in *Biology of Mammalian Fertilization and Implantation*, ed. K. S. Moghissi and E.S.E. Hafez. Courtesy of B. G. Brackett and Charles C Thomas, Publisher.

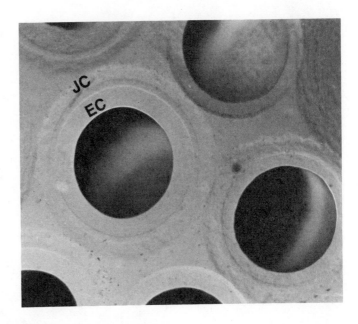

Figure 14-4 Mature eggs of *Xenopus*, an amphibian. A jelly coat *(JC)* is secreted over the egg coat *(EC)* immediately surrounding the eggs. The jelly coat is added as the eggs pass through the oviducts on their way to the exterior. ×15. Courtesy of W. F. Reynolds.

amphibians, also have relatively small eggs with limited amounts of yolk. Animals in which a long period of embryonic development is carried out with no outside source of nutrients, such as birds and reptiles, have large, yolky eggs that contain all the raw materials required for growth.

Surface Coats of the Egg The mode of future embryonic growth is also reflected in the type and complexity of external surface coats of the mature egg. If an egg cell remains protected within the female animal during fertilization and embryonic growth, as it does in mammals, the egg surface is covered by only a single external layer (as in Figure 14-3). Eggs that are released to the outside world for development have complex surface coats with several layers. These layers include materials that are secreted around the eggs during passage through ducts in the animal leading from the ovaries to the outside. The thick, gelatinous coat of frog eggs (Figure 14-4) or the albumen (white) and shell of a hen's egg are examples of these additional surface coats. The surface layers may consist of proteins, as in the albumen of a hen's egg, or glycoproteins (proteins linked to carbohydrate units), as in the gelatinous coat of frog eggs. Eggs released directly into relatively dry environments, such as the eggs of birds,

reptiles, and insects, have hard or leathery surface coats that protect the egg and developing embryo from drying and mechanical injury. In all cases, the surface coats protect the egg from bacterial and viral infections.

The Animal Egg at Maturity Oocytes in most animals become mature before completing meiosis. This maturation, as outlined above, consists of (1) transcription of messenger, transfer, and ribosomal RNA and synthesis of proteins and ribosomes; (2) yolk deposition; and (3) formation of the surface coats of the egg. When maturation is complete, most animal eggs enter a state of arrested activity without completing meiosis. A few remain in meiotic prophase; others proceed as far as metaphase I or metaphase II before entering arrest. At arrest, protein synthesis drops to unmeasurable levels, even though the cytoplasm is crowded with ribosomes, tRNA, and mRNA. Although large numbers of mitochondria persist in the cytoplasm, respiration stops almost completely.

Oocyte maturation includes: **(1) transcription of messenger, transfer, and ribosomal RNA and ribosomes, (2) yolk deposition, and (3) formation of the surface coats of the egg.**

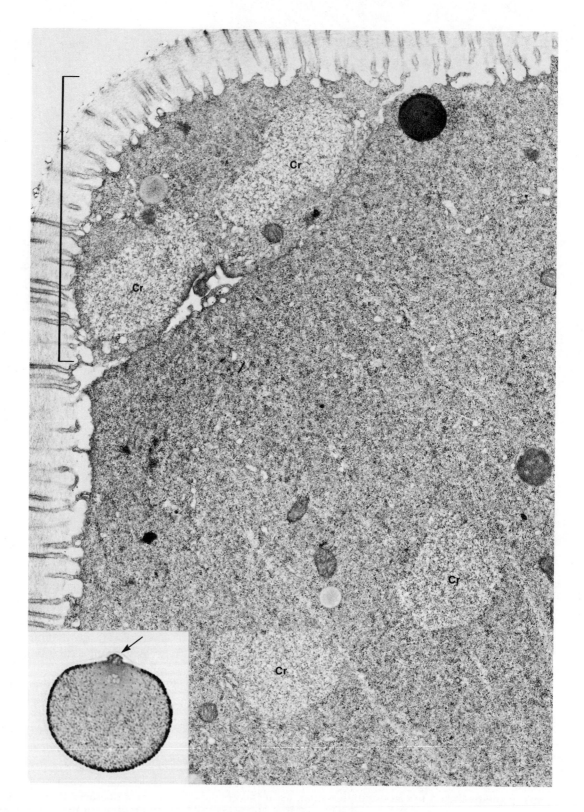

Figure 14-5 An egg of the surf clam *Spisula*. Its first polar body is indicated by the bracket in the electron micrograph and by the arrow in the light micrograph shown in the inset. The masses of chromatin *(Cr)* have been separated into the egg and polar body by the first meiotic division. × 19,500. Courtesy of F. J. Longo and Academic Press, Inc.

Depending on the time of meiotic arrest, one or more **polar bodies** may form on the surface of the maturing oocyte. These are small, abortive cells produced by the unequal division of the oocyte cytoplasm in meiosis (see Figures 11-7 and 14-5). The unequal divisions conserve most of the cytoplasm in the mature egg cell. If meiotic arrest occurs during prophase or metaphase I of meiosis, no polar bodies are formed. If the developing oocyte continues past anaphase I of meiosis, completing the first of the two meiotic divisions before arrest, the oocyte cytoplasm divides unequally, yielding one large product, the egg cell with the egg nucleus, and one small, nonfunctional cell, the first polar body with its nucleus. In a very few organisms, such as sea urchins, oocytes complete both meiotic divisions before maturing; in this case a second unequal division of the oocyte cytoplasm occurs, producing a second polar body at the surface of the mature egg. If the second division occurs, the first polar body either disintegrates, remains quiescent, or also divides, depending on the species. The result of meiotic arrest at egg maturity in different animals, according to the degree of progress through meiosis, is thus a single large cell—the egg—with no, one, two, or three polar bodies.

Polar bodies are small, abortive cells produced by unequal division of the egg cytoplasm in meiosis.

As a result of the patterns of synthesis and deposition of yolk, mRNA, and other materials, the cytoplasm of most animal eggs becomes subdivided into regions containing different components at maturity. In many eggs, this subdivision is most obvious as differences in the distribution of granules of pigment and yolk bodies in the egg cytoplasm. In mature frog eggs, for example, pigment granules are limited to one-half of the cytoplasm, distributed in a layer just under the egg surface (Figure 14-6a). Yolk bodies are concentrated in the opposite half of the egg. Eggs having a greater concentration of yolk at one end in this way are called **telolecithal** eggs (*telo* = end; *lekithos* = yolk). Birds, fish, reptiles, and other amphibians typically produce telolecithal eggs. Other distributions occur; insects and many other arthropods have yolk concentrated in a mass in the approximate center of the egg. Eggs of this type are called **centrolecithal**. Eggs with more or less evenly distributed yolk are called *isolecithal* (*iso* = equal).

The distribution of yolk in mature eggs modifies the pattern of cytoplasmic division after fertilization; yolky regions divide less rapidly than regions with less yolk.

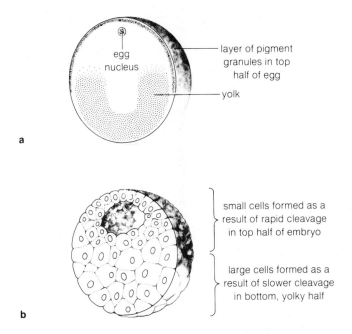

Figure 14-6 (**a**) Unequal distribution of pigment granules and yolk in frog eggs. (**b**) Pattern of cleavage in frog eggs. The presence of yolk causes cells in the bottom half to divide more slowly, producing larger cells in this location.

The distribution of yolk in the egg cytoplasm modifies the pattern of cell division during embryonic growth after fertilization. Regions of the fertilized egg containing greater concentrations of yolk generally divide more slowly, producing fewer and larger cells than less yolky regions. In frogs, the yolky bottom half of the fertilized egg divides more slowly than the top half containing less yolk; this pattern produces an early embryo with numerous small cells in the top half and a smaller number of large yolky cells in the bottom half (see Figure 14-6b). These differences are even more pronounced in the fertilized eggs of birds, which cleave only in a limited surface region containing little or no yolk (see Figure 14-7). New cell membranes formed by cleavage in these eggs typically extend only a short distance into the yolky regions (arrows, Figure 14-7).

The unequal distribution of yolk bodies, pigment granules, and other components in the mature egg is termed **polarity**. There is apparently always some degree of polarity in animal eggs, even in cases when the more visible components of the cytoplasm, such as yolk and pigment granules, are evenly distributed. For instance, the nucleus in the unfertilized eggs of most animals takes up a position near one end of the cell. This end of the egg,

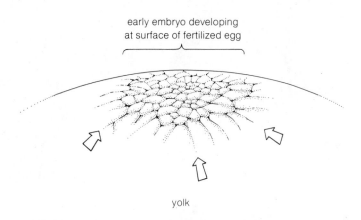

early embryo developing
at surface of fertilized egg

yolk

Figure 14-7 Early cleavage in a bird embryo. The cells formed by cleavage in these embryos are limited to a small region at the surface of the fertilized egg. The plasma membranes of the most recently formed cells are incomplete and extend only a short distance into the yolk (arrows).

called the **animal pole**, typically gives rise to surface structures and the anterior end of the embryo. When polar bodies are produced through meiotic divisions during the maturation of the egg, they are always formed at the animal pole of the egg and clearly mark this end. The opposite end of the egg, the **vegetal pole**, typically gives rise to internal structures of the embryos. Yolk, when unequally distributed in the egg cytoplasm, is most frequently concentrated in the vegetal half of the egg.

These arrangements of components in the egg are genetically controlled and undoubtedly reflect the activity of genes in the oocyte nucleus that function during egg maturation. Many experiments, carried out from the turn of the century to the present, indicate that the distribution of components in the egg involves more than the position of structural components such as the nucleus, yolk bodies, and pigment granules. These experiments have shown that developmental information in some form is also distributed differentially in the egg cytoplasm in fixed positions. (Two experiments demonstrating this fact are outlined in Supplement 14-2.) This information, possibly in the form of specific mRNA or protein molecules, influences the cytoplasm of different regions of the egg to develop into predetermined regions and parts of the embryo. Apparently, almost all animal eggs, with the possible exception of mammalian eggs, contain localized information of this sort that, to a greater or lesser extent depending on the species, fixes the developmental fate of regions of the egg cytoplasm.

Eggs containing cytoplasmic regions that are preprogrammed to develop into specific parts of the embryo are called **mosaic eggs**. If embryos derived from mosaic eggs are experimentally separated into individual cells at early stages of development, each cell typically develops only into the parts it would have formed in the intact embryo. Mosaic eggs of this type are found in molluscs (clams, oysters, squid, octopus, and relatives) and annelids (the segmented worms). Other animal eggs have much greater flexibility in the developmental fates of cytoplasmic regions. In the sea urchin, for example, any one of the cells of the early embryo, at the stage containing either two or four cells, can give rise to a small but complete embryo if separated from the rest. Eggs of this type, in which subregions of the cytoplasm give rise to complete embryos when compartmented into separate cells, are called **regulative eggs**. Vertebrate eggs, as well as the eggs of sea urchins and other echinoderms, are typically regulative in their developmental patterns.

The Time and Place of Oogenesis Oogenesis takes place at various stages in the life history of different animals. In frogs and salamanders, egg development occurs only in the ovaries of mature animals. In insects, however, oogenesis takes place before the female is fully mature, during an intermediate, larval stage. In mammals, oocytes begin development at an even earlier stage, during embryonic growth of the female. In humans, oocytes initiate maturation before birth and proceed as far as the transcription stage of meiotic prophase I. They remain arrested at this stage until the female enters puberty, when each month a single oocyte resumes meiosis under the influence of body hormones and matures. Final meiotic arrest comes at metaphase II of meiosis. The mature egg remains at metaphase II until fertilization, which must occur within about one day after the egg is released from the ovary. As a result of this pattern, human oocytes may remain arrested for 50 years or more between the initiation and completion of development.[1] (Supplement 14-1 gives further information about oogenesis and other stages of human development.)

---

[1]The long period of meiotic arrest at the synthesis stage of meiotic prophase I, during which homologous chromosomes are still paired, may be a cause of human birth defects that result from nondisjunction. In nondisjunction (see Supplement 13-2), one or more of the paired chromosomes fail to separate properly in meiosis; as a result, a gamete, and thus the embryo, receives too many or too few chromosomes. Supposedly, the longer the period of arrest at the synthesis stage, the greater the possibility that correct separation will not occur at ovulation. This relationship between meiotic arrest and nondisjunction might explain why the frequency of such birth defects as Down's syndrome (see p. 282), which results from nondisjunction, increases with age of the mother.

At maturity the animal egg is poised to enter development. The cytoplasm is packed with ribosomes and all the other components required for protein synthesis, and the nucleus is arrested at an intermediate stage of meiosis.

At maturity, the animal egg represents a system poised to enter development. The cytoplasm is packed with yolk bodies containing raw materials and an energy source, developmental information in the form of specific messenger RNA and protein molecules, and all the components required for nucleic acid and protein synthesis. The nucleus is arrested at different stages of meiosis depending on the species. All these systems await the triggering of development by the fertilizing sperm.

Sperm Formation in Animals: Spermatogenesis

The sperm cell has two primary functions in fertilization. One is to deliver the paternal set of chromosomes to the egg. The other is to break the arrest of the mature oocyte and trigger embryonic development. The changes occuring in the development of animal sperm, called **spermatogenesis**, create a cell type that is specialized to accomplish these ends.

The mature sperm cell is specialized to deliver the paternal set of chromosomes and break the arrest of the egg.

Sperm, like eggs, develop from cell lines that are set off very early in the life of the embryo. At some point in embryonic development, the cells destined to form sperm migrate to the male sex organs, the **testes**, where they increase in numbers through mitotic divisions. Eventually, these cells undergo meiosis and develop into mature sperm cells. During development, the cells destined to become sperm are called **spermatocytes**. In many species, spermatocytes are continually replaced as they develop into sperm, so that once spermatogenesis begins, it continues through the life of the animal. In human males, for example, sperm production begins at the onset of sexual maturity at ages 12 to 14 and normally continues throughout life.

Each of the four products of meiosis in males differentiates to produce a sperm cell. (In oogenesis, only one of the four meiotic products becomes a functional egg.) In spite of great morphological differences between different adult animals, the mature sperm cells of most species are basically similar in structure (Figure 14-8). With a few exceptions, most sperm are differentiated into three main body regions: a *head* containing the nucleus; a *middle piece* containing mitochondria and the base of the tail; and the *tail* itself. These structures are specialized to penetrate and fertilize the egg (the head), provide energy for movement (the middle piece), and generate movement (the tail).

The sperm head develops from the haploid nucleus resulting from meiosis. During this development, the chromatin of the nucleus packs tightly until it occupies only a small fraction of the volume of an interphase cell of the same species. During nuclear condensation in most species, the usual chromosomal proteins, the histone and nonhistone proteins (see p. 158), are replaced by smaller, even more basic proteins called **protamines**. All other nuclear substance is eliminated. By the time condensation is complete, the DNA and protein are packed into a structure organized so tightly and regularly that it approximates a crystal (Figure 14-9). These changes reduce the weight and volume of nuclear material to be transported by the mature sperm to the smallest possible levels. In various species, the nucleus becomes flattened, rounded, elongated, or hooked during the condensation process.

As the chromatin condenses, most of the cytoplasm surrounding the nucleus flows toward the end of the sperm that will develop into the tail. So much of the cytoplasm around the nucleus may be omitted that the nuclear envelope and plasma membrane may appear to fuse together, leaving almost no intervening space between them. All that remains in the sperm head as a cytoplasmic organelle at this time is the **acrosome** (see Figure 14-9), a dense, membrane-bound structure that fits like a cap over the tip of the nucleus. Biochemical analysis of the contents of the acrosome shows that it contains several enzymes that catalyze breakdown of the external coats of the egg during fertilization. In sperm development, the acrosome forms as a specialized secretion vesicle containing proteins synthesized in the cytoplasm and modified by the Golgi complex.

Other changes in the developing sperm cell provide the motile system that will transport the sperm nucleus to the egg. The tail, a typical eukaryotic flagellum with a 9 + 2 system of microtubules (see p. 64), arises from a centriole located at the base of the nucleus (see Figure 14-9). The mitochondria of the sperm cell migrate to the base of the developing tail and pack into a mass that forms the middle piece. Most of the remaining cytoplasm collects into a blisterlike outgrowth that pinches off, leaving the mature sperm cell with a greatly reduced cytoplasmic volume.

The mitochondria surrounding the base of the tail provide the ATP energy required to move the sperm flagellum. In many animals, the fluids released with the

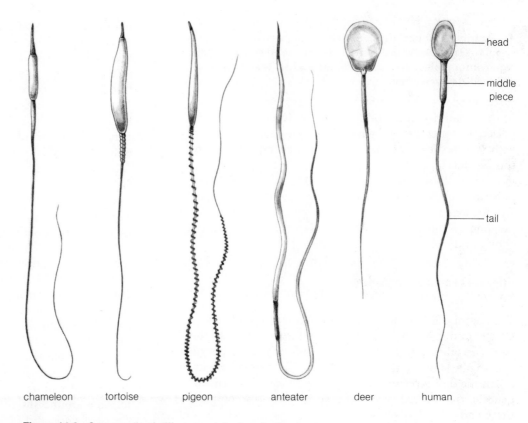

Figure 14-8 Sperm cells of different vertebrate animals.

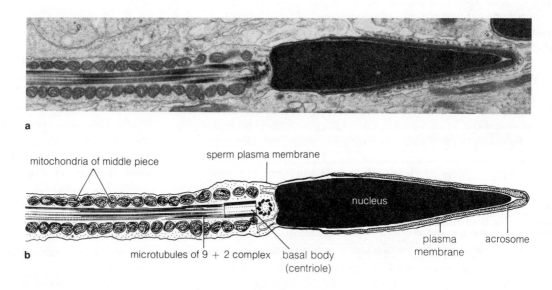

Figure 14-9 Structures of a mature sperm cell from a marmoset. ×13,700. Micrograph courtesy of J. B. Rattner and Academic Press, Inc.

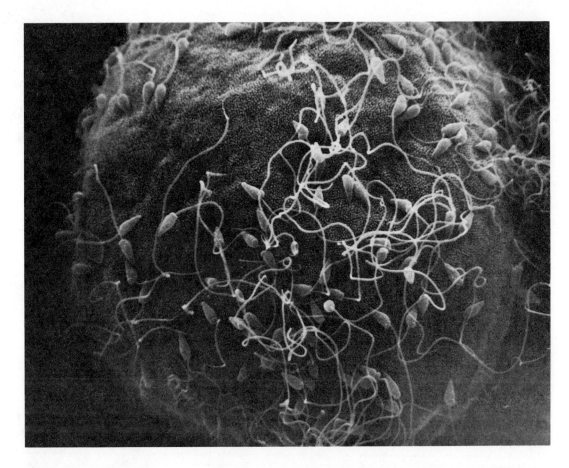

Figure 14-10 Attachment of sperm cells to the egg of the sea urchin *Strongylocentrotus*, viewed with the scanning electron microscope. ×2,300. Courtesy of M. J. Tegner. Copyright 1973 by the American Association for the Advancement of Science.

sperm contain substances that can be metabolized by the sperm to provide energy for movement. In humans, for example, the sperm fluid contains the sugar *fructose* as an energy source.

At maturity, the head of the sperm cell is streamlined and reduced to the smallest possible size; all it contains are the tightly condensed nucleus and the acrosome. Concentrated in the acrosome are enzymes that catalyze penetration of the egg coats. The remaining parts of the cell are limited to the middle piece, with its mitochondria generating energy for movement, and the motile tail.

The Meeting of Egg and Sperm: Fertilization

Fertilization, accomplished by the meeting and fusion of egg and sperm, initiates embryonic development and begins the entire train of genetically programmed events in

the life of the individual. Fertilization has two effects of immediate significance for embryonic development: (1) the egg and sperm nuclei fuse, restoring the chromosome number to the diploid value, and (2) the arrest of the egg is broken.

Fertilization breaks egg arrest and initiates protein synthesis, DNA replication, and completion of meiosis.

In most animals, movement of sperm cells to the egg is entirely unguided. With very few exceptions, no chemical attractants are released by the egg that direct movement of the sperm toward the egg surface. Thus contact of egg and sperm, if it occurs, is simply the result of random collisions. In sperm and eggs that are shed directly into water, as in the sea urchins and molluscs such as most clams and oysters, movement of the sperm tail is essential

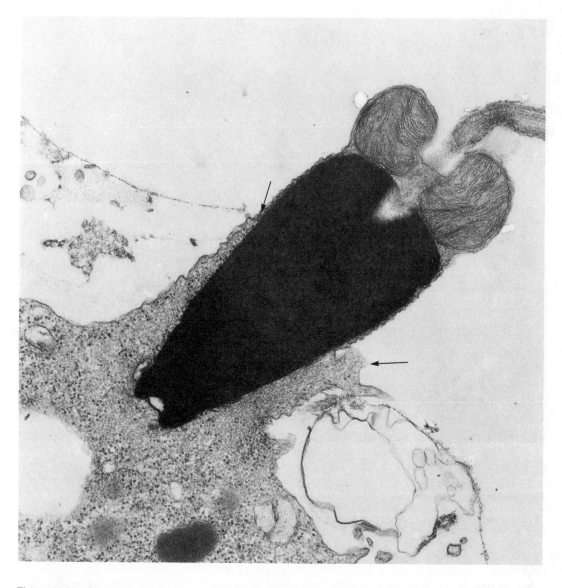

Figure 14-11 Fusion of egg and sperm membranes in fertilization of the sea urchin *Arbacia*. The egg cytoplasm (arrows) has begun to flow around the sperm nucleus. ×32,000. Courtesy of E. Anderson and The Rockefeller University Press.

for bringing eggs and sperm together. In animals that deposit sperm into the female reproductive tract, movement of sperm to the egg is often aided by contractions and motions of the female organs through which the sperm travel. In mammals, these movements are the primary agent bringing sperm to the egg; sperm motility apparently serves mainly to keep the sperm in suspension within the fluids of the reproductive tract. In humans and other mammals, the contractions of the female reproduc-

tive canal delivering the sperm to the egg are intensified by a hormone (a *prostaglandin*) secreted among the fluids ejaculated with the sperm by the male.

If egg and sperm are from the same species, their contact results in immediate attachment of the sperm head to the outer egg coats (Figure 14-10; see also Figure 14-3). This specific attachment is probably due to complementary binding sites on the surfaces of the sperm and egg coats that match only if the egg and sperm are from the

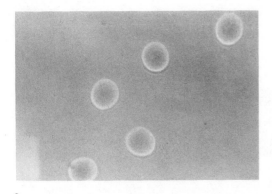
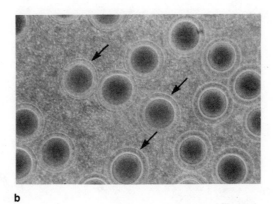

Figure 14-12 Elevation of the egg coat in response to fertilization in the sea urchin. (**a**) Before fertilization; (**b**) after fertilization. The arrows indicate the elevated egg coats. The space between the egg plasma membrane and the egg coat is filled with fluid and materials released from the egg cytoplasm.

same or closely related species. The binding holds the sperm in place and sets the stage for its penetration through the egg coats and fusion of the sperm and egg plasma membranes.

Within seconds after the sperm has attached to the egg, the membranes at the tip of the sperm separate, releasing the enzymes carried in the acrosome. These enzymes dissolve a narrow channel through the egg surface coats. As the channel forms, the sperm cell, still actively swimming, travels through it and approaches the outer surface membrane of the egg cell. As the egg and sperm cells touch, the plasma membranes of the two cells fuse (Figure 14-11) and the sperm nucleus enters and sinks into the egg cytoplasm. In some species, as in the sea urchin egg shown in Figure 14-11, the egg cytoplasm flows outward and actively surrounds and engulfs the fertilizing sperm in a process resembling phagocytosis (see p. 87).

Activation of the Egg Contact and fusion of the sperm and egg plasma membranes set off an almost instantaneous chain of reactions in the egg. One of the first of these changes prevents the entry of further sperm cells into the egg cytoplasm. This initial response, called the *block to polyspermy,* involves changes in the egg plasma membrane and surface coats that dislodge any additional sperm that may have attached. The block to polyspermy also involves modifications that alter the proteins or glycoproteins of the egg coats so that the enzymes released from the sperm acrosomes can no longer dissolve them. During these changes, the egg surface coat in most species detaches from the egg plasma membrane and raises so that it is now separated from the egg by an intervening

fluid-filled space (Figure 14-12). These elaborate measures ensure that only one set of paternal chromosomes is introduced into the egg at fertilization.

Other responses of the egg cytoplasm to fertilization include the initiation of protein synthesis, DNA replication, and gradually increasing RNA transcription. The energy required for this activity is provided by the oxidation of stored fuel molecules such as glucose, which begins almost immediately after fertilization and soon reaches the levels characteristic of actively synthesizing cells. The substances synthesized in quantity immediately after fertilization—DNA, the histone and nonhistone chromosomal proteins, and the microtubule proteins of the spindle—are the cellular products required for the rapid cell divisions of early embryonic development.

The total changes occurring in response to fertilization are called *egg activation.* It is likely that the fertilizing sperm cell merely triggers activation, since some animal eggs can be activated artificially by chemical treatment or sometimes simply by touching the egg surface with a needle. No molecules or information from the sperm seem to be required. In some animals, including insects and a few vertebrates, activation and normal development of the egg occur under natural conditions without any contact by sperm. Development of this type, called **parthenogenesis** (*parthenos* = virgin), occurs, for example, among bees and wasps, in which eggs laid without fertilization develop into haploid males. Fertilized bee and wasp eggs develop into females. Apparently, mechanical stimulation during passage of the unfertilized eggs through the oviducts triggers egg activation and parthenogenetic development of the haploid males. Parthenogenetic development has also

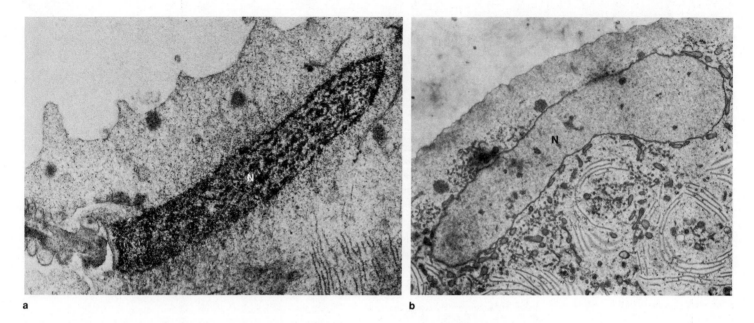

Figure 14-13 Successive stages in the return of sperm nuclei *(N)* to an interphase-like state after fertilization. **(a)** An early stage in the rat. ×23,400. Courtesy of L. Piko and Academic Press, Inc. **(b)** A later stage in the hamster. ×8,000. Courtesy of R. Yanagimachi and Academic Press, Inc.

been reported under natural conditions among domestic turkeys, in which a small percentage of eggs laid by virgin females are activated and enter development. In this case, diploid individuals are produced, apparently through fusion of two haploid cells at the initial stages of development.

Fusion of Egg and Sperm Nuclei After entry into the egg cytoplasm, the sperm nucleus loses its highly condensed appearance and returns to a state resembling an interphase nucleus (Figure 14-13). As part of these nuclear changes, the sperm nuclear proteins, the protamines, are replaced by histone and nonhistone proteins produced by the egg. While these changes in the sperm nucleus are taking place, the egg nucleus completes meiosis. The haploid egg and sperm nuclei then approach, and the chromosomes of the two nuclei join together, restoring the diploid condition. In most species, the resultant diploid nucleus, called the **zygote** nucleus, immediately enters the first mitotic division of embryonic development.

Fusion of egg and sperm nuclei restores the chromosomes to the diploid number.

Any additional structures carried into the egg by the fertilizing sperm, with the exception of the sperm cen-

trioles, eventually break down and do not contribute directly to the embryo. The sperm centrioles, upon entry into the egg cytoplasm, replicate and become aligned at opposite poles of the first mitotic spindle. (Most animal eggs lose their centrioles during maturation.) These centrioles replicate and divide normally during the subsequent mitotic divisions of the embryo. Hence the centrioles of animal embryos are normally derived from the fertilizing sperm. Other cytoplasmic organelles derived from the sperm cell break down and disappear. Thus all other major cytoplasmic organelles, such as mitochondria, are normally of maternal origin in animals.

EARLY DEVELOPMENTAL PATHWAYS IN ANIMAL EMBRYOS

After fertilization, the egg begins a series of rapid mitotic divisions. Because these early divisions take place without any overall increase in the size or mass of the embryo, they are called **cleavage** divisions. As a result of cleavage, the nuclei produced by each division are enclosed in successively smaller cells.

Cleavage divisions continue until the cytoplasm of the original fertilized egg is divided among hundreds to thousands of smaller cells. At this point, the embryo passes through two stages that are fundamental to the development of all animals: the **blastula** and the **gastrula.** The **blastula** (*blast* = bud or offshoot; *ula* = small) is simply a hollow ball of cells resulting from cleavage of the fertilized egg. The cells of the blastula then move and reorganize in highly ordered patterns to produce the second stage of development, the **gastrula** (*gaster* = gut or belly). Gastrulation lays down the primary cell layers of the embryo, including the beginnings of the surface skin and the digestive tract. By the time gastrulation is complete, the first hints of the adult animal can be detected, including identifiable dorsal (top), ventral (bottom), anterior (front), and posterior (rear) parts.

Blastula and Gastrula Formation

In different species cleavage of the fertilized egg produces a blastula containing from 100 to 60,000 or more individual cells. In most embryos, the blastula becomes hollow as it develops, forming a central, fluid-filled cavity called the **blastocoel** (*coel* = hollow; see Figure 14-14). Most higher animals complete the blastula stage within one to two days after fertilization.

Once the blastula forms, the cells of the embryo undergo a series of highly ordered movements that sort out the primary cell layers. Cell division continues as these movements take place. After completion of these migrations, which may involve both single cells and groups of cells, the embryo has a primary outer cell layer, the **ectoderm** (*ecto* = outside; *derm* = skin); an inner lining of cells, the **endoderm** (*endo* = inside), which forms the primary gut; and a third layer between these two, the **mesoderm** (*meso* = middle). All tissues and organs of the adult animal can be traced to beginnings in one of these three cell layers of the gastrula (see Table 14-1).

Gastrulation produces the primary cell layers of the embryo: ectoderm, mesoderm, and endoderm.

As the blastula develops into the gastrula, the embryonic cells differentiate to some degree: they become recognizably different in biochemistry and structure. The cells at this stage also become more limited in developmental potential than the fertilized egg from which they originated. At fertilization the egg cell is fully capable of developing into a complete embryo; a mesoderm cell, however, may develop into muscle or bone but not normally into outside skin or brain. These limits on developmental potential arise even though the cells of the embryo retain all the genetic information of the fertilized egg.

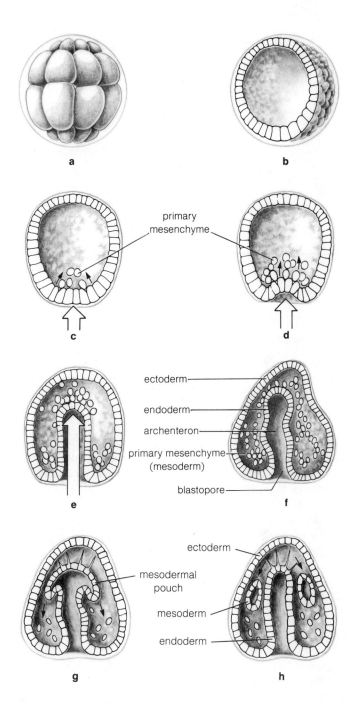

Figure 14-14   Cleavage and gastrulation in the sea urchin (see text).

Table 14-1   Origins of Adult Organs in the Primary Cell Layers

Ectoderm	Mesoderm	Endoderm
Skin, hair, feathers, scales, nails	Muscles	Inner lining of digestive system, including stomach and intestine
Entire nervous system	Blood	Liver
Lining of mouth and anus	Bone	Secretory cells of pancreas
	Reproductive organs	Lining of lungs
	Kidneys	
	Outer walls of stomach and intestine	

RNA and Protein Synthesis during Cleavage and Gastrulation

During the period of cleavage before gastrula formation, synthesis in the developing embryo is limited primarily to the molecules required for cell division. As gastrulation proceeds, RNA and protein synthesis in the embryo gradually increase and become more diversified. Ribosomal RNA synthesis, which often cannot be detected during cleavage and blastula formation, begins anew by the time gastrulation takes place. (In most embryos, ribosomes stored in the egg during oogenesis are used during the period before gastrulation.)

Messenger RNA transcription also changes as gastrulation proceeds. In earlier cleavage stages, the types of mRNA transcribed reflect the activity of the cells in division. In gastrulation, the spectrum of mRNAs transcribed broadens to include messengers for proteins characteristic of the primary tissues laid down. Transcription of transfer RNA remains at high levels throughout cleavage and gastrulation.

Major Patterns of Gastrulation

Cleavage and gastrulation follow different pathways in different major animal groups. Although different in detail, however, the major patterns of gastrulation in all animals include the same cellular mechanisms and produce the same differentiation of embryo tissues into ectoderm, mesoderm, and endoderm. These mechanisms of cleavage and gastrulation are illustrated in the following paragraphs by examples from the three best-studied animal groups: sea urchins, amphibians, and birds. Mammalian gastrulation, which resembles the pattern in birds, is described in Supplement 14-1.

Cleavage and Gastrulation in Sea Urchins   Yolk is relatively concentrated and distributed equally throughout sea urchin eggs. As a result, cleavage proceeds at approximately the same rate after fertilization in all regions of the developing embryo. Cleavage continues until a hollow fluid-filled ball containing about a thousand cells, the blastula, is formed (Figure 14-14a and b). Gastrulation begins as the wall at one side of the blastula flattens (large arrow, Figure 14-14c). The cells in this region become more elongated and cylindrical as flattening progresses. The flattened area then moves actively inward (Figure 14-14d and e), gradually pushing into the interior until the original blastula cavity is nearly eliminated (Figure 14-14f). Movement continues until the surface layer of cells is almost completely lined by an inner cell layer derived from the infolding wall. The inward movement, in effect much like pushing in the side of a hollow rubber ball, generates a new cavity in the embryo, the **archenteron** (arch = beginning, enteron = intestine or gut; see Figure 14-14f). Figure 14-15 shows a gastrulating sea urchin embryo. The archenteron communicates to the outside of the embryo through the **blastopore**, the opening produced by the inward movement of cells from the surface.

At this point the embryo has two complete cell layers—an outer layer, remaining from the original blastula surface, lined by a second, inner layer derived from the cells pushing inward to form the archenteron. These two layers make up the ectoderm and endoderm of the embryo. The middle cell layer, the mesoderm, is formed by cells that move from the endoderm of the developing gas-

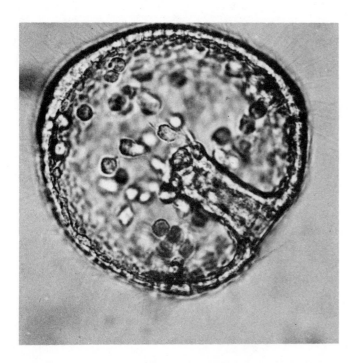

Figure 14-15 Light micrograph of a living sea urchin embryo undergoing gastrulation. × 500. Courtesy of G. Karp and Academic Press, Inc., from *Developmental Biology* 41:110 (1974).

trula. Some of these migrating cells, called **primary mesenchyme**, move singly to enter the gradually narrowing space between the endoderm and ectoderm as gastrulation proceeds (Figure 14-14*c* and *d*). The major source of mesoderm, however, arises from pouchlike invaginations of the endoderm that push into the space between ectoderm and endoderm (Figure 14-14*g* and *h*).

The migrations of all these cell types cease when they come into contact with other cells at their destinations in the embryo. Movement to a fixed destination in this way means that the migrating cells must slide over many other cells without stopping until they arrive at their permanent locations. The final locations are somehow recognized, and fixed cell-to-cell adhesions are made at these points. Trygvie Gustafson and Louis Wolpert, working at the University of Stockholm, have observed the details of this process in sea urchin gastrulation. In the early stages of gastrula formation in this organism, cells of the infolding blastula wall send out extensions that stretch across the blastocoel cavity and make contact with the inside wall of the ectoderm (arrows, Figure 14-14*f*). The extensions then move about on the inner ectoderm wall, as if searching for the correct locations for their final attachment. Finally, tight adhesions are made between the extensions and the

inner ectoderm wall at certain points and the extensions contract, pulling the infolding endoderm inward with them. This ability to move, to slide over some cells without adhering, and to recognize and make permanent attachments to others, called *selective adhesion*, is a basic property of animal embryonic cells.

As the ectoderm, mesoderm, and endoderm develop, the embryo lengthens into an ellipsoidal shape with the blastopore at one end. This elongation gives a recognizable shape to the embryo and establishes definite symmetry, such as dorsal and ventral sides. The organ systems of the adult then differentiate through further division, movements, and adhesions of the three basic cell layers. The outer skin arises from the ectoderm; muscles, digestive organs, and reproductive organs develop from the mesoderm; and the lining of the digestive tract forms from endoderm.

Gastrulation is accomplished by the cellular mechanisms of (1) cell movement, (2) cell division, and (3) selective cell adhesion.

The basic mechanisms producing the sea urchin gastrula thus include three activities of the embryonic cells: (1) motility through extension and contraction of parts of cells or movements of whole cells, (2) further mitotic divisions, and (3) establishment of selective adhesions between cells. These same mechanisms act in the gastrulation and further development of amphibian and bird embryos—and, in fact, all animal embryos.

Gastrulation in Amphibians The presence of large quantities of yolk concentrated toward one end of the egg cell in amphibians such as frogs and salamanders modifies the patterns of cleavage and gastrulation in these animals. The unfertilized frog egg (Figure 14-16*a*) is divided into top and bottom halves by the unequal distribution of yolk. The top half contains reduced amounts of yolk and is darkly pigmented; the bottom half is packed with yolk bodies and is light yellow in color. During cleavage, the darkly pigmented top half divides more rapidly than the yolky bottom half. As a result, cells in the pigmented half become smaller and more numerous than the cells in the yolky half (Figure 14-16*b*). Cleavage continues until the blastula contains about 15,000 cells. Within the ball of cells, the blastocoel cavity hollows out in the darkly pigmented top half of the embryo as development proceeds to this stage (Figure 14-16*c*).

The first signs of gastrulation appear at one side of the blastula in the pigment margin between the top and bottom halves of the embryo. The cells of this region elongate and move actively into the interior of the embryo. As

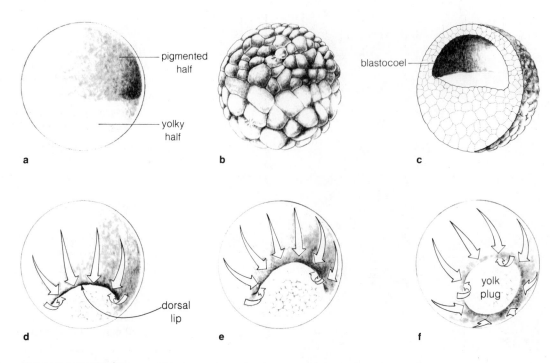

Figure 14-16 Cleavage and blastopore formation in an amphibian embryo (see text).

the cells move inward, they form a short linear depression in the surface along the border between the pigmented and yolky cells. The depression gradually extends laterally and begins to curve downward into a crescent at its sides. As the crescent develops, cells of the pigmented layer stream toward the depression, roll over its sides, and enter the interior of the embryo. The crescent-shaped depression gradually extends at its edges to form a complete circle, always with the pigmented cells migrating inwardly at its margins (Figure 14-16*d*, *e*, and *f*). The circle, when complete, forms the blastopore of the developing gastrula. By this time, the pigmented cell layer has expanded downward to cover the entire surface of the embryo. The yolky cells are completely enclosed by the movements and stretching of the pigmented layer and show only as a "plug" of cells that remains visible through the blastopore (see Figures 14-16*f* and 14-17).

The top margin of the blastopore, called the *dorsal lip* (see Figure 14-16*d*), is most active in the cell movements producing gastrulation in the frog. At the beginning of blastopore formation, the cells moving into the interior roll over the dorsal lip and slide upward and inward within the embryo (Figure 14-18*a* and *b*), forming an inner layer that lines the inside top half of the embryo. The first cells to enter contribute to endoderm formation and eventually give rise to the most anterior parts of the digestive tract.

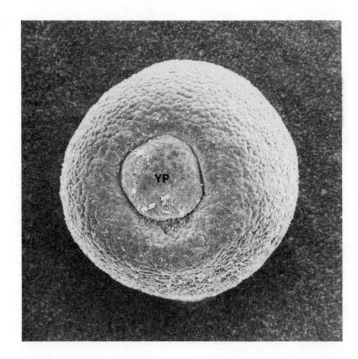

Figure 14-17 Early amphibian gastrula showing blastopore and yolk plug *(YP)*. Courtesy of P. B. Armstrong.

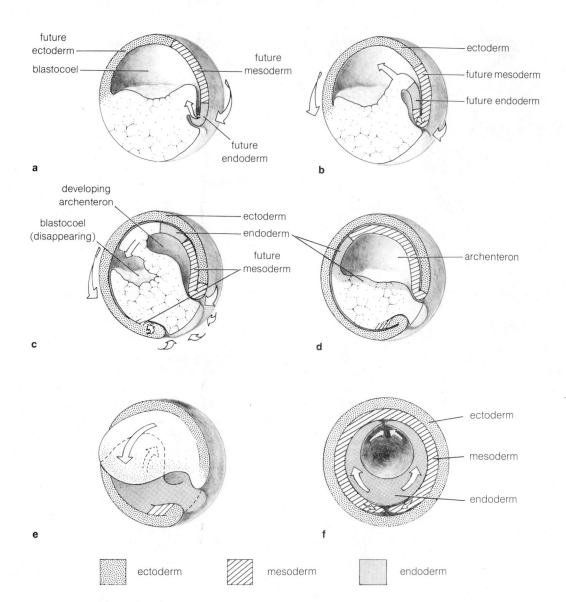

a

future ectoderm
blastocoel
future mesoderm
future endoderm

b

ectoderm
future mesoderm
future endoderm

c

developing archenteron
blastocoel (disappearing)
ectoderm
endoderm
future mesoderm

d

ectoderm
endoderm
archenteron

e

f

ectoderm
mesoderm
endoderm

ectoderm   mesoderm   endoderm

Figure 14-18 Gastrulation in an amphibian embryo (see text). In (**e**), the movements of the mesoderm (solid arrow) and endoderm (dotted arrow) are shown in three dimensions with the ectoderm cut away. The mesoderm grows and slides downward over the endoderm to form an almost complete layer between the ectoderm and endoderm. At the same time, growths of the endoderm slide upward under the advancing mesoderm, eventually meeting at the top to form a complete tube. These movements are shown in cross section in (**f**).

The following cells entering around the top and side margins of the developing blastopore are destined to form the mesoderm of the embryo. The pigmented cells remaining at the surface of the embryo form the ectoderm.

As the future mesoderm cells move inward around the edges of the blastopore, the surface depression forming the top margin of the blastopore gradually deepens (arrows, Figure 14-18b and c) and extends inward, even-

tually enlarging into the archenteron (Figure 14-18d). The archenteron at this stage has a roof of mesoderm that has migrated inward from the top half of the blastopore and a floor of yolk cells. The yolky floor is pushed downward as the cavity enlarges. At the innermost tip of the archenteron cavity, the yolk cells of the floor are continuous with the endoderm cells carried in by the first cell movements around the top margin of the blastopore. These yolk cells

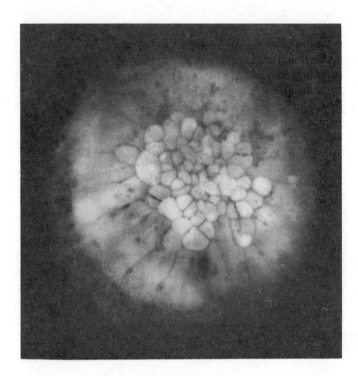

Figure 14-19 Layer of cells developing at the surface of a chick embryo to form the blastodisc. The developing blastodisc has been removed from the undivided yolky mass of the embryo. Courtesy of H. Eyal-Giladi and Academic Press, Inc., from *Developmental Biology* 49:321 (1976).

lining the archeteron floor also contribute to the formation of the endoderm. The archenteron continues to enlarge, gradually expanding at the expense of the blastocoel, which eventually disappears (Figure 14-18d).

Cells destined to become mesoderm continue to move from the surface of the embryo around the margins of the blastopore to the interior. Inside, they slide between the outer ectoderm and the developing endoderm cells. Eventually, the cells moving inward from the surface form an almost continuous sheet of mesoderm lying between the ectoderm and endoderm of the embryo.

The endoderm develops from cells derived from two parts of the embryo. Part of the endoderm, as noted, develops from the first cells to slide into the interior of the embryo as the blastopore forms. These cells form the endoderm at the anterior end of the embryo, opposite the blastopore. The remaining endoderm is formed from the yolky cells confined inside the embryo by blastopore formation. At first, these cells line only the floor of the archenteron cavity in the middle and posterior regions of the embryo. Toward the end of gastrulation, they move

upward along the sides of the archenteron cavity, eventually meeting at the top and fusing with the anterior endoderm to form a continuous inner layer lining the mesoderm and separating it from the archenteron cavity (Figure 14-18e and f). These movements complete the generation of the three primary tissues of the frog embryo. The cells of these layers continue to increase in number by further divisions as development proceeds.

Cleavage and Gastrulation in Birds The pattern of cleavage and gastrulation in birds is greatly modified by the presence and distribution of the yolk, which occupies almost the entire volume of the egg (see Figure 14-7). The portion of the cytoplasm that will divide and give rise to the primary tissues of the embryo is confined to a thin, disclike layer at the egg surface.

The early nuclear divisions following fertilization in the chick produce multiple nuclei, which gradually become completely separated by continuous plasma membranes except at the margins of the developing embryo (Figure 14-19). This process continues until a flattened layer of cells, about 2 millimeters in diameter, forms at the surface of the yolky mass. This layer, which is equivalent to the blastula of sea urchin and amphibian embryos, is called the **blastodisc**. When blastodisc formation is complete, the layer contains nearly 100,000 cells. Although the cells in the center of the layer are enclosed by complete plasma membranes, the cells at the margins are incompletely separated from the surrounding yolky regions, so that their cytoplasm is continuous with the yolk (see Figure 14-7).

The cells of the blastodisc then separate into top and bottom layers (Figure 14-20a). This separation produces the ectoderm, the endoderm, and a cavity enclosed between them: the blastocoel. The subsequent formation of mesoderm resembles blastopore formation in amphibian embryos. A longitudinal depression called the *primitive streak* forms along the outer surface of the ectoderm (Figure 14-20b). This depression corresponds to the dorsal lip of the blastopore in amphibians; as it gradually deepens, cells from the ectoderm roll over its margins and migrate downward into the blastocoel cavity to form the mesoderm (Figure 14-20c). These cells move laterally from the primitive streak until a continuous layer of mesoderm lines the space between the ectoderm and endoderm (Figure 14-20d). Some of the cells entering the primitive streak also contribute to the formation of the endoderm as well as the mesoderm.

Thus the pattern of gastrulation in amphibians and birds, although modified by the presence of yolk, produces the three primary cell layers by the same cellular

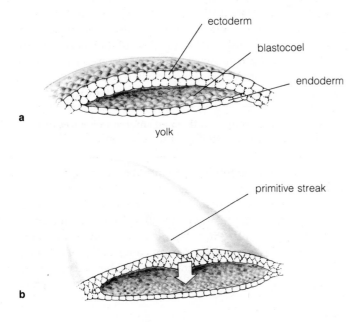

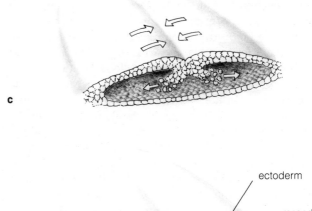

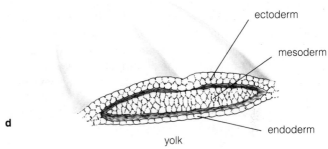

ectoderm

blastocoel

endoderm

a

yolk

c

primitive streak

b

ectoderm

mesoderm

d

endoderm

yolk

Figure 14-20 Gastrulation in the chick (see text).

mechanisms occurring in sea urchins: (1) cell movement, (2) further divisions, and (3) selective adhesion. (Cleavage and gastrulation in mammals, although unhindered by the presence of yolk, follow a pathway resembling the pattern in birds—see Supplement 14-1.)

Development of the organ systems of animals involves the cellular processes of cell movement, cell division, and selective cell adhesion plus two additional mechanisms: induction and differentiation.

The later development of animal embryos involves further interactions of cell movement, division, and selective adhesion plus two additional mechanisms that play an equally important role in embryonic development: *induction* and complete *differentiation* of cells to their final adult types. To illustrate how these mechanisms interact in the production of an adult organ system—especially induction and differentiation, which are perhaps the most interesting and least understood of the cellular activities in embryonic development—we will follow the events leading from gastrulation to the generation of the eye in vertebrate animals.

INDUCTION AND DIFFERENTIATION IN EYE DEVELOPMENT

Both the structural and biochemical changes occurring in eye development have been much studied by developmental biologists. (Information Box 14-1 outlines eye structure in vertebrate animals.) Although development of this organ system in amphibians is emphasized in the following discussion, the same processes occur as other organs develop in all vertebrate animals: (1) cell movement, (2) cell division, (3) selective cell adhesion, (4) induction, and (5) differentiation.

Development of the Amphibian Eye from the Gastrula

Eye development from the amphibian gastrula stage occurs in two major steps. First, the mesoderm induces the ectoderm of the gastrula to generate the beginnings of the central nervous system and brain. Then parts of the developing brain interact with additional ectoderm at the anterior end of the embryo to form the eye. In this second step, a series of back-and-forth inductions occurs between the developing brain and the ectoderm.

Step 1: Development of the Primary Nervous System As the ectoderm, mesoderm, and endoderm take form in amphibians, the embryo gradually elongates and forms anterior and posterior ends and dorsal, ventral, right, and left sides. The blastopore at this stage is positioned near the posterior end of the embryo. During gastrulation, the mesoderm rises inside the embryo and comes into close contact with the overlying ectoderm along the top of the embryo (see Figure 14-18*f*). In response, the ectoderm thickens and flattens into a broad band running longitudinally along the top of the gastrula (Figure 14-21). This flat, thickened layer of ectoderm, called the *neural plate*, is the forerunner of the central nervous system.

Various experiments have shown that the ectoderm along the dorsal margin of the amphibian embryo will not develop into the neural plate if the mesoderm is prevented from contacting it. The mesoderm is thus said to *induce* the overlying ectoderm to differentiate into the neural plate. In this induction, the underlying mesoderm either instructs the ectoderm how to develop into nerve tissue (via informational molecules or other signals) or it triggers a preprogrammed developmental pathway in the ectoderm itself. (These alternatives are discussed further in Chapter 16.) This induced change, the first in a series leading from ectoderm to eye formation, is typical of inductions in development, in which close contact between the inducing and induced cell layers is a constant feature.

In induction, one group of cells in an embryo influences or triggers another group to specialize or differentiate.

Once induced, the neural plate develops into the primary central nervous system through a series of contractions and movements of its cells (Figure 14-22). Cells along the centerline of the neural plate constrict at their top margins, forcing the plate to bend downward along the midline (Figure 14-22*a* and *b*). At the same time, cells along the edges of the plate contract at their bottom margins, raising elevated ridges along the sides of the neural plate

---

Information Box 14-1   Eye Structure

The structure of the eye is basically similar in fish, amphibians, reptiles, birds, and mammals (see the accompanying figure). The eye is a hollow ball filled with transparent fluid. Most of the eye wall is heavily pigmented and does not transmit light. But at one side, a clear area of the wall, the *cornea*, admits light into the eye. Just behind the cornea, a hard, crystallike *lens* within the eye focuses incoming light rays into an image on the opposite inside wall of the eye. In this region sensory cells of the *retina*, a light-sensitive layer, react to the patterns of light in the focused image. In response, the retinal cells send impulses through the *optic nerve* that connects the eye and the brain. In the brain, these impulses are translated into the perception of sight.

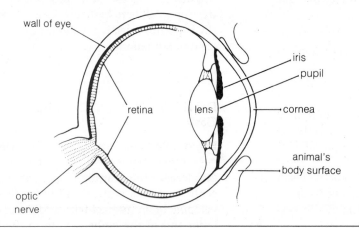

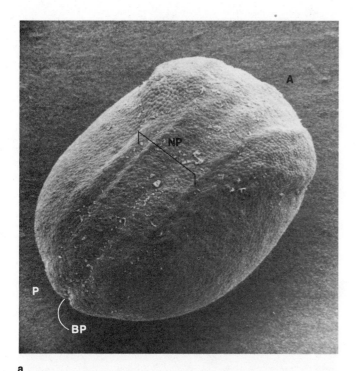

a

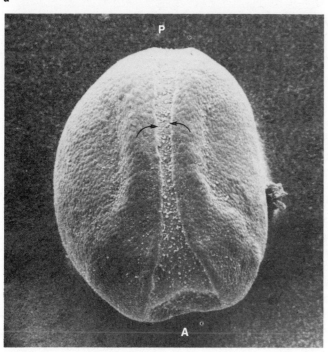

b

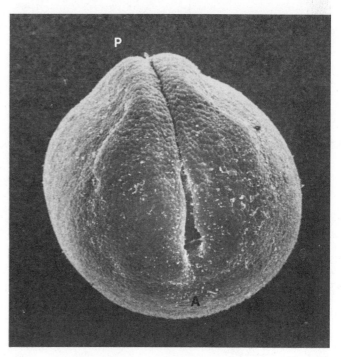

c

Figure 14-21 (**a–d**) Stages in neural plate formation in scanning electron microscope pictures of amphibian embryos (see text). Note the individual cells visible at the surfaces of the embryos. *A*, anterior end of embryo; *P*, posterior end; *NP*, neural plate; *BP*, blastopore (closing). Courtesy of P. B. Armstrong.

(Figure 14-22*b* and *c*). At the midline, the neural plate continues to sink downward, creating a deep central groove at the top of the embryo. As the groove deepens, the ridges at its sides extend into folds that move together and close over the center of the groove (Figure 14-22*c* and *d*). This movement and fusion converts the neural plate into a *neural tube* running the length of the embryo under the ectoderm. The cells along the midline then fuse tightly together and pinch off from the overlying ectoderm, freeing the neural tube to lie independently just under the dorsal surface of the embryo (Figure 14-22*e*). The central nervous system, including the brain and spinal cord, develops directly from this tube.

At the anterior end, the brain elaborates through the growth and enlargement of a series of hollow, pouchlike growths that swell outward from the tube (Figure 14-23). One paired set, called the *optic vesicles*, is directly involved in the further interactions with the ectoderm that form the eye.

Step 2: Development of the Eye Through Further Inductions The enlarging optic vesicles expand and move toward the surface of the embryo until they contact the ectoderm. Once an optic vesicle touches the underside of the ectoderm, an immediate series of reactions is triggered in each tissue (Figure 14-24). The outer surface of the optic

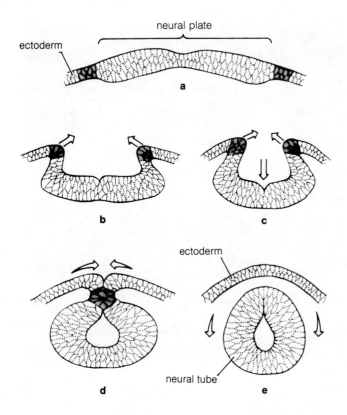

Figure 14-22 Cell movements converting the neural plate into the neural tube (see text). Part (**a**) is equivalent to Figure 14-21*a*; (**c**) is equivalent to Figure 14-21*b*; and (**d**) is equivalent to 14-21*c*.

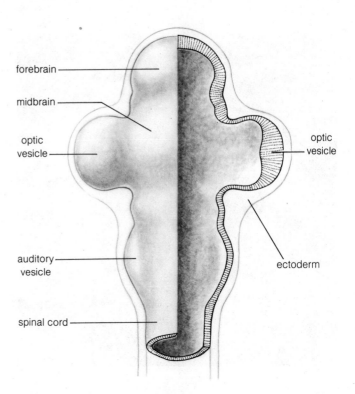

Figure 14-23 The neural tube enlarges at the anterior end into the vesicles of the brain. Eye structures develop through an interaction between the optic vesicles and the overlying ectoderm.

vesicle thickens and flattens at the region of contact with the ectoderm and then pushes inward, transforming the optic vesicle into a double-walled *optic cup* (Figure 14-24*a* to *c*). At the same time, the overlying ectoderm thickens into a disclike swelling (see Figure 14-24*c*). The thickened area of ectoderm, called the *lens placode*, subsequently undergoes a series of contractions and movements much like the transformation of ectoderm into the neural tube (Figure 14-24*d* to *f*). The center of the lens placode sinks inward toward the optic vesicle, and its edges rise and move toward the center to cover the deepening pit. Eventually they fuse together to close off the developing lens as a hollow ball of cells below the ectoderm surface. The ball of cells, called the *lens vesicle*, then pinches off from the ectoderm and descends to take up a position just inside the lip of the optic cup (Figure 14-24*g*). The ectoderm cells giving rise to the lens vesicle initially contain pigment granules. As the lens forms from the lens vesicle, these cells lose their granules and gradually become clear.

Formation of the lens vesicle from the ectoderm lying over the optic vesicle is the second major induction in the series leading to eye development. Contact between the optic vesicles and the overlying ectoderm is necessary for the lens to form. If optic vesicles of the brain are experimentally removed from embryos before lens formation, the ectoderm fails to develop lens placodes and vesicles. Moreover, transferring a removed optic vesicle to some other region of the embryo causes ectoderm in the new location to form a lens. For example, if an optic vesicle is severed and placed under the ectoderm in the belly region of a developing frog embryo, a normal lens placode and vesicle form at this point. Or, if the ectoderm over the optic vesicles is removed and ectoderm from elsewhere in the embryo is grafted in its place, a normal lens will develop in the grafted ectoderm, even though in its former location it would never differentiate into lens tissue.

The developing lens is responsible for the final major induction in eye development. As it develops, the lens increases in size and makes contact with the ectoderm, which has closed over it. The ectoderm cells in this region lose their pigment granules and become clear, forming the beginnings of the cornea. Eventually, the developing cor-

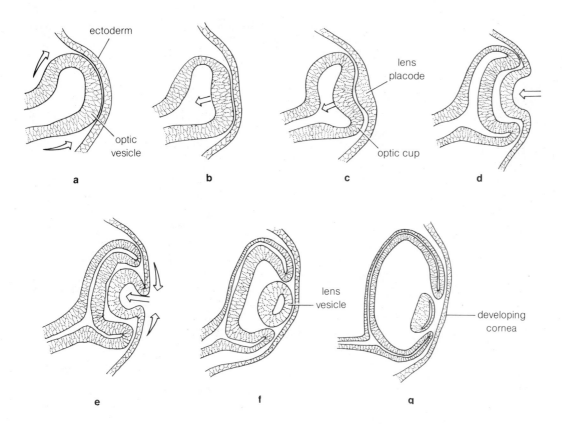

Figure 14-24 Stages in the formation of the eye (see text).

nea joins with the edges of the optic cup around the lens to complete the primary structures of the eye. Other cells contribute to accessory structures of the eye; for example, some of the outermost coats of reinforcing tissue in the wall of the eye and the muscles that move the eye are contributed by mesoderm cells.

Cell Differentiation in Eye Development

As the major eye structures appear, the cells forming them differentiate into fully specialized types. As the lens placode invaginates and becomes the lens vesicle, for example, the future lens cells synthesize large quantities of a fibrous protein called *crystallin* that hardens into glassy deposits. The lens vesicle cells gradually lengthen as they become packed with crystallin (Figure 14-25). Finally, the lens cells lose their nuclei and form an elastic, crystal-clear mass in the shape of a lens with curved front and back surfaces.

Cells differentiate by becoming specialized in structure and function.

The transformation of ectoderm into the lens provides an excellent example of biochemical differentiation of cell types. Ectoderm cells in the embryo that are not induced to form lens tissue characteristically synthesize a different protein, *keratin*, as the predominant cell product. Keratin is a component of surface structures such as feathers, scales, and horns. As a response to induction by the optic vesicle, the genes of the ectoderm cells coding for keratin proteins are permanently turned off and other genes coding for crystallin proteins are turned on. (The molecular mechanisms underlying this transcriptional regulation are discussed in Chapter 9.) These changes in biochemical differentiation occur even though the total genetic information of the cells remains the same.

The processes generating the eye in amphibians and other animals also operate in the development of all other organ systems in animal embryos. Besides the cellular

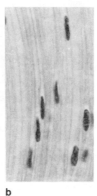

a   b

Figure 14-25 Developing lens cells of a chick. (**a**) A portion of the lens. ×150. (**b**) Some of the lens cells at higher magnification. These cells, already packed with crystallin protein, will eventually lose their nuclei and become completely transparent. ×960. Courtesy of J. Piatigorsky.

mechanisms of movement, division, and selective adhesion that take place in gastrulation, the development of organ systems, as we have seen, involves two further processes: induction and cell differentiation. These basic processes occur in the development of all animal embryos and, by interacting in various ways, account for all the complex organ systems of animals. How these processes work at the cellular and molecular level is of central interest to developmental biologists, and intensive research is currently being carried out in this area. This research, and the major conclusions that emerge from it, are the subjects of Chapter 16.

Questions

1. What is differentiation? Morphogenesis?

2. What is oogenesis? An oocyte? What substances are produced and stored in the oocyte cytoplasm during oogenesis? Where do these substances come from?

3. What is yolk? What is the relationship between the amount of yolk in an egg and the mode of embryonic development in a species?

4. What kinds of surface coats are formed around developing oocytes? How are the surface coats related to the pattern of embryonic development in a species?

5. What events accomplish oocyte maturation? Is meiosis always completed before maturation is complete? What are polar bodies? What is the relationship between polar body formation and meiosis in oocyte maturation?

6. Define telolecithal, centrolecithal, and isolecithal eggs. What effect does the distribution of yolk have on the pattern of cell division after fertilization of the egg?

7. What other elements in addition to yolk may be distributed unevenly in the egg cytoplasm? What significance does this distribution have for the pattern of embryonic development?

8. What are the animal and vegetal poles of an egg? What is the difference between mosaic and regulative eggs? How are mosaic and regulative development related to distribution of elements in the egg cytoplasm?

9. Review the systems and elements present in a mature egg cell.

10. What are the functions of sperm cells? How are sperm cells specialized to accomplish these functions?

11. What events occur in the nucleus during sperm maturation? In the cytoplasm? How are these events related to the function of sperm cells?

12. List the systems and elements present in a mature sperm cell and describe their functions.

13. What factors are important in the approach and attachment of sperm cells to the egg? How does the fertilizing sperm cell penetrate the egg surface coats?

14. What changes occur in the egg in response to fertilization?

15. How is the zygote nucleus formed?

16. What structures of the sperm contribute to the developing embryo?

17. What is cleavage?

18. Define blastula, blastocoel, gastrula, archenteron, blastopore, endoderm, ectoderm, and mesoderm.

19. What changes occur in RNA and protein synthesis during cleavage and gastrulation?

20. Compare the patterns of cleavage and gastrulation in the sea urchin, frog, and chick embryo. How are these patterns related to the distribution of yolk in the mature egg cell?

21. What cellular mechanisms accomplish gastrulation? What are the results of gastrulation?

22. Trace the development of the eye from the gastrula in a frog embryo. What cellular mechanisms are involved in eye formation?

23. What is induction? List the inductions occurring during eye formation in amphibians.

24. What cellular differentiations occur in the developing lens of the eye?

Suggestions for Further Reading

Balinsky, B. I. 1981. *An Introduction to Embryology.* 5th ed. Saunders, Philadelphia.

Berril, N. J., and G. Karp. 1981. *Development.* McGraw-Hill, New York.

Browder, L. W. 1980. *Developmental Biology.* Holt, Rinehart & Winston, New York.

Ebert, J. D., and T. S. Okada. 1979. *Mechanisms of Cell Change.* Wiley, New York.

Grant, P. 1979. *Biology of Developing Systems.* Holt, Rinehart & Winston, New York.

Lash, J., and J. R. Whittaker. 1974. *Concepts of Development.* Sinauer, Stamford, Connecticut.

Oppenheimer, S. B. 1980. *Introduction to Embryonic Development.* Allyn & Bacon, Boston.

Wolpert, L. 1978. "Pattern Formation in Biological Development." *Scientific American* 239:154–164.

SUPPLEMENT 14-1: HUMAN DEVELOPMENT

Embryonic development in humans follows the same major stages outlined in this chapter for other animals. In gametogenesis, mature eggs are formed in the female and sperm in the male. In fertilization, the union of sperm and egg cells restores the diploid number of chromosomes and initiates embryonic development. The embryo develops within the female's body; over a period of approximately 266 days the embryo grows from a single cell to a fetus weighing about 7 pounds. At birth, the infant is separated from the female and takes up a physically independent existence. Except for the developmental changes occurring at the onset of sexual maturity, growth after birth primarily involves an increase in size of the individual. There are no fundamental changes in the organ systems present.

Human Sexual Anatomy

Internal Sexual Anatomy in Females The internal sex organs of females (Figure 14-26) carry out a series of complex functions associated with each of the major stages of human development. Eggs are released from a pair of **ovaries;** one egg is released from one of the two ovaries each month. Once an egg is released, it is picked up by an **oviduct** (or *fallopian tube*) that leads from the ovary to the **uterus,** a saclike organ with thick, muscular walls in which most of embryonic development takes place. The lower end of the uterus forms an opening, the **cervix,** which extends into the **vagina,** a muscular canal that is continuous with the external genital organs.

Internal Sexual Anatomy in Males The internal sexual organs of human males (Figure 14-27) serve two major functions: the production of sperm and the secretion of fluids ejaculated with the sperm. Sperm cells produced in the paired **testes** are stored in the *epididymis,* a coiled tubule just above each testis. During intercourse, the sperm are transported at ejaculation from their storage place through paired tubes, the *deferent ducts.* These ducts meet in the region of the *prostate gland* to form a single duct leading to the exterior, the **urethra.** Several glands add fluids to the ejaculate, including a milky, viscous fluid secreted by the prostate gland and the *vesicular glands.* A clear lubricant is secreted before and during ejaculation by the twin *bulbourethral glands.* All these fluids, in combination with the sperm cells, make up the *semen* or *seminal fluid* ejaculated by the male. The sperm cells are activated by the seminal fluid and become motile at ejaculation.

Gametogenesis and the Menstrual Cycle in Females

Oocytes begin to develop in the ovaries before birth of the female. Development ceases by the eighth month of embryonic growth; at this time the oocytes are arrested at the transcription stage of meiotic prophase I. The oocytes remain stored in the ovaries until sexual maturity of the female at age 12 to 14. At this time, one oocyte breaks arrest each month and proceeds to a second arrest at metaphase II of meiosis. The mature oocyte is then released from the ovary in a process called **ovulation.** The ovulated egg is immediately picked up by the oviduct and transported to the uterus.

The release of an egg in ovulation is coordinated with changes in the uterus that prepare it to receive the egg if fertilization occurs. These changes, which make up the **menstrual cycle** (Figure 14-28), are keyed to the ovulation cycle and repeat approximately every 28 days. At the beginning of the menstrual cycle, the lining of the uterus is relatively thin. During the 10-day period before ovulation, the uterine lining gradually thickens, producing a layer, the *uterine mucosa* or **endometrium,** richly supplied with blood vessels and glands. Approximately 10 days after the lining begins to grow, ovulation occurs. The 10-day figure is only an average; in normal women, ovulation may sometimes take place from 4 to 16 days after growth of the lining begins. If fertilization does not take place, the uterine lining continues to thicken for another 12 to 18 days and then breaks down. Some of the lining is reabsorbed by the female at this time, but much of it is released to the exterior through the cervix and vagina as the *menstrual*

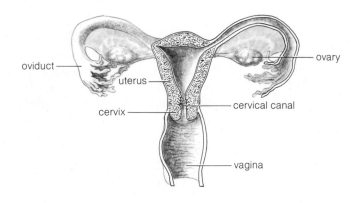

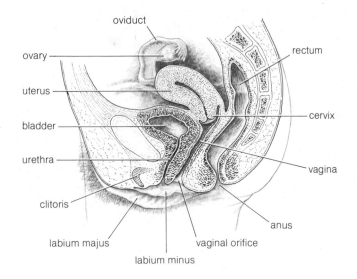

Figure 14-26 Internal sexual anatomy of the human female.

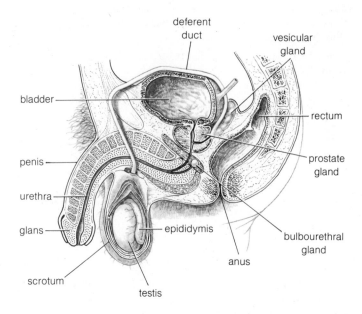

Figure 14-27 Internal sexual anatomy of the human male.

flow. Release of the old lining continues for 4 to 5 days until a new cycle begins in preparation for the next ovulation.

The changes in the uterus associated with the menstrual cycle are controlled by hormones secreted in various organs of the female, including the ovaries and several glands associated with the brain. At the beginning of the cycle, a segment of the brain called the *hypothalamus* stimulates the pituitary gland of the brain to secrete a hormone called **follicle-stimulating hormone (FSH).** In turn, FSH stimulates a single egg in the ovary to complete prophase I of meiosis and proceed to metaphase II, the stage at which it is ovulated to await fertilization. The ovary cells surrounding this egg grow into a blisterlike enlargement on the surface of the ovary. This fluid-filled enlargement containing the egg, which may become a centimeter or more in diameter, is called a **follicle** (see Figure 14-28).

As the follicle enlarges, it begins to produce several hormones of its own. These hormones, classified together as **estrogens,** stimulate the uterine lining to begin its cycle of growth. The estrogens also have a second effect; they inhibit the production of FSH by the pituitary. By about 8 days after the follicle begins to form, estrogen secretion has reached levels high enough to inhibit FSH production markedly. As a result, the concentration of FSH decreases until the end of the monthly cycle.

Reduction of FSH and the increase in the estrogens causes an abrupt rise in the secretion of a second pituitary hormone called **luteinizing hormone (LH).** The increase in LH, which stimulates release of the egg from the ovarian follicle, occurs about 10 days after the beginning of the cycle. After ovulation, LH secretion by the pituitary slows and by the end of the monthly cycle has dropped to its lowest level.

Besides initiating release of the egg from the follicle, LH also causes the follicle cells remaining at the surface of the ovary to grow into an enlarged, yellowish structure: the **corpus luteum** (*corpus* = body; *luteum* = yellow; see Figure 14-28). The corpus luteum acts as a temporary gland that continues to secrete estrogen for the remainder of the monthly cycle. A second hormone, **progesterone**, is also secreted by the corpus luteum. This hormone acts on the uterus to complete growth of the uterine lining and stimulates the breasts to form milk-producing glands. Because progesterone is active for a relatively short time during the monthly cycle, relatively little growth of the

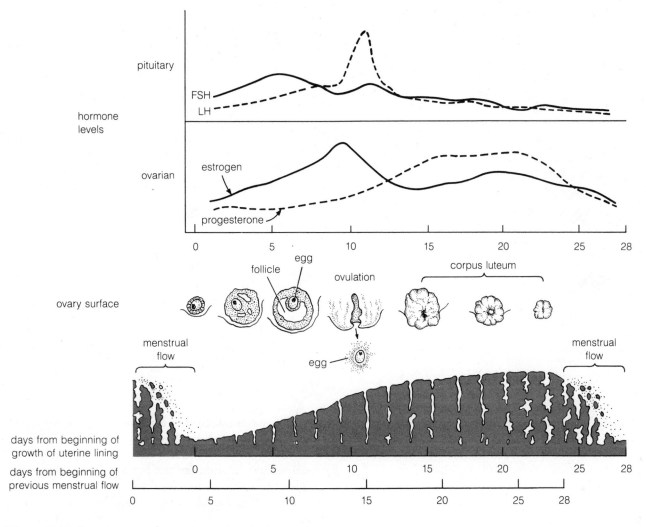

Figure 14-28 The menstrual cycle in human females (see text).

breasts occurs each month. The estrogen and progesterone secreted by the corpus luteum also inhibit FSH and LH secretion by the pituitary, probably through an effect on the hypothalamus.

If fertilization does not occur, the corpus luteum gradually begins to shrink. By about 11 days after ovulation, the reduction in the corpus luteum has progressed far enough to inhibit secretion of estrogen and progesterone. Without these hormones, the uterine lining begins to degenerate; by the twenty-fourth day after the beginning of the cycle, the lining breaks down and the menstrual flow begins.

When estrogen and progesterone secretion are reduced, FSH production is no longer inhibited and the pi-

tuitary begins to secrete FSH again. A new monthly cycle then begins, and again the various hormones act in concert to initiate growth of the uterine lining, ovulation, and menstrual flow.

For practical reasons the menstrual cycle is usually counted off from the beginning of the previous menstrual flow, since it is impossible to judge from external events when the uterine lining begins to grow in preparation for ovulation. On this basis (lower scale in Figure 14-28) ovulation usually takes place 14 days after the beginning of the last menstrual flow (but may vary from 8 to 20 days at times). The next menstrual flow begins in another 14 days if fertilization does not occur; this is 28 days after the beginning of the previous period of flow. These figures are

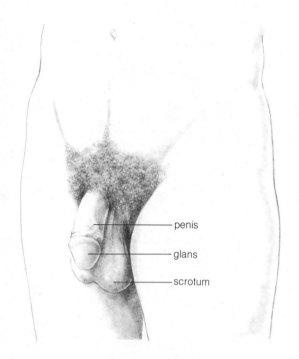

Figure 14-29 External sexual anatomy in the male.

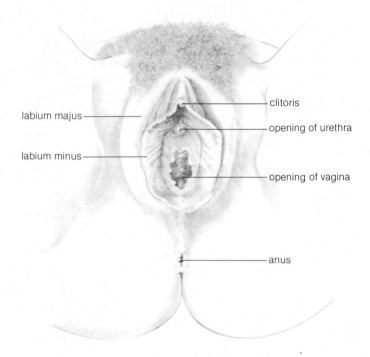

Figure 14-30 External sexual anatomy in the female.

only averages, however, and wide variations are not un-common. Emotional stress, infections, the common cold, or even worry about whether the menstrual flow will appear on schedule are sometimes responsible for significant delays in its appearance.

Copulation and Fertilization in Humans

Fertilization is accomplished in humans by the union of egg and sperm in an oviduct of the female. The sperm is delivered to this location by the activity of the external sex organs of the male and female in *copulation* and by movements of the uterus that accompany copulation and help to conduct the sperm to the oviducts.

External Sexual Anatomy of the Male Externally, sex organs in the male (Figure 14-29) consist of the *penis*, through which the duct transporting sperm from the testis to the anterior leads, and the *scrotum*, a baglike structure that houses the testes. Most of the interior of the penis is spongy tissue that, during sexual arousal, becomes filled with blood and causes erection of the penis. At its tip, the penis ends in a soft, caplike structure, the *glans*. Most of the nerve endings producing pleasurable sexual sensations

in the male are crowded into the glans and the region of the penis shaft just behind the glans.

External Sexual Anatomy of the Female The external female sex organs collectively form the *vulva* (Figure 14-30). Two fleshy folds, the *labia majora*, run from front to rear on either side of the opening to the vagina. Two smaller folds, the *labia minora*, lie just under the labia majora and are almost completely concealed by them. The labia minora immediately surround the opening of the vagina and at their tips, toward the front or anterior end of the vulva, join to partly cover a small bulblike swelling, the *clitoris*. The clitoris contains erectile tissue and has the same embryonic origins as the penis. Most of the nerve endings associated with pleasurable sexual sensations in females are concentrated in the clitoris and to a lesser extent in the labia minora and the opening of the vagina.

Transfer of Sperm to the Female When the male is sexually aroused, the internal cavities of the penis swell with blood, producing an erection in which the penis lengthens and enlarges to nearly twice its resting size. During continued sexual arousal, a clear lubricating fluid is released from the opening of the urethra and collects at the tip of the penis. Female sexual arousal results in enlargement and erection of the clitoris in a process analogous to erection of the penis in males. The labia minora and the vaginal wall also swell to some extent, and a clear lubricating fluid, secreted in the vagina and by glands that open near the clitoris, collects on the surfaces of the vulva.

Insertion of the penis into the vagina and the subsequent thrusting movements in copulation cause friction and sexual sensations leading to a physical and psychological climax in both male and female, the *orgasm*. In males, orgasm produces spasmodic contractions of the muscles surrounding the deferent ducts, prostrate, vesicular glands, and urethra, forcibly ejecting the semen into the vagina. Orgasm in females is accompanied by brief contractions of the vagina and prolonged contractions of the uterus, both of which serve to keep the sperm in suspension and conduct them from the site of ejaculation to the oviducts. (Female orgasm is not necessary for successful fertilization to occur, however.) Uterine contractions are also induced by a *prostaglandin* hormone that makes up part of the seminal fluids ejaculated by the male. The contractions and the random swimming movements of the sperm enable the sperm cells to reach the site of fertilization in the oviducts within minutes after their ejaculation into the vagina. Of the millions of sperm released into the vagina in a single ejaculation, only a few hundred actually reach the site of fertilization.

Fertilization The egg can be fertilized only during its passage through the third of the oviduct nearest the ovary. Since the egg is in this location for only 12 to 24 hours, it is only during this short period of the monthly cycle that fertilization can occur. Because sperm can survive in the female reproductive tract for only 2 days at best, it is not likely that intercourse will result in fertilization unless it takes place within a 3-day period: from about 2 days before to about 1 day after ovulation. If fertilized, the egg continues its passage through the oviduct and is released into the uterus to lodge in the uterine lining and continue developing.

To reach the egg the fertilizing sperm cell must penetrate a layer of cells derived from the ovary that surrounds the egg surface. This penetration is apparently aided by enzymes released from the acrosomes of the sperm invading the reproductive tract. As soon as a sperm reaches the egg surface and sperm and egg membranes fuse, the arrest of the egg is broken. The egg completes meiosis, and the chromosomes of the egg and sperm join on the spindle of the first mitotic division of the embryo to form the zygote.

Fertilization produces a major change in the balance of hormones secreted in the female that interrupts the

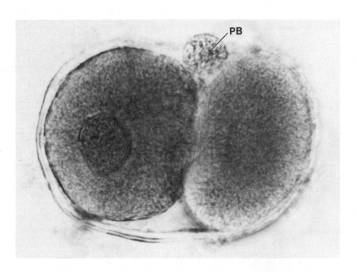

Figure 14-31 A human embryo at the two-cell stage. At this time, the embryo is in transit through the oviduct leading from an ovary to the uterus. A polar body *(PB)* is still present at one side of the embryo. ×1,500. Courtesy of A. T. Hertig, J. Rock, E. C. Adams, and W. J. Mulligan. The Carnegie Institution of Washington, Department of Embryology, Davis Division.

monthly cycle of menstruation. After fertilization, the egg continues its passage through the oviduct and is released into the uterus. In the uterus, it lodges or *implants* in the uterine lining and continues developing. As the embryo implants, cells of the developing placenta (see below) begin to secrete a hormone, *chorionic gonadotrophin*, which prevents the corpus luteum on the ovary from degenerating. As a result, the corpus luteum continues to secrete estrogen and progesterone, which prevents the uterine lining from breaking down at the end of the 28-day cycle as it normally would. Continued progesterone secretion by the corpus luteum also causes extensive development of the milk-producing glands of the breasts. Later in pregnancy the placenta too secretes progesterone, and dependence on the corpus luteum for this hormone gradually lessens.

Development of the Embryo and Placenta

Mitotic cleavage divisions proceed after fertilization until a ball of about 30 cells is formed. (Figure 14-31 shows the two-cell stage of a human embryo.) The ball then hollows out into the blastula, called a **blastocyst** in human development (Figure 14-32a). At one side, the blastocyst cells form a thicker layer. The innermost cells of this layer (shaded in Figure 14-32a) divide to form the embryo. The remaining cells, called the **trophoblast**, divide to produce the embryonic portion of the **placenta**, a structure that

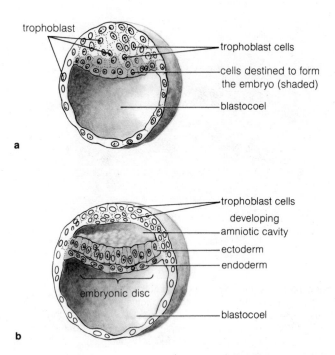

Figure 14-32 Development of ectoderm and endoderm in the embryonic disc stage of human embryos (see text). (**a**) Embryonic disc stage before formation of amniotic cavity. (**b**) After formation of the amniotic cavity.

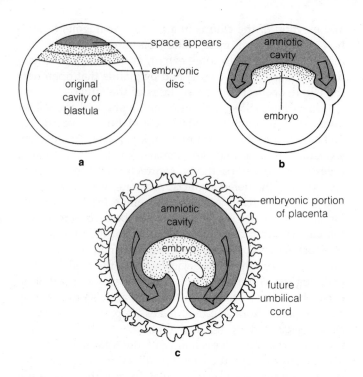

Figure 14-33 Development of the amniotic cavity surrounding the human embryo (see text).

conducts nutrients and wastes between mother and embryo. (Another part of the placental tissue is derived from the uterine lining of the mother.)

By this time, the blastocyst has reached the uterus. As contact with the uterine lining is made, the cells of the trophoblast divide and actively penetrate into the uterine lining, anchoring the embryo to the uterine wall. The embryo usually implants in the lining within 8 to 9 days after fertilization of the egg.

The cells destined to become the embryo now divide to produce two distinct cell layers (Figure 14-32b). The innermost layer of cells facing the blastula cavity develops into endoderm. The layer facing away from the large central cavity develops into ectoderm and mesoderm. At this two-layered stage, the embryo in humans and other mammals is called the **embryonic disc**.

As these primary layers develop, the cells of the blastula outside the embryonic disc continue to divide rapidly and grow into the uterine lining. The tissue formed from these invading cells eventually covers much of the uterine wall and fuses closely with the uterine lining to form the placenta of the developing embryonic system. Within the

placenta, the tissues growing outward from the embryo become so intimately mixed with the uterine lining that only the plasma membranes of adjacent cells separate the two systems. Across these intimate connections, blood nutrients and oxygen, but normally not blood cells, pass from the mother's circulation into the tissues of the embryo. Waste materials from the embryo pass in the opposite direction.

Formation of the Amniotic Cavity As the primary tissue layers differentiate in the embryonic disc, a space appears between the embryonic disc and the overlying trophoblast cells (Figure 14-33a; see also Figure 14-32b). This cavity gradually spreads outward, filling with fluid as it enlarges, until it completely surrounds the embryo (Figure 14-33b and c). Within the cavity, now called the **amniotic cavity**, the embryo lies suspended in fluid, connected with the placenta through a cord of tissue. Later, when the embryonic blood circulation develops, this connection between the embryo and placenta forms the *umbilical cord*, a tube-like organ containing several large blood vessels that conduct nutrients and waste materials between embryo and

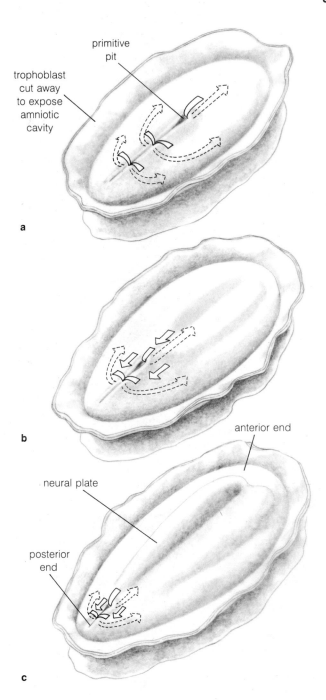

Figure 14-34 Cell movements forming the mesoderm and sorting out the three primary tissues in human embryos (see text). The neural plate has begun developing in (**b**) and (**c**). Adapted; original courtesy of H. Tuchmann-Duplessis and Springer-Verlag.

placenta. The fluid of the amniotic cavity insulates the embryo from mechanical shocks transmitted to the uterus from outside.

Development of Ectoderm and Mesoderm   The human embryo develops from the embryonic disc suspended within the amniotic cavity. The bottom layer of the two-layered disc has already differentiated as endoderm. The top layer is a composite tissue: cells destined to become mesoderm are at the center and future ectoderm is at the edges. Ectoderm and mesoderm differentiate from the top layer in a process resembling the movement of cells into the primitive streak in birds (Figure 14-34). The embryonic disc elongates, and a deep groove forms in its dorsal surface, running in the direction of elongation (Figure 14-34a). Cells of the top layer move downward into the groove and then spread laterally between the top and bottom cell layers of the developing embryo to form the beginnings of mesoderm. At one end of the groove, at the future anterior or head end of the embryo, the cells move rapidly from the surface into the developing mesoderm, forming a deep depression called the *primitive pit.* As cells flow into the pit, the depression moves toward the posterior end of the embryo, eliminating the groove as it progresses (Figure 14-34b and c). The cells remaining on the upper surface of the embryo form the ectoderm when this process is complete. As the primitive pit completes its rearward movement, therefore, the three primary germ layers of the embryo—ectoderm, mesoderm, and endoderm—are formed.

Further Growth and Birth

The primary nervous system then develops along a pattern similar to that described for amphibian embryos. A lengthwise groove appears in the ectoderm and deepens. As in neural tube formation in amphibians, the groove closes over and pinches off below the surface of the ectoderm as the neural tube (Figure 14-35a to d). The brain, spinal cord, and retina of the eyes develop from the neural tube. At the same time, within the embryo, the endoderm extends laterally and downward (Figure 14-35b and c), eventually fusing along its bottom margins to form a tube of endoderm running lengthwise inside the embryo (Figure 14-35d). This tube is the primitive gut of the embryo. The neural tube and the primitive gut are enclosed by lateral growths of the ectoderm and mesoderm, which gradually extend downward (Figure 14-35b to d) until the entire embryo is covered with a surface layer of ectoderm cells lined with mesoderm (Figure 14-35e).

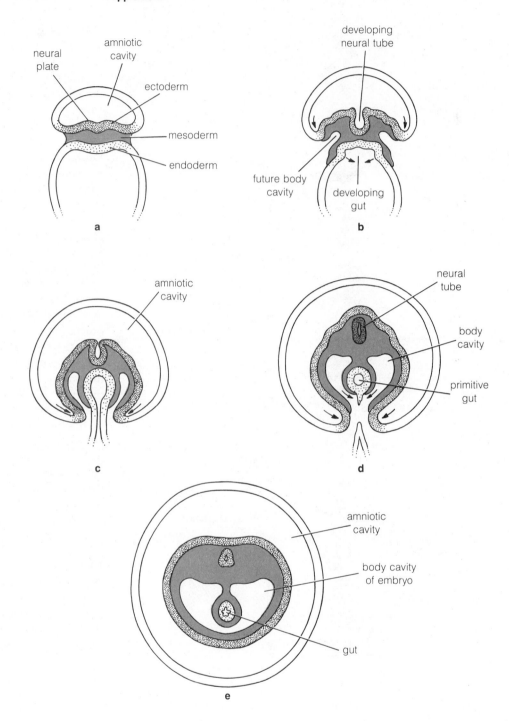

Figure 14-35 Steps in development of the neural tube and primitive gut, shown in diagrammatic cross sections of the embryo. The region connecting the embryo to the umbilical cord and placenta is not shown.

As in amphibian embryos, the ectoderm subsequently gives rise to the skin, parts of the eye, and all the surface structures of the embryo. The lining of the digestive system develops from the endoderm tube forming the primitive gut; organs of the digestive system, such as the liver and part of the pancreas, also originate from cells of the endoderm. Mesoderm tissue gives rise to the muscles, bone, connective tissues, and blood system of the embryo. As the embryonic blood system develops, it connects to the placenta through the vessels that run through the um-

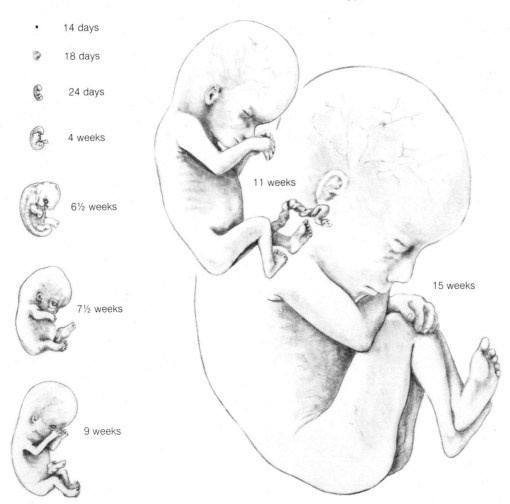

14 days

18 days

24 days

4 weeks

6½ weeks

7½ weeks

9 weeks

11 weeks

15 weeks

Figure 14-36 Growth of the human embryo and fetus from 14 days until birth (shown approximately life size).

bilical cord. In the placenta, the blood vessels of the embryo associate closely with vessels of the mother. Although the associations are intimate, allowing nutrients and wastes to pass over cell membranes between mother and embryo, the blood supplies of the two do not directly connect. In fact, the blood of the mother and embryo are often of different genetic types.

Within 60 days (8½ weeks) from fertilization, all the organ systems of the embryo have developed. (The human embryo is called a **fetus** after the eighth week of development.) From 60 days until birth at approximately 266 days, growth of the fetus occurs primarily through an increase in size and weight—from about 10 grams at 60 days to about 3 kilograms (7 pounds) as the fetus is born (Figure 14-36).

As the period of embryonic growth comes to a close, the beginning stages of childbirth are signaled by contractions of the uterus, first felt by the mother about 12 hours

preceding the actual time of birth. (The hormonal changes that initiate the birth process in humans are still unknown.) By the time the contractions begin, the fetus has normally turned so that its head points downward. The uterine contractions gradually increase in strength, forcing the fetus downward until its head passes into and dilates the cervix, the opening of the uterus into the vagina.

As the head of the fetus passes downward toward the cervix and vagina, the embryonic membranes holding the amniotic fluid around the embryo burst, releasing the "waters" that announce the onset of birth. Within 30 minutes to 1 hour after the head moves through the dilated cervix, the entire fetus is forced from the uterus through the vagina to the exterior. At this time, the umbilical cord, which still connects the infant to the placental tissues within the uterus, is usually cut and tied off by the birth attendant. Contractions of the uterus continue, soon expelling the placenta and any remnants of the umbilical

Table 14-2 Effectiveness of Popular Birth Control Methods

Method	Number of Pregnancies Expected in Each 100 Women Using the Method for One Year
Condom	5 to 15 pregnancies
Diaphragm alone	10 to 30 pregnancies
Diaphragm with spermicidal jelly	4 to 5 pregnancies
Rhythm method	25 to 35 pregnancies
Oral contraceptive pill	0 to 2 pregnancies
IUD	1 to 2 pregnancies

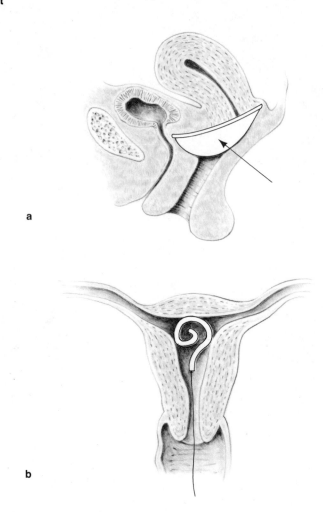

Figure 14-37 (**a**) A diaphragm (arrow) in place over the cervix. (**b**) An IUD placed inside the uterus.

cord and embryonic membranes as the ''afterbirth,'' usually within 15 minutes of the infant's birth. The short length of umbilical cord still attached to the infant dries and shrivels within a few days. Eventually it separates entirely and leaves a scar, the navel, to mark its former site of attachment during embryonic development.

Methods of Contraception or Birth Control

As human affairs go, there is often more interest in preventing fertilization and pregnancy than in accomplishing these ends. Various devices and techniques are used for this purpose with varying degrees of success (Table 14-2). These methods primarily include techniques for (1) preventing the sperm from reaching the site of fertilization, (2) preventing ovulation, or (3) interfering with implantation of the egg in the uterine lining if fertilization does occur.

The *condom*, a thin, close-fitting rubber sheath worn over the penis, is one of the traditional methods of preventing sperm from entering the vagina or uterus on ejaculation. Condoms may also provide a barrier to the transmission of disease between sexual partners. They are only moderately effective in preventing pregnancy, however, and may considerably reduce the sensations of sexual pleasure. The *diaphragm* is a cuplike rubber cap that fits closely over the cervix in females (Figure 14-37a). Used with a jelly or cream that kills sperm, diaphragms are somewhat more effective than condoms as a contraceptive. To be most effective, however, the diaphragm and spermicidal agent should be inserted from about an hour

to just before intercourse. Since it is frequently inconvenient to stop for this purpose, many sexual partners find using diaphragms unpleasant and rule out this method of birth control.

The remaining technique for preventing fertilization used with any frequency is the *rhythm method*, which consists of avoiding intercourse during the time of the month that fertilization can occur. This means abstaining from sex for a 3-day period—from about 2 days before to about 1 day after ovulation. Although it is apparently straightforward, the method is difficult to apply because of the unpredictability of the time of ovulation in many women. One of the few positive external indications of ovulation is a half-degree rise in body temperature that occurs on the morning of ovulation and persists until the menstrual flow begins. By keeping track of daily temperatures for a few months some women are able to predict when ovulation will occur each month. Avoiding intercourse for about 4 to 5 days before this date and 2 to 3 days after would, on the average, prevent sperm from reaching the egg at the right time. The averages in this case are none too dependable, however, and about one-third of the women practicing birth control by this method for a year can expect to become pregnant. The method carries with it the disadvantage of abstaining from intercourse for at least a week between menstrual cycles, which, for many couples, is difficult.

Some methods of contraception are designed to prevent ovulation or to prevent implantation of the egg if it is fertilized. The *oral contraceptive*, or *pill*, which contains a combination of synthetic hormones that prevents ovulation, is highly effective in preventing pregnancy if taken according to the prescribed schedule. Most pregnancies among women taking the pill apparently result from failure to take the pill on schedule and not from ineffectiveness of the pill itself. Some women, about one in four, experience unpleasant side effects from taking the pill, such as nausea, tenderness of the breasts, irritability, nervousness, or changes in skin color or texture. There have also been indications that in a very few women the pill may increase the incidence of blood clots in various parts of the body, sometimes with serious consequences. Clots appear normally in about 2 of every 100,000 women. Although studies indicating an increase in blood clots in women taking the pill are controversial, it seems that the pill may raise the incidence of clots to 3 or 4 out of 100,000 women. Even at these levels, however, the pill is still less of a health hazard than pregnancy and childbirth, which are much more likely to lead to physical impairment or death. (Complications related to pregnancy and childbirth cause about 15 deaths among every 100,000 women becoming pregnant.)

Contraceptive pills prevent conception through the effects of their content of estrogen and a synthetic hormone, **progestin,** that mimics the effects of progesterone. Because of its estrogen content, taking the pill just after menstrual flow stimulates growth of the uterine lining. Continued ingestion of the pill also inhibits secretion of FSH and LH by the pituitary, as do estrogen and progesterone during the normal cycle or during pregnancy. Because FSH and LH are inhibited early in the cycle, growth of an ovarian follicle and ovulation do not occur. When ingestion of the pill is temporarily stopped (after 20 to 21 days), the drop in concentration of estrogen and the progesterone-like hormone causes the uterine lining to degenerate and initiates the menstrual flow. Use of the pill thus causes the uterine lining to grow and be released in a monthly cycle mimicking the normal one. Since ovulation does not occur, however, fertilization and pregnancy are not possible.

Another technique for contraception much used until recently is insertion of an *intrauterine device* or *IUD*, a small metal or plastic coil that is placed in the uterus by a physician. Figure 14-37b shows an IUD inserted in the uterus. When in place, the IUD apparently prevents the implantation of fertilized eggs in the uterine lining. (How IUDs may prevent implantation is discussed in Supplement 9-1.) The primary source of pregnancies in women using the IUD seems to be displacement of the device from its correct position in the uterus; this may happen without warning or the user's awareness. A few women also experience unpleasant side effects from the IUD such as cramps, inflammation, uterine infections, or excessive menstrual bleeding. The plastic string left extending through the cervix into the vagina may also irritate the penis during intercourse.

Whatever the method of birth control, its effectiveness is improved if sex partners are highly motivated and careful in its use. The effectiveness of condoms, for example, is improved considerably if the penis is withdrawn immediately after ejaculation (before the seminal fluids have time to spread under the condom along the shaft of the penis) and not reinserted, with or without a condom, for a period of several hours. Similarly, high motivation in use of the rhythm method, which might require abstaining from intercourse for most of the month except for a few days just before and after the menstrual flow, considerably improves the percentage of success with this method. (Even in this case, some pregnancies are likely to occur.)

SUPPLEMENT 14-2:
EVIDENCE FOR THE DISTRIBUTION OF DEVELOPMENTAL INFORMATION IN THE EGG CYTOPLASM

The first experiments revealing the extent of cytoplasmically localized information in animal eggs were carried out in the early 1900s by the American investigator E. G. Conklin. In the organism he studied, an embryonic tunicate (a primitive member of the phylum Chordata), the egg cytoplasm is divided into regions with definite fates in the embryo. If Conklin separated the tunicate embryo into separate cells at the two-cell stage, for example, each cell

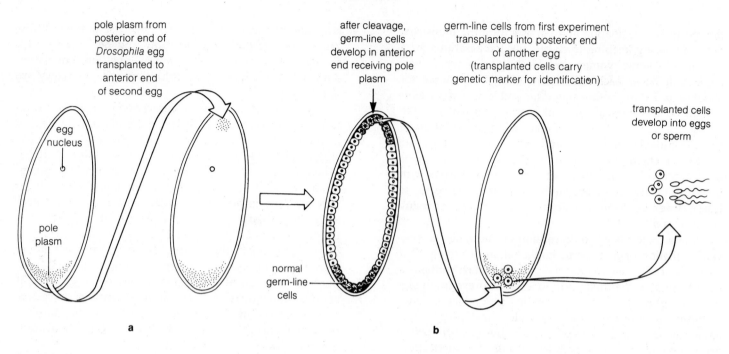

Figure 14-38 An experiment showing the predetermined fate of pole plasm cells in *Drosophila* (see text).

developed only into the parts it would have formed in the normal embryo; that is, each cell formed a half-embryo without the structures of the missing half. If the cells were separated at the four-cell stage, the same thing happened. Cells from the animal pole, for example, gave rise to ectodermally derived structures such as notochord, epidermis, and nervous tissue, but not to mesodermally derived cells such as muscle.

Conklin was able to show that the information causing the tunicate embryo cells to develop along predetermined pathways was localized in the egg cytoplasm. He used a centrifuge to displace parts of the cytoplasm to other regions of the egg. The degree of displacement was determined by noting the redistribution of pigment granules and yolk bodies. Conklin found that displacing parts of the cytoplasm also displaced the structures of the embryo formed from the cytoplasm. If a portion of the cytoplasm that normally occupied the animal pole of the egg and gave rise to nervous tissue was displaced to the vegetal half, for example, it still gave rise to nervous tissue in its new location after cleavage and further development of the egg.

Conklin's classic evidence for distribution of developmental information in different regions of the egg cytoplasm has been reinforced by the experiments of Anthony P. Mahowald and Karl Illmensee at the University of In-

diana. Mahowald and Illmensee followed the developmental fate of a region of the egg cytoplasm in insects that is predetermined to give rise to the **germ line** of the embryo. The germ line is the cell line that will eventually form sperm or egg cells when the organism becomes sexually mature. In insect eggs, the germ-line cytoplasm is concentrated at one end of the egg, where it can be readily identified under the light microscope by its tendency to color strongly with stains. Under the electron microscope, the same area proves to contain large numbers of granules of a type found only in this location. Because of its location at one end of the egg, the germ-line cytoplasm is termed *pole plasm;* the granules characteristically found in this region are termed *polar granules.*

Mahowald and Illmensee demonstrated the predetermined fate of the pole plasm in *Drosophila* by removing it from some eggs and injecting it into different regions of others (Figure 14-38a). They removed the pole plasm from the posterior end of the egg, where it normally occurs, and injected it into the anterior end of other eggs. As the injected eggs began to divide, cells that had received transplanted pole plasm developed into the germ-line type, even though germ-line cells would not normally form in this location. Subsequent experiments (Figure 14-38b) showed that cells from the anterior end of the embryo that had been induced to form germ-line cells by the transplanted pole plasm could actually produce functional gametes if they were placed in the developing sex organs of other embryos.

These experiments show that the pole plasm of *Drosophila* eggs contains information that can direct cells to develop into germ cells. This means that nuclei placed in this cytoplasm as the result of cleavage divisions are induced to regulate their genes toward final development and differentiation as gamete nuclei.

What is the nature of the information concentrated in the cytoplasm of insect and other animal eggs? In 1978, Mahowald isolated the polar granules from *Drosophila* eggs and discovered that they contain a basic protein that may be associated with storage of the information that directs differentiation of germ cells. The information may be coded in some way into the protein itself or in a nucleic acid molecule that becomes associated with the protein.

The differences in the development of plants and animals are as great as the differences in plant and animal life. In the midst of these differences, however, there are fundamental similarities. A complex, many-celled plant grows by mitotic cell divisions from a single original cell, the fertilized egg. In this growth, cells differentiate into highly specialized types with distinct structure and biochemical activity. As in animals, this differentiation is superimposed on cells that retain complete and identical copies of the basic genetic information of the individual.

A complex, many-celled plant grows by mitotic cell divisions from a single original cell, the fertilized egg.

Although cell division and differentiation occur in both plants and animals, several of the basic developmental processes of animals do not operate in plants and others are greatly reduced in importance. Most striking is the absence of cell movement and changes in selective cell adhesion. Once they are laid down by division, plant cells stay in place with relation to their neighbors. This characteristic of plant development reflects the presence of the cell walls that fix plant cells permanently in place. A second major dissimilarity in the developmental processes of plants is the limited importance of induction. Although developmental interactions equivalent to induction do occur in plants, they are infrequent in comparison to the central role played by inductive responses in animal development. This basic difference is related to the lack of motility and selective adhesion in developing plant cells. In animals, inductions primarily involve cell movements and the establishment of new contacts between the inducting cell type and the reacting tissue.

Thus only two basic developmental mechanisms, cell division and differentiation, are extensively shared between animals and plants. Moreover, plants have a third basic developmental process of their own: the amount and direction of cell growth or *elongation* after a cell is laid down by mitotic division. Differential cell elongation is an important part of all developmental interactions in plants, and it accounts for much of the final growth form taken by fully developed plant structures.

Plants develop primarily through three cellular mechanisms: cell division, cell elongation, and cell differentiation.

While the basic mechanisms of development are essentially the same in all plants, the details of growth from the fertilized egg to an adult plant vary widely in different plant groups. To study these mechanisms, we will consider development in the most familiar plant group: the flowering plants or *angiosperms*.

15

BASIC PROCESSES OF DEVELOPMENT IN PLANTS

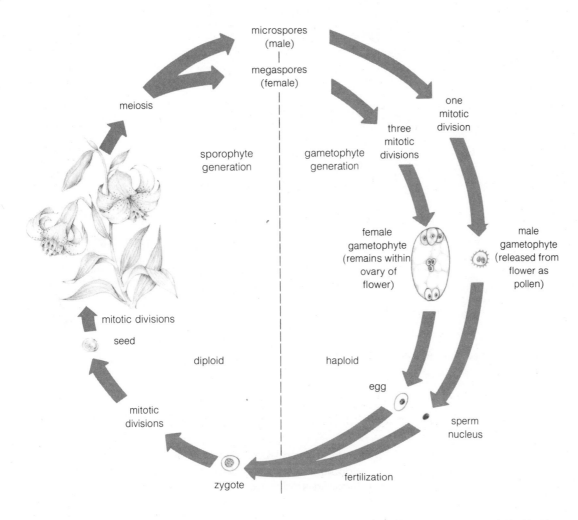

microspores (male)

megaspores (female)

meiosis

sporophyte generation

gametophyte generation

one mitotic division

three mitotic divisions

female gametophyte (remains within ovary of flower)

male gametophyte (released from flower as pollen)

mitotic divisions

seed

diploid

haploid

egg

mitotic divisions

sperm nucleus

zygote

fertilization

Figure 15-1 The life cycle of a flowering plant. The haploid, gametophyte generation (to the right of the dotted line) remains dependent on tissues of the sporophyte generation.

DEVELOPMENT IN THE FLOWERING PLANTS: AN OVERVIEW

As in animals, the process of development in plants begins with gamete formation and fertilization. However, these events occupy different positions in the lives of the plants emphasized in this chapter, the flowering trees and shrubs and the smaller flowering plants. In the flowering plants, as in all plants, the life cycle alternates between distinct haploid and diploid generations.

Fertilization in the flowering plants, as in animals, produces a diploid individual. These individuals, termed **sporophytes**, make up the predominant phase of the life cycle and are conspicuous as the flowers, trees, and shrubs of our environment (Figure 15-1). The sporophyte grows to maturity through mitotic divisions. At some point, meiosis occurs in certain cells of the sporophyte, giving rise to haploid products known as **spores**. As a result of the meiotic divisions, the spores produced have new combinations of genetic information that differ from those of the parent sporophyte. Spores grow by mitosis into a haploid phase of the life cycle termed the **gametophyte**. In the higher plants the gametophyte is a much reduced, microscopic plant located in the flower that remains dependent on the sporophyte for the nutrients required for growth.

In plant life cycles, sporophytes form diploid spores and gametophytes form haploid gametes.

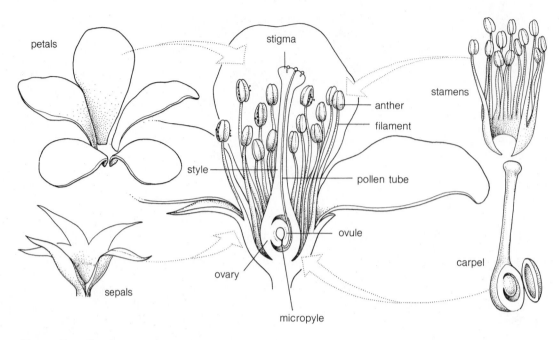

Figure 15-2 The parts of a flower. Redrawn from "Pollen" by Patrick Echlin. Copyright © 1968 by Scientific American, Inc. All rights reserved.

Gametophytes in flowering plants grow within the flower.

The haploid gametophytes are either male or female. In male gametophytes, cells differentiate into male gametes, the sperm cells, after two mitotic divisions. In female gametophytes, after three mitotic divisions of the gametophyte tissue, a single cell differentiates into the egg. Fusion of an egg and sperm in fertilization restores the diploid chromosome number and establishes the sporophyte phase of the life cycle.

The female gametophyte in the higher plants remains embedded within the tissues of the flower in the parent sporophyte and is fertilized in this location. After fertilization, the new sporophyte embryo grows by mitotic divisions in the same location until the cells that will form stems and roots have partially differentiated. At this time, protective coats are formed around the embryo and the new sporophyte stops developing and enters a dormant period as the **seed**. Eventually, seeds are released to the outside world and, if conditions are favorable, dormancy is broken and the embryo resumes development.

This chapter traces the major phases of development in the flowering plants through gamete formation, fertilization, and early development to produce the seed. Moreover, it describes seed germination and the further

developmental changes producing the primary organs and tissues of the adult plant. How cell division, elongation, and differentiation operate as basic processes in these major phases of plant development will become obvious as the description unfolds.

GAMETOGENESIS AND FERTILIZATION IN THE HIGHER PLANTS

Gametogenesis

Gametophytes and gametes in the higher plants are formed within two types of reproductive structures that differentiate from cells of the sporophyte plant stem. One type, the **stamen**, gives rise to male gametophytes; the second, the **carpel**, gives rise to female gametophytes. Stamens and carpels together form the reproductive parts of the flower (Figure 15-2). When fully developed, the stamens form stalklike structures with swellings at their tips. Within these swellings, called the **anthers** of the flower, the male gametophytes form. Carpels differentiate into an **ovary**, in which the female gametophytes develop, and an accessory structure called the **stigma**, which first receives the male gametes in fertilization. The stamens and carpels are surrounded by the outer, leaflike *petals* of the flower,

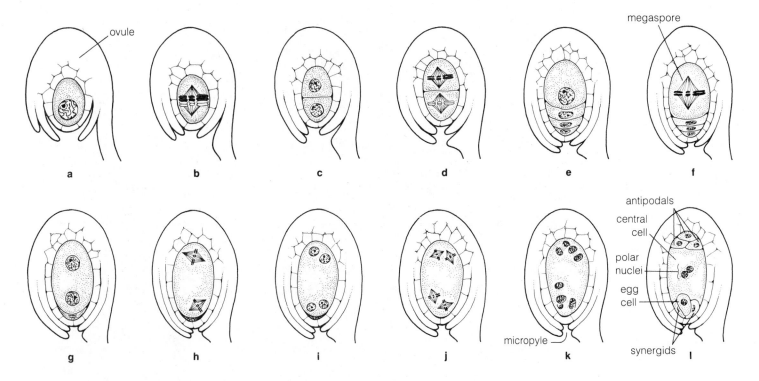

Figure 15-3 Development of the female gametophyte in a flowering plant. **(a)** to **(e)**: Meiosis produces four cells **(e)**, of which only one survives. **(f)** to **(k)**: Three successive mitotic divisions of the surviving cell produce eight nuclei, which become enclosed in seven cells: two synergids, the egg cell, a large central cell containing two nuclei, and three antipodals **(l)**.

which are frequently highly colored and showy, and the *sepals*, which are also leaflike but usually green in color and reduced in size. Although the petals and sepals are not directly involved in reproduction or gamete formation, they may still be important or even critical to fertilization: their function in some species is to attract animals that transfer pollen between flowers. The external form taken by the petals, sepals, stamens, and carpels varies widely in different species of flowering plants.

Growth of the Female Gametophyte and the Egg Cell
Female gametophytes develop within the ovaries of the flower in a bulblike swelling of ovary tissue called the **ovule** (see Figure 15-2). Within the ovule, one diploid cell enlarges and undergoes meiosis, producing four haploid products (Figure 15-3*a* to *e*). While the subsequent events differ in detail in the various flowering plants, all lead to the production of an egg cell and additional cells that give rise to accessory structures in the female gametophyte. In the best-known system, only one of the haploid products of meiosis survives, forming a single large cell, the **mega-**

spore. The haploid megaspore, still within the ovule, immediately undergoes a series of three sequential mitotic divisions (Figure 15-3*f* to *k*) to form the female gametophyte. At maturity the female gametophyte contains eight nuclei (Figure 15-3*l*).

Two of the nuclei of the female gametophyte become enclosed in cells called *synergids* that take up a position at the end of the female gametophyte. The synergids are the first cells to receive the sperm nuclei during fertilization. Another nucleus, which becomes the egg nucleus, is enclosed in a third cell that takes up a position just behind the synergids in most angiosperms (see Figure 15-3). This cell becomes the *egg cell* of the female gametophyte. Three more nuclei are enclosed in cells called *antipodals* that come to lie at the opposite end of the female gametophyte from the egg cell and synergids. The remaining two nuclei, called **polar nuclei,** share a common cytoplasm in a large *central cell* that makes up most of the volume of the female gametophyte. The polar nuclei eventually fuse with one nucleus of the male gametophyte during fertilization to form a nucleus containing three sets of chromosomes. This

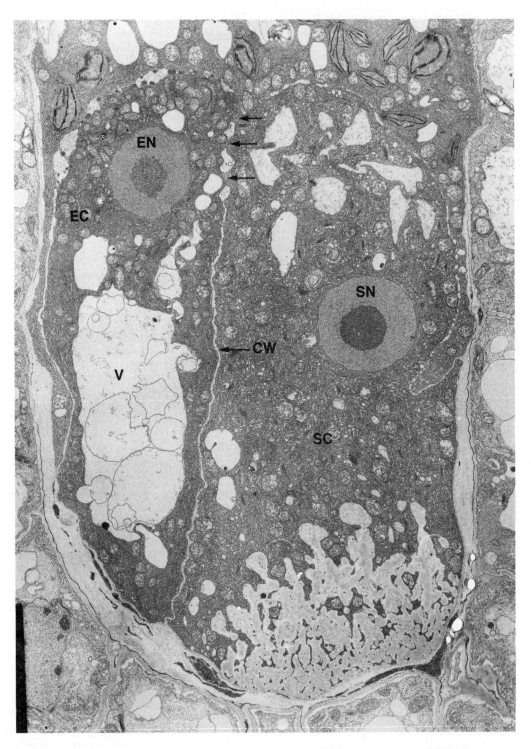

Figure 15-4 The egg cell and one of the synergids of the plant *Capsella* (shepherd's purse). The cell wall between the two cells is incomplete in the regions marked by the arrows. Note the large vacuole (*V*) in the egg cell. *EN*, egg nucleus; *EC*, egg cell cytoplasm; *CW*, cell wall; *SN*, synergid nucleus; *SC*, synergid cytoplasm. Courtesy of P. Schulz and Academic Press, Inc., from *Cell Division in Higher Plants*, M. M. Yoeman, ed., 1976.

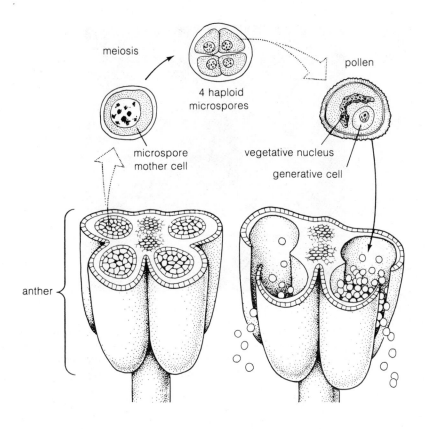

meiosis

4 haploid microspores

pollen

microspore mother cell

vegetative nucleus

generative cell

anther

Figure 15-5 Formation of the male gametophyte in a flowering plant. Cells in the anther undergo meiosis to produce four microspores (**a**); all these cells survive to form male gametes. Each microspore divides once mitotically to form the generative and vegetative cells of the male gametophyte (**b**). The generative cell will divide again during fertilization to form two sperm cells.

nucleus, which is thus a **triploid**, subsequently divides mitotically to form the **endosperm**, a triploid tissue that contributes nutritive substances to the developing embryo.

The female gametophyte at maturity thus contains eight nuclei enclosed in seven cells. All these cells become enclosed in cell walls. The walls between the egg cell and the synergids remain incomplete, however, so that extensive regions are separated only by adjacent plasma membranes (Figure 15-4).

The female gametophyte remains enclosed within the ovule, which eventually contributes the surface coats of the seed. As the gametophyte forms, the ovule develops a small opening, the **micropyle**, which communicates with the end of the gametophyte containing the synergids and the egg cell (see Figure 15-3).

The mature egg cell of flowering plants (Figure 15-4) is a relatively small cell, at least in comparison with the eggs of many animals. No reserve nutrients equivalent to yolk are present in the egg cell cytoplasm. The egg cytoplasm is packed with ribosomes, however, which are synthesized in large numbers during development of the female gametophyte. Presumably, the egg cell cytoplasm also contains developmental information in the form of stored mRNA and protein molecules. (Little is known of

the informational content of mature eggs in the angiosperms or other plants.) Mitochondria are also present, along with embryonic chloroplasts called **proplastids**. Much of the remaining cytoplasmic volume of the egg cell is occupied by one or more large vacuoles. At maturity the egg cell, the female gamete, remains as part of the gametophyte, awaiting fertilization within the female parts of the flower.

Formation of the Male Gametophyte Male gametophytes develop from slightly enlarged, diploid cells called *microsporocytes* within the anthers of the flower (Figure 15-5). These cells undergo meiosis, each producing four haploid **microspores**. Each of the four microspores resulting from meiotic division of a microsporocyte develops into a male gametophyte.

During this development each haploid microspore undergoes one mitotic division, producing a male gametophyte that contains two cells at maturity. The cytoplasmic division accompanying this mitosis is unequal and produces one large and one small cell. The smaller cell, called the **generative cell**, moves to the interior of the developing male gametophyte and becomes completely surrounded by the second, larger cell, called the **vegetative**

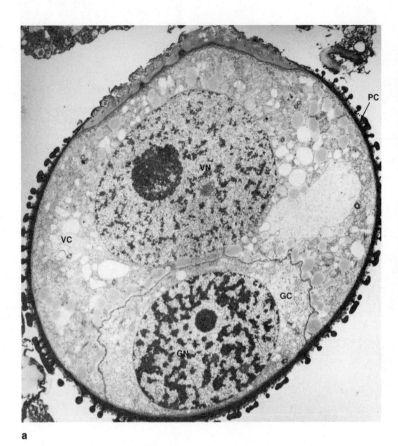

a

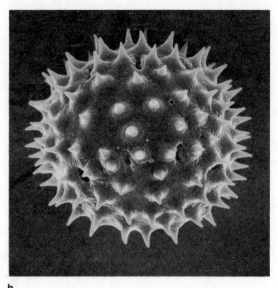

b

Figure 15-6 (**a**) A mature pollen grain of the lily as seen in thin section. *VC*, vegetative cell cytoplasm; *VN*, vegetative nucleus; *GC*, generative cell cytoplasm; *GN*, generative nucleus; *PC*, pollen coat. Courtesy of J. M. Sanger. (**b**) A scanning electron micrograph of a pollen grain. Courtesy of W. A. Jensen.

cell. During fertilization, the nucleus of the generative cell undergoes a second mitotic division to form two *sperm nuclei*. The vegetative nucleus, which undergoes no further divisions, directs growth of the pollen tube (see below) during fertilization. As its development becomes complete, the male gametophyte forms a hard, impermeable surface coat and enters a state of metabolic arrest. The mature male gametophyte, now known as a **pollen grain** (Figure 15-6), remains quiescent until it is released from the stamen to be carried by wind, water, insects, or other means to the female parts of a flower of the same species.

RNA synthesis during development of the male gametophyte produces quantities of mRNAs. Some of these code for proteins that are synthesized during gametogenesis, and some are stored in inactive form until fertilization. Among the mRNAs active during gametogenesis are several that direct the synthesis of *recognition proteins* that become embedded in the external coat surrounding the mature male gametophyte. These proteins function in recognition and attachment of the pollen grain to the stigma of a flower of the same species during fertilization. The recognition proteins are also responsible for the itching

and burning eyes, stuffy noses, and other allergic responses of humans who pick up pollen grains. The mRNAs stored in inactive form in the mature male gametophytes direct the synthesis of proteins involved in germination, growth, and penetration of the pollen into the tissues of the flower.

There is no indication that gametes in higher plants arise from separate and distinct germ-cell lines that are set aside early in embryonic development as in animals. Essentially any plant cell, if exposed to the correct conditions, appears to be able to divide to produce reproductive structures and gametes. For example, stem or leaf cuttings can be stimulated to generate complete plants that can form fully active and normal reproductive structures.

Fertilization

Fertilization in the higher plants, as in animals, restores the diploid number of chromosomes and initiates embryonic development. The first events in fertilization occur as a pollen grain contacts the stigma of a flower of the same species (Figure 15-7a). Almost immediately after contact,

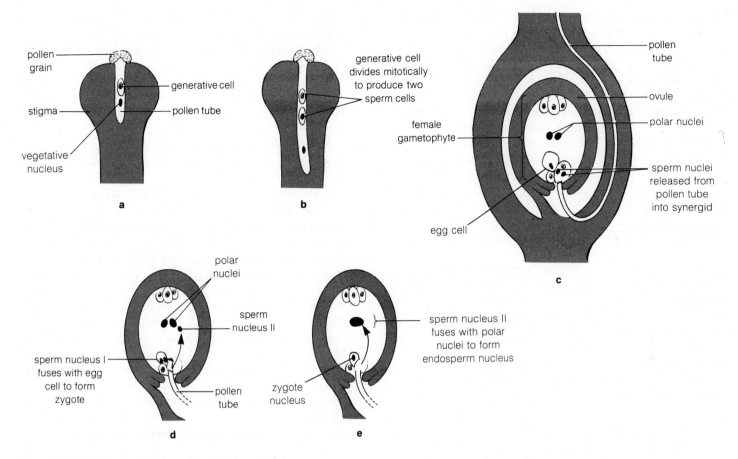

Figure 15-7 Fertilization in a flowering plant (see text).

the quiescent state of the pollen grain is broken. This change is detectable as an abrupt increase in respiration in the pollen grain. Protein synthesis also begins in the pollen grain within minutes of contact with the stigma. This protein synthesis evidently depends on mRNA molecules synthesized and stored in the cytoplasm of the male gametophyte during gametogenesis, because it will proceed even if RNA transcription is blocked at this stage by an inhibitor.

Soon after activation, the coat of the pollen grain splits, and a narrow outgrowth, the **pollen tube**, grows and penetrates into the tissues of the stigma (Figure 15-7a and b). The larger of the two cells in the pollen grain, the vegetative cell, forms the pollen tube. As the tube grows, the generative cell leaves the pollen grain and follows the advancing end of the tube as it penetrates into the stigma. This growth is accompanied by continued protein synthesis. Newly transcribed mRNAs also begin to appear at this

time; if RNA synthesis is experimentally inhibited during extension of the pollen tube, further elongation does not occur.

Activation of the pollen grain and growth of the pollen tube depend on the correct matching of recognition proteins on the surface coat of the pollen grain and the surface of the stigma. Unless the recognition proteins on the pollen and stigma are from the same species of flowering plant, activation and growth of the pollen grain do not occur. This adaptation prevents the germination of pollen grains from foreign species that might contact the stigma.

As the pollen tube penetrates more deeply into the tissues of the flower, the generative cell divides mitotically to produce two sperm cells (see Figure 15-7b). These cells collectively form the male gamete. The nucleus of one of the sperm cells will fuse with the egg nucleus to complete fertilization. The second will fuse with the two polar nuclei of the female gametophyte to form the primary nucleus of the endosperm. Although only one of the two nuclei produced by division of the generative nucleus actually fertilizes the egg cell, they are indistinguishable and both are usually identified as sperm nuclei. The sperm cells containing these nuclei are small, and their cytoplasm is reduced to a relatively thin layer surrounding the nuclei. No cell walls surround the sperm cells and, usually, no proplastids are present in the sperm cell cytoplasm.

The pollen tube continues its growth until it reaches the ovary and penetrates into the female gametophyte through the micropyle, the small opening at one end of the ovule tissues surrounding the gametophyte (Figure 15-7c). In most species the pollen tube first enters one of the two synergid cells at the end of the female gametophyte facing the micropyle. As the pollen tube enters the synergid, an opening appears at its tip and the two sperm cells are released (see Figure 15-7c). The plasma membrane and cytoplasm of the sperm cells are lost during this process, leaving the two sperm nuclei suspended in the synergid. The two sperm nuclei are then carried by streaming movements of the synergid cytoplasm to the vicinity of the egg and the large central cell housing the polar nuclei. One of the two sperm nuclei then penetrates into the egg cell and fuses with the egg nucleus (Figure 15-7d), presumably passing from the synergid to the egg cell through a region in which no walls separate the two cell types.

Fusion of the egg and sperm nuclei within the egg cell follows a pattern similar to animal fertilization. The nuclear envelopes of the two nuclei fuse together and become continuous. Moreover, the maternal and paternal chromosomes, at this point in the extended, interphase form, flow together to reestablish the diploid condition. During this process, DNA replication occurs and the diploid zygote nucleus enters its first mitotic division. The fertilized egg of plants is equivalent to the zygote of animals; all the tissues of the embryo and the mature plant arise directly from the fertilized egg by mitotic divisions. Directions for the growth of the plant are completely coded and programmed into the chromosomes of the zygote nucleus.

The second sperm nucleus fuses with the two polar nuclei of the central cell lying next to the egg nucleus (Figure 15-7e). Depending on the species, the polar nuclei may be separate at this stage, or they may already have fused into a single larger nucleus. The two polar nuclei and the second sperm nucleus fuse together into a single large nucleus with three complete sets of chromosomes. The triploid cell resulting from this fusion then enters a series of rapid mitotic divisions to produce the **endosperm** tissue. The endosperm surrounds the zygote and provides nu-

trients to the developing embryo, but it does not contribute cells directly to any part of the embryo or mature plant. Because two fusions of sperm nuclei occur in the gametophyte—one with the egg nucleus to form the zygote and one with the polar nuclei to produce the first endosperm cell—the entire process is called **double fertilization**.

In double fertilization, one sperm nucleus fertilizes the egg nucleus to form the zygote; the second fuses with the polar nuclei to form the triploid endosperm.

The release of the two sperm nuclei into the female gametophyte and double fertilization to produce both the zygote and endosperm take place only in the flowering plants. Although the coniferous trees, such as the pine and related species, form pollen and carry out fertilization by patterns similar to the flowering plants, only one sperm nucleus enters the female gametophyte and only the zygote nucleus is produced. No endosperm is formed. The lower plants do not form pollen but instead carry out fertilization by means of sperm cells that must swim through an aqueous medium to reach the female gamete. These sperm cells have flagella and in many respects resemble animal sperm cells. (A sperm cell of this type is shown in Figure 15-8.) In all cases, fertilization produces a diploid zygote nucleus that begins development to form the embryo.

EARLY DEVELOPMENT OF THE PLANT EMBRYO AND FORMATION OF THE SEED

Immediately after fertilization in the flowering plants the egg becomes active in protein synthesis. New ribosomes are assembled and packed into the egg cytoplasm, and other structures associated with cell synthesis, including the endoplasmic reticulum and Golgi complex, appear and become predominant. Further, the vacuole becomes somewhat smaller in most species, and any gaps in the cell wall are filled in. At the same time, the cell wall becomes thicker. These changes in the zygote cytoplasm and cell wall extend over a period of 1 to 2 days or more that pass before the first mitotic division of the zygote. During this delay in the division of the zygote, the endosperm divides rapidly, developing a tissue that gradually surrounds the zygote and, later, the developing embryo.

Cleavage and Early Embryonic Development

The initial divisions of the zygote produce successively smaller cells and are thus equivalent to the cleavage divi-

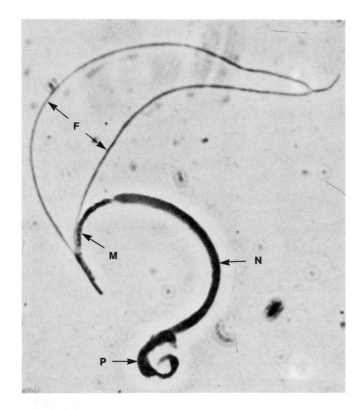

Figure 15-8 A sperm cell of the alga *Nitella*. *F*, flagella; *M*, mitochondria; *N*, nucleus; *P*, plastids. ×2,700. Courtesy of F. R. Turner and The Rockefeller University Press, from *Journal of Cellular Biology* 37:370 (1968).

Figure 15-9 The adult *Capsella* plant (shepherd's purse).

sions of early animal embryos. The pattern of these early cleavage divisions and the subsequent development of the embryo, which vary extensively in different flowering plants, have been traced most completely in *Capsella*, known commonly as shepherd's purse (Figure 15-9).

The fertilized egg of *Capsella* is distinctly polar (Figure 15-10), as are many animal eggs. The nucleus and most of the cytoplasm are crowded into one end of the egg. The opposite end is occupied by a large vacuole that restricts the cytoplasm at this end to a thin layer just inside the cell wall. The polarity of the fertilized egg is reflected in the first division, which is unequal and produces daughter cells of different sizes (Figure 15-11). Most of the cytoplasm of the zygote is concentrated in the smaller cell; the larger cell, since it receives the vacuole of the zygote, contains a relatively smaller amount of cytoplasm.

Only the smaller of these cells contributes directly to embryo formation. The larger cell, called the **basal cell**, divides two or three times to produce a short stalk, the **suspensor** (Figures 15-12 and 15-13a and b). The basal cell

and the suspensor retain numerous small openings (plasmodesmata) that serve as direct channels between the embryo at one end and the tissues of the parent plant at the other. As a result, nutrients from the parent tissue and endosperm are believed to be conducted from the basal cell through the suspensor and into the embryo in a pattern similar to the placental tissue of mammals (see Supplement 14-1).

The smaller cell formed from the first division of the zygote continues to divide until it forms a globular mass at the tip of the suspensor (Figure 15-13c and d). As the dividing mass reaches as few as eight cells, the first differentiation of cell types is noticeable. A highly ordered division occurs in such a way that each cell at the surface of the ball divides in a plane parallel to the surface (Figure 15-14a to c). The divisions produce a surface layer of cells that forms the primary "ectoderm" of the embryo, a surface tissue called the **protoderm** in plant embryology. In later development, this surface layer gives rise to the entire "skin" or outer **epidermis** of the plant. As the embryo

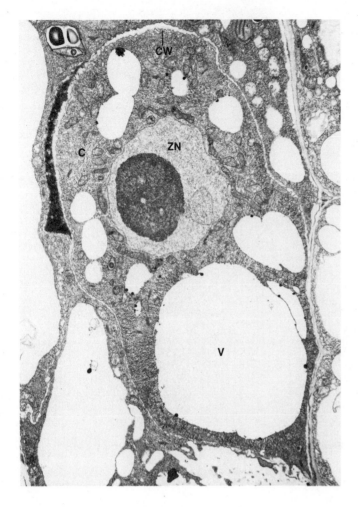

Figure 15-10 The *Capsella* zygote. *C*, cytoplasm; *ZN*, zygote nucleus; *V*, vacuole; *CW*, cell wall. Courtesy of P. Schulz.

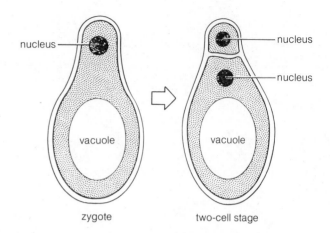

Figure 15-11 Zygote and two-cell stage of the *Capsella* embryo.

primary tissues of the later embryo and adult—epidermis, parenchyma, and vascular tissue.

Early development in the flowering plants establishes the three primary cell layers of the embryo: protoderm, ground meristem, and procambium.

The cells that develop into protoderm, ground meristem, and procambium differentiate in place, without changing positions, after they are laid down by cell division. Subsequent growth of the embryo continues through further division, elongation, and differentiation of cell lines derived from these primary layers. None of the active movements of single cells or cell groups that are so characteristic of animals contribute to the development of *Capsella* or other plant embryos.

As the *Capsella* embryo continues to divide after the globular mass of cells is formed, its shape begins to change. The protoderm and ground meristem cells at the tip of the embryo opposite the suspensor begin to divide more rapidly at two points, producing two protuberances that subsequently enlarge to give the embryo a heart-shaped appearance (Figure 15-13*e*). The developing plant at this stage is known as a *heart-stage embryo*. The protuberances continue to grow outward through rapid division and elongation, forming two thick, leaflike appendages called **cotyledons**. These structures serve as food storage organs in the seed.

As the cotyledons form, the protoderm, ground meristem, and procambial cells in the main body of the embryo also divide. The new cells produced by the division of these tissues elongate primarily along an axis running

grows, a central group of cells at the core of the globular mass differentiates into the **procambium**, which later gives rise to the **vascular system** of the plant. (The vascular system comprises the cells that transport water, minerals, and the products of photosynthesis between parts of the adult plant.) The procambium and protoderm are separated by a third cell layer called the **ground meristem** (*meristem* = group of dividing cells) that divides to form the **parenchyma** cells of the adult plant. Parenchyma cells in the adult remain relatively unspecialized and form a "filler" tissue that surrounds the vascular tissue. Thus the three cell layers of the embryo at the globular stage—protoderm, ground meristem, and procambium—remain in the same relative positions as they divide to form the three

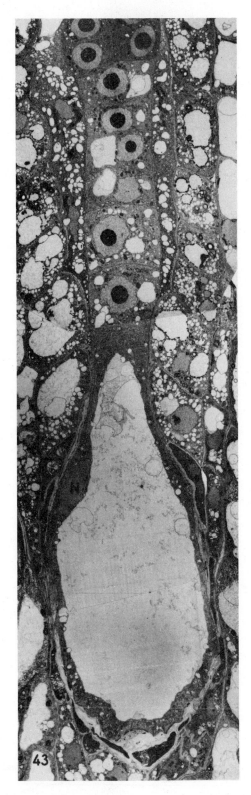

Figure 15-12 The basal cell, suspensor, and early embryo of *Capsella*. Courtesy of P. Schulz.

from the suspensor to the tip bearing the cotyledons. Division and elongation continue, gradually lengthening the embryo along this axis into the *torpedo stage* (Figure 15-13*f*). Within the torpedo-stage embryo, division and elongation of the procambium produce a band of cells that extends from the suspensor to the opposite tip between the cotyledons. At the two ends of the band, the procambial cells remain separated from the outer epidermis by an intervening layer of parenchymal cells.

While the embryo is lengthening to form the torpedo stage, cells at the tips of the embryo also divide rapidly, producing two cell masses from which the stem and root structures of the adult plant derive. At the tip opposite the suspensor (see Figure 15-13*g*), the cell mass formed by these divisions produces a swelling between the bases of the cotyledons. This swelling, the **shoot apical meristem** (*apical* = end), contains embryonic cells that divide continuously throughout the life of the adult plant to produce the stem, leaves, flowers, and all other structures above the soil. At the opposite end of the embryo, at the position of attachment to the suspensor, the second cell mass forms the **root apical meristem** (Figure 15-13*g*), a group of cells that produce the root structures of the mature plant. Both apical meristems are covered at their exterior surfaces by a layer of epidermis derived from the protoderm. Internally, they are continuous with the tips of the band of procambial cells that extends from end to end in the embryo.

As development reaches this stage, the *Capsella* embryo gradually bends double (see Figures 15-13 and 15-15). This bending, which shapes the embryo into the compact form found in the seed, occurs through the greater elongation of cells along one side of each cotyledon.

Seed Formation

Once the apical meristems have formed and the *Capsella* embryo has folded into its compact form, active cell division, elongation, and differentiation cease. The embryo then begins to dehydrate until as much as 85 percent of the cellular water has been removed. As a consequence of this dehydration, all metabolic activity, including respiration, comes to a halt and the embryo enters a state of suspended animation called *dormancy*. This change is accompanied by a loss of cellular vacuoles and conversion of the cytoplasm to a dense, compact form in which the individual cytoplasmic organelles become indistinguishable.

During the changes converting the embryo to the dormant state, the tissues of the ovule surrounding the embryo gradually harden and become impregnated with hydrophobic substances that prevent water movement between the enclosed embryo and its surroundings. This

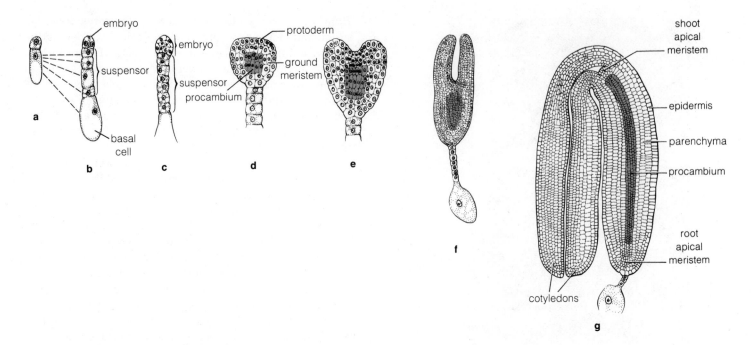

Figure 15-13 Developmental stages of the *Capsella* embryo. (**a**) to (**d**): Formation of the embryo and suspensor. (**e**) Heart-stage embryo. (**f**) Torpedo-stage embryo. (**g**) Mature embryo. *(a)–(e)* adapted from *Plant Morphology* by A. W. Haupt; copyright 1953 by McGraw-Hill Book Company; used with permission of A. W. Haupt and McGraw-Hill Book Company. *(f)–(g)* adapted courtesy of the Ohio Journal of Science.

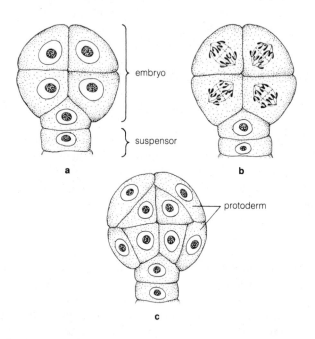

Figure 15-14 Cell divisions parallel to the surface (**a** to **c**) produce the protoderm, from which the epidermis of the adult plant develops (see text).

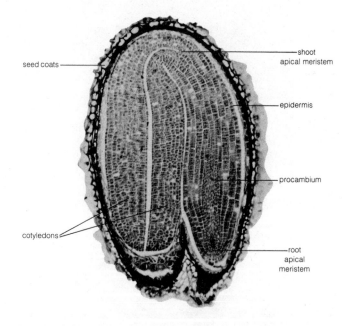

Figure 15-15 A *Capsella* embryo inside the developing seed. Courtesy of P. Schulz.

protective layer, derived from the ovule, forms the *seed coat* around the dormant embryo. The embryo, the seed coat, and any residual endosperm tissue enclosed with the embryo form the seed. Once the seed is mature, the embryo within remains dormant without further biochemical activity or growth until the seed germinates.

Plants with embryos and seeds of the type illustrated by *Capsella*, in which two cotyledons form during embryonic development, are called *dicotyledons* or **dicots**. Dicotyledonous seeds and development are typical of most flowering trees and shrubs, as well as food crops such as beans, peas, and peanuts. Many flowering plants, however, including the grasses and a few trees such as the palm follow a pattern of development in which only one cotyledon forms. These flowering plants are called *monocotyledons* or **monocots**. (Figure 15-16 shows the final structures in a monocot embryo encased in the seed coats.) Several important crop grasses, including wheat, rice, and corn, form monocotyledonous embryos and seeds.

Monocots form embryos with one cotyledon; dicots form embryos with two cotyledons.

Nutrients in the seeds of dicots are stored primarily in the cotyledons, which take up most of the space inside the seed; the endosperm in most dicots is depleted during embryonic development and is reduced or absent in the seed (as in Figure 15-15). In monocots, in contrast, the single cotyledon is frequently small and most of the seed's volume is occupied by endosperm, which persists as the major tissue storing nutrients until germination of the dormant embryo (as in Figure 15-16). In either case, the nutrients stored in the seed, in either the endosperm of the monocots or in the cotyledons of dicots, provide food not only for germinating plants but for much of the animal population of the world, including humans, as well. Food for humans and other animals is also supplied by *fruits,* which develop from tissues of the ovary surrounding maturing seeds.

Nutrients in monocot seeds are usually stored in the endosperm; nutrients in dicots are stored in the cotyledons.

In wheat, for example, about 80 percent of the seed volume is occupied by the endosperm. The endosperm material of this monocot is ground as flour after removal of the embryo (called the *wheat germ*) and the seed coat (called the *bran*). The flour derived from the endosperm, which contains about 75 percent of the protein of the wheat seed, provides bread for the world. The embryo or

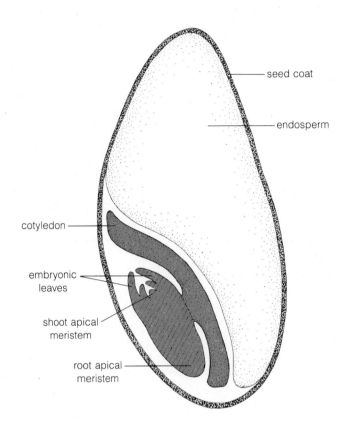

Figure 15-16 Structures in a monocot seed from wheat.

wheat germ is also eaten by humans or fed to livestock. The bran, although indigestible, is frequently included in the human diet as a source of bulk or "fiber." The endosperm of rice and corn are similarly used as food staples throughout the world. "Polishing" rice and wheat seeds for use as human food removes, along with the seed coats and the surface layers of the endosperm, many of the vitamins of the seeds.

Developmental Mechanisms During Growth of Plant Embryos

Development of the embryo from fertilization to the seed, as illustrated by *Capsella*, involves the three basic mechanisms of cell division, elongation, and differentiation. Once laid down by division, the cells retain their attachments to neighboring cells through their common cell walls and elongate and differentiate in this location.

The fact that plant cells remain in place once they are laid down gives cell division a special importance in plant development. Since there are no mass movements of cells

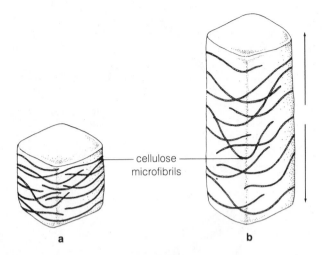

cellulose
microfibrils

a   b

Figure 15-17 Cellulose microfibrils running in hoops around the cell (**a**) restrict expansion in the direction in which the hoops run, but they allow the wall to extend in a direction perpendicular to the hoops (**b**). Only a portion of the microfibrils are shown in (**a**), and only the original fibers are shown in (**b**). As the wall expands, however, new microfibrils are laid down to fill the gaps created by elongation. Thus, as expansion is completed, the elongated wall in (**b**) becomes filled with cellulose microfibrils, many of them running in the direction of elongation.

or tissues in plant development, unlike animal development, much of embryonic development in plants is accomplished through localized changes in the *rate* and *direction* of cell division. In *Capsella*, for example, localized centers in which the rate of division increases produce the protuberances that form the apical meristems and cotyledons. Similarly, the direction of cell division is important in the production of structures such as the protoderm, which arises through a division of the cells at the surface of the embryo in a plane parallel to the surface (see Figure 15-14).

Both the rate and the direction of cell division are significant in plant development.

The mechanism of cell elongation is also of vital importance in the generation of embryonic structures in plants. The driving force for cell elongation is provided by water, which penetrates osmotically into plant cells in response to an increase in the concentrations of sugars, proteins, and other substances dissolved in the cytoplasm. As this force presses the cell contents against the walls, the

walls stretch to accommodate the increased internal pressure.

The direction of cell elongation is determined by the arrangement of cellulose microfibrils in embryonic cell walls.

Elongation is primarily unidirectional due to the pattern in which cellulose molecules (see p. 21) of the cell wall are laid down in the young or primary cell walls of the embryo. These molecules link together in groups of 30 to 40 to form composite cellulose fibers called **microfibrils** that run in roughly parallel fashion as hoops around the cell (Figure 15-17a). As a consequence, the microfibrils resist expansion of the wall in a direction parallel to the hoops, but allow the cell to elongate at right angles to this direction (Figure 15-17b). Unidirectional elongation in this way may proceed until the volume of the cell has increased by as much as 10 to 20 times. Most of this increase, as noted, is caused by the osmotic movement of water into the cell and is reflected in the growth of one or more large, water-filled vacuoles in the cytoplasm.

The pattern of cell differentiation in plants also reflects the fact that plant cells are immobile once they are formed by cell division. In plants, differentiation into one cell type or another is regulated largely by the *position* in which cells are laid down by division. For example, the location of cells at the tip of the embryo between the cotyledons is apparently the major factor responsible for their differentiation as the shoot apical meristem. The importance of cell position in plant development is related to the secretion and distribution of plant *hormones* or *growth factors*. Plant hormones become distributed in regions of developing embryos and plants and trigger the differentiation of cells in these regions into specific types. (Plant hormones and their effects on plant development are considered in further detail later in this chapter.)

Differentiation in plants depends on cell position in the embryo.

The basic cellular mechanisms operating in the development of plant embryos, although limited to cell division, elongation, and differentiation, achieve the same results as early animal development: the primary tissues from which the adult organ systems develop are established in the embryo. The further development of flowering plants after germination of the seed involves the same cellular mechanisms. Induction also occurs to a limited extent in the formation of some adult plant tissues and organs.

SEED GERMINATION AND DEVELOPMENT OF THE ADULT PLANT

Seed Germination

Seeds may remain dormant for years without any change in their capacity for germination and resumption of growth. The seeds of some plants have been successfully germinated after hundreds of years of storage. In one case, seeds of a small flowering plant, a lupine (*Lupinus arcticus*), found in lemming burrows buried deeply in frozen sediments believed to be 10,000 years old, were successfully germinated! The fact that seeds can remain dormant for such long periods without damage, with no measurable respiration or other metabolic activity, suggests that they are actually in a state approaching suspended life during the dormant stage.

Seeds end their dormancy and germinate if they are placed under the correct conditions of moisture and temperature. In many cases, the presence or absence of light also affects seed dormancy. All seeds require moisture to

---

Seed germination requires the correct conditions of moisture, temperature, and light.

---

germinate; if placed in water or a moist environment, seeds absorb water and swell to as much as twice their dormant size. Water absorption in many seeds requires abrasion or scarring of the seed coats. Unless the coats are damaged sufficiently for water to penetrate through them, as by mechanical abrasion, fire, or passage through the digestive tract of an animal, water absorption cannot occur. As water is absorbed, the cells of the embryo and endosperm lose their dense, compact appearance and the usual cytoplasmic organelles become visible again.

In many species hydration alone is not enough to break the dormancy. The seeds of some plants require exposure to temperatures near freezing followed by a return to warmer temperatures for germination to occur. Others simply require exposure to moderate temperatures. Other seeds, in addition to hydration and moderate temperatures, require exposure to light at certain wavelengths or periods of time or, in some cases, no light at all. These varying requirements probably represent adaptations ensuring that a germinating embryo meets conditions under which it can grow. The requirement for exposure to cold temperatures before germination, for example, may ensure that seeds will germinate in the spring instead of the late summer or fall following their formation. (What growth conditions might be favored by a requirement for the presence or absence of light?)

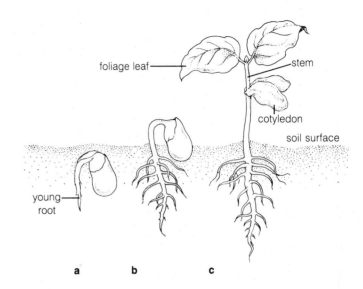

Figure 15-18 Germination of a bean seed, an example of a dicot. (**a**) Emergence of young root; (**b**) extension of cotyledons and shoot apex (both still within the seed coats) above the soil; (**c**) growth of the young stem and development of leaves. The cotyledons will eventually wither and fall from the plant.

The molecular effects of variations in light or temperature in producing germination are unknown. One current hypothesis proposes that an inhibiting chemical that prevents seeds from entering into protein synthesis is present during the dormant period. Exposure to the right conditions of moisture, temperature, and light, according to this hypothesis, triggers chemical events that break down or remove the inhibitor. *Abscisic acid*, a plant growth factor that often acts as an inhibitor, is present in seeds and is considered to be a good candidate for the hypothetical germination inhibitor.

When all the conditions required for germination are met, protein synthesis begins in the cells of the embryo. These early biochemical events probably depend on mRNA molecules stored in inactive form in the cytoplasm of the embryonic cells, since inhibitors of RNA transcription have no effect on the initiation of protein synthesis at this time. Once active enzymes appear and respiration increases to normal levels, the embryo begins to elongate and push outward through the seed coats. In most seeds, the first elongation occurs in the cells of the embryonic root behind the root apical meristem. (Figure 15-18*a*). This elongation pushes the root into the soil. Continued growth of the root depends on cell division in the root apical meristem, which begins soon after the germinating root pushes into the soil, and elongation of the newly formed cells behind the apical meristem. Thus the new root grows from the tip by constantly adding cells in the region of the apical meristem and by cell elongation just behind the tip.

Once the root has begun to push into the soil, the shoot at the opposite end of the embryo begins to elongate. Continued elongation and cell division carry the shoot apical meristem and the cotyledons into the air above the soil (Figure 15-18*b*). At this point, the shoot has fully germinated and any remnants of the seed coats fall away. The initial growth of the newly germinated shoot depends on nutrients stored in the cotyledons or endosperm. As leaves form at the shoot apex (Figure 15-18*c*) and photosynthesis begins, the nutrient source gradually shifts from the cotyledons or endosperm to the leaves.

Differentiation of Primary Tissues

The earliest growth of the new seedling is due primarily to the elongation of cells already present in the dormant embryo before seed germination. Once the new root and stem have escaped from the seed coats and begun to push into the soil and air, cell division at the root and shoot apex begins and rapidly adds new cells to the tips of these structures. Further growth of the seedling and plant is due to division and elongation of these newly formed cells.

The new cells are laid down, elongate, and differentiate in a similar pattern in both the root and the stem of the growing plant (Figure 15-19). At the tip of the root and stem is a short meristematic region in which active cell division continues. Just behind the tip, and extending for a short distance into the root or stem, is a *zone of elongation* in which cells stretch in the direction of growth. Just behind the zone of elongation is a *zone of differentiation* in which the elongated cells of the root or stem become specialized in structure and function. Typically the dividing, elongating, and differentiating cells of the root and stem extend back in regular rows or files from the regions of active division at the tips.

We will follow cell division, elongation, and differentiation in the stem to see how these mechanisms work in the development of the adult structures of a flowering plant. The three embryonic cell layers, protoderm, ground meristem, and procambium, persist at the tip of the stem (Figure 15-20), where they remain in their actively dividing, embryonic state throughout the life of the plant. From the cells laid down by these embryonic cell layers, the three primary tissues of the adult stem develop: epidermis, from the cells laid down by the protoderm; parenchyma, from the cells laid down by the ground meristem; and vascular tissue, from the cells laid down by the procambium.

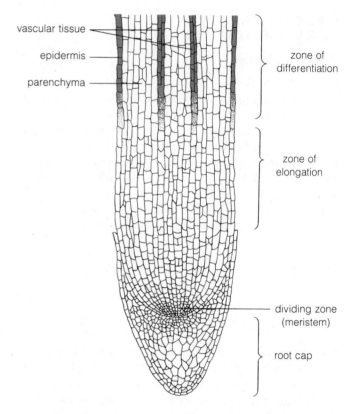

Figure 15-19 The zones of division, elongation, and differentiation in a growing root. The dividing cells at the tip add new cells both behind and in advance of the region of division (the meristem); the cells laid down in advance form the root cap, a structure that cushions the root tip as it extends into the soil.

In the development of adult structures, protoderm gives rise to epidermis, ground meristem to parenchyma, and procambium to vascular tissue.

Once laid down and left behind by the actively dividing cells at the stem tip, the newly formed cells stretch in length within the zone of elongation. This elongation, as noted, takes place in the direction of the long axis of the stem. After elongation, which occurs through the formation and enlargement of water-filled vacuoles within the developing cells, differentiation into the adult types begins. Epidermal cells form a waxy surface coat, the **cuticle**, that retards the evaporation of water from the stem. At intervals, epidermal cells differentiate into a **stoma** (plural = *stomata*), consisting of a small opening or *pore* surrounded by a pair of *guard cells* (Figure 15-21). During periods of rapid photosynthesis, the stomata, which also

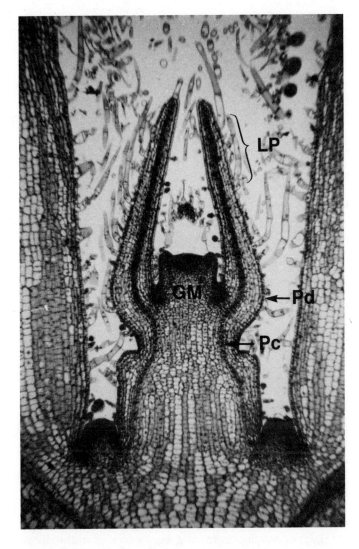

Figure 15-20 The protoderm *(Pd)*, ground meristem *(GM)*, and procambium *(Pc)* at the tip of a growing *Coleus* stem, which together make up the shoot apical meristem. These cells remain as actively dividing embryonic cells throughout the life of the plant. *LP*, leaf primordium.

occur in large numbers in the epidermis of leaves, open to permit the passage of oxygen and carbon dioxide between the plant interior and the surrounding atmosphere.

Within the stem the cells laid down by the ground meristem differentiate into parenchyma, a tissue characterized by large, irregularly shaped cells. Parenchyma fills in the spaces between the vascular tissue and the epidermis and also the center of the stem. The cells laid down by the procambium, which occur in separate strands that extend lengthwise through the stem, differentiate into

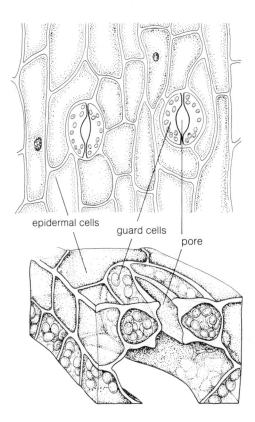

epidermal cells guard cells pore

Figure 15-21 Stomata in a plant epidermis. Each stoma consists of a pair of guard cells surrounding a pore.

two kinds of vascular tissue, **xylem** and **phloem,** after elongating.

Xylem is specialized to transport water absorbed in the roots to all parts of the plant. After elongating, the differentiating xylem cells develop thickened walls. This thickening, in many flowering plants, takes the form of regular spirals or circles of material laid down on the inside of the wall (Figure 15-22). As the walls thicken, the cells in them die and disintegrate, leaving the empty walls as a *xylem vessel element*. In the flowering plants, much of the end walls of the vessel elements is digested away, joining the elements of a row into a *vessel tube*. Openings called *pits* also appear in the side walls of the xylem vessels.

Xylem conducts water from the roots to stems, leaves, and flowers. Phloem conducts the products of photosynthesis from leaves to all parts of the plant.

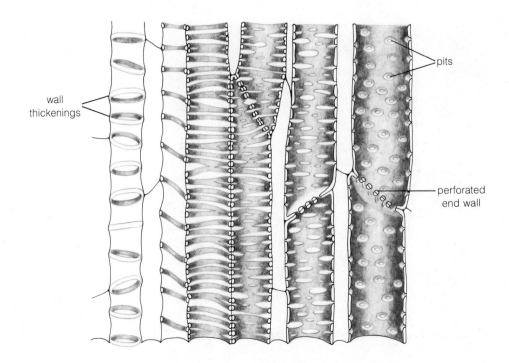

wall thickenings

pits

perforated end wall

Figure 15-22 Mature xylem vessel elements showing the spiral or circular wall thickenings and pits (see text).

Phloem is specialized to carry the products of photosynthesis to all parts of the plant. Conducting elements of the phloem differentiate from pairs of cells aligned side by side. In one cell of the pair, the cytoplasm thins out and becomes reduced to little more than a watery solution containing strands of a protein called *P-protein* that extend from end to end of the cell. The end walls become perforated, forming a structure called the *sieve plate* (Figure 15-23). As these changes are completed, the nucleus disintegrates, completing differentiation of this member of the pair into a *sieve cell*, the primary conductive element of the phloem. The other cell of the pair retains its nucleus and cytoplasm and also retains direct connection to its adjacent sieve cell via numerous plasmodesmata. This cell, which presumably directs the activities of its sieve cell, is called a *companion cell*. Often sieve cells line up end to end in the phloem of flowering plants to form *sieve tubes*. The adjacent sieve cells in these tubes make connections through the openings in the sieve plates, producing a continuous cytoplasm running through the phloem.

Cells in the developing vascular bundles located on the side toward the epidermis typically differentiate into phloem whereas cells located toward the interior develop into xylem. Differentiation of phloem and xylem is similar in monocots and dicots with one major exception. In monocots, the bundles developing from the strands of cells laid down by the procambium are distributed randomly throughout the stem (Figure 15-24*a*). In dicots, the bundles are arranged in a circle forming an interrupted cylinder of vascular elements inside the stem (Figure 15-24*b*).

Leaves develop from outgrowths formed in the dividing region at the tip of the stem. At the tip, cells of the ground meristem and procambium just below the protoderm begin to divide in a direction parallel to the surface of the plant. This development produces a protuberance called a *leaf primordium* at the growing stem tip (see Figure 15-20). The primordium then extends through further divisions of the protoderm, ground meristem, and procambium and eventually forms a small leaflike blade at the stem tip. At different points around the growing stem tip, the leaf primordia appear in definite patterns that produce the arrangement of leaves typical in different plant species. How the pattern is controlled is unknown. As a growing leaf primordium extends, groups of cells located along the central axis and other points within the developing leaf continue to divide in different directions. It is the direction of these divisions and the subsequent elongations that determines the shape of the leaf.

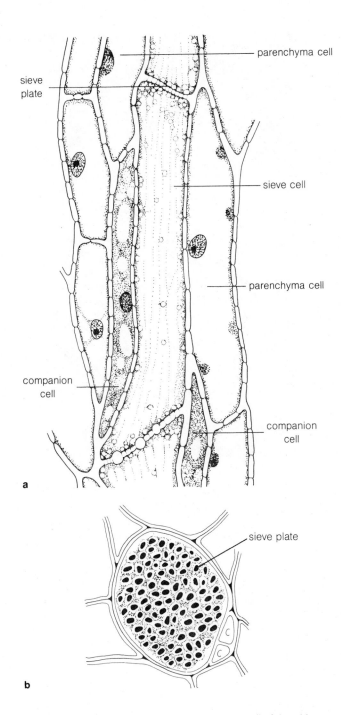

sieve plate

parenchyma cell

sieve cell

parenchyma cell

companion cell

companion cell

a

sieve plate

C

C

b

Figure 15-23 (a) A sieve cell and companion cell of the phloem. (b) The sieve plate of a sieve cell.

Within the leaf blade, cells differentiate after elongation into the tissues of the leaf (Figure 15-25). Cells derived from the protoderm form a waxy surface cuticle and differentiate into an epidermis covering the entire leaf. At intervals on the lower leaf surface, the epidermis is interrupted by stomata consisting of a central pore surrounded by guard cells as in the stem. Inside the epidermis, cells laid down by the ground meristem differentiate into parenchyma of two types that forms the primary photosynthetic tissue of the leaf. Toward the upper surface of the leaf, the parenchyma cells are columnar and closely spaced in even rows lying perpendicular to the leaf surface. These cells form the *palisade parenchyma* of the leaf interior. Toward the bottom surface of the leaf, the cells are more irregular in shape and arranged in a pattern that leaves many open spaces that communicate with the exterior through the stomata. This layer forms the *spongy parenchyma* of the leaf interior. Strands of cells laid down by the procambium differentiate into vascular bundles that extend throughout the leaf; xylem elements are located toward the upper leaf surface, phloem toward the lower surface. These vascular bundles, since they form as outgrowths of the procambium of the stem tip, make direct connections to the xylem and phloem elements of the stem.

The primary photosynthetic tissue in leaves is parenchyma.

In dicots, cell division, elongation, and differentiation cease as the leaf assumes its final size and shape. No embryonic dividing cells persist at any points within the dicot blade. In many monocots such as the grasses, in contrast, embryonic cells persist at the base of the blade. If the leaf is cut or broken by a grazing animal or mowing human, the persistent embryonic cells are stimulated to divide and the leaf resumes growth at its base.

At some point in the growth of a plant, primordia at the tip of the stem develop into flowers instead of leaves, often in response to changes in day length. While the outgrowths destined to become flowers develop initially in patterns much like leaf primordia, later development varies greatly and produces an almost infinite range of flower types in different angiosperms.

Development of Secondary Tissues

In some plants, such as the small wildflowers that grow, bloom, and die each year, the primary epidermal, parenchymal, and vascular tissues that differentiate from cells laid down at the tips of roots and stems are the only tissues of the plant. Cell division is thus restricted to the root

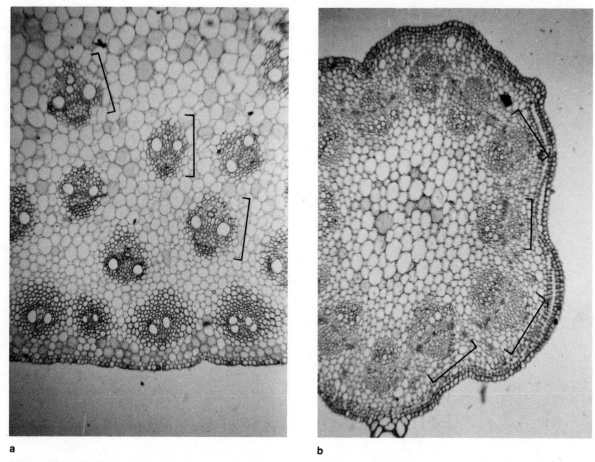

a

b

Figure 15-24 (**a**) The random distribution of vascular bundles (brackets) in a monocot stem (from corn). (**b**) The circular arrangement of vascular bundles (brackets) in a dicot stem from *Haemanthus*.

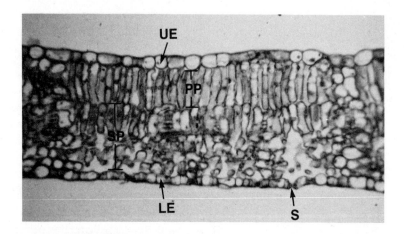

Figure 15-25 The differentiated cells in a *Syringa* leaf blade. Photosynthesis occurs primarily in the parenchymal cells in the leaf interior. *UE*, upper epidermis; *PP*, palisade parenchyma; *SP*, spongy parenchyma; *LE*, lower epidermis; *S*, a stomate.

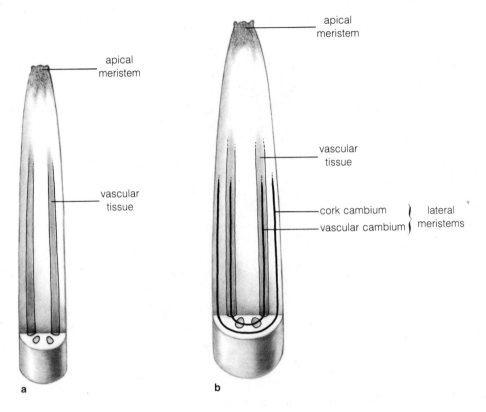

apical
meristem

vascular
tissue

apical
meristem

vascular
tissue

cork cambium
vascular cambium

} lateral
meristems

a

b

Figure 15-26 (**a**) The primary meristem of a young stem. (**b**) The lateral meristems that add secondary growth to the stem.

and stem tips, which are called the *primary meristems* of the plant (Figure 15-26a). Other plants, primarily dicots such as the perennial trees and shrubs, develop additional regions of dividing cells that are distributed lengthwise within stems and roots and extend throughout the length of the plant (Figure 15-26b). These lengthwise dividing regions, called *lateral meristems*, add secondary tissues to the plant. While the tissues laid down by the primary meristems contribute largely to the length of stems and roots, the activity of the lateral meristems adds to the width or thickness of these structures.

Secondary tissues are produced through the activity of lateral meristems, which increase the thickness of stems and roots.

The growth of secondary tissues in dicot stems stands as a good example of this pattern of development. The secondary growth of vascular tissues is of special interest in this development because it provides one of the rare examples of induction in plants.

Dicots undergoing secondary growth develop two lateral meristems in the stem. One, called the *vascular cambium*, develops from embryonic cells originally laid down by the procambium that persist in the vascular bundles.

Vascular cambium gives rise to additional xylem and phloem tissues; cork cambium gives rise to the outer layers of the bark.

This cambial layer divides to give rise to the secondary xylem and phloem of the stem. The second lateral meristem, the *cork cambium,* arises from a layer of parenchymal cells lying just inside the epidermis. When fully developed, the vascular and cork cambiums form complete, concentric cylinders of actively dividing cells within the stem (see Figure 15-26b).

The vascular cambium develops from strands of embryonic cells that persist between the phloem and xylem of the primary vascular bundles in a layer parallel to the surface of the stem (Figure 15-27a). Just before secondary growth is ready to begin, the cells in these persistent layers of procambium induce the parenchymal cells lateral to

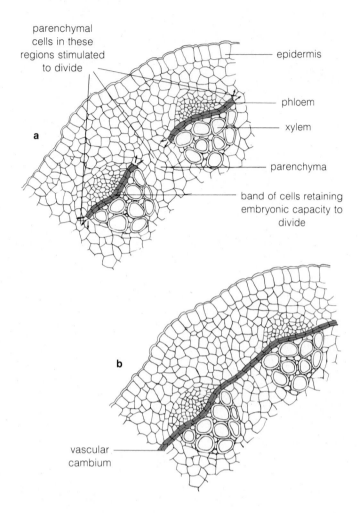

parenchymal
cells in these
regions stimulated
to divide

epidermis

phloem

xylem

parenchyma

band of cells retaining
embryonic capacity to
divide

a

b

vascular
cambium

Figure 15-27 Development of the vascular cambium. (**a**) Residual embryonic cells between the phloem and xylem induce neighboring parenchyma cells to divide and differentiate into embryonic, dividing cells (small arrows). (**b**) The process, when complete, forms a complete cylinder of dividing tissue, the vascular cambium, around the stem. These cells divide to give rise to phloem (toward the surface of the stem) and xylem (toward the interior).

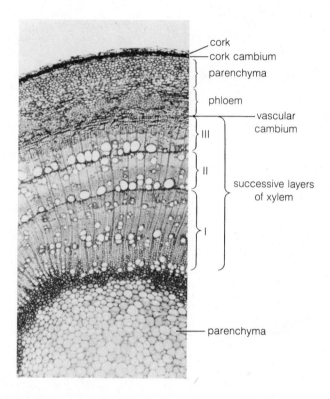

cork
cork cambium
parenchyma

phloem

vascular
cambium

III

II

successive layers
of xylem

I

parenchyma

Figure 15-28 The growth rings produced by annual activity of the vascular cambium. As each successive ring of xylem is laid down, the phloem is pushed outward. Courtesy of W. A. Jensen.

them to divide. The small cells resulting from this division remain active in division. As they form, they stimulate additional parenchymal cells beside them to divide until the layer of dividing cells closes into a complete cylinder around the stem (Figure 15-27b). The cylinder of dividing cells, which extends the length of the stem containing primary vascular bundles, forms the vascular cambium. Since the persistent, dividing layer of the vascular bundles is necessary for the adjacent parenchymal cells to differentiate into vascular cambium, information in some form flows from the vascular bundle to "instruct" or trigger de-

velopment of the adjacent parenchyma into cambium. The change thus stands as an example of an induction in plant development. As in animal inductions, the nature of the information flowing from the inducing to the induced cells is unknown (see Chapter 16).

After it is fully formed, the vascular cambium continues to divide. Since the plane of cellular divisions runs parallel to the surface of the stem, consecutive layers of cells are laid down to the outside and inside of the vascular cambium. As in the development of the primary tissues, the cells on the side of the vascular cambium toward the surface of the plant develop into phloem. The cells laid down on the side facing the center of the stem develop into xylem (Figure 15-28). The combined division and differentiation on both sides of the vascular cambium forms a ring of new growth extending around the stem. In most dicots, the vascular cambium undergoes its most rapid divisions in the spring and summer of each year, producing the *growth rings* visible in cut stems and trunks. The growth rings (except for the most recently formed ring) consist entirely of xylem; the phloem of previous years is crushed by the outward growth of new phloem from the vascular cambium and disappears.

The cells in the cylinder of cork cambium also divide in a plane parallel to the surface of the stem. In this case, divisions produce new cells only on the side of the cambium toward the stem surface. These cells develop deposits of a fatty substance, *suberin*, that makes them impermeable to water. As they grow outward in a solid layer around the stem, they break and separate the original epidermis. This outer layer, broken on its surface but continuous underneath, forms the *cork* layer of the bark. The innermost layer of the bark, separated from the surface cork layer by the cork cambium, is formed by the phloem and any remaining parenchymal cells that persist between the phloem and the cork cambium (Figure 15-29).

The Effects of Hormones on Plant Development

The formation of primary and secondary tissues in the flowering plants demonstrates the importance of *position* in the plant as the primary factor determining whether plant cells differentiate into epidermis, parenchyma, vascular tissues, or cork. Cell position, in turn, depends on the effects of a small group of plant *hormones* or *growth factors*.

The importance of cell position in plant development depends on local differences in concentration of the plant hormones.

Five kinds of hormones have been described in plants: **auxin,** the **cytokinins** and **gibberellins, abscisic acid,** and **ethylene** (Figure 15-30). Although certain parts of plants synthesize one or more of these hormones in greatest quantity, as in the increased auxin production by growing stem tips, apparently any cell or group of cells is capable of synthesizing them almost anywhere in a plant under the appropriate conditions. After synthesis in various parts of the plant, the hormones travel through the vascular system to promote or inhibit cell division, elongation, differentiation, and other effects. (Some of the known or proposed effects of the five plant hormones are summarized in Table 15-1.) Evidently the final effects on a plant cell's elongation and differentiation depend on how much of each hormone reaches the position occupied by the cell. The combined action of the hormones thus triggers cell differentiation toward specific cell types such as epidermis, xylem, phloem, parenchyma, or leaf cells.

The combined effect of the hormones, and the importance of their concentrations, is nicely illustrated by a classic experiment carried out by Folke Skoog and Carlos O. Miller at the University of Wisconsin. Skoog and Miller removed a block of parenchyma cells from the interior of

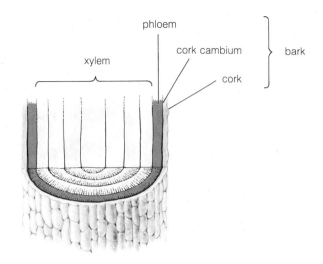

Figure 15-29 The bark formed at the surface of the stem consists primarily of an inner layer of phloem and an outer layer of cork cells separated by the cork cambium. The woody tissue inside the bark consists of xylem cells.

a tobacco plant and placed them in a nutrient medium containing auxin and cytokinin at different relative concentrations (Figure 15-31). With cytokinin at intermediate levels and auxin at low concentrations, the parenchyma cells were stimulated to divide without differentiating. This division produced a growing mass of essentially embryonic cells called a *callus* (Figure 15-31a). If cytokinin was raised to high levels in relationship to auxin, cells in the callus differentiated into the structures of stems, and normal stems and leaves gradually grew from the callus (Figure 15-31b). Changing the concentration of the hormones to the reverse ratio, with auxin high and cytokinin low, caused normal roots to differentiate and grow from the callus (Figure 15-31c).

Since the five plant hormones interact in this complex way to produce one or more combined effects, it is difficult to identify the activity of each in promoting or inhibiting cell division, growth, and differentiation. The situation is further complicated by the fact that one hormone, in arriving at a point in a plant, may stimulate or inhibit the synthesis of the other four types. As a consequence, the individual effects of each hormone are controversial, and it is frequently uncertain whether a given stimulation or inhibition is a primary effect of a hormone or a side effect caused by the stimulation of other hormones. Of the five hormone types, auxin, the cytokinins, and the gibberellins seem to be most important in regulating cell division, elongation, and differentiation.

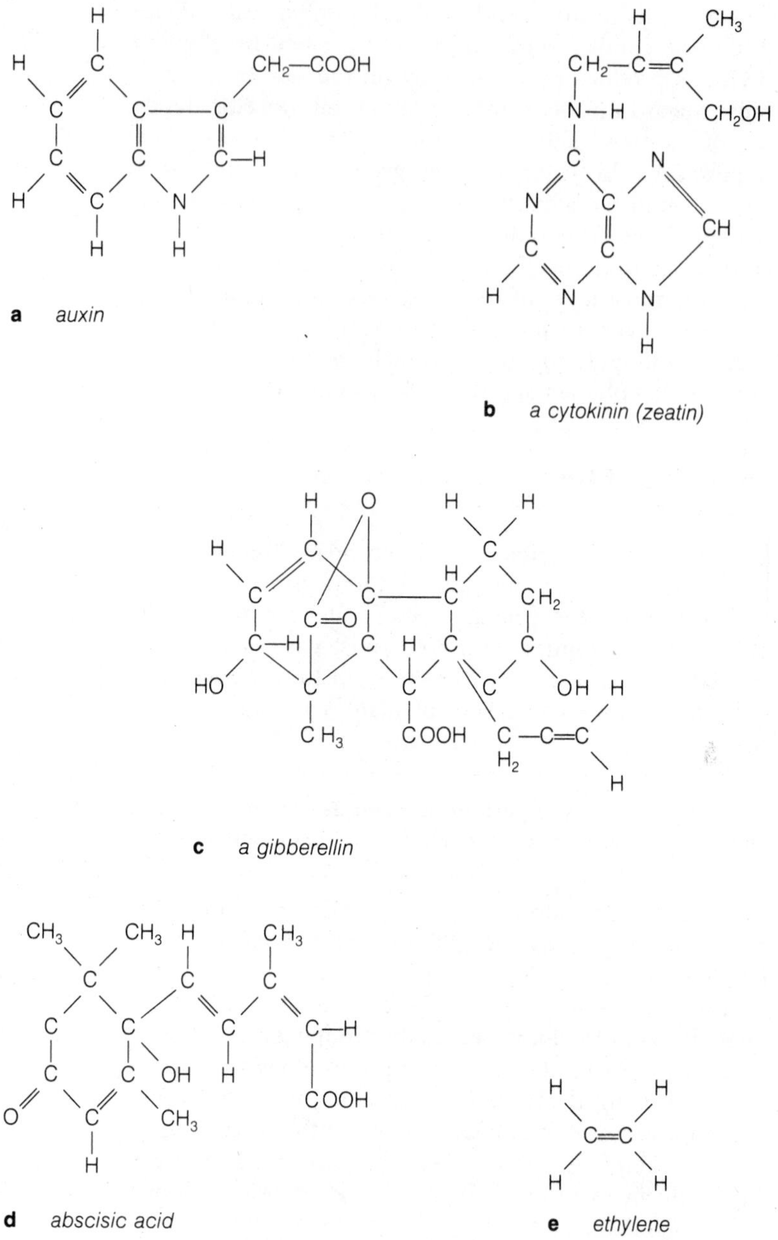

a *auxin*

b *a cytokinin (zeatin)*

c *a gibberellin*

d *abscisic acid*

e *ethylene*

Figure 15-30 The plant hormones or growth factors.

The Five Hormone Types

Auxin The first plant hormone to be discovered and described was auxin. Its effects on growing plant stems were first studied by Charles Darwin and his son Francis in 1881. Darwin and his son noted that stems, which normally curve toward a light source, failed to do so if the tips were covered. This effect was later found to be due to the activity of auxin by Frits Went in 1926. Went, who was a graduate student in Holland at the time, carried out the experiment diagrammed in Figure 15-32 with growing oat shoots. Curvature toward light occurs in the zone of elongation below the tip of the stem (Figure 15-32a). If the tip was cut off, no curvature occurred and the stem failed to respond to light directed from one side (Figure 15-32b). Went then placed the removed stem tip on a piece of gelatin (Figure 15-32c). After allowing the tip to remain in this

Table 15-1   Plant Hormones and Their Effects

Hormone	Stimulatory Effects	Inhibitory Effects	Interactions with Other Hormones
Auxin	elongation of stems and roots	retards elongation at high concentrations	+ gibberellins, promotes lateral growth of stems
	elasticity of cell walls	retards root growth at high concentrations	+ gibberellins and cytokinins, promotes fruit enlargement
	root branching	retards abscission of leaves, flowers, fruits	
	fruit development	retards flowering	stimulates ethylene formation
	ethylene formation		
	responses to gravity and light		
	development of vascular tissue		
	cambial activity		
Cytokinins	cell division	inhibits formation of lateral roots	+ auxin, promotes organ formation
	fruit enlargement	delays senescence	+ auxin and gibberellins, promotes fruit enlargement
	formation and growth of lateral buds and flowers	can inhibit elongation	can reverse elongation effect of auxin
	expansion of cotyledons and leaves		
	development of chlorophyll and grana in chloroplasts		
	radial enlargement of stems and roots		
Gibberellins	cell division in stem tips		+ auxin, promotes growth of lateral meristems
	fruit enlargement		+ auxin and cytokinin, promotes fruit enlargement
	elongation and growth of stems		
	germination of dormant seeds and buds		reverses effects of abscisic acid
	flowering		
Abscisic acid	abscission of leaves and fruits	promotes dormancy in buds and seeds	+ ethylene, causes senescence
	maintenance of internal cell pressure to retard wilting	inhibits growth	can reverse effects of auxin, cytokinins, and gibberellins
Ethylene	ripening of fruit	retards elongation of stems and roots	+ abscisic acid, causes senescence
	root branching	inhibits flower formation in most species	may inhibit auxin transport
	radial enlargement of stems and roots		
	stimulates own synthesis		

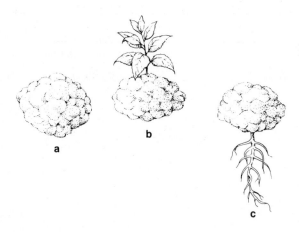

Figure 15-31 The Skoog and Miller experiments demonstrating how changes in the relative concentrations of auxin and gibberellin affect differentiation. (**a**) The callus formed with cytokinin at intermediate concentration and auxin at low concentration. (**b**) The differentiation of stems in solutions containing cytokinin at high concentrations with respect to auxin. (**c**) The differentiation of roots in solutions containing auxin at high concentrations and cytokinin at low concentrations.

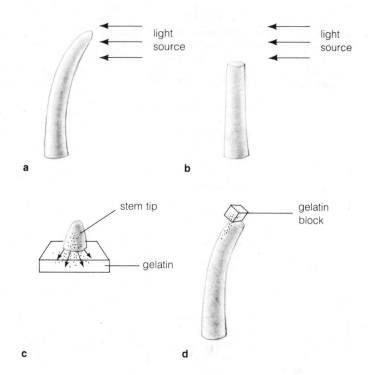

Figure 15-32 Went's experiment demonstrating the effects of auxin on elongation (see text).

position for a time, he removed the tip and cut the gelatin into small blocks. Went found that placing a block of gelatin on one side of a cut stem tip caused the same side of the stem to elongate. As a result, the stem curved away from the side on which the gelatin block was placed (Figure 15-32*d*). It was obvious that a substance diffused from the cut tip into the gelatin; when the gelatin was placed on one side of the cut stem, the substance diffused into the stem and promoted elongation of the cells on the same side. Went called the growth substance auxin (from the Greek *auxin* = to increase). It was later discovered to be *indoleacetic acid*, the molecule shown in Figure 15-30*a*. Research has shown that auxin enhances elongation by increasing the elasticity of cell walls and making them more extensible.

The positive response of growing stems to light is called positive *phototropism* (*tropos* = turn). Auxin also regulates the negative response of stems to gravity. If a growing stem or shoot is placed horizontally, the stem will respond by curving upward, away from the center of the earth. This response is due to the accumulation of auxin on the lower side of the stem, causing greater elongation of the cells on this side. A negative response to gravity of this type is called negative *geotropism*. Auxin also regulates the positive geotropism of roots: when placed horizontally, a growing plant root will curve downward toward the center of the earth. In this case, evidently, the effects are indirect. Although auxin accumulates on the lower side of

the root, as in a stem held horizontally, its presence there stimulates the production of ethylene, which overrides the auxin effect and inhibits elongation on this side. Elongation proceeds normally on the upper side, and the root curves downward.

Auxin has a variety of additional effects apart from its ability to promote stem elongation and regulate the responses of plants to light and gravity. Its primary source in plants is in growing stem tips; from there it diffuses to other parts of the plant in a gradient running from high concentrations at the stem tips to low at the roots. The particular auxin concentration at a given point in the plant, in combination with the concentrations of the other hormones present, produces the varied effects noted in Table 15-1. Although auxin promotes growth as a natural hormone in plants, the same hormone and artificial auxins such as 2,4-D (2,4-dichlorophenoxyacetic acid) are potent plant killers when applied at high concentrations. They are particularly effective as killers of broad-leaved plants, such as lawn weeds, and are much used for this purpose. (The narrow blades of grass plants absorb much less of the applied auxin and are relatively unaffected.) Synthetic or natural auxins are also applied commercially to retard fruit drop and the sprouting of potatoes or to promote the growth of roots from leaf or stem cuttings. Generally, the natural and synthetic auxins used for these purposes are relatively nontoxic to humans and other animals.

Cytokinins The hormones classed as cytokinins were discovered through the development of methods to grow plant cells in tissue culture. Early attempts using auxin in combination with mineral nutrients were unsuccessful until coconut milk (a form of liquid endosperm) was added to the culture medium. The active ingredient in the coconut milk was later discovered to be a cytokinin. Cytokinins of various kinds were subsequently found to be present as growth hormones in all plants. Although other regulatory effects are noted, their primary role is in the enhancement of cell division. Hundreds of different cytokinins have been discovered, but individual plant species usually synthesize less than ten of them in detectable quantities. Curiously, many of the cytokinins are chemical derivatives of adenosine, a base–sugar complex occurring in nucleic acids. Cytokinins are especially abundant in young leaves, fruits, and seeds and in the root tips of plants.

Gibberellins The plant hormones known as gibberellins were discovered in Japan through a study of diseased rice seedlings that grew taller than normal plants. The tall plants, which were called *bakanae* ("foolish seedlings") by the Japanese because they usually became topheavy and collapsed, were found to be infected by a fungus, *Gibberella fujikuroi*. The fungus produced a growth-promoting substance that was subsequently isolated, identified, and called gibberellin. Later, gibberellins were found to be natural growth hormones produced by all higher plants and many fungi. About 50 different gibberellins are now known in different plants and fungi.

The gibberellins, by stimulating both cell division and elongation, promote plant growth. When applied externally, the hormone can convert dwarf plants to normal size and stimulate certain normal plants to grow to unusual dimensions. Cabbage plants, for example, which rarely grow to more than a foot in height, can be stimulated by applied gibberellins to grow more than six feet tall! In the usual amounts naturally present in plants, the gibberellins stimulate normal patterns of cell division and elongation, the growth of fruits and dormant buds, and the germination of seeds. They are produced in quantity in developing seeds and then diffuse to promote the growth of the surrounding ovary tissue into fruit. Gibberellins are also synthesized in young leaves and in roots. Their effect on fruits is utilized commercially to increase the size of seedless grapes and oranges, which otherwise develop relatively slowly.

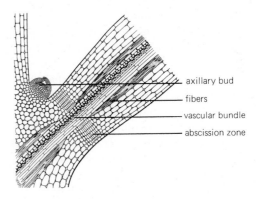

Figure 15-33 An abscission zone formed through the effects of abscisic acid. The layer of thin-walled cells in the abscission zone eventually breaks, releasing the structure outside the zone from the stem. Abscission zones are responsible for the release of ripened fruit and leaves from stems in the fall.

Abscisic acid This hormone was discovered in the 1960s through the study of dormant buds and seeds. Abscisic acid was found to be present in quantity in these dormant plant structures, in which it appears to inhibit growth by counteracting the effects of auxin and the cytokinins and gibberellins. Abscisic acid also promotes the formation of the *abscission zone* (Figure 15-33), a crosswise layer of thin-walled cells that breaks to cause the drop of leaves and fruits.

Abscisic acid is evidently the primary inhibitor of seed growth; seeds will not germinate unless their internal concentration of the hormone is reduced or eliminated. The soaking or washing provided by spring rains or human planters is frequently sufficient to remove the hormone from seeds. Conversely, added abscisic acid can retard or inhibit germination. The natural sites of abscisic acid production in plants are still unknown.

Ethylene Unusual among the plant hormones because it is a gas at growth temperatures, ethylene was discovered through its effect in stimulating fruits to ripen. Smoke produced by burning organic material had been used for centuries to promote the ripening of fruits. Later, ripe fruits themselves were found to emit a gas that causes other fruits to ripen. An agricultural report at the turn of the century, for example, mentioned that oranges and bananas should not be stored together, since a substance released by the oranges would cause the bananas to ripen too quickly and spoil. In the 1930s, the substance involved in promoting fruit to ripen in all these instances was discovered to be ethylene. With a molecular weight of only 28, ethylene is easily the smallest molecule forming a hormone yet discovered in either plants or animals.

Within the stems and roots of plants ethylene retards elongation and stimulates cells to grow in the opposite direction—that is, to become larger in radial diameter. This has the effect of thickening stems and roots in regions of high ethylene concentration. The hormone is synthesized in response to stress or wounding and, by increasing the diameter of stems and roots, strengthens the plant in the stressed region. Ethylene can be synthesized by cells essentially anywhere in a plant. Since auxin too stimulates ethylene synthesis, the pattern of growth in a given plant region may depend on the final ratio of auxin to ethylene, one hormone counterbalancing the effect of the other. Ethylene is used commercially to ripen fruits that are shipped green to retard spoilage, and also to promote flowering in pineapples.

The Molecular Activity of Plant Hormones The action of plant hormones at the molecular level is largely unknown. In contrast to the hormones of animals, which have specific effects on their target cells, plant hormones have general effects on almost all cells of the plant body. Among these general cellular effects are enhancement or inhibition of RNA transcription and protein synthesis—auxin, the cytokinins, and the gibberellins enhance these molecular activities whereas abscisic acid and ethylene inhibit them. This enhancement or inhibition of RNA and protein synthesis may well involve the control of specific mRNAs or protein molecules such as enzymes that carry out regulatory effects of the hormones. A few such cases have been described. The gibberellins, for example, which promote seed germination, specifically enhance the synthesis of an amylase enzyme that breaks down the starch molecules stored in endosperm and makes the glucose derived from them available for use by the germinating embryo.

Three hormones—auxin, cytokinin, and gibberellin—are primarily stimulatory in their effects on plant growth. Two—abscisic acid and ethylene—are primarily inhibitory.

Plant hormones act at the molecular level by stimulating or inhibiting RNA transcription and protein synthesis and by modifying transport through the plasma membrane.

Moreover, the plant hormones probably modify the transport of ions through the plasma membrane. Auxin, for example, promotes the expulsion of H^+ from the cytoplasm into the cell wall, an activity related to cell elongation. The increase in H^+ ion concentration activates enzymes that are active in acid solutions but relatively inactive at the neutral pH characteristic of nongrowing cell

walls. The activated enzymes break chemical bonds that hold cellulose microfibrils rigidly in place in the walls, freeing the microfibrils and allowing the walls to extend. Through these molecular effects, and others that no doubt remain to be discovered, the five hormones interact to regulate cell division, elongation, and differentiation in plants.

Growth and differentiation of plants from the fertilized egg to the adult, as in animals, occurs without changes in the genetic information of developing cells.

As in animals, differentiation into the structures of the mature plant occurs without any change in the total genetic information of the developing cells; all the nuclei of the adult retain the full complement of genes for that organism. The evidence for this basic fact of development is even stronger in plants than in animals, because almost any nucleated plant cell, if removed from the adult and cultured under the right conditions, can be induced to divide and regenerate an entire new plant.

Although they are imperfectly understood, the mechanisms of differentiation in the cell nucleus are believed to be the same in both plants and animals. In some manner, the activity of genes is controlled by the histone and nonhistone chromosomal proteins, which turn genes on and off in a programmed sequence as development proceeds (see Chapter 9). If traced to its ultimate source in the cell, this control by chromosomal proteins eventually rests on the activity of genes that control the synthesis and modification of the histone and nonhistone proteins. Thus, in both plants and animals, development is ultimately controlled by the genes of the nucleus. Whether a fertilized egg grows into a pine tree or a human is a matter of the number and types of genes present in the egg and the program for development they bear.

Questions

1. What mechanisms of cellular development are shared by animals and plants? What developmental mechanisms of animals do not occur in plants or are of reduced importance? How are these similarities and differences related to the fact that plant cells are stationary?

2. How do generations alternate in the flowering plants? What is a sporophyte? A gametophyte? Where do mitosis, meiosis, and fertilization occur in the life cycles of flowering plants?

3. Define flower, petal, carpel, sepal, stamen, anther, ovary, and stigma. Which of these structures are associated with the male part of the flower? With the female part? How does each part function in the flower?

4. Trace the development of the megaspore and female gametophyte. Where do these events occur in the flower?

5. Define synergid, antipodal, egg cell, central cell, polar nuclei, ovule, and micropyle.

6. Trace the development of the microspore and male gametophyte. Where do these events occur in the flower?

7. Define vegetative cell, generative cell, sperm nuclei, pollen grain, and recognition protein. How does each of these structures function in fertilization?

8. Outline the events of fertilization. What does *double fertilization* mean? As fertilization becomes complete, what structures in the flower are haploid? Diploid? Triploid?

9. What is the endosperm? What parts of the male gametophyte contribute directly to the embryo after fertilization?

10. What is the basal cell? What is the suspensor? How are these structures formed? How do they function during embryonic development?

11. Define protoderm, procambium, and ground meristem. To what tissues do these embryonic cells give rise during later stages of development?

12. How do cell differentiation, elongation, and differentiation operate during growth of the *Capsella* embryo?

13. Define heart stage, torpedo stage, cotyledons, shoot apical meristem, and root apical meristem. What is a meristem?

14. What is a seed? Trace the events in seed formation in *Capsella*. Define endosperm, seed coats, and fruit. What is the origin of each of these structures?

15. Define monocot and dicot. How do monocotyledonous and dicotyledonous seeds differ?

16. What effects do the rate and direction of cell division have in plant development?

17. How is the arrangement of cellulose microfibrils in cell walls related to the direction of cell elongation in plant development?

18. What factors operate to break the dormancy of seeds? What events occur during seed germination?

19. What are the zones of elongation and differentiation in growing stems and roots?

20. Trace the events in the development of primary phloem and xylem from the time cells are first laid down at the stem tip. How does the development of vascular tissue differ in monocots and dicots?

21. Define phloem, xylem, vessel element, vessel tube, sieve cell, sieve plate, sieve tube, and companion cell. What is the function of each of these elements in vascular tissue?

22. Trace the development of a leaf at the stem tip. Define epidermis, palisade parenchyma, and spongy parenchyma with respect to leaf structure.

23. How does secondary vascular tissue develop in dicots? What are lateral meristems? How do lateral meristems increase the diameter of stems in dicots?

24. Define vascular cambium and cork cambium. Trace the steps occurring in development of the vascular cambium.

25. What is bark? What are the components of bark? What tissues give rise to bark in dicots?

26. Identify the five plant hormones and list their major developmental effects.

27. How do changes in the relative concentrations of the plant hormones affect plant cell division, elongation, and differentiation?

28. How is auxin related to the responses of plants to light and gravity? Define phototropism and geotropism.

29. How do concentration gradients of the plant hormones affect plant development? What is the relationship between hormone concentration and cell position in plant development?

30. Outline an experiment demonstrating how plant hormones interact to direct cell differentiation.

Suggestions for Further Reading

Browder, L. W. 1980. *Developmental Biology.* Saunders College/ Holt, Rinehart & Winston, Philadelphia.

Jacobs, W. P. 1979. *Plant Hormones and Plant Development.* Cambridge University Press, New York.

Jensen, W. A., and F. B. Salisbury. 1972. *Botany, an Ecological Approach.* Wadsworth, Belmont, California.

Rost, T. L., et al. 1979. *Botany.* Wiley, New York.

Salisbury, F. B., and C. W. Ross. 1978. *Plant Physiology.* 2nd ed. Wadsworth, Belmont, California.

Yeoman, M. M. 1976. *Cell Division in Higher Plants.* Academic Press, New York.

The processes of animal and plant development described in the preceding chapters of this unit illustrate how the basic mechanisms of embryonic growth operate in the two kingdoms of living organisms. Animals develop through the coordinated activities of cells in division, movement, selective adhesion, induction, and differentiation. Plants share two of these activities, cell division and differentiation, as primary developmental mechanisms, and also rely on directional cell elongation for their development. Induction occurs only to a limited extent in plants. In both animals and plants, these processes convert the fertilized egg into the adult through a series of elegantly timed and programmed steps.

The processes of development convert the fertilized egg into the adult through a series of elegantly timed and programmed steps.

The basis for this seeming miracle of development is only partially known despite more than a century of intensive scientific effort. Only one of the developmental mechanisms, cell movement, is reasonably well understood today. There are still fundamental gaps in our knowledge of the cellular and molecular basis for cell division, selective adhesion, elongation, and differentiation. The mechanisms underlying induction remain as baffling today as at the time of their first discovery early in this century.

Interest in these developmental processes extends far beyond the field of embryology, because development touches all branches of biology and has practical significance beyond the world of science. Complete understanding of the processes of development, if this is ever possible, would almost certainly suggest the means for correcting or preventing birth defects in humans. It is also possible that regeneration of lost or damaged organs and tissues, or even whole limbs, could be induced if we fully understood the developmental mechanisms producing them in the embryo in the first place. (In contrast to many other animals, humans have very limited abilities to regenerate lost parts.) And if we could work out the mechanisms of differentiation, perhaps we could learn how to control cancer cells, which override the regulation imposed in development and revert to uncontrolled movement and growth.

This chapter outlines the present state of our knowledge of the cellular and molecular basis of the developmental mechanisms. (The probable basis for cell elongation in plants is described in Chapter 15.) The extent of the problems remaining to be solved in developmental biology will become obvious as the discussion unfolds.

16

THE CELLULAR AND MOLECULAR BASIS OF DEVELOPMENT

CELL DIVISION IN ANIMAL AND PLANT DEVELOPMENT

Cell Division in Development

The events underlying cell division have been actively researched for nearly a century. While there are still many questions concerning the molecular basis of mitotic cell division, the cellular mechanisms involved have been understood since about 1900 (see Chapter 10). Two features of mitotic cell division are currently of special interest in developmental biology: its *direction* and *rate*. The first of these involves the position taken by the plane of cytoplasmic division that separates the parent cell into daughter cells after mitotic division of the nucleus. This plane determines the direction in which new cells are laid down by division. This feature of cell division is precisely controlled in both plants and animals and, particularly in early embryos, is one of the most significant factors determining the outcome of embryonic development. The second feature of major interest is the mechanism controlling the *rate* at which cells proceed through the cell cycle in different parts of the embryo. In both plants and animals, changes in the frequency of division significantly affect the final shape and size of the organ systems of the developing embryo.

The plane of cell division influences the distribution of cytoplasmic information and also the size and shape of embryonic structures in both plant and animal development.

Control of the Direction of Cytoplasmic Division

The Cleavage Furrow in Animals Cytoplasmic division in animals is governed by the cleavage furrow, a depression that forms around the margins of the cell at the surface and deepens until the parent cell cytoplasm has been pinched off into two parts, each containing a daughter nucleus (see Figure 10-15). The location of the furrow is critical to embryonic development in many animals, especially in the stages immediately following fertilization of the egg. In these animals the egg cytoplasm is organized into separate regions that give rise after cleavage to specific parts of the embryo (see p. 296). The early cleavage divisions, which proceed in a coordinated manner, cut the cytoplasm of the fertilized egg into daughter cells that have highly specialized developmental fates. In some animals, for example, the cells eventually giving rise to gametes are compartmented off during the first or second division of the

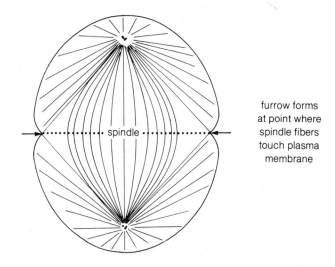

furrow forms at point where spindle fibers touch plasma membrane

Figure 16-1 The furrow (arrows) dividing the cytoplasm in animal cells penetrates into the cell along a plane that follows the spindle midpoint (dotted line).

fertilized egg. If the plane of initial cell divisions is experimentally disturbed in these highly organized eggs, the embryo fails to develop normally.

The cellular mechanisms underlying cleavage in animal embryos, as in most cell movements, depend on the coordinated activities of microtubules and microfilaments. The exact position of the furrow is determined by the location of the spindle, a collection of microtubules that separates the chromosomes in mitosis (see p. 203 for details). The deepening furrow that divides the cytoplasm between daughter cells always penetrates through the widest part of the spindle at its midpoint, following a plane at right angles to the spindle axis (Figure 16-1). Disturbing the position of the spindle in the early stages of mitotic division by centrifuging or pushing the spindle out of place with a microneedle displaces the cleavage furrow to the new location of the spindle midpoint. The relationship between the spindle and the cleavage furrow may depend on long microtubules that radiate from the two ends of the spindle and approach the plasma membrane opposite the spindle midpoint. Interaction between these microtubules and the cytoplasmic region just under the plasma membrane may cause the cleavage furrow to form in this position.

Once the position of the furrow is fixed, the furrow itself forms and deepens through the activity of microfilaments. In 1968, Thomas E. Schroeder of the University of Washington determined that a band of microfilaments extends entirely around the dividing cell at the advancing edge of the furrow (see Figure 10-16). In a mechanism resembling the tightening of a drawstring around the cell, these microfilaments slide over each other to tighten the band and deepen the furrow.

Thus the direction of cell division in animals depends on the coordinated activities of microtubules and microfilaments. Microtubules of the spindle divide the chromosomes and, through an unknown mechanism, interact with the cell boundary to determine the plane to be followed by the cleavage furrow. Once the plane of cytoplasmic division is determined, microfilaments form and deepen the cleavage furrow. Knowing the roles of microtubules and microfilaments in nuclear and cytoplasmic division does not tell the whole story, however. The cellular and molecular events that position the spindle initially, and thus determine the plane of furrowing later in division, remain completely unknown.

Control of Cytoplasmic Division in Plants When plant cells divide, a new cell wall called the **cell plate** grows between the daughter cells resulting from mitosis. The cell plate separating daughter cells is laid down in a plane through the spindle midpoint at right angles to the long axis of the spindle. Its position is thus analogous to the path of the furrow in dividing animal cells (see Figure 10-18 and p. 203).

Since plant cells remain in position once they are laid down by division, the plane of cytoplasmic division is as significant to the outcome of development and as precisely controlled as in animals. One example of this control, described in Chapter 15, is the first division of the fertilized egg in flowering plants (see Figure 15-11). This division produces two cells of unequal size that have widely different fates in embryonic development. The smaller of the two products gives rise to the embryo; the larger gives rise to the suspensor, a supportive structure that acts as a plant "placenta" and supplies nutrients to the developing embryo.

The plane of cell division is fixed in both animals and plants by the position of the mitotic spindle.

As in animals, disturbing the spindle position experimentally before or during formation of the cell plate moves the plane of division to the new position along with the spindle. The connection between the plane of cytoplasmic division and the spindle midpoint seems more direct in plants because the new cell wall is first laid down within the spindle itself and then extends outward to meet the side walls—in reverse of the direction of furrowing in animal cells. As in animals, the cellular and molecular factors that position the spindle initially and thus fix the plane of cytoplasmic division remain unknown.

Cell Cycle Controls in Embryonic Growth

The rate of cell division changes in different tissues as development proceeds in animal and plant embryos. In animals, cell divisions are most rapid as the fertilized egg begins cleavage. During this period, the early embryonic cells spend almost no time in interphase and enter the next cycle of DNA replication and mitosis as soon as the previous mitotic division ends. As the embryo develops and cells differentiate, the time spent in interphase increases and varies in length in different cell types. Some cells, when fully differentiated, remain permanently fixed in interphase and enter into no further replications or divisions.

The size and shape of embryonic structures in both plants and animals is determined also by the rate of cell division.

In the development of some organs in animals, the final form taken by embryonic structures depends on localized regions in which cell division proceeds at higher rates than the surrounding tissues. In the growth of mammalian salivary glands, for example, active cell division at restricted points produces lobelike growths (Figure 16-2a); these regions subsequently break into smaller centers to produce the treelike tissue of the gland (Figure 16-2b).

The fact that plant cells remain in position makes the rate of cell division of special importance in plant development. We have seen that in dicots the cotyledons are generated by two centers of rapid cell division that grow outward from the tip of the embryo (see p. 338). Similarly, localized regions of active cell division at the tips of roots and stems extend these structures and give rise to accessory structures such as branches, leaves, and flowers.

The mechanisms controlling the cell cycle are not well understood in either plants or animals (for details, see Chapter 10). Whatever form these controls take, the first events probably occur just before DNA replication; from this point onward most cells are fixed in the pathway leading to mitosis. Research in this area suggests that the initiation of division is controlled by genes programmed to turn on at specific times in the life cycle of the cell. The activity of these genes, in turn, is modified by factors both in the cytoplasm and outside the cell.

In plants, for example, plant growth hormones are known to modify the rate of cell division. One group of plant hormones in particular, the cytokinins (see p. 355),

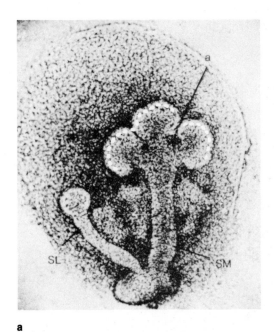
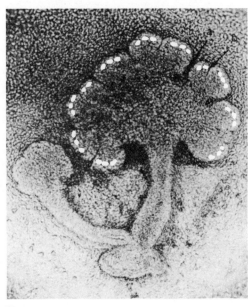

a

b

Figure 16-2 The activity of localized regions in which cell division is more active (dotted lines) produces tree-like lobes within a mouse salivary gland. (**a**) Early stage. (**b**) Later stage in which more lobes have developed. *a* marks the same developing cleft in both pictures; *b* marks a new cleft that has appeared in the right-hand picture. SM, developing submandibular gland; SL, developing sublingual gland. Courtesy of N. K. Wessels.

increases the rate of cell division if present in high concentration in subregions of the embryo. Presumably these hormones act directly or indirectly on the chromosomes of embryonic cells by triggering the genes responsible for initiating DNA synthesis and mitosis.

The factors controlling the rate of cell division are of general interest among biologists and medical scientists as well as embryologists because cells in malignant tumors typically lose their controls and divide rapidly. The growths then block the blood supply to regions of the body and interfere with other vital functions. In this sense, cancer cells return to the rapidly dividing, embryonic condition. The current research attempting to discover the normal controls of the rate of cell division may lead to methods for reestablishing these controls to stop the malignant growth of cancer cells. (The characteristics of cancer cells are outlined further in Supplement 16-1.)

CELL MOVEMENT IN ANIMAL DEVELOPMENT

Cells move actively during the embryonic growth of animals, both singly and as sheets or groups. Moreover, embryonic cells undergo changes in shape that generate

movements such as the infolding of embryonic surface layers to produce endoderm or mesoderm. Recent studies of these cells have revealed that motility of both kinds, both whole cell movements and shape changes, depend on the activities of microtubules and microfilaments.

Embryonic cell movements depend on the activity of microtubules and microfilaments.

Microtubules and Microfilaments in Coordinated Cell Movements

The formation of the neural tube and lens vesicle in amphibians (see Chapter 14) are among the best-studied examples of the coordinated movements of cell layers in generating embryonic structures. The cell movements occurring in the generation of both these organs are quite similar. First the ectoderm thickens and flattens in the region where the neural tube or lens vesicle will form. The flattened area then folds inward, or invaginates, to form the tube or vesicle. The first of these movements, the flattening and thickening of the ectoderm, is due to the activities of microtubules; microfilaments are responsible for the subsequent invagination of the thickened layers.

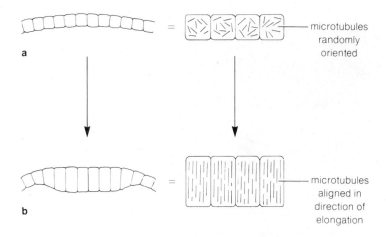

microtubules randomly oriented

microtubules aligned in direction of elongation

Figure 16-3 (**a**) and (**b**) Flattening of the ectoderm to produce the neural plate. Cells of the ectoderm change from cubelike to a tall, elongated shape during neural plate formation.

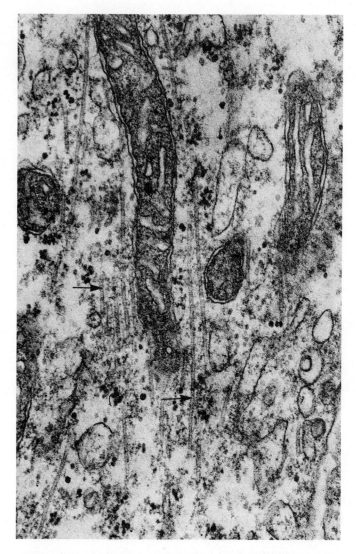

Figure 16-4 Parallel alignment of microtubules (arrows) in the direction of elongation in a flattening ectoderm cell. ×65,000. Courtesy of M. B. Burnside.

Microtubules and Flattening of the Ectoderm When ectoderm flattens and thickens before the nerve tube and lens vesicle form, the cells in the ectoderm layer change from a cubelike shape to tall columns (Figure 16-3). In 1964, Breck Byers and Keith R. Porter, then at Harvard University, discovered that the cytoplasm of these elongating cells contains masses of microtubules aligned in the direction of elongation (Figure 16-4). These microtubules are believed to slide actively over each other, producing a force that pushes the ends of the cells farther apart and elongating them. Other investigators have since tested this idea by injecting colchicine (a chemical that interferes with the assembly of microtubules) into developing embryos. The colchicine disrupts the microtubules and inhibits the cell elongation that produces the flattening of the neural plate region of the ectoderm. This is precisely the result to be expected if this cell movement results from the activity of microtubules.

Microfilaments and Invagination of the Ectoderm After the ectoderm thickens, it folds inward to form the neural tube or lens vesicle. This inward folding of the ectoderm results from a change in cell shape from columnar to wedgelike (Figure 16-5). As the tops of the cells in the ectoderm constrict and narrow, the entire cell layer is forced inward as if a drawstring is being tightened along the top surface of the ectoderm. In 1970, Schroeder found that each wedge-shaped cell contains a group of microfilaments arranged in a circle at the narrow tip of the wedge

(Figure 16-6). Cytochalasin B, a drug that interferes with the activity of microfilaments, dispersed the microfilament circle at the top of the cells and stopped the invagination of the ectoderm. Evidently the microfilaments of the circle slide over each other, tightening the circle and narrowing the top of the cell.

Thus, thickened through the activity of microtubules, the ectoderm of the developing neural tube or lens vesicle folds inward through the activity of microfilaments. After the initial investigations of Byers, Porter, and Schroeder pointed the way, many additional embryonic movements

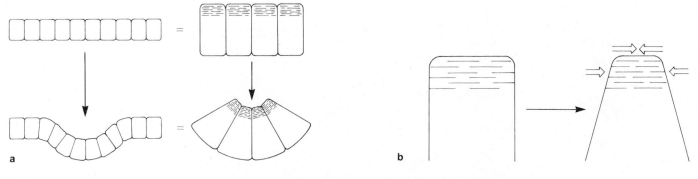

Figure 16-5 Inward folding of ectoderm in neural tube and lens vesicle formation. (**a**) The ectoderm cells change from columnar to wedge-shaped during the invagination. The shape change results from contraction of a microfilament band at the tops of the cells (see text). (**b**) How sliding microfilaments progressively overlap to produce the contraction.

of cell layers were discovered to depend on the activities of microtubules and microfilaments acting either singly or in coordination.

Microtubules and Microfilaments in Individual Cell Movements

Many embryonic movements involve movements of single cells. Among the most spectacular examples of these individual movements are the migrations of a cell type called *neural crest cells* in vertebrate embryos such as amphibians, birds, and humans. During formation of the neural tube these cells arise from the region where the closing tops of the tube join the ectoderm (Figure 16-7). As the neural tube closes, the neural crest cells are released into the body cavity. From the body cavity they then migrate over relatively long distances, following specific routes to reach distant points in the developing embryo. Some migrate to the head, where they form nerves communicating between the brain and parts of the head; others contribute to parts of the brain itself or to cartilage of facial structures. Others migrate to form nerve cells leading from the spinal cord to body structures or to form nerves of the developing gut. Still others move to the skin, where they form pigment cells.

Individual cell movements of this type also involve the activities of microtubules and microfilaments. Cells grown outside the body in a tissue culture medium illustrate how both structures coordinate to produce movement. A cell forms an attachment to the substrate, in this case the glass or plastic surface of the culture vessel (Figure 16-8a), and moves forward by elongating from the point of attachment (Figure 16-8b). A new attachment is then made at the advancing tip (Figure 16-8c). The cell subsequently contracts, producing tension until the rearmost attachment is broken (Figure 16-8d). The front attachment is then used as the base for another elongation that extends the cell body in the direction of movement (Figure 16-8e).

Electron microscopy of such cells shows that the cytoplasm contains both microfilaments and microtubules (Figure 16-9). The microfilaments form a network at the cell surface; the microtubules lie in deeper layers of the cytoplasm, where they are oriented more or less in the direction of movement. Although the results are somewhat controversial, treatments with colchicine and cytochalasin B indicate that microtubules and microfilaments coordinate to produce the movement. Colchicine interferes with the elongation of the cell in a forward direction from an attachment point (the movement shown in Figure 16-8b), indicating that this part of the movement is produced by sliding microtubules. Cytochalasin B, in contrast, inhibits the contraction that places tension on the cell and breaks the rearmost attachment (as in Figure 16-8d). Therefore this segment of the movement is apparently due to sliding microfilaments.

This pattern of movement observed in cultured cells, involving attachment, elongation, and contraction, is also believed to occur inside embryos. Moreover, single embryonic cells have also been observed to move by simply extending lobes of cytoplasm and flowing actively into them. This pattern of freely flowing motion depends primarily on the activities of microfilament networks girdling the cells just under the cell surface. These microfilaments, by sliding over each other, are considered to constrict the cell and squeeze the fluid cytoplasm into the advancing lobe, much like squeezing a tube of toothpaste.

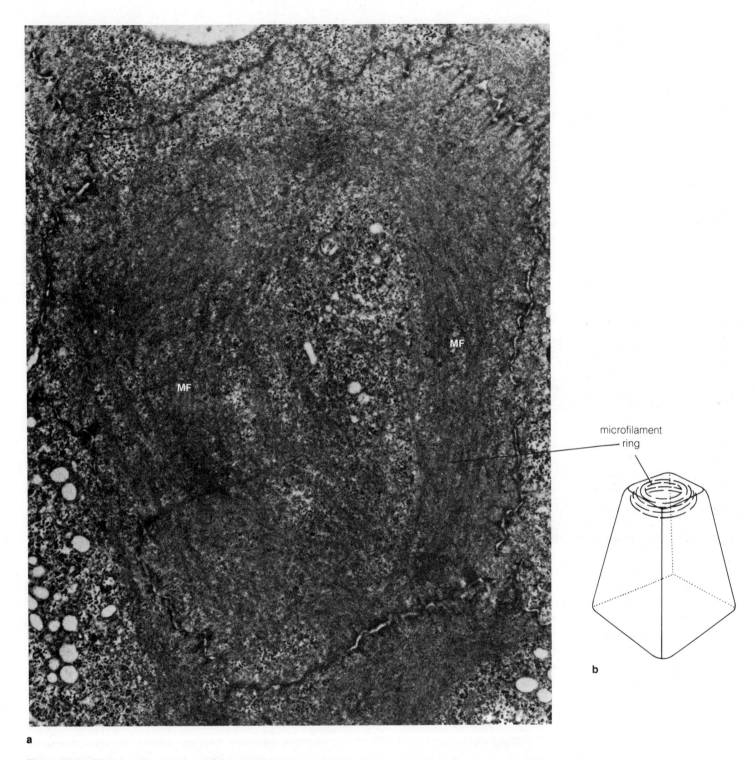

microfilament
ring

b

a

Figure 16-6 The microfilament ring *(MF)* constricting the tops of the cells in the ectoderm invaginating to form the neural tube in the amphibian *Xenopus.* ×20,500. Courtesy of P. C. Baker, T. E. Schroeder, and Academic Press, Inc.

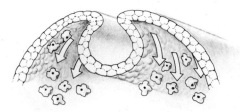

Figure 16-7 Neural crest cells arise from the top margins of the developing neural tube and migrate relatively long distances to reach their final locations in the embryo.

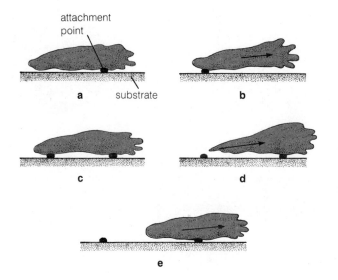

Figure 16-8 One pattern by which single cells move in tissue culture vessels and, presumably, also in embryos (see text).

Plant cells, as noted in Chapter 15, do not move either as individuals or as cell layers during embryonic development. Whole plant regions may undergo movements, however, as in the curvature of stems away from the center of the earth or toward light. Movements of this kind in plants are produced by unequal cell elongation on two sides of the stem instead of microtubule or microfilament activity. As the cells on the side opposite the direction of curvature become longer, they push the stem away from gravity or toward light. These movements are controlled through differences in the concentrations of plant growth hormones on the two sides of the stem (for details, see p. 352).

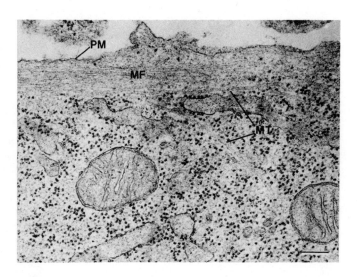

Figure 16-9 Microtubules *(MT)* and microfilaments *(MF)* in the cytoplasm of a cell migrating in the pattern shown in Figure 16-8. Courtesy of G. L. Nicolson, from *Biochimica et Biophysica Acta* 457:57 (1976).

SELECTIVE CELL ADHESION IN ANIMAL DEVELOPMENT

Cell movement in animal development is closely related to selective cell adhesion. As development proceeds in animals, some cells remain fixed in position by cell-to-cell adhesions. Others break their adhesions and move. Even during movement, cells progress by making and breaking adhesions. The development of these adhesions is selective; embryonic cells may make temporary adhesions at some points while in motion and later, after recognizing their final stations in the embryo, make permanently fixed attachments (as in primary mesenchyme movement and adhesion in sea urchin embryos—see p. 305). The final, cell-to-cell adhesions hold the embryo in its correct shape and form. Once made, these final attachments are maintained throughout the adult life of the organism; only a few normal cells in adults, such as the white blood cells (leukocytes) of vertebrate animals, retain the capacity to break their adhesions and move.

Cells make temporary adhesions as they move and make permanent adhesions as they reach their final locations in developing embryos.

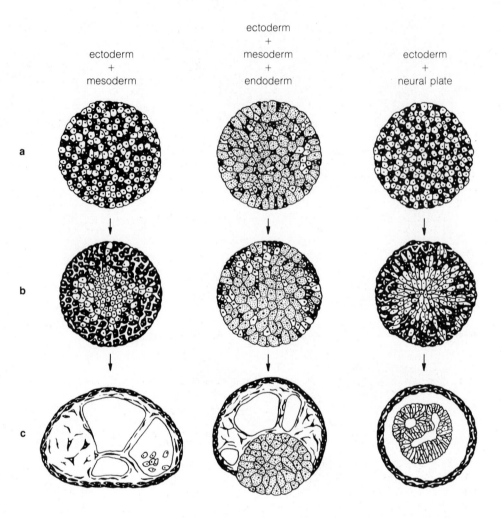

ectoderm
+
mesoderm

ectoderm
+
mesoderm
+
endoderm

ectoderm
+
neural plate

a

b

c

Figure 16-10 Cell migrations sorting out tissue layers in segments removed from amphibian embryos. (**a**) Ball containing randomly mixed cells; (**b**) and (**c**) sorting out to form gastrula-like arrangements in which the cells occupy the relative positions they would normally occupy in the embryo. Redrawn from an original courtesy of J. Holtfreter.

The selective nature of cell adhesions was first demonstrated in a classic experiment of developmental biology carried out in 1955 by Johannes Holtfreter and his colleagues at the University of Rochester. Holtfreter dissected a living frog embryo into separate tissue slices of ectoderm, mesoderm, and endoderm and then placed the slices together in various combinations in a culture medium. The cells could be followed in a microscope because frog ectoderm, mesoderm, and endoderm are pigmented differently. Initially the cells moved from the layers to form a ball in which the cell types were mixed together more or less randomly (Figure 16-10a). After a few days, the cells sorted themselves out and moved into gastrula-like arrangements resembling their normal locations in the embryo (Figure 16-10b and c). For example, the cells in slices of ectoderm and mesoderm, when placed together, sorted out into a ball with ectoderm on the outside and mesoderm on the inside (first vertical column, Figure 16-10). In

the most remarkable experiment of this type, Holtfreter arranged a sandwich of layers lined up against totally incorrect neighbors: mesoderm and endoderm constituted the "bread" on the outside of the sandwich, and ectoderm was the "meat" inside. In this case the cells broke loose from the sandwich and sorted themselves into a gastrula-like arrangement with endoderm on the inside, mesoderm in the middle, and ectoderm covering most of the outside surface (second vertical column in Figure 16-10).

Other experiments by Holtfreter's group showed that cells of the same primary tissue can also sort into aggregates in which their proper relationships with neighbors are reestablished. In one of these experiments, ectodermal cells destined to form a neural tube were mixed with ectodermal cells from another region that would normally give rise to the epidermis or outer skin of the embryo. These cells, when mixed together, sorted out into a ball with the future epidermal cells on the outside and the future neural tube cells on the inside—just as they would be located in a normal embryo (third vertical column, Figure 16-10).

These results suggested that cells contain recognition and aggregation factors, probably located on the cell surface, that enable them to identify their correct neighbors in the embryo and, once in position, to adhere and remain in place. Later experiments by others showed that this is in fact the case and, moreover, that the recognition molecules are probably glycoproteins.

Embryonic cells probably contain cell surface recognition and adhesion glycoproteins that enable them to make temporary and permanent cell adhesions.

Aron Moscona and his coworkers at the University of Chicago found that cells separated from the retina of the chick eye can sort out and reform into a retinal tissue when mixed with other cell types. Further, the cells reformed in patterns similar to the results obtained by Holtfreter with frog tissues. This ability to reaggregate was destroyed if the surfaces of the cells were treated with trypsin, an enzyme that breaks down proteins. The ability of the trypsin-treated cells to reaggregate was restored if a factor released from the surfaces of normal retinal cells was added to the culture medium. This factor, when isolated and purified, was identified as a glycoprotein. Moscona's findings are supported by other experiments using enzymes to probe the ability of cells to recognize each other and adhere in specific combinations. Generally this ability is destroyed by exposing cells to enzymes that catalyze the breakdown of either protein or carbohydrate groups at the cell surface.

Thus the factors giving embryonic cells the ability to recognize each other and specifically adhere are evidently glycoprotein molecules at the cell surface (see p. 89). These molecules consist of a protein base, embedded in the plasma membrane, bearing a carbohydrate "antenna" that extends outward from the cell surface (Figure 16-11). Since the carbohydrate portion can be made in an almost endless variety of types, depending on the particular sugars present and the patterns in which they are combined, glycoproteins provide a way for each cell type to label its surface with specific recognition and adhesion factors. Presumably these factors are made by each cell type at certain stages in embryonic development in accord with timed directions coded into the genes. Depending on the glycoprotein types present, individual cells may adhere to other cells or release from their adhesions to migrate and seek new combinations. Under this mechanism, for example, all ectoderm cells would possess the same recognition and

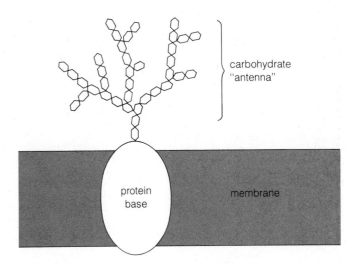

Figure 16-11 A membrane glycoprotein molecule with its carbohydrate "antenna" directed away from the cell surface. Differences in the monosaccharide subunits of the antenna unit provide each cell type with specific recognition and aggregation groups.

aggregation glycoproteins at their surfaces early in gastrulation (Figure 16-12a). Later, when the neural tube begins to develop, cells destined to become parts of the neural tube would lose surface glycoproteins of the ectoderm type and synthesize new glycoproteins that (1) identify them as neural tube cells and (2) allow them to adhere to other neural tube cells with the same surface markers, but not to ectoderm cells (Figure 16-12b). Glycoprotein molecules at cell surfaces probably also mark the routes to be followed by cells, such as the neural crest cells, that migrate over long distances in the embryo.

Migrating cells probably follow "directions" coded into cell surface glycoproteins in their movements within developing embryos.

Selective cell adhesion, like cell division, is of general interest in biology and medical science because cancer cells characteristically break loose from their specific cell adhesions to lodge and grow elsewhere in the body where their presence blocks vital functions. (For details, see Supplement 16-1.) Research into the basic embryonic mechanism allowing normal cells to form and keep adhesions may not only provide clues to the reasons why cancer cells break these adhesions but also suggest methods for preventing the migration of malignant cells.

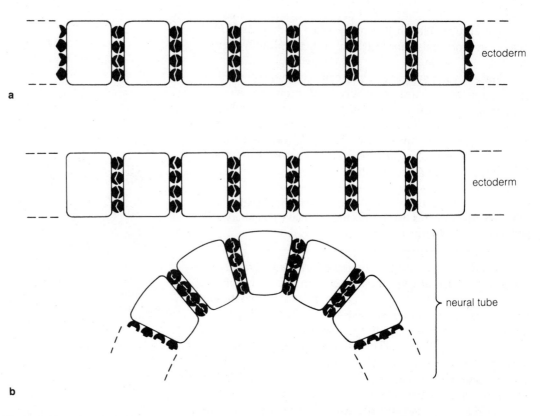

Figure 16-12 Specific adhesions of cells by surface recognition and binding groups. (**a**) Early in gastrulation, all ectoderm cells contain identical surface recognition groups. (**b**) Later, cells of the developing neural tube lose their ectoderm groups and develop new recognition and adhesion groups characteristic of the neural tube.

INDUCTION IN EMBRYONIC DEVELOPMENT

Embryonic induction is a process in which one tissue stimulates another to develop along a prescribed pathway. Inductions occur repeatedly in the pathways developing the organ systems of animals. The formation of the neural tube and eye in amphibians, one of the foremost examples of induction in animal embryos, is described in Chapter 14. In this series of inductions, the mesoderm first induces the overlying ectoderm to differentiate into the neural tube. The brain, developing from the anterior end of the tube, then induces ectoderm in the head region to differentiate into the lens vesicle. Once formed, the lens vesicle in turn induces the overlying ectoderm to differentiate into the cornea.

Inductions also take place in plant development, although to a much more limited extent than in animals. Chapter 15 outlines one induction occurring in the devel-

opment of the vascular cambrium leading to secondary growth in dicot stems: residual embryonic cells within the vascular bundles laid down by the first year's growth induce nearby cells of the parenchyma to divide and form actively dividing cambial tissue (see Figure 15-27).

Close proximity between the inducing cells and the responding tissue is a constant feature of the process in these and all other known inductions in animal and plant embryos. If close association between the two cell types is prevented experimentally, induction is blocked and normal development does not occur.

Although the fact of induction has been amply demonstrated by experiments with a wide variety of animal embryos, the cellular and molecular mechanisms underlying the process, and even some points concerning its fundamental nature, are still unclear. These uncertainties can be summarized in three questions. First, does the inducing tissue tell the responding cells how to develop or does induction merely trigger a preprogrammed sequence in the induced cells? The two alternatives to this question are sometimes described as the **instructive** ("do this") versus the **permissive** ("go ahead") hypotheses of induction. Second, does the close proximity between the inducing and responding cells observed as a constant feature of the process mean that actual cell-to-cell contact is required for induction to take place? This question is closely related to the third question: Does instruction or permission travel from the inducing to the responding cells in the form of informational molecules, or is contact between the two cell types sufficient to pass the information?

In instructive induction, the inducer "tells" the reactant what to do; in permissive induction, the inducer triggers a preprogrammed pathway in the reactant.

Perhaps the easiest way to summarize the progress toward answering these questions is to trace briefly the major developments in research into embryonic induction. Induction was first discovered early in this century by a German scientist, Hans Spemann, one of the founders of modern developmental biology. Spemann studied the developmental processes leading to nerve tube formation in amphibians; his experiments were the first to demonstrate that close association with the underlying mesoderm is necessary for the ectoderm along the top margin of frog embryos to develop into the neural plate. Spemann assumed that the ectoderm receives information from the mesoderm—possibly in the form of molecules transferred from the inducing cells—that induces it to develop into the neural plate. Thus he assumed that induction is instructive and that an inducing molecule passes from the inducer to the responding tissue.

Spemann's hypothesis touched off a long search for the inducing molecules that has extended, without success, to the present day. A wide variety of substances extracted from embryos have been thought at one time or another to be inducer molecules until further work showed them to be unrelated to the process. A complicating factor in this work is the fact that many substances, some of them artificial products never found in embryos, can mimic inducers and apparently instruct certain reactant cells to develop along their normal pathways. These difficulties gradually led to the hypothesis that perhaps there are no inducer substances after all. It may be that the signals for induction are passed from the inducer to the responding tissue in some other manner, possibly by direct cell-to-cell contact. In this case, the instructions would be passed in the form of surface groups on the inducing cells that are read by the responding cells during the period of contact. Which of these alternative hypotheses is correct—that is, whether induction involves the passage of infor-

mational molecules or cell-to-cell contact (or both)—remains a matter of extensive controversy and research in developmental biology.

Induction is probably both instructive and permissive. That is, the inducing tissue chooses one of several pathways that are preprogrammed into the reactant tissue.

Is induction permissive or instructive? Recent work suggests that neither alternative is exclusively correct and that inductions probably involve both instructive and permissive elements. Some of the best evidence for this conclusion comes from experiments following the induction of ectoderm by mesoderm to form surface structures such as hair or feathers. If mesoderm derived from the leg region of a chick embryo is grafted under ectoderm from the wing region, it induces the wing ectoderm to form leg feathers instead of wing feathers (Figure 16-13a). Mesoderm from a region of the embryo that forms feathers, if placed under ectoderm that remains featherless, induces the normally featherless ectoderm to grow feathers (Figure 16-13b). If placed under ectoderm from a normally feathered region, mesoderm from a leg region that forms scales will induce scales to form in the normally feathered ectoderm (Figure 16-13c). In this system, the mesoderm obviously "tells" the ectoderm whether to form wing or leg feathers, or whether to form feathers or scales, and is instructive to this extent.

Other experiments using mesoderm and ectoderm derived from different species show that this same system also contains permissive elements; that is, the reacting ectoderm cells already possess several preprogrammed pathways that lead either to leg feathers or wing feathers or scales. Mesoderm derived from the mouse, for example, if placed under chick ectoderm, induces the chick ectoderm to form feathers (Figure 16-14). This result shows that the chick ectoderm cells already know how to make feathers; it is unlikely, after all, that mouse mesoderm can relay instructions for manufacturing feathers instead of hair. The reverse combinations also work: mesoderm derived from a chick embryo can induce mouse mesoderm to form hair.

Thus the mesoderm–ectoderm systems studied in these experiments probably involve both instructive and permissive processes. They are instructive to the extent that mesoderm chooses which developmental pathway the reacting ectoderm will follow; they are permissive to the extent that the different pathways are already programmed into the ectoderm. Most embryonic inductions probably involve both instructive and permissive elements in this way.

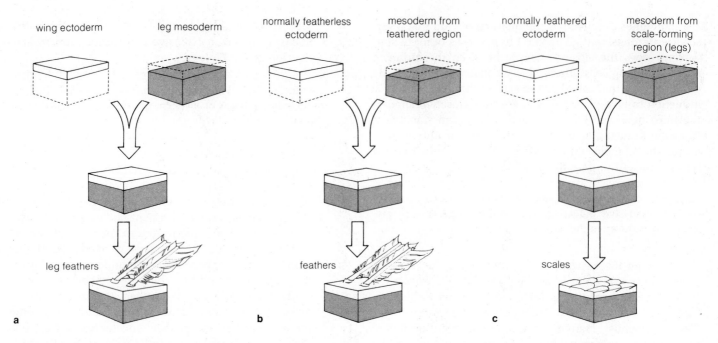

wing ectoderm   leg mesoderm

normally featherless ectoderm   mesoderm from feathered region

normally feathered ectoderm   mesoderm from scale-forming region (legs)

leg feathers

feathers

scales

a   b   c

Figure 16-13 Induction of ectodermal structures by mesoderm (see text).

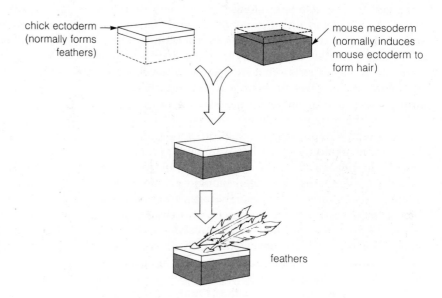

chick ectoderm (normally forms feathers)

mouse mesoderm (normally induces mouse ectoderm to form hair)

feathers

Figure 16-14 Induction of chick ectoderm by mouse mesoderm (see text).

CELL DIFFERENTIATION

The result of embryonic development in plants and animals is the adult organism, composed of cells that are specialized in structure and biochemistry to carry out the different tasks necessary for life. This specialization is accomplished by differentiation of basic embryonic cells into the various cell types of the adult. As developing cells make the transition to fully differentiated types, their developmental potential and biochemical activity become restricted. Embryonic epidermal cells in mammals, for example, may give rise through division and differentiation to skin, hair, nerve, and lens cells—all recognizably different cell types. Once these cells begin to differentiate, however, they become more restricted in potential. Cells of the developing lens give rise only to lens cells through division and further differentiation; they no longer have the developmental potential to differentiate into other types, such as hair cells. At the same time, their biochemical activity becomes restricted to the production of relatively few molecules. Lens cells, for example, concentrate their synthetic machinery on the production of the lens protein, crystallin, which makes up 80 to 90 percent of the total protein synthesized. Other proteins characteristic of epidermal cells, such as keratin, are not assembled in lens cells.

The Characteristics of Cell Differentiation

Differentiation occurs in the development of both plants and animals without any addition or subtraction of genes. With few exceptions, all the cells of an individual retain the same basic genetic information, no matter how restricted or specialized they become as development proceeds. Thus the changes that occur in differentiation are superimposed on the same basic directions contained in the genes of all the organism's cells. Somehow, in differentiation, certain directions are permanently filed away while others are retained in active form for immediate use.

Differentiation occurs without changes in the genetic information of cell nuclei.

Several experiments have demonstrated that the full complement of genes is retained in differentiated cells. In animals, evidence for this conclusion was developed in the 1960s by J. B. Gurdon of Cambridge University through his experiments with nuclear transplantation (Figure 16-15). In his experiments, oocytes of the frog *Xenopus* were irradiated with ultraviolet light to destroy the egg nucleus; the ultraviolet light also activated the egg cyto-

plasm and stimulated respiration and protein synthesis to begin. Nuclei were then removed from a fully differentiated *Xenopus* tissue (from the layer of cells lining the intestine) and injected into the activated egg cytoplasm. A significant percentage of the eggs containing the transplanted intestinal cell nuclei subsequently developed normally into tadpoles and adult frogs, confirming that the nuclei in the fully differentiated intestinal cells retain their full genetic capacity.

Equivalent experiments have also been carried out with plants. During the 1950s, F. C. Steward and his associates at Cornell University developed the experimental techniques in which plant cells are cultured to form callus tissue (see Figure 15-31a). In some of their experiments, cells separated from a callus grown from carrot phloem cells entered a series of divisions to form embryo-like structures (Figure 16-16). Growth of these "embryoids" in a medium containing the correct ratios of the plant hormones auxin and cytokinin led to the generation of normal carrot plants. Subsequent experiments have produced complete plants from differentiated cells of buds, leaves, and stems of tobacco, petunia, and asparagus plants.

Although these experiments demonstrate that the nuclei of fully differentiated cells retain their full complement of genes and the capacity to dedifferentiate to an embryonic state, cells retained in position in adult plants and animals do not normally do so. The differentiated state is normally permanent, and the genetic controls imposed during development remain in effect throughout the life of the adult in both dividing and nondividing tissues.

The Molecular Basis of Differentiation

At the molecular level, differentiation ultimately involves regulation of the activity of genes in RNA transcription. There is ample evidence that transcription is in fact regulated during development. Some of the best evidence comes from Gurdon's nuclear transplantation experiments. In 1969, Gurdon transplanted nuclei from amphibian embryos at the gastrula stage into activated eggs in which the egg nucleus had been destroyed by ultraviolet light. He then followed the levels of RNA synthesis occurring in the transplanted nuclei. When the nuclei were transferred, they were rapidly transcribing RNA of all types. Within 20 minutes after injection into the eggs, the gastrula nuclei stopped all detectable RNA transcription. Genes in the transplanted nuclei were then turned on again according to the normal schedule for RNA synthesis in the embryo. As the egg receiving the transplanted nucleus reached the late cleavage stages before blastula formation, transfer RNA synthesis began; at late gastrulation,

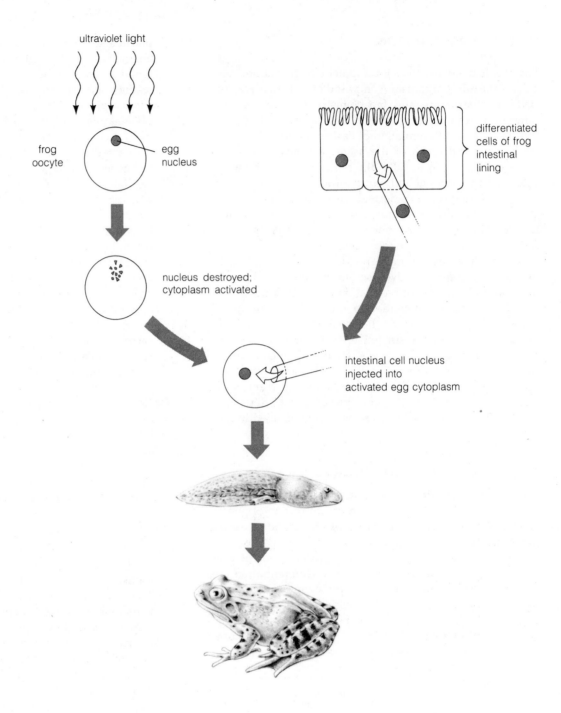

ultraviolet light

frog
oocyte

egg
nucleus

nucleus destroyed;
cytoplasm activated

differentiated
cells of frog
intestinal
lining

intestinal cell nucleus
injected into
activated egg cytoplasm

Figure 16-15 Gurdon's nuclear transplantation experiment (see text).

ribosomal RNA transcription could be detected. These experiments clearly demonstrate that genes can be turned on and off in an ordered sequence and, incidentally, that cytoplasmic factors can change the activity of genes in the nucleus. Exposing the late gastrula-stage nucleus to the egg cytoplasm almost immediately turned off the genes of the nucleus and imposed the normal embryonic schedule on them.

Another experimental system, using chromosomes of the larvae of flies such as *Drosophila*, has given direct visual evidence of transcriptional regulation. The chromosomes used in this system (Figure 16-17) are formed in

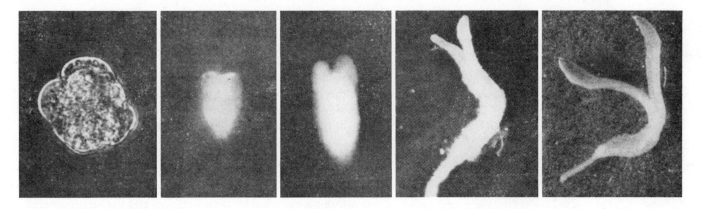

Figure 16-16 Formation of embryo-like structures ("embryoids") from carrot root cells grown in tissue culture. Subsequent growth of embryoids in culture fluids containing the correct ratios of the plant growth hormones auxin and cytokinin produced normal plants. Courtesy of F. C. Steward, from *Science 143*:20 (1964).

nuclei in certain tissues of fly larvae, such as the salivary glands, that undergo repeated replication without mitosis or cytoplasmic division. The repeated replication produces chromosomes that contain thousands of DNA molecules, all lined up side by side in a long cable of chromatin fibers.

These **polytene** (*poly* = many; *tene* = thread) chromosomes become thick enough to be visible in interphase cells under the light microscope. Each chromosome appears as a cross-banded structure in which dark crossbands alternate with light interband spaces. Various lines of evidence have shown that the dark bands contain the DNA of a single gene repeated many times and condensed into the tightly folded structures that form the bands. At points along the polytene chromosomes, one or more bands unfold to form a structure known as a *puff* (arrow, Figure 16-17). These puffs have been shown to represent the many copies of the gene of a band, all uncoiled and single genes active in the synthesis of mRNA. Inactive genes remain coiled into the tightly packed bands. Thus inactive genes appear as bands and active genes as puffs. (Figure 16-18 shows the relationship between bands and puffs.)

Tracing the appearance of polytene chromosomes under the light microscope shows that as development proceeds, puffs appear and disappear in definite patterns (Figure 16-19). Since the puffs are sites of mRNA synthesis, this evidence demonstrates directly that the transcription of individual genes is regulated during development and, moreover, that genes are turned on and off in sequence as differentiation proceeds.

The molecular mechanism directly turning genes on and off in cell differentiation probably depends on the histone and nonhistone chromosomal proteins. Some of the experiments supporting this conclusion, and the hypotheses derived from them, are discussed briefly in Chapter 9. Briefly, these experiments reveal that the histones, when added to DNA in a cell-free system containing all the factors required for RNA transcription, inhibit RNA synthesis in a general way; there is no selectivity for specific genes.

Transcriptional controls are imposed by the histone and nonhistone proteins of the chromosomes.

The nonhistone chromosomal proteins, in contrast, control transcription in a specific way. When added to a DNA–histone combination in a cell-free system in which RNA synthesis has been generally inhibited by the presence of the histones, the nonhistones can activate specific genes and open them for RNA transcription. The genes turned on are in this case related to the cell type from which the nonhistone proteins have been extracted. Nonhistone proteins isolated from liver cells, for example, activate genes typical of liver cells when added to DNA and histones in a cell-free system (see Figure 10-18). These experiments, therefore, indicate that the histones have a general effect in controlling the availability of large blocks of genes but do not specify which genes within a block are to be transcribed. Specific control is exerted by the nonhistone proteins, which are able to bind to specific genes within the available blocks and open them to the RNA polymerase enzymes that transcribe RNA.

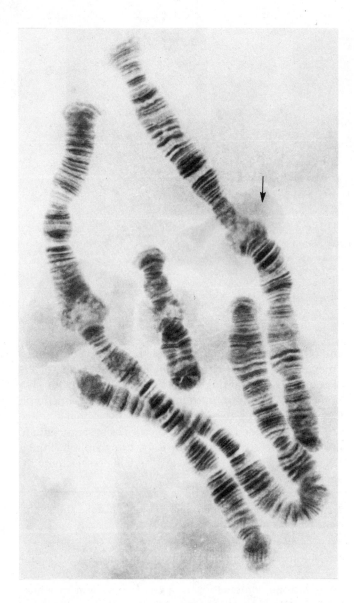

Figure 16-17 Polytene chromosomes from a salivary gland cell of a larva of the fly *Chironomus*. Several large puffs are visible in the chromosomes (one of these is indicated by the arrow). Courtesy of W. Beermann and Springer-Verlag, from *Protoplasmatalogia* 6D (1962).

These transcriptional controls interact in a highly ordered way to produce the changes we recognize as differentiation in embryonic development. Whatever the combinations of genes operating from the time of egg formation and fertilization through the rest of development, the program for the controls rests ultimately on the information coded into the nucleus and stored in the cytoplasm of the fertilized egg. This information determines which of the uncounted species of plants or animals the egg will develop into and, within the species, determines all distinguishing characteristics such as sex, size, and color. Even the time of sexual maturity, middle age, and old age—although modified by circumstances and the environment—is programmed into the fertilized egg.

The cellular mechanisms discussed in this unit are responsible for the growth and development of all eukaryotic organisms. Animals, as we have seen, develop through the interactions of five basic mechanisms: cell division, movement, selective adhesion, induction, and differentiation. Plants share cell division and differentiation and, to a limited extent, induction with animals, but they add another process, cell elongation, as a major developmental mechanism of their own. These cellular processes, working together, faithfully reproduce the members of each successive generation of sexually reproducing organisms.

Questions

1. What basic developmental mechanisms operate in animals? In plants? Give one specific example illustrating each mechanism in animals and plants.

2. How is the plane of cytoplasmic division important in animal development? In plant development? How does the spindle influence the plane of cytoplasmic division?

3. How are the activities of microtubules and microfilaments coordinated in cell division in animals?

4. How is the rate of cell division important in animal development? In plant development?

5. What kinds of cellular movements occur in animal development? How do microtubules and microfilaments operate to produce these movements? How is the sliding action of microtubules and microfilaments related to these movements?

6. Outline the roles of microfilaments and microtubules in the development of the neural tube in amphibians. What parts of these developmental movements would be inhibited by the drug colchicine? By cytochalasin B?

7. How do microtubules and microfilaments coordinate in the movements of single cells?

8. Outline one hypothesis that may account for the ability of migrating cells to find their way in developing embryos. Are such cell movements limited entirely to embryonic cells?

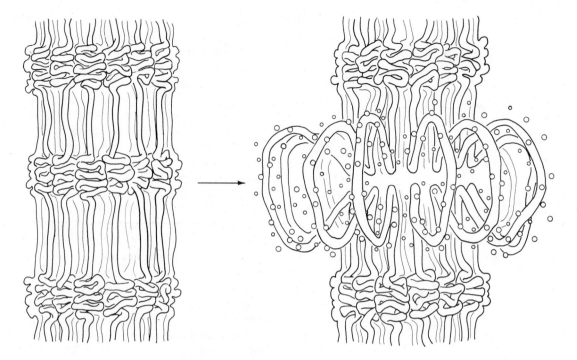

Figure 16-18 How the chromatin fibers of the bands unwind to form puffs active in RNA transcription in polytene chromosomes.

9. What is the relationship between cell adhesion and cell movement in animal development?

10. Outline an experiment demonstrating the ability of embryonic cells to make selective adhesions.

11. Define glycoproteins. Outline two lines of experimental evidence supporting the hypothesis that surface glycoproteins may be responsible for cell adhesion.

12. Define induction. What is the difference between permissive and instructive induction?

13. Chick leg mesoderm taken from a region that normally forms scales, if placed under wing ectoderm, induces the wing ectoderm to form scales. In what way is this induction instructive? In what way is it permissive?

14. What evidence indicates that the differentiated cells of animals retain all their genes? What about the differentiated cells of plants?

15. What is transcriptional regulation? How is it related to differentiation?

16. Outline one experiment showing that transcriptional regulation occurs as part of development.

17. What are polytene chromosomes? Define band and puff. How does the observation of polytene chromosomes support the hypothesis that transcriptional regulation occurs as part of development?

18. How are the histone and nonhistone proteins believed to control transcription?

Suggestions for Further Reading

Browder, L. W. 1980. *Developmental Biology.* Saunders College/ Holt, Rinehart & Winston, Philadelphia.

DeRobertis, E. M., and J. B. Gurdon. 1979. "Gene Transplantation and the Analysis of Development." *Scientific American* 241:74–82.

Grant, P. 1978. *Biology of Developing Systems.* Holt, Rinehart & Winston, New York.

Maclean, N. 1977. *The Differentiation of Cells.* Edward Arnold, London.

Oppenheimer, S. B. 1980. *Introduction to Embryonic Development.* Allyn & Bacon, Boston.

Stein, G. S., J. S. Stein, and L. J. Kleinsmith. 1975. "Chromosomal Proteins and Gene Regulation." *Scientific American* 232:46–57.

Wessells, N. K. 1977. *Tissue Interactions and Development.* Benjamin, Menlo Park, California.

Wolpert, L. 1978. "Pattern Formation in Biological Development." *Scientific American* 239:154–164.

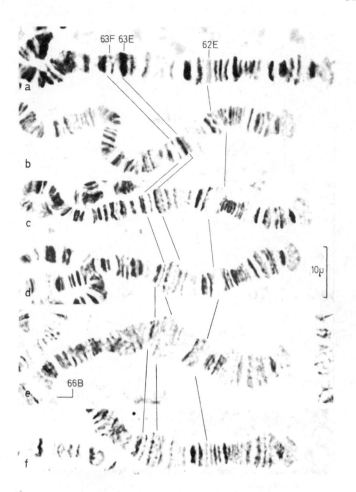

Figure 16-19 The appearance and disappearance of puffs at different points in a polytene chromosome as development proceeds in the fly *Drosophila*. Band 63F forms no puffs during successive developmental stages in the sequence shown in (**a** to **f**). Band 63 enlarges and forms a puff in (**c**), (**d**), and (**e**) and retracts slightly in (**f**). Band 62E puffs in (**b**) and (**c**) and retracts in (**d** to **f**). The lines connect the same bands in (**a** to **f**). Courtesy of M. Ashburner and Springer-Verlag, from *Chromosoma* 38:255 (1972).

SUPPLEMENT 16-1:
CANCER: THE REVERSAL OF DEVELOPMENT

Once development is complete, differentiated cells normally remain as specialized types. Cell division in normal tissues either does not occur at all or it is limited to the amount required to replace cells lost by normal wear and tear. In some individuals, however, groups of cells may partially dedifferentiate, initiate more rapid division, and break loose and move to new locations within the body. As they grow, these cells form cell masses called **tumors**. Tumors are common in vertebrate animals and plants; for unknown reasons, they are infrequent in invertebrates.

As tumors grow they may block the blood supply to normal tissues or organs, interfere with the function of nearby structures, or cause pain or inflammation by displacing normal tissues. Since some large tumors are insufficiently supplied with blood, cell masses within them die and produce lesions that will not heal. Tumors with these characteristics are *malignant;* unless eliminated or reduced in size they lead to impairment or death of the individual carrying them. Not all tumors are malignant, however. Nonmalignant or *benign* tumors, as they are called, include small, slowly growing cell masses such as warts and moles that do not become large enough to impair body functions.

Malignant tumors share several common characteristics. Apart from their characteristic uncontrolled cell division and partial dedifferentiation, malignant tumor cells are *invasive.* They tend to detach easily from the tumor masses, move through spaces between normal cells, and spread to other parts of the body. Frequently the invasive characteristic of tumor cells leads to their penetration through the walls of small blood vessels and into the bloodstream. This entry into the bloodstream, called **metastasis**, carries the tumor cells to all parts of the body, particularly the lungs, where they may lodge to initiate additional growths. Another common characteristic of malignant tumor cells related to their invasiveness is an alteration of the cell surface. Some of the cell surface groups associated with normal cell recognition and adhesion are lost, and new surface groups may appear.

Since dedifferentiation of tumor cells is typically only partial, the malignant cells retain certain characteristics of their tissue of origin. As a result, tumors are as varied in cell type as the tissues of the body. Although the ability to make certain proteins and other molecules characteristic of the tissue of origin is frequently lost, others, such as pigments, hormones, and structural proteins or enzymes, are still produced. According to these retained characteristics, tumors may be identified as **carcinomas** derived from epithelial tissues, **sarcomas** from connective tissue, **melanomas** from pigment cells, **lymphomas** and **leukemias** from blood cells, and **teratomas** from the primary germ cells that normally give rise to eggs and sperm.

Although the molecular and cellular events that change normal cells into cancer cells are unknown, they are believed to involve changes in the DNA sequences of genes or changes in the chromosomal proteins that regulate them. Certainly many of the agents known to promote cancer are also **mutagenic**—that is, they can induce alterations in the nucleotide sequences of DNA. These agents include certain kinds of irradiation, such as x rays or ultraviolet light, and a variety of mutagenic chemicals. The genetic changes induced by these substances may alter or destroy the normal regulatory mechanisms that control cell division, motility, or the recognition and adhesion groups at the cell surface that keep normal cells in place in their tissues.

A variety of additional agents have been identified as cancer-causing or **carcinogenic** because of the increased incidence of cancerous tumors among groups of people exposed to them, such as smokers or workers in certain industries. These agents include benzene, vinyl chloride, asbestos fibers, coal dust, soot, the tars of smoke, compounds of cadmium, and nitrosamines. When such agents are suspected to cause cancer, they are tested by exposing rats or mice or cultured cells to them. Although the production of cancer in test animals does not prove that the agents cause cancer in humans, the chances are good because the metabolism and reactions of different mammalian species are basically similar.

Viruses too have been found to cause certain kinds of cancer in birds and mice. Although virus particles have also been found in human tumor cells, such as those of cervical tumors of the uterus, it has not been possible to determine whether the viruses actually cause the tumor to develop or are simply present by coincidence. Cells of human cervical tumors, for example, frequently contain particles of a virus known as *herpes* that commonly infects humans. So many persons are infected with herpes, however, that the virus is frequently found in both normal and tumor cells. Thus it is difficult to establish a direct cause-and-effect relationship between the virus and cancer. The situation is further complicated by the fact that humans cannot be experimentally infected with viruses or cancer cells to test the possible connections between a given virus and cancer types.

Effective treatment for cancer depends on early recognition and surgical removal of a malignant tumor. Tumor cells can also be killed by freezing or by radiation such as x rays if the tumor is localized and accessible. If extensive growth and metastasis has occurred, so that these treatments are no longer effective, chemicals that kill or reduce the growth of the tumor are applied. This technique, called **chemotherapy**, involves the use of chemicals that interfere with DNA, RNA, protein synthesis, or respiration in the tumor cells. Although these agents also affect normal cells, they often have greater effects on tumors because DNA replication, RNA and protein synthesis, and general cell metabolism generally proceed at greater rates in cancer cells. Thus malignant cells are more likely to incorporate large quantities of the chemicals and be killed. Chemotherapy therefore amounts to establishing a balance that kills the tumor but not the patient. Despite the difficulty of establishing this balance, chemotherapy has been effective in curing or reducing the growth of a variety of cancers, including leukemia, Hodgkin's disease, and, to some extent, breast and intestinal tumors.

The Diversity of Life

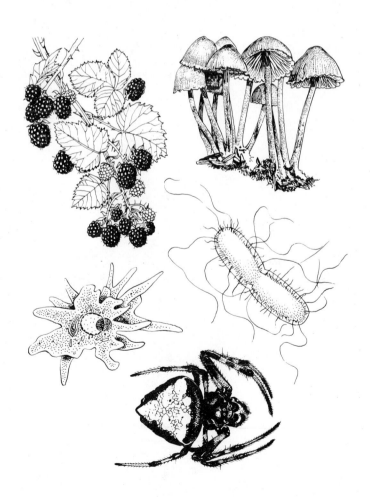

17

CLASSIFICATION AND THE DIVERSITY OF LIFE

The study and classification of the organisms of the earth have been central human interests since ancient times. One of the earliest schemes for classifying animals into logical groups according to similarities and differences was proposed by Aristotle no less than 2000 years ago. Since that time, many modifications and new proposals for classifying living things have been advanced. None of these systems, however, has ever succeeded in assigning all animals, plants, and microorganisms to unambiguous places in a completely logical hierarchy. As a result, efforts to develop an all-inclusive and perfect classification system continue to the present day.

The driving force behind this effort stems from several sources. Biologists classify living things to make information about the world's organisms easier to store, retrieve, and communicate. Further, classification of organisms past and present often reveals relationships between different ancient and modern groups while suggesting pathways through which the present-day species might have evolved. In biochemical and physiological work, moreover, the known characteristics of different plant and animal groups often become a basis for further hypotheses and experiments. Finally, biologists classify organisms simply through curiosity and the desire to learn and identify as many plants and animals as possible.

SYSTEMATIC CLASSIFICATION

The aim of contemporary biological classification (also called **systematics** or **taxonomy,** from *taxis* = arrangement) is to arrange living organisms into categories that are based not on artificial criteria but on natural similarities and differences. In the past, organisms were often classified according to artificial systems; during the Middle Ages, for example, plants were often grouped according to their uses as medicines. Utilitarian systems of this type, however, assign organisms to categories that are not likely to reflect their relationships in nature. The use of natural criteria in classification—items such as coloration, geographic distribution, and anatomical and physiological characteristics—is complemented by an attitude among biologists that a classification system should also be based on evolutionary pathways. That is, organisms placed in the same category should share the same evolutionary origins.

In practice, the goal of producing natural rather than artificial systems is met by using as many separate characteristics as possible when describing organisms for classification. As a result, all possible categories of information are used, including morphology, gross and microscopic anatomy, biochemistry, embryology, molecular biology, behavior, and distribution. If the number of different characteristics used for identification is sufficiently large, the classification system is likely to reveal evolutionary lineages as a matter of course.

The Basic Unit of Classification: The Species

All classification systems are built on the **species** as a basic unit. Most of us understand intuitively what a species is. That is, we all recognize that cats and dogs, for example, are separate and distinct kinds of animals, that any cat chosen at random will possess certain features found in all other cats, and that normally cats interbreed only among themselves and not with dogs or other animals. Trivial as they might seem, these common observations also form the basis for the most widely accepted scientific meaning of the word species. In these terms, a biological species is defined as a group of organisms that resemble each other more closely than the organisms of any other group and, moreover, differ from the organisms of any other group in at least one clearly defined characteristic. Ideally, the members of a species interbreed only among themselves to produce fertile offspring.

A species is a group of organisms that resemble each other more closely than the organisms of any other group and differ from the organisms of any other group in at least one clearly defined characteristic. Ideally, the members of a species interbreed only among themselves to produce fertile offspring.

This definition of a species can be applied to most organisms of the world as the basic unit of classification on which higher and more inclusive categories are built. There are difficulties in applying the concept to all living things, however, and some groups cannot be clearly identified as separate and distinct species according to the basic parts of the definition. Some groups of animals or plants merge so subtly with one another through a continuous series of intermediate forms, for example, that no clear boundary allows separate species to be identified. Other organisms reproduce primarily or exclusively by asexual means, making the test for successful interbreeding difficult if not impossible. Further, members of different species can sometimes interbreed successfully to produce fertile offspring, making the interbreeding test meaningless in this case.

In spite of these difficulties with exceptional groups, the species concept and definition can be applied with success to most organisms and, for the most part, the species serves as a workable starting point for classification. For the exceptions—when different organisms cannot be separated into distinct species because intermediate types merge between them or when tests of breeding success are impossible—the taxonomist simply sets arbitrary limits to the species with the hope that further work with additional characteristics will reveal the natural dividing lines.

How Species Are Described and Named

Until the time of Charles Darwin, the originator of the modern theory of evolution, most scientists and laymen alike believed that living organisms were fixed entities existing in exactly the forms placed on earth by God during creation. Change or evolution of a species over time was not considered possible. This attitude was reflected in the way in which a species was described: by using single examples that were considered to be the absolute standard for the group. Although differences among the individuals of a species were known to occur, these variations were thought to be imperfections—that is, failures to achieve the ideal species type. Darwin's careful observations of nature, however, established that variability is the rule among individuals. The members of a species collectively show such a wide range of characteristics, in fact, that it is impossible to describe them in terms of a single ideal type. In our own species, for example, different individuals vary over wide ranges according to skin, eye, and hair color, height, weight, and a host of other characteristics. There is no ideal or absolute type.

As a consequence of these observations, biologists realized that the variations in different characteristics must be included in order to describe a species accurately. This realization led to the description of species collectively as **populations** rather than as ideal individuals, an approach that underlies modern classification. Studying populations rather than single individuals is also the fundamental approach of contemporary evolution and ecology (discussed in Units Seven and Eight).

Species are described by averaging the collective characteristics of a population of individuals.

After they have been described in terms of the characteristics of populations, species are named according to a system first used extensively in the eighteenth century by the Swedish naturalist Carolus Linnaeus. Linnaeus gave each species a double name with two parts roughly analogous to our first and last names. Our own species, for example, is given the two-word identifier *Homo sapiens* (Latin for "wise man"). *Homo*, the first part of the name (equivalent to a last name like *Smith*), indicates the **genus** to which a species belongs. (A genus is a group of closely related species.) The second word of the name, *sapiens* (equivalent to a first name like *John*), identifies a particular species within a genus. This two-word **binomial nomenclature,** as it is called, is used universally to describe a species. The same two-word names are used for the same species in all languages and all parts of the world.

The names used for species follow rules set up by international congresses that meet every few years. Usually derived from Latin, the names either reflect structural, geographical, or historical information about the species or commemorate famous scientists, important personages, or the discoverer of the species.[1]

Groups Above the Species

The present classification system arranges species into higher groupings that are progressively more inclusive. Ideally, all the organisms placed together in each group should have common evolutionary origins. Species that are closely related are grouped together in a **genus** (plural = *genera*). Genera with similar characteristics and origins are grouped into **families**. In turn, families are grouped into **orders**, orders into **classes**, and classes into **phyla** (singular = *phylum*). **Division** is used instead of phylum for the major groups of the plant kingdom. Each of these groups can be subdivided if the organisms included are sufficiently varied (see Table 17-1). Our own species looks like this when classified into all its major categories and subdivisions:

KINGDOM: *Animalia* (animals)
Subkingdom: *Metazoa* (many-celled animals)
  PHYLUM: *Chordata* (axial body support, dorsal nerve cord, gill slits at some stage of life)
  Subphylum: *Vertebrata* (jaws and paired limbs)
  Superclass: *Tetrapoda* (two pairs of limbs, lungs, bony skeleton)
    CLASS: *Mammalia* (hairy skin, warm-blooded, young suckled)
    Subclass: *Theria* (fetus nourished through placenta or in pouch)
    Infraclass: *Eutheria* (fetus nourished through placenta)

ORDER: *Primates* (each limb with five distinct fingers or toes bearing nails)
Suborder: *Anthropoidea* (upright stance, enlarged cranium)
Superfamily: *Hominoidea* (tailless anthropoidea with no cheek pouches)
  FAMILY: *Homidea* (large brain, flat face, reduced brow ridges, small canine teeth)
  GENUS: *Homo* (teeth evenly sized, grinding molars)
    SPECIES: *sapiens* (toolmakers, thinkers, speakers and organizers)

THE MAJOR KINGDOMS OF LIFE

Many proposals have been advanced for dividing the world's organisms into the most inclusive categories: the **kingdoms.** Traditionally, living things were lumped together as either animals or plants because all known organisms seemed to fall clearly into one of the two categories. By the 1800s, however, investigations into less well known life forms, particularly microorganisms, revealed other groups that could not be clearly identified as either plant or animal. As these organisms were discovered and described, it became obvious that the old two-kingdom concept was inadequate. As a result, the number of major kingdoms has gradually expanded from two to as many as five in modern systems.

Although several systems are currently in use, many biologists now favor a five-kingdom system proposed in 1969 by R. H. Whittaker of Cornell University. Whittaker's system divides past and present organisms according to several broad criteria: whether prokaryotic or eukaryotic, whether unicellular or multicellular, and whether nutrition is by photosynthesis, ingestion, or absorption of organic matter from the surroundings. The five kingdoms set up according to these criteria are clear and distinct: the *Monera* (bacteria and blue-green algae), the *Protista* (protozoa and certain unicellular algae), the *Fungi* (fungi), the *Plantae* (plants), and the *Animalia* (animals). The five kingdoms and their proposed evolutionary relationships are diagrammed in Figure 17-1.

The Kingdom Monera

The kingdom Monera includes the prokaryotic organisms of the earth: the bacteria and blue-green algae. Prokaryotes have relatively simple cell organization with a single surface membrane, the plasma membrane, and limited de-

---

[1]Sometimes taxonomists slip names based on less serious criteria past the international congresses. One invented the genus name *Zyzzyx* for a group of wasps for obvious reasons but also to have the distinction of naming the last genus listed in classification categories. Another named a series of beetles with words created by scrambling the letters in his mistress's name. Still another named a fly in which males have a red-tipped abdomen with the species identifier *balsafyr*.

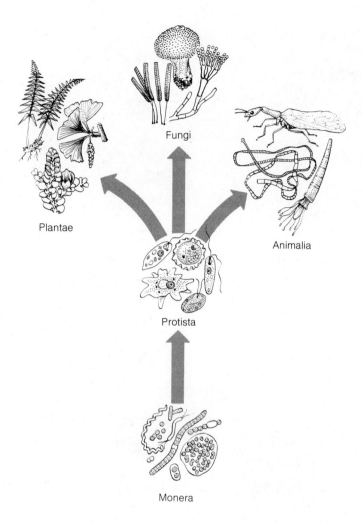

Figure 17-1 Evolutionary lineages among the five kingdoms of organisms.

velopment of internal membranes. No mitochondria, chloroplasts, endoplasmic reticulum, or Golgi complexes are present in the cytoplasm, and no membranes separate the nucleoid from the cytoplasm. The DNA of the nucleus is ''naked''—that is, uncomplexed with histone or nonhistone proteins—and no nucleolus is present. Prokaryotic cells typically have a cell wall surrounding the cell outside the plasma membrane. The Monera are either single-celled or colonial; if colonial, there is virtually no specialization or division of labor among cells of the colonies.

The Monera are all prokaryotic, single-celled organisms.

Classifying the Monera Within the Monera, several characteristics are used to subdivide the kingdom into phyla and lower groups. One of the most important characteristics is the composition of the cell wall. Most of the Monera show distinctive cell wall components and structures that clearly set them off from other members of the kingdom. Other major characteristics include the type of motility and the mode of cellular nutrition.

According to these characteristics, the Monera are subdivided into several phyla. The blue-green algae (phylum Cyanophyta; Figure 17-2) differ from the remaining groups of the Monera in cell wall constituents and in their photosynthetic mechanism, which uses chlorophyll *a*, evolves oxygen as a by-product, and otherwise closely matches photosynthesis in the higher plants (Chapter 6). The remaining phyla of the kingdom Monera, all bacteria,

Table 17-1	Groups Used in Classifying Animals and Plants	
Animals	Plants
Kingdom	*Kingdom*
Subkingdom	Subkingdom
Phylum	*Division*
Subphylum	Subdivision
Superclass	
	Class
Class	Subclass
Subclass	
Infraclass	*Order*
Superorder	Suborder
Order	*Family*
Suborder	Subfamily
Superfamily	Tribe
	Subtribe
Family	
Subfamily	*Genus*
Supergenus	Subgenus
	Section
Genus	Subsection
Subgenus	
	Species
Species	Subspecies
Subspecies	Variety
	Form

Figure 17-2 A blue-green alga, *Spirulina*. ×37,400. Courtesy of N. J. Lang.

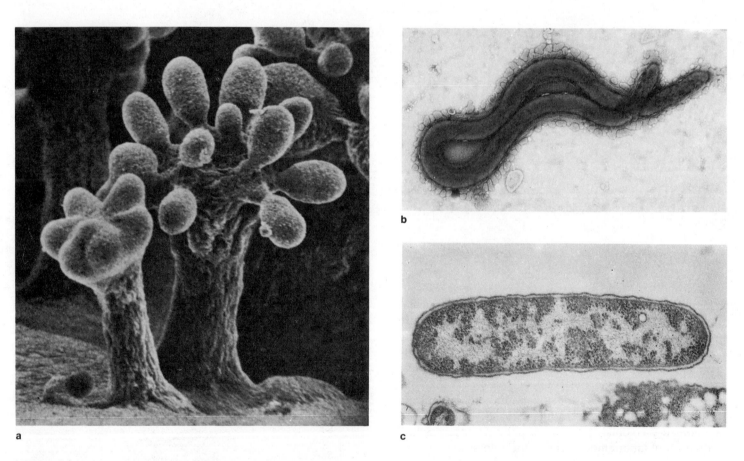

a

b

c

Figure 17-3 Representative bacteria. (**a**) Scanning electron micrograph of the reproductive structures of a myxobacterium. ×600. Courtesy of J. Pangborn. (**b**) A spirochete. ×57,000. Courtesy of M. A. Listgarten and the American Society for Microbiology. (**c**) A eubacterium, *Pseudomonas*. ×12,700. Courtesy of J. Pangborn.

are classified primarily according to differences in cell wall structure and type of movement. Members of the phylum Myxobacteriae (Figure 17-3a) have thin, flexible cell walls and move by gliding over a substrate. The Spirochaetae (Figure 17-3b) have thin cell walls and helical shapes; these bacteria move by means of a set of fibrils embedded in the flexible cell wall. Members of the Eubacteriae (Figure 17-3c), the largest phylum in the kingdom, have thick walls and are either nonmotile or move by means of one or more thin, hairlike fibers, the bacterial flagella (see Figure 4-7), that extend outward through the cell wall. The Eubacteriae include all the photosynthetic bacteria. The two groups of photosynthetic bacteria within the Eubacteriae, the green and purple bacteria, use **bacteriochlorophylls** rather than chlorophylls as their major photosynthetic pigments. In contrast to the Cyanophyta or blue-green algae they possess only part of the photosynthetic mechanisms of the plants. In particular, the photosynthetic bacteria cannot use water as a source of electrons for photosynthesis and do not release oxygen as a by-product.

Viruses and the Kingdom Monera   Viruses in the free form (see Figure 4-22) are noncellular complexes consisting of a nucleic acid core (either DNA or RNA) surrounded by a protective coat of protein. Some of the more complex viruses infecting animal cells also have a lipid layer surrounding the protein coat. Viruses are about 20 to 100 nanometers long; at this size, they are of molecular rather than cellular dimensions. The smallest viruses are smaller than ribosomes or the largest protein molecules. With these characteristics of structure and size, viruses do not fit conveniently into the Monera or any other kingdom of living organisms.

Viruses differ from all living organisms in other fundamental ways. None of the systems characteristic of life, such as motility, synthesis, growth, or oxidative reactions producing ATP, are present or active in virus particles. The genes coded into viral nucleic acids are replicated or transcribed only if the virus infects a host cell. These characteristics suggest that virus particles may be the evolutionary remains of a group of parasitic organisms, probably prokaryotic like the Monera, that have lost all cell structure except the list of coded directions for making their nucleic acid molecule and its protective protein coat. (For details, see p. 69.)

Viruses are alive only when they take over the chemical machinery of a host cell.

The Kingdom Protista

The organisms of the kingdom Protista—and all other living things except the Monera—have eukaryotic cell structure. A true nucleus with nucleolus and nuclear envelope is present, and the cytoplasm contains mitochondria. In some Protista and the Plantae, the cytoplasm also contains chloroplasts. Flagella, when present, contain the 9 + 2 system of microtubules characteristic of eukaryotic flagella or a variant of this system. Cell division occurs by mitosis or meiosis.

The Protista are eukaryotic organisms, primarily single-celled, that live by photosynthesis, absorption, or ingestion.

The eukaryotes of the kingdom Protista are either single-celled or colonial. If colonial, there is practically no cell specialization in the colonies. The Protista include two major groups: the single-celled algae and the protozoa. The Protista are divided into these two groups by mode of nutrition. The algal protistans carry out photosynthesis using chlorophyll a and other pigments in a complete mechanism that uses water as electron donor and releases oxygen as a by-product. The protozoa, by contrast, are nonphotosynthetic and live by ingesting organic matter or absorbing it from their surroundings. Almost all protozoa are motile at some stage of their life cycles.

The algal protistans (Figure 17-4) are further subdivided according to their photosynthetic pigments, the chemical components of their cell walls, and the types of organic molecules they store as food reserves. Among the important algal phyla of the kingdom Protista are the Euglenophyta (Figure 17-4a), single-celled algae with one or two flagella and chloroplasts. The photosynthetic pigments present in the Euglenophyta, chlorophylls a and b and several carotenoids, resemble the pigments of the green algae and higher plants. Some Euglenophyta have evolved into forms that have lost their pigments and are colorless; these organisms feed by ingesting organic matter. The Chrysophyta (the diatoms; Figure 17-4b and d) have cell walls consisting of two segments that overlap like the lid and bottom of a candy box; photosynthetic pigments include chlorophyll a and an abundance of carotenoids, which mask the chlorophyll and give the Chrysophyta their characteristic golden color. The final important phylum of protistan algae, the Pyrrophyta or dinoflagellates (Figure 17-4c), are golden-brown algae in which the chlorophyll a and c pigments present are masked by yellow and brownish carotenoids. The Pyrrophyta typically have two flagella and complex cell walls formed from several overlapping plates. Although separated from the surrounding cytoplasm by a nuclear envelope, the Pyrro-

a

b

c

d

phyta nucleus contains chromosomes that closely resemble prokaryote nucleoids (Figure 17-5). None of the chromosomal proteins typical of eukaryotes—the histone and nonhistone proteins—are present. These structural resemblances to the prokaryotes have led to the hypothesis that the Pyrrophyta are descendants of very primitive eukaryotes that once formed a link between the prokaryotes and eukaryotes.

The protozoan members of the kingdom Protista (Figure 17-6) are subdivided into phyla primarily on the basis of their mode of locomotion. The Zoomastigina (Figure 17-6a) move by means of one or more flagella. The Sarcodina (Figure 17-6b) move by extending footlike lobes of cytoplasm called **pseudopodia**. The Sporozoa (Figure 17-6c) are parasitic protozoa that are either nonmotile or move by gliding over the substrate. The final phylum of protozoa, the Ciliophora (Figure 17-6d), move by means of numerous short flagella called **cilia**. The Ciliophora are also distinguished from all other organisms by having two different kinds of nuclei in each cell: a **micronucleus**, which serves as the storage site for genetic information and functions only in mitosis and meiosis, and a **macronucleus**, which is the site of RNA transcription for cell synthesis and growth.

The Kingdom Fungi

The members of the kingdom Fungi are nonmotile organisms that live by absorbing organic matter from their surroundings. Although often plantlike in appearance, they have no photosynthetic pigments. The basic body structure of the Fungi is a microscopic filament called the **hypha** (plural = *hyphae*; from *hypha* = web). These filaments (Figure 17-7), although of small diameter, may be many millimeters or even centimeters in length. In many fungi, the hyphal filaments contain large numbers of nuclei, all suspended in a common cytoplasm without intervening cell walls or membranes. The walls of the hyphal tube may contain either cellulose or a hard polysaccharide material, **chitin,** or both substances. (Chitin also occurs as a protective external covering in animals such as insects.) Hyphae branch and interweave in masses to form the body structure of the fungus: the **mycelium** (*mykes* = fungus).

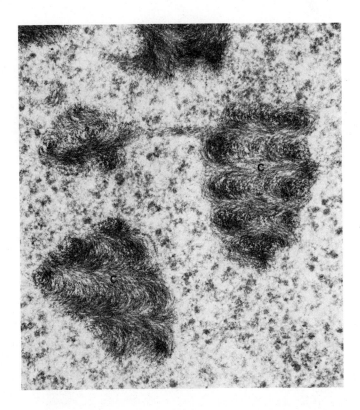

Figure 17-5 Nuclear chromatin bodies (C) of the dinoflagellate *Gyrodinium.* The bodies, which are the individual chromosomes of the dinoflagellate, closely resemble bacterial nucleoids in structure. ×47,500. Courtesy of D. F. Kubai and The Rockefeller University Press.

The Fungi are many-celled organisms that live by absorbing organic matter from their surroundings.

There is little differentiation or specialization of the mycelial filaments except for reproductive structures. Typically the main body mass of the fungus is embedded directly in the organic matter being absorbed as food; only the reproductive structures are exposed. The familiar

Figure 17-4 Representative algae of the kingdom Protista. (**a**) Euglenophyta *(Colacium).* N, nucleus; C, chloroplasts; M, mitochondria; V, vacuole; F, flagella (in cross section). ×10,000. Courtesy of P. Krugens. (**b**) Chrysophyta *(Navicula);* a dividing cell. N, nucleus; Nu, nucleolus; C, chloroplast; M, mitochondrion; R, ribosomes; Go, Golgi complex. ×30,000. Courtesy of M. L. Chiappino and B. E. Volcani. (**c**) Pyrrophyta; the dinoflagellate *Peridinium.* ×1,600. Courtesy of M. Ricard. (**d**) Chrysophyta; the shell of a diatom as seen in the scanning electron microscope. ×3,700. Courtesy of H. Paerl.

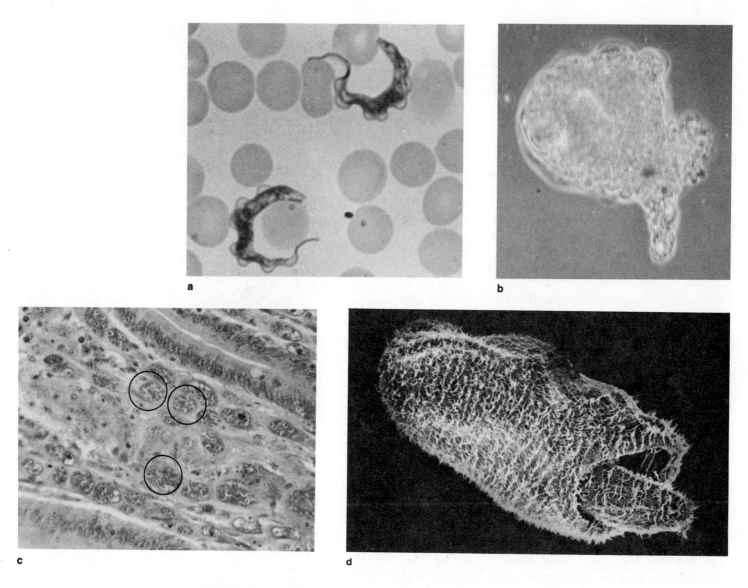

Figure 17-6 Representative protozoans of the kingdom Protista. (**a**) Zoomastigina; *Trypanosoma equiperdum*, a blood parasite of horses and donkeys. ×1,700. (**b**) Sarcodina; *Amoeba proteus.* ×350. (**c**) Sporozoa; *Eimeria tenella* cells (circled) living in the tissue of chick intestine. (**d**) Ciliophora. The ciliate shown in this scanning electron micrograph, *Woodruffia*, is ingesting another ciliate, *Paramecium.* ×300. *(d) courtesy of T. K. Golder.*

mushrooms and toadstools, for example, are only the reproductive structures of a much larger mycelial mass embedded in organic matter in the soil. The different groups of Fungi are classified on the basis of the morphology and type of reproductive structures produced by the mycelium. The kingdom Fungi includes the yeasts, slime molds, other molds, and mushrooms. (Figure 17-8 shows some representative forms.)

The Kingdom Plantae

The organisms classified together into the kingdom Plantae are all many-celled forms that live by photosynthesis. This mode of nutrition is associated with several characteristics common to all plants. Every plant has many cells that contain chloroplasts and carry out photosynthesis using chlorophyll *a* in combination with other chlorophylls and carotenoids as light-absorbing pigments. All plants are nonmotile except for the flagellated male gametes or sperm cells that occur in many types. This immobility is reflected in the structure of plant cells, which have rigid cell walls that support individual cells and larger parts of the plant.

The Plantae are many-celled organisms that live by photosynthesis.

The bodies of almost all plants are differentiated into three major regions that perform functions necessary for efficient photosynthesis: (1) leaves or leaflike structures specialized as photosynthetic organs, (2) stems or stemlike structures that hold the photosynthetic organs in positions favorable to receive light, and (3) roots or holdfasts that anchor the plant firmly to the substrate. In more advanced plants, roots are also specialized for absorbing water and minerals from the soil. In these higher plants the roots, stems, and leaves are connected by vascular tissues that transport water and the products of photosynthesis. Since plants are immobile, their bodies are structured to provide maximum contact with the environment. Roots and stems are highly branched, for example, and leaves or leaflike structures are flattened to expose the greatest possible surface area to the sun.

The pattern of reproduction in plants reflects their immobile life-style. The male gametes produced in the reproductive organs of plants are either actively motile and capable of bridging the gap between parents by swimming or, in the most highly evolved plants, specialized as pollen to be carried passively between plants by water, winds, insects, or other means.

The structures of plants reflect their immobile life-style: plant parts are shaped and arranged to provide the maximum surface area in contact with the environment.

Classifying the Plantae

Photosynthetic Pigments and Products Photosynthetic pigments and food storage molecules are much used in classifying the algae of the kingdom Plantae, which are distinguished from the protistan algae by their pigments and their primarily many-celled structure. Various algal plants have different combinations of chlorophylls *a*, *b*, and *c* together with carotenoids and sometimes other pigments. In the different groups of algal plants, food storage molecules, which are usually starches of various kinds, are arranged in different ways into granules of characteristic

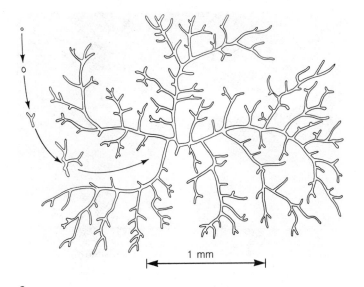

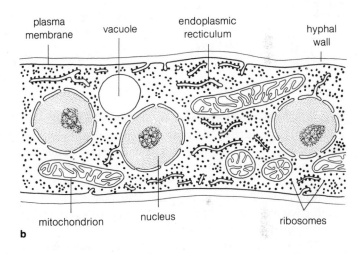

Figure 17-7 Hyphae of a fungus. (**a**) A web of hyphae forming a mycelium. (**b**) Structures present in a fungal hypha. Redrawn from *The Biology of Fungi*, courtesy of C. T. Ingold and Hutchinson Publishing Group Limited.

size and shape. The distribution and kinds of other biochemical compounds, including sugars, fats, oils, alcohols, and proteins, also distinguish subgroups within both the algal and higher plants. Classification according to all these criteria is frequently termed *biochemical taxonomy.*

Vascular Tissues Plants fall naturally into one of two major groups according to the presence or absence of specialized conductive elements: the vascular tissues. One group, including the algal plants and the bryophytes (the liverworts and mosses), has no vascular tissue. The remaining group,

a

b

Figure 17-8 Representative fungi. (**a**) An edible mushroom, *Agaricus*. (**b**) A morel, *Morchella*. (**c**) A slime mold, *Didymium*, growing on agar in a culture dish. Photographs *(a)* and *(b)* courtesy of V. Duran; *(c)* courtesy of O. R. Collins.

which includes the ferns and seed plants, has two types of internal conductive elements: **xylem** and **phloem**. Xylem (see Figure 15-22) is specialized to transport water and inorganic minerals from roots to all parts of the plant. Phloem (see Figure 15-23) transports the products of photosynthesis to all parts of the plant. The plants with these conductive elements, the ferns and seed plants, are collectively termed the **vascular plants.**

The development of vascular tissues allowed the ancestors of higher plants to leave the water and take up a fully terrestrial existence. The nonvascular plants, the algae and bryophytes, are limited to aquatic or very moist terrestrial environments.

Reproductive Patterns and Structures The morphology and function of reproductive organs provides one of the main criteria used to classify the major groups within the vascular plants. Besides showing clear and distinct differences among the various groups of vascular plants, the reproductive parts provide another advantage to taxonomists: they remain relatively constant in structure when plants grow in different environments. In contrast, the vegetative parts of vascular plants—the roots, stems, and leaves—may vary greatly in size, pattern, and morphology under different environmental conditions, making them less useful as criteria for classification.

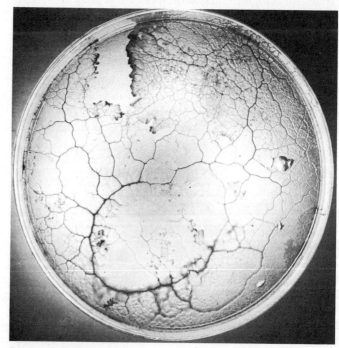

c

Understanding reproductive structures in plants requires some knowledge of an important feature of plant life cycles: the alternation between haploid and diploid phases in successive generations (see p. 329 and Figure 15-1). In this alternation, cells of the diploid generation, called the **sporophyte**, undergo meiosis to produce haploid spores.[2] The spores germinate and grow by mitotic division to produce the alternate, haploid generation of the life cycle, called the **gametophyte**. At some point, cells of the gametophyte differentiate as male and female gametes; fusion of these gametes reestablishes the diploid sporophyte generation.

Plants undergo an alternation between diploid sporophyte and haploid gametophyte generations; the relative prominence of the sporophyte and gametophyte generations varies from division to division in plants.

In some plant groups, the two generations have equal duration in the life cycle: both gametophytes and sporophytes grow as separate, conspicuous plants of equal size and complexity. In other groups, either the gametophyte or sporophyte may form the dominant phase of the life cycle. In either case, the plants of the two generations may be identical in outward appearance or they may differ.

Since variations of these patterns occur among the algal plants, the life cycle is another valuable key to identifying and classifying these groups. In bryophytes, the gametophyte generation is typically dominant and conspicuous whereas the sporophyte is reduced and short-lived. Among the vascular plants, the gametophyte generation is reduced and the dominant, conspicuous phase of the life cycle is the diploid sporophyte. In the most highly evolved plants, the seed plants, which make up the common trees, shrubs, and smaller plants of our environment, the haploid gametophyte is reduced to a microscopic structure containing only a few cells that remains dependent on the sporophyte for its nutrients.

Various features of the life cycle and reproductive organs of the ferns and seed plants are used in their classification. In ferns, both the sporophyte and gametophyte stages live as separate, independent plants (Figure 17-9). The fern gametophyte produces flagellated male gametes that must swim through a watery medium to reach the egg. Thus the ferns, although supplied with vascular tissues and other adaptations that make them successful in

inhabiting terrestrial environments, still require water for union of gametes in reproduction.

The remaining vascular plants, the seed plants, share two special adaptations in their life cycles (see Figure 15-1) that set them apart from the ferns and from all other plants: **pollen** and **seeds**. Pollen, a type of male gametophyte, is resistant to drying and can be transported through the air to female gametophytes without damage. At no time in the life cycle are male gametes required to swim through an external medium to reach the egg. Consequently, seed plants do not require an aqueous environment for fusion of gametes.

The second major adaptation found only in seed plants is the seed itself, a capsule containing an embryonic sporophyte arrested at an intermediate stage of development (see Figure 15-16). In the seed, the embryo is protected from damage by hard, impermeable surface coats. As seeds, embryonic plants can be distributed by wind, water, or animals and can remain dormant for years until favorable conditions for growth are met. The development of the seed and pollen enabled the seed plants to exploit terrestrial environments to the fullest.

Seed plants are divided into *angiosperms* and *gymnosperms* according to differences in their reproductive structures. In the gymnosperms, which include the pines, firs, cedars, and other primarily evergreen plants, reproductive structures are formed in a stack of modified, scalelike leaves arranged in a **cone** (Figure 17-10). The name gymnosperm (*gymno* = naked; *sperm* = seed) refers to the fact that seeds are borne in a relatively exposed location on the surfaces of the scales of a cone. In the angiosperms (*angio* = vessel), seeds are formed in a protected location (an **ovary**) within a specialized structure of the reproductive organ of this class: the **flower** (see Figure 15-2). The angiosperms include all the small annual flowers and the flowering shrubs, vines, and trees.

Within the angiosperms, which comprise the most diverse group of plants, subdivision into orders, families, genera, and species depends primarily on the internal and external structure of flowers, the arrangement of flowers on the plant, and the form taken by seeds and fruits. The flower structures used in angiosperm classification include the sepals and petals, which occur in various numbers in different angiosperm groups. In some species, sepals and petals are reduced to hairs or scales. Vein pattern within sepals and petals is also much used in classification. Carpels show great variation in the number of ovules carried, their position within the ovaries, and the location of the micropyle, the small opening into the ovule (see Figure 15-3). In some angiosperm groups, flowers are bisexual and contain both male and female reproductive parts; in

---

[2]The cells of haploid individuals contain one copy of each chromosome of the set characteristic of the species; diploids have two copies of each chromosome. (For further information, see Information Box 10-3.)

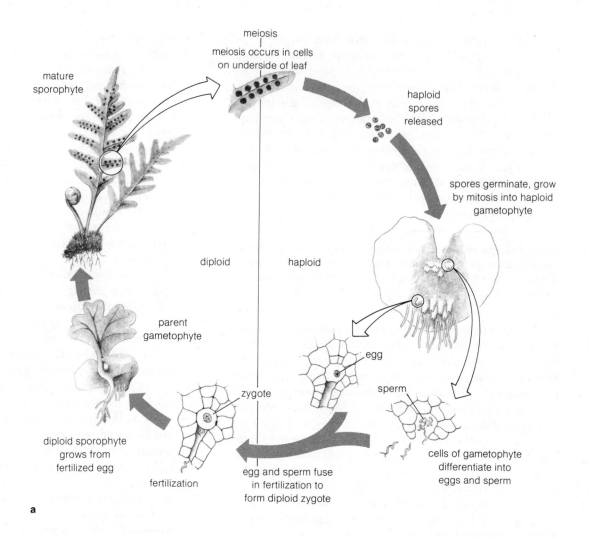

meiosis

meiosis occurs in cells on underside of leaf

mature sporophyte

haploid spores released

spores germinate, grow by mitosis into haploid gametophyte

diploid   haploid

parent gametophyte

egg

sperm

zygote

diploid sporophyte grows from fertilized egg

fertilization

egg and sperm fuse in fertilization to form diploid zygote

cells of gametophyte differentiate into eggs and sperm

a

b

c

Figure 17-9 Ferns. (**a**) Alternation of generations in the life cycle of a fern. (**b**) Frond of a fern sporophyte. Spores are produced in the spotlike outgrowths on the frond surfaces. (**c**) Fern gametophytes. ×5.

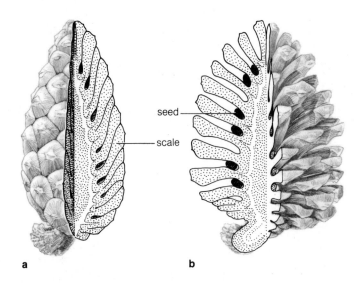

a   b

Figure 17-10 Gymnosperm cones. (**a**) Developing embryos protected within the tightly closed immature cone. (**b**) Mature seeds exposed in an open cone.

Table 17-2	Characters Distinguishing Dicots from Monocots
Dicots	Monocots
Two cotyledons in the embryo	One cotyledon in the embryo
Leaf veins in a network	Leaf veins parallel
Root system with a taproot or large primary root with branch roots	Root system without a principal taproot; roots about equal in size
Stem with vascular bundles arranged in ring or cylinder	Stem with vascular bundles arranged randomly

others, individual flowers may lack either stamens or carpels and are unisexual.

Flower morphology is especially valuable as a basis for classifying angiosperms because variations in flower parts closely reflect the evolutionary pathways followed by the different angiosperm groups. As a result, classification based on flower form and structure produces a natural system that reveals relationships and evolutionary pathways among these plants. The reproductive parts of the gymnosperms are of equivalent value in classification.

Embryology The pattern of embryonic growth followed by plants, although it is used less often than morphology as a basis for classification, is sometimes valuable as an indicator of lineages and relationships. Within the flowering plants, one of the primary subdivisions is established by the number of cotyledons (see Figure 15-13) formed on the embryo. **Monocot** embryos have one cotyledon; **dicot** embryos have two. Other characteristics of the flowering plants, such as distribution and pattern of xylem and phloem in stems, roots, and leaves, also show similarities and differences among lines following the monocot–dicot division (see Table 17-2). Presence or absence of endosperm (see p. 336), differences in developmental stages of the embryo, and variations in the position and form of the embryo within the developing seed are also useful classification keys.

Other Criteria for Plant Classification Depending on the group of plants being classified, information from other souces may also be used. Two sources are of particular value in revealing evolutionary relationships: chromosome analysis and geographic distribution. Changes in chromosome number through doubling of basic diploid or haploid sets, and modifications in individual chromosomes within doubled or tripled sets, have been frequent sources of evolutionary change in plants. Frequently these changes can be traced, and relationships and lineages established, by comparing the number and types of chromosomes visible in cells of different species during mitosis and meiosis. Chromosome analysis is also useful if hybrids can be formed between the different species being investigated. In these hybrids, the degree of meiotic pairing between chromosomes from the different parent species is often a good indicator of the closeness of their evolutionary relationships. Classification through comparisons of geographic distribution is often successful because plants with common evolutionary ancestors tend to inhabit similar regions of the earth.

Classification using as many information sources as possible permits plants to be grouped in divisions (the equivalents of phyla in the plant kingdom) that are clearly distinct and do not overlap significantly (Figure 17-11). Information from the same sources is used to establish the lower categories of the plant kingdom: the classes, orders, families, genera, and species within the major divisions.

The Major Divisions of the Kingdom Plantae The first four divisions of the kingdom Plantae, the Rhodophyta (red algae), Phaeophyta (brown algae), Chlorophyta (green algae), and Bryophyta (liverworts and mosses) are nonvascular plants. The remaining divisions are all distinguished by the presence of vascular tissues, the conductive

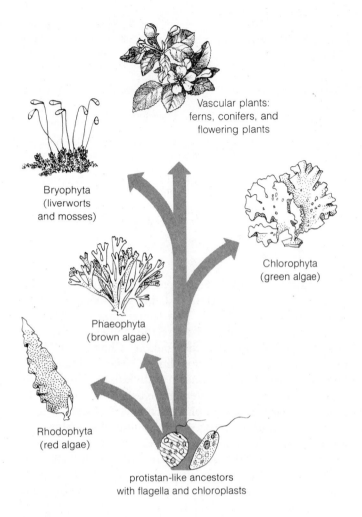

Vascular plants:
ferns, conifers, and
flowering plants

Bryophyta
(liverworts
and mosses)

Chlorophyta
(green algae)

Phaeophyta
(brown algae)

Rhodophyta
(red algae)

protistan-like ancestors
with flagella and chloroplasts

Figure 17-11 The major plant divisions arranged according to evolutionary lineages.

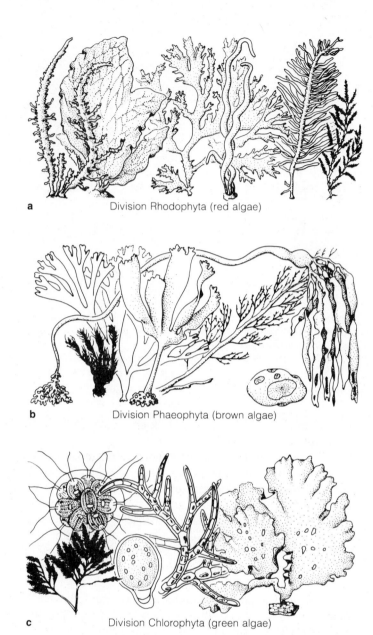

a   Division Rhodophyta (red algae)

b   Division Phaeophyta (brown algae)

c   Division Chlorophyta (green algae)

Figure 17-12 The algae of the kingdom Plantae.

elements of the xylem and phloem. The aerial surfaces of vascular plants are covered with a waxy cuticle that restricts water loss into the air. The sporophyte stage of the life cycle is dominant in these plants, which comprise the most successful and prominent land plants: the ferns, the gymnosperms, and the angiosperms.

Division Rhodophyta (Red Algae) These primarily marine algae (Figure 17-12a) have two classes of photosynthetic pigments—*phycocyanin* (a blue pigment) and *phycoerythrin* (a red pigment)—in addition to chlorophyll *a* and the carotenoids. The cell walls of these plants often contain calcium carbonate, giving them a corallike appearance. No cells are flagellated at any stage.

Division Phaeophyta (Brown Algae) The algae of this division (Figure 17-12b), which consists almost entirely of marine seaweeds, have a brown carotenoid, *fucoxanthin*, in addition to other carotenoids and chlorophyll *a* and *c*. Photosynthetic products are stored as *laminarin*, a nonstarch compound. The cell walls of the brown algae contain a characteristic pectinlike substance, *algin*. The gametes of brown algae have flagella and are motile; all other stages

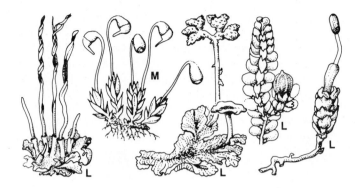

Figure 17-13 Representative liverworts *(L)* and mosses *(M)* of the Bryophyta.

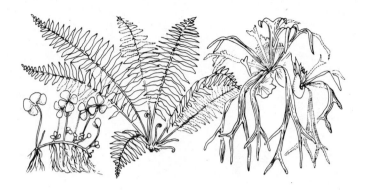

Figure 17-14 Representative fern species.

of the life cycle are nonmotile. Brown algae vary from microscopic size to the largest marine algae, the giant kelps.

Division Chlorophyta (Green Algae) The green algae (Figure 17-12c), which grow in both freshwater and marine environments, have chlorophyll *a* and *b* in addition to yellow carotenoids, all in proportions resembling the higher plants. The Chlorophyta range in size from microscopic plants to individuals 30 centimeters or more in length. Different species of the green algae consist of single cells, small colonies of cells, or larger filamentous or leaflike plants. The gametes of this division are flagellated and motile; many of the microscopic green algae also possess flagella and are motile forms.

Division Bryophyta (Liverworts and Mosses) The mosses are familiar to us as green, downy mats that cover trees, stones, and soil surfaces in moist and shaded environments. Liverworts resemble mosses in size and distribution, but generally they grow as flatter, more leaflike plants. Although the bryophytes (Figure 17-13) are almost exclusively terrestrial, they lack vascular tissues and do not possess true roots, stems, or leaves containing xylem and phloem cells as do the vascular plants. Rootlike structures called **rhizoids** anchor the bryophytes to the soil or substrate, however, and outgrowths resembling leaves are formed in aerial parts of the plant. The external surfaces of the bryophytes are covered with a waxy cuticle that retards water loss as in the vascular plants.

The conspicuous bryophytes represent the haploid gametophyte stage of the life cycle; the sporophyte generation is reduced to a small plant that in most species remains dependent on the gametophyte. The male gametes of the bryophytes are flagellated and motile and require a watery medium to reach and fertilize the egg.

Division Pterophyta (Ferns) Although they are widely distributed as modern plants, the ferns (Figure 17-14; see also Figure 17-9) are an ancient group that flourished on the earth as dominant plants some 300 to 400 million years ago. Most ferns grow as an underground stem or rhizome from which leaves extend upward out of the soil. Thus the ferns familiar to us consist only of leaves; the stems lie out of sight beneath the ground. Although true roots are present, they are usually short, single structures that penetrate only a few centimeters into the soil without branching into the elaborate networks characteristic of the seed plants. Typically the parts of the fern plant above the soil, the leaves or *fronds* as they are called, are subdivided into a stemlike central axis and numerous small leaflets called *pinnae* (see Figure 17-9b). Although a frond resembles a stem bearing small leaves on either side, the entire structure is actually a single leaf. All parts of the fern plant, stem, leaves, and roots, have vascular elements formed from xylem and phloem.

The conspicuous plants with fronds visible above the soil are the sporophytes of the ferns. The gametophytes, which are separate, free-living plants, are small, leaflike structures only a centimeter or less in diameter. The fern gametophytes form flagellated, swimming male gametes as in the lower plants (see Figure 17-9a).

The remaining vascular plants all bear seeds. Although relative newcomers in the evolutionary sense, the seed plants evolved to dominate the land and gradually replaced the ferns and other more primitive types that were once the most numerous land plants. The seed plants include two important divisions: the Coniferophyta (the conifers) and the Anthophyta (the flowering plants). The Coniferophyta make up the most important phylum of the gymnosperms. The Anthophyta comprise the angiosperms.

Figure 17-15 Representative Coniferophyta (conifers).

Division Coniferophyta (Conifers) The conifers (Figure 17-15) are familiar to us as the pines, firs, spruces, hemlocks, and larches of the forests that cover much of North America. The shrubs and trees of this division (also called Pinophyta) typically have small, needlelike leaves and produce their gametes in cones. (Figure 17-16 outlines the life cycle of a typical conifer.) Unlike the cycads and ginkgoes, no freely swimming sperm cells are formed at any point in the life cycle in the conifers. In these plants, pollen germinates to form a pollen tube that carries the sperm nucleus to the egg as in the flowering plants. In contrast to the flowering plants, only one sperm nucleus is functional and fertilization is single instead of double (see below). Following fertilization, the embryo grows inside the female cone to form a seed that is released at maturity.

Division Anthophyta (Flowering Plants) In this division (also called Magnoliophyta; Figure 17-17) the male and female gametophytes are formed inside the structures of the flower (see Figure 15-1). Fertilization, as in the conifers, is by pollen; no free-swimming sperm cells develop at any time. Typically two functional sperm nuclei are released from the pollen tube as it germinates and grows into the female parts of the flower. One of these nuclei fuses with the egg nucleus to form the zygote; the second fuses with two accessory nuclei of the female gametophyte to form the **endosperm** (see Figure 15-7). Fertilization in the flowering plants is thus *double* instead of single as in the conifers. Once fertilized within the ovules of the flower, the zygote develops into a seed in the same location.

Trends in Plant Evolution The classification of plants and the arrangement of plant divisions into a natural hierarchy reveals several overall trends in plant evolution. The most

striking of these is associated with the invasion of dry-land habitats by the more highly evolved plants. The most primitive plants, the algae and bryophytes, are limited to aquatic or very moist environments. Several developments made the most highly evolved plants, the vascular plants, successful in dry-land habitats:

1. Conductive cells specialized for rapid transport of water from the soil to parts of the plant suspended in the air

2. An impermeable surface coat (the waxy cuticle) that retards water loss from aerial plant parts (also present in the bryophytes)

3. In the most highly evolved vascular plants, the development of nonswimming male gametes (enclosed in pollen) that do not require a watery medium to reach the egg

4. Protection of the female gametophyte within structures of the sporophyte

5. Development of the seed, which protects the embryo from damage due to water loss and allows distribution of the species in dry environments

Plants show an overall evolutionary trend toward adaptations to a terrestrial habitat.

The Kingdom Animalia

The major characteristics distinguishing the kingdom Animalia from other groups are associated largely with the mode of nutrition. All animals live by eating or absorbing nutrients from other organisms, primarily plants and other animals. In most animals, this feeding pattern is associated with a motile life-style: most animals move actively to find and capture food. This food-getting movement may involve motions of body parts or the animal's entire body. The few exceptions to this motile pattern are among the parasites—animals that absorb food directly from their hosts and do not move during much of their life cycles.

The Animalia are many-celled organisms that live by ingesting organic matter.

The active motility associated with the feeding habits of animals is reflected in a number of body specializations that we recognize as typical of the group. Except in the most primitive forms, animals have nervous and muscular systems that permit them to recognize food organisms and to produce coordinated movements for their capture and ingestion. Ingested food is broken down into small, easily

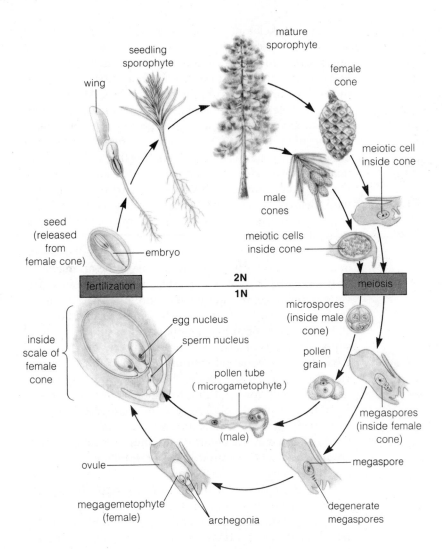

Figure 17-16 The life cycle of a conifer, the common pine.

absorbed organic and inorganic molecules by specialized digestive cells or systems. The digested nutrients are carried to the animal's body cells, and wastes are removed, by a circulatory system. Being self-contained, these systems do not interfere with the requirement for motility. Even the form taken by reproductive systems in many animals permit young to be produced with virtually no interference with the feeding habits of the adult.

The structural characteristics of animals reflect their motile lifestyle.

Within the animal kingdom, the problems of structuring muscular, nervous, digestive, circulatory, and reproductive systems to meet the demand for motility and

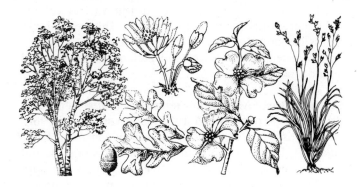

Figure 17-17 Representative Anthophyta (flowering plants).

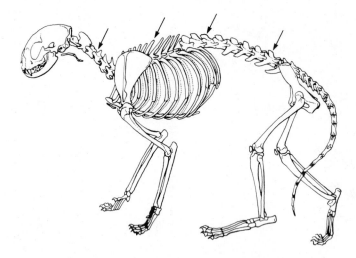

Figure 17-18 The axial support of the vertebrates is structured from a column of individual bony, jointed elements: the vertebrae (arrows).

efficiency in feeding have met with many solutions in evolution. These differing solutions provide the primary basis for classifying animals into phyla and lower subdivisions within the kingdom Animalia.

Classifying the Animalia Primary among the characteristics used in classifying animals are differences in the skeletal system, the type of body cavity in which the internal organs are suspended, the presence or absence of repeating body units called **segments**, the type of body symmetry, the digestive system present, and the pattern of embryonic development of these systems. Grouping animals according to these characteristics reveals an evolution toward more specialized and efficient ways of meeting the requirements for a motile life-style. This general pattern is elaborated with side branches representing different but often equally successful patterns of feeding and reproduction.

Skeletal Systems Most animals have skeletal systems consisting of rigid or semirigid elements that support the softer body tissues. The skeletal units may consist of hard, inflexible elements such as bone or the shells of crabs and lobsters. Less rigid elements, such as cartilage, which support tissues of the ears and the tip of the nose in humans, also serve as skeletal units. In some animals, body cavities filled with fluid under pressure support the softer body tissues. The legs of spiders, for example, are supported and extended by internal fluid pressure.

Animal skeletons may be internal or external. Animals with internal skeletons have supportive elements on the interior that form the central axis of the main body and body parts. Muscles and other soft body tissues surround the skeletal units and form the body surface. Animals with external skeletons have support elements located at their body surfaces. In most animals with this type of skeletal organization, the surface units are structured like a suit of armor linked together by flexible joints that permit the animal to move. Muscles lining the inside surfaces of the skeletal units connect and move one part of the skeleton with respect to another. Animals such as the insects, crabs, and lobsters are the prime examples of this type of skeletal organization.

Animals are frequently grouped for convenience into one of two major categories according to the arrangement of the skeletal system: vertebrates and invertebrates. The **vertebrates** have an internal skeleton organized around a central, axial support running the length of the body. In all but the most primitive members of this group, the axial support consists of a series of repeated bony units forming a backbone or vertebral column (Figure 17-18). The individual units of the backbone, the *vertebrae*, line up end to end in a single row and link together by flexible joints to form the vertebral column. All the vertebrates share a basic body structure built around the axial backbone: a head is located at the anterior end and, in most vertebrate groups, two sets of paired limbs or appendages are connected laterally to the backbone.

The animals of the second major group, the **invertebrates**, have no backbone or other axial body support. The skeletal system of these organisms may be internal or external, or it may contain elements of both types. There is great variability in the number and types of limbs or appendages among the invertebrates; moreover, no basic body plan is shared by all members of the group. Different invertebrate phyla range from relatively primitive animals to others as complex and specialized as the vertebrates.

Among the invertebrates there are many side branches from the evolutionary tree (see Figure 17-23). Four of these branches represent major evolutionary trends, each with a characteristic skeletal type. The *annelids* (earthworms, leeches, and related forms) have bodies supported primarily by internal spaces containing fluid under pressure. The *arthropods* (including lobsters, crabs, insects, and spiders) have hard external skeletons as their primary body support. In the *echinoderms* (the starfish, sea urchins, and related forms), the primary body support is also at the surface of the organism, in this case just within the skin. Most of the animals of the remaining major invertebrate group, the *molluscs* (clams, oysters, snails, squids, and octopuses), have a characteristic shell that supports the soft body parts. The shell is external in clams and oysters but internal in squids. In octopuses, the shell has been lost in evolution.

Type of Body Cavity The most primitive animals have no internal body cavity. Internal tissues and organs in these species are embedded in a gelatinous mass containing loosely organized cells (Figure 17-19*a*). These organisms, which include sponges, coelenterates, and flatworms (see Figure 17-23), are called the **acoelomate** (*a* = no; *coelom* = cavity) animals. All other animals have an internal cavity in which the digestive system and other organs are suspended. In the primitive representatives of this group, the internal cavity is only partially lined with cells derived from embryonic mesoderm (see p. 303), so that in some places ectoderm or endoderm tissue directly faces the body cavity (Figure 17-19*b*). Of these animals, called the **pseudocoelomate** phyla, only the nematodes (roundworms) are abundant and widely distributed. The remaining animals have body cavities completely lined with mesoderm (Figure 17-19*c*). These animals, the **coelomate** phyla, include the most highly evolved invertebrates (the annelid worms, arthropods, molluscs, and echinoderms) and all the vertebrates.

Segmentation The more primitive animals, including sponges, jellyfish, flatworms, and roundworms, show no internal or external division of the body into repeating units. In many of the more advanced animal phyla, the body is formed from more or less identical units or **segments** that are repeated lengthwise. (Figure 17-20 shows the repetition of body parts in a conspicuously segmented animal: the leech.) The repeated segments may be completely identical, as in the posterior segments of the leech or earthworm, or they may be individually modified to carry internal or external structures that vary from segment to segment as in the insects (Figure 17-21). In some animals, among them the vertebrates, segmentation is limited to a few internal body parts such as the vertebrae and ribs. In these animals segmentation is not conspicuously visible on the body surface.

Bilateral symmetry is characteristic of animals that are highly motile; radial symmetry is typical of animals with reduced capacity for movement.

Body Symmetry Most animals show one of two primary types of body symmetry (Figure 17-22): bilateral or radial. Animals that move actively generally have distinct right

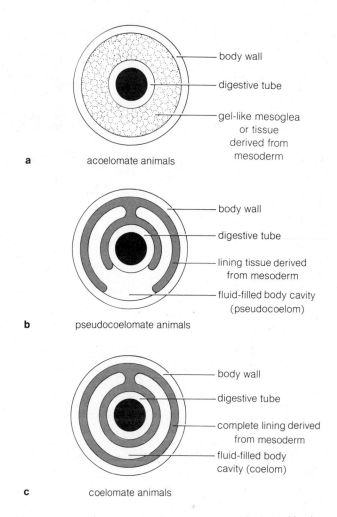

a   acoelomate animals

- body wall
- digestive tube
- gel-like mesoglea or tissue derived from mesoderm

b   pseudocoelomate animals

- body wall
- digestive tube
- lining tissue derived from mesoderm
- fluid-filled body cavity (pseudocoelom)

c   coelomate animals

- body wall
- digestive tube
- complete lining derived from mesoderm
- fluid-filled body cavity (coelom)

Figure 17-19 The arrangement of mesoderm and body cavities in the acoelomate (**a**), pseudocoelomate (**b**), and coelomate (**c**) phyla (see text).

and left sides and also a clearly defined front or anterior end containing the organs of major senses such as touch, sight, and smell. These animals, which are said to have **bilateral symmetry** (Figure 17-22*a*), also have recognizable top (dorsal), bottom (ventral), and rear (posterior) ends. In the other major body pattern, called **radial symmetry** (Figure 17-22*b*), body parts are arranged in a circle around a center or axis passing through the mouth. These organisms have either top and bottom or anterior and posterior ends but no recognizable right or left sides. Radial symmetry is typical of animals with reduced ability to move (such as the starfish) or animals that are anchored to a substrate for much of their lives. Animals that are carried about passively by water currents frequently show radial symmetry.

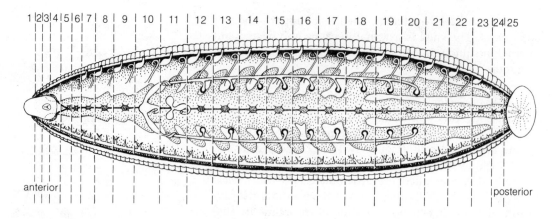

Figure 17-20 Segmentation of internal and external body structure in a leech. Redrawn from *General Zoology* by T. I. Storer, R. L. Usinger, R. C. Stebbins, and J. W. Nybakken. Copyright 1972 by McGraw-Hill, Inc. Used with permission of McGraw-Hill Book Company.

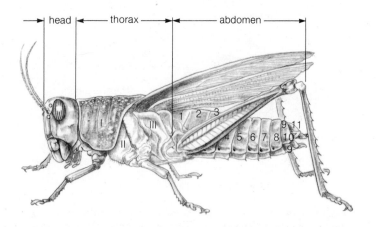

Figure 17-21 External segmentation in an insect, the grasshopper. Some segments carry appendages such as antennae, legs, and wings. Redrawn from *General Zoology* by T. I. Storer, R. L. Usinger, R. C. Stebbins, and J. W. Nybakken. Copyright 1972 by McGraw-Hill, Inc. Used with permission of McGraw-Hill Book Company.

Digestive System The most primitive animals, the sponges, have a central body cavity that functions only in the capture of food. Digestion is *intracellular;* that is, it is carried out entirely within the cells of the body wall; pieces of food are engulfed whole by these cells and broken down within them by digestive enzymes. No enzymes or other digestive substances are secreted into the body cavity. All other animals have a specific tubelike cavity within the body that is specialized for digestion. A series of digestive organs secretes enzymes and digestive substances such as

acids into the tube. Most of the digestive processes occur in the tube and not in the cells lining the digestive tract; thus digestion is *extracellular.* In primitive animals, the digestive tube has a single opening that serves as both mouth and anus; more advanced animals have a digestive tube with separate openings for the mouth and anus.

Pattern of Embryonic Development The types and stages of embryonic development in animals are frequently the most important characteristics used for classification and establishing evolutionary lineages. Often animals reveal similarities in embryonic stages that are completely lost and undetectable in adult forms.

One developmental characteristic that has been useful in establishing fundamental evolutionary relationships among the animal phyla is the relationship between the mouth and the embryonic blastopore (see p. 304). This relationship has revealed common lineages within two major groups of phyla within the animal kingdom. One group includes the annelids, arthropods, and molluscs; the other includes the echinoderms and the chordates, the phylum comprising the vertebrates.

In the annelid–arthropod–mollusc group, the mouth develops in the region near the embryonic blastopore. In the echinoderm–chordate group, the mouth forms in a different region of the embryo completely separate from the blastopore. For this reason, the annelid–arthropod–mollusc phyla are sometimes termed the **protostome** phyla (*proto* = first; *stome* = mouth) whereas the echinoderm–chordate phyla are the **deuterostome** phyla (*deutero* = second). Still other embryonic developmental patterns separate the two groups of phyla. Among these characteristic features are

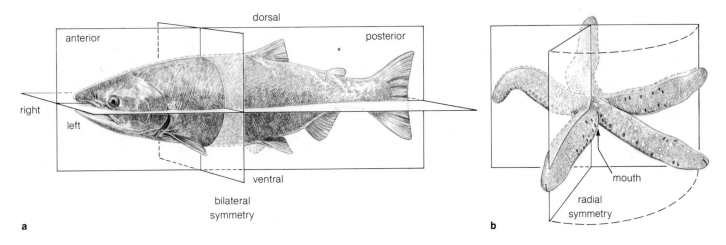

right
anterior
dorsal
left
ventral
posterior
bilateral
symmetry

a

mouth
radial
symmetry

b

Figure 17-22 The two primary types of body symmetry in animals. (**a**) Bilateral symmetry. Organisms with this symmetry have anterior, posterior, dorsal, ventral, and left and right sides. (**b**) Radial symmetry. Organisms with this pattern have body parts arranged in a circle with its center passing through the mouth. These organisms have either dorsal and ventral ends, or anterior and posterior ends, but no right and left sides. Redrawn from *General Zoology* by T. I. Storer, R. L. Usinger, R. C. Stebbins, and J. W. Nybakken. Copyright 1972 by McGraw-Hill, Inc. Used with permission of McGraw-Hill Book Company.

details of the cleavage pattern, the degree to which specific regions of the egg cytoplasm are fixed into determined development pathways (see p. 296), and also the form taken by swimming larvae, when present.

These distinctive developmental patterns are of great interest because they suggest that the protostome and deuterostome phyla separated in very ancient times as two major branches of the animal kingdom. One branch led to the modern annelid–arthropod–mollusc group and the other to the echinoderm–chordate group of phyla to which we belong. While most people would not expect or perhaps appreciate comparison with the starfish and its kin, the possibilities are good that we are more closely related to the echinoderms than any other phylum of the animal kingdom.

Other Criteria for Classification Studies of animal behavior have become increasingly important in recent years as an aid to animal classification. Species can often be identified more clearly by behavioral activities such as courtship rituals and songs (in birds, frogs, toads, and crickets, for example) than any other characteristic. Other behavioral criteria involve the patterns and materials used in nest building or the form in which eggs are laid. Differences in coloration, types of parasites carried, geographic distribution, and chromosome types and number are also used in classification. Biochemical and physiological factors such as temperature tolerance and the types of proteins, lipids, and pigment molecules produced are still other classification criteria.

Information from all these sources allows taxonomists to classify the animals of the world into a number of distinct phyla (Figure 17-23) and establishes the probable evolutionary lineages of these groups. Within the phyla, similarities and differences in the same characteristics are used to establish the classes, orders, families, genera, and species as well as the subdivisions of these categories.

Outline of the Kingdom Animalia

Phylum Porifera (Sponges) The animals of this aquatic group (Figure 17-24) have bodies that may be radially symmetric or formless. Although sponges are anchored to the substrate and thus immobile, flagellated cells within them beat actively to move water carrying food particles through an extensive system of internal channels. Water enters these channels through pores located at various points on the body surface, and it exits through a large main pore or channel. Organic matter carried into the sponge is engulfed by cells lining the internal channels and digested intracellularly.

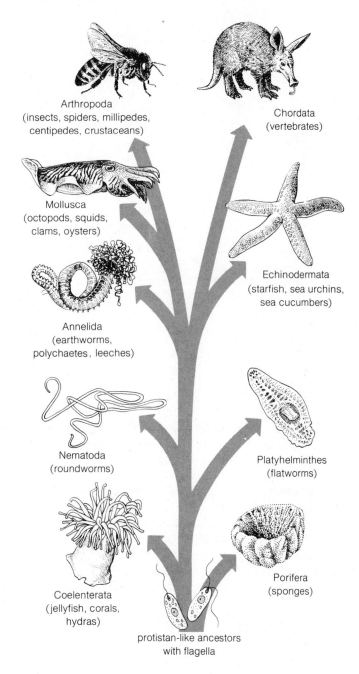

Figure 17-23 Evolutionary lineages and the relationships among the major phyla of the animal kingdom.

Arthropoda
(insects, spiders, millipedes,
centipedes, crustaceans)

Chordata
(vertebrates)

Mollusca
(octopods, squids,
clams, oysters)

Echinodermata
(starfish, sea urchins,
sea cucumbers)

Annelida
(earthworms,
polychaetes, leeches)

Nematoda
(roundworms)

Platyhelminthes
(flatworms)

Coelenterata
(jellyfish, corals,
hydras)

Porifera
(sponges)

protistan-like ancestors
with flagella

The body walls of sponges are built up from an outer layer of flattened cells that faces the exterior, an ill-defined middle layer composed of wandering cells, and an inner layer of flagellated cells that surrounds the central channels and cavities. The walls are supported by a flexible

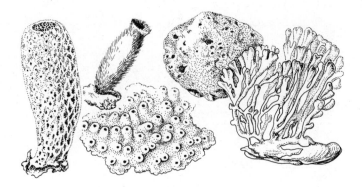

Figure 17-24 Phylum Porifera (sponges).

network of protein fibers or by hard, calcium- or silica-containing elements called **spicules**. The flexible network of protein fibers, when cleaned of cellular material, is sold commercially as natural bath or cleaning sponges. The Porifera have no coelom or nervous system; the muscular system is limited to a few primitive contractile cells.

Phylum Coelenterata (Jellyfishes, Corals, Sea Anemones, Hydras)
This phylum, also called the Cnidaria (Figure 17-25), comprises a large group of freshwater and marine species that share a radially symmetric body plan. At the center of symmetry is the mouth, which leads to an internal digestive cavity. Surrounding the mouth is a circle of tentacles covered by specialized stinging cells called *cnidoblasts*. The body wall is formed by two distinct layers: an outer epidermis and an inner *gastrodermis* separated by an intervening noncellular layer of gelatinous material called the *mesoglea* (see Figure 17-19). Although the mesoglea is primarily noncellular, it contains scattered wandering cells.

The body organization of the coelenterates is simple. The digestive cavity is blind, and the same opening serves both as mouth and anus. Digestion is both extracellular, through the activity of enzymes secreted into the central digestive cavity, and intracellular, through the breakdown of food particles engulfed by the endodermal cells. Although there is no coelom or organized nervous or muscular systems, cells of the epidermis and gastrodermis are specialized for sensory and motor tasks. Cells of the epidermis are specialized for sensory or contractile roles; cells of the gastrodermis are specialized for digestion and contraction. The coelenterates include three major classes, all shown in Figure 17-25: the Hydrozoa (hydras), Scyphozoa (jellyfishes), and the Anthozoa (corals and sea anemones).

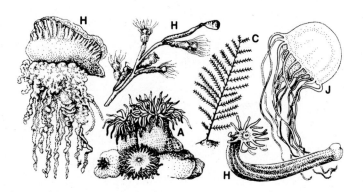

Figure 17-25 Phylum Coelenterata. Representatives of the classes Hydrozoa or hydras *(H)*, Scyphozoa or jellyfish *(J)*, and Anthozoa or corals *(C)* and sea anemones *(A)* are shown in the figure.

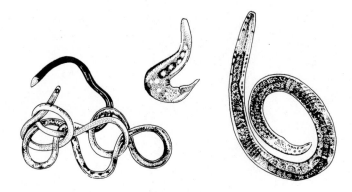

Figure 17-27 Roundworms (class Nematoda) of the phylum Aschelminthes.

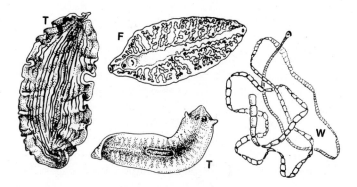

Figure 17-26 The phylum Platyhelminthes (flatworms). Representatives of the classes Turbellaria *(T)*, Trematoda (flukes, *F*), and Cestoda (tapeworms, *W*) are shown in the figure.

Phylum Platyhelminthes (Flatworms) The flatworms (Figure 17-26), as the name suggests, have conspicuously flattened, ribbonlike bodies. The digestive system, when present (some parasitic flatworms have no digestive cavity), has a mouth but no anus. In contrast to the Porifera and Coelenterata, the flatworms possess three distinct body layers derived from ectoderm, mesoderm, and endoderm. Although there is no coelom or circulatory system, a nervous system consisting of a primitive brain or **ganglion** and one to three longitudinal nerve cords is present. A definite muscular system, derived in embryonic development from the mesoderm as in higher animals, is also present. Different flatworm species live in both freshwater and saltwater habitats or are parasites of other animals.

Phylum Aschelminthes (Roundworms) The members of this phylum (Figure 17-27) are bilaterally symmetric animals with bodies that are wormlike and cylindrical. All have a complete digestive system; mouth and anus are located at opposite ends of the body and are connected by a more or less straight digestive tube. The nervous system consists of a ringlike ganglion at the anterior end of the animal that communicates with the body through a number of longitudinal nerve cords. All the organisms of this group are protostomes; that is, the blastopore of the embryo becomes the mouth of the adult. The body cavity, since it is incompletely lined with mesoderm (see Figure 17-19), is a pseudocoelom.

The most important class of Aschelminthes, the Nematoda or roundworms, are abundant as both free-living and parasitic species in almost every freshwater and terrestrial habitat. The nematodes typically have long, cylindrical bodies that are tapered and pointed at both ends. Only one longitudinal layer of muscles lies inside the body wall; contraction of these muscles causes seemingly uncoordinated whiplike flexures of the entire worm. No cilia are present at any point on the surface of the body; instead the surface is covered by a resistant cuticle.

Phylum Annelida (Segmented Worms) The annelid worms (Figure 17-28) show the protostome pattern of embryonic development, in which the mouth arises from the blastopore. In this respect, they resemble the Aschelminthes. They differ from the roundworms, however, in possessing a true coelom: a body cavity completely lined and separated by mesoderm from ectodermal and endodermal structures (see Figure 17-19). Apart from these characteristics, the Annelida have segmented bodies in which a unit structural plan is repeated lengthwise along the body. The

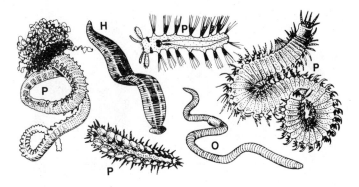

Figure 17-28 Representative annelids, including the classes Oligo-chaeta *(O)*, Hirudinea *(H)*, and Polychaeta *(P)*.

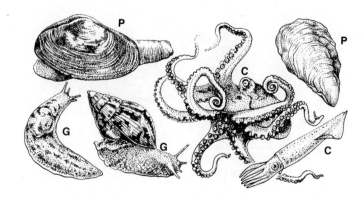

Figure 17-30 Representative molluscs from the classes Pelecypoda *(P)*, Gastropoda *(G)*, and Cephalopoda *(C)*.

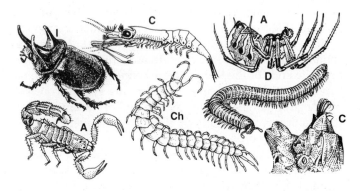

Figure 17-29 Members of the phylum Arthropoda, including Crusta-cea *(C)*, Arachnida *(A)*, Chilopoda *(Ch)*, Diplopoda *(D)*, and Insecta *(I)*.

Phylum Arthropoda (Insects, Lobsters, Crabs, Spiders, Mil-lipedes, Centipedes) The animals of this phylum (Figure 17-29), like the annelids, are segmented. In the arthropods, however, many of the segments fuse together to form larger, composite structures. The arthropod body is bilaterally symmetric and covered with a hard external skeleton that is jointed like a suit of armor. At least some of the segments in most arthropods bear jointed appendages, including legs, jaws, and antennae. Movement is accomplished by a complex muscular system that moves each segment or joint with respect to the next. The nervous system of the arthropods closely resembles the annelid plan and consists of a major anterior ganglion with a ventral, segmented nerve cord that expands in each segment into smaller ganglia carrying lateral nerves. The digestive system is complete; there are separate openings for mouth and anus. The circulatory system is open; that is, vessels leading from the heart empty into large body spaces that bathe structures and return blood to the heart. Like the annelids and molluscs, the arthropods follow the proto-stome pattern of embryonic development and possess a true coelom. The coelom in arthropods, however, is much reduced and limited to regions that house the gonads. The arthropods are highly successful animals that are widely distributed; in terms of numbers of species they are easily the most abundant animal phylum on earth. There are five important classes of arthropods (Figure 17-29): the Crus-tacea (crabs, lobsters, and their kin), the Arachnida (spiders, ticks, mites, and scorpions), the Chilopoda (centipedes), the Diplopoda (millipedes), and the Insecta (insects).

segmented bodies of the annelids are bilaterally symmetric and elongated; the members of this phylum in addition possess a complete digestive system with separate openings for mouth and anus. The nervous system consists of a large anterior ganglion or brain with a ventral nerve cord extending the length of the body. The ventral nerve cord carries minor ganglia and lateral nerves in each segment. The body plan also comprises a complex muscular system, with several layers of muscles providing movement in a variety of directions, and a circulatory system that includes vessels leading to and from a series of hearts. The three classes of the phylum Annelida include the Oligo-chaeta (the earthworms and their relatives), the Hirudinea (leeches), and the Polychaeta (clamworms and other segmented marine worms).

Phylum Mollusca (Clams, Oysters, Snails, Octopuses, Squids)
These animals (Figure 17-30), which share a protostome pattern of development with the annelids and arthropods, have bilaterally symmetric, soft bodies covered by an enveloping fold of tissue called the *mantle*. An external or internal shell is secreted by the mantle in many molluscs. The body is usually divided into a fleshy, muscular *foot*, which may be subdivided into tentacles, and a soft *visceral mass* containing most of the body organs. Some molluscs have in addition a clearly defined head. A closed circulatory system, with continuous blood vessels leading to and from the heart, is present. The nervous system of the molluscs consists of three major ganglia located in the head, foot, and visceral regions, all interconnected by nerve cords. A true coelom is present but much reduced. The digestive system of the molluscs is complete; there are separate openings for mouth and anus. Molluscs of various kinds inhabit freshwater, marine, and terrestrial habitats. Of the classes, three are important and widespread: the Pelecypoda (clams, mussels, oysters, and scallops), the Gastropoda (snails, slugs, abalones, and others), and the Cephalopoda (squids and octopuses).

Phylum Echinodermata (Starfish, Sea Urchins, Sea Cucumbers)
The echinoderms (Figure 17-31) share with the remaining phylum of animals, the chordates, the deuterostome pattern of development: formation of the mouth is unrelated to the embryonic blastopore. The Echinodermata are radially symmetric marine animals that in most species have hard, calcium-containing plates or spines embedded in their skin as supportive elements. The digestive system is complete; there are mouth and anal openings on opposite sides of the body. A true coelom is present that extends throughout the body and forms the internal cavity in which the digestive, reproductive, and other organ systems are suspended. Echinoderms are unique in possessing an unusual *water-vascular* system consisting of a central ring and a series of radial canals (Figure 17-32). The canals have short lateral branches ending in numerous *tube feet* that extend to the outside through the body wall. The system is filled with water that enters the canals through a sievelike opening, the *madreporite*, at the surface of the animal. Echinoderms move by a combination of changes in water pressure and muscular contractions in the tube feet. Of the several classes in this phylum, three contain animals that are widely distributed in the oceans of the world: the Asteroidea (starfish), the Echinoidea (sea urchins and sand dollars), and the Holothuroidea (sea cucumbers).

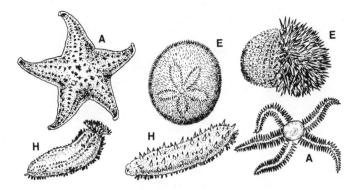

Figure 17-31 The phylum Echinodermata, including the classes Asteroidea or starfish *(A)*, Echinoidea or sea urchins and sand dollars *(E)*, and Holothuroidea or sea cucumbers *(H)*.

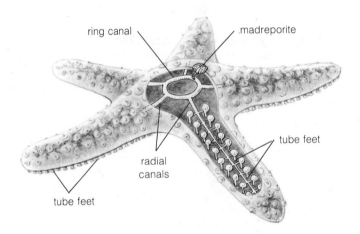

Figure 17-32 The water-vascular system of a starfish (see text). A portion of the body wall is cut away to show the ring and radial canals; other internal organs are not shown. Portions of the stomach are encircled by the ring canal.

Phylum Chordata All the members of this phylum, to which our own species belongs, share several characteristics in common (see Figure 17-33). All possess at some time in their life cycles (1) a *notochord*, an elastic rod of tissue forming an axial body support, (2) a hollow *dorsal nerve cord*, lying just above the notochord, that communicates between the brain and the posterior regions of the body, and (3) slitlike openings, the *gill slits*, that lead from the anterior portion of the digestive tract (the pharynx) to the exterior surface of the animal. The chordates include

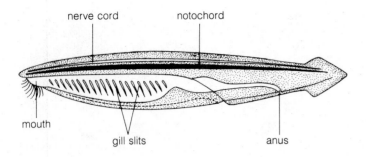

nerve cord | notochord

mouth

gill slits | anus

Figure 17-33 All chordates share several characteristics in common: the arrangement of the notochord, dorsal nerve cord, and gill slits in *Amphioxus*, a cephalochordate.

three major subphyla: the Cephalochordata (lancelets or amphioxi), the Urochordata (tunicates or sea squirts), and the Vertebrata (the vertebrate group to which we belong).

Subphylum Cephalochordata (Lancelets) The Cephalochordata (Figure 17-33) are fishlike chordates with the three chor-

date characteristics clearly developed: notochord, hollow dorsal nerve cord, and gill slits. These small chordates, which average only a few centimeters in length, live partly buried in the bottom sediments of marine environments. Oxygen is obtained from water taken in the mouth and expelled through the gill slits; food particles taken in with the water are retained and enter the digestive system.

Subphylum Urochordata (Tunicates) The adult animals of this subphylum seem to have little resemblance to the chordates (Figure 17-34a). These animals, which live attached to a substrate as adults, have a tough tunic or mantle that covers the body surface. The internal organs lie in a cavity within the tunic that opens to the exterior through siphons. Primitive chordate characteristics are revealed by the larvae of the urochordates, however, which have a notochord, hollow dorsal nerve cord, and gill slits. (Figure 17-34b shows several stages in the conversion of the larval to the adult form.)

Subphylum Vertebrata In this subphylum (Figure 17-35) the notochord is replaced to a greater or lesser extent by a

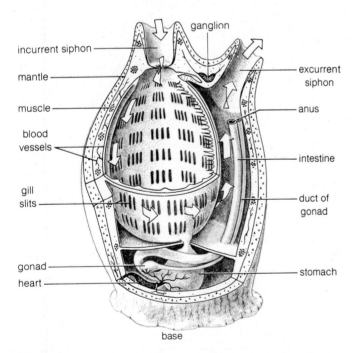

incurrent siphon — mantle — muscle — blood vessels — gill slits — gonad — heart — ganglion — excurrent siphon — anus — intestine — duct of gonad — stomach — base

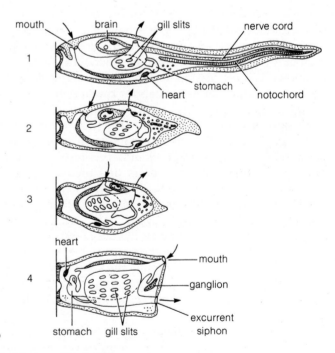

mouth — brain — gill slits — nerve cord — stomach — heart — notochord — heart — mouth — ganglion — excurrent siphon — stomach — gill slits

b

Figure 17-34 A tunicate, a urochordate that seems totally unlike the vertebrates in adult characteristics (**a**), proves to have a notochord, a primitive vertebrate structure, during embryonic development. (**b**) Successive stages (1–4) in the development of the adult form. Redrawn from *General Zoology* by T. I. Storer, R. L. Usinger, R. C. Stebbins, and J. W. Nybakken. Copyright 1972 by McGraw-Hill, Inc. Used with permission of McGraw-Hill Book Company.

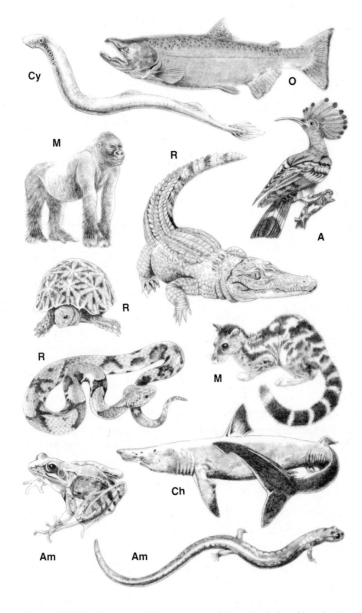

the pharynx in humans, for example, is a modified gill slit.) All vertebrates are bilaterally symmetric animals with a distinct head containing a complex brain and the major sense organs. In the more advanced members of the phylum, the body bears two pairs of limbs or appendages. There are seven classes of living vertebrates: the Cyclostomata (hagfish and lamprey eels), Chondrichthyes (sharks and rays), Osteichthyes (bony fishes), Amphibia (frogs, toads, and salamanders), Reptilia (turtles, snakes, lizards, and alligators), Aves (birds), and Mammalia (mammals). The major characteristics of these vertebrate classes are summarized in Table 17-3.

Trends in Animal Evolution   Although there are various exceptions and reversals of the overall direction of animal evolution, several major trends emerge when the kingdom is traced from the most primitive to the most advanced phyla. These trends reveal:

1. A gradual increase in cell specialization and interdependence

2. A gradual increase in body size and complexity

3. An increase in mobility and with it, greater complexity of muscular and nervous systems

4. Development of increasingly complex, internal circulatory systems that bring water, oxygen, and nutrients to all body parts and carry away wastes

5. Development of bilateral symmetry and, with it, concentration of major sense organs and functions at the anterior end of the body

All these developments lead to greater efficiency and success in feeding and reproduction. The evolutionary mechanisms responsible for these changes in animals, and for the overall trends in the evolution of plants and other organisms, are described in the next unit of this book.

Questions

1. Why is the classification of living things considered important?

2. What is the goal of classification?

3. What is a species? Why is it difficult to define a species?

4. Why is a rigid "ideal type" description of a species unworkable? Why are species now described in terms of populations?

5. What is binomial nomenclature?

6. List the groups larger than the species.

7. Explain the criteria used to define the five kingdoms of living organisms.

8. Where would you place viruses in the five kingdoms? Why?

Figure 17-35   The seven living classes of the subphylum Vertebrata: the Cyclostomata *(Cy)*, Chondrichthyes *(Ch)*, Osteichthyes *(O)*, Amphibia *(Am)*, Reptilia *(R)*, Aves *(Av)*, and Mammalia *(M)*.

column of jointed, bony elements: the vertebrae. The nerve cord lies just above the vertebral column. Gill slits are present in the more primitive chordates; in the more highly evolved members of the phylum the gill slits are modified into other structures. (The eustacian tube that communicates between the internal parts of the ear and

Table 17-3 Vertebrate Classes and Their Characteristics

Class	Respiration	Circulation	External Body Covering
Cyclostomata	Gills	Two-chambered heart	Skin with mucous glands
Chondrichthyes	Gills	Two-chambered heart	Skin with mucous glands and scales
Osteichthyes	Gills covered by operculum	Two-chambered heart	Skin with mucous glands and scales
Amphibia	Lungs, gills in larvae and some adults; also through skin	Three-chambered heart	Skin with mucous glands
Reptilia	Lungs	Incompletely four-chambered heart	Skin scales and few glands
Aves	Lungs	Four-chambered heart	Skin with feathers
Mammalia	Lungs	Four-chambered heart	Skin with hair

9. Outline the differences between the Monera and the Protista. Why are the blue-green algae placed in the Monera and not in the Protista or Plantae?

10. Outline the differences between the Fungi and Plantae.

11. What characteristics are used in classifying the Plantae? Why are reproductive structures used extensively in plant classification?

12. In which plant divisions does fertilization take place by motile sperm?

13. Which plant divisions have vascular tissues? Which have leaves, stems, and roots?

14. Outline the alternation of generations in plants.

15. What are gymnosperms and angiosperms? Which plant divisions are included in these groups?

16. What major evolutionary developments allowed plants to live successfully on land?

17. What criteria are used in classifying animals? How are these characteristics related to the mode of animal nutrition?

18. Compare the digestive systems of the animals of the major animal phyla.

19. Compare the skeletal systems of animals in the major animal phyla. What is the difference between vertebrates and invertebrates?

20. What kind of body cavity is found in each animal phylum? Define acoelomate, pseudocoelomate, and coelomate.

21. List the *protostome* and *deuterostome* phyla. What do the two terms mean?

22. Define segmentation. What animal phyla show segmentation?

23. What major evolutionary trends are evident in the animal kingdom? Which of these trends were important in the adaptation of animals to living on the land?

Suggestions for Further Reading

Buchsbaum, R. 1948. *Animals Without Backbones*. 2nd ed. University of Chicago Press, Chicago.

Curtis, H. 1968. *Marvelous Animals: An Introduction to the Protozoa*. Natural History Press, New York.

Jahn, L. T., and F. F. Jahn. 1949. *How to Know the Protozoa*. William C. Brown, Dubuque, Iowa.

Jensen, W. A., and F. B. Salisbury. 1972. *Botany: An Ecological Approach*. Wadsworth, Belmont, California.

Lerwill, C. J. 1971. *An Introduction to the Classification of Animals*. Constable, London.

Keeton, W. T. 1980. *Biological Science*. 3rd ed. Norton, New York.

Margulis, L., and K. Schwartz. 1980. *Phyla of the Five Kingdoms*. Freeman, San Francisco.

Rost, T. L., et al. 1979. *Botany*. Wiley, New York.

Stanier, R. Y., M. Douderoff, and E. A. Alderberg. 1976. *The Microbial World*. 4th ed. Prentice-Hall, Englewood Cliffs, New Jersey.

Whittaker, R. H. 1969. "New Concepts of Kingdoms of Organisms." *Science* 163:150–160.

Feeding	Reproduction	Habitat
Mouth sucking or biting	External fertilization	Marine and freshwater (lampreys); marine (hagfish)
Biting jaws with teeth	Internal fertilization	Marine; very few freshwater
Mouth with teeth	Usually external fertilization	Marine and freshwater
Mouth without prominent teeth; tongue used to catch prey	External or internal fertilization	Freshwater, terrestrial
Mouth with teeth; beak	Internal fertilization	Terrestrial primarily; some freshwater, marine
Beak	Internal fertilization	Terrestrial
Mouth with teeth modified according to diet	Internal fertilization	Terrestrial; some marine, some freshwater

Evolution

18

EVOLUTION: THE EVIDENCE AND THE DARWIN–WALLACE THEORY

Even the briefest survey of the earth's organisms makes it obvious that living things exist in a seemingly endless variety of types and kinds. How did these organisms arise, and why are they present today in their numbers and obvious relationships? Where did we ourselves come from?

These questions about the variety and origins of the life around us, and about our own origins, have been a major preoccupation of humans since they first began to wonder about themselves and their surroundings. Attempts to explain the beginnings and variety of life go back to ancient times and appear among our earliest written works.

The earliest formal explanation, and the one most accepted in Western thought for many centuries, was special creation: the belief that the present collection of living things was created in its entirety by a supernatural being and placed on the earth in the form in which it exists today. In its most rigid form, this doctrine, still very much alive, states that the species on earth have not changed in the past and are not changing today.

Despite its importance in molding popular attitudes about our origins, there have always been difficulties with this belief. The ancient Greeks unearthed and described the fossil remains of organisms that were unknown as living forms. By the end of the eighteenth century enough of these remains of unknown creatures had been cataloged to make it obvious that many kinds of plants and animals had lived long ago and become extinct. Moreover, careful study of the fossils sometimes revealed sequences of apparently related species running in a progression from ancient to more modern forms.

These observations suggested to the most advanced thinkers that, rather than remaining the same, the living things of the earth have been in a constant state of change, undergoing **evolution** from one set of forms into another.

The living organisms of the earth have been in a constant state of change since the origin of life on our planet.

By the mid-1800s, these ideas culminated in the work of Charles Darwin and Alfred Russel Wallace. Through careful observation of living creatures past and present, Darwin and Wallace reinforced the conclusion that evolution has occurred and, moreover, proposed a mechanism to account for it. Their theory laid the groundwork for contemporary ideas about how these changes take place.

The Darwin–Wallace theory of evolution and the modern theory based on it are described in this unit. In this chapter, we will survey the kinds of evidence establishing that evolution has occurred in the past and is still taking place today and will consider the original theory proposed by Darwin and Wallace to account for evolutionary change. The evidence for evolution stems not only from the study of fossils but from several additional sources: comparisons of the anatomy, development, and biochemistry of living organisms; comparisons of the sequences of DNA and protein molecules of living species; and observations of evolutionary change within recorded history or in contemporary organisms.

THE FOSSIL EVIDENCE FOR EVOLUTION

Fossils and How They Are Dated

Fossils represent parts, casts, or impressions of organisms that have been preserved and protected from decay or breakdown for millions or hundreds of millions of years. (Figure 18-1 shows a representative selection of fossils.) Although soft-bodied organisms without hard tissues or elements were sometimes fossilized, usually only supportive or skeletal parts—the bones, teeth, or shells of animals, for example, or the surface coats or cell walls of plants—were preserved. Nevertheless, impressions of soft tissues sometimes persist through deposits that remain around the bones or other hard parts. (Figure 18-1c shows a fossil of this type, in which the outline of softer tissues can be traced around some of the bones.)

In some cases, the process of fossilization replaced all the organic material with mineral matter, so that none of the original tissue remains at all. When such fossils were formed, bones and softer parts were surrounded by material that gradually compacted and hardened into rock. Later the bones and any remaining tissue dissolved away, leaving hollow spaces in the rock that filled with other minerals, gradually forming a recognizable cast of the fossil. Fossils of any kind were most likely to form when animals or plants were buried soon after death. Prompt burial reduces the chance of decay by microorganisms, oxidation by exposure to air, or dissolution by flowing water.

Fossils are found in rocks formed from layers of sediment that compacted and hardened after millions of years of compression under the surface of the earth. Since these rock layers formed one on top of another, deep layers are generally older than shallow layers. As a result, fossils can

a

b

c

Figure 18-1 Some representative fossils. (**a**) A trilobite, a fossil ancestor of the arthropods; some fossil marine shells are also present. (**b**) A fossil fern. (**c**) A segment of the vertebrae and ribs of a dinosaur showing traces of softer tissues. Courtesy of The American Museum of Natural History.

Table 18-1 Radioisotopes Used in Dating Rocks

Isotope	Half-life (Years)	Product
Rubidium-87	500,000,000,000	Strontium-87
Uranium-238	4,510,000,000	Lead-206
Potassium-40	1,280,000,000	Argon-40
Uranium-235	700,000,000	Lead-207
Carbon-14	5,730	Nitrogen-14

often be placed in a relative time sequence with forms found in deeper layers preceding those in the layers closer to the surface.

Originally, fossils found in layers of sedimentary rock were dated by estimating the rate at which the mud, clay, or sandy sediments containing them hardened into rock. In recent times these estimates have been reinforced by a method known as **radioactive dating**. Many rocks contain radioactive isotopes of elements such as uranium, thorium, potassium, and rubidium. As the rocks age, the radioactive isotopes within them break down into other elements at a steady rate (Table 18-1). By measuring the relative amounts of the original isotope and its breakdown products left in the rock, and comparing this ratio with the known rate of decay of the isotope, the absolute age of the rocks can be determined. It takes 700 million years, for example, for half of the uranium-235 (^{235}U) originally present in rock to decay into lead-207 (^{207}Pb). If ^{235}U and ^{207}Pb are found in equal quantities in a rock sample, the rock can therefore be dated as 700 million years old. Other ratios give other ages—for example, the presence of ^{235}U and ^{207}Pb in a 2:1 ratio would indicate that a rock sample has progressed half the way to the midpoint or ½ × 700 million = 350 million years. The method is capable of giving results accurate to about ±2 percent.

Radioactive dating works best with volcanic rocks that formed by cooling from the molten state rather than rocks compacted from sediment. Even so, fossil-containing sedimentary rocks can still be dated indirectly if they lie at the same levels as volcanic rock. More recent fossils can also be dated directly by measuring their residual content of carbon-14, an isotope that is absorbed from the environment and incorporated into body molecules by all living organisms.

All these dating methods have allowed many fossils and the sedimentary rocks containing them to be put in place on an absolute **geological time scale** that measures the eras, periods, and epochs of the earth in terms of millions of years (Figure 18-2). The oldest known fossils of many-celled organisms are between 600 and 700 million years old and date from the late Precambrian period of the geological time scale. Although few fossil animals are known from this period (some are detected only by the tracks they left in sediments), fossils are relatively plentiful in sediments laid down from the Cambrian period onward.

Horse Fossils and Evolution

One of the most complete lineages of fossil animals, assembled through comparisons and dating of thousands of specimens, is the line leading from ancient ancestors to the modern horse. This well-documented story is a classic demonstration of evolution through fossils. The compilation began in Darwin's day, and by 1900 the ancestral lineage of the horse was essentially complete. Discoveries since that time have reinforced the findings of the earlier fossil hunters. Besides providing evidence for evolution, the horse lineage is also of general value in revealing overall patterns common to the evolution of many plants and animals.

The Time Scale of Horse Evolution Horses are comparative latecomers in the evolutionary history of plants and animals: the first ancestors of modern horses appear in rocks of the Eocene epoch, laid down some 50 million years ago. By that time, vertebrate animals had already been in existence for 450 million years. The age of reptiles in the Mesozoic era had come and gone and the mammalian line, which dates from some 200 million years ago, was well established. The conifers and flowering plants, which have their origins in the period from 140 to 180 million years ago, were also abundant.

The oldest fossils in the horse lineage date from about 50 million years ago.

The earliest fossils assigned to the family tree of horses would not be considered horselike by modern standards. The discovery of thousands of fossils of intermediate age and morphology leaves no doubt, however, that direct relationships exist between the earliest fossil horses and modern horses. These fossils form an unbroken series leading, with several important side branches, from the

Era	Period	Epoch	Major Events	Millions of Years Ago
Cenozoic	Quaternary	Recent	Historic time (last 10,000 years)	0
		Pleistocene	Ice ages; humans appear	2.5
	Tertiary	Pliocene / Miocene	Apelike ancestors of humans appear	7 / 26
		Oligocene / Eocene / Paleocene	Origins of most modern mammals	38 / 54 / 65
Mesozoic	Cretaceous		Flowering plants dominant; dinosaurs become extinct	136
	Jurassic		Flowering plants, birds, and mammals appear; dinosaurs dominant	190
	Triassic		Reptiles increase; mammallike reptiles appear; gymnosperms dominant	225
Paleozoic	Permian		Origins of most modern insect orders	280
	Pennsylvanian		Origins of reptiles; amphibians dominant	325
	Mississippian		Land plants dominant	345
	Devonian		Bony fishes dominant; first seed plants; first amphibians and insects	395
	Silurian		First land plants	430
	Ordovician		First vertebrates	500
	Cambrian		Origins of most invertebrates	570
	Precambrian		Prokaryotes and eukaryotic algae dominant; origins of first invertebrates	

Figure 18-2 The geological time scale.

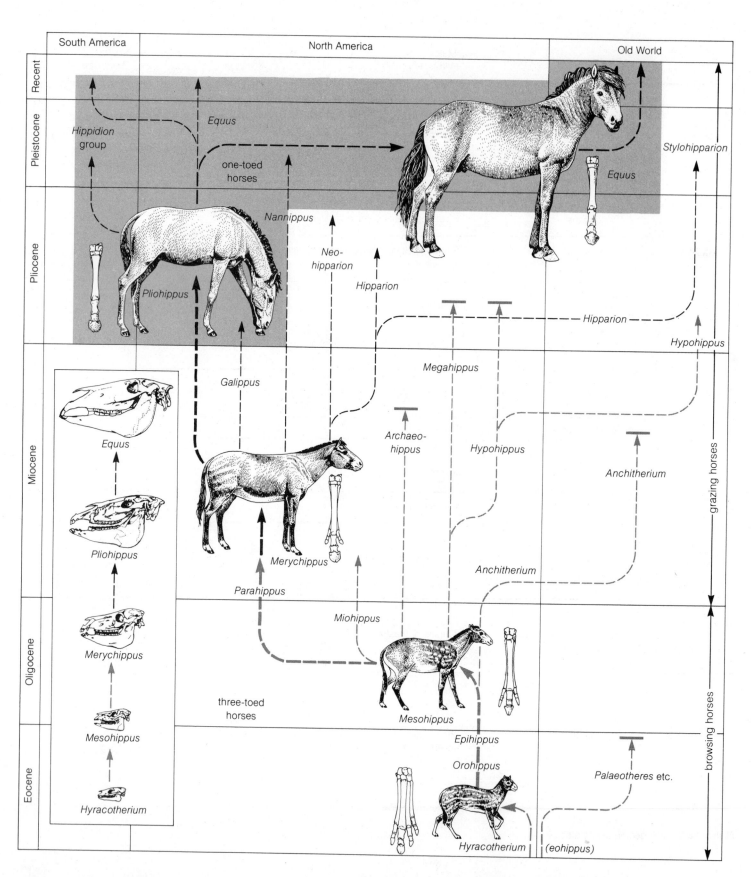

South America　　　　North America　　　　Old World

Recent

Pleistocene

Pliocene

Miocene

Oligocene

Eocene

Hippidion group

Equus

one-toed horses

Nannippus

Neo-hipparion

Hipparion

Stylohipparion

Equus

Pliohippus

Hipparion

Galippus

Megahippus

Hypohippus

Equus

Pliohippus

Merychippus

Archaeo-hippus

Hypohippus

Anchitherium

Merychippus

Parahippus

Miohippus

Anchitherium

three-toed horses

Mesohippus

Mesohippus

Epihippus

Orohippus

Palaeotheres etc.

Hyracotherium

Hyracotherium (eohippus)

grazing horses

browsing horses

Figure 18-3 The lineage of horse evolution. Redrawn from *Horses: The Story of the Horse Family in the Modern World and through Sixty Million Years of History* by George Gaylord Simpson. Copyright ©1951 by Oxford University Press, Inc.; renewed by G.G. Simpson 1979. Reprinted by permission.

first to the modern horse (Figure 18-3). Although only fossilized bones remain of these ancestral horses, much of the body form, including sometimes the outlines of softer body tissues, can be reconstructed by noting the size and arrangement of surface patterns on the fossil bones marking the places where muscles and other body parts were attached.

The Horse Lineage

The Eocene Horse: Hyracotherium When first discovered, the earliest fossil horse, called *Hyracotherium* (see Figure 18-3), was not recognized as belonging to the horse lineage.[1] Instead the earliest horse was initially classified as a relative of modern African animals known as conies or rock rabbits, a group classified in an order called the Hyracoidea. This early horse lived in the Eocene epoch and was a small animal not much more than 45 to 50 centimeters tall, with forelegs and hind legs bearing toes instead of a single hoof. Each toe, four on the front and three on the hind feet, ended in a small hoof; most of the body weight was carried on a doglike pad on the bottom of the foot behind the toes. The rest of the animal was also more canine than equine: an arched back, a head with a relatively short snout, and a long tail.

The skull of *Hyracotherium* reveals much about the probable life-style of this fossil horse. The teeth included front incisors and side molars with smooth, rounded surface projections or cusps more like our own teeth than those of modern horses. These teeth were suited for browsing on soft plants or fruits, but not for grinding tough and abrasive grasses. Casts made of the skull cavity reveal that the brain was relatively small—about half the size of the brain of modern horses in relation to overall body size. Leg and body form indicate that, although small, *Hyracotherium* was probably a swift runner, no doubt as fast as modern horses or dogs. Abundant fossil remains from all parts of the northern hemisphere indicate

---

[1]Fossils are most often named as a genus or family rather than a species. The fossil genus *Hyracotherium* was for many years known by the popular name *eohippus*, meaning "dawn horse."

that the little "dawn horses" were widespread and successful animals.

When *Hyracotherium* first appeared about 54 million years ago, the landmasses of present-day Europe and North America were still part of the same ancient continent. At about the time different *Hyracotherium* species became plentiful, movements of the earth's crust separated the European and American landmasses, leading to different fates for the descendants of these earliest horses. Of the two lines, only the North American *Hyracotherium* can be traced through time to modern horses.

Later horse fossils from the Eocene epoch, ranging in age from about 50 to 40 million years, show few differences from *Hyracotherium*. In these horses, called *Orohippus* and *Epihippus*, most of the changes are in the side teeth, which have sharper crests. These crested side molars were more efficient in cutting the stems and leaves of vegetation and allowed the late Eocene horses to browse on tougher plants than could *Hyracotherium*. However, these early molars are still not as highly specialized for grinding and cutting as the side teeth of modern horses.

The Oligocene Horse: Mesohippus Fossil horses from the Oligocene, the epoch following the Eocene, show clear relationships to *Hyracotherium* and its immediate descendants and also to later horses more nearly like the modern horse. The fossil horses of this period, called *Mesohippus* (see Figure 18-3), which flourished 36 million years ago, were taller than *Hyracotherium*. Individuals averaged around 50 to 55 centimeters tall and had only three toes on the front and hind feet. The skull was more horselike, with a longer snout and eyes set farther toward the rear. Casts of the skull show that the brain of *Mesohippus* was larger, approaching three-quarters the brain volume of modern horses with respect to body size. The teeth of *Mesohippus* and other horses of the Oligocene showed further modifications toward still sharper molar crests. In external appearance *Mesohippus* was probably much more like modern horses than *Hyracotherium*, except for the three-toed feet, an arched back, and relatively small stature.

The Miocene Horse: Merychippus Fossil horses from the Miocene epoch were taller than *Mesohippus* and, except for their three-toed feet, were probably close to modern horses in external appearance. These animals, classified together in the genus *Merychippus* (see Figure 18-3), averaged about 90 centimeters in height. Although they were three-toed, the center toe of each foot was much larger than the others and hooflike. The center hoof supported the weight of the animal; the doglike pad on the bottom of the foot was gone and the animal essentially stood on its toes. The bones of the lower limbs were fused together into single, solid structures as they are in modern horses. (Other mammals, such as humans, retain two parallel bones in the lower leg and forelimb.) The side teeth were longer, and tall, sharp ridges of enamel were supported by a softer cement that filled in the depressions between the ridges at the tops of the teeth. The eyes were set well back in the head along the line from the muzzle to the rear of the skull, in a position much like that of a modern horse.

These changes in the horse line kept pace with major alterations in climate and vegetation that took place during this period. *Hyracotherium* lived in the forests that covered much of North America during Eocene times. As the Eocene gave way to the Oligocene epoch, the climate became drier and much of the forest yielded to open grasslands which continued to spread in the Miocene. *Merychippus* flourished during the Miocene and gave rise to many evolutionary side branches, all of which became extinct except the single line leading to the modern horse. These animals were successful partly because of their sharp, ridged teeth, which allowed them to feed on the tough grasses that became widespread during the Miocene epoch.

The Pliocene Horse: Pliohippus The horses of the Pliocene epoch were almost like modern horses in appearance. Each leg of these fossil horses bears a large central toe forming a single hoof; the side toes are represented only by small, residual bones. The primary and representative horse from this period, *Pliohippus* (see Figure 18-3), stood about two-thirds as tall as modern horses and had the rigid, slightly depressed backbone typical of modern horses. Except for minor differences, the skull of *Pliohippus* has the same shape and proportions as the modern horse, with the eyes placed well back on the head.

The Modern Horse: Equus After the Pleistocene epoch began, the descendants of *Pliohippus* gave rise to the modern horse. The changes noted in fossils from this period, dating back to about a million years ago, include a further increase in size to the dimensions of contemporary horses and, moreover, complete reduction of the residual side toes to thin strips of bone beneath the skin at the sides of the legs.

Both *Equus* and its immediate predecessor, *Pliohippus*, thrived in North America. As *Equus* appeared, the European branches of the horse family, which were descendants of the Miocene horse *Merychippus*, died out, leaving Europe without horses for the third or fourth time since the Eocene. *Equus*, a highly successful traveler, reinvaded

Asia and Europe across land bridges existing at the time and also ranged southward into South America. In addition to the modern horse, *Equus caballus,* evolution in the genus also gave rise to the zebras and donkeys. For unknown reasons, the horse line later died out in North and South America some time after the arrival of the first Indians on our continent about 50,000 years ago. By the time the Europeans arrived, horses were unknown. Thus all the domestic and wild horses now in the Americas are Asian and European descendants of *Equus caballus* introduced to this hemisphere by colonizing Spanish, English, and French soldiers and settlers.

Overall Patterns in Horse Evolution The series of fossils tracing out the development of horses from Eocene to recent times provides clear evidence for evolution. It also illustrates several patterns that are common to many of the animal and plant groups for which the fossil record is reasonably complete. The fossil record shows above all that individual species come and go, appearing and becoming extinct in the course of time, and in the process undergo change as one species gives way to the next. Where the changes are gradual, as they are in the horse lineage, it is possible to trace the rate of change of certain characteristics. Horse teeth may be regarded as changing rapidly, in evolutionary terms, during the millions of years of the Eocene epoch whereas the brain changed more slowly during this time. During the Oligocene, in contrast, the brain developed rapidly but evolution of the teeth slowed. The same patterns are noted also for entire species: some evolve gradually, over millions of years, as in the horse lineage, and some appear suddenly in the fossil record as if created by major and rapid evolutionary changes.

Another important evolutionary characteristic is illustrated by the horse lineage: the change from older to newer types most often follows a branching pathway in which some branches persist and others become extinct. Thus evolution rarely proceeds in a straight-line pattern with one species giving rise to only a single new species as its descendant. This characteristic is clearly demonstrated by the patterns of horse evolution in Figure 18-3. The heavy dashed line follows the lineage leading from *Hyracotherium* to the modern horse. The lighter dashed

Fossils in the horse lineage demonstrate three basic evolutionary patterns: (1) species appear, change with time, and become extinct; (2) the rate of change varies with time; and (3) evolution frequently follows a branching pathway.

lines show the many other evolutionary branches that developed from *Hyracotherium* and its descendants.

These patterns—the appearance and disappearance of new species, variation in the rate of change, and branching in evolutionary lines—are common to most of the known lineages with relatively complete fossil series. Another trend observed in the horse and frequently in other lineages is the general tendency toward greater size. Many animals and plants have evolved toward organisms of larger dimensions; their maximum size is dictated by such factors as motility, body support, nutrient circulation, temperature regulation, and ability to gather food. Another trend noted in the horse line and often observed elsewhere is a tendency toward increasing specialization. In the horse line, specialization is reflected primarily in the digestive system, which became modified toward high efficiency in chewing and digestion of grasses, and in the system for locomotion, which developed toward efficient production of high running speeds over flat terrain. There are exceptions to these general trends, however. In some fossil lines (including some branches leading from *Hyracotherium*) descendants became smaller; in others, evolutionary development led to less specialized forms.

Many evolutionary lines show an overall tendency toward the development of greater size and specialization.

The Adequacy of the Fossil Record

Although the fossil record is invaluable in providing evidence for evolution and tracing the lineages of certain plants and animals, it is rarely as complete for most lines as it is for horses. This shortcoming derives from two major difficulties with the fossil record: **bias** and **incompleteness**.

The fossil record is biased because only certain kinds of animals and plants living in favorable locales are likely to be preserved as fossils. Only the harder parts of organisms (bones, teeth, shells, hard surface coverings) are apt to survive intact. As a result, animals consisting entirely of soft parts, such as worms and most microorganisms, are rarely fossilized. Some modern groups of soft-bodied organisms, such as the presently ubiquitous nematode worms, have never been discovered as fossils.

The fossil record is incomplete and shows bias toward organisms with hard or bony parts that lived in moist or marine environments.

Among the organisms actually fossilized, marine species are best represented. Of these, bottom dwellers or burrowing forms living in shallow water have been most frequently preserved. Among land animals and plants, species dwelling in boggy, wet areas are most likely to have died in places where they could become promptly buried and fossilized. Organisms perishing in drier, more exposed locations evidently decay, dry out, and disintegrate quickly. These factors bias the fossil record toward animals with bony parts or shells that lived in moist or marine environments. Animals and plants with body parts hard enough to satisfy these conditions evidently first appeared in quantity in Cambrian times, 600 million years ago.

The bias of the fossil record contributes directly to its second great weakness, its incompleteness. Since few organisms are fossilized and surviving fossils are often deeply buried and difficult to find, the fossil record will never be complete. George Gaylord Simpson, formerly of Harvard University, has estimated that between a minimum of 50 million and maximum of 4 billion different species of organisms have lived on earth at one time or another, with 500 million as a reasonable guess. Of these, approximately 1 million are contemporary and have been described and cataloged as living species. Another 90,000 have been discovered as fossils of extinct forms. If Simpson's estimates are reasonably correct, the species known as either fossils or living forms amount to only 0.2 percent of the grand total. Thus the record of fossils, even when combined with living forms, provides only a very incomplete picture of the evolution of life. Even so, the fossil record provides good evidence that the plants, animals, and microorganisms on earth have changed and evolved with time.

OTHER EVIDENCE FOR EVOLUTION

Anatomical Comparisons

Evidence for evolution has also been developed through comparisons of the anatomy of living and extinct forms. These comparisons reveal relationships that cannot be reasonably explained except by evolution from common ancestors. The forelimbs of different animals, for example, are modified for functions as diverse as walking or running (in dogs), swimming (in whales), flying (in birds), and reaching and grasping (in humans). Despite these widely different functions, the arrangement of bones within the forelimbs in these animals is strikingly similar (Figure 18-4). The bones are so similar, in fact, that there is little doubt that all evolved from common ancestors with forelimbs consisting of the same bony elements. The same

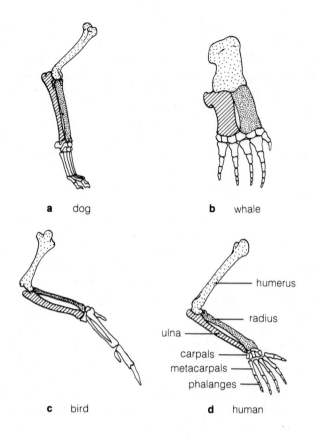

a dog   **b** whale

humerus
radius
ulna
carpals
metacarpals
phalanges

c bird   **d** human

Figure 18-4 An anatomical comparison of the bones of vertebrate forelimbs adapted for widely different functions.

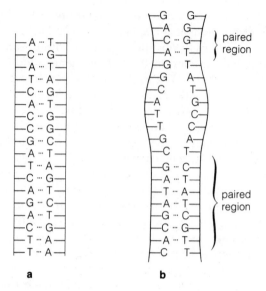

a   **b**

Figure 18-5 (**a**) Completely complementary DNA chains. (**b**) DNA chains from different organisms with complementary and noncomplementary regions (see text).

kinds of relationships can also be discerned through comparisons of the embryonic development of different plant and animal groups (see, for example, p. 400).

Molecular Comparisons

Similar conclusions have been reached by extending anatomical comparisons to the molecular level. These studies have produced evidence for evolution, and also for evolutionary relationships, by detecting similarities in the sequences of the DNA and protein molecules of different species.

Similarities in the nucleotide sequences of DNA molecules from different species are determined primarily by a method known as *DNA hybridization.* This method depends on the fact that DNA molecules consist of a double helix in which two nucleotide chains pair together. In order to pair, the two nucleotide chains must be *complementary:* an adenine in one chain must be matched with a

thymine, and a guanine with a cytosine, at the same positions in the opposite chain (Figure 18-5*a*). If single DNA nucleotide chains from different species are mixed together, they can pair and wind into "hybrid" double helices only in regions where they contain complementary sequences (Figure 18-5*b*). Therefore the extent to which they pair indicates the proportion of coding sequences that are the same in different species.

Application of the hybridization technique to humans and other primates (Table 18-2) reveals that 80 percent or more of DNA molecules of the various species shown are similar enough to form hybrid double helices. (Human and chimpanzee DNA are so similar in sequence that the hybridization technique cannot detect any differences between them!) The technique is also useful as a measure of the evolutionary *distance* between species—that is, how closely or distantly they may be related. Among primates, as noted, sequence similarities are found in 80 percent or more of the DNA in the different species tested. We share only 21 percent of our DNA sequences with mice, however, and only 10 percent with chickens. Thus we are clearly more closely related in evolutionary lineages with monkeys, apes, and gibbons than with mice, and more closely related to mice than chickens, at least on the basis of DNA sequence similarity. This conclusion is also supported by other comparisons such as gross anatomy, patterns of embryological development, and fossil lineages.

Table 18-2 Sequence Similarity Between the DNA of Humans and Other Organisms

Primates	
Human	100%
Chimpanzee	100
Gibbon	94
Rhesus monkey	88
Capuchin monkey	83
Tarsier	65
Slow loris	58
Galago	58
Lemur	47
Other Animals	
Tree shrew	28
Mouse	21
Hedgehog	19
Chicken	10

Table adapted from T. Dobzhansky, F. J. Ayala, G. L. Stebbins, and J. W. Valentine, *Evolution*. W. H. Freeman, San Francisco, 1977.

Protein sequencing provides evidence of a similar kind. In this case, a protein type found in all species of interest is extracted, purified, and sequenced. The resulting amino acid sequences give a direct measure of the structural similarity of the proteins in each species studied. Proteins frequently used for sequence comparisons include *cytochrome c*, one of the carriers of the electron transport systems of all eukaryotic organisms; *hemoglobin*, the blood protein of vertebrate animals; *myoglobin*, a muscle protein of higher vertebrates; *immunoglobin*, an antibody protein found in many animals; and *carbonic anhydrase*, an enzyme that converts CO_2 into carbonic acid in animals.

The sequences of cytochrome *c* molecules from humans and chimpanzees, for example, are identical; human and rhesus monkey cytochrome *c*'s differ in only one out of 104 amino acids. Other proteins show similar relationships: human and chimpanzee hemoglobin molecules are identical, and the carbonic anhydrase molecules of humans and chimpanzees differ in only one amino acid out of 115 sequenced in the molecules. Once again, similarities

this close in the sequence of protein molecules cannot be explained reasonably unless it is assumed that humans and other primates evolved from a common ancestor.

Since comparisons of amino acid sequence also provide a measure of the evolutionary distance between different species, they too can be used to establish probable evolutionary lineages. The method is especially valuable when a protein such as cytochrome *c*, which occurs in all eukaryotes, is used for the comparison. (Table 18-3 shows the numbers of amino acids differing between the cytochrome *c* molecules of humans and a variety of other species.)

EVIDENCE FROM CONTEMPORARY TIMES

The evidence for evolution developed from the study of fossils, comparative anatomy, and molecular sequences is complemented by examples of evolutionary change within recorded history. The evidence showing evolution in contemporary times refutes the idea that species are unchanging and have existed in their present forms since the origin of life. Some species, such as the passenger pigeon, have become extinct within the last century; many others are in danger of becoming so. Other organisms have become sufficiently modified within recorded history to show that species, rather than being fixed entities, undergo evolutionary change from one form to another. Some of these modern examples affirm that the changes occurring in evolution vary in rate and, moreover, that evolutionary lines may branch into many separate pathways as change takes place.

Among the best-documented modern-day evidence for evolution is the change in color observed in peppered moth populations in industrialized areas of England. These alterations in color confirm that a species may change with time, and they also reveal something about the mechanisms that bring about this change. A second study, of geographic variation in the English sparrows of North America, confirms that a species may branch or radiate along several different evolutionary pathways at the same time.

Color Changes in the Peppered Moth

The peppered moth, *Biston betularia* (Figure 18-6), is abundant in wooded areas in Great Britain and continental Europe. Before the 1840s, most members of this species were gray or off-white in color, with darker markings resembling pepper spots on the wings. Although uniformly dark individuals occurred, they were so rare before 1850 that

Table 18-3 Amino Acid Differences Between Cytochrome c of Humans and Other Organisms

Species Pair	Number of Amino Acid Differences
Human–rhesus monkey	1
Human–horse	12
Human–cattle, sheep	10
Human–dog	11
Human–rabbit	9
Human–chicken, turkey	13
Human–pigeon	12
Human–snapping turtle	15
Human–rattlesnake	14
Human–bullfrog	18
Human–tuna fish	21
Human–dogfish	24
Human–fruit fly	29
Human–screwworm fly	27
Human–silkworm moth	31
Human–wheat	43
Human–*Neurospora*	48
Fruit fly–screwworm fly	2
Fruit fly–silkworm moth	15
Fruit fly–tobacco hornworm moth	14
Fruit fly–dogfish	26
Fruit fly–pigeon	25
Fruit fly–wheat	47

Adapted from V. Grant, *Organismic Evolution.* W. H. Freeman, San Francisco, 1977.

the capture of a single black peppered moth in 1848 was an event worth publishing. Black moths began to increase in number after 1850, however; by 1895, 98 percent of the peppered moths captured in the English Midlands were uniformly black in color whereas the light gray, speckled variety was becoming rare. An equivalent change occurred during the same period in the Ruhr Valley in Germany. Both areas, the English Midlands and Ruhr Valley, were heavily industrialized between 1850 and 1900, a period of about 50 generations for the peppered moth.

In the 1930s an investigator studying the changes in peppered moth populations, E. B. Ford of Oxford University, linked the darker color to the blackening of trees, leaves, and rocks by soot in the industrial areas. The darker color, Ford suggested, was an evolutionary response in the moths to the requirement for darker camouflage against the sooty trees and rocks. In an unpolluted environment, light-colored moths with the speckled pattern are almost perfectly concealed whereas black moths stand out and are easy prey (Figure 18-6a). In a polluted, sooty environment, however, the black color offers better camouflage whereas pale, speckled moths stand out clearly against the dark background (Figure 18-6b). The change in color was not restricted to the peppered moth; more than 200 species of insects were noted to have evolved toward darker forms in response to the general blackening of the environment by soot.

Variability in populations and the effects of predators in the environment were significant factors in the evolution of the peppered moth toward darker color.

Ford's hypothesis was carefully tested during the early 1950s by another investigator from Oxford, H.B.D. Kettlewell. Kettlewell marked and released both dark and speckled peppered moths in Dorset, England, an unpolluted area, and noted the survival rate of the two colors against predation by birds. In the unpolluted environment, birds captured and ate about six times as many dark moths as speckled ones. In a polluted area near Birmingham, where trees were blackened by soot, speckled moths were captured and eaten about three times more often than dark moths.

From behind a blind, Kettlewell observed birds in the two environments and noted that they searched the trunks looking for moths. The birds—including robins, flycatchers, nuthatches, and thrushes—took the contrasting individuals most often and overlooked the camouflaged variety. Thus the immediate environmental factor favoring the survival of individuals of certain colors was predation by

Figure 18-6 (**a**) Dark and light forms of the peppered moth on a tree trunk in an area unpolluted by soot. The light form is indicated by a bracket. (**b**) Dark and light peppered moths on a tree trunk blackened by pollution. The dark form is indicated by a bracket. From the experiments of H.B.D. Kettlewell, University of Oxford.

birds; effectively camouflaged individuals were more likely to be missed by the predators and survive.

Genetic studies showed that the dark color was produced by the dominant allele of a single gene. Individuals carrying this allele as either homozygotes or heterozygotes (see pp. 250–251) were dark; homozygous recessives were speckled, light-colored moths. In polluted environments, predation by birds greatly reduced the number of moths carrying the allele for lighter, speckled color and increased the proportion of individuals with the allele for dark color. As a result, the dark moths were more numerous at mating time and hence more likely to pass on their alleles to the next generation.

The study of peppered moths confirms that the average characteristics of a species can change with time.

The study of color change in peppered moths confirms that a species can evolve from one form to another

with time. The study also illustrates two details of the evolutionary process that are crucial in producing change: variability and environmental selection. **Variability** existed in the population of peppered moths in the 1840s: the allele for black color, although rare, was already present as variability in the moth population before the industrialization of Europe in the 1850s. This alternate allele arose through chance **mutations** in the hereditary molecules of the moths. Mutations occur at a low but constant frequency in all natural populations and provide a source of variability that may be of survival value if environmental conditions change, as they did for the peppered moth after 1850.

The second detail of the evolutionary process revealed by the peppered moth study is the effect of the environment in selecting one allele over another. To result in evolution, selection must enhance or reduce the likelihood that a certain genetic type will survive and be more successful than other types in reproducing and passing its

Figure 18-7 Color differences due to evolutionary change in English sparrows collected from Death Valley (top row across page) and Oakland (bottom row across page) in California. Note that the feather color on top of the head (**a**) and on the breast (**b**) is lighter in the Death Valley specimens. Courtesy of R. F. Johnston and R. K. Selander.

alleles to the next generation. Reproduction is therefore nonrandom. That is, certain genotypes, as a result of selection, are more likely than others to survive and appear in the next generation. Environmental selection resulting in nonrandom reproduction may take many forms, including predators (such as the effects of birds preying on peppered moths), resistance to disease, competition for living space and food, or attractiveness to potential mates. How these factors operate in evolution is described further in Chapter 19.

As a result of selection, some genotypes are more likely than others to appear in the next generation.

Changes in English Sparrow Populations

A few English sparrows (*Passer domesticus*; Figure 18-7) were introduced from Europe into eastern North America in 1852. Descendants of these birds colonized the entire

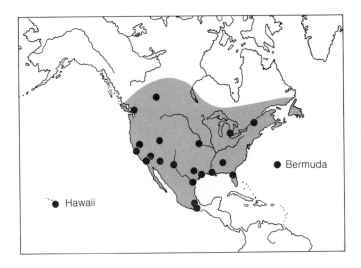

Figure 18-8 The distribution of English sparrows in North America (shaded area) and points at which samples were taken (dots) in the Johnston and Selander study. Courtesy of R. F. Johnston and R. K. Selander. Copyright 1964 by the American Association for the Advancement of Science.

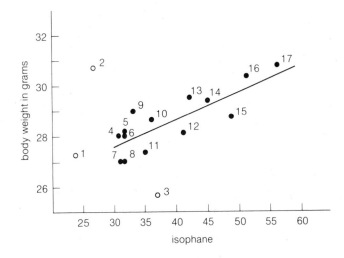

Figure 18-9 Average body weights of adult male English sparrows plotted against isophanes of the localities from which the birds were collected. (Isophanes, calculated from latitude, longitude, and altitude, reflect climate; higher isophanes generally indicate cooler climates.) Localities: (1) Oaxaca City, Mexico; (2) Progreso, Texas; (3) Mexico City, Mexico; (4) Houston, Texas; (5) Los Angeles, California; (6) Austin, Texas; (7) Death Valley, California; (8) Phoenix, Arizona; (9) Baton Rouge, Louisiana; (10) Sacramento, California; (11) Oakland, California; (12) Las Cruces, New Mexico; (13) Lawrence, Kansas; (14) Vancouver, British Columbia; (15) Salt Lake City, Utah; (16) Montreal, Quebec; (17) Edmonton, Alberta. Courtesy of R. F. Johnston and R. K. Selander. Copyright 1964 by the American Association for the Advancement of Science.

North American continent, reaching most areas by about 1900. English sparrows do not migrate in the cold season, and they remain in a relatively restricted territory during their lifetime. Therefore their spread across the continent in 100 or so generations resulted from reproduction and gradual dispersal of offspring into new areas near the regions occupied by the parent birds the year before.

In the range now inhabited by the species, from northern Canada to Mexico and from the east to the west coast, environmental conditions vary over wide extremes. Birds from the limits of this range (Figure 18-8) were compared in 1964 by Richard F. Johnston of the University of Kansas and Robert K. Selander of the University of Texas. Their study revealed that evolutionary change has taken place in different directions in various parts of the range.

Sparrows collected in the arid Southwest, from southern California to central Texas, were pale in color; birds of lightest color lived around Death Valley, California, and Phoenix, Arizona. Birds found in the rainy Northwest, around Vancouver, British Columbia, were much more darkly pigmented. This variation in body color from lighter pigments in arid regions to darker coloration in moist climates is an evolutionary trend frequently noted in other warm-blooded species (see Gloger's rule, p. 471). Johnston and Selander noted other differences in coloration in sparrows from various parts of the North American range. Birds in Louisiana and Mexico, for example, had a conspicuous yellow color on the posterior underside of the body; this color was absent from sparrows sampled in other regions.

Among the other characteristics measured by Johnston and Selander was body weight, which also varied among the sparrows along a north–south distribution (Figure 18-9). Birds in Mexico were lightest in weight, averaging about 27 grams; in the most northerly regions of Canada body weight averaged 31 grams. This weight distribution also follows an evolutionary pattern frequently noted for warm-blooded animals, called Bergman's rule, in which larger animals are found in colder climates (see p. 471).

Thus various branches of the North American English sparrow population have evolved in different directions that depend on environmental factors in different parts of the continent. This branched evolution is reflected in variability in the population as a whole; variability also occurs regularly within the populations sampled in restricted areas. Figure 18-10 shows the range of variability in bill length in individuals sampled in Oakland, California, and

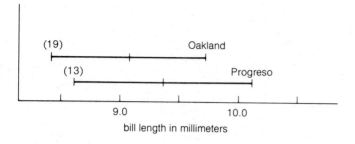

Figure 18-10 Range of variability in bill length of sparrows captured in Oakland, California, and Progreso, Texas. The length of the line shows the range of the variability; the vertical line in the center indicates the average bill length. Courtesy of R. F. Johnston and R. K. Selander. Copyright 1964 by the American Association for the Advancement of Science.

Progreso, Texas. Bill length is different in the two locations, averaging about 9 millimeters in Oakland and 9.5 millimeters in Progreso. Among the birds from either location, variability in bill length is also obvious—running from a minimum of 8.4 to a maximum of 9.7 millimeters in Oakland and between limits of 8.6 and 10.1 millimeters in Progreso.

The differences among the various sparrow groups, although relatively small, are significant and surprising considering the short time, on the evolutionary scale, since the birds first colonized North America. Continued change along these lines, in response to the environmental conditions experienced by birds in different parts of the country, will probably lead to more pronounced differences in sparrows, perhaps even to branching of the sparrow population into several separate species. How this might come about is described in Chapter 20.

Other instances of evolution within recent history have been recorded. Bananas, for example, were first introduced into the Hawaiian Islands by Polynesians about a thousand years ago. After introduction of the banana, a group of native moths normally feeding on palms in Hawaii evolved into forms feeding on bananas. This evolution has continued until there are presently five different moth species that eat nothing but the banana. (See also the recent evolution of hawthorne flies, p. 465.)

Still other contemporary examples are provided by the development of resistance to antibiotics by various disease-causing bacteria. Antibiotics such as penicillin, streptomycin, and tetracycline were developed and administered to control disease bacteria during the 1940s and 1950s. At first these antibiotics were effective at low concentrations in killing or retarding the growth of bacteria. However, the effects of each antibiotic altered according to a consistent pattern in the years following its introduction. Gradually, greater and greater concentrations of the antibiotic were required to produce the same controlling effect. Eventually strains of the bacteria appeared that were completely resistant to the antibiotic.

Experiments with *Escherichia coli*, a nonpathogenic bacterium living in the large intestine of humans, illustrate how the resistance develops. Although *E. coli* is killed readily by streptomycin, a rare mutation appearing in as few as one out of one billion cells allows *E. coli* to survive in the presence of the drug. Even though the mutation is rare, the fact that a single bacterial cell can produce 10 billion descendants in a 24-hour period greatly increases the chance that a mutant form resistant to streptomycin will appear and become established. Indeed, some *E. coli* strains have appeared that cannot survive unless the antibiotic is present!

Resistance of insects to insecticides follows a similar pattern. The pesticide DDT, developed during the early 1940s, was highly effective against houseflies at low concentrations when it was first applied. By 1947, however, resistant populations had become established; these gradually spread until the effectiveness of DDT was greatly reduced and successively higher concentrations were required to produce a killing effect. The resistant forms are selected from the constantly appearing mutants in the pest populations by the applied antibiotics or insecticides, just as black peppered moths are selected by the effects of bird predators that eat light moths exposed against a dark background.

These contemporary examples confirm and extend the conclusions about the evolutionary process drawn from the fossil record and from anatomical and molecular studies. Species do change with time, evolving from one form to others; in doing so, some new species appear and others vanish. As evolution continues, the products may branch along several different evolutionary pathways at any one time.

DARWIN, WALLACE, AND THE THEORY OF EVOLUTION

The evidence described to this point shows that evolutionary changes have taken place among species both past and present and, moreover, that such changes continue today. The mechanisms underlying evolutionary change are less obvious, however, and are still the subject of some controversy among scientists. This controversy surrounds only the details of the process: all scientists agree that evolution has taken place and that species change with time.

Figure 18-11 Charles Darwin as a young man. The portrait was painted in 1840, a few years after his return from the *Beagle* expedition. By kind permission of the President and Council of the Royal College of Surgeons of England.

Figure 18-12 Alfred Russel Wallace in 1853. Courtesy of the American Museum of Natural History.

Charles Darwin and Alfred Russel Wallace

The groundwork for the modern theory of evolution was laid by Charles Darwin and Alfred Russel Wallace (Figures 18-11 and 18-12) and published in 1859. Darwin was educated to be an Anglican priest but was actually a naturalist by inclination and profession. In 1831, at the age of 22, he signed on the British ship *Beagle* as resident naturalist on a cruise officially undertaken to chart coasts and harbors for use by British vessels around the world. Even though the post was offered without pay, Darwin had several competitors and had to prove to the captain that he would be a tolerable companion for the voyage, which was planned to extend for several years. On the voyage, which lasted until 1836, Darwin observed nature in the ports and islands the *Beagle* visited, carefully cataloging everything that captured his attention as he went. While the ship called at South American ports, Darwin traveled overland on horseback and had an opportunity to observe extensive beds of fossils exposed by erosion. Among the fossils, he saw skeletons obviously mammalian in form but of creatures no longer known on the earth. Yet these fossils were similar enough to living South American mammals to suggest that the living species may have been derived from these now-extinct forms. Darwin also made extensive observations of the plants and animals on the Galápagos Islands, where he saw species that were related to animals and plants on the nearby South American mainland. The island species were different from their mainland counterparts in many details, however; differences could be noted even between members of the same species living on different islands of the group. These islands, and the animals and plants peculiar to them, were to be particularly important to Darwin's theory of evolution. By the time his travels on the *Beagle* ended, Darwin, originally a believer in special creation, was convinced that species of animals and plants do not remain constant with the passage of time but change from one form into another.

On his return to England in 1836, Darwin began to write down his thoughts on how evolutionary change in living organisms might occur. He also gathered evidence from the comparative anatomy and embryology of living species and experimented with the effects of artificial selection on pigeons. It was obvious from these experiments that human intervention too, by selecting unusual variants or "sports" among the offspring of domesticated animals for breeding, could bring about change. Darwin was also heavily influenced at this time by an essay on human population growth written by the English clergyman Thomas Malthus. Malthus pointed out that humans have so great a potential for growth that if all our offspring survived, we would soon outstrip the available resources and living space of the world. Malthus demonstrated in his essay that this outcome is prevented by environmental factors such as disease, starvation, and war, which keep the human population within reasonable limits. Darwin realized that although all organisms have the same potential for unlimited growth, environmental factors intervene in all cases to reduce the number of offspring that actually survive. In 1838 the idea came to him that this influence of environment on the potentially unlimited nature of population growth produces evolutionary change. With this realization, the major parts of his theory of evolution fell into place. Darwin was extremely cautious in expounding his ideas, however, and continued to add observations and perfect his concept until 1844, when he wrote a long essay detailing his views on evolution.

All organisms have the potential for unlimited growth. They are kept from growing to unlimited numbers by environmental factors.

Since Darwin considered his work incomplete, he merely discussed his essay with friends and colleagues rather than publishing it. His painstaking efforts to document his views would no doubt have been extended considerably except for the work of another naturalist, Alfred Russel Wallace. By coming up independently with the same ideas, Wallace forced Darwin to publish.

Wallace was as impulsive as Darwin was deliberate. His interest in natural history had carried him to such out-of-the-way places as the Amazon and the East Indies. By 1855 he had published a paper, sent from Indonesia, revealing his belief that species can evolve with time. Wallace continued his intensive collection and observation in Indonesia until 1858, when an attack of malarial fever forced him to bed for a week. With little else to do but think about his work, and relieved of his pressing daily schedule, Wallace suddenly saw the interrelationships in his findings. He too had read Malthus' essay and was strongly influenced by it; he quickly realized that evolution results from the effects of the environment on organisms that produce more offspring than can possibly survive. Unlike Darwin, however, he immediately set his thoughts on paper. Within three days he had outlined his theory of evolution and sent it to Darwin for his opinion, unaware that Darwin had come to the same conclusions years before. Wallace's paper on evolution was a shock to Darwin, who now realized that his work of many years had been independently discovered by someone else and was written in a form ready for immediate publication.

Darwin's first reaction was to concede credit to Wallace for deducing the mechanism of evolution. At this point, friends and colleagues of the two naturalists stepped in and convinced them that the fairest procedure was to announce their conclusions jointly. Accordingly Darwin prepared his own results in brief form for publication and the two papers were read at a meeting of the Linnean Society of London on July 1, 1858. Subsequently Darwin condensed his planned longer work into a much shorter version: his classic *On the Origin of Species by Natural Selection,* published in 1859.

Publication of the theory produced a tidal wave of controversy and criticism. The public was outraged. Not only did the theory directly contradict the biblical teaching of special creation but proposed that we have ancestors in common with the apes. Some scientific criticism came from other naturalists who were disturbed by gaps in the Darwin–Wallace theory. Although later discoveries eventually answered the major scientific objections, controversy on religious grounds has continued to the present day.

The Darwin–Wallace Theory

The evolutionary theory outlined by Darwin and Wallace in 1858 contained the following propositions:

1. Species produce more offspring than can in fact survive. If all the offspring of animals and plants lived to adulthood and reproduced, there would be no living space left on the earth within a few generations. The fact that this does not occur shows that environmental factors limit the numbers of individuals that actually survive.

2. Among the offspring of a species small variations exist, so that virtually none are exactly alike. Some of these varied individuals have characteristics that make them more successful than the others in their environment; others have differences that reduce their ability to grow and compete for space and food.

3. The environmental factors that reduce the survival of offspring tend to eliminate individuals with less favorable variations. Individuals with favorable characteristics are more likely to survive.

4. The surviving individuals live to adulthood and reproduce. These individuals pass on their more favorable characteristics to the next generation.

5. As a result, the next generation has different average characteristics. Over many generations, this process, termed **natural selection**, results in gradual evolution toward greater adaptation to the environment. Eventually the changes are extensive enough to produce new species.

The theory of evolution presented by Darwin and Wallace is perhaps the most fundamental and important deduction of biology. No idea has had a greater effect on the development of the science. Nevertheless, two gaps in the theory were to lead to a half-century of intensive controversy among scientists. One was that the theory did not explain the *source* of variation noted among the members of a species. What mechanism produces the stable variations that can be selected and passed on to offspring? The second weakness involved the *transmission* of characters from parents to offspring. In Darwin and Wallace's day, inheritance was generally believed to ''blend'' the characteristics of the parents. Such blending, it seemed, would always dilute the favorable characteristics and produce intermediate types; there was hence no possibility of producing the highly adapted species observed in nature. How were the favorable characteristics of the survivors, once selected for, preserved intact and passed on to the next generation?

Answers to these criticisms had to await the findings of genetics. Although Mendel was a contemporary of Darwin and Wallace, and discovered genes and their manner of transmission from parents to offspring during the 1860s, another 40 years were to pass before this information was recognized or understood by biologists. With the rapid development of genetics after 1900, the major gaps in evolutionary theory were filled in. Scientists realized that gene mutation is the ultimate source of the variability observed in nature. Moreover, the existence of genes and alleles as hereditary units that are passed unchanged from parents to offspring fully explained the persistence in later generations of the traits created by mutation and selected by the environment.

The new information from genetics therefore allowed the process of evolution to be traced in terms of genes and their functions. With this information, the basic cellular mechanisms producing variability and transmitting characteristics from parents to offspring could be explained and understood. These advances in the understanding of evolutionary mechanisms have been complemented in recent years by work in environmental biology (ecology), which has shed new light on how species interact with their environment to undergo evolution. These modern additions to the theory of evolution, and the mechanism of evolution as now understood in the light of these additions, are the subjects of the next chapter.

Questions

1. What are fossils? How were they formed? What kinds of animals are most frequently found as fossils? What conditions favor the formation of fossils?

2. How are fossils dated?

3. Trace the major developments in the evolutionary line leading from *Hyracotherium* to modern horses. What environmental conditions contributed to these developments?

4. What evolutionary trends and patterns are evident in the horse lineage?

5. What are the limitations of the fossil record?

6. How are anatomical comparisons used as evidence for evolution? Why do you suppose that horses have four legs, rather than six like insects?

7. How do molecular studies provide evidence for evolution? How are these studies used to establish evolutionary lineages?

8. What environmental conditions were responsible for the shift in color in the peppered moth from speckled gray to black? What parts of the evolutionary mechanism are illustrated by the color change in peppered moths?

9. What variations were noted among sparrow populations in the United States? How do these variations provide evidence that evolution has occurred in these birds?

10. What evolutionary mechanism underlies the development of resistance to antibiotics in bacteria?

11. Can you think of any further examples of present-day evolution not covered in the text? Can you think of any new species that have evolved or become extinct within recorded history?

12. Do you regard the development of new breeds of domestic animals and plants as an example of evolution? Why?

13. Do you think that humans are the most highly evolved animals? Why?

14. Do you think that a group of living beings on some other planet, if they exist, would be exactly like those on earth? Why?

15. What observations changed Darwin's belief in special creation to a conviction that evolution has occurred?

16. What influence did Malthus have on Darwin's thinking?

17. Outline the main points of the Darwin–Wallace theory.

18. What is natural selection?

19. What major parts of the evolutionary process were not explained by the Darwin–Wallace theory?

20. Do you think that the theory of evolution denies the existence of a supreme being? Why?

Suggestions for Further Reading

Ayala, F. J., and J. W. Valentine. 1979. *Evolving.* Benjamin/Cummings, Menlo Park, California.

Bishop, J. A., and L. M. Cook. 1975. "Moths, Melanism and Clean Air." *Scientific American* 232:90–99.

Futuyma, D. J. 1979. *Evolutionary Biology.* Sinauer Associates, Sunderland, Massachusetts.

Gould, S. J. 1977. *Ontogeny and Phylogeny.* Harvard University Press, Cambridge.

Johnston, R. F., and R. K. Selander. 1964. "House Sparrows: Rapid Evolution of Races in North America." *Science* 144:548–550.

Kettlewell, H.B.D. 1959. "Darwin's Missing Evidence." *Scientific American* 200:48–53.

Moorehead, A. 1969. *Darwin and the "Beagle."* Harper & Row, New York.

Patterson, C. 1978. *Evolution.* Cornell University Press, Ithaca, New York.

Scientific American 239, September 1978 (Evolution Issue).

Simpson, G. G. 1951. *Horses.* Oxford University Press, New York.

Smith, J. M. 1975. *The Theory of Evolution.* Penguin, Baltimore.

The findings of genetics filled in the gaps in the Darwin–Wallace theory and made it clear that (1) mutations are the source of the variability acted upon by natural selection and (2) hereditary information is maintained and passed from generation to generation in the form of specific units, the genes, rather than through blending of parental traits. The research following the rediscovery of Mendel's findings in 1900 allowed the basic mechanism of evolution to be worked out in the light of these findings: how the variability produced by mutations appears and is maintained in populations, and how the environment acts upon this variability to produce evolutionary change.

These basic parts of the evolutionary mechanism—the production and maintenance of variability and the interaction between variability and the environment to produce evolutionary change—are described in this chapter. The following chapter shows how the evolutionary changes arising from these sources may become extensive enough to produce the central result of evolution: the creation of new species.

VARIABILITY: ITS PRODUCTION AND MAINTENANCE

Producing Variability: Mutations

Variability is present in all species. Among humans, variability is reflected in the common observation that "no two individuals are alike." Although Darwin observed and carefully cataloged this natural variability in humans and other species, he was unable to account for its source. We now know that all inheritable variations arise ultimately through mutations in DNA, the molecule that stores genetic information.

All inheritable variation arises ultimately from mutations. Any change in the nucleotide sequence of DNA is a mutation.

A mutation may be defined simply as any change in the sequence of nucleotides in a gene. For a mutation to be significant to the evolutionary process, it must occur in the DNA of reproductive cells—that is, in cells that give rise to new individuals. In animals and plants that reproduce sexually, the important cells are those giving rise to gametes, the eggs and sperm. Mutations in the DNA of these cells are incorporated into the nucleus of the zygote at fertilization and, through the mitotic divisions taking place during development, are duplicated and passed on to all the body cells of the new organism. As a result of the mutations, the organism may be more or less likely to

19

THE MECHANISM OF EVOLUTIONARY CHANGE

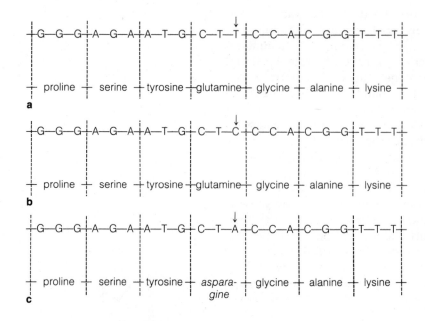

Figure 19-1 The effect of mutations on proteins encoded in a gene. (**a**) The gene sequence and the corresponding sequence of amino acids in a protein. (**b**) A mutation changing thymine to cytosine at the position marked by the arrow would cause no change in the amino acid sequence since the DNA code words CTT and CTC both code for glutamine. (**c**) A change from thymine to adenine at the same position would cause substitution of asparagine for glutamine.

pass on its mutated genes to its own offspring. If the mutations impart greater reproductive success to the individual, its offspring will be more numerous, thereby increasing the proportion of individuals carrying the mutated genes in the next generation. Mutations that occur in body cells that do not give rise to new individuals (somatic cells) have virtually no effect on the process of evolution.

The Nature of Mutations

Gene Mutations Mutations that occur within the length of DNA coding for a single function such as an mRNA or rRNA molecule are called **gene mutations.** Most common are changes in single bases of a DNA sequence, involving the substitution of one base for another or the addition or deletion of single bases. (The molecular mechanisms producing such changes are described in Supplement 19-1.) Mutations of this type may cause the substitution of one amino acid for another in the protein encoded in the mutated gene (Figure 19-1*a* and *b*). Not all mutations involving single base changes cause amino acid substitutions, however, because many of the amino acids are specified by more than one three-letter code word in DNA. Muta-

tion of the code word CTT to CTC, for example, would cause no change in the amino acid chain of protein coded for by a gene because both CTT and CTC are code words for the amino acid glutamine (see Figure 19-1*a* and *b*). A mutation of CTT to CTA, however, would cause asparagine to be substituted for glutamine (Figure 19-1*c*). If an amino acid substitution does occur, the mutation creates a new **allele** of the gene.

Gene mutations that involve single nucleotide changes are sometimes called **point mutations.** Gene mutations may also involve longer segments of DNA including several nucleotides instead of single bases. Most commonly these longer changes involve deletions in which a sequence of the code involving several bases is lost. If the deletion includes more than just a few nucleotides, such losses are usually lethal to an organism.

Although most mutations are deleterious, some provide an advantage to the individual carrying them.

Most gene mutations, whether they involve single nucleotides or several, are deleterious or, at best, bring no advantage to the individuals carrying them. This is be-

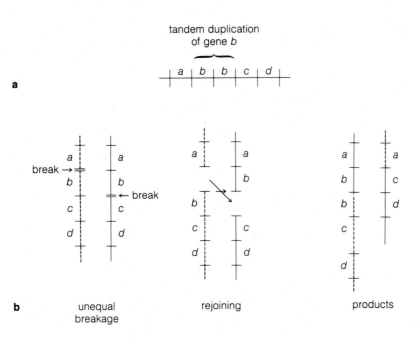

Figure 19-2 Tandem duplications of genes (**a**) may arise through unequal crossing over in meiosis (**b**).

cause most gene mutations ultimately show up as changes in structural and enzymatic proteins. Changing these molecules at random, as mutations do, is likely to upset the complex balance that produces the coordinated biochemical activity of an organism. A good analogy would be to note the effects of inducing "mutations" in a complex electronic device such as a computer by changing the values of resistors, capacitors, and other components at random. Most of the random changes would interfere with performance or, at best, would have no effect. A few mutations out of thousands or millions might, however, increase the speed, efficiency, or capacity of the computer—or, in the case of mutations in nature, the reproductive success of an organism.

These predictions are supported by experimental evidence. A test of mutations detected in the X chromosome of *Drosophila melanogaster* showed that 90 percent reduced the general viability and reproductive success of the flies carrying them; the remaining 10 percent were favorable. Another study, in barley, showed even fewer favorable mutations—in this species, only 0.1 to 0.2 percent of the mutations increased the viability of the plants. Among the recorded examples of single mutations leading to greater viability are the changes producing resistance to poisons or antibiotics among bacterial, insect, and other pest species. Increased resistance in these species can frequently be traced to single gene mutations.

Occasionally gene mutations involve duplications of an entire gene, so that offspring receive two copies of the gene that are repeated in tandem in the chromosome (Figure 19-2a). Such changes may arise through rare mistakes in the breakage and exchange mechanism producing recombination in meiosis (see Figure 11-9). In these cases, the breakage and exchange is unequal and one chromosome receives extra DNA (Figure 19-2b). Gene duplication by such mechanisms has probably been an important source of genetic change in evolution. The hemoglobin molecules of lower vertebrates, for example, are relatively simple proteins consisting of polypeptides transcribed from a single gene. The more complex and efficient hemoglobin molecules of higher vertebrates contain polypeptides transcribed from two different genes. These genes, which are similar in sequence and arranged in tandem on the chromosomes, probably arose through a duplication resulting from unequal crossing over during meiosis.

Chromosome Mutations Extensive changes involving the DNA of several genes are called **chromosome mutations**.

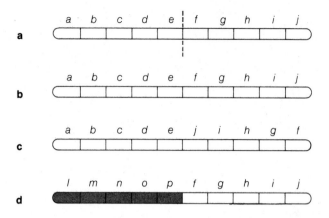

Figure 19-3 Chromosome mutations. Suppose that the DNA of a chromosome breaks at the position shown by the dotted lines in **a**. The broken segment may attach (**b**) in the same position, (**c**) in reversed position, or (**d**) to the end of a different chromosome.

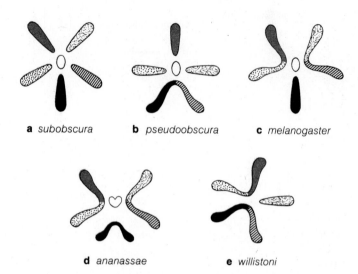

Figure 19-4 Chromosome mutations in five *Drosophila* species. The ancestral chromosome probably resembled the arrangement present in *D. subobscura* (**a**); a variety of breaks and reattachment of chromosome parts led to the rearrangements present in the other four species (**b–e**). Chromosome regions shown with the same shading contain the same genes. Courtesy of F. J. Ayala, from T. Dobzhansky, F. J. Ayala, G. L. Stebbins, and J. W. Valentine, *Evolution*. W. H. Freeman, San Francisco, 1977.

Most of these changes result from breaks introduced into the sugar–phosphate backbone chains of DNA (Figure 19-3). These breaks separate a DNA molecule into two or more pieces, which may subsequently reattach in the same position (Figure 19-3a), attach in reversed position (Figure 19-3b), attach to the DNA molecule of a different chromosome (Figure 19-3c), or be lost entirely. All these changes, which take place in nature, can be induced by various chemicals found in the environment or by radiation such as x rays.

Deletions of chromosome segments are usually lethal at an early stage of development to any individuals carrying them. As a result, this type of chromosome mutation has little significance as a source of variability for evolution. Reversed attachment of a broken segment (Figure 19-3b), however, or the attachment of a broken segment to the end of a different chromosome (Figure 19-3c), may have less drastic effects. One effect may be to prevent recombination of alleles within reversed or translocated segments through interference with chromosome pairing in meiosis. Or such rearrangements might affect the mechanisms regulating the transcription of genes within the altered segments. While most of these rearrangements are expected to have deleterious effects, as in the case of chromosome mutations, some are favorable and increase the reproductive success of their carriers.

These chromosome mutations, which are frequently encountered in nature, are believed to be an important source of variation for evolutionary change. (Figure 19-4 shows the chromosome rearrangements observed between several closely related *Drosophila* species.) Chromosome rearrangements are especially common in plants; some animals, such as scorpions, grasshoppers, cockroaches, and certain species of flies, also show a large proportion of individuals with altered chromosomes. Chromosome rearrangements occur in humans as well. Down's syndrome (see p. 282), a mutation with a deleterious effect in humans, involves addition of an entire extra chromosome. In a few animal and plant species, chromosome mutations are so frequent that it is difficult to decide which chromosome arrangement is the normal **wild type** for the species.

There is another type of extensive chromosome change that may act to promote variability. In many plants, and to a lesser extent in animals, the chromosome number sometimes doubles when a cell fails to divide after replicating its DNA. If doubling of this type occurs in the divisions leading to gametes or in the first division of the fertilized egg, a line of individuals called **polyploids** may be produced with two, three, four, or more times the normal number of chromosomes. This change, usually lethal in animals, is tolerated by many plants and often leads to individuals of greater size and vigor.

Why polyploids are less viable in animals is uncertain. Attempts to produce animal polyploids by artificial techniques in the laboratory are rarely successful. Since most of the induced animal polyploids die as embryos, developmental difficulties may be responsible. Animal development, which is characteristically longer and more complex than plant development, depends on a precise balance and interaction between many genes; polyploidy probably upsets this balance and leads to lethal developmental abnormalities.

Polyploids are produced by changes that multiply the chromosome number of a species.

Polyploids can be artificially induced by exposing cells to colchicine, a drug that destroys the spindle and prevents division of the chromosomes. At telophase, the nuclear envelope reforms around the entire mass of chromosomes, producing a nucleus with twice the normal chromosome number. Artificial chromosome doubling by this method has been used with success by plant breeders to produce hardier and more productive lines of ornamental and crop plants, including wheat, corn, bananas, and tobacco.

Mutation Rates   Mutations of all kinds occur at low but steady and predictable rates—from the rarest at one out of 1 billion to the most frequent at one out of 10,000 copies of a gene (see Table 19-1). Average rates fall in the vicinity of about one in 100,000 gene copies. This means that for each gene with this average mutation rate in a diploid species, about one out of every 50,000 individuals would carry a newly mutated form of that gene. (The number is 50,000 instead of 100,000 since diploids carry two copies of each gene.) While these mutation rates seem low, the large number of genes in individuals and the large number of individuals in many species assure that new mutations will appear frequently. Assuming that humans have about 30,000 different genes, for example, the average rate of one mutated gene out of 100,000 copies indicates that each person is expected to carry about $30,000 \times 2 \times 1/100,000 = 0.6$ new mutations. (Here 30,000 is multiplied by 2 since we are diploid and carry two copies of each gene.) In other words, each of us has about 1 chance in 2 of harboring a mutation not found in our parents. Over the human population as a whole, which includes about 4 billion people, 0.6×4 billion = 2.4 billion new mutations are expected to appear each generation.

These mutations are the raw material for the mechanism of evolution. While many or even most will be unfavorable, some will increase the chance that their carriers will leave offspring in the next generation. The chance that new mutations will be favorable is enhanced in changing

Table 19-1 Mutation Rates of Some Human Genes	
Character Caused by Mutated Allele	Number of Mutants per Million Gametes
Achondroplasia (dwarfing)	6 to 13
Aniridia (absence of iris in eye)	2.9 to 2.6
Retinoblastoma (tumor of retina)	6 to 7
Muscular dystrophy	43 to 92
Hemophilia (failure of blood to clot)	22 to 32
Tuberous sclerosis (brain tumors; progressive mental deterioration)	6 to 10.5

From F. Vogel and R. Rathenberg, *Advances in Human Genetics* 5:223 (1974); courtesy of F. Vogel and Plenum Publishing Corporation.

environments or among individuals colonizing a new habitat. In these cases the variability supplied by the mutations may provide types that are better suited to survive under the new conditions.

Maintaining Variability: Diploidy and Sexual Reproduction

Since the time of Darwin and Wallace, evolutionary theorists have argued about how variability is maintained in a species once it has been generated by mutations. According to evolutionary theory, environmental factors select favorable mutations and eliminate those that are less useful. Some evolutionists claim that if this line of reasoning is carried to its logical end, the genotypes in a species would gradually become almost completely uniform as the most favored and efficient type. No variation among individuals would be likely except for new alleles entering through mutation.

Yet observation of species in nature shows that variability is the rule rather than the exception. Somehow great stores of variability are maintained in the form of different genetic alleles. Further, calculations show that among eukaryotic organisms the amount of variability from one generation to the next is too great to be accounted for by new mutations alone. How is this additional variability generated and maintained?

The maintenance of this additional variability depends primarily on two features common to the life history of almost all higher organisms: **diploidy** and **sexual reproduction.** (For a discussion of diploidy see Information Box

10-3.) During sexual reproduction, alleles in diploids *recombine* to produce new combinations for environmental testing.

Diploidy and Variability The most important single characteristic that promotes the maintenance of variability in higher organisms is the condition of diploidy. Except for prokaryotes and a few eukaryotic species such as fungi and mosses and certain algae, all organisms spend a major part of their life cycles as diploids. In this condition, each gene is present in two copies (except for genes carried on the sex chromosomes in XY or XO individuals—see pp. 263–269). Because each gene is present in two copies, one or both genes of a pair may carry a mutation. If both genes of a pair are in the mutant form, and if the mutation is unfavorable in the environment, selection against the allele will be as complete as it would be in a haploid. However, if the mutant allele is carried in an individual as a recessive along with a more favorable dominant allele of the same gene, the mutation may be sheltered from the effects of selection. This protection of alleles in diploid individuals carrying one normal and one mutant allele reduces the rate at which unfavorable alleles are eliminated from populations and maintains the store of variability produced by mutations.

Diploidy, by allowing heterozygotes to carry recessive alleles without detriment, reduces the rate at which unfavorable alleles are eliminated by selection and maintains variability by mutation.

The human disease sickle-cell anemia (see p. 281) provides an excellent illustration of this point. This disease is caused by a recessive, mutant allele of a gene for hemoglobin protein. As diploids, all humans carry two copies of the gene for the protein in question. Individuals with two copies of the mutant allele (homozygous recessives) produce a defective S form of hemoglobin (HbS) that leads to breakdown of red blood cells and causes extreme deficiencies in circulation of the blood. These persons are so greatly reduced in viability that they rarely live long enough to reproduce and pass their mutation on to the next generation. Thus one would predict that the allele would soon be reduced to very low levels of much less than 1 percent through mutation. Yet the mutant allele for HbS in nonmalarial areas of the world is present in 2 or 3 percent of the individuals tested. (In malarial areas as many as 40 percent of the population carry the allele.)

Maintenance of the allele at this level depends on the fact that heterozygotes—that is, individuals with one normal and one mutant allele—produce enough normal hemoglobin to survive and reproduce successfully. As a result, the mutant allele is sheltered from selection in these heterozygotes and persists at higher levels than expected from mutation alone.

Sickle-cell anemia also illustrates how the variability "protected" by the diploid condition may provide an advantage in changing environments. Among humans living in nonmalarial areas, the mutant allele causing sickle-cell anemia offers no advantage to heterozygotes. In fact, the allele is a disadvantage because it reduces the reproductive potential of heterozygous carriers; some of the offspring of two heterozygotes who mate are likely to be homozygous recessives who cannot survive. But in malarial regions, heterozygotes possessing the recessive allele have an advantage. They are less susceptible to malaria than noncarriers of the allele because the organism causing the disease, a protozoan blood parasite, has only a limited ability to grow in blood cells containing HbS. (The red blood cells of heterozygotes carry HbS as well as the normal, nonmutant hemoglobin.) As a result, heterozygotes have a much lower malarial infection rate both as children and adults; even when the disease is contracted, its effects are greatly reduced in severity. If the world environment changed so that malaria spread out of control, the reservoir of variability maintained in heterozygous carriers of the HbS allele might become an important factor in the preservation of our species.

Haploid organisms such as bacteria and the blue-green algae are able to maintain variability through mutations alone because generations come and go more rapidly than substantial environmental changes. The life span of a generation of bacteria may occupy only 20 minutes, for example. In such organisms, a single cell carrying a new mutation could produce 36 generations, or $2^{36} = 700$ billion individuals, in only 12 hours under the most favorable conditions. By this means, a favorable mutation could spread rapidly in the face of changing environmental conditions whereas unfavorable ones could be promptly eliminated.

Maintaining variability through mutations in haploids is possible only in simple organisms with very short generation times. The longer generations characteristic of complex plants and animals favored the appearance of a new mechanism for maintaining variability; this mechanism was supplied by the evolution of diploidy. The fact that almost all higher organisms have diploid phases in at least part of their life cycles shows that this condition arose very early in the evolution of life, probably at about the time of the first eukaryotic cells.

Sexual Reproduction and Variability Diploidy allows less favorable mutations to persist in a species as a reservoir of variability that may be of use if environmental conditions change. The second feature of the reproductive cycles of higher organisms, sexual reproduction, provides a means for mixing this store of mutations into a wide variety of new combinations for testing by the environment.

This mixing of the mutant and "normal" alleles of a diploid individual is accomplished by the mechanisms of meiosis, gamete formation, and fertilization. In meiosis (see Chapter 11), the two representatives of each chromosome pair align together, thereby bringing the two alleles of each gene into close proximity. The paired chromosomes may then trade segments through the mechanism of breakage and exchange (see Figure 11-9). As a result of the exchange, the chromosomes, while retaining the same genes in the same order, have new combinations of alleles. Further generation of new combinations occurs in the divisions of meiosis and during the fusion of sperm and egg nuclei in fertilization. During the meiotic divisions, the two members of each chromosome pair are separated and delivered in random combinations to the haploid gamete nuclei. At fertilization, chance dictates the pair among the millions or billions of gametes that actually fuse to form the zygote.

The total variability that can be generated in a species that has numerous chromosome pairs is huge. Measurements of heterozygosity have shown that in humans, for example, about 6.7 percent of the genes present in an individual are likely to be in the heterozygous condition. Assuming that humans have about 30,000 genes means that each person would have about 2000 genes with different alleles on the two chromosomes of the pairs carrying them. These heterozygous alleles could recombine to produce about 2^{2000} or 10^{600} different gametes. This number is much larger than the total number of atoms in the universe!

Sexual reproduction and recombination in diploids greatly increase variability by mixing alleles into new combinations.

Sexual reproduction and recombination thus provide the advantage of generating an almost infinite variety of genetic types to preserve the species in the face of a changing environment. But there are also disadvantages to the sexual pattern of reproduction. Since only females bear offspring, the number of individuals that directly contribute new individuals amounts to only one-half of the species (assuming that males and females occur in equal numbers). Moreover, the rapid and extensive mixing of alleles through recombination may break up favorable combinations as well as generate new ones. As a result, the species may never become as highly specialized and successful in a stable environment as it might without sexual reproduction and recombination.

These disadvantages have evidently been significant enough to produce some higher eukaryotes that have lost the ability to reproduce sexually either partially or completely. Both plant and animal species provide instances of this development. Aphids, for example, reproduce partially by parthenogenesis, a mechanism by which females reproduce additional females without fertilization by a male. These examples are the exception rather than the rule, however. For most species, the environment is sufficiently unstable to make the advantage of the variability generated by sexual reproduction and recombination outweigh the disadvantages. Thus the sexual pattern of reproduction is selected and maintained in evolution.

The total variability generated by mutation and maintained by diploidy and sexual reproduction, which provides the raw material for evolution, is continually tested by the environment. As a result of environmental conditions, some mutations are selected to be passed on in greater numbers to the next generation.

The variability produced and maintained by mutation, sexual reproduction, and recombination is the raw material of evolution.

VARIABILITY AND THE ENVIRONMENT: NATURAL SELECTION

Environmental factors have different effects on the individuals of a species containing a variety of genetic types. These factors reduce the reproductive potential of some individuals and increase the chance that others will grow to maturity and leave offspring. Because the more successful individuals have greater reproductive potential, their alleles are passed on to the next generation in greater numbers than those of less successful individuals. This differential transmission of alleles alters the genetic makeup of the next generation toward an average distribution that is more successful in reproducing in the environment. The choice of reproducers of the next generation by environmental forces constitutes natural selection. The end result, the altered genetic makeup of the succeeding generation, is evolution.

Natural selection is the choice of the parents of the next generation by the forces of the environment.

Environmental Conditions Active in Natural Selection

Any environmental condition that affects an individual's ability to reproduce is a force in natural selection. These conditions fall into two broad classes: **density-independent** factors and **density-dependent** factors. Density-independent means that the effects of the environmental conditions have no relationship to the number of individuals per unit area. Density-independent conditions are primarily inanimate or **abiotic** forces of the environment such as the climatic conditions of temperature, moisture, and wind or natural catastrophes such as volcanic eruptions, fires, floods, and earthquakes. These physical factors reduce the reproductive potential of individuals whether their density is high or low.

Density-dependent factors, the second class of environmental factors, proportionately reduce the reproductive potential of individuals as their number per unit area increases. The density-dependent factors affecting the reproductive potential of individuals are primarily **biotic**— that is, they result from the activities of living organisms. Most important among these biotic factors are **predation**, the use of one group of organisms by another as a food source, and **competition,** the attempt by individuals of the same or different groups to use the same limited resource. Among the resources frequently under competition are such vital needs as food, water, mates, and living and nesting space. Both competition and predation are highly effective as selective forces that reduce the reproductive potential of individuals and alter the genetic makeup of succeeding generations.

Density-Independent Factors   Climatic factors are the abiotic environmental forces that are least subject to intensification by the presence of other organisms. Climatic factors important in natural selection include temperature, amount and type of precipitation, humidity, atmospheric pressure, wind, and daily and seasonal variations in light. Various individuals have differing ranges of tolerance for these natural conditions. A population of nonmigratory, nonhibernating birds of North America such as English sparrows, for example, survive best at temperatures between 20 and 24°C (68 and 75°F). Temperatures above and below this optimal range are tolerated, but with increasing impairment of function until extremes of about −20 and 42°C (0 and 110°F) are reached. Beyond these limits, virtually no individuals can live to reproduce. At any given temperature between these extremes, some individuals function better than others and have greater reproductive potential.

Selection by climatic factors becomes more intensive as the extremes are approached. An example illustrating this principle was carefully cataloged in 1898 by H. C. Bumpus at Brown University. Following a severe winter storm, Bumpus noted that many sparrows were stunned by the cold. Of 136 birds brought into his laboratory, 72 revived and 64 died. Measurements of the survivors and nonsurvivors revealed that survival was correlated with various types of physical characteristics such as wing length, body weight, length of bill, width of skull, and length of bones in the wings and legs. Bumpus concluded that survivors generally tended to be animals with body dimensions closer to the average values. Recent, more detailed evaluation of Bumpus's data shows that among males the slightly larger individuals were better able to survive; among females, individuals close to the average fared better. Selection by extreme cold in this case tended to eliminate individuals far from the norm, with more deaths among birds of smaller size, and tended to enhance the reproductive potential of average and slightly larger individuals.

The presence or absence of additional individuals among the English sparrows subjected to extreme cold would probably have had little effect on the outcome. Other abiotic factors with equivalent density-independent effects—apart from natural catastrophes like earthquakes, fires, and floods—include more gradual changes such as the elevation of mountain ranges or islands, formation or recession of seas and lakes, changes in the course of rivers and streams, and the advance or retreat of glaciers and layers of ice.

Although the effects of these abiotic factors are primarily density-independent, the distinction is not always complete. For example, the ability of animals to find a suitable shelter during a natural catastrophe such as a fire or flood might be affected by other individuals also seeking shelter. In this case the effects of fire or flood might depend to some extent on the number of individuals per unit area. (For additional details on density-independent factors and their environmental effects, see Chapter 22.)

Density-Dependent Factors   The selective factors in the environment that result from the presence of other organisms, such as competition and predation, are almost exclusively density-dependent in their effects. Competition, one of the most important of these factors, immediately brings to mind the vision of animals locked in deadly combat over a resource wanted by both. Direct competition of this type, called **interference competition,** is in reality less frequent than competition through the effort of some individuals to absorb nutrients more efficiently from the environment than other individuals. By this means, the more competitive types simply grow faster and crowd out the less able. Competition of this type, called **exploitation competition**, may even occur without face-to-face meetings of the competing individuals—as in the use of a plant resource as food by animals that feed at different times of day.

Species occupying the same environment actually tend to develop forms and life-styles that reduce competition of either kind. Contest or exploitation competition is normally intensive only when a new environment is colonized or when an extensive change drastically upsets conditions in a previously inhabited region. Under normal conditions, species tend to evolve modifications in behavior, structure, or other characteristics that allow resources, nutrients, and living space to be divided with reduced competition. David Lack of Oxford University, in his study of two closely related water birds, the cormorant (*Phalacrocorax carbo*) and the shag (*P. aristotelis*), noted that on first inspection the birds seemed highly competitive because they hunted for fish in the same water areas. Closer study revealed, however, that the cormorant feeds on bottom-dwelling fish while the shag takes species that swim near the surface. Thus the two species of birds have evolved behavior patterns that reduce or avoid competition.

Different species occupying the same environment tend to evolve forms and life-styles that reduce direct competition.

Predation involves any consumption of one organism (the **prey**) by another (the **predator**). Although predation may involve the death of the prey species, as in the capture and consumption of a fly by a spider, it may also involve slow consumption of prey by predator, as in the grazing of plants by cattle, deer, and insects or the infection of host species by parasites or disease organisms. In this case, predation may simply reduce the reproductive capacity of susceptible individuals rather than killing them outright.

Predator and prey species affect each other in a reciprocal relationship: the survival of prey depends on the numbers and skill of the predators; the survival of predators depends on the abundance and evasiveness of the prey. Among caribou herds in Canada and Alaska, for example, predation by wolves acts as a force selecting for reproduction caribou that have the speed and stamina to outrun their predators. At the same time, the ability of the surviving caribou to run at high speed for extended periods acts as a force selecting wolves with the same qualities and, moreover, hunting skills that lead to entrapment and capture of their prey. The linkage of two species in this pattern, through back-and-forth selective effects, is termed **coevolution.** (Further details of density-dependent factors and their effects are covered in Chapter 22.)

Competition and predation are probably the principal forces reducing the number of individuals that survive and grow to reproductive age. All forms of selection, however, whether density-dependent, density-independent, biotic, or abiotic, act in the same manner. Through selection, the probability that carriers of favorable mutations will survive to reproduce is increased. Their greater reproductive potential increases the frequency of their alleles in the next generation; these altered genetic frequencies amount to evolution.

Populations: The Units of Evolution

The effects of selection are felt by *individuals*. Through selection, the potential of individuals to reproduce is enhanced or reduced, depending on their favorable or unfavorable characteristics. Single individuals cannot evolve, however. An individual represents a fixed combination of genes and alleles. No matter how intense the effects of selection, an individual's genetic combination remains the same throughout its life. For this reason, evolutionary change cannot be detected in single individuals.

Moreover, the study of single individuals provides no clues to the possible direction of evolution. Even if they are favorable, the alleles carried by an individual may perish with it should it happen to die before mating or be unsuccessful in finding a mate. Or an individual with unfavorable alleles may, by lucky accident, mate successfully and pass its alleles to the next generation. Because chance may operate in this way, the future evolutionary effects of a single individual, no matter how favorable or unfavorable its alleles, cannot be predicted.

If evolution cannot be detected or predicted by studying single individuals, how can evolutionary change be followed? The answer lies in the findings of genetics, which reveal that the members of a group of interbreeding organisms share a common collection of genes and alleles. Discovery of this common **gene pool,** as it is called (see also p. 271), provides a method for detecting and predicting evolutionary change. By studying a group of interbreeding individuals sharing a common gene pool rather than single individuals evolutionists can use statistics to predict and evaluate average trends. Although the results obtained cannot predict the fate of single individuals, they are valid for the group as a whole.

Individuals are the units of selection; populations are the units of evolution.

For the purposes of study, then, the unit of evolution

is defined as a group of interbreeding individuals. The word **population** is used in a special sense by evolutionists to describe this group. The opportunity for individuals to interbreed defines the limits of an evolutionary population. Individuals of the same species that are unable to breed with members of a population because of barriers of time, space, or physical obstacles are considered outside the population or members of a different population. All the individuals of a population are thus potential mates and all share the same gene pool.

A population is a group of interbreeding individuals sharing the same gene pool.

Evolution is a change in the frequency of alleles in the gene pool of a population with the passage of time.

The population concept can also be applied to organisms such as bacteria and plants that reproduce primarily by asexual means with only occasional interbreeding. In this case, all the asexual descendants of a single individual, because they are genetically identical, form a **clone.** Since the members of a clone are genetically identical (unless a mutation occurs), they are not affected differently by environmental forces and cannot evolve. Thus a clone is equivalent to an individual as far as evolution is concerned. A population of organisms that reproduce primarily by asexual means usually contains several clones, however, each with a different genetic makeup. These will react differently to the selective forces of the environment; in fact, they react as individuals do in sexually reproducing populations. Even in bacteria, moreover, some form of sexual reproduction eventually takes place between members of different clones. Therefore the outcome is the same: some clones will react favorably to the selective forces of the environment and hence will be more likely to reproduce and pass their particular combination of alleles to the next generation. In such populations, the *clone* is the unit of selection instead of the individual. The unit of evolution, however, remains the population. Its shared gene pool of variability is the raw material for evolutionary change.

Populations may be small, containing less than a hundred individuals, or they may comprise billions of individuals or even the entire species. Usually a species contains many populations, each more or less distinct with little or no interbreeding between them. When studied in terms of populations, the effect of natural selection, and thus evolution, is defined simply as a detectable change in the frequency of alleles in the gene pool of a population with the passage of time.

The Hardy–Weinberg Principle The detection of changes in populations undergoing evolution is based on a mathematical relationship discovered independently in 1908 by G. H. Hardy, an English mathematician, and W. Weinberg, a German physician. This relationship, now called the **Hardy–Weinberg principle,** provides a standard for evolutionary comparisons by showing how the proportions of alleles will stabilize in hypothetical populations undergoing no evolutionary change. Comparing the actual proportions of alleles observed in real populations with the proportions predicted by the Hardy–Weinberg principle then gives a quantitative measure of the extent and type of evolutionary change taking place.

The Hardy–Weinberg principle provides a baseline for measuring evolutionary change.

Because the Hardy–Weinberg principle describes expected allele and genotype frequencies in populations undergoing no evolutionary change, five assumptions must be made: (1) mating is completely random in the populations; that is, there is no bias for or against any genetic types in choice of mates; (2) selection does not occur; (3) mutation does not occur; (4) no individuals immigrate into or emigrate from the population; and (5) the population must be large, so that the average values for the frequency of alleles are unaffected by the luck of single individuals in survival and mate selection. The Hardy–Weinberg principle states that in such a population the frequencies of the alleles of a gene will remain the same from generation to generation, and, moreover, that after one generation the frequencies of genotypes arising from these alleles will stabilize and will also remain the same.

How does the principle operate in practice? Suppose that a gene in a diploid population invading a new area has two alternate alleles, *A* and *a*. The possible genotypes with respect to these alleles in the populations are *AA*, *Aa*, and *aa*. The frequency of allele *A* among individuals in the population is equivalent to

$$\text{Frequency of } A \text{ allele} = p = \frac{\text{all } AA \text{ individuals} + \frac{1}{2} \text{ of } Aa \text{ individuals}}{\text{total number in population}} \quad (19\text{-}1)$$

The frequency of *a* allele is equivalent to

$$\text{Frequency of } a \text{ allele} = q = \frac{\text{all } aa \text{ individuals} + \frac{1}{2} \text{ of } Aa \text{ individuals}}{\text{total number in population}} \quad (19\text{-}2)$$

Traditionally the frequencies of the *A* and *a* alleles in the population are designated *p* and *q* as in Equations 19-1 and 19-2.

The Hardy–Weinberg principle states that the frequency of alleles in large, randomly mating populations remains constant in the absence of selection, mutation, immigration and emigration (gene flow), and genetic drift. After one generation, the frequencies of genotypes also remain constant.

These frequencies mean that if all the chromosomes carrying the gene in question in the population were pooled together, some proportion *p* would carry the *A* allele and some proportion *q* would carry the *a* allele. These proportions, or frequencies, add up to 1. That is, if one-tenth or 0.1 of the chromosomes carry the *A* allele, nine-tenths or 0.9 must carry the *a* allele.

Gametes receive only one chromosome of each pair. For the alleles in question, the gametes of the population therefore have *p* chances of receiving allele *A* and *q* chances of receiving allele *a*. All the sperm cells produced in the population will therefore carry the *A* and *a* alleles in the proportions *p* and *q* and, similarly, all the eggs will carry those alleles in the same proportions. The frequencies of genotypes expected from matings of these gametes can be predicted by constructing a Punnett square (see Figure 19-5). Adding the values of the squares in Figure 19-5 shows that the genotypes expected in the next generation will be equivalent to $p^2(AA) + 2pq(Aa) + q^2(aa)$. The same result is obtained by algebraic multiplication of the frequencies of alleles in the gametes (see Supplement 12-1):

$$(p + q)(p + q) = p^2 + 2pq + q^2 \qquad (19\text{-}3)$$

The Hardy–Weinberg principle states that genotypes of large populations will remain constant in these frequencies, $p^2(AA) + 2pq(Aa) + q^2(aa)$, in all subsequent generations if mating is random and if selection, mutation, and emigration and immigration are absent. (A change in allele frequencies due to emigration or immigration is termed **gene flow**.)

As an example, consider an initial population of 100,000 individuals in which 10,000 have the *AA* genotype, 20,000 have the *Aa* genotype, and 70,000 have the *aa* genotype. The frequency of the *A* allele is

$$\text{Frequency of } A \text{ allele} = p = \frac{10{,}000 + \tfrac{1}{2}(20{,}000)}{100{,}000} = \frac{20{,}000}{100{,}000} = 0.2$$

frequencies of alleles in sperm \ frequencies of alleles in eggs	$p(A)$	$q(a)$
$p(A)$	$p^2(AA)$	$pq(Aa)$
$q(a)$	$pq(Aa)$	$q^2(aa)$

frequencies of genotypes

Figure 19-5 Genotypes expected from a mating involving egg and sperm cells containing the allele *A* at frequency *p* and allele *a* at frequency *q* (see text).

The frequency of the *a* allele is

$$\text{Frequency of } a \text{ allele} = q = \frac{70{,}000 + \tfrac{1}{2}(20{,}000)}{100{,}000} = \frac{80{,}000}{100{,}000} = 0.8$$

In the gametes produced by this population, sperm and egg cells will contain the two alleles in the frequencies *p* and *q*. As a result, the frequencies of the genotypes in the next generation will be $p^2(AA) + 2pq(Aa) + q^2(aa)$, or $0.04AA + 0.32Aa + 0.64aa$. If this generation too has 100,000 individuals, 4000 would have the *AA* genotype, 32,000 the *Aa* genotype, and 64,000 the *aa* genotype.

Calculating the frequency of the *A* allele (*p*) in this population gives

$$p = \frac{4000 + \tfrac{1}{2}(32{,}000)}{100{,}000} = \frac{20{,}000}{100{,}000} = 0.2$$

The frequency of the *a* allele (*q*) is

$$q = \frac{64{,}000 + \tfrac{1}{2}(32{,}000)}{100{,}000} = \frac{80{,}000}{100{,}000} = 0.8$$

These frequencies, and the frequencies of the gametes expected from this population, have thus remained fixed at $p = 0.2$ and $q = 0.8$. They will remain at these levels in all succeeding generations as long as mating is random and there is no selection, mutation, and gene flow through emigration or immigration. The proportions of *AA*, *Aa*, and *aa* individuals, established at 0.04*AA*, 0.32*Aa*, and 0.64*aa* after the first generation, will also remain fixed at these levels in all succeeeding generations. Once the genotypic equilibrium has been established after the first generation, the frequencies of individual alleles can be determined simply by taking the square root of the frequencies of the homozygotes $0.04AA + 0.32Aa + 0.64aa$:

$$p = \sqrt{p^2} = \sqrt{0.04} = 0.2$$
$$q = \sqrt{q^2} = \sqrt{0.64} = 0.8$$

Hardy–Weinberg and Evolutionary Change The Hardy–Weinberg principle describes a static population—a population in which no evolutionary change occurs through mutation, selection by the environment, gene flow, or luck in survival or mate selection. As such, the equilibrium probably applies to no real population in nature. Nevertheless the principle is of fundamental value because it provides a baseline for determining the extent, rate, and sometimes the source of evolutionary change in real populations. Usually the method is applied by calculating the genotypes expected for the next generation according to the Hardy–Weinberg principle. The actual numbers and types of individuals are then determined and the expected and actual results are compared. (The chi square method, described in Supplement 12-1, is often used to evaluate the significance of the differences between actual and expected results.) The degree of difference gives a measure of the extent and rate of evolution taking place in a population.

Table 19-2 shows how a population departs from the Hardy–Weinberg principle if selection against one of the homozygous genotypes, *aa*, is complete—that is, if all *aa* individuals die before reaching reproductive age (as individuals homozygous for the HbS or sickling hemoglobin mutation might in a human population). In the initial population, the frequencies of *A* and *a* are equal: $p = 0.5$ and $q = 0.5$. After the hundredth generation, $p = 0.99$ and $q = 0.01$. The population has therefore evolved, since evolution is defined as a change in gene frequencies with time. The cause of the change is the selective force that causes all *aa* individuals to die before reaching reproductive age.

The Effects of Chance: Genetic Drift Some individuals in populations survive and mate, or die and fail to reproduce, because of good or bad luck. As a result, they affect the frequencies of alleles in the next generation whether their own alleles are favorable or unfavorable. Such changes, due to the effects of chance and not natural selection, are called **genetic drift**.

Genetic drift is the change in the frequencies of alleles due to chance.

Small populations are more susceptible to genetic drift than large ones. This tendency can easily be understood by comparing the effects of chance removal of one individual in populations of small and large size. Imagine that a population of 12 individuals has recently colonized an island. The alleles *A* and *a* are present in equal frequencies ($p = 0.5$, $q = 0.5$), with three *AA*, six *Aa*, and three *aa* individuals in the population. By chance, one *AA* individual does not find a mate. The proportions of alleles in the gametes will then be

Proportion of *A* alleles in gametes =
$$p = \frac{(3 - 1) + \frac{1}{2}(6)}{11} = \frac{2 + \frac{1}{2}(6)}{11} = 0.45$$

Proportion of *a* alleles in gametes $= q = \dfrac{3 + \frac{1}{2}(6)}{11} = 0.55$

Thus chance removal of one individual in this case causes a significant shift in the frequencies of the two alleles—from 0.5 to 0.45 for *p* and from 0.5 to 0.55 for *q*, a difference of about 4 percent. Substituting these values for *p* and *q* in the expression $p^2(AA) + 2pq(AA) + q^2(aa)$ reveals a significant shift in the frequencies of genotypes—from the $0.25AA$, $0.5Aa$, and $0.25aa$ of the initial population to altered levels of $0.2AA$, $0.49Aa$, and $0.3aa$ in the next generation.

Now suppose that the initial colonizing population is ten times larger, with 120 individuals instead of 12. If one *AA* individual in this population does not find a mate, the resulting shift in allelic frequencies is much smaller, about 0.5 percent:

Proportion of *AA* alleles in gametes =
$$p = \frac{29 + \frac{1}{2}(60)}{119} = \frac{59}{119} = 0.495$$

Proportion of *aa* alleles in gametes =
$$q = \frac{30 + \frac{1}{2}(60)}{119} = \frac{60}{119} = 0.504$$

The resulting shift in genotypes in the next generation too will be much smaller.

Changes due to the effects of chance in small populations are called genetic drift because they cause gene frequencies to shift aimlessly from their equilibrium values. Since the direction of such shifts is likely to change in each generation, the gene frequencies wander back and forth across their original average values for the population. The smaller the population, the more pronounced the wandering. This pattern is in contrast to the effects of natural selection, which cause allelic frequencies to shift from the values predicted by the Hardy–Weinberg principle in def-

Table 19-2 Departure from Hardy–Weinberg Equilibrium If Selection Against the Homozygous Recessive Is Complete

Generation	Frequency (%) of Dominant Homozygote ($AA = p^2$)	Frequency (%) of Heterozygote ($Aa = 2pq$)	Frequency (%) of Recessive (Lethal) Homozygote ($aa = q^2$)
1	25.00	50.00	25.00
2	44.44	44.44	11.11
3	56.25	37.50	6.25
4	64.00	32.00	4.00
5	69.44	27.78	2.78
9	81.00	18.00	1.00
10	82.64	16.53	0.83
20	90.70	9.07	0.23
30	93.65	6.24	0.10
40	95.18	4.76	0.06
50	96.12	3.84	0.04
100	98.03	1.96	0.01

inite and steady patterns (see below) as the generations pass.

Genetic drift has been repeatedly demonstrated in laboratory populations of *Drosophila* grown under identical conditions. One such experiment, carried out by Sewell Wright at the University of Chicago, used flies with a mutation for body bristles (called *forked*) mixed with flies having normal, wild-type bristles. At the beginning of the experiment, 96 populations were started with four males and four females of each genetic type. From the offspring of the first generation, four males and four females were chosen at random to start the next generation. By the time 16 generations had passed, chance in the selection of the four males and females used to start each subsequent generation led to elimination of the mutant in 41 populations and elimination of wild-type flies in 29 populations. Both alleles were still present in the remaining 26 populations. Since the populations were grown under identical conditions, the variation noted after 16 generations was due to chance or genetic drift. Small, natural populations occupying separate but nearby regions sharing the same environmental conditions show the same apparently haphazard

variations apparently due to genetic drift. Cypress trees in California, for example, frequently exist in small, isolated populations. Although the climatic and topographical conditions acting as selective forces in many of the populations are apparently similar, each cypress population has its own distinct genetic characteristics.

The Results of Natural Selection

The Hardy–Weinberg principle makes it possible to distinguish between the influence of genetic drift and natural selection on the numbers and kinds of alleles passed from one generation to the next. Calculations using the principle have revealed that natural selection may have **directional**, **divergent**, or **stabilizing** effects on the frequency of alleles in a population.

Directional Selection Evolution is *directional* when the average distribution of alleles shifts progressively to a new position as one generation is replaced by another. Consider a typical curve representing the distribution of body

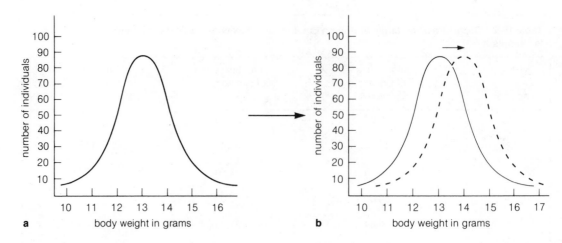

Figure 19-6 The effects of directional selection (see text).

sizes in a population (Figure 19-6*a*). Evolutionary change in the population toward larger body size would be reflected in a directional shift in the average toward large body size (Figure 19-6*b*). The total range of body sizes, however, might remain as broad as before. In such a case, the distribution curve retains the same height and shape as it shifts to the new position. Directional selection usually takes place in response to directional changes in the environment: toward colder or warmer climate, increasing or decreasing moisture, or changes in the number and types of predators. Directional selection may also take place in response to colonization of a new region or habitat by an emigrating population.

There are many examples of directional evolution in nature. The directional changes in body size, head length, and leg structure in horses, the change in moths toward darker colors in industrial regions, and the development of resistance to antibiotics in bacteria have already been described. In our own species, comparison of the cranial capacity of our fossil ancestors shows a steady progression from a cranial volume of about 700 cc to the present average of about 1400 cc.

One of the most interesting patterns of directional selection, the development of certain types of *mimicry*, occurs in response to the reciprocal evolutionary relationship between predators and their prey. As predators evolve toward greater skill in capture of their food species, the prey species frequently undergoes directional change toward forms that are more successful at hiding or escape. These changes include mimicry, in which an edible species gradually develops features that imitate the behavior or external appearance of another poisonous or foul-tasting species.

Figure 19-7 Mimicry due to directional selection in butterflies. The foul-tasting monarch butterfly (**top**) is mimicked by the edible viceroy (**below**). Specimens courtesy of A. M. Shapiro.

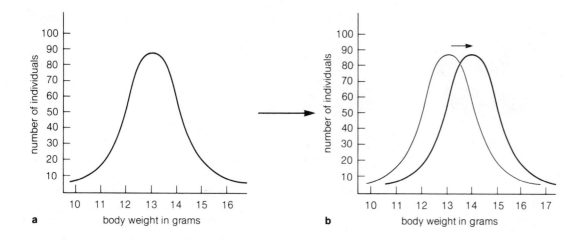

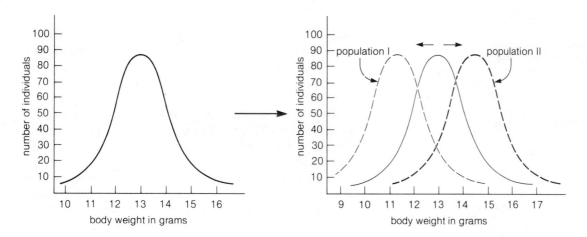

Figure 19-8 The effects of divergent selection (see text).

One of the best known examples of mimicry is the imitation of the monarch butterfly (Figure 19-7a) by the viceroy butterfly (Figure 19-7b). In regions where both butterflies occur together, birds avoid the edible viceroy because of its association with the noxious monarch. In this case, the pressures of selection led to directional evolution of the viceroy toward a coloration that resembles the monarch.

Divergent Selection Evolution is *divergent* when two or more parts of a population evolve in different directions. A distribution curve reflecting such a change would shift from one peak to two or more (Figure 19-8). Divergent evolution may take place in large populations that occupy ranges wide enough to include significantly different en-

vironments, such as the population of English sparrows in North America (see Chapter 18). In the wide range covered by the population, the differing environmental conditions select for different alleles in separate areas.

Examples of divergent selection are abundant among natural populations. The directional development of the modern horse from its evolutionary ancestor, *Hyracotherium*, has been emphasized as an instance of directional evolution. The same ancestor, however, through the characteristic branching pattern followed by evolution, also gave rise to lines that led under different environmental conditions to the tapirs and rhinoceroses. The human species also provides examples of the effects of divergent selection, as in the development of greater or lesser de-

Figure 19-9 Mimics among female *Papilio dardanus* butterflies. (**a**) Appearance of *Papilio* females in regions where Danaidae do not occur. (**b**) A noxious species of Danaidae butterflies. (**c**) An imperfect *Papilio* mimic from an area where Danaidae butterflies occur in limited numbers. (**d**) A *Papilio* mimic from an area where Danaidae butterflies are abundant. Parts *(a)*, *(c)*, and *(d)* drawn from originals courtesy of A. C. Clarke, P. M. Sheppard, and *Heredity*; part *(b)* drawn from original courtesy of Cambridge University Press.

grees of skin pigmentation in response to localized differences in the amount of solar radiation.

Another example of mimicry among butterflies, studied by A. C. Clarke and P. M. Sheppard of Liverpool University, shows how varying environmental conditions over the range occupied by a species can lead to divergent evolution. The butterfly *Papilio dardanus* (Figure 19-9), an edible species for birds, lives in various localities in central and southern Africa. Many of the regions occupied by the *Papilio* butterflies are shared with species of another family of butterflies, the Danaidae, all of which are noxious and avoided by birds. In Madagascar and parts of Ethiopia, where the Danaidae butterflies are absent, no mimicry is observed (Figure 19-9*a* and *b*). In these areas, both male and female *Papilio* have yellow and black wings with tails. Where the Danaidae occur in relatively low numbers, as in Kenya and Tanzania, a few imperfect mimics occur as variations among the *Papilio* females (Figure 19-9*c*). In other areas of Africa, where the Danaidae are abundant, variations in wing markings and shape appear among *Pap-*

ilio females that closely mimic the tailless, noxious Danaidae species (Figure 19-9*d*).

Genetic tests by Clarke and Sheppard showed that the differences in wing color and shape were produced by several alleles of two different genes in the *Papilio* butterflies. In regions shared with the noxious Danaidae butterflies, birds capture *Papilio* butterflies of "normal" appearance more frequently while avoiding the forms with the alleles producing characteristics that resemble the noxious Danaidae. This pattern of predation increases the probability that bearers of the mutant alleles resembling the Danaidae will survive to mate, thereby increasing the proportion of individuals bearing these alleles in the next generation. In regions where no Danaidae exist, the alleles producing Danaidae mimics are not favored and hence the "normal" forms predominate. Thus evolution in the *Papilio dardanus* butterflies, in response to predation by birds, follows several divergent pathways depending on the presence and abundance of Danaidae butterflies in the locale.

Stabilizing Selection   Evolution is *stabilizing* when the more extreme types are eliminated from the distribution of types in a population and individuals closest to the norm or average are preserved as breeders for the next generation. As a result, the curve representing the distribution of types in the population is gradually narrowed to a sharp peak (Figure 19-10) as time passes. The survival of sparrows with body size closest to average values in response to cold shock during a storm (p. 438) provides a typical example of stabilizing evolution. This pattern is less significant than directional or divergent evolution as a source of change because it reduces variability and tends to fix the population into a narrow range of types.

Adaptation and Fitness

The evolutionary changes brought about by directional, divergent, and stabilizing selection produce populations containing individuals that are better suited to their environments and have better chances of finding mates and reproducing. The individuals in the populations are said to be *adapted* to their environment; any structural or behavioral modification resulting from selection that increases their success in survival and reproduction is called an **adaptation**. The wings of birds, the webbed feet of ducks, the mimics among butterflies, and the opposable thumbs of humans are all adaptations resulting from natural selection of favorable types. In the evolutionary sense, the success of such adaptations is measured by the degree of success given to their bearers in survival and

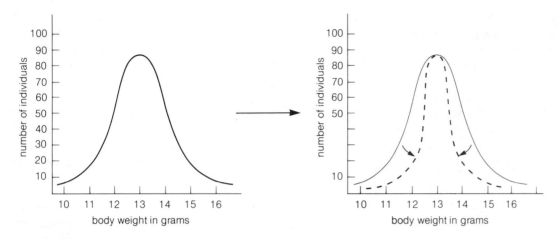

Figure 19-10 The effects of stabilizing selection (see text).

reproduction. In evolutionary terms, the degree of reproductive success is a measure of the **fitness** of a genetic type.

An adaptation is any structural, biochemical, or behavioral modification resulting from selection that increases the reproductive potential of an individual.

The adaptive changes in a population brought about by selection are not necessarily in a "better" direction from the standpoint of human values. The eye, for example, is commonly regarded as one of the greatest accomplishments of evolution. Yet certain cavefish have evolved into forms without eyes because they provide no advantage in the complete darkness of their environment and possibly are more subject to injury and infection than other body parts. Increased complexity too is commonly regarded as a "goal" of evolution. Yet parasites such as the tapeworms inhabiting the intestinal tracts of vertebrates have become reduced through selection to little more than a highly efficient reproductive system; because all other functions are provided by the unwilling host, they are rudimentary or lost in the parasite.

Fitness is a measure of reproductive potential.

It is also important to note that adaptations are not purposeful, no matter how deliberate they may seem from the human point of view. The development of nearly perfect monarch mimics by viceroy butterflies does not result from a purposeful evolutionary effort on the part of viceroys to imitate monarchs. This, like every other adaptation, is simply the result of the reproduction of certain genetic types and the elimination of others due to the effects of natural selection.

The Evolution of Neutral Characteristics

Evolutionists since Darwin's time have puzzled over the selection of apparently neutral or nonadaptive characteristics in animals and plants. Almost all species show these characters; in humans, for example, it is difficult to understand how the eyelid fold producing the eye shape of the Oriental races might have been selected as a character increasing reproductive potential. Among rhinoceroses, the Indian species has one horn and the African species two. Such character differences appear to be of neutral value and neither help nor hinder the reproductive success of their bearers. How do they become established and persist in evolution?

Questions over this issue multiplied into a major controversy when it was discovered that such apparently neutral characteristics also appear in great numbers at the molecular level. All the proteins investigated by sequencing and other techniques have proved to exist in a bewildering multitude of forms usually differing by one amino acid substitution. Many of the substitutions, which persist in evolution, appear to be in nonfunctional "filler" regions of the proteins. As such, they have no detectable effect on the activity of the protein as an enzymatic, structural, or other functional element in the organism. Of the many amino acid differences found in the hemoglobin molecule of humans (see Supplement 19-1), for example, at least 43 appear to be without significant effects on the function of the blood protein.

The unexpectedly large number of apparently neutral molecular mutants has polarized contemporary evolutionary theorists into two camps. One group, led by M. Kimura of The National Institute of Genetics in Japan, and variously called the "neutralists," "nonclassicists," or "non-Darwinians," argues that neutral characteristics occur simply by chance. Since they are neutral, and bring neither advantage nor disadvantage to their carriers, they are not acted upon by selective forces and thus become established or are lost by genetic drift. According to this school of thought, most of the molecular and anatomical characteristics of living beings actually arise in this way and are fixed or lost simply by chance. Mutations that are either advantageous or disadvantageous (most will be disadvantageous), and hence acted upon by selection, are in a minority. Thus, according to the neutralists, evolution is primarily the result of chance rather than natural selection.

The opposite camp, called the "selectionists," "classical," or "Darwinian" group, argues that the neutral or nonadaptive character of molecular and anatomical mutations is only apparent. While this school does not deny that neutral mutations do occur and persist by chance, its proponents claim that most of the amino acid substitutions in molecules, and the apparently nonadaptive structural characteristics such as the extra horn of African rhinoceroses, do in fact bring slight advantages or disadvantages to their carriers and thus are acted upon by selection. Even though their advantages or disadvantages are so slight that they remain undetected by present methods, they do exist. Eventually the forces of selection will either eliminate these characteristics or fix them as adaptations in a population. Thus, according to the selectionists, evolution is primarily the result of selection, and chance plays a relatively minor role.

The two schools of evolutionary thought therefore differ in the relative importance given to selection and chance in evolution. Both sides agree that both processes contribute to evolutionary change; the question is, how much does each contribute? While the controversy is impossible to resolve at the present time, there are several lines of evidence indicating that the selectionists have the edge in the argument. Most of this evidence stems from the fact that when many apparently neutral mutations are analyzed in sufficient detail, they prove to respond to selection—indicating that they do in fact have subtle effects on their carriers. Drosophila pseudoobscura and D. persimilis, for example, two fly species that have been separate for about a million generations, share 39 mutations causing amino acid substitutions that have no detectable effects on the proteins carrying them. Even though the two species have been evolutionarily separate for such an extended period,

the frequencies of 35 of the mutant alleles have been maintained at almost the same levels in both species. This pattern indicates that the mutant alleles have subtle effects and are maintained by selection. If the alleles were truly without effects on their carriers, so that their presence or absence would make no difference, genetic drift would be expected to have produced significant variations in the frequencies of the alleles by this time. A variety of additional alleles from Drosophila and other species, originally thought to be neutral in adaptive value, have since been found to respond to selection and therefore to have small but significant effects after all.

Thus while some alleles undoubtedly become established and persist in populations by chance, the majority probably have some adaptive significance and are maintained by selection. Within this distribution the relative contributions of chance and selection will be influenced by population size: in small populations the effects of chance will be increased; in large populations, where genetic drift is less likely, the effects of selection will be increased.

It is also likely that at least some nonadaptive characteristics are maintained in populations because the genes controlling them also control other functions with definite survival value to the species. The number and arrangement of bristles on the head of Drosophila flies, for example, probably have little significance for the survival of individuals. However, the developmental pattern giving rise to the bristle arrangement also controls development of the simple eyes of Drosophila—a characteristic that does have definite significance for survival of individual flies. Many genes studied in detail in Drosophila, corn, and other species have multiple effects of this kind; some characteristics that are apparently neutral are linked to others that do have selective value. This reflects the fact that the biochemical pathways controlled by genes in embryonic development affect a variety of adult characters. If one function controlled by a gene is beneficial, the entire group of effects will be selected for, including any side effects that we may recognize and describe as nonadaptive.

Neutral characteristics may persist because of chance or because the genes controlling them also control other structures or functions with survival value.

Thus the myriad adaptations of living organisms can be explained by the theory of evolution. These adaptations arise primarily through the selective effects of environmental factors that increase the likelihood that individuals with favorable alleles will survive to reproduce. The increased reproductive potential of individuals selected by the environment because they carry more favorable alleles results in greater levels of these alleles in the next generation. The resulting change in the frequency of these alleles in the next generation produces adaptations and accomplishes evolution. Some alleles, especially in small populations, also fluctuate or persist through the effects of chance. The next chapter outlines how these changes may become great enough to lead to the creation of new species.

Questions

1. What is a gene mutation? A point mutation? A chromosome mutation?

2. Do all gene mutations result in changes in the amino acid sequences of proteins? Why?

3. Why are most mutations likely to be deleterious?

4. What is a polyploid? How are polyploids artificially induced?

5. Why is diploidy important in the maintenance of variability?

6. How do deleterious alleles such as the one causing sickle-cell anemia persist in the human population?

7. How does sexual reproduction add to the variability of a species?

8. What are the disadvantages of sexual reproduction?

9. Why might variability provide an advantage in times of adverse conditions?

10. What are abiotic and biotic environmental factors? List examples of each. Which of your examples are density-dependent? Which are density-independent? What do these terms mean?

11. Does competition necessarily involve direct confrontation or combat? What kinds of competition may occur?

12. What is a population from the evolutionary standpoint? Explain why individuals are the units of selection whereas populations are the units of evolution.

13. What is a gene pool? How are gene pools significant in evolution?

14. What is a clone? How do clones resemble individuals?

15. What is the Hardy–Weinberg principle? What conditions must be met for this concept to be valid?

16. The Hardy–Weinberg principle describes a nonevolving population. Why are the characteristics of nonevolving populations of value to evolutionary studies?

17. Define gene flow and genetic drift. How is genetic drift important in evolution?

18. What are directional, divergent, and stabilizing selection? Give one example of each.

19. What is an adaptation? How do adaptations arise in evolution?

20. What is fitness?

21. What are neutral or nonadaptive characteristics? What processes might account for their appearance and persistence in evolution? What is the difference between the "neutralist" and "selectionist" viewpoints of the evolutionary mechanism?

Suggestions for Further Reading

Ayala, F. J., and J. W. Valentine. 1979. *Evolving*. Benjamin/Cummings, Menlo Park, California.

Dobzhansky, T., et al. 1977. *Evolution*. Freeman, San Francisco.

Futuyma, D. J. 1979. *Evolutionary Biology*. Sinauer Associates, Sunderland, Massachusetts.

Grant, J. 1977. *Organismic Evolution*. Freeman, San Francisco.

Kimura, M. 1979. "The Neutral Theory of Molecular Evolution." *Scientific American* 241:98–126.

Lack, D. 1947. *Darwin's Finches*. Cambridge University Press, New York.

Scientific American 239, September 1978 (Evolution Issue).

Stebbins, G. L. 1971. *Processes of Organic Evolution*. 2nd ed. Prentice-Hall, Englewood Cliffs, New Jersey.

SUPPLEMENT 19-1: MORE ON GENE MUTATIONS

The nucleotide base changes that cause point mutations in genes may occur in two ways: transition or transversion. A **transition** occurs when one purine is replaced by another (a replacement of adenine by guanine or guanine by adenine) or when one pyrimidine is replaced by another (cytosine by thymine or thymine by cytosine; see Figure 19-11a). Transitions have the effect of substituting a different base pair for the original one. **Transversions**, the second type of nucleotide change, occur when a purine is replaced by a pyrimidine or a pyrimidine is replaced by a purine (a replacement of either cytosine or thymine by guanine or adenine or vice versa; Figure 19-11b). In transversions, the same base pair is retained but the purine and pyrimidine bases of the pair change sides in the double helix.

Both changes, transitions and transversions, may arise from a variety of sources. Rarely, a nucleotide base may take up an altered distribution of its atoms in space that permits unusual base pairs to form (Figure 19-12). These changes, for example, would permit adenine to pair with cytosine instead of its normal pairing partner thymine (as shown in Figure 19-12) during replication. Changes of this type, called tautomeric shift, occur naturally in the nucleotide bases at a low frequency.

Other changes may be induced in the nucleotide bases of DNA by the action of chemicals such as nitrous acid, hydroxylamine, and mustard gas. Nitrous acid, for example, converts adenine to a modified base, *hypoxanthine*, by removing an amino group (Figure 19-13a). At a

Figure 19-11 (**a**) A transition—a mutation that has the effect of substituting another base pair for the correct one. (**b**) A transversion—a mutation in which the correct base pair is retained but the purine and pyrimidine bases of the pair exchange places in the double helix.

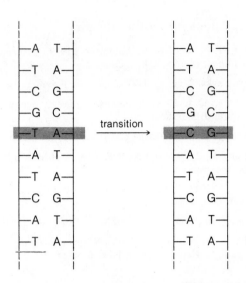

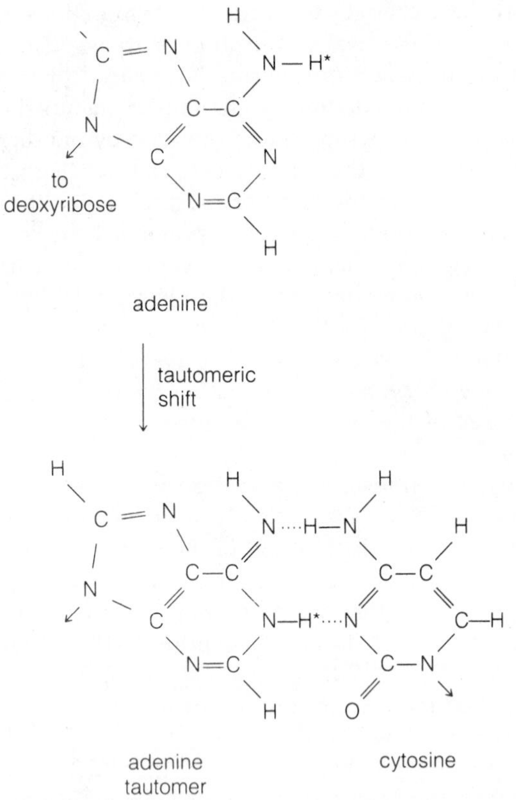

Figure 19-12 An altered distribution of the atoms in adenine leading to pairing with cytosine instead of its normal pairing partner, thymine.

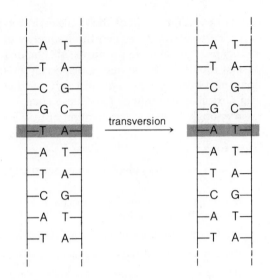

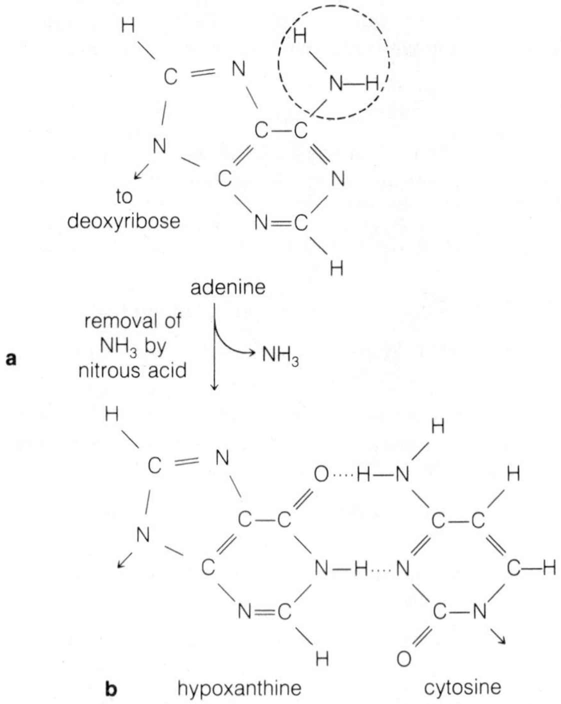

Figure 19-13 (**a**) Conversion of adenine to hypoxanthine by nitrous acid. (**b**) Hypoxanthine pairs with cytosine instead of thymine. This change would produce a transition in the DNA chain.

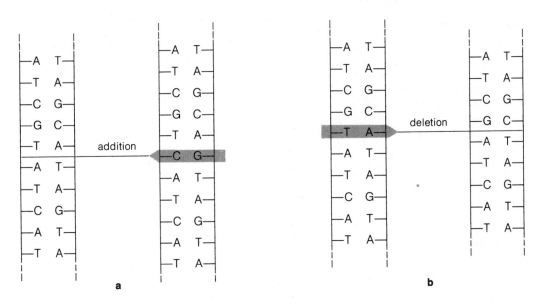

Figure 19-14 Mutations through additions and deletions of single base pairs: (**a**) an addition; (**b**) a deletion.

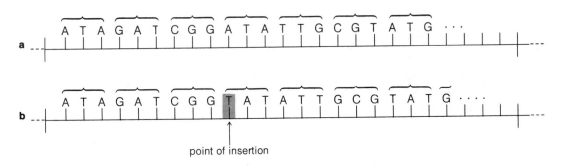

point of insertion

Figure 19-15 A change in reading frame of a DNA sequence because of an addition. (**a**) The unmutated gene; triplets are read in sets of three as shown. (**b**) Insertion of a base pair at the point shown causes a change in reading frame. All the triplets are read incorrectly after the point of insertion.

subsequent replication, instead of pairing with thymine, as the adenine normally would, hypoxanthine pairs with cytosine (Figure 19-13b). As a result, C is substituted for the T that would normally be placed in this position in the newly synthesized chain. Thus an A–C base pair is converted to a G–C pair in the DNA in subsequent generations. Hydroxylamine has the reverse effect: it induces chemical alterations that change a G–C base pair to an A–C pair.

Tautomeric shifts and chemical alterations of the DNA bases can also be induced by *ionizing radiations* such as x rays, gamma rays, and neutrons. These forms of radiation are called "ionizing" because they cause ejection of an electron from one of the DNA bases or from molecules in the medium surrounding the DNA. This ejection creates a chemically reactive ionic group that can alter the bases or cause breaks in the DNA backbone chain. Ultraviolet light, a nonionizing form of radiation, can also alter DNA—in this case by inducing chemical cross-links to form between pyrimidine bases in the DNA chain. These links, by causing the bases to pair incorrectly, can lead to both transitions and transversions.

Depending on their position in the DNA code for a protein or polypeptide, these sequence alterations may change a coding triplet from one amino acid to another, or they may change a coding triplet into a terminator codon.

These changes may have various effects on the polypeptide coded for by the transcription unit. If a change in amino acid sequence occurs in the active site of an enzyme, the protein may be altered so extensively that its catalytic activity is destroyed. Changes of amino acids in other parts of the polypeptide chain of an enzyme may have little or no effect. Changes from one codon to another specifying the same amino acid would have no effect on the protein encoded in a gene.

Gene mutations also arise through additions or deletions of one or more nucleotide bases in a DNA chain (Figure 19-14, page 451). These alterations may be caused by acridine dyes, for example, which can remove from one to 20 or more adjacent nucleotides from a chain. Such deletions and insertions cause a change in **reading frame** of the DNA code. From the position of the insertion or deletion to the end of the gene, the second or third base of each triplet will now be read as the first base of each triplet (Figure 19-15, page 451). Single deletions or insertions, especially if located near the beginning of the gene, cause extreme changes in the amino acid sequence of the encoded protein, usually with complete loss of activity. Insertions or deletions near the end of the gene, or additional insertions or deletions that combine to reestablish the reading frame within a few triplets, have less drastic effects.

Mutations noted in the genes coding for hemoglobin in humans give an idea of the extent of such changes. Of the 169 mutations reported in different regions of the two genes coding for the polypeptides of this protein, 161 can be accounted for by single base changes. One mutation would require a change of two bases within the coding triplet affected. Another mutation involves a change in a terminator codon that results in the addition of 31 extra amino acids to the polypeptide. The rest of the changes are deletions that have various effects ranging from amino acid changes to removal of one to five amino acids from the polypeptide chains.

The process of evolutionary change eventually leads to the appearance of new species. Among organisms that reproduce sexually, a change of this magnitude places an extra requirement on the evolutionary mechanism. This requirement is connected with the major characteristic that defines a species: Members of a species breed only among themselves to produce fertile offspring; they do not breed successfully with other species. This means that, among the adaptations that produce a new species, a change must occur that prevents members of the evolving species from interbreeding with members of its parent species. Otherwise the unique collection of alleles that sets apart the emerging species from its parent population would be lost in the mixing and recombination that occurs in sexual reproduction. Thus the establishment of a new species amounts to the development of a block to crossbreeding, a development termed **reproductive isolation** by evolutionists.

The appearance of new species requires reproductive isolation of the emerging species from its parent population.

It is easy to imagine how reproductive isolation develops in a species that evolves into a different type over a long period of time. In this case, the changes in gene frequencies that accumulate over many generations due to mutations, recombination, genetic drift, and selection gradually produce a new form. This new form, moreover, is so different from its ancestral populations that interbreeding between the new and old forms of the species, if they could be brought together, would be unsuccessful. This pathway, which is easily observed in fossil lineages such as the line leading from ancient to modern horses, has undoubtedly occurred frequently to produce new species in evolution. This pattern of speciation, in which one species gradually evolves into another, does not increase the total number of species, however, since each species changes only from an older to a newer form.

For the number of species to increase, a single species must split into two or more daughter species that exist at the same time. If they appear at the same time, however, what prevents crossbreeding and keeps the unique gene combinations of the emerging species from mixing and being swamped out? Two patterns of change are thought to block crossbreeding: geographic speciation and sympatric speciation. **Geographic** or **allopatric speciation** (*allo* = different; *patrio* = country) involves the appearance of physical barriers, such as the upheaval of a mountain range, as initial blocks to crossbreeding. The barrier splits a population and keeps individuals of the separated groups apart until differences in reproductive patterns evolve that

20

THE ORIGIN AND MULTIPLICATION OF SPECIES

are extensive enough to prevent successful crossbreeding if they mix again. In the second pathway, called **sympatric speciation** (*sym* = same; *patria* = country), new species evolve within a population occupying a single geographic region; in this case there is no separation of individuals by geographic barriers. In this pathway, a sudden, extensive change takes place, such as the formation of polyploids, that almost instantly establishes reproductive isolation of individuals and their descendants within the population.

The evolutionary changes that bring about reproductive isolation are also likely to include alterations in the way the emerging species occupy and use the resources of their environment. This pattern of environmental utilization, called the **ecological niche** of a species, involves the species' habitat, the way food is obtained, the items used as food, the manner in which the species interacts with other species in the environment, waste disposal, and all other functions of life peculiar to the species. Because the ecological niche includes both the environment and the manner in which the species uses it, it is unlikely that any two species on earth can occupy exactly the same niche even if they live in the same geographic region. Since the occupancy of an ecological niche depends on the activity of a large number of genes and alleles in an organism, it is as susceptible to genetic change as reproductive behavior and function. When a population splits into new and distinct species through geographic isolation, therefore, it is also likely to diverge into a separate ecological niche.

The ecological niche is the manner in which a species occupies and uses its environment.

This change in the ecological niche may be as vital to species survival as reproductive isolation if two populations, recently evolved as separate species, become sympatric again. Occupying different ecological niches reduces direct competition between species and allows them to survive and reproduce in the same environment. (Further details of the ecological niche and its evolutionary effects are presented in Chapter 23.)

Occupying different ecological niches reduces direct competition between species and allows them to survive and reproduce in the same environment.

The observation of present-day species and the fossil record both show that the multiplication of species has taken place repeatedly in evolution. The 600 to 700 different species of *Drosophila* living on the Hawaiian Islands today, for example, descended from a very few *Drosophila*, probably a single species, that first colonized the islands some 10 million years ago.

THE MECHANISM OF GEOGRAPHIC SPECIATION

Geographic speciation is considered to take place in several clearly defined steps: (1) A natural event separates a single population into two or more parts. (Remember that a population consists of a group of interbreeding individuals sharing the same gene pool.) The separating event may be a cataclysm such as a volcanic eruption, or it may be a less dramatic but more frequent happening such as colonization of a new and distant habitat by a few individuals that migrate permanently from a home population. (2) Because of the geographic isolation, interbreeding between the separated populations is greatly reduced or eliminated. (3) As a result, the gene pools of the separated populations evolve independently in response to differences in mutations, recombination, genetic drift, and selection in the different regions. (4) If the populations remain isolated long enough, changes in their gene pools produce changes in behavior, structure, and physiology that reduce the likelihood of interbreeding if members of the two populations meet again.

The key factor in speciation by geographic isolation is the separation of a single population into two or more populations that are prevented from crossbreeding by geographic barriers.

At this point, the isolated populations have begun to emerge as separate species. If natural events should bring them together again, members of the two species may occupy the same geographic region without interbreeding. In fact, for reasons outlined below, natural selection tends to speed and perfect the evolution of reproductive differences between the emerging species if they are brought together again before reproductive isolation is complete. Since occupation of a separate ecological niche has survival value to the emerging species, natural selection also tends to increase the differences in the way the new species occupy and use the resources of their environment.

Details of the Mechanism

Factors Producing Geographic Separation Any external event or situation that fragments a single population into two isolated groups can set the mechanism of geographic speciation into operation. For land dwellers, formation of a new riverbed or invasion of the land by an arm of the sea have frequently caused population splits in the past. The narrow neck of land presently connecting North and South America, for example, has submerged and risen from the sea to fragment populations many times in evolutionary history. The advance of glaciers, the creation of mountain barriers by uplifts in the earth's crust, volcanic activity—all have separated single populations into isolated groups. Even changes in vegetation may separate parts of a population of animals. For example, a continuous forest covering a wide area and containing a single population of forest-dwelling animals may be split in two by erosion, fire, or the conversion of intervening segments into grasslands or desert (a common event in history). For water-dwelling forms, division of a river into two arms or separation of a lake or sea into two parts by uplift of the land or silting may effectively separate previously single populations.

Geographic barriers may be surprisingly narrow and yet effective. Even organisms as mobile as birds often hesitate to cross water barriers as narrow as a few kilometers. For sedentary species such as plants or aquatic animals that remain fixed to the substrate, a few meters of intervening uninhabitable space may completely prevent crossbreeding between adjacent populations.

Populations can also become segregated by the occasional migration of individuals from the home range into a new environment. This mechanism of geographic separation has often operated in the case of islands lying at some distance from a larger landmass or in habitable regions separated by a wide river, lake, or desert. A few individuals may cross the intervening water or land barriers and colonize the new area, but there is no regular flow of crossbreeders between the new and old populations.

Genetic Divergence After Geographic Separation All these mechanisms of geographic separation greatly reduce breeding between the isolated populations and set the stage for the next major step in the multiplication of species: divergence of the gene pools of the separated populations. This step is a natural consequence of differences in mutation, recombination, genetic drift, and selection in the different groups.

Gene pools diverge as a natural consequence of differences in selection and genetic drift in separated populations.

Since no two environments are likely to be identical, the effects of natural selection in different regions cause changes in the average frequency of alleles in the separated gene pools. Some combinations of alleles, less successful in one area, give their bearers greater reproductive success in a different area and are selected for. The more different the conditions existing in the regions occupied by the separated populations, the more extensive the genetic divergence.

Even if the separated populations were to inhabit identical environments, which is unlikely, genetic drift is still likely to lead to divergence of their gene pools. The chance that the same genetic types will find mates, or that the same mutations will occur in the same sequence and at the same time in separated populations is essentially zero. The possibility that recombination will produce the same gene combinations in separated populations is equally remote. Thus isolated populations are likely to diverge genetically even if they should occupy identical environments. Small populations are expected to be most susceptible to changes in gene frequencies because of genetic drift (see p. 442).

Another chance factor contributing to the genetic divergence of small colonizing populations has been termed the **founder effect** by Ernst Mayr of Harvard University. Since a small colonizing population is likely to carry only a fraction of the alleles of their former population, their descendants will probably differ significantly in average allele frequencies. The destruction of most of the individuals in a population separated by natural or other causes is likely to have similar effects. In this case, the population dies back to a few individuals that survive to serve as breeders. This version of the founder effect is likely to have occurred, for example, among northern elephant seals decimated by hunting. The present population of elephant seals, which amounts to about 30,000 individuals, is descended from some 20 seals that were the sole survivors left by hunters in the 1890s.

Development of Reproductive Differences Some of the genetic changes occurring in separated populations affect reproductive morphology, behavior, and physiology. These changes occur simply because the populations are separate and free to evolve in different directions—not because of any evolutionary advantage gained from the development of reproductive isolation at this point. Reproductive functions are particularly susceptible to change because in most plants and animals the structural, behavioral, and biochemical activities related to reproduction are complex· and controlled by large numbers of genes and alleles working together in delicate balance. Because so many genes are involved, it is likely that at least some will be affected by selection, mutation, and recombination. As a consequence, the separated populations are almost certain to develop distinct reproductive patterns.

Further Developments Producing New Species Isolated populations turn into separate species as a natural consequence of further divergent evolution of reproductive activities. At some point, the reproductive differences produced by divergent evolution are great enough to prevent or greatly reduce the possibility of successful crossbreeding if the two populations later occupy the same geographic region. The populations are now established as separate species. If the new species occupy the same environment at this point, natural selection will reinforce the reproductive barriers between them.

Separate populations are converted into separate species through the development of reproductive isolation.

An experiment by Karl F. Koopman of the American Museum of Natural History demonstrates how reproductive isolation is reinforced in this way. Koopman mixed together populations of two fruit fly species, *Drosophila pseudoobscura* and *D. persimilis*. Reproductive barriers between these closely related species are incomplete; if temperatures are held at low levels (16°C), limited crossbreeding occurs and hybrids form. The fact that limited crossbreeding occurs means that some flies of the two species carry alleles that permit hybridization under these conditions. Since the hybrids formed are less successful than either parent species in competing for the resources of the environment, however, they tend to be eliminated. Eliminating the hybrids, the offspring of parents with alleles that permit crossbreeding between the two species, then lowers the frequency of these alleles in the next generation. This process reduces the amount of hybridization that can occur and reinforces reproductive barriers between the two species. Koopman tested this outcome by removing hybrids before they could reproduce, thus increasing the selection against them to 100 percent. As expected, the number of hybrids formed in each successive generation steadily decreased.

Natural selection tends to increase both reproductive isolation and niche specialization if newly separated species are brought together again.

Natural selection also tends to separate the ecological niches of diverging species if they come to occupy the same geographic region again. Where two closely related warbler species, *Dendroica pinus* (pine warbler) and *D. dominica* (yellow-throated warbler), occupy totally separate geographic regions, for example, their beak lengths are similar (Figure 20-1a). Where the two species live together in the same region, their beak lengths are distinctly different (Figure 20-1b). The longer beak length of *D. dominica* allows it to live on insects hidden deeply in pine cones; the same insects remain inaccessible to *D. pinus*, which feeds more successfully on other insects that live in shal-

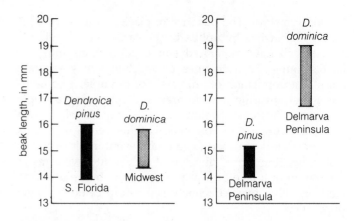

Figure 20-1 Where the two warbler species *Dendroica pinus* and *D. dominica* occupy separate geographic regions (**a**), their beak lengths are similar. In populations of the two species occupying the same geographic region (**b**), beak length is distinctly different. The difference in beak length in sympatric populations allows individuals of the two species to occupy different ecological niches. The Delmarva Peninsula is the land mass represented by Delaware, Maryland, and Virginia. Redrawn from an original courtesy of R. W. Ficken.

low recesses in the bark of pine trees. This feeding difference, promoted by the divergence in beak length, reduces competition between the two species and allows them to coexist in the same region. Since the differences have this effect, they are selected in the regions occupied jointly by the two species. (For further details of the environmental effects reinforcing niche separation, see Chapter 23.)

Reproductive Barriers Between Species The study of populations that have recently separated into distinct and different species reveals that newly developed barriers to interbreeding may involve many parts of the reproductive system (see Table 20-1). Some barriers may prevent mating because individuals derived from the different populations now reach reproductive maturity at different seasons of the year. In some cases reproductive behavior, including the elaborate courtship rituals of animals, may be so modified that individuals do not recognize each other as potential mates. Another barrier in plants and in animals such as insects is change in reproductive structures so that fertilization is mechanically impossible. Flower size, for example, may change so much that an insect carrying pollen between the flowers of one species can no longer enter the flowers of the second species. These barriers, which prevent crosses between species by preventing mating, are called *prezygotic barriers* because eggs and sperm from the diverging species cannot unite to form a zygote.

Table 20-1   Possible Barriers to Crossbreeding Between Different Species

1. Prezygotic barriers (which prevent fertilization from taking place)
   a. Behavioral differences that prevent mate recognition
   b. Occupancy of different habitats
   c. Reproductive activity in different seasons
   d. Differences in reproductive structures that prevent copulation or fertilization

2. Postzygotic barriers (which prevent development if fertilization takes place)
   a. Death of zygote immediately after fertilization because of incompatibility of sperm and eggs
   b. Death of embryo later in development because of genetic imbalances
   c. Reduced hybrid fitness to reproduce
   d. Hybrid sterility

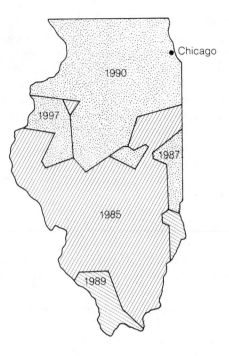

Figure 20-2   The patterns of emergence of cicadas in Illinois. The cross-hatched areas represent 13-year cicadas, and the stippled areas signify 17-year cicadas; the dates indicate the next expected year of emergence. The emergence pattern establishes an effective seasonal reproductive barrier between the populations. Redrawn from an original courtesy of the Trustees of the British Museum (Natural History).

Seasonal differences in sexual activity are important prezygotic barriers in both animals and plants. *Pinus radiata* and *P. attenuata,* for example, two closely related pine species that grow in the same geographic regions in coastal California, are effectively isolated reproductively because *P. radiata* releases pollen and carries out fertilization about 6 weeks before *P. attenuata*. Similarly, among cicadas, a group of insects that spends its larval stage underground, adult reproductive forms emerge at different times after spending either 13 or 17 years as larvae. Within a given region, broods are out of synchrony (see Figure 20-2) and crossbreeding between them is therefore impossible. Such seasonal barriers may be effective within much shorter time intervals. *Drosophila pseudoobscura* and *D. persimilis* breed at the same season of the year, but they are reproductively isolated because *D. persimilis* is sexually active in the mornings and *D. pseudoobscura* in the evenings.

Other prezygotic barriers appear more frequently in either plants or animals. Structural differences causing mechanical difficulties in the exchange of gametes are most important in plants. Two closely related species of sage plants, *Salvia mellifera* and *Salvia apiana*, are effectively isolated by the size and arrangement of their flower parts and by the sizes of the bees that visit their flowers and carry pollen between them (Figure 20-3). The flowers of *S. mellifera* are arranged so that small bees can enter and carry pollen from plant to plant. The flowers of *S. apiana* have long stamens and an arrangement of petals that prevents access by all but the largest bees, such as bumblebees and carpenter bees. The difference is effective in blocking cross-pollination between the two species.

Differences in courtship behavior are effective prezygotic barriers to crossing between related animal species, particularly among groups such as birds that carry out elaborate courtship rituals. Mallard and pintail ducks, which have become separate species in relatively recent times, provide an excellent example of the effectiveness of behavioral isolation. The two species are still capable of interbreeding and producing viable hybrids if they are brought together in captivity. In their natural environments, however, extensive differences in courtship and nesting habits greatly reduce the frequency of hybrids and keep the species almost completely separate, even though the two kinds of ducks live together in thousands of lakes, ponds, and streams in North America.

Other reproductive barriers are *postzygotic;* that is, they prevent the development of an embryo or individual if members of two diverging species mate and fertilization is successful. The embryo may die soon after fertilization or during development. Of, if development is successful, the hybrid individual may be sterile or so inferior that it is unlikely to reach reproductive maturity.

Imperfect growth or death of embryos appear most frequently among animals as postzygotic blocks to crossbreeding between related species. These blocks reflect the pattern of embryonic development in animals, which is complex and susceptible to the effects of genetic changes. Few animal hybrids, in fact, are capable of progressing through gastrulation even if cross-fertilization between species is accomplished. In plants, crosses between species often produce viable individuals that are hardy and grow to maturity. Difficulties in pairing and separation of the chromosomes during meiosis, however, usually make them sterile as adults.

Reproductive isolation is so important to the maintenance of separate species that evolution usually leads to multiple barriers. That is, the effects of two or more prezygotic and postzygotic barriers combine to make crossbreeding nearly impossible. For example, cross-fertilization

---

Species are usually separated by a variety of prezygotic and postzygotic reproductive barriers.

---

between the two related fruit fly species *Drosophila pseudoobscura* and *D. persimilis* is greatly reduced by differences in courtship displays and the periods of the day in which adults are sexually active. If cross-fertilization does occur, hybrids either die or are greatly reduced in viability and reproductive capacity. The combined effects of these isolating mechanisms make reproductive isolation of the two species essentially complete.

Geographic isolation of populations, followed by divergent evolution of the separated gene pools to produce reproductive isolation, has taken place literally millions of times in evolutionary history. The variety and number of new species created by these mechanisms, which continue to operate today, are so extensive that we will probably never be able to catalog the living species completely, let alone the many that have lived in times past.

A Classic Example of Geographic Speciation: Darwin's Finches

The Galápagos Islands, located some 900 kilometers off the west coast of South America, are distant enough from each other and from the mainland that they are rarely reached by migrating land animals or by windblown or water-dispersed plants. Each island of the group has its own distinct topography and climate. Since the time about 3 million years ago when the islands first emerged from the sea through volcanic activity, various species of plants and animals have colonized them. The successful colo-

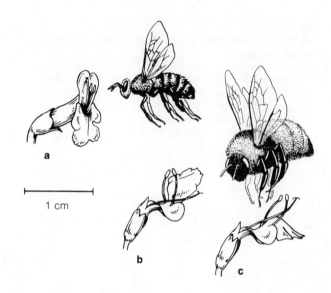

Figure 20-3 Reproductive isolation in flowers of two sage species: *Salvia mellifera* (**a**) and *S. apiana* (**b**) and (**c**). Only small bees can enter the *S. mellifera* flowers and pick up pollen; only large bees are able to open the *S. apiana* petals to enter these flowers. Entry of large bees opens the *S. apiana* flowers by tripping open the large, leaflike extension of the flower, shown in the untripped (**b**) and tripped (**c**) positions. Redrawn from an original courtesy of V. Grant.

nists, often only a few individuals of each species, have evolved on the islands into a variety of new species unknown on the nearby mainland.

The best known of these Galápagos Islands species are a group of finches, probably all derived from a single colonizing species that migrated there hundreds of thousands of years ago from the South American mainland. These birds were first studied by Darwin when he visited the island during his cruise on the *Beagle*. Later, David Lack and others examined the Galápagos finches in detail and fully documented their pattern of evolution on the islands as the classic example of speciation through geographic isolation.

According to Lack's analysis, one of the large central islands of the Galápagos (see Figure 20-4) was probably the first to be colonized by the birds. A small number of finches, possibly only a single pair, arrived from the mainland. The birds and their descendants survived in the new environment and, cut off from interbreeding with the mainland population, soon evolved in response to the environmental conditions on their island. Genetic drift and the founder effect also speeded evolution away from the parent finch population of the mainland.

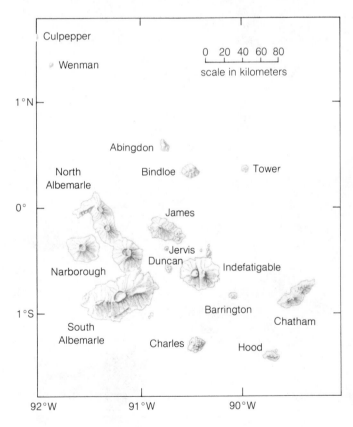

Figure 20-4 The Galápagos Islands. The photograph shows South Plaza Island, which lies off the east coast of Indefatigable. The sparse *Opuntia* cactus seen in the background covers the entire island. Besides the iguana lizards seen here, South Plaza supports swallow-tailed gulls, blue-footed boobies, red-tailed tropic birds, and three species of finches. Photograph by Larry Keenan.

As a result of these pressures, the colonizing finches gradually evolved into a new species that was adapted to the special demands of feeding and reproduction in the island environment. As the species on the main island evolved, the basic process repeated itself. A few of these birds eventually colonized one of the outlying, smaller islands, where they founded another population. This group crossbred only rarely with the main island species. Because of genetic drift, the founder effect, and the new conditions encountered on the smaller island, this new population also diverged genetically, eventually producing a species different from both the parent island species and the ancestral finches of the mainland. This process of colonizing new islands repeated itself again and again until all the Galápagos Islands were occupied by finch populations, each evolving in distinct and new directions.

Eventually some of the emerging finch species from the outlying islands migrated back to the main island and to the other islands of the group. These birds were able to maintain themselves as distinct species because of the reproductive barriers evolved during their period of separation. Since niche differences evolved also, the different species were able to survive on the same island with minimal competition. Selection then intensified and reinforced the reproductive barriers and niche separation between the species. In time, many of the islands were populated by several distinct finch species through this process; one island is now occupied by ten separate species.

The four genera and fourteen species of finches presently living on the islands, although probably all derived from a common ancestral species, show many behavioral and morphological differences reflecting their ecological specialization (see Figure 20-5 and Table 20-2). Other organisms have shown the same pattern of speciation in response to geographic isolation on the Galápagos Islands. Distinctive species of plants belonging to the Compositae (sunflower) family are common on the islands; some of these are large trees unknown anywhere else in the world. Many of the islands have their own characteristic Compositae species. Reptiles, including lizards and tortoises, also exist on the islands in distinctive species related to but different from any known species on the South American continent.

Quantum Speciation

Within populations newly separated by geographic barriers, the changes creating reproductive isolation may be slow and gradual. Sometimes, however, reproductive isolation may appear through sudden genetic changes that convert a separate population into a new species in a relatively quick jump. Evolution of a new species in this way,

Figure 20-5 The finches of the Galápagos Islands. Courtesy of The American Museum of Natural History.

Table 20-2 Adaptations in Darwin's Finches

Bill Shape	Species	Bill Type	Feeding Habit
	Certhidia olivacea	Probing bill	Insect eater (trees)
	Camarhynchus pallidus	Probing bill	Insect eater; uses twig or cactus spine to probe insects from crevices
	Camarhynchus heliobates	Grasping bill	Insect eater (trees)
	Camarhynchus crossirostris	Crushing bill	Cactus seed eater
	Geospiza magnirostris	Crushing bill	Seed and nut eater (ground)

by sudden, extensive genetic changes, is called **quantum speciation.**

New species may arise either gradually or in sudden jumps.

Several mechanisms may account for quantum speciation. One is the formation of polyploids, which is particularly important as a source of new species in plants. Another is hybridization, which although rarely successful, sometimes leads to the sudden appearance of a new form that is fertile and reproductively isolated from either of its parent species. The rapid appearance of new species by this pathway is restricted almost entirely to plants. Still another mechanism involving chromosome changes that may produce new species rapidly is the rearrangement of

chromosome segments through breakage and rejoining in various ways (see Supplement 19-1). The rearrangements, which can appear suddenly through the effects of natural radiation and chemicals in the environment, may limit fertile crosses to the individuals carrying them. Finally, new species may appear rapidly through mutations in single genes that regulate major steps in embryonic development or critical phases of courtship physiology or behavior. This mechanism may have been most significant as a source of quantum speciation in animals.

Polyploids Although new species are suspected to have arisen by the formation of polyploids in a few animal groups such as insects and freshwater snails, animal polyploids are rarely viable because of disturbances to the processes of embryonic development. In plants, however,

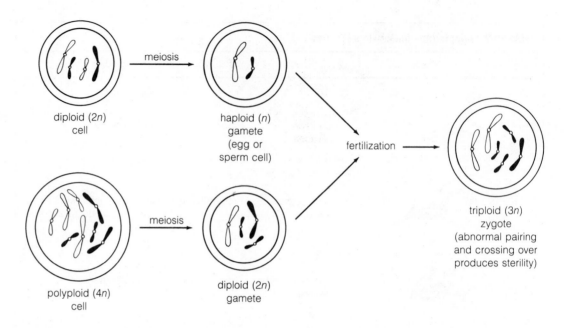

diploid (2n)
cell

meiosis

haploid (n)
gamete
(egg or
sperm cell)

fertilization

triploid (3n)
zygote
(abnormal pairing
and crossing over
produces sterility)

polyploid (4n)
cell

meiosis

diploid (2n)
gamete

Figure 20-6 Crosses between diploid and polyploid individuals with twice the number of chromosomes in diploids produces triploid individuals. Although triploid plants may be hardy individuals, they are usually sterile due to difficulties in pairing and crossing over between the chromosomes during meiosis.

polyploids are frequently vigorous individuals that can interbreed among themselves and are fully fertile. Usually these polyploids are different enough to be reproductively isolated from their parent species from the start, since a cross with one of the diploid parental species would produce a sterile triploid individual (Figure 20-6) that cannot carry out meiosis successfully. This reproductive isolation eliminates interbreeding between the polyploid and its parent species.

Thus the polyploid is an "instant species" that is free to evolve further along its own pathway without genetic disturbance from crossbreeding with its parent species. As many as 95 percent of the extant fern species, and about half of the flowering plants, probably arose through the appearance of polyploids. Many productive food crops are polyploids derived from ancestral species with smaller chromosome numbers. The potato is a 48-chromosome polyploid derived from an ancestral species with 12 chromosomes; cotton is cultivated both as a 26-chromosome diploid and as a 52-chromosome polyploid.

Hybrids Occasionally new species arise suddenly through matings between individuals of two different but related species. Because hybrids are usually sterile, the origin of

viable and reproductively successful species by this means is rare in nature.

Sterility in hybrids usually results from faulty pairing of chromosomes during meiosis. This form of sterility in plant hybrids is sometimes circumvented, however, through the simultaneous production of polyploids. Doubling each of the two separate chromosome sets gives each chromosome a pairing partner in meiosis and leads to perfect separation of the chromosomes in the meiotic divisions (Figure 20-7). The gametes formed after meiosis are viable and can combine successfully to produce hybrid, polyploid individuals. Modern cultivated wheat has evolved from an ancestral species with 14 chromosomes through hybridization and chromosome doubling in this pattern (Figure 20-8). A famous modern example, created by the Russian scientist G. D. Karpechenko, is the "radocabbage." The radocabbage was produced by a cross between a radish and a cabbage, combined with treatment by the drug colchicine to generate a fertile polyploid. (Colchicine produces polyploids by destroying the spindle and preventing chromatids from separating—see p. 202.) Although it is a fertile, viable species, the radocabbage has neither the large head of the cabbage nor the fleshy root of the radish and thus has no value as a potential food crop. Since polyploidy is usually incompatible with the complexities of animal de-

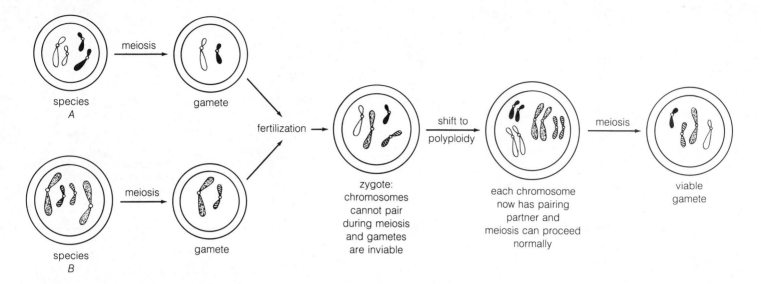

Figure 20-7 Formation of fertile offspring by hybridization followed by a shift to polyploidy. The hybrid *AB* is sterile because neither chromosome has a pairing partner and meiosis cannot proceed. A shift to polyploidy, however, creates pairing partners for each chromosome. Meiosis can now proceed normally and produce viable gametes.

velopment, the equivalent production of fertile animal hybrids through polyploidy is generally unworkable.

Chromosome Changes and Gene Mutations Individuals carrying chromosome rearrangement frequently produce sterile offspring if they mate with individuals of the parent population without the rearrangement. Sterility in the offspring of such matings usually results from the failure of the rearranged chromosomes to pair properly with the unchanged chromosomes of the parent population during meiotic prophase. Breakage and exchange between the improperly paired chromosomes produces tangles that fail to separate in the subsequent meiotic divisions, leading to gametes with too few or too many chromosomes. Individuals carrying the *same* chromosome rearrangement, however, can often mate and produce fertile offspring in which the chromosomes can pair and meiosis proceeds normally. Since the individuals carrying the same rearrangement can breed successfully among themselves but not with the parent population they are established as instant species. Chromosome rearrangements of this type are considered to have produced new species rapidly in a wide variety of plants and animals, including insects such as grasshoppers and flies, various rodent species, and plants such as *Clarkia*, a genus of small flowering plants resident in California. Among two species of *Clarkia*, for example, *C. lingulata* and *C. biloba*, three rearrangements of chromosome segments and one chromosome that has split into two separate parts in *C. biloba* produce enough pairing and mechanical difficulties in meiosis to make the offspring of crosses between them sterile, even though the two species are otherwise close genetically.

Mutations in single genes can conceivably lead to rapid speciation in animals if they affect important steps in embryonic development, critical parts of metabolic reactions such as the ability to feed on specific plant or animal hosts, or genetically controlled elements of courtship behavior. For example, the males and females of many animals find each other and mate through the action of chemical attractants called **pheromones** that are unique to each species (Figure 20-9). The chemical differences between the pheromones of related species are frequently so small that single gene mutations could easily account for them and thus radically alter reproductive behavior.

Gradual versus Quantum Speciation The relative importance of gradual and quantum changes in the evolution of new species has been the subject of considerable debate among evolutionary theorists. Much of this argument stems from differences in the interpretation of the fossil

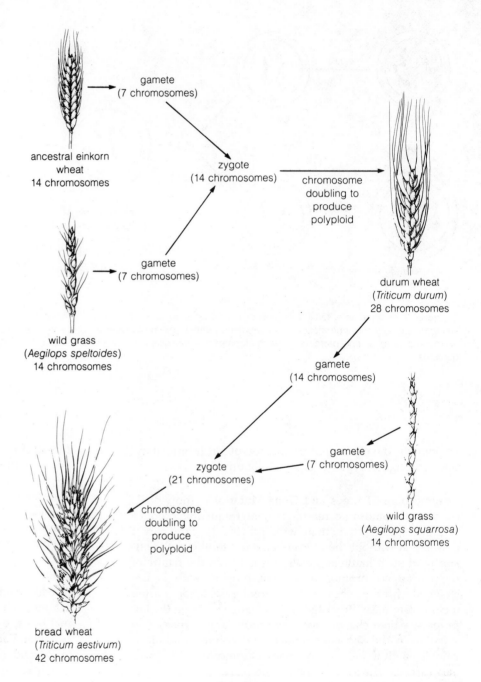

Figure 20-8 The evolution of bread wheat through hybridization followed by shifts to polyploidy. Einkorn wheat, which still grows in the Near East, was the first cultivated species. Hybridization of this species with the wild grass *Aegilops speltoides* followed by polyploidization produced durum wheat *(Triticum durum)*, which is now cultivated for use in spaghetti, macaroni, and other forms of pasta. A second hybridization between durum wheat and another wild grass, *A. squarrosa*, followed by another chromosome doubling, produced bread wheat, *Triticum aestivum*.

Labels within figure:

ancestral einkorn wheat 14 chromosomes
gamete (7 chromosomes)
wild grass (*Aegilops speltoides*) 14 chromosomes
gamete (7 chromosomes)
zygote (14 chromosomes)
chromosome doubling to produce polyploid
durum wheat (*Triticum durum*) 28 chromosomes
gamete (14 chromosomes)
wild grass (*Aegilops squarrosa*) 14 chromosomes
gamete (7 chromosomes)
zygote (21 chromosomes)
chromosome doubling to produce polyploid
bread wheat (*Triticum aestivum*) 42 chromosomes

record. New species, and even new genera and higher groups such as families and orders, characteristically appear suddenly in the fossil record without intermediate forms persisting as evidence of gradual change. These episodes of apparently rapid change are followed by long periods, sometimes amounting to millions of years, in which many species appear hardly to change at all. This pattern has been interpreted by some evolutionists, most notably Niles Eldredge of the American Museum of Natural History and Stephen Jay Gould of Harvard University, to mean that species appear primarily through sudden quantum jumps. Others maintain that the sudden appearance of new forms in the fossil record simply reflects the fact that the record is so incomplete that most intermediate forms have been lost. The evolution of new species, according to these theorists, has been primarily gradual.

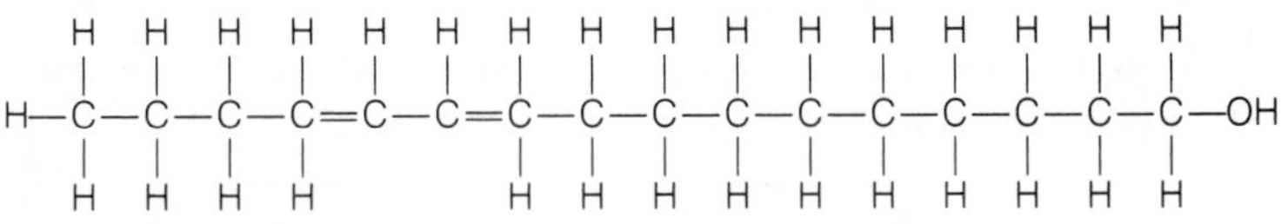

silkworm moth

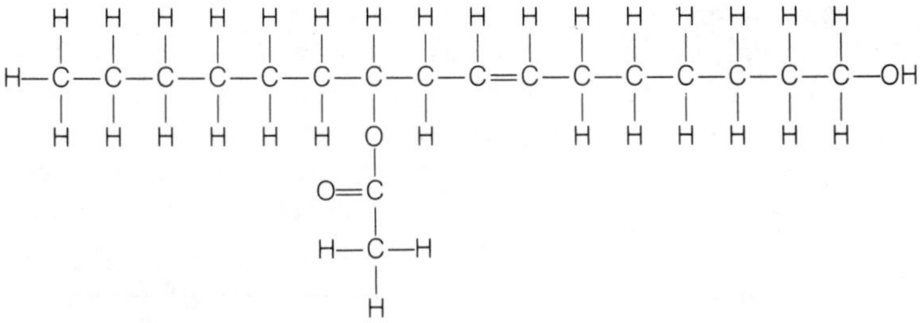

gypsy moth

Figure 20-9 The pheromones of two insect species, the silkworm and gypsy moths.

Given the incomplete nature of the fossil record, it is impossible to decide which group is nearer the truth. All that can presently be said with certainty is that both gradual and quantum speciation have undoubtedly been important in the origin of new species in evolution.

SYMPATRIC SPECIATION

The separation of populations by geographic barriers is probably the primary route by which species split and multiply in evolution. Even so, speciation almost certainly occurs too within a population occupying a single geographic region. In order for this pattern of speciation to occur, a genetic change must take place that achieves almost immediate reproductive isolation or allows a subgroup within a population to occupy a new ecological niche. This new niche must be so separate from the niche of the main population that there is virtually no contact between individuals in the two niches even though they are located in the same geographic region.

Sympatric speciation is a form of quantum speciation.

Since the change leading to reproductive isolation or niche separation must be rapid and essentially complete from the outset, sympatric speciation represents a form of quantum speciation. As such, it requires rapid and extensive genetic changes including polyploidy, hybridization, formation of chromosome rearrangements that prevent crossbreeding, or mutations in genes that control key steps in development, polyploidy, or reproductive behavior.

Once the new population is separated from the parent stock, it is free to evolve separately in response to differences in genetic drift and selection in its own subgroup. Among the expected changes are adaptations that reinforce the separation from the main population by additional prezygotic and postzygotic reproductive barriers or increase specialization for occupancy of a new niche.

Sympatric speciation provides another source of intensive controversy among contemporary evolutionists. Many theorists question whether the special requirements for sympatric speciation could be met frequently enough for this pathway to be a major source of new species in evolution. Others maintain that it does occur frequently. According to this interpretation, sympatric speciation may be responsible for some of the rapid increases in species in regions such as Hawaii, where *Drosophila* species have multiplied at great rates within a restricted geographic location.

Whatever the outcome of this controversy, there is at least limited evidence that sympatric speciation has probably occurred in some locales. The classic example is the hawthorne fly, *Rhagoletis*, which originally infested only native hawthorne fruits in the northwestern United States. In 1864, a subgroup of the hawthorne fly population suddenly switched to cultivated apple trees planted near the native hawthornes and became established as a serious pest of this fruit. In 1960, a subgroup of the apple-infesting race shifted to adjacent orchards planted with cherry trees in Wisconsin and is now established as a pest of these trees. Since cherries ripen before apples, the apple and cherry races of the hawthorne fly are effectively isolated. As a consequence, the adult forms of the flies emerge and breed at different times in the summer. These changes in the flies apparently depend on mutations in single genes that control the biochemistry of taste and digestion, leading to a sudden change in host preference. Sympatric speciation may frequently involve parasite–host relationships of this type, in which mutations lead to infestation of a new animal or plant host by a parasite. Once in the new host, the population subgroup is effectively isolated and free to evolve independently. There is then little chance that its unique combination of alleles will be lost due to crossbreeding with its parent population.

Although some new species apparently arise through sympatric speciation, the primary route of species multiplication has probably been through separation of populations into subgroups that are initially prevented from crossbreeding by geographic barriers. This route for speciation, common to plants and animals alike, involves splitting of a parent population into subpopulations inhabiting separate geographic regions. The next step is divergent evolution of the separated populations because of differences in natural selection, mutations, recombination, and genetic drift. The final stage is the development of reproductive barriers between the populations as a result of divergent evolution. It is this development of reproductive barriers that establishes the populations as separate species.

THE ORIGINS OF HIGHER TAXONOMIC GROUPS

How do higher taxonomic groups—genera, families, orders, classes, and phyla—originate? This question stands as one of the major unresolved problems of evolution. As in most past and present controversies in the field, there are two extreme views concerning the origins of higher taxonomic groups. Both sides of the issue have their adherents. One view, probably shared by the majority, is that higher groups evolve by the same mechanisms producing species: primarily geographic isolation, with either gradual genetic changes or quantum jumps producing new forms that are reproductively isolated from their parent populations. The opposite view, vigorously advanced by other evolutionists, is that the evolution of higher taxonomic groups proceeds by special mechanisms that do not occur in the evolution of species. The primary observation underlying this view is the fact that orders, classes, and phyla typically appear suddenly in the fossil record without intermediate forms surviving to link them directly to other groups.

Higher taxonomic groups probably arise through the same evolutionary mechanisms as species.

Despite the objections raised through this observation it seems most likely that the genera and higher taxonomic groups of the various kingdoms have in fact appeared through the same mechanisms that give rise to new species. It is improbable that hidden and as yet unknown mechanisms are responsible for producing the higher groups. Darwin's finches provide an especially well-documented example of the evolution of higher taxonomic groups through an extension of the same geographic mechanism giving rise to new species. The 14 species of finches inhabiting the Galápagos Islands, all probably derived from a single ancestral species, now fall into at least four reasonably distinct genera. All four genera can be accounted for most simply by assuming that geographic isolation of small finch populations on the islands of the Galápagos group led to evolution of differences significant enough to define separate genera. A similarly well-documented study of honeycreepers in Hawaii (Figure 20-10) indicates that four genera, containing 24 species of birds comprising a separate family (Drepanididae), evolved from a single original species through geographic isolation in the Hawaiian Islands.

Within this pattern of evolution of species and higher groups through similar mechanisms there may be special circumstances that lead to the rapid appearance of new groups showing only limited resemblance to their ancestors and to each other (Figure 20-11a). One is the development of a major new adaptation that allows species to invade an entirely new and extensive habitat. The development of lungs by a group of fishes in Devonian times allowed vertebrate animals to invade the land and led to evolutionary lines of new forms. These lines evolved so rapidly under the influence of the new and radically different environment that the evolutionary products of these lines, the amphibians, reptiles, birds, and mammals, show only distant relationships to each other and their aquatic ancestors. The differences are great enough to justify placing the products in categories as high as classes.

The rapid appearance of higher taxonomic groups may follow the appearance of an adaptation that allows invasion of a radically different habitat. Sweeping changes in the environment may lead to the same result.

Another special circumstance leading to the same result is sweeping climatic or geographic changes that open extensive new environments—such as the emergence of a

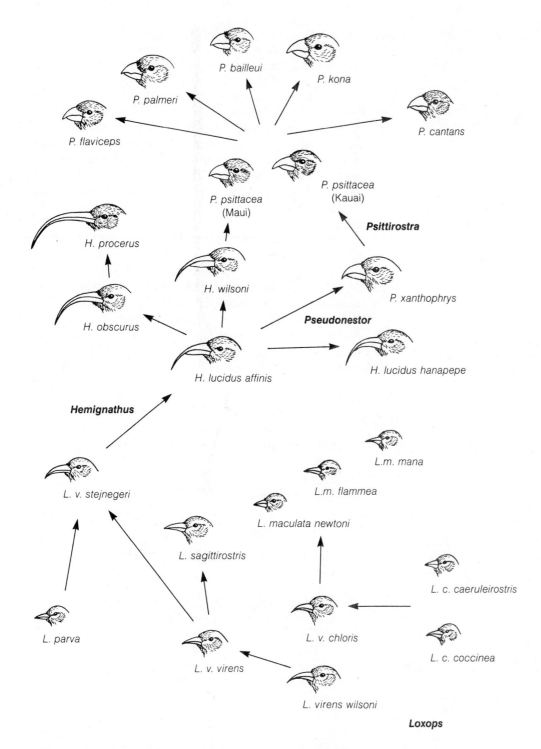

Figure 20-10 The evolution of four genera of present-day Hawaiian honeycreepers from an original *Loxops* species that first colonized the islands. The arrows show the probable evolutionary lineages leading to the modern species. Courtesy of F. J. Ayala, from T. Dobzhansky, F. J. Ayala, G. L. Stebbins, and J. W. Valentine, *Evolution.* W. H. Freeman, San Francisco, 1977.

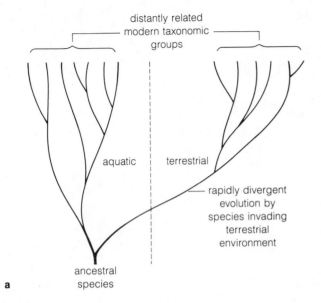

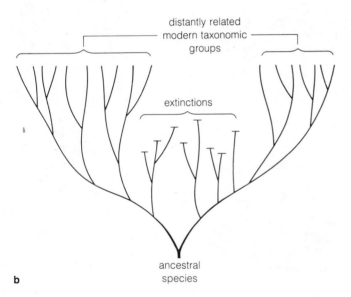

Figure 20-11 Two circumstances in evolutionary lineages that lead to gaps between the higher taxonomic groups. (**a**) A branch of an evolutionary line diverges rapidly through the appearance of an adaptation that permits the invasion of a new and radically different habitat—in this example, a terrestrial habitat invaded by a line descended from aquatic ancestors. (**b**) A series of extinctions eliminates species that would otherwise persist to fill the taxonomic gap between modern groups. The gap in this instance would appear absolute if none of the extinct species had formed fossils.

landmass from the sea, the splitting or collision of continents, recession of glaciers, and major changes in ocean currents. All these changes, which have actually occurred in the history of the earth, would open new environments of vast size and promote rapid evolution of forms that might show little relationship to each other or their ancestors. The last ice age, for example, ended only 10,000 to 20,000 years ago, a relatively short period on the geological time scale. The exposure of vast land areas formerly covered by ice for colonization led to the rapid evolution of new plant and animal species. Many of these species are so distinctive that they fall naturally into new genera, families, and orders.

There is a final circumstance that may lead to the elimination of intermediate forms and appearance of gaps in the fossil record: the pattern of extinctions (see Figure 20-11*b*). Rapid evolution in response to changing conditions may be accompanied by such rapid extinctions that intermediate forms do not persist long enough and in sufficient numbers to leave fossils. Moreover, the same climatic and geographical changes opening new environments may close old ones, as in a landmass that rises from the sea to create a terrestrial environment. As the new land environment is created, the old aquatic environment is eliminated—along with most or all of its former inhabitants. In this mass extinction a wide variety of intermediate and linking forms may be lost without a trace.

Mass extinctions without extensive fossilization may produce wide gaps in the fossil record.

EXTINCTION AND THE MECHANISMS OF SPECIATION

We rightfully regard extinction as a tragic loss of species that will never be seen again on the face of the earth. But extinctions are the rule rather than the exception in evolution. In fact, extinctions are as much a part of the evolutionary process as the creation of new species. The vast majority of species that once lived on the earth are now extinct; on the average, terrestrial species persist only some 50,000 years and marine species about twice as long. These extinctions take two forms. A species may disappear simply by evolving and changing into another species. Extinctions of this type, which account for 5 to 20 percent of the recorded disappearances of species, are sometimes called *pseudoextinctions*. The remaining disappearances are true extinctions in which all members of a species die without leaving direct descendants.

Extinctions are the rule, rather than the exception, in evolution.

Why do species become extinct? The answer is that evolution has no foresight. The adaptations that make species successful are selected by past and present environmental conditions, not future ones. Whether the same adaptations will provide reproductive success in a future environment is a matter of chance. In a sense, the more highly a species is adapted and specialized for survival in a present environment, the less chance there is that the same species will be able to survive in a changed environment. The probability of survival is affected by the rate of change of environmental conditions. If environmental changes proceed slowly, new adaptations may be selected and appear rapidly enough for a species to survive. If the change is too rapid, it may outstrip the rate of evolutionary change and lead to extinction. In times of most rapid change, the only species to survive are those with preexisting adaptations that, by chance, suit them to the new environment.

The environmental factors leading to extinction may be abiotic or biotic. Changes in average temperature or rainfall, submergence or emergence of continents, volcanic eruptions, changes in level, salinity, and currents in the ocean, elimination or creation of lakes and streams, collision or separation of continents—all are abiotic factors that may lead to extinctions, especially if the change is rapid. Among biotic factors are competition, predation, parasitism, and disease, which if intensified or prolonged may lead to the elimination of entire species. Elimination is especially likely where a new species is accidentally or deliberately introduced into an area populated by other species occupying similar ecological niches. If the introduced species is a highly efficient competitor or predator, the native species may be wiped out before niche differences have time to evolve.

Since the ability to develop new adaptations is crucial to survival, the total variability stored in the gene pool of a species is one of the factors determining whether it persists or becomes extinct. If the total variability is great, there is a better chance that new combinations of alleles will produce types with adaptations that may happen to suit them to the new environment. This chance will be greatest in diploid species with many individuals possessing a large proportion of genes in the heterozygous condition—that is, possessing different alleles of a gene on the two chromosomes of a homologous pair.

The list of examples of past extinctions is almost endless. Within recent history, the passenger pigeon has become extinct and a variety of species, such as the condor

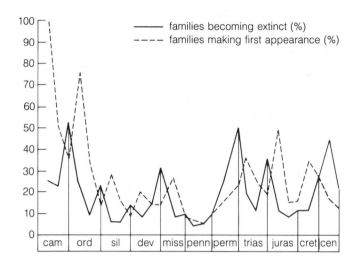

Figure 20-12 Mass extinctions occurred at or near the boundaries between the major periods of the geological time scale. Each period of mass extinction was followed, however, by a burst in the development of new species. From an original courtesy of N. D. Newell, from *Special Papers* 89:63 (1967) of the Geological Society of America.

and certain whales, are in danger of becoming so. Others, such as the dinosaurs, sabertooth cats, ground sloths, mastodons, mammoths, and North American horses, became extinct in prehistoric times and are now known only as fossils. Some of these extinctions are scattered examples involving the disappearance of species here and there among others that survive. Others are mass extinctions in which all the species as well as higher taxonomic groups in a wide geographic area are eliminated. While the major climatic changes leading to mass extinctions can sometimes be traced, the factors causing individual extinctions, such as the disappearance of horses in the Americas or the sabertooth cat, are much more difficult to determine from the fossil or geological record.

The record of mass extinctions shows a direct relationship to periods of rapid evolution of new species and higher taxonomic groups (Figure 20-12). Each of these major extinctions occurs at or near the boundary of a period in the geological time scale; that is, the close of the Cambrian, Ordovician, and the remaining geological periods is marked by mass extinctions amounting to as much as 50 percent or more of the known living forms (Figure 20-12a). Each period of mass extinction vacated major habitats, however, and opened the way for the rapid evolution of new forms to fill the emptied space. As a result, an almost equivalent number of new species and higher taxonomic groups appears after each mass extinction (Figure 20-12b).

Mass extinctions are followed by rapid and extensive appearance of new species in the fossil record.

Thus extinctions fit into the evolutionary process by opening opportunities for the rapid growth and diversification of new species and higher taxonomic groups.

TRENDS AND PATTERNS IN EVOLUTION

The mechanisms of evolution and speciation described in the chapters of this unit have taken place continuously since the origin of life on earth. Although the products of evolution are almost infinitely varied, there are trends and patterns common to many evolutionary lines.

Some of the trends have already been mentioned. Among the most basic is the development of *diploidy* as the dominant life phase of most animals and plants. The selective advantages of diploidy have already been described: the diploid condition allows recessive genes to be carried without injury to heterozygotes and provides a store of variability that makes a species more flexible in the face of a changing environment. Recombination of the dominant and recessive alleles of diploids in meiosis and sexual reproduction continually presents new combinations of these alleles to the changing environment, greatly increasing the potential of diploid organisms for genetic flexibility and change.

Another striking trend of evolution is toward increasing *complexity* in living organisms. From the simple, single-celled forms of the earliest known life, evolution has led toward organisms of many cells and extreme complexity. With the increase in complexity has come a general increase in the amount of information stored in DNA in cell nuclei. The evolutionary advantages of complexity are not immediately obvious, because we are often led to believe that simple solutions are best. Survival in environments of great potential but harsh conditions, such as those encountered by aquatic organisms when they first invaded the land, apparently requires solutions of greater and greater complexity, such as adaptations for efficiency in skeletal support, respiration, resistance to drying, and food gathering.

Life has evolved toward forms of greater complexity and size.

Increasing complexity is reflected in the frequent evolutionary trend toward greater *size*. From the microscopic or even molecular dimensions of the first life forms, evolution has led to the present distribution of organisms that are immense by comparison. Besides the requirement for greater body size to house specialized organs of greater complexity, selection pressures for increasing size involve greater physical strength and weight, an advantage for survival of both predator and prey; heat conservation and resistance to extreme temperature change in both warm-blooded and cold-blooded animals; competition for available sunlight in plants; and resistance to drying in land-dwelling plants and animals. These selection pressures have led some organisms to evolve toward the greatest possible size in each environment: elephants among land animals, sequoia trees among land plants, and whales and squid among aquatic animals.

It must be emphasized that none of these trends is absolutely uniform and universally found in all evolving lines. Even within lines that show an overall trend toward greater size and complexity, there are individual groups that reverse the trend and evolve toward smaller size or greater simplicity. Although the general trend among land plants has been toward greater size and complexity, individual groups, such as many of the flowering plants, have evolved toward smaller overall size and greater simplicity in the number and arrangement of flower parts. Among animals, parasitic species such as tapeworms lack many of the body systems of their free-living ancestral species and are more simple by comparison.

Evolution also follows patterns that have been repeated frequently in the history of life. The most important of these patterns, and one that is found almost universally, is the characteristic branching of evolutionary lines. This typical branching pattern of evolution, called **adaptive radiation** by evolutionists, is a response to the survival advantages of greater niche specialization in both new and old environments. In new environments particularly, a colonizing species tends to split rapidly into a variety of new types inhabiting every possible ecological niche. The varied Galápagos finches, all probably descended from a single species colonizing the islands, provide a classic example of adaptive radiation.

Organisms in similar environments tend to evolve along similar pathways.

Two additional patterns, called **convergence** and **parallelism**, although found less often in evolutionary lines than adaptive radiation, are still observed frequently in fossil and living species. Convergence and parallelism refer to the tendency of organisms to evolve along similar pathways in similar environments. Whales and fishes, for example, have developed equivalent body forms and appendages in response to the similar selective pressures of their aquatic environment. If the organisms are only distantly related, as in the case of whales and fishes, the pattern is called convergence; if the organisms are closely related, the pattern is termed parallelism. An often quoted example of parallelism is the fact that in the early Mesozoic several different groups of reptiles evolved simultaneously toward warm-bloodedness, producing several major lines of warm-blooded animals that evolved independently but along the same pathways. Since convergence and parallelism differ only in the degree of relatedness of groups evolving along similar lines, the distinction between the two terms is sometimes arbitrary.

Parallelism among warm-blooded animals has been studied extensively. From this research, several so-called rules of evolution have become established. Although there are many exceptions to these generalizations, parallel patterns do occur in many warm-blooded animals. The most famous of these is **Bergman's rule**, which formalizes the observation that warm-blooded animals tend to be larger in cold regions. The selective advantage responsible for this commonly observed evolutionary pattern is believed to be the relationship between surface area and volume as size increases. Large organisms have a lower surface/volume ratio and expose less surface to cold per unit of body volume, reducing heat loss as a result.

Another common response of warm-blooded animals to the selective pressures of survival in cold environments is a tendency toward reduction in the size of body appendages. This "rule," called **Allen's rule**, is really an extension of Bergman's rule because decreasing appendage size also decreases the surface/volume ratio of an organism. Allen's rule is reflected in the reduced wing and leg length of warm-blooded animals in cold climates; small noses in human races living in extremely cold climates is also said to be a response to the same selective forces.

A final "rule" of parallelism, **Gloger's rule**, notes that in warm-blooded animals black surface pigments are reduced in warm, dry regions whereas brown surface pigments are reduced in cold, moist climates. The English sparrows studied by Johnston and Selander (see Chapter 18) have evolved lighter body colors in the warm, dry desert climate of the American Southwest. Although the selective advantages of these patterns in color response are not clear, Gloger's rule apparently operates frequently among warm-blooded animals. Even some cold-blooded forms such as insects obey Gloger's rule. (This is an example of convergence rather than parallelism, however, since insects are only distantly related to warm-blooded animals.)

All these trends and patterns show that, far from being a random process, evolution proceeds in definite directions in response to the demands and pressures of the environment. As a result, overall trends such as diploidy and the tendency toward greater size and complexity can be discerned in the evolution of plants and animals. And, in similar environments, evolution often follows similar and parallel pathways. It is important to note that these overall trends and patterns do not represent *goals* in evolution; they are simply common adaptations developed in response to selection by similar past and present forces of the environment.

THE THEORY OF EVOLUTION TODAY: A SUMMARY

According to the contemporary theory of evolution, evolutionary change is controlled by a series of basic mechanisms in nature. The raw material of evolution is supplied by random *mutations* in the genetic material of a population. These mutations, in diploid, sexually reproducing organisms, are mixed into new combinations by *sexual reproduction* and *recombination.* These mechanisms—mutation, sexual reproduction, and recombination—produce *variability* among the individuals of a population.

This variability is subjected to *selection* in the environment. Individuals with characteristics making them more successful in using the resources of the environment are more likely to survive and reproduce. Others with less favorable characteristics are less likely to reproduce. Because individuals carrying favorable characteristics are more likely to reproduce, the genetic alleles controlling these characteristics are passed in greater frequency to the next generation. The resulting change in the frequencies of alleles in the next generation constitutes evolution. Gene frequencies also change due to the effects of chance or genetic drift, especially in small populations.

Although natural selection acts on individuals, the outcome of evolution cannot be traced for individuals because single organisms carry the same combinations of genes and alleles for their lifetimes and cannot themselves evolve. Moreover, an individual with many favorable alleles may happen not to find a suitable mate and reproduce. The outcome and trend of evolution can, however, be traced through time for *groups* of interbreeding individuals. This group, the *population,* can be analyzed statistically and its evolution predicted in terms of average numbers. Although the unit of selection is the individual, the unit of evolution is the population.

New species may evolve either through rapid genetic changes in individuals living in the same geographic region *(sympatric speciation)*, or from the geographic isolation of populations of a species (geographic speciation). In sympatric speciation, the sudden appearance of reproductive or niche isolation separates populations effectively enough to allow their evolution into separate species. In geographic speciation, the isolated populations, subjected to different pressures of mutation, recombination, and natural selection, gradually evolve along different pathways. If the populations are separated long enough, these differing selective pressures will bring about differences in their reproductive functions and activities. These differences, if they become extensive enough to prevent successful interbreeding, then establish the populations as new and different species.

While all evolutionists agree on the fact of evolution and the basic parts of the theory as outlined here, there is still considerable controversy about details of the theory. The relative importance of natural selection and chance in evolution, whether new species arise primarily through gradual changes or in sudden quantum jumps, the relative importance of geographic and sympatric speciation—these are the most hotly debated issues among evolutionists today.

Questions

1. How does the evolution of new species by splitting differ from the change of one species into another?

2. What is the significance of reproductive isolation in the formation of new species?

3. What is an ecological niche? What is its significance to the evolution of new species?

4. What steps occur in geographic speciation?

5. List environmental changes that may lead to geographic separation of populations.

6. What processes lead to genetic divergence of populations once they are separated? Why does this change occur?

7. What is the *founder effect* in the generation of new species?

8. What is the difference between prezygotic and postzygotic barriers to reproduction? Which is more economical to the emerging species as far as utilization of environmental resources is concerned?

9. Describe these barriers to reproduction: behavioral, seasonal, structural, reduced hybrid viability, hybrid sterility.

10. How are reproductive barriers intensified if species with incomplete barriers to reproduction are brought together?

11. What is the difference between gradual and quantum speciation? What mechanisms may produce quantum speciation?

12. What is sympatric speciation? How is it related to quantum speciation?

13. What special circumstances may underlie the sudden appearance of higher taxonomic groups in evolution?

14. Why do extinctions occur? How have extinctions been an important factor in the evolution of new species?

15. Define adaptive radiation, convergence, and parallelism.

16. What are Bergman's, Allen's, and Gloger's rules? Are these evolutionary patterns really absolute rules? Explain your answer.

17. The proposal that the species on earth were specially created by a supreme being may be said to be a valid idea but not a valid scientific hypothesis. What does this statement mean?

Suggestions for Further Reading

Ayala, F. J., and J. W. Valentine. 1979. *Evolving*. Benjamin/Cummings, Menlo Park, California.

Dobzhansky, T., et al. 1977. *Evolution*. Freeman, San Francisco.

Futuyma, D. J. 1979. *Evolutionary Biology*. Sinauer Associates, Sunderland, Massachusetts.

Gould, S. J. 1977. *Ontogeny and Phylogeny*. Harvard University Press, Cambridge, Massachusetts.

Grant, V. 1977. *Organismic Evolution*. Freeman, San Francisco.

Mayr, E. 1970. *Populations, Species, and Evolution*. Harvard University Press, Cambridge, Massachusetts.

Scientific American 239, September 1978 (Evolution Issue).

The earth is estimated to be about 4½ billion years old. It was then, 4½ billion years ago, that our planet condensed out of the primordial matter and began its long transition into the environment we know today. Since the oldest known fossils of bacteria-like cells, discovered in deposits at North Pole in remote northern Australia, are estimated to be approximately 3½ billion years old, life must have originated on our planet at some time during the first billion years of its existence.

Very few fossils remain to tell us of the characteristics of the cellular life of 3½ billion years ago; there is just enough information to indicate that the earliest cells were probably bacteria-like prokaryotes (Figure 21-1). Nothing at all exists to inform us about the earlier period, between 4½ and 3½ billion years ago, when the transition from nonliving to living matter took place. No intermediate forms survive, and there is no readable fossil record of this period. Thus we are left with hypothesis, speculation, and conjecture to unravel the events of this time.

It is not necessary to assume that life appeared spontaneously on earth through the chance interactions of nonliving matter. The first life may have appeared through divine intervention, as many religions teach us. Alternatively, life may have been carried to earth on particles from outer space or on machines delivered here by civilizations based elsewhere in the universe. Divine intervention is not within the domains of science, however, and speculation about the origin of life by this route is the concern of theologians, not biologists. Assuming that life was introduced from somewhere else in the universe merely begs the question because we are then left with the difficulty of reconstructing the origin of life on some other planet. From the scientific point of view, the assumption that life on earth originated here through inanimate chemical processes is the only productive and reasonable approach to follow and test. In fact, a number of working hypotheses based on this assumption have even been tested successfully in the laboratory.

From the scientific point of view, the assumption that life on earth originated through inanimate chemical processes is the only productive and reasonable approach to follow.

Discussion of the evolutionary origins of life requires a special definition of "life." Probably the first spark of life appeared in collections of molecules much simpler than even the most simple cells we know today. Present-day cells, even the most rudimentary ones, have a complex of components: a boundary membrane separating the cell interior from the exterior; one or more nucleic acid coding molecules; a translation system, including the various

21

THE ORIGINS OF CELLULAR LIFE

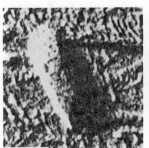

Figure 21-1 Fossils of bacteria-like prokaryotes exposed in 3-billion-year-old rocks by polishing. Courtesy of E. S. Barghoorn. Copyright 1966 by the American Association for the Advancement of Science.

RNAs and ribosomes, capable of converting the coded information into biological molecules; and a metabolic system providing the energy needed to carry out cellular activities. Because these systems are so complex, it is hardly conceivable that life suddenly appeared at this level without being preceded by less organized forms. Therefore the transition from nonliving matter to the first cells was probably gradual. It is unlikely that a sudden event caused cellular life in all its present-day complexity to appear at one instant.

Presumably the transition from nonliving to living matter was gradual; no sudden event caused cells to appear all at one instant.

The problem, then, is to decide the minimum level of organization and activity that is required for a collection of interacting molecules to be considered alive. Some investigators claim that very simple groups of molecules are alive if they can use an energy source to carry out a single, continuous energy-requiring reaction. Others insist that life also involves the ability to grow and reproduce in kind.

Because the latter, more complex requirements fit our present-day concept of life, we will use this level of complexity to define precellular life. Collections of molecules will thus be considered to have had the first spark of life if they were able to (1) use either light or chemical energy to drive internal reactions requiring energy, (2) increase in mass by controlled synthesis, and (3) reproduce into additional collections of matter of the same kind. The ability to reproduce in kind is considered to carry with it a requirement for an information-coding system and a system for translating the coded information into finished mole-

cules. At this minimum level, life is relatively complex but still much simpler than the simplest cells we know of.

STAGES IN THE EVOLUTION OF LIFE

Although the process was probably continuous, the first evolution of life can be divided for convenience into four successive stages. The first was the formation of the earth and its initial atmosphere. This first stage provided the inorganic raw materials for the evolution of life and set up the conditions under which they could interact. In the second stage, complex organic molecules were produced by inanimate forces such as lightning or ultraviolet radiation acting on the inorganic chemicals in the environment. This hypothetical stage has been successfully duplicated in the laboratory. In the third stage, the newly produced organic molecules collected by chance into clusters capable of chemical interaction with the environment. Experiments testing ideas dealing with this stage also have met with some success. In the fourth and final stage, some of these clusters were successful in converting the energy of complex molecules absorbed from the environment into useful chemical energy; some of this energy was used to synthesize other complex molecules such as proteins and nucleic acids. Gradually, in this stage, the coding function of the nucleic acids became established and related to the sequences of amino acids in proteins. This development guided synthesis and reproduction in the primitive molecular assemblies. At this level of organization, life appeared in the assemblies. Evolution then changed from chemical to organic: natural selection of favorable mutations in the coding system became the basis for further evolutionary development and change. Although the development of the coding relationship between nucleic acids and proteins is the most difficult part of the entire process of evolution to deal with experimentally, it too has been partially reconstructed and tested with some success.

Given similar conditions and sufficient time, it is probable that life could evolve elsewhere in the universe.

Once life began in the molecular assemblies and the basis for organic evolution was established, natural selection led to the appearance of cells equivalent to the most primitive prokaryotes known today. Presumably this level was reached at or before the time the North Pole deposits were laid down in Australia some 3½ billion years ago. Interactions between these early prokaryotes and further

Figure 21-2 A cosmic cloud of gas and dust (the Horsehead Nebula in Orion) in space some 1300 light-years from Earth. Palomar Observatory photograph.

selection eventually established the lines leading to eukaryotic cells, a development that may have required another 1 to 1½ billion years.

To study the evolution of life by the scientific method we must assume that all the processes occurring in these successive stages were inanimate and took place by chance.[1] Given similar conditions and sufficient time, it is probable that a similar process could occur again and that life has evolved or is evolving now at other locations in the universe.

---

[1]Some theoreticians think that the chemical processes giving rise to the first life did not occur by chance. According to their view, the particular collection of inorganic molecules assembled together when the earth first appeared, the conditions of pressure, temperature, and so forth existing at the time, and the types of energy input could lead only to the chemical results actually obtained. Thus the results were not random but fixed as chemical probabilities when the earth first condensed.

The First Stage: The Origin of the Earth and its Primitive Atmosphere

The contemporary view of the earth's origins is that the sun and planets of our solar system and all the stars and other bodies of the universe condensed out of cosmic clouds of gas and dust particles. The composition of these clouds, which still persist in the universe today (Figure 21-2), has been studied by analyzing the light transmitted or reflected by them. Most of the matter of the clouds is hydrogen gas at extremely low concentrations; lesser amounts of nitrogen, helium, and neon are also present. Other elements and compounds are also suspended in the clouds as solid particles—including metallic iron and nickel; the silicates, oxides, sulfides, and carbides of these and other metals; inorganic and organic carbon compounds; ammonia; and frozen water (see Table 21-1).

The sun and the planets of the solar system probably condensed out of cosmic clouds of dust.

According to the condensation hypothesis, stars, suns, and planetary systems are continually forming from clouds of gas and dust and disintegrating into dust again. Our own solar system condensed from one of these large clouds of dust. Most of the cloud condensed rapidly around a single center, causing high pressure and heat to develop in the interior and setting off a thermonuclear release of energy; this release created the sun. Smaller centers of condensation produced the planets.

As the condensing planets formed, their temperatures increased due to the effects of solar heating, gravitation, and internal pressure. Although never reaching the temperature of the sun, the heat inside the condensing earth became high enough to melt the collected materials. In the melted form, the heavy, metallic elements settled to form the core of the earth. The lighter materials, such as the silicates and carbides of these metals, rose to the surface where they cooled and solidified to form the rocks and particles of the surface crust.

The original atmosphere of the earth probably contained large quantities of hydrogen, nitrogen, and water vapor originating from the cosmic cloud. As the atmosphere and surface cooled, much of the water vapor condensed into droplets and rained down on the dust and rocks of the crust. Eventually, after years of torrential rains, water collected into the rivers, lakes, and seas of the primitive earth.

Some of the water was retained as vapor in the atmosphere. Other gases, such as hydrogen, nitrogen, and carbon dioxide, interacted with each other and with carbides, nitrides, and sulfides in the crust to produce methane, ammonia, and hydrogen sulfide. Carbon dioxide may also have been expelled from the interior of the earth along with other gases by erupting volcanoes. Free oxygen is not believed to have been present in the atmosphere in significant amounts because it would have reacted quickly with particles and rocks of the crust to form oxides.

The fact that the oldest rocks of the earth's crust, now deeply buried but originally exposed to the atmosphere, contain reduced rather than oxidized substances supports these ideas about the characteristics of the earth's primitive atmosphere. Moreover, analysis of the light transmitted and reflected by the atmospheres of the largest planets, Jupiter and Saturn, which are believed to have retained their primitive complement of gases almost intact, reveals the same strongly nonoxidizing character. Free oxygen is absent, and NH_3, CH_4, and vaporized H_2O are prominent among the gases present.

The nonoxidizing quality of the primordial atmosphere was a necessary condition for the evolution of life.

Table 21-1   Atoms, Molecules, or Chemical Groups Detected in Cosmic Clouds or Outer Space

Atom, Molecule, or Radical	Symbol
Hydrogen atom	H
Hydroxyl radical	$OH^\bullet$
Ammonia	NH_3
Water	H_2O
Formaldehyde	HCHO
Carbon monoxide	CO
Cyanogen radical	$CN^\bullet$
Hydrogen cyanide	HCN
Cyanoacetylene	HC_2CN
Methyl alcohol	CH_3OH
Formic acid	HCOOH
Carbon monosulfide	CS
Formamide	$HCONH_2$
Silicon oxide	SiO
Carbonyl sulfide	OCS
Acetonitrile	CH_3CN
Isocyanic acid	HNCO
Hydrogen isocyanide	HNC
Methylacetylene	CH_3C_2H
Acetaldehyde	CH_3CHO
Thioformaldehyde	HCHS
Hydrogen sulfide	H_2S
Methylene imine	H_2CNH

Adapted from S. W. Fox, *Molecular and Cellular Biochemistry* 3:129 (1974); courtesy of S. W. Fox and *MCB*.

The absence of oxygen and the presence of hydrogen, methane, ammonia, and water vapor gave the atmosphere of the primitive earth a nonoxidizing rather than an oxidizing quality. This nonoxidizing quality was fundamental

Table 21-2   Natural Sources of Energy on the Earth

Source	Energy (cal/cm²/yr)
Sun (total radiation including ultraviolet)	260,000
Ultraviolet light	4,000
Electrical discharges	4
Shock waves	1.1
Radioactivity (to 1 km depth)	0.8
Volcanoes	0.13
Cosmic rays	0.0015

Adapted from S. L. Miller, H. C. Urey, and J. Oro, *Journal of Molecular Evolution* 9:59 (1976), with permission of Springer-Verlag.

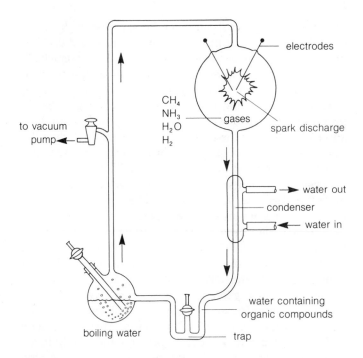

Figure 21-3   The Miller apparatus demonstrating that organic molecules can be produced spontaneously in a primitive atmosphere (see text). Courtesy of S. L. Miller. Copyright 1955 by the American Chemical Society.

to the next stage of evolution: the appearance of complex organic molecules through the action of natural energy sources on the inorganic matter of the crust and atmosphere.

The Second Stage: The Spontaneous Production of Organic Molecules

The hydrogen, nitrogen, methane, ammonia, water vapor, and other gases of the primordial atmosphere, and the same gases dissolved into bodies of water, were exposed to continual inputs of energy from a number of natural sources (Table 21-2). One source, then as now, was sunlight. Besides the visible light from the sun, ultraviolet light in greater quantities than today reached the lower atmosphere and acted on surface chemicals and waters. (At present, most of the ultraviolet light approaching the earth is absorbed by oxygen and ozone in the outer atmosphere and never reaches lower levels or the surface.) Another energy source was provided by heat from absorbed light and volcanic activity. Electrical discharges during the violent rainstorms of the period also supplied energy to the atmosphere and surface of the earth.

In 1924, the Russian biochemist A. I. Oparin proposed that the action of these energy sources on the inorganic matter of the earth would have caused complex organic molecules to form. The same idea was advanced independently a few years later by an English geneticist, J.B.S. Haldane. Both Oparin and Haldane reasoned that a great variety of organic chemicals would have been produced and would have accumulated in the absence of oxidation and decay by microorganisms, the chief routes by

which organic matter is broken down in the present-day environment. Haldane thought that the concentration of these organic substances in the seas might even have reached the consistency of a "hot, dilute soup."

The Oparin–Haldane proposal received direct support in 1953 when Stanley L. Miller developed an apparatus at the University of Chicago to test the effect of electrical discharges on a simulated primitive atmosphere. In the Miller apparatus (Figure 21-3), water vapor, methane, ammonia, and hydrogen flowed continuously through a chamber exposed to repeated sparking from electrodes. Below the chamber, the water vapor and any organic chemicals produced were cooled, condensed, and trapped at a low point in the tubing. Operating the apparatus for one week yielded a surprising variety of organic chemicals, including urea, several amino acids, and lactic, formic, and acetic acids. Thus organic compounds could indeed have been produced by the action of an energy source on the gases of the primitive atmosphere as Oparin and Haldane proposed.

After Miller's pioneering experiments showed the way, a great many additional experiments demonstrated that other organic molecules can be synthesized by varying the gases present in the starting mixture. Adding hydrogen cyanide, which is readily produced by interactions between methane and nitrogen, produced additional amino acids and the purine and pyrimidine building blocks of the nucleic acids. Adding formaldehyde, another gas readily produced by reactions between the gases of the primitive atmosphere, led to the production of sugars, including the ribose and deoxyribose sugars of DNA and RNA. Other variations produced the subunits of lipids. A key feature in these experiments is the absence of oxygen in the simulated atmosphere. If oxygen is added to the mixture of gases, virtually no organic molecules are produced.

Thus it has been possible to demonstrate that the building blocks of all the major biological molecules could have been synthesized on the primitive earth. Analysis of meteorites and interstellar dust clouds also supports the hypothesis that organic molecules could have been synthesized spontaneously. Organic molecules found in a meteorite impacting in 1969 near Murchison, Australia, include 6 of the 20 amino acids found in proteins and 12 additional nonbiological amino acids. Purines and pyrimidines too have been found in trace quantities among the organic molecules of the meteorite. These organic substances and the organic molecules detected in interstellar dust clouds were presumably synthesized by the same inanimate, spontaneous mechanisms proposed in the Oparin–Haldane hypothesis.

Even proteinlike chains of amino acids have been synthesized in laboratory experiments approximating the primitive environment. S. W. Fox of the University of Miami produced proteinlike molecules by heating dried mixtures of amino acids to 160 to 210°C for several hours. He termed these molecules **proteinoids**. Fox and others have also formed nucleic acids by heating mixtures of nucleotides and phosphates to about 60°C and holding them at this temperature for some time.

All these experiments establish that a wide variety of complex organic substances could have arisen spontaneously over the millions of years between the formation of the earth and the first appearance of living matter. Accumulation of these compounds, which probably included all the major molecules found in living organisms, provided the raw materials for the third stage in the evolution of life: the collection of organic matter into assemblies capable of interacting with each other and the environment.

Interactions between the earth's chemicals and natural energy inputs probably resulted in the synthesis of all the major organic molecules of living organisms.

The Third Stage: The Spontaneous Collection of Organic Molecules into Functional Assemblies

As the concentration of organic substances increased in the environment, certain types of molecules assembled spontaneously into aggregates of various kinds. As a preliminary to this assembly, they may have been concentrated somewhat by the evaporation of water from lakes, inland seas, and tidal flats. These more concentrated solutions of organic matter may then have followed several possible routes of assembly into functional units. Some of these routes have been shown to be possible under conditions imitating the primitive environment.

Oparin proposed that a mechanism known as **coacervate** formation was crucial in the assembly process. Coacervates form when protein molecules in solution separate into concentrated droplets as a result of attractions between charged and polar groups on the surfaces of different polypeptide chains. If many of these polar groups face the surfaces of the protein droplet, the surrounding water molecules tend to form a film several molecules thick around the coacervate. This film of water molecules gives the boundary between the droplet and the surrounding medium many of the properties of a membrane. As a result, coacervates may concentrate molecules from the surrounding medium or they may shrink or swell in response to changes in inside or outside concentrations as do living cells.

A similar process of droplet formation has been studied by Fox and his colleagues. They showed that if solutions of proteinoids are heated in water and then allowed to cool, small spherical particles containing the proteinoids separate out of solution. These small particles (Figure 21-4a), termed **microspheres** by Fox, are similar to coacervates in activity and can take in various substances from the surrounding medium. Some of the microspheres in Fox's experiments were also able to speed the rate of various biochemical reactions (see Table 21-3), including the hydrolysis of ATP to ADP. Other microspheres were observed to bud, split, or fragment in a process superficially similar to cell division (see Figure 21-4b).

Other assembly mechanisms may have also been important. J. D. Bernal of the University of London has proposed that molecules may have collected into aggregates by absorption on particles of clay in mud around tidal flats and river mouths. Clay beds of this kind might have reached extensive size and provided large sun-warmed areas in which life might slowly have evolved. Some clays containing metallic compounds also act as catalysts; they can enhance the breakdown and conversion of organic

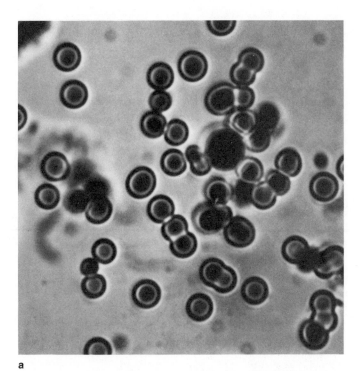

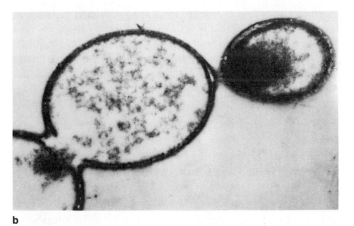

Figure 21-4 Fox's microspheres. (**a**) A collection of microspheres made by cooling a heated solution of proteinoids (light micrograph). ×1000. (**b**) Microspheres fragmenting in a process superficially similar to cell division (electron micrograph). ×11,000. Courtesy of S. W. Fox from *Molecular Evolution and the Origin of Life* by S. W. Fox and K. Dose.

molecules into other substances and thus carry out limited chemical interactions.

Another possible route of aggregation is the formation of films or particles by many types of lipid molecules.

Table 21-3   Catalytic and Related Activities of Microspheres

Type of Reaction	Substrate Broken Down by Microspheres
Hydrolysis	ATP Nitrophenyl acetate Nitrophenyl phosphate
Decarboxylation	Glucaronic acid Pyruvic acid Oxaloacetic acid
Amination	Glutamic acid
Reduction–oxidation	H_2O_2
Synthesis (with added ATP)	Nucleic acids Peptides

Adapted from S. W. Fox, *Molecular and Cellular Biochemistry* 3:129 (1974); courtesy of S. W. Fox and *Journal of Molecular and Cellular Biochemistry*.

When placed in water, these lipid molecules spontaneously form layers two molecules thick called **bilayers** that have many of the properties of cell membranes (see Figure 2-16 and p. 26). Often such lipid bilayers suspended in water round up into closed vesicles consisting of a saclike continuous "membrane" enclosing a central space (Figure 21-5). Other organic molecules, including proteinlike polypeptides, nucleotides, and nucleic acids, could have been absorbed into the lipid bilayers or trapped in the space inside the closed vesicles.

Supposedly, one or more of these assembly processes—the formation of coacervates, microspheres, clay beds, or lipid bilayers—took place repeatedly in the primitive environment. Sooner or later the fourth stage in the evolution of life, life itself, appeared within some of these assemblies of protein, nucleic acid, lipid, and carbohydrate molecules.

The Fourth Stage: The Development of Life in the Primitive Assemblies

We have defined precellular life as the stage at which the primitive molecular assemblies could (1) use an energy source to drive energy-requiring reactions to completion, (2) increase in mass by controlled synthesis, and (3) reproduce additional assemblies like themselves. These steps toward life have proved to be much more difficult to reconstruct and test in the laboratory. Nevertheless, some

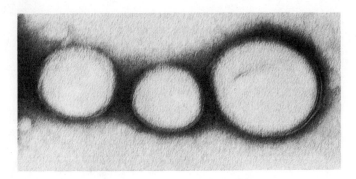

Figure 21-5 Vesicles produced from phospholipids synthesized under primitive earth conditions. Electron micrograph of negatively stained preparation. Courtesy of W. R. Hargreaves, from *Nature* 266:78 (1977).

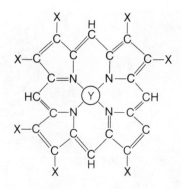

Figure 21-6 A porphyrin ring—a pigment molecule that could have been synthesized in quantity under primitive earth conditions. In present-day organisms, porphyrin rings form the active chemical groups of chlorophyll, hemoglobin, and the cytochromes. The porphyrin rings of these substances differ in the substitution of chemical groups at the positions marked with an X and the central metal ion marked by a Y. (Y is a magnesium ion in chlorophyll and an iron ion in hemoglobin and the cytochromes.)

ideas about them have been developed and a few have even been tested experimentally.

Energy to drive reactions "uphill" was probably first obtained by breaking down molecules absorbed from the surrounding medium, including ATP and the other high-energy nucleoside triphosphates CTP, GTP, UTP, and TTP. Some of the reaction mechanisms might also have removed electrons from other organic molecules—and, in so doing, released usable energy.

As these energy-releasing systems developed, synthetic pathways also increased in complexity. N. H. Horowitz has suggested a pattern by which complex synthetic pathways might have appeared. Suppose that a modern pathway in cells makes a required substance such as an amino acid. This pathway begins with a simple inorganic substance A and leads through the steps $A \rightarrow B \rightarrow C \rightarrow D$ to produce the amino acid. Initially the amino acid was abundant in the environment and was absorbed directly for use in the molecular cluster. Later, as the amino acid became depleted, chemical selection favored clusters that could make the amino acid from substance D, a slightly less complex organic molecule still found in abundance in the environment. As D became exhausted, selection favored assemblies developing a pathway in which the even more simple substance C could be absorbed and used to make D. This process continued until the entire pathway leading from the simple substance A to the amino acid was established.

All these developing reaction systems yielding and utilizing energy were probably speeded initially by the general catalytic activities of coacervates, lipid vesicles, microspheres, or the clays absorbing organic molecules. Gradually, catalytic functions were taken over by more specialized molecules with increased specificity, leading eventually to enzymes.

Some ideas about the possible steps in the appearance of enzymes have emerged from the study of the catalytic properties of substances like iron. Metallic iron, for example, can act as a catalyst to increase the breakdown of hydrogen peroxide (H_2O_2) to water and oxygen. Combining iron into iron oxide further increases the rate of the reaction. If iron is bound into a complex organic structure known as a porphyrin ring (see Figure 21-6), its catalytic ability is increased about a thousand times. Although complex, this ring structure could have been synthesized spontaneously in the primitive environment. Finally, if combined with the correct protein, the catalytic properties of iron are increased millions of times. Thus the catalysts of precellular life probably first evolved from small inorganic and organic molecules and, later, increased in activity by combination with more complex organic molecules and polypeptides.

The first of these polypeptides may have simply provided a stable framework anchoring an inorganic catalyst to the primitive molecular assemblies. Many amino acids, however, particularly those with acidic or basic properties, also have catalytic activity; thus some of the catalyst–polypeptide complexes would have been more efficient in speeding reactions than the inorganic catalyst alone. With the development of a coding system directing the synthesis of proteins, the catalytic polypeptides took on greater specificity and reproducibility until the first enzymes appeared.

The beginnings of the information system may have developed as an offshoot from the use of the nucleotides as an energy source. Occasionally these nucleotides probably assembled spontaneously into nucleic acids, as the experiments simulating primitive-earth conditions have shown. Some combinations of nucleotides in the nucleic acids formed in this way may have favored the absorption of certain amino acids into the assemblies. As a result, the sequence of amino acids in the polypeptides made in the assemblies became less random. The first vestiges of directed synthesis had appeared. With time, and with the appearance of additional spontaneous interactions between nucleic acids and proteins, enough specificity developed to lay down the beginnings of the genetic code.

The nucleotide triphosphates, probably first important in energy transfer, combined later into nucleic acids and took on a coding function.

Fox has reported that microspheres containing both proteinoids and nucleic acids are able to carry out the first steps in this process. In Fox's experiments some correlation was detected between the type of nucleic acid included in a microsphere and the type of amino acid absorbed. Inclusion of a poly-A nucleic acid (a nucleic acid molecule containing only one type of base, adenine, repeated in the sequence AAA . . . and so on) favored the absorption of the amino acid lysine into the microspheres. Including poly-C favored the absorption of proline into the microspheres. Similarly, poly-G nucleic acid molecules favored the uptake of glycine and poly-U acid favored phenylalanine. These results are of special interest because AAA is the code word for lysine, CCC is the code word for proline, GGG is the code word for glycine, and UUU is the code word for phenylalanine in the genetic code!

Once the coding relationships between the nucleic acids and proteins appeared, the way was open for the transition from chemical to organic evolution.

Once the coding system appeared and the relationship between the nucleic acid code words and the amino acids became fixed, the way was open for the change from chemical evolution to organic evolution. Mutations in the sequence of the coding nucleic acids would then cause changes in the proteins of the assemblies. Some of these changes would be favorable and increase the ability of the assembly to compete for organic molecules in the medium. Other favorable mutations would allow the assemblies to

manufacture more of their required organic substances for themselves. By this stage, the assemblies would probably also have acquired surface membranes. These could have arisen either during the initial aggregate formation or later as an adaptation that improved the ability of the assemblies to survive and compete. At this stage, these first living beings would have assumed the crucial characteristics of cellular life: a coding system, a system capable of translating the coded information into specific proteins, and systems coupling energy-yielding to energy-requiring reactions, all surrounded by a surface membrane.

FROM PROKARYOTIC TO EUKARYOTIC CELLS

The first primitive cells probably resembled the most primitive prokaryotic cells known today. Three sequential events occurring at this time were critical for the later appearance of eukaryotes. One was the development of photosynthesis using water as a raw material and releasing oxygen to the atmosphere as a by-product. This development set up conditions for the second event critical to the later evolution of more advanced cells: the conversion of the atmosphere, through the oxygen released from photosynthesis, from a nonoxidizing to an oxidizing medium. This change made possible the third event: the appearance of cells using oxygen in their energy-releasing reactions.

The Appearance of Photosynthesis, Oxygen, and Aerobes

While the functions of life were developing, the primitive molecular assemblies were dependent on absorbed organic substances as an energy source. The most efficient energy-yielding reactions probably involved breakdown of ATP or removal of electrons from absorbed fuel molecules. Because there was no free oxygen in the environment, the assemblies probably used inorganic compounds such as sulfates or nitrates as final electron acceptors for the reactions removing high-energy electrons from fuel molecules. Organisms using molecules other than oxygen as final electron acceptors are termed **anaerobes**. As the anaerobes became more widely distributed, the supply of organic fuel substances dwindled until the lakes and seas more closely resembled those of today. The dwindling fuel supply would certainly have led to extinction of the new life except for the appearance of photosynthesis among the early cells. Since the appearance of photosynthesis allowed some of the primitive cells to use light instead of absorbed organic substances as their energy source, it liberated them from dependence on the supply of fuel molecules from the environment. At the same time, moreover, nonphotosynthetic cells were able to switch to using the photosynthetic cells as a source of organic fuel molecules.

Photosynthesis as we now know it has a number of significant steps (Figure 21-7). In the first step, electrons are released by a donor substance (Figure 21-7a) and passed to a pigment molecule. The pigment molecule then absorbs light energy; this energy is used to raise the electrons derived from the donor substance to a high-energy form (Figure 21-7b). The electrons, raised in energy level by this mechanism, then power the chemical work of the cell. Thus, in photosynthesis, the energy of the electrons entering chemical reactions originates from sunlight rather than from complex organic molecules.

The earliest photosynthetic reactions may have involved pigment molecules, such as porphyrins, that were synthesized spontaneously and included by chance in molecular aggregates. As with the other biochemical activities of emerging life, light absorption and the excitation of electrons to high energy levels became regulated as enzymes appeared in the molecular assemblies.

The earliest photosynthetic pathways were forced to use an initial donor substance such as H_2S that releases electrons at relatively high levels, as the photosynthetic bacteria still do today. Eventually mutations appeared that allowed water, which releases very low energy electrons, to be used as the electron donor for photosynthesis. This adaptation established the pathway for photosynthesis used in the modern blue-green algae and the eukaryotic plants, in which water is split as a source of electrons and oxygen is released as a by-product to the atmosphere.

The development of photosynthesis changed the earth's atmosphere from nonoxidizing to oxidizing and set up the conditions necessary for the appearance of eukaryotes.

As a consequence of this new adaptation, which appeared in ancestral prokaryotes giving rise to the blue-green algae, oxygen was released to the atmosphere in ever-increasing quantities, gradually changing it from a nonoxidizing to an oxidizing character and leading to the balance of gases present in our environment today. Release of oxygen into the atmosphere also formed an ozone layer in the outer reaches of the atmosphere. This ozone layer protected the cellular life developing on the earth's surface from the injurious mutations and other chemical effects of ultraviolet light.

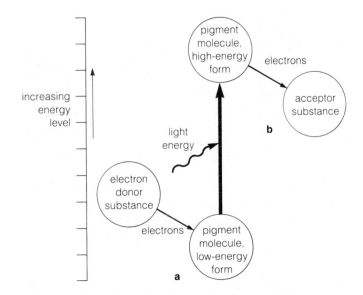

Figure 21-7 The overall steps of photosynthesis. (**a**) Electrons released by a donor substance are picked up by a pigment molecule, which absorbs light and raises the electrons to higher energy levels. (**b**) These high-energy electrons are then released from the pigment molecule and used to reduce an acceptor substance, which is converted into a complex high-energy molecule in the process. In eukaryotic plants, the ultimate acceptor substance is CO_2, which is converted into units of carbohydrate by the reaction.

The gradual increase in the levels of oxygen in the atmosphere set the stage for another biochemical development that was to be of critical importance for the later evolution of more complex cells. This was a series of mutations that allowed the newly abundant oxygen to be used as the final acceptor for the electrons removed in oxidative reactions. Since oxygen accepts electrons at very low energy levels, much more energy could be tapped from the electrons removed from fuel substances and used to power cellular activities (Figure 21-8). The prokaryotic cells using this complete oxidative pathway were the first **aerobes** to appear on the earth.

Thus the evolutionary developments of this period include three major events that were of supreme importance for the later development of eukaryotes: (1) development of photosynthesis and adaptation of some of the primitive photosynthetic prokaryotes to the use of H_2O, giving rise to the forerunners of the blue-green algae; (2) release of oxygen to the atmosphere, converting its character from reducing to oxidizing; and (3) evolution of bacteria using oxygen as final acceptor for the electrons removed during cellular oxidations. By this time, the change in the atmosphere and the nearly total depletion of organic molecules

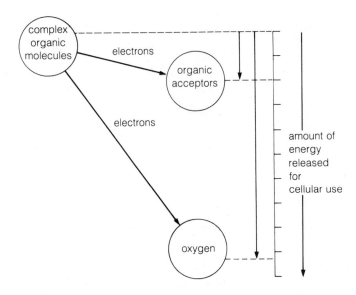

Figure 21-8 The advantage of using oxygen as final electron acceptor for electrons removed during cellular oxidations. Oxygen accepts electrons at very low energy levels and permits more of their energy to be tapped off for cellular use.

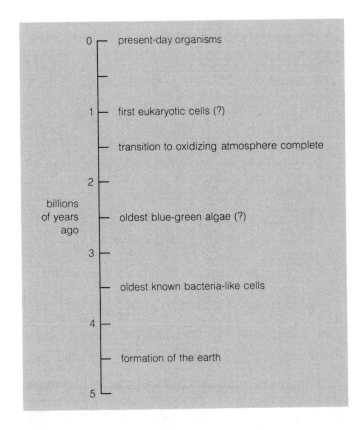

Figure 21-9 The time scale of cellular evolution (see text).

in the rivers, lakes, and seas altered the earth to conditions approaching the present-day environment.

The fossil and geological record suggests the probable time scale of these events (Figure 21-9). The oldest known bacteria-like cells date back about 3.5 billion years. Evolution of the blue-green algae from these early bacteria may have required another 500 million years. This conclusion is based on discoveries of limestone-containing deposits called **stromatolites** in rocks laid down 1.6 to 2.7 billion years ago. Stromatolites are good indicators of the presence of blue-green algae; in fact, they are still formed by certain kinds of blue-green algae today (Figure 21-10). The time required for the blue-green algae to release oxygen in abundance to the atmosphere has been estimated from the degree of oxidation of iron-containing sediments. Sediments in nonoxidized form persist in layers deposited as recently as 1.8 billion years ago. These finally give way to oxidized "red beds" about 1.5 billion years old. This pattern indicates that the change to an oxidizing atmosphere, begun nearly 3 billion years in the past, probably occupied 1.5 billion years and became complete not much more than 1.5 billion years ago.

The Appearance of Eukaryotic Cells

Eukaryotic cells undoubtedly developed from one or more of the different prokaryotic cells that became abundant some 1.5 billion years ago. This process, which included evolution of new structural features and organelles such as mitochondria, chloroplasts, and the nuclear envelope, may have taken place via several routes. One possibility is that many of the membranous structures characteristic of eukaryotes may have arisen through invaginations of the surface or plasma membranes of prokaryotic cells (Figure 21-11). In prokaryotes, the enzymes and biochemical activities associated with oxidation and photosynthesis are linked to the plasma membrane (see p. 141). Possibly, through invaginations of the plasma membrane (Figure 21-11a), portions of the membrane containing these enzymes and biochemical activities extended into the cytoplasm and pinched off (Figure 21-11b), giving rise to organelles that gradually evolved into mitochondria and chloroplasts.

The nuclear envelope membranes may have developed by the same process—that is, through invaginations of the plasma membrane that extended inward and gradually surrounded the nucleus (Figure 21-12). Similar invaginations may also have given rise to the membranes of the rough and smooth endoplasmic reticulum.

A different hypothesis about the possible origins of mitochondria and chloroplasts is based on the obvious resemblances between these organelles and prokaryotes.

a
b

Figure 21-10 Stromatolites (arrows)—deposits laid down by blue-green algae both in primitive times and to-day. (**a**) Fossil stromatolites embedded in rock. (**b**) Living stromatolites photographed under the ice in an oasis lake in the Antarctic. (*a*) Courtesy of Biology Media. (*b*) Photograph by F. G. Love and L. Hoare; courtesy of B. C. Parker and G. M. Simmons, Jr., from *Trends in Biochemical Sciences* 6 (1981).

According to this idea, advanced in its most recent and complete form by Lynn Margulis of Boston University, mitochondria and chloroplasts evolved from ancient pro-karyotes that were originally ingested as food particles. Instead of breaking down, the ingested cells persisted as functional units in the cytoplasm of the feeding cells. Mutations increasing the interdependence between the ingested prokaryotes and their host cells eventually led to their conversion into chloroplasts and mitochondria.

Margulis proposes that in the development of mitochondria, the evolution of prokaryotes proceeded to the point where complete photosynthesis was common and oxygen was present in large quantities in the atmosphere. Among these prokaryotes were some nonphotosynthetic types that had developed the capacity to engulf other cells through invaginations of the plasma membrane. These cells were thus feeding types that lived by oxidizing the organic molecules picked up by this means. Some of these nonphotosynthetic feeding cells were capable of using oxygen as final electron acceptor and were thus aerobic. Others, still limited in their oxidative reactions to using organic substances that accept electrons at intermediate energy levels, were anaerobic.

Interactions among these cells led to the evolution of mitochondria. This development began when groups of the nonphotosynthetic, anaerobic cells ingested nonphotosynthetic aerobic cells in large numbers. Instead of breaking down, some of the ingested aerobes persisted intact in the cytoplasm of the capturing cells and continued to respire aerobically. As a result, the cytoplasm of the host anaerobes, formerly limited to the use of organic molecules as final electron acceptors, became the residence of an aerobe capable of carrying out the much more efficient transfer of electrons to oxygen.

The new association would have brought advantages to both the host cell and the ingested prokaryote. Some of the chemical energy produced by the ingested aerobe would diffuse into the cytoplasm of the anaerobe, thereby benefiting the host cell. In turn the host cell, which was probably highly efficient in the capture and ingestion of organic matter, would supply the aerobe living in its cytoplasm with the foodstuffs needed for survival. As this relationship developed, mutations gradually increased the interdependence between the host and ingested cells. Among these mutations were the loss of many motile and synthetic functions of the ingested aerobe, since these activities could be supplied by the host. As the ingested aerobes became more specialized as energy-converting organelles, they eventually completed their transition into mitochondria. All that remains today of the nucleus of the ancestral parasites that gave rise to mitochondria are small circles of DNA that code for a few limited and essential functions that evidently cannot be supplied by their "hosts" (see Supplement 9-2).

This new cell type, a former anaerobe containing an aerobe in its cytoplasm, is thought to have given rise to the eukaryotic cell line. Other features of eukaryotic cells, including the nuclear envelope and endoplasmic reticu-

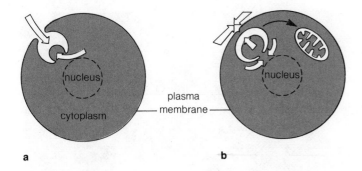

Figure 21-11 A hypothetical route by which mitochondria may have formed: through invagination of the plasma membrane of a prokaryote. (**a**) Invagination of the plasma membrane (which contains oxidative enzymes in prokaryotes) to form a cytoplasmic vesicle. (**b**) Separation and gradual conversion of the vesicle into a mitochondrion (see text).

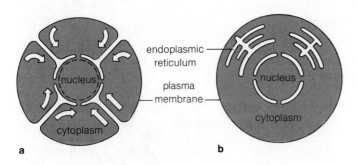

Figure 21-12 The hypothetical origin of the nuclear envelope of eukaryotic cells through invaginations of the plasma membrane. Fragments destined to form endoplasmic reticulum could have originated through the same route.

lum, gradually appeared in the developing eukaryotes, possibly through invaginations of the plasma membrane as proposed in the alternative hypothesis.

Some time after the evolution of eukaryotic cells to this level, Margulis proposes, a second interaction of the same type led to the development of chloroplasts. In this case, some of the feeding cells already containing mitochondria ingested prokaryotes capable of photosynthesis. Some of the photosynthetic cells persisted in the cytoplasm of the ingesting cells and supplied their hosts with the ability to use light as a source of energy. As this relationship developed, the ingested prokaryotes gradually evolved into chloroplasts by the same process of gradually increasing interdependence between the host cells and their cytoplasmic residents. These primitive eukaryotes,

containing both mitochondria and chloroplasts, founded the cell lines leading to the eukaryotic algae and plants.

Modern examples of relationships between ingested and host cells provide evidence that this route of origin could indeed have taken place. Many kinds of animal cells, including representatives of eight major phyla, ingest algal cells or chloroplasts and retain them as beneficial cytoplasmic residents. One of the most interesting examples, described by Robert Trench at Oxford University, involves a group of marine snails that contain chloroplasts derived from algae used as food. Initially the snails hatch and develop as embryos without chloroplasts. As soon as they reach a larval feeding stage, however, chloroplasts from algae eaten by the young snails are taken up by cells lining the gut. These chloroplasts persist in the gut cells as the snails grow to the adult form and, moreover, continue to carry out photosynthesis in their new location (Figure 21-13). Trench has shown by the use of radioactive labels that molecules synthesized by the ingested chloroplasts diffuse into the surrounding cytoplasm and are used as fuel substances by the host snail. Individual chloroplasts may remain active in the gut cells for months. Thus the routes envisioned by Margulis for the origin of mitochondria and chloroplasts can be demonstrated to operate today.

As these membranous organelles and structures developed in the cytoplasm of the primitive eukaryotes, other major adaptations appeared that completed the conversion of eukaryotic cellular life to contemporary forms. Among these adaptations were increases in the complexity of chromosomes and the development of microtubules to regulate cell division and motility. Later evolution led to the aggregation of cells into colonies and, eventually, to the specialization of cells and the appearance of many-celled organisms of greater and greater complexity, including our own form of life.

Evolution of eukaryotes from prokaryotes may have required as much as 2 billion years.

How long did it take for the first eukaryotes to appear? The time required for evolution of eukaryotes from prokaryotes has been difficult to establish because microscopic structures such as mitochondria, chloroplasts, and even nuclei are usually poorly preserved in fossil cells. Fossils dating from as long ago as 1.5 billion years have been claimed by some investigators to be eukaryotic; others fix the earliest eukaryotic cells as 1 billion years old. If we assume that the first fully eukaryotic cells appeared some time between these limits of 1.5 and 1 billion years ago and that the oldest prokaryotes date from 3½ billion

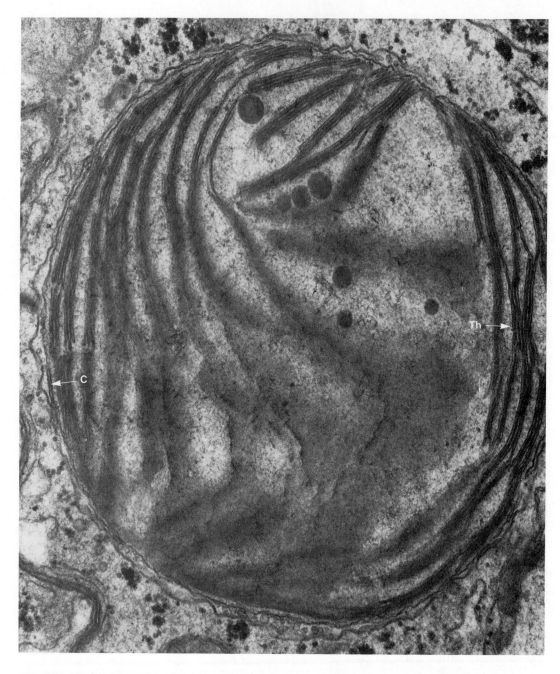

Figure 21-13 An active chloroplast in the cytoplasm of a digestive cell of the snail *Elysia*. *C*, chloroplast boundary membranes; *Th*, thylakoid membranes. ×53,000. Courtesy of R. K. Trench.

years in the past, then as much as 2 billion years was required for evolution of eukaryotes from prokaryotes (see Figure 21-9). This interval is greater than the total period required for the entire evolution of prokaryotes from the earth's origins. Considering the complexities of eukaryotic cells and the advances they represent over the prokaryotes, it is entirely possible that their evolution may have taken this long.

The chemical and biological events leading from the inanimate earth to the first eukaryotic cells entail so many hypothetical steps that the entire process may seem unlikely to have occurred. The total time involved, some 3½ billion years, is so great, however, that even events of very low probability are likely to have happened more than once. Given the time span of these events, as George Wald of Harvard University has put it, "the impossible becomes possible, the possible probable, and the probable virtually certain. One has only to wait: time itself performs the miracles."[2]

Questions

1. Why is it necessary to assume for scientific purposes that life originated through inanimate chemical processes?

2. Do you think that it would be scientifically correct either to deny or affirm divine intervention in the origin of life? Why?

3. How would you define life as applied to a bacterium? A eukaryotic cell? A human? A tree? Are these definitions necessarily the same?

4. Summarize the stages in the evolution of life.

5. What is the condensation hypothesis for the origin of the sun and planets of our solar system?

6. Why was the absence of oxygen among the gases of the primitive atmosphere important to the evolution of life? What evidence indicates that the atmosphere of the primitive earth was actually nonoxidizing?

7. What energy sources are believed to have acted on the chemicals of the primitive environment to produce complex organic molecules? What evidence indicates that complex substances may in fact have been produced in this way? What kinds of molecules have been synthesized in systems simulating the primitive environment?

8. What are coacervates and microspheres? What roles might clays and lipids have played in the assembly of molecules in the primitive earth? What is a proteinoid?

9. What importance did molecules like ATP, GTP, UTP, TTP, and CTP have in the evolution of life?

10. Which is probably older, anaerobic life or aerobic life?

11. What was the importance of photosynthesis in the evolution of life?

12. Outline two hypotheses that account for the origins of mitochondria and chloroplasts.

13. What environmental conditions were necessary for the appearance of eukaryotes?

14. Could eukaryotic cells have survived if they had been introduced into the primitive earth from a spaceship before prokaryotic life had evolved here?

---

[2]G. Wald, "The Origin of Life," *Scientific American* 191:45 (1954).

Suggestions for Further Reading

Bernal, J. D. 1967. *The Origin of Life.* World, New York.

Calvin, M. 1969. *Chemical Evolution.* Oxford University Press, New York.

Day, W. 1979. *Genesis on Planet Earth.* House of Talos Publishers, East Lansing, Michigan.

Dickerson, R. E. 1978. "Chemical Evolution and the Origin of Life." *Scientific American* 239:70–86.

Fox, S. W., and K. Dose. 1977. *Molecular Evolution and the Origin of Life.* 2nd ed. Marcel Dekker, New York.

Keosian, J. 1964. *The Origin of Life.* Reinhold, New York.

Margulis, L. 1970. *Origins of Eukaryotic Cells.* Yale University Press, New Haven.

Ponnamperuma, C. 1972. *The Origins of Life.* Thames & Hudson, London.

Schopf, J. W. 1978. "The Evolution of the Earliest Cells." *Scientific American* 239:111–138.

Wolfe, S. L. 1981. *Biology of the Cell.* 2nd ed. Wadsworth, Belmont, California.

Ecology

22

POPULATIONS AND THEIR ENVIRONMENT

The plants, animals, and microorganisms of the world, the result of evolution to the present time, live in a state of constant interaction with each other and their environment. In some parts of the world the interaction maintains the organisms of the region in a fairly constant balance of numbers and kinds. In others, it produces change as individuals of a species increase or decrease in number or are replaced by other types. The science of ecology (*oikos* = house, living place) seeks to understand these interactions: the forces of the environment that affect living organisms and the changes in the environment brought about by the presence of life.

These problems are studied in ecology at several levels of increasing breadth and complexity. At the most basic level, ecologists study the numbers and kinds of individuals and their distribution in defined regions of the world. At this level, ecology closely resembles natural history. Although much has been learned about investigations at this level, the study of individuals reveals little about overall changes or trends brought about by interactions between organisms and their environment. These interactions and changes are studied instead in terms of **populations**. As in evolution, a population is defined as a group of individuals of the same species existing in the environmental region under study. As such, the populations under investigation may vary from a few individuals in a highly localized region to all the members of a species distributed throughout the entire world. Studies made at the level of a population, as in evolution, allow changes in numbers and distribution to be evaluated statistically, yielding results that are valid for the population as a whole. Of particular interest in ecological studies are the environmental factors that regulate population size, growth, and distribution—factors such as climate and soils, the availability of living space and nutrients, and the effects of other living organisms.

The populations studied in ecology are groups of individuals of the same species occurring within a region under investigation.

At still higher and more inclusive levels, the interactions between organisms and their environment are studied in the **community**, which includes all the populations living in a given region. At this level ecologists are interested in the places occupied by populations within the community, their activities, and their interactions in these places. The interactions of interest at this level include the use of some populations by others as sources of food, the number of populations and the factors controlling this number, the changes occurring in the populations of a community with time, and the communities that are more or less typical of certain regions of the earth.

Finally, at the highest and most inclusive level, the **ecosystem**, ecologists investigate the sources and flow of energy and the cycles of matter between the living and nonliving worlds as well as the living communities themselves. Ecosystems thus encompass all the physical, chemical, and biological factors of a defined region. The ecosystems under study, depending on the limits set by the investigator, may be as small as a single leaf or as large as the earth itself.

Whatever the level under study, investigators attempt to answer the fundamental questions of ecology. Why do species occur in one location and not in another? What determines the abundance of organisms in regions where they do occur? What factors regulate the growth of populations and limit their total size? How are the populations of a community integrated with each other and with their environment? How does energy flow and matter cycle between the living and nonliving parts of ecosystems? And finally, perhaps of greatest significance to human endeavors, what is the effect of organisms, including our own species, on the environment? Work at this level has such significance for our present and future world that ecology in common usage has come to mean concern and care for the environment.

The three chapters of this unit summarize the results of efforts to answer these and other questions about ecology at the level of individuals, populations, communities, and ecosystems. This chapter concentrates on ecology at the individual and population level and discusses the factors regulating population growth, size, and distribution. The structure, activities, and changes that occur in communities are described in Chapter 23. Chapter 24 considers the integration of the physical, chemical, and biological factors of organisms and their environment in ecosystems, as well as the impact of our own species on the communities and ecosystems of the world.

POPULATIONS AND POPULATION GROWTH

The Population Concept in Ecology

The population studied in ecology consists of a group of individuals potentially capable of interbreeding and sharing the same gene pool. As such, the individuals of a population are members of the same species. While this definition is the same as the one used in evolution, the populations studied in ecology are usually defined for convenience by the boundaries set by the investigator rather than the limits of interbreeding. Thus the populations studied in ecology are most frequently viewed as groups of individuals of the same species occurring within a region under investigation.

Individuals have fixed characteristics such as sex and dates of birth and death. Populations have group characteristics that individuals do not possess, such as growth rates, distribution and density, and rates of emigration and immigration. These additional characteristics allow overall patterns of dispersion, activity, and growth of populations to be traced and related to environmental conditions. Since these environmental relationships and growth trends are of central interest in ecology, it is the population, rather than the individual, that is the fundamental unit investigated in ecological studies.

Population Growth

One of the most significant characteristics studied by ecologists is the rate of population growth and the environmental factors that determine it. At any time, populations may be growing, diminishing, or remaining stable. These trends can be visualized if the numbers of individuals in a population are counted at intervals and the two types of information, numbers and time, are plotted against each other (see Figure 22-1). Plots of this type take characteristic forms for populations that are growing, stable, or decreasing in numbers.

Exponential Growth If environmental conditions are favorable, all populations have the capacity for rapid growth. A favorite example is the housefly. A female housefly lays about 120 eggs. If all offspring survive and live to reproductive age, the first generation derived from a single mating pair will number about 120 individuals. Assuming that half these individuals are females, that all females bear young at the same rate, and that all young survive, the individuals in the second generation will number $60 \times 120 = 7200$. In the next generation, if all offspring survive, the numbers expected are $3600 \times 120 = 432,000$. As the numbers increase, the potential for reproduction increases. By the fifth generation, the population would number 1.5 billion individuals. By the seventh generation, equivalent to one year of reproduction for houseflies, the flies derived from the single original pair would number 5.5 trillion individuals.

Rapid, uncontrolled growth of this type is called **exponential growth**. If individuals are plotted against time,

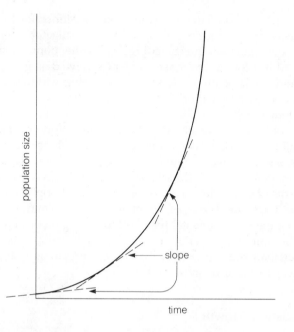

Figure 22-1 Exponential population growth. The slope of the curve at any point (dashed line) shows how fast the population is increasing in size at that instant.

a typical exponential curve is obtained in which the line slopes upward at a constantly steeper angle from the horizontal (see Figure 22-1). Curves of this type are also

---

An exponential growth curve slopes upward at a constantly steeper angle.

---

called *J-shaped curves* in population studies. Note that the slope of the curve at any point shows how fast population size is increasing at that time. More precisely, the slope reflects the *rate of increase* in the number of individuals in the population at a given time.

In ecological studies the rate at which a population increases over any given time interval Δt (read "change in time") is usually calculated from data collected on (1) birth rate in the population, (2) death rate, and (3) the total population size. If this information is obtained, the rate of increase in number of individuals over the time interval Δt, for exponentially growing populations without immigration or emigration, is equivalent to

Rate of increase in individuals =

$\qquad$ (birth rate) − (death rate) × (population size)　(22-1)

or

$$\frac{\Delta N}{\Delta t} = (b - d)N \qquad (22\text{-}2)$$

where ΔN is the change in number of individuals, Δt is the time interval chosen, b is the birth rate, d is the death rate, and N is the population size. As long as b and d remain constant and the quantity $(b - d)$ is a positive number, the population will grow exponentially. Conversely, if $(b - d)$ is negative, the population will decline.

Imagine a human population of 1 million persons in which 50 children are born each year for every 1000 individuals so that the birth rate is 50/1000 or 0.05. During the same time interval, 20 persons out of each 1000 individuals die, giving an average death rate of 20/1000 or 0.02. The quantity $(b - d)$ in this case is $0.05 - 0.02 = 0.03$ and is positive. The equation for growth rate then gives

Rate of increase in individuals = $\dfrac{\Delta N}{\Delta t} = (b - d)N =$

$0.03 \times 1,000,000 = 30,000$ per year　(22-3)

Thus the rate of increase for the population at this time is 30,000 persons per year. In the next year, even with the same birth and death rates, the rate of increase will be larger because N will be 1,030,000 instead of 1 million:

Rate of increase = $0.03 \times 1,030,000 = 30,900$ persons per year

Thus as long as $(b - d)$ is a positive number, the rate of increase will be larger each year and the population will grow exponentially.

Calculations of this type assume that the rates of immigration and emigration of individuals to and from the population, which would also affect population growth, are equal and can thus be ignored. This assumption is often made in ecological studies because the rates of immigration and emigration in natural populations are impossible to measure.

Equation 22-2 allows us to calculate the rate of increase in number of individuals over some time interval Δt, such as hours, days, weeks, or years. (In the human population example given above, the time interval Δt was 1 year.) If the growth rate at any instant in time is desired, essentially the same equation is used. Here Δt is vanishingly small and is designated dt instead of Δt. The growth in numbers during this instant of time is similarly dN instead of ΔN:

$$\frac{dN}{dt} = (b - d)N \qquad (22\text{-}4)$$

Because it allows calculation at any instant in time, the quantity dN/dt is called the **instantaneous growth rate**.

The quantity $(b - d)$ is of such importance in evaluating population growth that it is given the special designation r:

$$r = (b - d) \qquad (22\text{-}5)$$

The factor r, called the **intrinsic rate of population increase**, represents the potential rate of increase when there is no limit to population growth due to a lack of food, space, or other requirements of survival and reproduction. Substituting r for $(b - d)$ in Equations 22-2 and 22-4 gives

$$\frac{\Delta N}{\Delta t} = rN \qquad (22\text{-}6)$$

for the rate of increase in individuals for intervals of time Δt and

$$\frac{dN}{dt} = rN \qquad (22\text{-}7)$$

for the instantaneous growth rate of populations.

Logistic Growth It is obvious that no real population could continue to grow indefinitely at an exponential rate. Whether houseflies or humans, the numbers would soon be so vast that the weight of the population would be greater than that of the earth itself. In nature, although newly established populations may grow exponentially for a time, the rate of increase eventually slows and stops or even reverses. Thus populations never realize their innate capacity for unrestrained growth. Slowing and stopping of the population growth rate is caused by a decrease in the birth rate or an increase in the death rate, or both, until the quantity $(b - d)$ or r becomes zero or negative.

No populations in nature can continue to grow indefinitely at an exponential rate.

The various factors regulating populations and keeping their growth from approaching their ultimate potential are of two basic types: density-independent and density-dependent. **Density-independent** factors reduce population growth without regard to the number of individuals per unit area (see also p. 438). Most of the factors in this category are purely physical effects of the environment, such as climatic conditions or natural catastrophes like earthquakes, floods, fire, or volcanic eruptions. These physical factors kill individuals or reduce their reproduc-

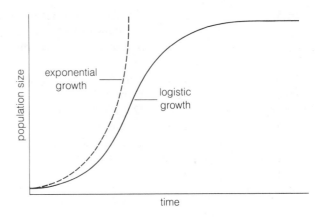

Figure 22-2 Logistic population growth. The initial part of the curve is almost exponential. But as the population grows, density-dependent factors have an increasingly greater effect and the rate of growth slows and eventually levels off.

tive capacity whether population density within the affected region is high or low.

While density-independent factors do reduce population growth, the most significant controls fall into the second basic category. In this case the effects of the regulatory factor intensify as the number of individuals per unit area increases. These **density-dependent** controls, as they are called (see also p. 438), include the effects of predators and competition for food, raw materials, mates, and living space. For individuals of the population, the density-dependent factors add up to starvation, disease, or being eaten—conditions that become more severe and more frequent as population size increases.

Density-dependent factors operate in such a way that, when population density is very low, their effects are hardly felt and population growth is almost exponential (initial part of the curve in Figure 22-2). At these levels, the individuals per unit area are so few that there are more than enough resources to go around and virtually no competition takes place. Similarly, predators may not be active in hunting members of the population because the prey are too few to make their presence known. As population density increases, however, limitations in the supply of food, living space, and other factors needed for survival result in competition that gradually intensifies as the number of individuals per unit area grows. Population growth may also attract predators, which increase in numbers or activity as the population size increases. As these factors begin to affect the population, the rate of population growth slows from the exponential phase and the slope of the growth curve gradually becomes less steep (middle portion of the curve in Figure 22-2). These density-dependent factors gradually increase in intensity until population growth is brought to a stop or even reversed (final portion of the curve in Figure 22-2).

The effects of density-dependent factors in regulating population growth can be approximated by multiplying the right-hand side of Equation 22-7 by (1 − N/K), where K is the **carrying capacity** of the environment. The carrying capacity is defined as the maximum number of individuals that the region occupied by the population can support over long periods of time:

$$\frac{dN}{dt} = (b - d)N(1 - \frac{N}{K}) = rN(1 - \frac{N}{K}) \qquad (22\text{-}8)$$

At very low population densities, N/K is almost zero and the factor (1 − N/K) is almost equal to 1. Population growth at this point is essentially exponential and unrestrained. As the population size increases, however, the value of N becomes larger and the entire quantity (1 − N/K) becomes a fraction of gradually decreasing size. Ultimately, as N becomes equal to K (or, in other words, as the number of individuals in the population matches the carrying capacity of the region), the value of (1 − N/K) becomes zero and population growth stops. Equation 22-8, called the *logistic equation*, thus produces the smooth S-shaped or sigmoid curve shown in Figure 22-2. Population growth following this pattern is called **logistic growth**.

Populations grown under ideal laboratory conditions often closely follow the path predicted by the logistic equation. A culture of yeast cells, for example, grown under carefully controlled conditions with a constant but limited food supply, increased almost exponentially at first and then gradually slowed and reached a stable maximum level (Figure 22-3). The growth curve produced by the yeast culture, representing the number of individuals counted at 1-hour time intervals, was sigmoid and closely matched the predictions of the logistic equation with r = 0.53 and K = 665 individuals. (In this case K represented the carrying capacity of the volume of nutrient medium used.)

Natural populations, however, grow less frequently in patterns matching the predictions of the logistic equation. This is because birth rate, death rate, and K, the carrying capacity of the environment, are rarely constants as they are assumed to be in the logistic equation. Moreover, the logistic equation assumes that all individuals have the same reproductive potential and are affected equally by density-dependent regulatory factors such as competition and predation. Since all the conditions assumed in the logistic equation are rarely met in nature, natural populations usually vary from logistic growth to a greater or lesser extent. Nevertheless, the logistic equation is valu-

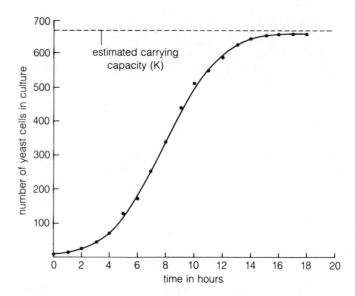

Figure 22-3 Growth of yeast cells in a laboratory culture. The pattern of growth closely matches a logistic curve. Data from R. Pearl, *Quantitative Review of Biology* 2:532 (1927).

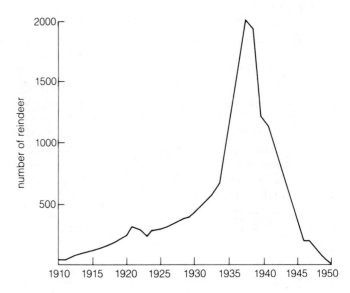

Figure 22-4 Population crash following a period of exponential growth in a reindeer herd introduced on St. Paul Island, one of the Pribilof Islands near the Alaskan coast. Data from U. B. Scheffer, *Science Monthly* 73:356 (1951).

able to ecological studies because it provides a baseline for comparing the growth of natural populations with ideal conditions.

The logistic equation provides a baseline for comparing the growth of populations in nature.

One important natural variation from the middle and final phases of an ideal logistic curve is observed in populations that grow rapidly and almost exponentially and then "crash." This pattern of growth is illustrated by reindeer on St. Paul Island, one of the Pribilof Islands near the Alaskan coast. A population of 25 reindeer introduced on the 45-square-mile island in 1911 grew exponentially to a herd of more than 2000 animals in 1938 and then crashed to a population that numbered only 8 reindeer by 1950 (Figure 22-4). This pattern of growth is also frequently observed in pest species such as insects and weedy plants invading a new region or a habitat recently disturbed by a natural disaster such as a flood or volcanic eruption. Typically the populations with this growth pattern enter a favorable environment with an initially abundant food supply and few or no predators, reproduce rapidly to levels that outstrip the carrying capacity of the environment, and then suddenly die out.

Other important variations occur in the final phases of population growth. Although much of the initial and intermediate growth patterns follow the logistic curve, populations may begin to fluctuate in numbers as they approach the carrying capacity rather than stabilizing at this point as predicted by the logistic equation. The fluctuations around the carrying capacity may be random (Figure 22-5a), or they may occur in regular cycles (Figure 22-5b). In stable populations, the variations remain small and the population level remains close to the carrying capacity of the environment. Such populations usually contain individuals that are highly adapted and successful competitors. Unstable populations show much wider fluctuations that, in some cases, extend to the zero level and result in extinction.

The lynx and snowshoe hare populations in the Canadian Arctic (Figure 22-5b) provide one of the classic examples of fluctuating populations. In this case, the fluctuations vary in a regular cyclic pattern: both the hare and lynx populations cycle at approximately 10-year intervals. In most cycles, peaks in the lynx population follow peaks in the hare population. While many hypotheses have been advanced to account for regular cycles of this type, the most widely accepted models propose that the oscillations involve a *time lag* in the reciprocal effects between predator and prey populations.

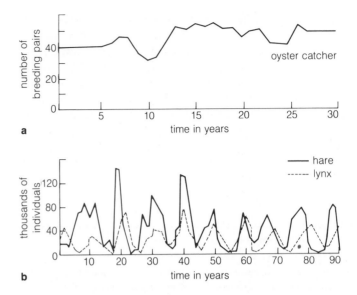

Figure 22-5 Two populations showing typical patterns of fluctuation around the carrying capacity. (**a**) A population of shorebirds, the oyster catcher, in Great Britain, showing more or less random fluctuations; (**b**) regular, cyclic variations in hare and lynx populations in Canada. Data for (*a*) from D. Lack, *Journal of Animal Ecology* 38:211 (1969); data for (*b*) from D. A. MacLulich, *University of Toronto Studies, Biology Series* 43 (1937).

In the hare–lynx interaction, the regular cycles are believed to depend on both reciprocal interactions between the hares and their prey (the plants they feed on) and between the hares and the lynx. According to this idea, the hares grow rapidly in times of good food supply. Eventually they reach levels in which they equal the carrying capacity of their environment, determined primarily by the supply of plants on which they feed. Population growth does not decline immediately; even though adults are poorly fed at these levels, reproduction still occurs. Thus there is a time lag in the effect of the dwindling food supply, and the hare population continues to grow past the carrying capacity for a time. The high-density hare population eventually exhausts the food supply, however, and a rapid dieback begins.

The lynx population follows a cycle that follows slightly behind the hare population due to the same time-lag effects. As the hare population grows, the lynx preying on the hares also grow in numbers in response to the increases in their food supply. Eventually the lynx population reaches such high densities that the lynx overeat the hares, adding to the effects of the dwindling plant supply eaten by the hares themselves. The result is a rapid and sudden crash of the hare population. Since the lynx adults are still capable of reproduction, there is a time lag in the effect of the crash of the hare population, and the lynx population continues to grow for a brief period. The inevitable decline in the lynx population soon follows, however, as the adults and offspring finally eliminate most of their food supply. The small numbers of lynx and hares surviving after the crash give the plant cover a chance to regrow; following recovery of the plants, the lynx–hare cycle is ready to repeat. Regular cyclic variations in population numbers of this kind may also be keyed to changes in the seasons or, in the case of microorganisms that reproduce in minutes or hours, to daily fluctuations in environmental factors such as light or temperature.

K Selection and r Selection The patterns noted in these variations from the logistic growth curve have led to the concept that two different growth and reproductive strategies have developed through natural selection in response to the regulatory factors of the environment: the r *strategy* and the K *strategy*.

The first strategy has been selected in response to environmental conditions that are unstable and fluctuate widely enough to pose a constant threat to survival, as in regions subject to frequent floods or fires. Selection of this type is called **r selection**, and the pattern of reproduction and growth selected in response is the r strategy. The "r" in these terms refers to the r of the logistic equation (Equation 22-8), which represents the intrinsic rate of increase of a population. Populations evolving the r strategy in response to r selection typically grow rapidly by producing large numbers of offspring when conditions are favorable. Such "big bang" populations disperse and fill the environment quickly when other populations are dislodged by fluctuating conditions. Typically nutrients and living space are temporarily plentiful for r strategists, and competition and predation are low. The population lasts only as long as the environmental disturbance persists. It then dies back, leaving seeds, spores, or dormant forms to provide the basis for another phase of rapid growth when favorable conditions appear again. The growth pattern of r strategists thus resembles the curve shown in Figure 22-4, in which a rapid, almost exponential phase is followed by a crash. Species using the r strategy are typically of small size and have short life spans and generation times. Selection of r strategists has thus produced species that invest most of their resources in reproduction rather than in highly adapted individuals that can compete and survive effectively for long periods of time.

The r strategists invest their resources in reproductive capacity rather than adult survival.

The opposite strategy has been selected in response to environmental conditions that remain fairly constant and stable for long periods. Selection by these conditions is called **K selection**. The corresponding growth and reproductive pattern is termed the K strategy since populations growing in stable environments tend to reach levels at or near the carrying capacity of the environment (K in Equation 22-8) and remain more or less constant at this size. Species evolving the K strategy in response to K selection tend to produce highly adapted individuals that are good competitors for the resources of the environment. These resources are limited since the populations inhabiting the region live at or near the carrying capacity. The growth of K strategists follows pathways more closely resembling the logistic curve, with random or cyclic variations at levels near the carrying capacity. Generation times and life spans of K strategists are typically longer than those of r strategists, and individuals are frequently of larger body size. Small numbers of offspring are produced by individual K strategists; developmental periods, moreover, are usually longer than in r strategists. Selection of K strategists thus leads to species that invest their resources in adult survival rather than reproductive capacity. Table 22-1 summarizes the characteristics of r and K selection and strategies.

The K strategists invest their resources in adaptations making adults competitive rather than in high reproductive rates.

Many insects, microorganisms, and weed plants are typically r strategists and survive through high reproductive rates in temporarily favorable environments. Other species, such as birds and many mammals, are K strategists that survive as highly competitive adults. Many organisms also adapt at levels between these extremes by striking an evolutionary balance between r and K strategies.

Survivorship Curves The various strategies evolved in different species in response to r and K selection are reflected in the average life expectancy of individuals in populations. The effects of these strategies on life expectancy can be visualized by plotting the number of individuals surviving against age, usually in percentage of total life span (see Figure 22-6). Plots of this kind show patterns associated with r and K selection and the strategies that are a compromise between the two extremes.

Populations of species that have evolved the r strategy with high mortality among offspring, for example, typically have survivorship curves with an initial sharp drop followed by an extended, gradually leveling line (curve a, Figure 22-6). Individuals in these populations have a high

Table 22-1   Characteristics of Population Growth Under *r* and *K* Selection

Characteristic	*r* Selection	*K* Selection
Reproduction	Early in life cycle	Delayed
Reproductive pattern	Single "big bang"	Continued through much of life
Number of offspring	Many, small	Few, larger
Development	Rapid	Slow
Body size	Small	Larger
Length of life	Short; usually less than 1 year	Longer
Population size	Highly variable; often below carrying capacity	Fairly constant; usually at or near carrying capacity
Competition	Variable; often not pronounced	Usually intensive

Adapted from E. R. Pianka, *Evolutionary Ecology*, 2nd ed. (New York: Harper & Row, 1978).

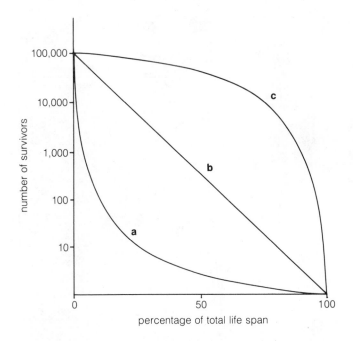

Figure 22-6   Survivorship curves. (**a**) Population with high mortality rate of young. (**b**) Population with equal mortality rate at any phase of the life span. (**c**) Population with low mortality rate of young and long life span, in which most individuals live to an age near the maximum life expectancy. Note that the vertical scale is plotted in logarithmic units: each higher division is greater by a factor of 10.

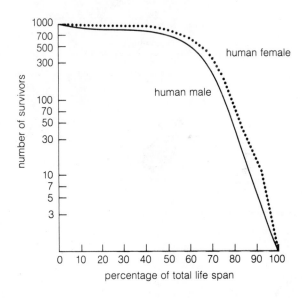

Figure 22-7   Survivorship curve for the human population based on data collected during the years 1940–1941. Data from L. I. Dublin et al., *Length of Life* (New York: Ronald Press, 1949).

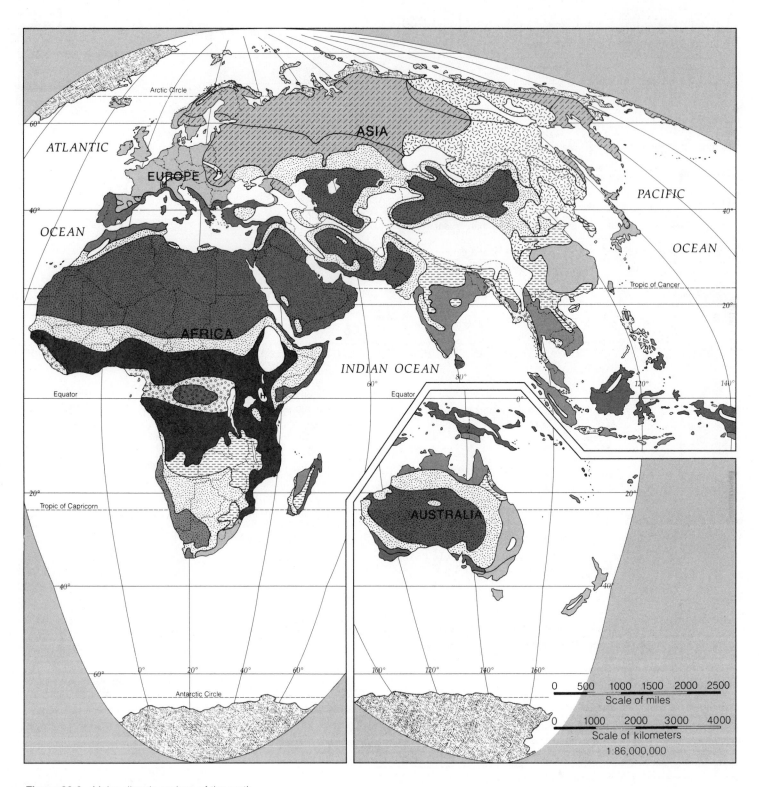

Figure 22-8   Major climatic regions of the earth.

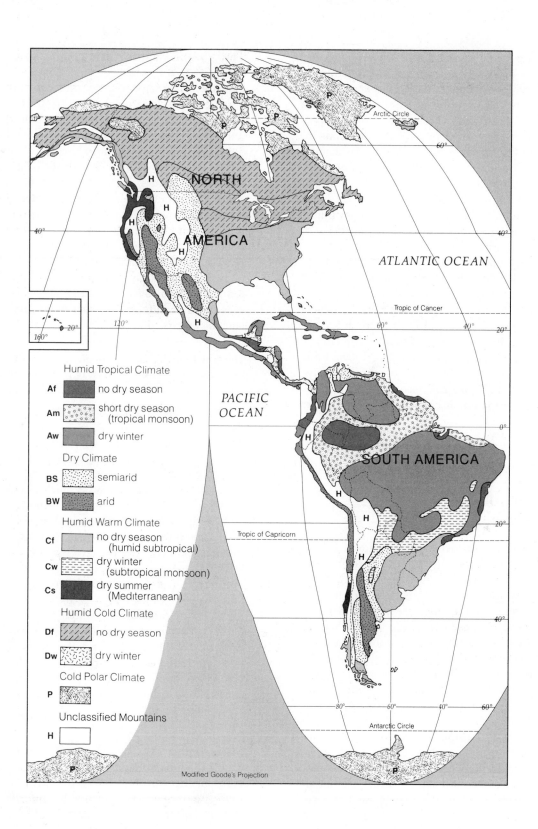

Humid Tropical Climate

Af no dry season

Am short dry season (tropical monsoon)

Aw dry winter

Dry Climate

BS semiarid

BW arid

Humid Warm Climate

Cf no dry season (humid subtropical)

Cw dry winter (subtropical monsoon)

Cs dry summer (Mediterranean)

Humid Cold Climate

Df no dry season

Dw dry winter

Cold Polar Climate

P

Unclassified Mountains

H

NORTH AMERICA

ATLANTIC OCEAN

PACIFIC OCEAN

SOUTH AMERICA

Arctic Circle

Tropic of Cancer

Tropic of Capricorn

Antarctic Circle

Modified Goode's Projection

probability of dying while young and a low average life expectancy. Curves of this type are typical of organisms such as fish, marine invertebrates, flatworm parasites, insects, and many plants that produce thousands or millions of offspring, each with little individual chance of survival.

A second type of curve is produced by populations in which the probability of elimination is approximately equal at any age. This pattern occurs in organisms with growth strategies that reflect a balance between K and r selection (curve b in Figure 22-6). In the type of plot used in Figure 22-6, in which the vertical axis is logarithmic (each successive division increases by a tenfold factor) and the horizontal axis is arithmetic, this pattern produces a straight line. Curves of this sort are typical of some microorganisms and birds, among them common songbirds such as the American robin.

The final curve is typical of K strategists in which there is low mortality of young and most individuals approach the maximum life expectancy. Survivorship curves of this type (curve c, Figure 22-6) remain almost level until late in the life span, when they dip down precipitously. The human population (Figure 22-7, see p. 497) and populations of many other mammals have survivorship curves of this type.

THE FACTORS REGULATING POPULATION GROWTH

Both density-dependent and density-independent factors restrict the growth of populations and keep them from reaching their full reproductive potential. Among these factors, density-dependent controls are most effective in population regulation since they increase in intensity with population size. Most of these density-dependent factors are primarily **biotic**—that is, they result from the activity of living organisms. These density-dependent biotic factors include competition, which occurs when two populations attempt to use the same limited resource, and predation, which involves the use of one organism by another as a food source. Density-independent population controls are primarily the **abiotic** factors of the environment, including natural catastrophes, climate, and differences in soils. The distinctions between density-dependent, density-independent, biotic, and abiotic factors are not always clear. The effects of a climatic change such as a severe frost, for example, may depend to some degree on the ability of animals to find shelter, which, in turn, may depend on population density.

The Abiotic Factors: Climate and Soils

Climate   The major weather patterns of the earth produce a variety of climates ranging from the humid tropics through temperate zones and deserts to the cold and barren poles (Figure 22-8, see pp. 498–499). The climatic conditions within these regions affect the ability of populations to survive, grow, and reproduce. The amount of light available to plants, depending on conditions such as day length and cloud cover, affects the rates of photosynthesis and plant growth and hence the growth of animal populations dependent on plants for food. Low temperatures reduce the growth rates of plants and also the growth and reproduction of animals that cannot regulate their body temperatures. High temperatures or arid conditions also reduce survival by evaporating the water required by all organisms for life or by destroying vital body molecules such as enzymatic proteins.

The growth of populations depends directly on the conditions of light, moisture, and temperature produced by climate.

Each species population survives, grows, and reproduces by means of adaptations evolved in response to selection by the climatic conditions of a region. Some organisms, including many insects and spiders and the higher plants, survive the drying effects of arid climates by adaptations such as waxy surface cuticles that retard moisture losses. Other species, among them millipedes and centipedes, avoid desiccation by behavioral responses such as hiding in shaded crevices or underground during the day. Cold climates are survived by adaptations such as thick fur, seasonal migrations, loss of leaves, or winter dormancy. These evolutionary responses allow populations to survive extremes of high and low temperature, humidity or moisture, day length, and wind. There are limits, however, to the ability of each species to withstand extreme conditions. Beyond these limits, climatic conditions are lethal and no individuals of the species can survive. The distribution of species populations therefore corresponds more or less closely to the major climatic types (see Figure 23-21).

Microclimate   The broad climatic types shown in Figure 22-8, called **macroclimates**, occur over large geographic regions of the world. Although the average climate within these major regions is reasonably uniform, local conditions of temperature, humidity, and wind may differ considerably from the average values. These local variations within major climatic regions are called **microclimates**.

Microclimates are produced by local variations in light, moisture, and temperature.

The soil surface itself is a microclimate. Studies of temperature differences measured at intervals of a few centimeters above and below the surface of the soil show that daytime temperatures are highest at the surface level and drop off at points both above and below the surface level. These conditions are reversed at night, when the soil surface cools more rapidly than the underlying soil layers or the air above the surface. An inclination of the surface with respect to the sun also alters local climatic conditions. Slopes facing south in the northern hemisphere receive more sunlight and are warmer and drier than north-facing slopes. Plant cover also affects local conditions by shading the surface and breaking the force of the wind. Under the plant cover, temperature and wind velocities are lower, and humidity higher, than in regions receiving direct sunlight above the plants.

These local variations in microclimate may produce strikingly different living conditions for populations in the region. The conditions on the underside of a leaf, where temperatures are reduced and humidity is increased by water evaporating from the leaf surface, are entirely different from a nearby rock baking in the direct rays of the sun. These local variations in microclimate limit the growth and distribution of populations: an aphid population, for example, might thrive under the leaf but die within seconds if exposed on the rock.

Detailed measurements show that microclimates exist within aquatic habitats also. Because of its high specific heat, water warms more slowly during the day, and cools more slowly at night, than nearby points on land. As a result, aquatic microclimates are cooler during the day and warmer at night than nearby land microclimates. Moreover, bodies of water modify nearby terrestrial microclimates by increasing the local humidity and lowering temperatures through evaporation from the water surface.

The microclimates of lakes deeper than about 5 meters often vary in a regular cycle with changes in the seasons. During summer the surface heats, producing a layer of warm water that "floats" on the cooler, more dense water lying underneath. (Water reaches its maximum density at 4°C.) Currents created by wind blowing over the lake surface circulate water within the warmer layer, creating a surface layer that does not mix readily with the cooler deep layers (Figure 22-9a). This surface layer remains distinct until fall, when the surface gradually cools. When the surface and underlying layers reach the same temperature, the distinct layers are lost and the circulation created by

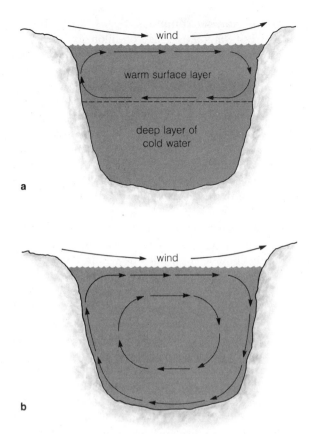

Figure 22-9 The fall turnover in deep lakes. In summer, solar heating creates a layer of warm water at the surface (**a**). The warm water floats on top of the colder water lying at deeper levels; water circulates within the warm layer at the surface but does not mix to any extent with the underlying cold water. As the surface layer cools to the same temperature as the deeper water in the fall, layering is lost and the entire lake circulates as a single unit (**b**).

the wind involves the entire lake (Figure 22-9b). As a result, deep waters are brought to the surface and mixing is complete. This complete mixing is called the *fall turnover*. A second turnover may occur in spring, when cold water released by melting ice displaces warmer water at the bottom.

These differences in temperature and water circulation shape the growth of populations living within the lake, particularly in temperate zones. During summer, organisms grow more rapidly in the surface layer, where temperatures are warmer, light is more intense, and oxygen is abundant. As the organic litter from these populations sinks to the deeper cold layers and decomposes, it adds nutrients to the deeper layers but removes oxygen.

As a result, as summer continues, deeper layers become nutrient-rich but stagnant. This condition limits population growth and survival to the warmer layer at the surface, except for a few types, such as bacteria and certain insect larvae, that can tolerate the low-oxygen conditions underneath the surface layer. The limitation of most populations to the surface continues until the fall turnover, when oxygen is carried to the depths and nutrients are brought to the surface. Frequently the mixing of oxygen and nutrients throughout the lake during the fall turnover greatly increases the growth of aquatic algae and plants and the animals that feed on them, producing a "bloom" of life in the lake at this time.

Thus both macroclimates and microclimates restrict survival, growth, and reproduction of living organisms to regions where climatic conditions lie within the limits tolerated by different populations. Where climatic conditions are optimal, growth is limited primarily by biotic factors such as competition and predation. Under climatic conditions approaching the tolerance limits of a population, the levels of temperature, humidity, and wind may instead become the primary factors limiting population growth.

Soils The soils of a region, which have a direct influence on the kinds of plant and animal populations existing there, are created by a combination of weathering of parent rock and the activities of living organisms. Weathering, through exposure to moisture and cycles of freezing and thawing, splits bedrock into smaller fragments. Moreover, water dissolves minerals from the bedrock—either directly, as in the case of soluble minerals such as calcium sulfate ($CaSO_4$), or through the action of acids such as carbonic acid (H_2CO_3), formed when atmospheric CO_2 dissolves in water. Bedrock has also been fragmented into soil-sized particles through the abrasive action of slowly moving glaciers.

Soils are produced by the action of weather and living organisms on the bedrock of a region.

Living organisms contribute to soil formation through two major routes: plant growth and burrowing animals. Plant growth adds to the fragmentation of rock and soil as plant roots grow into crevices and expand. Burrowing animals such as earthworms and rodents also contribute to pulverization and mixing of the soil. Both routes add organic material to the soil through the accumulation of litter from dead plants and animals.

The joint action of weathering and the activities of living organisms produce a distribution of organic matter and rock particles that is found in soils all over the world. Organic matter is at greatest concentration at the surface; at successively deeper layers it gradually gives way to particles derived from the parent rock. Eventually, usually at depths of a meter or so, the soil consists almost entirely of rock particles. This layer of rock particles extends down to the parent bedrock underlying the soil.

Water percolating from above causes further stratification of materials in the soil and produces what is known as a *soil profile* (Figure 22-10). Organic litter and largely undecomposed organic matter form the *O horizon* at the top of the soil. Below this layer, the *A horizon* forms a *humus* layer rich in decomposing organic matter mixed with fine rock particles. Water percolating downward through the A horizon dissolves organic chemicals and mineral salts from the humus and descends to the next lower layer, the *B horizon*, which consists primarily of particles derived from the parent bedrock. Frequently much of the organic and mineral matter dissolved from the O and A horizons precipitates and accumulates in the B horizon, producing a region enriched in nutrients within this soil layer. The bottom soil layer, the *C horizon*, contains rock particles of various sizes that are largely unweathered; this layer is altered little by material leaching downward from above. Bedrock underlies the C horizon. Each of these layers may be subdivided to various degrees depending on the patterns of weathering and organic deposition in the region.

The sizes of the weathered and unaltered rock particles directly influence the ability of soils to support populations of living organisms. Particles with dimensions between 2.0 and 0.02 millimeter are called *sands*. Particles of this size pack in loose aggregations that readily admit oxygen but have little ability to hold water. As the particle size diminishes through *silts*, with diameters between 0.02 and 0.002 millimeter, and *clays*, with particles below 0.002 millimeter, the capacity to hold water increases but aeration becomes poorer. Soils with the most favorable characteristics for plant growth contain a mixture of the three types—sufficient sand or larger silt particles to provide aeration and enough clay to prevent water from draining too rapidly.

The plant populations growing in a region are highly dependent on the soil type. Water must be available in the soil, as well as sufficient oxygen for roots to survive below the soil surface. The inorganic mineral salts essential for plant growth must also be available among the chemicals dissolved from the parent rock. Some soil types contain abundant supplies of these minerals; others, such as soils derived from granite and sandstone, are often deficient in the inorganic chemicals needed by plants. A few soils,

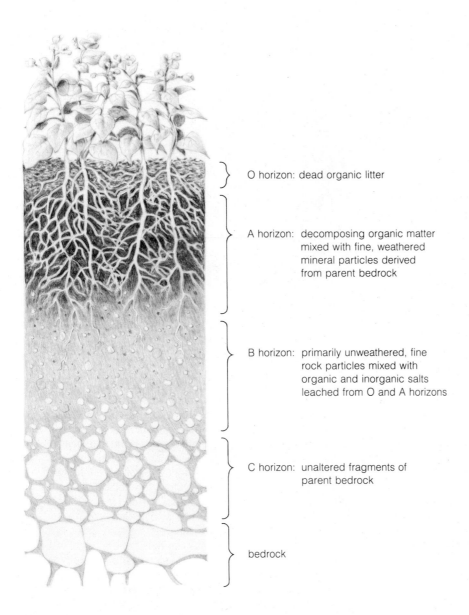

O horizon: dead organic litter

A horizon: decomposing organic matter mixed with fine, weathered mineral particles derived from parent bedrock

B horizon: primarily unweathered, fine rock particles mixed with organic and inorganic salts leached from O and A horizons

C horizon: unaltered fragments of parent bedrock

bedrock

Figure 22-10 Soil profile showing the O, A, B, and C horizons (see text).

such as those derived from serpentine rocks, contain concentrations of minerals that are toxic to plants. (Serpentine rock contains Mg^{2+} in concentrations high enough to retard plant growth.)

Regions with soils capable of supporting abundant plant life also support a variety of grazing animal populations that feed on plants and, in turn, predators that feed on the grazers. Both plant and animal populations may be severely limited, however, in regions containing unfavorable soils such as types derived from serpentine, granite, or sandstone bedrock.

The Major Biotic Factors: Competition and Predation

The major biotic factors regulating population growth, **competition** and **predation**, in contrast to climate and soils, are primarily density-dependent in their effects on population survival and growth: they increase in their effectiveness as population density increases. Competition occurs when different individuals attempt to use the same limited resource. Predation involves the use of the individuals of one population by another as a food source. Of the

Information Box 22-1:

Symbiosis: Mutualism and Commensalism

Competition and predation reduce the ability of individuals to survive and reproduce. Two other interactions between populations, **mutualism** and **commensalism**, benefit one or both participants rather than reducing their survival and reproduction. These interactions, collectively called **symbiosis**, have the effect of increasing the carrying capacity of the environment or expanding the ecological niche of one or both of the interacting populations.

There are many natural examples of mutualism, which benefits both populations in the symbiotic interaction. One of the most interesting is the relationship between termites and a group of flagellate protozoans found nowhere else but in the termite digestive tract (see Figure 4-2c). Although termites ingest wood, they do not possess the enzymes (cellulases) necessary to break down cellulose, the primary organic component of woody tissues. That capacity is supplied by the flagellate protozoans, which secrete enough of the cellulase enzymes to digest wood fibers for both themselves and their termite hosts. A similar mutualistic relationship allows animals such as cattle and deer to digest the cellulose of grasses and other plants eaten as food. These animals have a pouchlike cavity leading from the stomach, the *rumen,* which houses bacteria that can secrete cellulose-digesting enzymes. The association provides the animals with the ability to utilize cellulose as an energy source while giving the bacteria a constant food supply and what amounts to a natural cultural vessel with conditions maintained at optimal levels for bacterial growth. This association is vital not only to the ruminants themselves but also to the animals, including humans, that depend on cattle, sheep, deer, and the like as a food source.

Plants too have developed mutualistic relationships that benefit both participants in the interaction. A large and important family of plants, which includes legumes such as the garden pea, house bacteria in their roots that can convert inorganic nitrogen into forms that can be directly absorbed and used by the plant tissues (for details, see p. 551). Without the association, nitrogen would be unavailable to the plants and hence to the animals and other organisms that depend on them as a food source. Another example of mutualism in plants is the association between green algae and fungi in the composite structures known as *lichens.* Although lichens look like individual plants, each consists of two separate species: an alga, which carries out photosynthesis and supplies organic nutrients to the association, and a fungus, which grows as a leaflike structure that provides support to the algae, traps water,

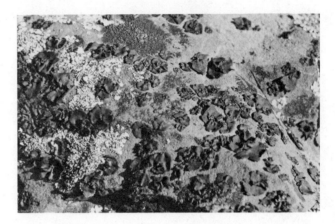

Several types of lichens growing on a rock surface. Lichens are composite organisms consisting of a mutualistic association between an alga and a fungus.

and secretes acids that dissolve inorganic minerals from the bare rock surfaces on which lichens grow.

There are also many examples of commensalism in nature. The nesting of birds in trees, which benefits birds without harming the trees, is one; another is the relationship between some species of small marine fishes and sea anemones. The fishes use the tentacles of the anemones for protection and concealment. (Stinging cells of the anemones immobilize intruders but not the commensal fish.) The anemones are evidently neither benefited nor injured by the relationship.

Mutualism and commensalism thus have a beneficial effect on the survival and reproduction of the populations engaging in these relationships. Since they are positive rather than negative interactions between species, and their effects are less spectacular, these symbiotic relationships are sometimes regarded as being insignificant among the factors affecting population growth. Yet without two positive results of symbiosis alone—nitrogen fixation and the digestion of cellulose—much of the present collection of populations of the world could not survive. Without the nitrogen fixed into organic compounds by the bacterium–legume association, the raw materials for nitrogen-containing molecules such as the nucleic acids DNA and RNA would be unavailable to many of the plants and animals of the earth. And without the symbiotic relationships that allow cellulose to be digested in ruminant animals such as cattle and deer, much of the most abundant organic molecule would be eliminated as a direct or indirect food supply to the world community.

biotic factors regulating populations, competition and predation have the greatest effect as population controls. (Additional biotic factors affecting population growth, **mutualism** and **commensalism**, are outlined in Information Box 22-1.)

The biotic factors regulating growth are primarily density-dependent in their effects.

Competition The common resources needed by competing populations may be either abiotic or biotic elements of the environment. Among plants, abiotic resources such as light, water, inorganic nutrients, and living space are the most frequent objects of competition. Animals commonly compete for food, water, mates, or living space in the form of feeding, nesting, refuge, or hiding sites. The competition for these resources may take the form of direct attempts to prevent another individual or population from reaching the resource. Although **interference competition** of this type is more typical of animals, it may also occur among plants. Black walnut trees and certain sagebrush species, for example, release toxic chemicals that prevent other plants from germinating or growing nearby.

Interference competition is the attempt to prevent another individual or population from using a resource. Exploitation competition is the attempt to use a resource faster or more efficiently than another individual or population.

Competition may be more subtle, however, and simply involve an effort to use a resource faster or more efficiently than other individuals or populations. Competition of this type, called **exploitation competition**, may occur without direct confrontation between individuals. Exploitation competition for water might occur, for example, between one population of animals that uses a water hole during the day and another that visits it at night.

The ecological niche is a population's way of utilizing resources and functioning in the environment.

Competition may occur when the niches of two populations overlap.

If time and energy that otherwise might be used in growth and reproduction are spent in competition, so that the birth rate is reduced or the death rate is increased, interference and exploitation competition may serve as effective population controls. Since populations compete for limited resources, they grow at or near the carrying capacity of the environment. As a result, competing populations are usually subject to K selection and frequently show the characteristics of K-selected organisms (see Table 22-1).

Competition and the Ecological Niche Competition and its effects are most easily understood in terms of the **ecological niche** of populations (see also Chapter 20). The ecological niche represents a population's pattern of utilizing resources and functioning in its environment—including the type and sizes of food eaten, time in which feeding occurs, ranges of oxygen, salinity, carbon dioxide, temperature, light, altitude, and moisture tolerated, and courting and mating seasons. The ecological niche of a population differs from its **habitat**, which is restricted to the physical space it occupies. (One investigator has compared the habitat of a population to its address and its niche to its profession.) Since the ecological niche represents a population's way of using resources and functioning in its environment, competition between populations may occur whenever the niches of two populations coincide or overlap.

Niche overlap can be represented graphically if the same niche characteristics of two populations—such as the size of prey taken, the range of temperatures tolerated, or the daily period of feeding activity—are plotted and compared (Figure 22-11). Competition may occur if the niches plotted in this way for two populations overlap at any point (shaded region in Figure 22-11a). The degree of overlap provides an idea of the extent of potential competition: the greater the degree of overlap, the more intense competition between the populations may be. Conversely, the amount of nonoverlapping regions in the two niches gives an idea of the opportunities for avoiding the negative effects of competition. If the nonoverlapping regions are sufficiently large and the populations can restrict their activities to these regions (called **niche refuges**), survival without competition is possible. If a population restricts its activities to only part of its niche as a result of competition, the niche actually occupied (called its **realized niche**; see Figure 22-11b) becomes smaller than the niche it would occupy in the absence of competition (the **fundamental niche**).

Plotting the ecological niches of separate populations thus provides a measure of the expected degree of competition between them. In most cases, the populations plotted in this way are from different species and the competition expected is **interspecific competition**—that is, competition between individuals of different species.

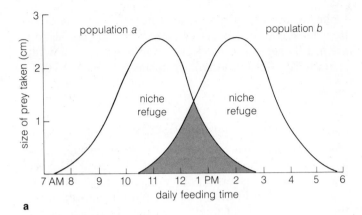

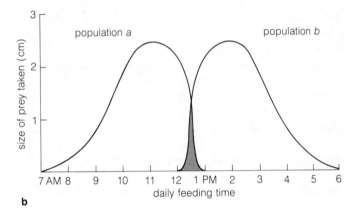

Figure 22-11 (**a**) Niche overlap (shaded) in two bird populations feeding on insects in a forest. The greater the degree of overlap, the greater the potential for competition between the two populations. The nonoverlapping portions of the niches, called *niche refuges*, provide opportunities for feeding without competition. (**b**) Reduction of competition by restriction of feeding activities to the niche refuges (see text).

Niche plots can also be made for the individuals of a single population. In this case, each individual's pattern of resource use is plotted (Figure 22-12). Although individual variations in patterns of resource use would produce some differences, most plots would overlap extensively, especially in the center of the distribution. (Note that the sum of the individual niche plots in Figure 22-12 is equivalent to the ecological niche of the population.) As a result, few individuals in the population could find niche refuges if the resources plotted are required for survival but limited in supply. The degree of overlap between the niche plots for individuals in a population thus indicates that competition between individuals of the same population, or **in-**

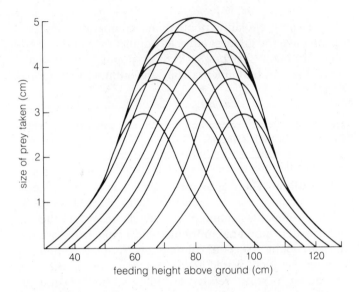

Figure 22-12 Niche plots for individuals of the same population. Since overlap is almost complete, competition is expected to be intensive (see text).

traspecific competition, as it is called, is likely to be the most intensive competition of all.

The Effects of Competition The effects of competition on populations have been studied in both controlled laboratory experiments and natural environments. Laboratory experiments have shown that competition between two populations may lead to either coexistence or extinction of one population. In most laboratory experiments, extinction is the usual outcome.

In one classic example of this type, G. F. Gause, a Russian ecologist, grew two protozoan species, *Paramecium caudatum* and *P. aurelia*, together in culture vessels under controlled conditions with a constant but limited food supply. Niche overlap for the two species in this artificial environment was essentially complete and competition for food was intense. Under these conditions, *P. aurelia* was the superior competitor and in all experiments the population levels of *P. caudatum* fell steadily until all the *caudatum* individuals died out.

A variety of laboratory experiments yielding the same result led to the development of an ecological principle known as **competitive exclusion**. This concept states that when the niches of competing populations overlap completely or nearly completely, competition will lead eventually to extinction of one of the two populations. Whether this concept, developed from laboratory studies of populations in greatly simplified environments, applies to natural populations has been hotly debated by ecologists for many years.

Competitive exclusion may occur when the niche overlap of two populations is almost complete.

A few laboratory experiments have shown that different populations can coexist if their niche overlap is reduced sufficiently to provide niche refuges. In fact, Gause's classic experiments with protozoa also demonstrated this outcome. Gause found that when *P. bursaria* was used as the species competing with *P. aurelia* instead of *P. caudatum*, both populations survived. *Paramecium aurelia* feeds near the top of the culture vessel whereas *P. bursaria* feeds at bottom levels avoided by *aurelia*. This difference in habit was enough to provide niche refuges, and survival, for both populations.

Studies of populations in their natural environments reveal that, in contrast to the most frequent outcome of laboratory experiments, competition usually leads to coexistence rather than extinction. This difference probably depends on the greater complexity of natural environments, which offer so extensive a variety of resources and conditions that niche refuges are much more easily found in nature than in the laboratory.

The complexity of natural environments is such that populations which at first seem to occupy the same niche prove, on detailed examination, actually to occupy separate niches within the same environment. Populations of five different warbler species living in New England forests, for example, seem to feed and nest in trees in the same positions and manner. On this basis, the populations should be subject to competitive exclusion if the exclusion principle applies to natural populations as well as laboratory cultures. Detailed investigations by Robert H. MacArthur of Princeton University, however, revealed that each warbler population actually feeds at different positions and levels (Figure 22-13) and has different mating and nesting periods in the forest. These niche differences are sufficient to allow the warbler populations to survive and coexist.

A similar study by David Lack of Oxford University has already been mentioned in the discussion of evolution (Chapter 19). Lack studied two bird species of the genus *Phalacrocorax*, the cormorant and the shag, that seemed to feed on ocean fish in the same ecological niche. Close examination of the feeding habits of the two species, however, revealed that the cormorant feeds on fish that live near the bottom in shallow water whereas the shag feeds on fish that swim near the surface in water at some distance from the shore. Reproductive habits of the two populations also differ enough to provide separate nesting space for the two populations: the cormorant nests on broad ledges on cliff faces whereas the shag uses narrow ledges. Most studies of natural populations reveal the same pattern. Populations inhabiting the same geographic regions coexist through niche differences that reduce the effects of competition.

The evolutionary processes of natural selection tend to reinforce niche separation and reduce competition. Imagine two species with niches that overlap to some extent but leave extensive regions of nonoverlap that provide niche refuges (as in Figure 22-11a). Individuals with characteristics that place them in the regions of overlap will suffer the most intensive effects of competition and be less likely to survive and leave offspring. Other individuals, with characteristics that place them in the nonoverlapping parts of the two niches, will be more likely to survive and reproduce. Consequently, individuals with characteristics placing them in the region of overlap will tend to disappear from succeeding generations whereas individuals in the niche refuges will tend to increase in numbers. As a result, the niches of the populations will shift toward the nonoverlapping portions and the region of overlap will be omitted (as in Figure 22-11b).

This evolutionary process stemming from competition, called **niche specialization**, has been demonstrated repeatedly in studies of natural populations. Two of the Galápagos finches, for example, *Geospiza fuliginosa* and *G. fortis*, have similar beak sizes and feed on similar food sources when they occur on different islands (Figure 22-14a and b). Where the populations of the two species live on the same island, however, competition and natural selection have led to the evolution of beak sizes (Figure 22-14c) and feeding behavior different enough to produce niche specialization and reduce competition.

Coexistence in natural environments depends on the avoidance of competitive exclusion by niche specialization.

Thus the coexistence frequently observed in natural environments depends on niche specialization, which, in turn, results from the effects of competition and natural selection. Extinction due to competitive exclusion is therefore probably infrequent in nature unless the environment of a region is too simple and uniform to offer sufficient opportunities for niche specialization or, alternatively, is so unstable that change occurs too rapidly for niche specialization to evolve. The recorded cases of competitive exclusion in nature, as a result, are limited almost entirely to

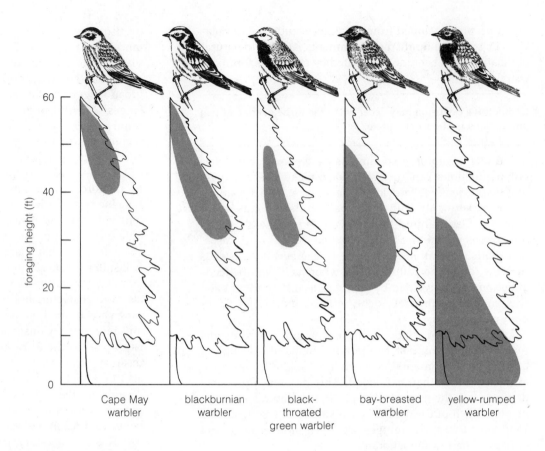

Figure 22-13 Locations at which five different warbler species (genus *Dendroica*) feed in Maine spruce trees. Data from R. H. MacArthur, *Ecology* 39:599 (1958). Adapted from R. E. Ricklefs, *Ecology*, 2nd ed., Chiron Press, 1979.

situations in which natural disasters or human intervention have disturbed the environment more rapidly than populations can respond by developing niche specializations. Competitive exclusion has occurred, for example, as the result of the introduction by humans of a foreign species into a new region. Since the introduced species lacks niche specializations corresponding to the new region, its niche often overlaps extensively or completely with one or more native populations. If the introduced species is more efficient than the native populations in the use of a common resource, its presence may result in extinction of the native species before natural selection can lead to sufficient niche specialization to allow coexistence and survival. The Hawaiian Islands, for example, were once home to a variety of bird species that occurred nowhere else in the world. The introduction by humans of highly competitive species with overlapping niches, such as the English sparrow and starling, led to rapid extinction of many of

the native birds. Similarly, introduction of the dingo dog and European fox has resulted in extinction of native species such as the Tasmanian wolf from many regions in Australia.

Territoriality Competition in many animals is expressed as an effort by an individual to defend a region used for feeding or mating against the intrusion of other individuals of the same or different populations. This form of competition, called **territoriality**, occurs among a few insects, some fish, many mammals, and most birds. The males of many bird species, such as the redwinged blackbird, establish territories just before mating season that are defended vigorously against other males. Many fights ensue, and some males are eliminated entirely from regions usable as nesting and feeding grounds before final territories are established. Once territories are formed, fighting diminishes and the males maintain their territories primarily

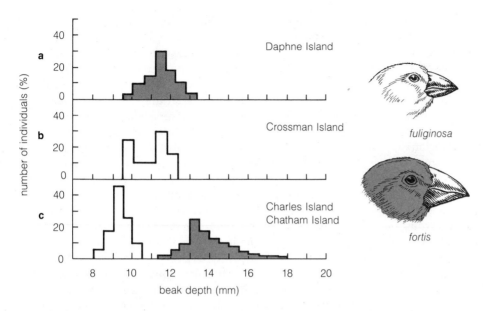

Figure 22-14 Beak sizes in two species of ground finches (*Geospiza*) on several islands of the Galápagos group. Where populations of the two species occur together, as on Charles and Chatham Islands (bottom row), selection has led to evolution of different beak sizes and niche specialization (see text). Shaded bird and plot *g. fortis*; unshaded bird and plots *g. fuliginosa*. Adapted from R. E. Ricklefs, *Ecology*, 2nd ed., Chiron Press, 1979.

by perching in conspicuous locations and singing. Efforts are still made, however, to prevent males of the same species, or birds of other species, from entering the territory. (Some birds, such as mockingbirds, will even attempt to defend their territories against animals as large as humans by swooping and calling loudly.) Once the males pair off with females and the nesting season is over, defense of the territory gradually diminishes and finally ceases. Many other animals set up temporary or permanent territories by similar patterns of defense.

Territoriality is the defense of a region used for feeding or mating against the intrusion of other individuals.

By increasing the space between individuals in a region, territoriality effectively reduces population density and places an upper limit on population growth. The area of the territory depends on the body size of the territorial individuals and the supply of food or other resources in the region. Song sparrows in Ohio, where food supplies are relatively scarce, defend individual territories averaging 2700 square meters in area; in the salt marshes of San Francisco Bay, where food is more abundant, the territory for song sparrows is reduced to an average of 400 square meters.

Territoriality reduces population density.

Competition in its various forms thus provides a biotic factor regulating population growth. While there is no doubt that competition can limit population size, there is extensive debate among ecologists concerning its importance compared to other population controls such as predation. Because selection leads to niche specialization and reduction of competition between populations, some ecologists maintain that competition will be at minimum levels in well-established communities. Therefore, according to this way of thinking, competition is likely to be relatively unimportant in controlling population growth. Others argue that even with niche specialization, enough niche overlap will exist to produce significant levels of competition and hence reduction in the survival and reproduction of individuals. Thus, according to this argument, competition for limited resources remains an effective population control even in established communities. The problems in deciding between these alternatives are compounded by the fact that competition is difficult to study in natural environments. Isolation of populations to test growth in the absence of competition, for example, often disturbs the environment so extensively that the results obtained are impossible to evaluate.

Many other questions about competition and its effects remain to be answered by ecologists. How much niche overlap is possible before competitive exclusion occurs and with it the elimination of one competing population? Does competitive exclusion in fact occur in undisturbed environments? Does niche overlap always lead to niche specialization? Is there a limit to the number of niches that can develop? Is there an ultimate limit therefore to the number of different species that can inhabit a region? These and other questions concerning competition are today the subjects of intensive ecological research.

Predation Predation includes any consumption of one living organism by another—the capture of a deer by a wolf, an insect by a spider, the ingestion of grasses by a browsing moose, flies by a venus fly trap, acorns by a squirrel, or the attack of animals and plants by disease organisms and parasites. Predation of one animal by another usually results in the sudden death of the prey. However, disease organisms such as bacteria and parasites such as the roundworm and flatworm parasites of animals frequently "nibble" at their hosts, sapping vitality and reducing reproduction but not necessarily causing immediate death. The animals of various kinds that graze on plants, called **herbivores**, resemble parasites in that they frequently nibble on the plants they feed on without killing them outright.

Predation involves any consumption of one organism by another.

The Effects of Predation Whether it involves sudden death of the prey or gradual reduction of vigor and health, predation effectively limits the growth and reproduction of populations. Like competition, predation may control population growth by increasing the death rate or by reducing the birth rate of the prey population. The death rate is increased by the outright killing or slow death of prey; the birth rate is reduced not only through impairment of the health of the prey but also through the loss of energy that must be directed from reproduction to avoiding predators or developing defensive measures.

One of the best examples of the regulation of population growth by a predator comes from a study of moose on Isle Royale, a large island in northern Lake Superior. Moose first migrated to the island and established a population not long after 1900. They flourished on the island, and by the 1930s the population included as many as 3000 individuals—an average population density of almost 15 moose per square mile. This situation was changed in the 1940s, however, by the appearance on the island of a

moose predator, the timber wolf. The wolves preyed on the moose and gradually reduced the herd to about 600 individuals. The moose population is now apparently held stable at these numbers by the activity of the predators.

Other studies have revealed the effects of predation on population growth through the removal of predators. On the Kaibab Plateau of Arizona, deer herds were once preyed upon by wolves, mountain lions, and coyotes. In 1906, the region was designated as a game preserve and the natural predators of the region were systematically eliminated. Following removal of the predators, the deer population gradually increased in density, growing from an estimated 4000 to between 60,000 and 70,000 individuals by 1923. At this point the deer outgrew their food supply, and a population crash followed. Only 20,000 deer survived by 1931; by 1939, only 10,000 remained. The results of predator removal in this case suggest that the predators originally regulated the deer at relatively low population densities that were more compatible with the resources of the region than the high densities achieved after predator removal.

Although predators can maintain the prey population at stable levels, their presence can also produce regular oscillations in which the numbers of predators and prey rise and fall at fairly constant intervals. While such oscillations are usually identified with animal predators and prey, as in the snowshoe hare and lynx populations described earlier in this chapter, the interactions between herbivores and the plants they feed on can also produce cyclic oscillations in numbers. (Part of the lynx–hare cycle, in fact, is due to overgrazing by the hares; see p. 495.) The cyclic variations in the numbers of lemmings, small rodents that live in the tundra in the far northern regions of North America and Eurasia, have been linked to a regular pattern of overgrazing on tundra plants, followed by a crash in the lemming population. Then recovery of the plants provides food for another increase in the lemming population, and the cycle repeats.

Regulation of prey populations by predators has been used with some success to control both animal and plant pest species. Scale insects on orange trees in Southern California have been brought under control by the introduction of ladybird beetles that prey on the scale pests. Similarly, the growth of Klamath weed, an introduced plant with toxic effects on cattle, was reduced by an estimated 99 percent by another introduced species, a beetle that feeds on the weed.

Evolutionary Responses of Predators and Prey Predation exerts a potent selective force as well as a control of population growth. Predation removes from the prey populations in-

dividuals that are less successful in avoiding capture or ingestion; other individuals, with adaptations making them more difficult to catch or consume, survive and give rise to the next generation. Thus selection tends constantly to increase the ability of a prey population to elude its predator. At the same time, the predators that are more successful in capturing and consuming prey are more likely to survive and leave offspring. Thus the predator population tends to evolve toward individuals more skillful at catching or ingesting their prey.

Natural selection tends to increase not only the efficiency of predators in capturing prey but also the ability of prey to elude their predators.

Natural selection due to the pressures of predation has produced an impressive battery of avoidance and defensive adaptations in prey species. Animal prey have responded to predation with such adaptations as the quills of porcupines, the protective camouflage of zebras, dead-leaf butterflies and walking-stick insects, the ink cloud released as a protective screen by squid and octopus, the swift and evasive running of antelope and deer, and the repellent spray of skunks (Figure 22-15). Less spectacular, but just as effective, are defensive measures such as increased size, as in the rhinoceros and elephants, and the production of reproductive adults in large numbers at intervals. The emergence of cicadas, example, spaced at intervals of many years, occurs in such numbers that

Figure 22-15 Adaptations of prey species that provide defense against their predators.

predators are unlikely to be able to eat all the available adults.

Plants too have developed adaptations that reduce their palatability and availability to their herbivore predators. The spines, hairs, barbs, thorns, and hooks of stems and leaves and the hard shells of some seeds are typical adaptations that reduce the effectiveness of herbivores in

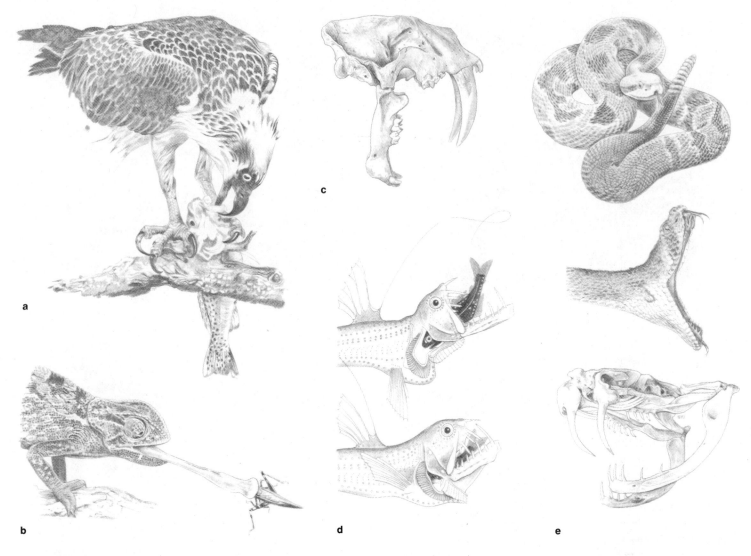

Figure 22-16 Adaptations that increase the efficiency of predators in capturing prey.

consuming plants. Some plants have evolved a variety of chemical defenses that make them distasteful or toxic to animals. Like the cicadas, other plants have responded to predators by reproduction in huge numbers at irregular intervals: beech trees, for example, release large numbers of seeds at irregular intervals of several years. The intervening years of low seed production keep the population of predators at levels low enough to ensure that not enough seed-eaters will be present to eat all the seeds in years of heavy production. While these defensive measures reduce the death toll due to predation, they also require of the prey energy and resources that otherwise might be used for reproduction.

The intense evolutionary interaction between predators and prey has also led to a wide variety of adaptations improving the prowess of predators. The complex webs of spiders, the sticky tongues and accurate aim of toads and frogs, the teeth, talons, and swiftness of cats, the camouflage coloration of rattlesnakes and tigers, which makes them almost invisible in their natural habitats—all come readily to mind among the many striking predator adaptations (Figure 22-16). Parasites and disease organisms too evolve in response to their effects on prey species. Since a predator acting as a parasite or infectious disease organism can only exist as long as its victim lives, selection in this case leads toward predators that consume their prey slowly enough to allow their prey just to survive.

Predation thus represents another biotic factor keeping the level of population growth below the carrying capacity of the environment. In contrast to competition, the effects of predation are more easily measured and detected. As a consequence, predation is somewhat less controversial among ecologists as a factor in the regulation of population growth. Nevertheless, many questions remain to be answered about predation. The most important of these concern the specific factors producing cyclic variations in predator and prey population size and the extent to which predators can stabilize prey populations at numbers that avoid explosions and subsequent crashes.

Population Regulation: A Summary

The density-independent and density-dependent factors outlined in this chapter—climate, soils, competition, and predation—work together to regulate the growth of populations. Of the various controls, only the density-dependent factors, competition and predation, are expected to regulate populations at stable levels at or near the carrying capacity of the environment. While the density-independent factors of climate and soils can reduce the growth of populations below their ultimate potentials, they do not control populations over long periods of time at a stable equilibrium.

The populations of a region, held in check by these environmental controls, form a complex, interdependent community of organisms. The interrelationships of the populations in communities, the sources of the energy and raw materials used by these populations, the disposition of community wastes, and the changes of communities with time are described in the next chapter of this unit.

Questions

1. What is a population? What population characteristics and properties are studied in ecology?

2. What is *exponential* population growth? Plot the growth of a population initially containing 1000 individuals evenly distributed between males and females in which each female has one offspring per year.

3. Calculate the yearly rate of increase for a human population of 100,000 individuals in which the birth rate is 70 per every 1000 persons and the death rate is 10 per every 1000 individuals. (*Hint*: Use Equation 22-2.)

4. What is *logistic* population growth? How do density-dependent controls convert population growth from an exponential to a logistic curve?

5. When do density-dependent controls begin to slow the rate of population growth? In what ways does population growth vary from the predictions of the logistic equation?

6. What is K and r selection? What strategies do populations develop in response to K and r selection?

7. What are survivorship curves? How are survivorship curves related to K and r selection?

8. What kinds of density-dependent factors regulate population growth?

9. In what ways does climate affect population growth? Can climate have density-dependent effects?

10. What is microclimate? How might microclimate differ in the canopy, shrub layer, and ground level in a forest?

11. What causes the fall turnover in deep lakes? How do the conditions before and after the turnover affect population growth?

12. How are soils produced from the parent rock? What are soil horizons? How are soil horizons produced?

13. What are sands, silts, and clays? How do soil types affect population growth?

14. What is competition? What kinds of competition occur between populations? How does competition affect population growth?

15. What is an ecological niche? What is niche overlap? How is niche overlap related to competition?

16. What is a niche refuge? A fundamental niche? A realized niche? How is niche overlap related to population survival?

17. Explain the principle of competitive exclusion. How do natural populations inhabiting the same geographic region avoid extinction by competitive exclusion?

18. How does natural selection reduce competition? What is niche specialization?

19. What is territoriality? How does territoriality limit population growth?

20. What is predation? What kinds of predation may occur? How does predation affect population growth?

21. Can predators ever have beneficial effects on their prey populations? How?

22. What effects do the defensive measures evolved by prey species have on their population growth?

Suggestions for Further Reading

Krebs, C. J. 1978. *Ecology: The Experimental Analysis of Distribution and Abundance*. Harper & Row, New York.

McNaughton, S. J., and L. L. Wolf. 1979. *General Ecology*. 2nd ed. Holt, Rinehart & Winston, New York.

Pianka, E. R. 1978. *Evolutionary Ecology*. 2nd ed. Harper & Row, New York.

Ricklefs, R. E. 1979. *Ecology*. 2nd ed. Chiron Press, New York.

Whittaker, R. H. 1975. *Communities and Ecosystems*. 2nd ed. Macmillan, New York.

23

THE INTERACTIONS OF POPULATIONS IN COMMUNITIES

The populations occupying a region form a **community**, an assemblage of living organisms interacting with each other and their environment. The interaction gives communities features that individual populations do not have, just as populations have properties not possessed by individuals. These properties and characteristics include *community structure*, the arrangement of populations in space and the pathways of energy flow between them; *community stability*, the degree to which the populations of a community change in numbers and types with time; *community succession*, the sequences in which populations appear and disappear in changing communities; and *community climax*, a final and stable assembly of populations that may form in a particular geographic region as a result of community succession.

Communities comprise the living populations of a region and their interactions.

The boundaries of the communities studied in ecology may be fixed by natural lines or barriers setting apart continents, mountains, plains, islands, streams, lakes, oceans, or other geographic features of the earth. Or community boundaries may be defined to suit the requirements of an investigation, as in a forest or field plot, a segment of a lake or ocean, a rotting log, or even a drop of water. In either case, ecologists attempt to include as many of the factors affecting communities as possible in the effort to understand what holds them together and what makes them work. This chapter outlines the results of these ecological studies of communities and describes, in a supplement, the major climax communities of the world.

COMMUNITY STRUCTURE

The Arrangement of Populations Within Communities

Each of the populations forming a community occupies a horizontal and vertical position in space. The horizontal distribution of populations in a community may be *continuous*, with the individuals of a species spread more or less evenly throughout the community, or *patchy*, with the individuals of a species collected in isolated aggregations. A plant species such as bluestem grass, for example, may be distributed continuously in a grassland community (Figure 23-1a). The same grass species may be dispersed in a patchy distribution in another community dominated by small trees and shrubs (Figure 23-1b). Detailed studies of the horizontal structure of communities have revealed that

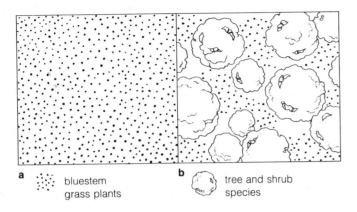

a :·:· bluestem
grass plants

b ⊙ tree and shrub
species

Figure 23-1 Continuous (**a**) and patchy (**b**) distribution of bluestem grass in grassland and woodland communities.

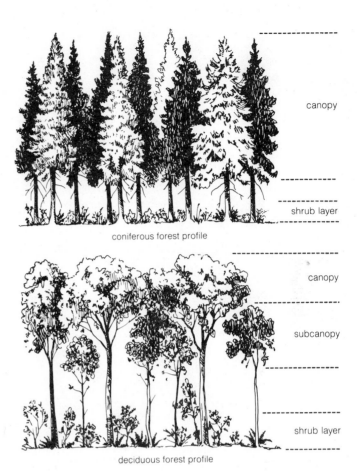

coniferous forest profile

canopy

shrub layer

canopy

subcanopy

shrub layer

deciduous forest profile

Figure 23-2 Vertical stratification in coniferous and deciduous forests.

populations tend to be dispersed in patchy rather than continuous distributions.

Studies of the vertical structure of communities show that populations also tend to be unevenly spaced or *stratified* at different heights above or below the ground or water level (Figure 23-2). In a forest, for example, plants of different heights and sizes form vertical layers or *strata* running from algae at the surface of the soil through mosses, herbs, shrubs, small trees at intermediate levels, to the tallest trees of the topmost canopy. The roots or holdfasts of these plants form similarly stratified layers extending below the surface of the soil. Each of these layers of vegetation above and below the soil surface has its own characteristic assemblage of animal life—including forms such as burrowing mammals below the surface, birds at each level above the soil, and insect, mite, and crustacean populations of different kinds at levels both above and below the ground. Aquatic habitats reveal a similar stratification. In marine communities, green algae grow in the zones near the surface, where light intensity is highest; brown algae grow in intermediate zones; red algae are found at the deepest layers penetrated by light. As in terrestrial environments, these stratified zones of algae tend to have their own complement of animals, including crustaceans, molluscs, and annelids that live in each stratum.

The horizontal distribution of populations in a community may be continuous or patchy; their vertical distribution tends to be stratified.

Most of the patchy horizontal distribution and vertical stratification of plants, animals, and other forms within communities reflects differences in ecological niches and the effects of competition, predation, and symbiosis in restricting niche size and producing niche specialization (see p. 507). A community is thus structured as a collection of different ecological niches set up in horizontal and vertical arrangements by abiotic environmental conditions and interactions between the community populations.

Communities are structured collections of niches maintained by biotic and abiotic environmental conditions.

Studies of the distribution of populations in communities reveal that certain species frequently occur together. Beech trees are often found in mixed stands with maples in deciduous forests in midwestern and eastern states; in drier forests, oaks, hickory trees, and flowering dogwood often grow together. These recurring combinations have led to the proposition, made originally by Frederick E. Clements of the Carnegie Institute, that communities are

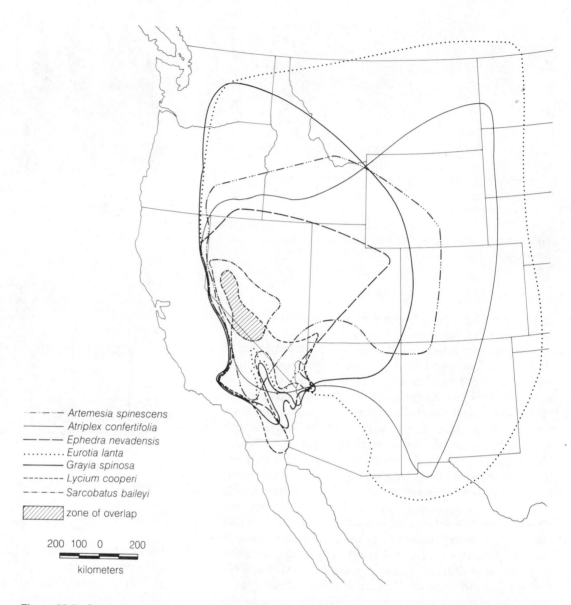

Legend:

- - - - - Artemesia spinescens
──────── Atriplex confertifolia
- - - - - Ephedra nevadensis
· · · · · · · Eurotia lanta
──────── Grayia spinosa
- - - - - - Lycium cooperi
- · - · - · Sarcobatus baileyi

▨ zone of overlap

200 100 0   200

kilometers

Figure 23-3 Distributions of plant species that grow in the shadscale community (shaded) in western Nevada. The distribution of each species extends well beyond the shadscale community and overlaps with other communities (see text). From data in W. D. Billings, *American Midl. Nat. 48*:87 (1949).

natural units representing a sort of "superorganism" in which certain species depend on each other in subtle ways for their existence and always grow together. This proposal has resulted in one of the major unresolved controversies in community ecology. At one extreme is Clements' idea that the populations of communities occur in constant combinations. At the opposite extreme is the viewpoint that the presence of each population in a community is determined strictly by local conditions of soil, moisture,

temperature, and other abiotic factors and biotic factors such as competition and predation. According to this view, the frequent presence of two species together in communities is nothing more than a result of the coincidence that their habitats and ecological niches overlap to a greater or lesser extent.

Investigations of community structure suggest that the distribution of each species is in fact largely independent and determined by local biotic and abiotic conditions.

If communities are natural units, and certain species are closely interdependent, then the distributions of the species in a linked unit should be almost identical and none should extend significantly beyond the borders of their community. In most cases, examinations of the actual distributions of species show that the opposite is true. Almost none of the distributions are identical or even close; most extend beyond the boundaries of a given community. Figure 23-3, for example, shows the distributions of the plant species forming the *shadscale* desert community of western Nevada. The distribution of each species extends well beyond the borders of the shadscale community (shaded zone) and overlaps with other communities located as far away as southern Canada and northern Mexico. These and other observations of the same kind support the idea that the occurrence of species together in communities results largely from the chance that their distributions and niches overlap.

The occurrence of populations in communities depends largely on a common zone of overlap in their ranges of distribution.

Studies of the distribution of communities along gradients of moisture, temperature, or elevation also argue against the idea that the species of communities occur in distinct and repeated combinations. Plots of the plants occurring along a gradually increasing slope (Figure 23-4*a*), for example, show that the species and communities overlap and gradually intergrade from one to the next. There are no abrupt changes or boundary lines as might be expected if certain species tended to associate exclusively together. Plots of the species occurring along gradients of moisture or temperature show the same gradual change in species (Figure 23-4*b* and *c*). Gradual changes in community structure along environmental gradients of this type are called **ecoclines**.

The studies of community structure described in this section therefore demonstrate that the populations of communities tend to be distributed in patchy distributions horizontally and also to be stratified vertically. Although some species frequently occur together in communities, their individual distributions are usually different and extend well beyond the borders of the community. Therefore the location of species together in communities depends primarily on a common zone of niche overlap in some part of their total range.

Energy Flow in Communities: Trophic Levels and Food Webs

Much of a community's structure is determined by predation—in other words, by who eats whom in the com-

munity. Maintenance of this feeding structure depends on the amount of energy available to the predators in the form of the complex body molecules of the prey populations. Since this is the case, much can be learned about community structure by tracing the flow of energy through communities in terms of the organic molecules produced by one population and consumed by another.

The Levels of Energy Flow

Primary Production in Green Plants Energy enters most communities through the activity of green plants in absorbing sunlight and converting it to the chemical energy of complex organic molecules (see Figure 23-5 and Chapter 6). In their role as the organisms that capture and convert the energy of sunlight, green plants are called the **primary producers** of communities. In various terrestrial communities, the primary producers may be trees, shrubs, grasses, mosses, lichens, or photosynthetic bacteria. In aquatic communities, microscopic floating algae called **phytoplankton** (see Figure 23-6) or rooted aquatic plants are the major primary producers.

Primary producers convert the energy of sunlight into the chemical energy of organic molecules.

The primary producers convert only about 1 percent or less of the sunlight falling on them into the chemical energy of complex carbohydrate, lipid, and protein molecules of plant tissues. The actual amount of light converted into the energy of organic matter in different regions of the earth depends on several factors. One is the availability of the necessary sunlight. In some regions—landmasses almost continuously covered by clouds, for example, or the deeper levels of lakes, rivers, and oceans—the small amount of light reaching plants limits primary productivity. Other factors, such as low temperatures or the scarcity of water or inorganic nutrients, can also limit productivity. The primary producers of arctic communities are limited by low temperatures as well as low light intensities; in deserts, productivity is limited by the scarcity of water even though light intensities are high. The productivity of the oceans is reduced by low supplies of inorganic nutrients as well as insufficient light. Table 23-1 summarizes the primary productivity of different major communities.

Not all of the light energy converted into chemical energy by the primary producers becomes available as an energy source for predators in the community. Part of the organic matter produced is used by the plants themselves in respiration and is lost to the community. Depending on

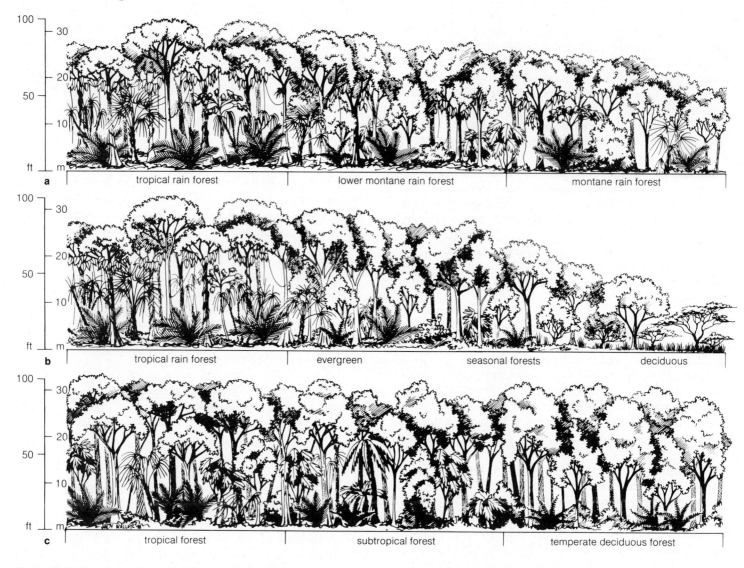

Figure 23-4 Overlap and intergrading of populations and communities along gradients of elevation (**a**), moisture (**b**), and temperature (**c**).

the community and the kinds of plants present, from 30 to 75 percent of the captured light energy is lost in this way. The total production before losses to respiration is the **gross primary production**. The remaining chemical energy, which forms the **net primary production** of the plants, is stored in the plant tissues—some to be consumed by herbivores and some to be added to the decaying organic matter forming the litter or **detritus** of the community (see Figure 23-5).

Net primary production is the chemical energy remaining in the organic molecules of primary producers after losses to respiration.

Energy Flow to the Primary Consumers Herbivores obtain the energy required for their community interactions by feeding directly on the chemical substances stored in the tissues of the primary producers. In this role, the herbivores

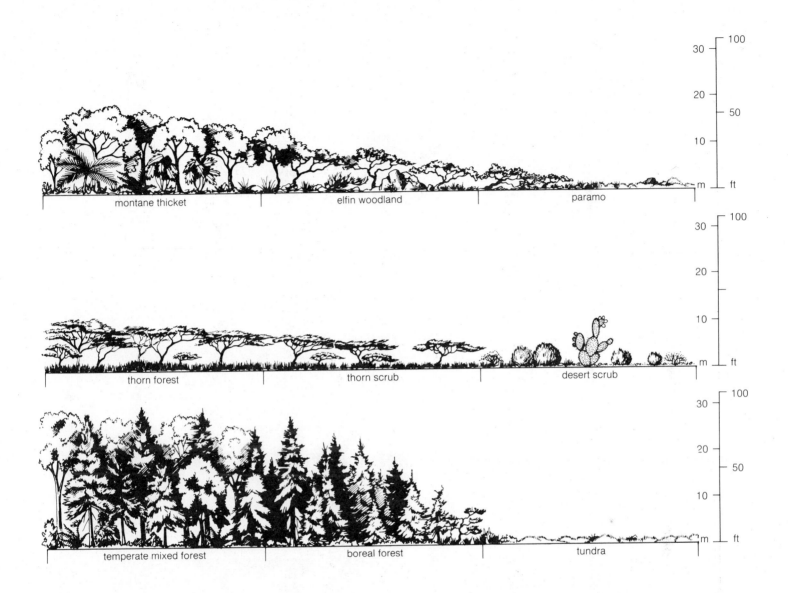

act as the **primary consumers** of the community. In terrestrial communities, the main herbivores are grazing animals such as rabbits, deer, cattle, or insects that feed on leafy plants, grasses, or mosses. In aquatic communities, the primary consumers are usually microscopic floating animals and larvae, called **zooplankton** (Figure 23-7), that feed on phytoplankton.

Depending on the community, from 5 to 20 percent of the energy stored in the primary producers is consumed and used by the primary consumers (Figures 23-5 and 23-8); the remainder eventually adds to the decaying organic matter of the detritus. Within the primary consumers from 30 to as much as 90 percent of the energy taken up is used in respiration and lost as heat; the remainder is stored in their tissues. This energy, stored as complex organic molecules in the primary consumers, has the same fate as the energy stored in the plant tissues of the primary producers. A part is consumed by carnivorous animals

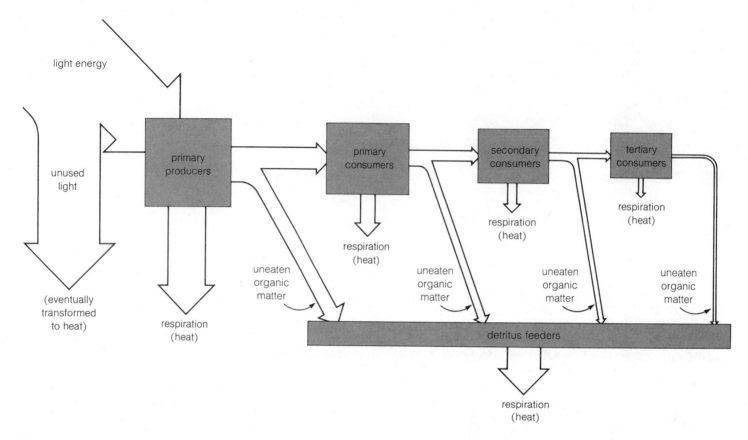

Figure 23-5 The general pattern of energy flow through the primary producers, consumers, and detritus feeders of a community (see text). The size of the boxes gives a relative estimate of the total weight or biomass of the populations at each level.

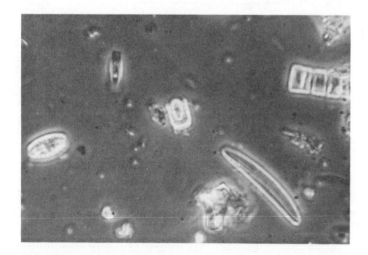

Figure 23-6 Microscopic plants of freshwater phytoplankton, the primary producers of many aquatic communities.

Figure 23-7 Zooplankton, the small floating or swimming primary consumers of aquatic communities.

Table 23-1 Net Primary Production for Different Major Communities

Community	Net Primary Productivity per Unit Area (g/m²/yr) Normal Range	Mean	World Net Primary Production (10⁹t/yr)
Tropical rain forest	1000–3500	2200	37.4
Tropical seasonal forest	1000–2500	1600	12.0
Temperate evergreen forest	600–2500	1300	6.5
Temperate deciduous forest	600–2500	1200	8.4
Boreal forest	400–2000	800	9.6
Woodland and shrubland	250–1200	700	6.0
Savanna	200–2000	900	13.5
Temperate grassland	200–1500	600	5.4
Tundra and alpine	10–400	140	1.1
Desert and semidesert scrub	10–250	90	1.6
Extreme desert, rock, sand, ice	0–10	3	0.07
Cultivated land	100–3500	650	9.1
Swamp and marsh	800–3500	2000	4.0
Lake and stream	100–1500	250	0.5
Total continental		773	115
Open ocean	2–400	125	41.5
Upwelling zones	400–1000	500	0.2
Continental shelf	200–600	360	9.6
Algal beds and reefs	500–4000	2500	1.6
Estuaries	200–3500	1500	2.1
Total marine		152	55.0
Full Total		333	170

Adapted from R. H. Whittaker, *Communities and Ecosystems*, 2nd ed. New York: Macmillan Publishing Company, 1975.

feeding on the herbivores, and a part is added to the decaying organic matter of the detritus.

Secondary and Tertiary Consumers   At the next level a group of carnivores of the community called the **secondary consumers** lives by eating the herbivorous primary consumers. Secondary consumers in terrestrial communities include animals such as insects and other invertebrates, owls, foxes, cats of various kinds, wolves, and to some extent humans. In aquatic environments, animals such as

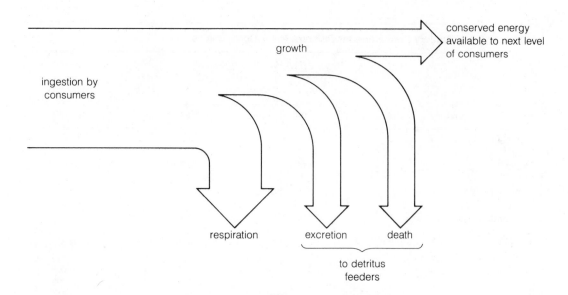

Figure 23-8 Details of the energy disposition at the level of a consumer in a community.

trout, bass, cod, and whales fill the same role; microscopic forms, including protozoa, are also important secondary consumers in some aquatic communities. In this role, aquatic and terrestrial carnivores manage to take in about 25 to 30 percent of the energy stored in the primary consumers; the remainder adds to the decaying matter of the detritus (see Figure 23-5). Secondary consumers pay a high price in respiration for their hunting and capturing role as carnivores: as much as 60 to 90 percent of the food they take in is used to provide energy for their activities and is lost as heat.

The secondary consumers may themselves be consumed by yet another level of larger carnivores such as wolves, lions, and cougars on land or sharks, tuna, and barracuda in aquatic communities. These animals act as **tertiary consumers**; that is, they feed on the secondary consumers. As in all the other levels of producers and consumers in the community, the secondary consumers may simply die and add their molecules to the detritus.

Consumers rarely exist in communities past the secondary or tertiary levels. There is simply not enough energy left. Since from 25 to 90 percent of the energy captured at each successive level is lost as heat, only the remainder is left as usable energy for the next level. And since herbivores and carnivores never succeed in completely consuming all the individuals they prey on, some are left to die and add their stored organic molecules to the detritus. As a rule of thumb, ecologists estimate that only 10 percent of the energy stored at one level is actually passed on to the next level in the energy pathway. That is, for every 100 calories of light energy converted to chemical energy by the primary producers, only 10 calories is likely to be taken up and stored in the chemical molecules of the primary consumers (Figure 23-9). Only 1 calorie of this energy, on the average, finds its way to the secondary consumers to form the complex molecules of the organisms at this level. This leaves only 0.1 calorie for tertiary consumers at the next level. As a result, there is little energy left to support a significantly large population of consumers past the tertiary level.

Only a fraction of the energy at each level in the producer → consumer → decomposer pathway is passed on to the next level.

The Final Users of Energy: The Detritus Feeders Because herbivores and carnivores are rarely 100 percent effective in eating the level just before them in the energy chain, many plants, animals, and microorganisms simply die to add their energy, in the form of complex organic molecules, to the detritus of the community. Detritus, which is still rich in stored energy, includes the dead bodies or body parts of animals and the dead leaves, stems, and roots of plants. Detritus collects at the ground level in terrestrial communities and on the bottom in aquatic communities. The organisms feeding on it, such as bacteria, fungi, protozoa, termites and other insects, mites, millipedes, and small aquatic and terrestrial worms, are called **detritus feeders**.

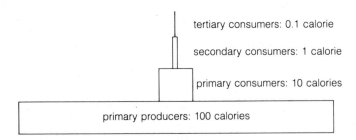

Figure 23-9 Average energy conserved at successive levels in the energy pathways of a community.

At this level (see Figure 23-5) most of the remaining energy of the organic matter is used for respiration and lost in the form of heat. Some decaying organic matter is also simply oxidized by the molecular oxygen of the atmosphere—either slowly, over the ages, or rapidly, as in the case of fire sweeping through the litter of a forest floor. Eventually, most of the chemical energy of the organic molecules remaining in the community as detritus is converted to heat by one of these two routes.

Depending on the community, more or less of the energy captured by the primary producers may go directly to the detritus feeders. In some terrestrial environments, as in a mature forest (Figure 23-10), comparatively little of the vegetation, only 5 to 7 percent, is eaten by herbivores. The remaining 93 to 95 percent falls to the forest floor and is consumed by detritus feeders or fire. The portion of primary productivity adding to the detritus in other terrestrial communities is similar. Marine herbivores (Figure 23-11) are somewhat more efficient and harvest about 40 percent of the primary productivity; the remaining 60 percent of the stored products of the primary producers falls to the bottom to be consumed by detritus feeders.

Trophic Levels, Food Chains, and Food Webs The various levels of producers and consumers in a community are identified as **trophic levels** (*trophikos* = nourishing). The primary producers form one trophic level, the primary consumers the next, and so on. Eventually the detritus feeders form the last trophic level.

A more or less straight-line pathway of energy flow through a series of trophic levels—as in the plant → herbivore → carnivore pathway shown in Figure 23-10—is called a **food chain**. Energy flow through the trophic levels of natural communities rarely follows such direct pathways, however. In nature, the path of energy flow is complicated by the fact that many organisms feed at more than one trophic level. Foxes and snakes, for example, may feed on fruits as well as small rodents and thus act as herbivores as well as carnivores. Some animals may even act as a herbivore, carnivore, and detritus feeder. (Have you ever eaten a bacon, lettuce, and tomato sandwich with cheese?) For this reason, plotting out energy flow in natural ecosystems often reveals a complex, interconnected pattern that ecologists call a **food web** (Figure 23-12).

The flow of energy in most communities follows a complex branched pathway called a food web.

Energy Flow at Silver Springs A classic study of energy flow in natural communities was made in 1957 at Silver Springs, Florida, by Howard T. Odum, then at the University of North Carolina. Odum measured the total solar energy falling on the community, the amount absorbed and converted to chemical energy by aquatic plants, and the pathways followed by the captured energy as it flowed through consumers and detritus feeders (Figure 23-13). Odum also sampled and calculated the total dry weight, called the **biomass**, of the organisms existing at each trophic level. He found that the plants forming the primary producers of the community were able to convert 1.2 percent of the total light falling on the community into the chemical energy of organic compounds. Respiration in the primary producers tapped off 58 percent of this captured energy, leaving less than half as net primary production. The amount remaining as net primary production thus represented only 0.5 percent of the total solar energy falling on the community. At the next trophic level, the herbivores of the community took up 38 percent of the energy stored in the primary producers. Of this energy, 56 percent was tapped off in respiration to power the activities of the herbivores, leaving 44 percent as energy potentially available to secondary consumers. Because of losses to respiration, the energy conserved by the herbivores represents only 0.08 percent of the total solar energy falling on the community.

At the next level, secondary consumers took up 26 percent of the energy stored in the herbivores. Of this energy, about 82 percent was used to meet the respiratory needs of the secondary consumers. As a result, only 18 percent of the energy taken in by the secondary consumers remained available for tertiary consumers—a figure amounting to only 0.004 percent of the total light energy entering the community. The final level of tertiary consumers conserved energy at about the same efficiency levels; they managed to store only 0.00035 percent of the total solar energy entering the community (see Figure 23-14).

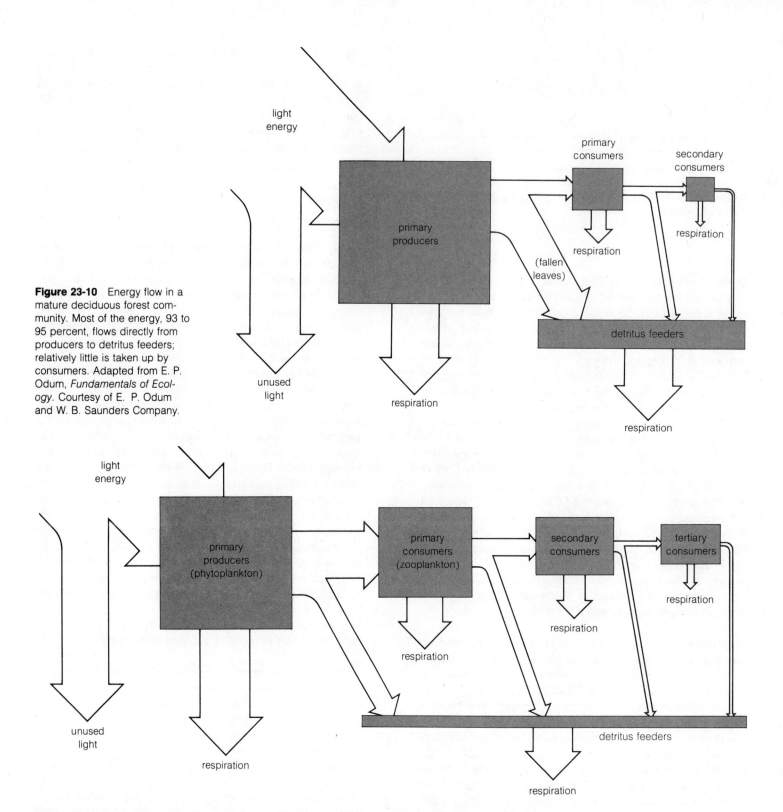

Figure 23-10 Energy flow in a mature deciduous forest community. Most of the energy, 93 to 95 percent, flows directly from producers to detritus feeders; relatively little is taken up by consumers. Adapted from E. P. Odum, *Fundamentals of Ecology.* Courtesy of E. P. Odum and W. B. Saunders Company.

Figure 23-11 Energy flow in a marine community. More of the energy flows through producers to consumers than in forests and relatively less to detritus feeders. Adapted from E. P. Odum, *Fundamentals of Ecology.* Courtesy of E. P. Odum and W. B. Saunders Company.

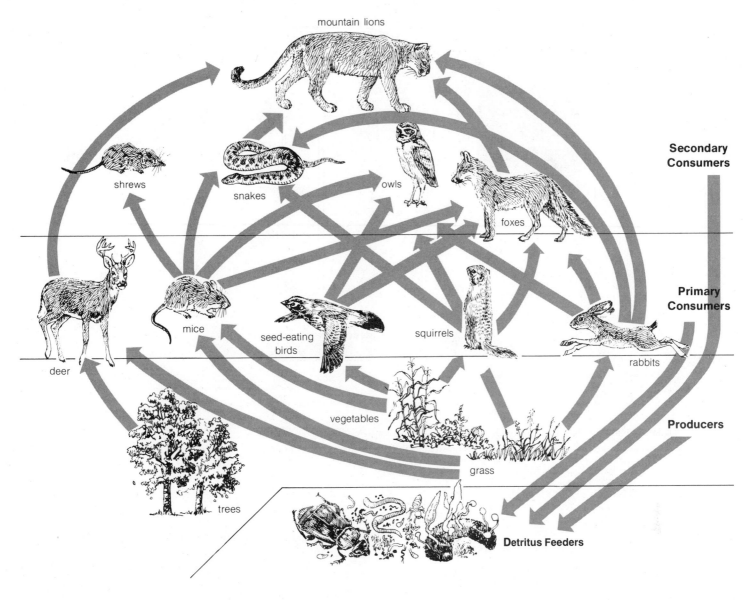

Figure 23-12 A simplified terrestrial food web.

At each trophic level, the uneaten organic matter entered the detritus pathway. For the Silver Springs community, the total energy entering this pathway amounted to 57 percent of the net primary production. Most of the energy taken in by the detritus feeders, about 91 percent, was used in respiration at this trophic level. Another 28 percent of the energy captured in net primary production simply washed downstream to be added to the detritus of other communities. (A small amount of energy also entered the Silver Springs community by the same route:

through leaf litter falling into the spring from surrounding terrestrial plants; see Figure 23-13.)

The food chains and webs observed at Silver Springs illustrate the basic pattern of energy flow in communities. The energy flows in one direction only. In this case it enters as sunlight and exits as heat lost in respiration. At each successive level, the amount remaining available to the community as useful energy becomes smaller and smaller, until at the last level in the producer → consumer pathway only a small fraction is conserved. Eventually

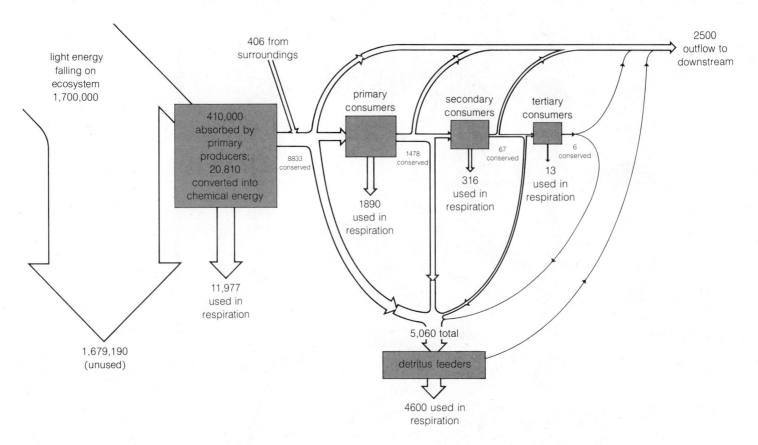

Figure 23-13 Energy flow in the Silver Springs community as plotted by H. T. Odum (see text). Energy totals in each pathway are given in kilocalories per square meter per year (kcal/m²/yr). Redrawn from H. T. Odum, *Ecological Monographs* 27:55–112 (1957). Courtesy of H. T. Odum. Copyright 1957 by the Ecological Society of America.

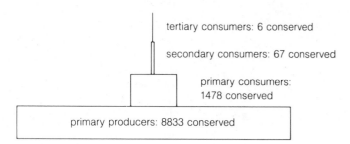

Figure 23-14 An energy pyramid for the Silver Springs community showing the relative amount of energy (in kcal/m²/yr) conserved at each trophic level.

even this fraction is converted to heat and lost to the community through the activities of the detritus feeders. The reduced energy available at each successive level in the producer–consumer food chain is reflected in the amount of biomass supported at these levels, which becomes progressively smaller. (Note the width of the boxes representing the biomass of each of these trophic levels in Figure 23-14.)

It is obvious from this study that the energy potentially available to organisms in a food chain depends on how close their trophic level is to the primary producers.

The greatest efficiency in use of the solar energy captured by a community is attained by consumption of the primary producers.

At the herbivore level, losses to respiration have occurred only in the primary producers, and comparatively large amounts of energy are still conserved in the form of or-

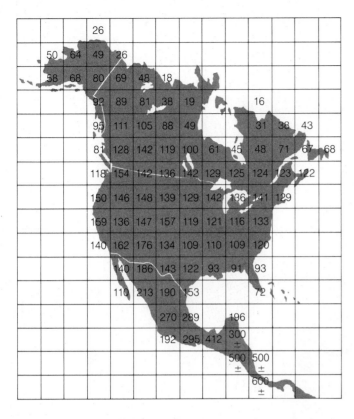

Figure 23-15 The numbers of bird species breeding in areas of equivalent size in North and Central America. Note the marked increase in species from north to south. From data presented in R. H. MacArthur, *Biology Journal of the Linnean Society* 1:19 (1969). Copyright Linnean Society of London.

Community Diversity

The energy available to communities supports the food chains and webs linking living organisms into community structure. One of the most perplexing questions about this structure concerns **community diversity**—that is, the number of different populations living in the community. Some communities are highly diverse, with many different kinds of species living at each trophic level; in others, there are relatively few species in the entire community. What factors control these differences? Community diversity rests on more complex causes than energy availability alone, because some communities that are highly efficient in primary and secondary productivity, such as freshwater lake and salt marsh communities, are often low in species diversity.

Species diversity in many communities varies in regular patterns with changes in temperature, moisture, elevation, and topography. Communities living in warm, moist, tropical climates near the equator are much more diverse than communities in the cold and dry climates toward the poles. This difference, easily the most striking noted in community diversity, occurs in plants as well as animals. Four-acre plots in temperate forests in the United States, for example, typically contain only 10 to 15 different tree species whereas equivalent plots in the tropical rain forests of Malaya may contain more than 200. Animals such as birds, mammals, and insects show similar increases in species diversity from the poles to the tropics (Figure 23-15). Diversity increases also with greater topographical relief: communities in mountainous regions such as the Rockies and the Appalachians typically contain a greater diversity of plant and animal species than flatter regions (Figure 23-16). Other more or less regular gradients in species diversity are also noted. Diversity generally increases from the source to the mouth of rivers and streams, for example, and increases also with the area of continents, islands, and peninsulas.

Community diversity frequently increases with increased temperature, moisture, topographical relief, elevation, and area of landmasses, peninsulas, and islands.

Some of these factors are more easily related to species diversity than others. Topographical relief obviously provides variety in the physical environment and affords greater opportunities for niche specialization. Similarly, large landmasses would be expected to provide more habitat complexity and increased opportunities for the establishment of specialized niches in food chains and webs. The relationship of temperature and moisture to diversity

ganic molecules. By the time energy reaches secondary or tertiary consumers, much of it has been lost through respiration and inefficiency in feeding at intervening levels. This fact has special significance for our own species, since we can enter communities at different trophic levels. In countries where vegetable matter forms the major part of the diet, the energy captured by plants is used with much greater efficiency since less is lost to respiration by intervening levels of consumers. In our own American community, by eating beef, pork, and lamb as diet staples we act as secondary rather than primary consumers. Thus our efficiency in absorbing the energy captured by primary producers is greatly reduced. From this pattern we can predict that, as the world's population increases and food energy becomes more scarce and expensive, more Americans will be forced to become accustomed to a meatless diet as primary rather than secondary or tertiary consumers in the world community.

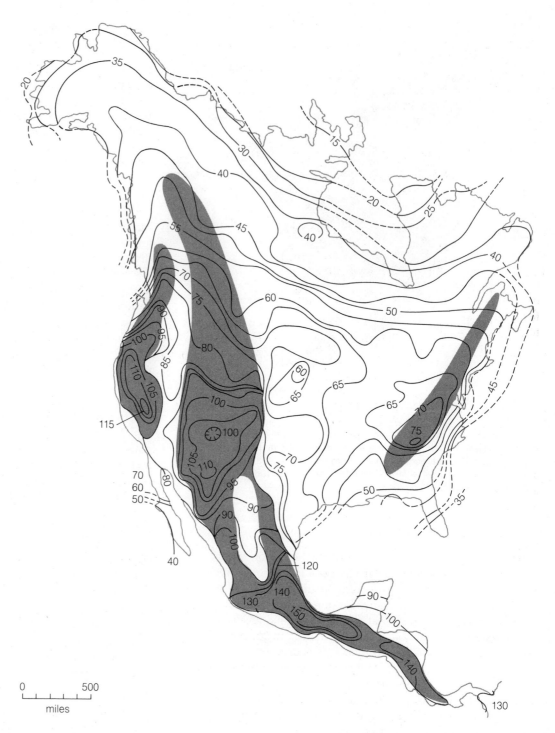

Figure 23-16 The number of mammalian species occurring in North and Central America. Note the increase in species in mountainous regions (shaded). From an original courtesy of G. G. Simpson.

0   500

miles

is less clear, however. Although increased temperature and moisture levels are expected to favor the overall growth of both plants and animals, at least up to a point, there is no obvious reason why increases in these factors should lead to a greater number of species instead of more individuals or larger individuals of a few species.

The difficulty in relating species diversity directly to increased temperature and moisture has led to a search for other factors that might be responsible for the greater number of species noted in some regions, particularly in the tropics. One idea is that predation and competition are more intense in the tropics, leading to greater niche specialization and, with it, more species. A number of experiments support the notion that predation, at least, can lead to greater species diversity. In one of these experiments, Robert Paine of the University of Washington removed all individuals of a starfish (*Pisaster*) population that acted as major predators on mussels, barnacles, and limpets along a rocky coast in Washington. After removal of the predator, diversity in the rocky coastal community fell from 15 species to 8. Without the predator, a mussel (*Mytilus*) gradually increased in numbers and crowded many of the original species from rocks in the community.

Related to the hypothesis that predation and competition increase diversity is another idea originally advanced by Alfred Russel Wallace, a founder with Charles Darwin of the theory of evolution. According to this idea, the tropics are more diverse in species composition because they have existed longer than temperate regions without serious environmental disturbances. As a result, the selective forces of competition and predation have had more time to produce greater niche specialization and niche spaces for more species. Temperate regions, in contrast, have been devastated in recent evolutionary time by the glaciers that advanced over the continents from the north and eliminated all life.

The idea that evolutionary time itself promotes species diversity is sometimes called the *historical* hypothesis. Another hypothesis frequently advanced in opposition, called the *equilibrium* hypothesis, has emerged from the study of speciation on islands. According to this concept, advanced by Robert H. MacArthur of Princeton University and Edward O. Wilson of Harvard University, evolutionary time is relatively unimportant. Species diversity on islands is determined instead by an interplay between two major factors: (1) the rate of immigration of new species from the mainland or other sources and (2) the rate of extinction on the island. These rates in turn are affected by two other elements: distance and area. The distance separating the island from the mainland or source of new species affects the rate of immigration, since migrating species have a greater chance of reaching islands that are close to the mainland. The area of the island too affects the rate of extinction, since the small populations maintained on smaller islands will undergo greater fluctuations in numbers due to chance factors such as genetic drift and have a greater probability of becoming extinct. According to MacArthur and Wilson each island quickly reaches a level of species diversity representing a balance between the rate of immigration and the rate of extinction. Species diversity increases as island area increases and distance from the mainland decreases. Supporters of the equilibrium hypothesis claim that it applies to all communities. In their

view, every community can be considered as an island separated by greater or lesser distances from other communities that act as the source of migrating species.

The equilibrium hypothesis has been supported by experiments carried out by Wilson and by Daniel S. Simberloff of Florida State University in which all animal species on small islands off the Florida Keys were eliminated by fumigation. Within a few years, however, recolonization from the mainland or other islands returned the islands to their original levels of species diversity. Moreover, the maximum level of diversity on each island was related to island size and distance from colonization sources as proposed in the hypothesis. The findings thus indicate that, on these small islands at least, diversity depends on relative rates of immigration and extinction rather than evolutionary time.

Whether the hypothesis applies as well to the larger and more complex communities of mainland regions is a subject of considerable controversy among ecologists at the present time. It seems likely that all the factors proposed as sources of diversity—topographical relief, the area of islands, peninsulas, and continents, temperature, moisture, predation, competition, time, and the equilibrium between immigration and extinction—probably contribute in different degrees to the number of species populations living in each community. Establishing which factor is most influential in determining the diversity of given communities, particularly in large, complex communities such as the tropical rain forests, presents one of the most challenging problems facing ecologists.

SUCCESSION, CLIMAX, AND STABILITY

The various terrestrial and aquatic communities of the world exist in a constant state of change. Not only does the number of individuals in community populations rise and fall and fluctuate in density but populations themselves come and go as some become established through immigration and others become extinct. These changes may be rapid enough to detect on a yearly, seasonal, or even daily basis, or they may proceed so slowly that significant alterations in populations may require hundreds or thousands of years. The relative rate of change in the size of the populations of a community, and in the types of populations present, is a measure of **community stability**.

Aside from regular fluctuations in population numbers resulting from seasonal changes in climate, the most striking alterations in communities occur in regions where the environment has been disturbed by human intervention or natural disasters. In such regions, **succession** occurs as populations of plants, animals, fungi, and other forms replace each other in sequence. Succession also takes place more or less rapidly as pioneer species establish the first communities on previously barren regions such as sand dunes, lava flows, land newly exposed by the retreat of a glacier, or a recently formed body of water.

In many parts of the world succession may lead to stable combinations of populations, the **climax communities,** that are characteristic of the region. If left undisturbed, land in the plains regions of the central United States tends to form grasslands; shrubs and trees in the plains are limited to moist pockets and belts along streams and rivers. Other types of climax communities appear characteristically on undisturbed land in other regions, such as the tundra of the far north or the deciduous forests of the northeastern United States. As climax communities become established, changes in population diversity and types gradually slow so that significant alterations take place only over hundreds or thousands of years.

Community Succession

Succession takes place whenever the organisms of a community cause extensive changes in the environment through their presence and growth. As the populations alter the environment, conditions favorable to the growth of additional organisms gradually arise. Depending on such factors as the chance combinations of new species that colonize the region, their relative ability to compete, and the effects of predation, the early populations are then replaced by others. This environmental change and population replacement continues until a combination of species appears that grows without extensively altering the environment. At this point a relatively stable community, the climax community, is established. Although some of the species populations may overlap and appear in successive stages, usually none persist throughout an entire succession. Thus a climax community is usually different in species makeup from the pioneer community that first colonizes a region.

Community succession takes place when the organisms in an ecosystem cause extensive changes in the environment through their presence and growth.

Although there are differences and frequent exceptions in the details of succession, the process of environmental modification and replacement of one species by another often has common elements in different regions. In most successions, net primary production gradually in-

creases and the plants of the community increase in size and numbers (biomass). The increase in biomass is reflected in the detritus level of the community: it too gradually rises and supports increasing numbers of detritus feeders as succession proceeds. The activities of the living organisms lead to a progressive development of the soil, which typically increases in depth, organic content, and stratification into different horizons (see Figure 22-15). In many successions, both the horizontal and the vertical structure of the community increase in complexity as larger plants become established, producing a greater diversity of microclimates within the community. The increased structural complexity of the community forms a variety of smaller habitats that provides diverse niches for additional plants and animals. As a result, climax communities frequently include a greater variety of species than the pioneer communities first inhabiting a region. (These and other common elements in successions are summarized in Table 23-2.)

Successions occurring in regions previously unoccupied by living organisms, such as bare rock surfaces, are called **primary successions**. **Secondary successions** take place where an established community has been destroyed by natural disasters or human intervention. Since the soils necessary for the growth of larger and more complex plant species take many years to develop, primary successions proceed relatively slowly. Secondary successions, in contrast, usually run their course more rapidly because the soil is already established.

Primary Succession: Some Examples

Forest Succession on Isle Royale During the early part of this century, William S. Cooper of the University of Minnesota traced out the probable succession of communities leading from bare rock to the climax community on Isle Royale in northern Lake Superior (Figure 23-17). According to Cooper's analysis, exposed rocks on the island first become inhabited by lichens and mosses, small pioneer producers that can survive in harsh conditions. The pioneers gradually fill in the rock surface, depositing organic matter as they grow. Eventually, in a process that may require many years, enough organic matter and trapped soil collects on the rocks to allow rooting of small herbs and shrubs, such as cinquefoil, bluebell, yarrow, blueberry, bearberry, and juniper. The continued growth of these plants, particularly the shrubs, produces a low, compact layer of vegetation called the *heath mat*.

Soon after the heath mat has developed, the first forest trees take root. Important among these early trees on Isle Royale are jack pine, black spruce, and occasionally aspen.

Table 23-2   Trends Frequently Observed in Successions

Community Characteristic	Trend Toward Climax
Net productivity	Increases
Total organic matter (biomass)	Increases
Soil depth and stratification	Increases
Horizontal and vertical structure	Becomes more complex
Niches	Greater specialization
Energy pathways	From simple food chains to complex food webs
Detritus feeders	Greater importance
Organism size	Greater
Life cycles	More complex
Selective forces	From r selection to K selection

Courtesy of E. P. Odum. Adapted from E. P. Odum, *Science 164*:262(1969); copyright 1969 by the American Association for the Advancement of Science.

These trees gradually increase in height and density until a more or less complete cover is formed. During the process, the smaller plants of the heath mat are pried loose or crowded out by the growing forest. The jack pine/black spruce forest creates the conditions of shade and humidity required for sprouting of the climax community of trees, a mixture dominated by balsam fir, paper birch, and white spruce. As these trees grow up through the canopy of pine and spruce, they eventually crowd them out to establish the climax community.

Each group of plants in the succession worked out by Cooper is accompanied by a characteristic group of animal consumers. Although these animals were not cataloged by Cooper, a typical succession in the northern region studied would probably include small insects and spiders as the primary and secondary consumers among the first animal inhabitants. Later, as the heath mat formed, larger consumers such as mice and a variety of birds would appear. As the pine/spruce forest invaded the heath mat, the number and kinds of birds, mammals, insects, and other species would increase and begin to include larger species such as hawks, owls, and, depending on the location, rabbits, deer, lynx, and bobcats. On present-day Isle Royale, the largest animals living in or near the climax forest community include moose and wolves in a balanced predator–prey system.

Small Lakes in Forested Regions   Succession in small lakes (Figure 23-18) usually begins with the microscopic phytoplankton floating near the surface of the lake. As these small photosynthesizers become more abundant and primary production increases, the first consumers of the zooplankton became established. Detritus from the dead bodies of these organisms gradually sinks and accumulates at the bottom of the lake, mixing with sediment and silt washing in from surrounding land areas. As this material continues to accumulate on the lake bottom, it gradually raises the bottom until the water is shallow enough to permit growth of larger submerged plants such as musk grass and floating plants such as water lilies. Colonization by these plants first takes place along the shore, where the water is shallow. As filling of the bottom continues, growth extends toward the center of the lake and more of the surface becomes occupied by floating plants. Growth of these plants and the continued accumulation of silt and dead organic matter fills in the bottom until the water becomes so shallow that plants which root on the bottom but extend above the surface, including rushes, sedges, and cattails, become abundant. Like the completely aquatic species, these plants grow first near the shoreline and gradually extend toward the center. As they become established and the bottom fills in, the lake gradually changes into a bog or swamp.

| exposed rocks | lichens and mosses | small herbs and shrubs | heath mat | jack pine, black spruce, and aspen | balsam fir, paper birch, and white spruce climax community |

Figure 23-17 A primary succession from a pioneer community toward forest on Isle Royale in northern Lake Superior (see text). Each community is named for the dominant plant species.

Continued accumulation of sediment and organic matter eventually raises the shorelines so much that land plants such as grasses, alders, and willow shrubs colonize these regions. The zone occupied by these plants then extends inward until the entire lake is filled in and completely occupied by land plants. Finally, the shrubs are replaced by trees characteristic of the surrounding forests, such as pines, maples, beeches, or elms, and all traces of the former lake disappear. (In drier regions, grasses or shrubs may persist as the final community.)

Secondary Succession In regions where an established community has been destroyed, succession progresses more rapidly because soils capable of supporting complex plant life are already in existence. Secondary successions of this type have been studied extensively in farmland that has been cleared, cultivated for a time, and then abandoned.

Secondary successions on abandoned farmland in Long Island, New York, studied by G. M. Woodwell and others at the Brookhaven National Laboratory, begin with crabgrass and other annual weeds (Figure 23-19). These colonizers are replaced after about 2 years by perennial grasses such as broomsedge, which form a wild meadow that persists from 15 to 20 years until a variety of shrubs, among them blueberries and huckleberries, supplant the grasses. The shrubs, in turn, are replaced after another 15 to 20 years by pines that take root and grow up through the shrubs. Oaks then colonize the region, forming a mixed oak/pine stand that reaches full development within 50 years after abandonment of the fields. Over the years, scarlet and white oaks gradu-

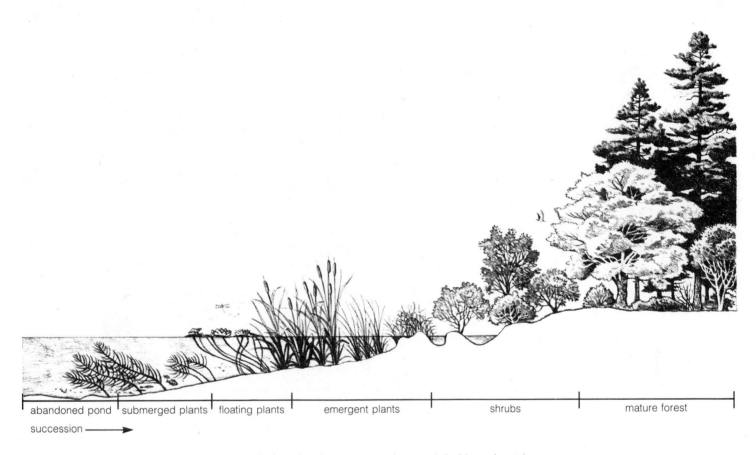

Figure 23-18 Primary succession in a shallow lake from the pioneer community toward deciduous forest (see text).

ally crowd out the pines, which cannot germinate and grow succesfully under the shade of the oak canopy. Eventually, after intervals amounting to perhaps 200 years, the forest becomes an almost pure stand of oaks. The outcome of secondary successions of this type in other forested regions of the United States, which may lead to evergreen, oak/hickory, beech/maple, or other combinations of trees in the mature community, depends on local moisture, temperature, soils, and topography.

The sequence of plants growing in secondary successions is accompanied by animals and other consumers that move in and occupy the microenvironments provided by the plants. As the environment becomes more complex and stratified through the growth of shrubs and trees, the complement of primary and secondary consumers also passes through a succession from annual weed through meadow, shrubland, and forest types. (Figure 23-20 shows the succession of bird species occurring during a transition from abandoned land to hardwood forest in the southeastern United States.)

Climax Communities and Community Stability

The mature communities of primary producers, consumers, and detritus feeders established by succession are relatively stable and undergo change only slowly, at rates measured in hundreds or thousands of years. The particular combinations of species populations of climax communities, as noted, depend on local conditions of moisture, temperature, soils, and topography. Although local conditions produce variations in the species populations present, the final

---

The combination of species occurring in a stable climax community depends on local conditions of climate, soils, and topography.

---

overall type of climax community can be predicted within broad limits. That is, grasslands of various kinds can be expected as the typical climax community in arid regions of the central plains of the United States; deciduous or coniferous forest can be expected in eastern regions depending on the elevation. These major climax communities typical of

canopy

lower
canopy trees

understory
trees

tall shrub
understory

low shrub
ground layer

annual
weeds

perennial weeds
and grasses

shrubs

young pine forest

mature oak forest

succession ⟶

Figure 23-19 Secondary succession in abandoned farmland from grasses and annual weeds toward deciduous forest (see text).

broad regions of the continents are called **biomes**. (Supplement 23-1 outlines the species composition of the world's
major biomes.)

What makes climax communities stable? How do they
persist and compensate for environmental disturbances? Ap-
preciating the magnitude of these questions requires an understanding of stability in terms of the individuals,
populations, and energy balance of a community. The overall structure of a stable community, in terms of the species
populations present, remains essentially the same from one
year to the next. Although daily or seasonal fluctuations
may occur in the numbers of individuals in the different
species populations of a stable community, these average out
over the years to a constant level. Thus, for these populations, the rates of births and deaths of individuals average
out at equivalent levels over long periods of time: individuals are born and die, but the community remains the same.
The total production of biomass in the community, as a consequence, is also stable. This means that, averaged over
many years, energy gains due to primary production are

balanced by losses due to respiration in consumers and detritus feeders.

**Stable communities retain the same average structure, species di-
versity, and biomass for long periods of time.**

Although several hypotheses have been advanced to
account for community stability, none has proved adequate.
One idea, proposed in its most complete form by Robert
MacArthur, is that community stability depends on species
diversity. The more complex a community is in terms of the
number of different species present, the more stable it is.
The major factor underlying stability in this model is the
complexity of trophic levels and food webs, which allows
many opportunities for checks and balances to compensate
for environmental disturbances. When one source of nour-
ishment becomes temporarily short in supply in a complex,
highly interlinked food web, consumers can survive by
switching to something else.

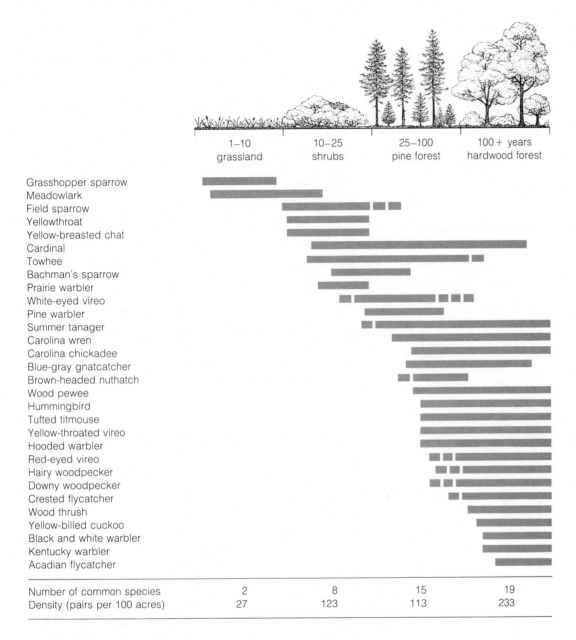

	1–10 grassland	10–25 shrubs	25–100 pine forest	100 + years hardwood forest
Grasshopper sparrow				
Meadowlark				
Field sparrow				
Yellowthroat				
Yellow-breasted chat				
Cardinal				
Towhee				
Bachman's sparrow				
Prairie warbler				
White-eyed vireo				
Pine warbler				
Summer tanager				
Carolina wren				
Carolina chickadee				
Blue-gray gnatcatcher				
Brown-headed nuthatch				
Wood pewee				
Hummingbird				
Tufted titmouse				
Yellow-throated vireo				
Hooded warbler				
Red-eyed vireo				
Hairy woodpecker				
Downy woodpecker				
Crested flycatcher				
Wood thrush				
Yellow-billed cuckoo				
Black and white warbler				
Kentucky warbler				
Acadian flycatcher				
Number of common species	2	8	15	19
Density (pairs per 100 acres)	27	123	113	233

Figure 23-20 The bird species occurring in a secondary succession in abandoned farmland in the southeastern United States. Note that the number or diversity of the bird species increases toward the climax community. Adapted from E. P. Odum, *Fundamentals of Ecology*. Courtesy of E. P. Odum and W. B. Saunders Company.

The diversity model for community stability was developed partly from observations of laboratory cultures. In these studies simple communities containing only two or three species (as in Gause's *Paramecium* experiments outlined in Chapter 22) proved to be highly unstable. Adding species to form a more complex community would sometimes produce a stable culture. Moreover, outbreaks of pest species such as insects appeared to be more common in temperate forest communities, which have relatively few species, than in the highly complex tropical rain forests. Reducing complexity by removing predators has also led to community instability with wide fluctuations of species populations. Removal of predators in the Kaibab Plateau of Arizona, for example, led to instability in the deer population, which subsequently grew to large numbers and then crashed (see p. 510).

Other observations, however, contradict the diversity hypothesis. One study of moth and butterfly populations in Canadian forests showed that populations of these species were in fact more unstable in communities with complex food webs than in more simple communities. The insect and other animal populations of cultivated fields, which are simple communities reduced to a single primary producer, are often noted to be more stable than expected from the diversity hypothesis. Moreover, stability decreases with increasing diversity in computer models that simulate natural communities.

The inadequacies of the diversity hypothesis revealed by such observations have led to the advancement of other ideas to explain community stability. Some ecologists argue that community stability depends on the stability of the physical environment. (Stable environments produce stable communities.) Others insist that stability depends on the degree to which populations evolve adaptations that allow them to coexist, such as niche specializations reducing competition and beneficial symbiotic relationships. Neither of these hypotheses is widely accepted or completely supported by evidence, however, and it is clear that none of the models provide the complete answers to the questions concerning stability. As in the case of community diversity, it seems most likely that community stability will be found to depend on a multitude of factors: species diversity, environmental stability, evolutionary adaptation, and others yet to be discovered. Certainly the factors underlying community structure, succession, diversity, and stability are well worth investigating on practical as well as scientific grounds. A complete understanding of community ecology would not only help us to evaluate the impact of human disturbances on communities but would also suggest means to allow the coexistence of human and natural communities throughout the world.

Questions

1. What is an ecological community? What characteristics of communities are studied in ecology?

2. What is community structure? Define patchy and continuous distributions of populations in communities. How is a forest community structured vertically?

3. What are primary producers? What is the difference between gross and net primary production?

4. What are primary consumers? Secondary consumers? Detritus feeders? What is biomass?

5. What happens to the energy "lost" at each trophic level of a community? How much energy is lost on the average at each trophic level?

6. Define a food chain and a food web. Why do food chains rarely extend beyond three or four trophic levels?

7. What is community diversity? What regular patterns are observed in community diversity in relation to latitude, topography, temperature and moisture, and island area?

8. Explain the *historical* and *equilibrium* hypotheses for community diversity. What evidence supports the equilibrium hypothesis?

9. What is community succession? List the processes frequently observed as common elements in successions in different regions.

10. What is the difference between primary and secondary succession? Which usually proceeds more rapidly? Why?

11. Trace the overall pattern of succession expected in a small lake. Trace the succession expected in an abandoned field surrounded by deciduous forest.

12. What is a climax community? What is community stability? What factors are believed to underlie community stability?

Suggestions for Further Reading

Barbour, M. G., J. H. Burk, and J. D. Pitts. 1980. *Terrestrial Plant Ecology.* Benjamin/Cummings, Menlo Park, California.

Krebs, C. J. 1978. *Ecology: The Experimental Analysis of Distribution and Abundance.* Harper & Row, New York.

McNaughton, S. J., and L. L. Wolf. 1979. *General Ecology.* 2nd ed. Holt, Rinehart & Winston, New York.

Pianka, E. R. 1978. *Evolutionary Ecology.* 2nd ed. Harper & Row, New York.

Ricklefs, R. E. 1979. *Ecology.* 2nd ed. Chiron Press, New York.

Whittaker, R. H. 1975. *Communities and Ecosystems.* 2nd ed. Macmillan, New York.

SUPPLEMENT 23-1: MAJOR BIOMES OF THE WORLD

The climax communities found in various regions of the world depend on local conditions of temperature, moisture, and topography. Since these environmental factors merge continuously, and the species occurring in different communities overlap broadly, there are usually no sharp boundaries separating one climax community from another. Even so, it is still possible to distinguish overall community types that reflect average conditions extending over large areas of the continents, such as the deciduous forests of the eastern United States or the grasslands of the central plains. While somewhat arbitrary, these major community types, called *biomes,* are useful in providing a broad characterization of the species occurring in large-scale habitats. In the broadest and most general classification, these world biomes include seven major terrestrial types: *tropical forest, temperate deciduous forest, coniferous forest, tundra, shrublands, grasslands,* and *desert.* There are five major aquatic biomes: *freshwater lakes and streams, ocean shoreline, coral reef, ocean shelf,* and *deep ocean.*

Terrestrial Biomes

Much variation occurs within the major terrestrial biomes—tropical forest, temperate deciduous forest, coniferous forest, tundra, shrubland, grassland, and desert—depending on local moisture, temperature, and topography. As a result, each of the major terrestrial biomes, if reviewed in greater detail than the map shown in Figure 23-21, would reveal many subdivisions with local communities diverging to a greater or lesser extent from the overall type represented by the major biome for the region. And, over much of the world, particularly in regions once occupied by grasslands and temperate deciduous forest, most of the natural climax communities have been replaced by farms, cities, and highways.

Tropical Forest   Tropical forest (Figure 23-22) occurs in humid regions near the equator where temperatures and rainfall are high throughout the year. The plant and animal life of these forests are the richest and most varied of any terrestrial biome on earth. Hundreds of trees, generally tall with evergreen leaves, form a dense canopy that spreads 25 to 45 meters or more above the forest floor. Since little light penetrates to the ground level, much of the life of tropical forests is concentrated in the canopy; relatively few plants grow at or near ground level. Long climbers called lianas send shoots from the forest floor into the canopy; other plants such as orchids, cactuses, and ferns grow entirely in the canopy without rooting in the soil. Since so few plants grow in the permanent twilight under the canopy, the forest floor is relatively open. The thick jungle growth usually associated with tropical forests is limited primarily to the edges of open areas such as streambanks and clearings.

Much of the animal life too is concentrated in the canopy, where both vertebrate and invertebrate species are abundant. Insects, including mosquitos, flies, and butterflies, occur in large numbers in the canopy. Other insects, including gigantic beetles and roaches, walk on the forest floor. Vertebrates include insect-eating and fruit-eating bats, rodents, many birds (among them colorful species such as parrots and toucans), and a variety of primates. In Africa and Asia, the tropical forest primates include monkeys, chimpanzees, gorillas, orangutans, gibbons, and langurs; in South America, monkeys, marmosets, and cap-uchins are common. Predator species are typically small animals such as ocelots and bush dogs that prey primarily on the rodents in the tropical forest biome.

In tropical regions with less rainfall, the evergreen tropical forest gives way to trees that lose their leaves in dry seasons or to shrublands or grasslands. The canopy of the tropical forests in these regions is more open than the evergreen tropical forests, and shrubs or grasses are frequently abundant on the forest floor.

Temperate Deciduous Forest   Temperate deciduous forest (Figure 23-23) once covered eastern North America, much of Europe from the Atlantic to the Ural Mountains of Russia, and large parts of China. The major characteristic of this biome is a rich variety of deciduous trees—including beeches, maples, birch, ash, and basswood in moist regions and oak and hickory where the forest is drier. Elm, cottonwood, willow, and sycamore trees are common along streams and rivers. Some evergreens also occur among the deciduous trees of this biome, particularly at higher elevations or as succession stages in burned areas. Frequently a lush understory of shrubs and herbs grows among the trees. Deciduous forest supports a variety of insects, small mammals such as shrews, mice, squirrels, chipmunks, and rabbits, and larger animals including foxes, wildcats, whitetailed deer, mountain lions, wolves, and black bears. Ruffed grouse, ovenbirds, and a variety of owls are but a few of the many bird species.

Coniferous Forest   Coniferous forest (Figure 23-24) grows in a broad band across North America and Eurasia between the tundra in the north and the deciduous forests or grasslands to the south. In North America, the trees living in the cool, moist climate of this biome include white and black spruce mixed with balsam fir trees in the north. Toward the southwestern regions of the coniferous forest, ponderosa pines, Douglas fir, white fir, and redwoods become dominant types. Near the Great Lakes, the spruce and fir of northern regions give way to pines and hemlock. Pines, particularly the jack pine, and deciduous birches occur in large numbers in successional communities growing in regions where the climax community has been destroyed by fire, a common occurrence in coniferous forests. Lichens, mosses, shrubs such as huckleberries, and small flowering plants grow under the trees. Animals of this biome include a variety of insects such as beetles, wasps, mosquitos, and other biting flies. A major pest species, the spruce budworm, often causes extensive damage to coniferous forests. Other animals include nuthatches, juncos, jays, ravens, and warblers as typical birds; mammals range from squirrels and other small rodents to wolves, lynx, deer, elk, and moose.

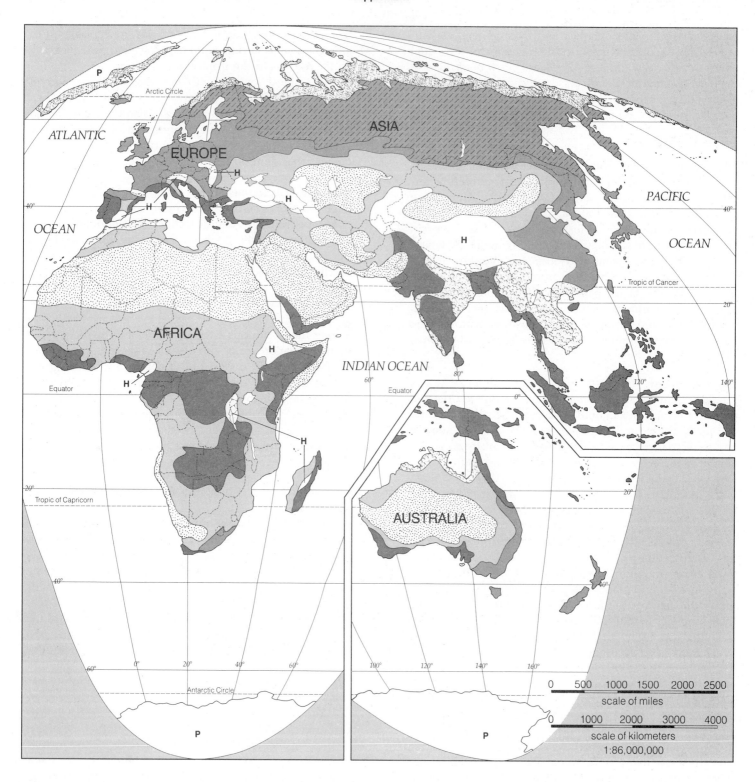

Figure 23-21 The major terrestrial biomes of the world.

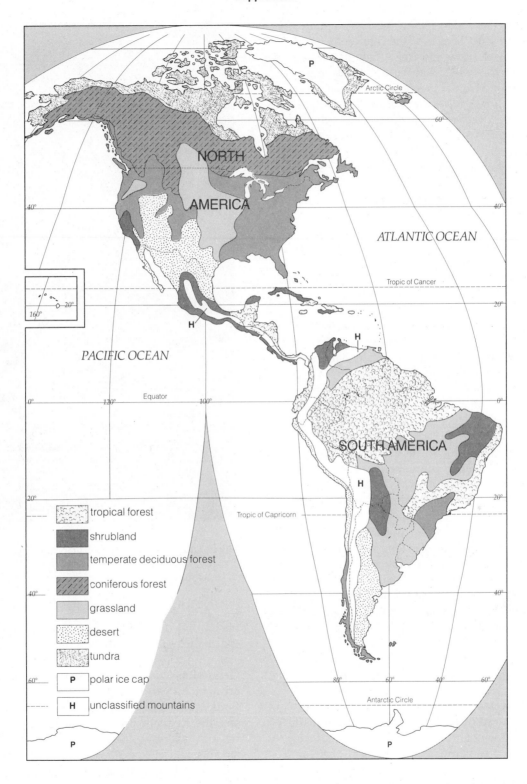

tropical forest

shrubland

temperate deciduous forest

coniferous forest

grassland

desert

tundra

P polar ice cap

H unclassified mountains

Tundra Tundra (Figure 23-25) fills in the high latitudes between the northernmost fringes of the coniferous forest and the arctic wastes of snow and ice. Although average precipitation is comparatively low at these altitudes, much soil moisture is retained by a permanently frozen layer called *permafrost* lying only a few feet below the surface. This frozen layer effectively seals off the topsoil and prevents downward percolation of water from the surface. Many areas, in fact, are boggy during the summer growing season, and the region is dotted with countless small lakes and ponds.

Vegetation in the tundra is limited by cold and the short growing season to a low growth of lichens, mosses, grasses, and sedges in boggy regions. In better-drained areas, other plants such as dwarf willows, huckleberries, bilberries, and flowering herbs grow in addition in the tundra. A similar plant community, called *alpine tundra*, occurs in more southerly latitudes at high elevations in the Rockies and Sierra Nevada. Vegetation in alpine tundra resembles tundra in better-drained regions of the far north, but there is a greater proportion of small shrubs and mountain wildflowers among the lichens, mosses, sedges, and grasses in alpine tundra.

Animals of the northern tundra include insects (mosquitos and other biting flies are common) and mammals ranging from small rodents such as voles and lemmings to arctic foxes, wolves, musk-oxen, caribou, and brown bears. Birds of the northern tundra include plovers, horned larks, and snowy owls among other species. Alpine tundra supports a wider variety of insects, birds, and mammals such as marmots, mountain goats and sheep, deer, and black or grizzly bears.

Shrublands Shrublands (Figure 23-26), including the *chaparral* of California coastal mountains and the Sierra foothills, grow in temperate or warm climates with a summer dry season. The growth includes small trees and shrubs such as live oak, manzanita, deerbrush, and buckbrush. Many of the plants typical of shrublands, particularly California chaparral, contain small, leathery leaves with high concentrations of aromatic oils that burn readily. As a result, shrublands are susceptible to fires that burn rapidly through the dense brush of this biome. Most shrubland plants are tolerant to damage by fire; the aboveground growth destroyed by flames is quickly regenerated by undamaged rootstock or from fire-resistant seeds that are actually stimulated to germinate by the heat of fires. Summer fires in California, where many homes have been built in the chaparral, cause extensive property damage each year. Animals of the chaparral include insects, lizards, snakes, small birds, and small mammals such as rab-

Figure 23-22 The tropical forest biome (in Peru). Courtesy of C. A. Toft.

bits, wood rats, and chipmunks. Mule deer migrate into the chaparral during the winter rainy seasons.

Grasslands Grasslands (Figure 23-27) occur over huge areas of the world in North and South America, eastern Europe and Asia, Africa, and Australia. Moderate rainfall and the frequent fires that sweep the grasslands restrict trees to sparsely scattered growths or to the edges of streams and rivers. Grasses ranging from low, clumped varieties a few centimeters in height to species taller than 1 meter are the predominant vegetative type. In North America, grasslands occur in most of the central United States between the Rocky Mountains and the Mississippi

Figure 23-23 Temperate deciduous forest in the Appalachians.

Figure 23-24 Coniferous forest in Idaho.

Figure 23-25 Arctic tundra. Gazing caribou are visible in the distance. Courtesy of the U.S. Department of the Interior, Bureau of Land Management, Alaska Office.

Figure 23-26 Shrublands (California chapparal).

Figure 23-27   Grasslands in the Western U.S.

Figure 23-28   Desert in Nevada.

River, eastward in Illinois and Indiana, and also in regions of California and Washington. Toward the eastern limits of the grasslands, where rainfall is greatest, taller grasses are most common; shorter varieties including buffalo grass dominate the drier western limits. Grasshoppers, ground squirrels, prairie dogs, gophers, jackrabbits, coyotes, a limited variety of birds, and large herbivores such as pronghorn antelope and bison are (or were) the conspicuous animals of the North American grassland biome. The soil of the grasslands, typically rich in organic matter, has provided some of the most productive farmland of the world, particularly when domestic grasses such as corn and wheat are planted as major crops.

Grasslands in the more tropical regions of Africa, Australia, and South America include widely spaced trees and are called *savannas*. The savannas of each continent contain characteristic collections of animals—such as the kangaroos of Australian savannas and the overwhelming variety of species living in the African savanna, including wildebeests, dik-diks, gazelles, kudus, springboks, sable antelopes, zebras, and impalas, all preyed upon by cheetahs, lions, leopards, hyenas, and jackals.

Desert Desert (Figure 23-28) occurs where precipitation is very low, less than about 20 centimeters a year. In North America deserts are limited primarily to the southwest and the high plateau between the Sierra Nevada and Rocky Mountains. The desert biome is characterized by sparsely distributed low shrubs and clumps of grass with much bare earth between them. In the higher, cooler deserts of the plateau between the Sierra and the Rockies sagebrush is the dominant plant. Toward the south, sagebrush gives way to creosote bush and cactus species of various kinds including prickly pear and saguaro. Characteristic desert animals include a rich variety of lizards and snakes, locusts and other insects, and arachnids such as scorpions and spiders. Mammals are often abundant, especially small rodents such as pocket mice, hamsters, and kangaroo rats. Bats and insectivorous birds such as cactus wrens and sage sparrows may also be present. Many of the desert animals survive the hot, dry climate by restricting their activities to nocturnal, early morning, or evening periods.

Aquatic Biomes

The species living in the major aquatic biomes of the world, including freshwater lakes and streams, ocean shoreline, and the marine biomes—coral reef, ocean shelf, and deep ocean—are as varied as those of terrestrial biomes. Variety of species types is especially marked in the coral reef biome, which in many ways is the "tropical forest" of aquatic communities. Like terrestrial communi-

ties, the density and variety of populations in aquatic biomes depends on several environmental factors, among them the amount of sunlight penetrating into the community, temperature, rate of water movement, water depth, salinity, and availability of inorganic nutrients. Within the major aquatic biomes local variations in these factors produce communities that differ to some degree from the major biome type for that region. Thus the aquatic biomes, like their terrestrial counterparts, appear as a mosaic of subtypes if local regions are examined in detail.

Freshwater Lakes and Streams The kinds of animals, plants, and other organisms occurring in freshwater lakes and streams are highly dependent on water depth, inorganic nutrients, temperature, and, in streams, the rate of water movement. Primary producers in lakes consist of floating, planktonic species (including green algae, diatoms, and blue-green algae) and rooted, aquatic plants that grow in shallow water. Plankton forms the major primary producer of deep lakes; in shallow lakes, bogs, and ponds, both plankton and rooted aquatic plants carry out primary production. Animals include microscopic crustaceans of the zooplankton near the surface and a variety of larger crustaceans, worms, and insect larvae that live on the bottom. The fish present in North American lakes reflect depth and water temperature: trout species are found in deep, cold lakes, smallmouth bass, perch, and pike species in warmer lakes of intermediate depth, and largemouth bass, catfish, and carp in warmer shallow lakes.

Floating plankton is sparse in freshwater streams; primary production is carried out instead by green and blue-green algae growing in a film on rocks in streams with rapidly moving water or by rooted aquatic plants in slower streams and pools. Much energy input also enters streams as detritus washed from the surrounding land. Animal life includes insect larvae, burrowing worms, clams and mussels, and crayfish on the bottom; a variety of fish including trout and smallmouth bass are present in faster, cooler waters and carp, largemouth bass, and catfish in slower streams and pools.

Ocean Shoreline Ocean shoreline (Figure 23-29) is subject to violent disturbances by wave action and, along the zone nearest the land, to alternate immersion and exposure to the air as waves advance and recede and the tide rises and falls. In the shallowest regions, primary producers are limited to microscopic green and blue-green algae that form a film on rocks and sand or silt particles. In deeper parts of shorelines that are almost continuously immersed, larger species of green, brown, and some red algae carry

Figure 23-29 California ocean shoreline in a rocky region at low tide. Surf clams are clustered in pockets on the rocks in the foreground; the more distant rocks are covered with brown algae. A pair of oyster catchers and a seagull are searching for food among the rocks.

out primary production. Animal life of the shorelines consists of a wide variety of snails, clams, limpets, barnacles, crabs, sea urchins, and sea stars on the rocks, sand, or mud of the bottom; protozoa, shrimp, and fish represent swimming species. Fish along shorelines are limited to a relatively few species of surf and rockfish.

Coral Reef Coral reef biomes (Figure 23-30) occur in tropical oceans as fringes or reefs surrounding islands or lying just offshore along continental coasts; they may also occur as *atolls*, ringlike reef formations that enclose a central shallow lagoon. Corals (see p. 402) are coelenterates that secrete shells of calcium carbonate. Individuals gradually build one shell on another to form the main structure of reefs and atolls. Primary production is supplied by photosynthetic dinoflagellates and green and red algae that grow within the coral structures and, moreover, by a variety of sea grasses that grow in pockets around the coral. The complex shapes taken by different corals provide habitats for a rich variety of animal life, including worms and molluscs of different kinds growing within the coral mass, sea anemones, sponges, and sea plumes attached to the coral, and crabs, snails, and echinoderms moving over the coral surfaces. The fish swimming around coral, many of them highly colorful, form one of the most diverse fish communities of aquatic habitats.

Ocean Shelf Ocean shelf fills in a transition region between the shoreline biome and the deepwater biome of the open ocean. Water is shallow enough in the continental shelves for light to penetrate to the bottom; in these regions, collections of green, brown, and red algae grow as primary producers in dense beds. Primary production is also carried out by planktonic dinoflagellates and diatoms near the surface. The algal beds on the bottom are often rich with animal life, including snails, clams, worms, crustaceans, and echinoderms that live on the bottom, and a variety of fish, including flounders, groupers, sea bass, cod, and herring, that swim between the bottom and the surface.

Deep Ocean Deep ocean occurs as the biome in regions extending from the outermost edges of the continental shelves to the deepest trenches of the ocean floor. Because

Figure 23-30 Coral reef near St. Croix Island. The tall, cylindrical outgrowths are pillar corals (*Dendrogyra cylindricus*). Photograph by S. K. Webster, Monterey Bay Aquarium and Biological Photo Service.

light does not penetrate to the bottom in the deep ocean biome, primary producers are limited to microscopic or near-microscopic green algae, diatoms, and dinoflagellates that float as part of the plankton near the surface. Mixed with the photosynthetic forms are microscopic herbivores, including protozoans such as radiolarians and foraminifera, that feed on the microscopic plants of the plankton. Detritus from the plankton sinks to the bottom, where a variety of animals live in the darkness and immense pressure of the deep ocean floor. Among these bottom-dwellers are sea whips, sea fans, sea lilies, brittle stars, clams, snails, and fish that often take bizarre forms. Other swimming animals, such as tuna, sailfish, sharks, and whales, range over the deep ocean biome. These animals feed near the surface as herbivores or carnivores in food chains of both the deep ocean and continental shelf communities.

When the study of organisms and their environment is expanded from the living community to include the ultimate sources and disposition of the physical factors necessary for life—energy and inorganic nutrients—the entire combination of organisms and physical factors is termed an **ecosystem**. Ecosystems, like populations and communities, have characteristics that are unique to their level of organization. Of primary interest are the flow of energy from the sun to living communities, the gradual conversion of energy from useful forms into heat that cannot be utilized by living organisms, and the cycling of elements such as carbon, nitrogen, and phosphorus between inorganic molecules of the environment and the organic molecules of life. Like the communities they contain, the boundaries of ecosystems under study may be as small as a leaf or as large as the entire earth. The earthly ecosystem, which comprises all the organisms of the entire world, the total energy flow of the earth, and the earthly cycle of inorganic nutrients, is called the **ecosphere** or **biosphere**.

The integration of physical and biological elements in ecosystems is described in this final chapter of the ecology unit. The chapter also summarizes the application of community and ecosystem ecology to the human population and outlines the major ecological problems that must be solved if our population is to survive.

24

ECOSYSTEMS AND HUMAN ECOLOGY

ECOSYSTEM ECOLOGY

The Energy Component of Ecosystems

The energy powering the activities of ecosystems flows in what is essentially a one-way direction. As chemical or

---

The energy powering the activities of ecosystems flows in a one-way direction.

---

biological change takes place anywhere in an ecosystem, some energy is always converted to heat and lost to the system. For this reason, the activities of ecosystems can continue only if an input of usable energy from outside is constantly available. The outside energy source for the largest ecosystem, the **ecosphere**, is the sun. Within the ecosphere, smaller ecosystems may depend either directly or indirectly on sunlight for their energy supply. Most ecosystems are directly dependent on sunlight and are based on green plants that absorb the energy of sunlight and convert it to chemical energy through photosynthesis.

The energy for almost all ecosystems originates in the sun. Solar energy is captured in ecosystems by photosynthesis in primary producers.

Almost all the ecosystems that do not depend directly on sunlight use organic matter built up in other sunlight-dependent ecosystems as an energy source. In a forest litter ecosystem, for example, the organic matter of fallen leaves is oxidized as an energy source. If traced back to its ultimate origins, however, the energy for this ecosystem too comes from the sun. The only exceptions to the general dependence on sunlight as the ultimate energy source are ecosystems based on bacteria that can obtain energy by breaking down inorganic substances such as H_2S. The energy these bacteria are able to obtain by processes of this type, called **chemosynthesis**, represents such a tiny fraction of the total earthly energy flow that it can essentially be ignored.

Energy Flow and the Thermodynamic Laws Energy flow in ecosystems obeys the same two basic thermodynamic laws governing energy changes in any physical or chemical system (see p. 36). According to the first law, energy (or "matter-energy") can neither be created nor destroyed. In an ecosystem, this means that the energy flowing into the ecosystem must eventually be released to its surroundings in exactly the same quantity. For the largest ecosystem, the ecosphere, the first law affirms that the energy flowing to the earth from the sun is ultimately released in equal quantities from the earth to outer space.

The second law of thermodynamics states that chemical or physical changes can take place only if the total entropy, or disorder, of a system increases. For ecosystems to run, this law requires that a price in the form of entropy must be paid for each change, activity, or reaction anywhere in the system. The entropy toll is eventually paid out as heat, which radiates from the ecosystem and is lost as a potential energy source. In terms that relate more directly to the activities of living organisms, this entropy change involves a conversion of radiant energy from the shorter wavelengths of visible light, which contain sufficient energy to drive the reactions of photosynthesis, to the longer wavelengths of infrared or heat radiation, which individually contain less energy and are generally unusable as a direct energy source for photosynthesis or other activities of life. To make up for energy lost to entropy, and to allow continued operation of ecosystems, energy must be continually supplied from outside. Payment of the entropy toll is the primary reason why energy flows in a one-way direction through ecosystems and does not cycle to be used again.

Energy Flow and the Sun The sun is a thermonuclear reactor that converts matter into energy. Inside the sun, pressures and temperatures are high enough to fuse hydrogen nuclei into heavier helium nuclei. As this fusion takes place, some of the mass of the hydrogen is converted into energy, which radiates outward from the sun in all directions:

> The sun liberates about 100,000,000,000,000,000,-000,000,000 (or 10^{26}) joules of energy per second. If we could completely harness this energy, each person on earth each second would have for his own personal use over 70,000 times the annual power consumption of the United States. The sun uses up about 4.2 million tons (4.1 billion kilograms) of its mass every second in producing this enormous amount of energy. We need not worry about the sun running out of fuel, however. In the normal life cycle of stars, the sun is entering middle age. It has probably been in existence for at least 6 billion years, and there is enough hydrogen left to keep it going for at least another 5 billion years.[1]

The energy released from the sun is emitted as radiant energy at wavelengths ranging from relatively low-energy radio waves through visible and ultraviolet light to very short, high-energy x rays and gamma rays. Of these wavelengths, only light in the visible wavelengths is used in significant quantities as an energy source by the earth's ecosystems.

Only one fifty-millionth of the total light energy radiated by the sun falls on the earth and its atmosphere. Moreover, of this fraction (Figure 24-1) about 30 percent is reflected back to space by clouds and dust particles in the atmosphere or bright regions of the earth's surface such as sand, snow, rocks, or bodies of water. Of the unreflected light, almost all is absorbed in the atmosphere, in bodies of water, or at the earth's surface. Only a tiny fraction, 0.0023 percent, is absorbed by plants and converted into the chemical energy of organic molecules.

Most of the ultraviolet light striking the atmosphere is absorbed by an **ozone** layer consisting of oxygen molecules containing three (O_3) rather than two (O_2) oxygen atoms. This absorption protects the earth's ecosystems because ultraviolet light, through its ability to induce mutations in nucleic acid coding molecules, is highly injurious to living organisms. There is now concern that the ozone layer is being destroyed through the effects of oxides of nitrogen (NO and NO_2) released in the exhaust of jet aircraft and

---

[1]G. T. Miller, Jr., *Living in the Environment*, 1st ed. (Belmont, California: Wadsworth, 1975).

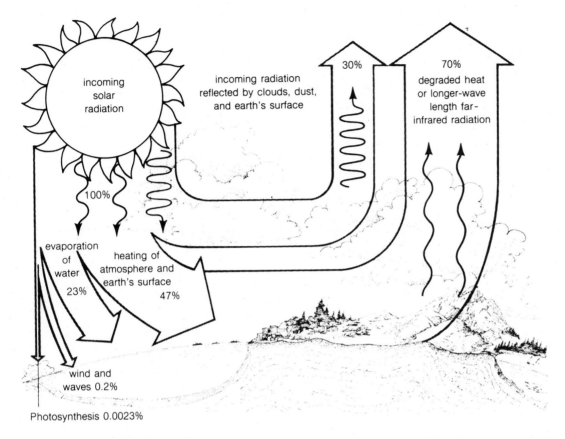

Figure 24-1 The pathways followed by solar energy striking the earth and its atmosphere. Only 0.0023 percent of this energy is absorbed in photosynthesis and converted to the chemical energy of organic molecules.

the fluorocarbon compounds used as propellants in aerosol cans and refrigerants. One of the many injurious effects likely to arise from this ozone depletion is an increase in the incidence of human skin cancer due to the effects of the greater quantities of ultraviolet light reaching the earth's surface.

Most of the light absorbed in the atmosphere, at the surface of the earth, and within bodies of water is converted directly into heat at this stage and becomes unavailable as a usable energy source for ecosystems. Even so, this heat energy contributes indirectly to ecosystems by creating the conditions of temperature necessary for life to exist. Moreover, solar heating of land and air masses causes evaporation of water and produces air currents that interact to create the humidity, cloud cover, and precipitation typical of different regions of the earth.

In contrast to energy, matter cycles in the world ecosystem.

Chemical Cycles of the Ecosphere

The organic molecules of ecosystems are built up from inorganic molecules such as water and carbon dioxide obtained from the environment. Within smaller ecosystems of the ecosphere these inorganic substances may either flow or cycle through the living community. In a river or stream ecosystem, for example, some inorganic matter may enter from upstream, some may remain to cycle between the living community and the atmosphere, water, and bottom sediment of the ecosystem, and some may be lost with the outflow. The matter of the largest ecosystem, however, the ecosphere, remains intact and cycles back and forth between the living community and the nonliving environment rather than flowing through as energy does. Thus, in the ecosphere as a whole, there is no net input of matter from outer space (except for the negligible quantities entering in meteorites) and almost no loss. Ecologists call these major cycles of the ecosphere **biogeochemical cycles** (*bio* = life; *geo* = earth).

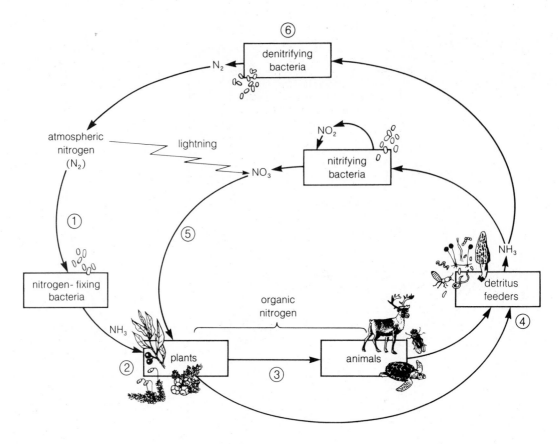

Figure 24-2 The nitrogen cycle (see text).

More than 40 chemical elements, some in pure form and some in combination with other elements, are known to be important in the major nutrient cycles of the ecosphere. Some of these elements, called **macronutrients,** cycle in relatively large quantities. These include the elements occurring in highest concentrations in protoplasm: carbon, oxygen, hydrogen, nitrogen, phosphorus, potassium, calcium, magnesium, and sulfur. Others, the **micronutrients,** shuttle in much smaller quantities or in trace amounts. Among these are the elements iron, sodium, chlorine, copper, zinc, boron, vanadium, and cobalt.

In all cases, these elements cycle between large natural reservoirs and their temporary residence within living organisms. There are three of these natural reservoirs: (1) the atmosphere surrounding the earth, (2) the bodies of water of the earth, and (3) the sedimentary rocks and soil of the earth's crust. Most of the gaseous macronutrients, including hydrogen, oxygen, and nitrogen, are pooled in both the atmosphere and in solution in bodies of water. Carbon, in the form of carbon dioxide, is pooled

in both of these natural reservoirs and in carbonate-containing sedimentary rocks as well. Other elements, including the macronutrients sulfur, potassium, phosphorus, calcium, and magnesium and all the micronutrients, are pooled in sedimentary rocks of the earth's crust, from which they are slowly released by weathering and erosion and to which they slowly return through the gradual deposition of new rock. Three of these biogeochemical cycles, the nitrogen, carbon, and phosphorus cycles, are outlined as examples in this section.

The atmosphere, bodies of water, and the sedimentary rocks of the earth form the natural reservoirs of the matter cycling in ecosystems.

The Nitrogen Cycle   Nitrogen is an essential element for all living organisms. Its primary chemical roles in protoplasm are in the amino ($-NH_3$) group of the amino acids

Figure 24-3 Nodules (arrows) containing nitrogen-fixing bacteria in the roots of a clover plant. Photograph by Carolina Biological Supply Company.

of proteins and in the purine and pyrimidine bases of nucleotides and nucleic acids. The major reservoir of this element, the atmosphere, contains nearly 80 percent nitrogen by volume. However, no plants or animals can use this gaseous nitrogen directly as a building block for their nucleic acids and proteins.

Nitrogen is made available to living organisms primarily through the activity of nitrogen-fixing bacteria.

Instead, nitrogen cycles from the atmosphere primarily through *fixation* by microorganisms (see step 1 of Figure 24-2). Among the most important nitrogen-fixers in terrestrial environments are almost all species of the bacterial genus *Rhizobium*. These bacteria live in the roots of plants including some 13,000 species of legumes, among them peas, beans, alfalfa, clover, and soybeans (Figure 24-3). *Rhizobium* carry out fixation by adding electrons and hydrogen ions to gaseous nitrogen to form ammonia (NH$_3$):

$$N_2 + 6H^+ + 12ATP \rightarrow 2NH_3 + 12ADP + 12PO_4^{2-} \quad (24\text{-}1)$$

The reaction, an "uphill" one, requires the breakdown of comparatively large quantities of ATP. In the host plants, the ammonia yielded in this reaction is used in step 2 of Figure 24-2 to produce amino acids, nucleotides, and eventually proteins and nucleic acids through a variety of biochemical reactions. Nitrogen fixation through the process shown in Reaction 24-1 is also carried out in both terrestrial and aquatic environments by free-living bacteria such as *Azotobacter* and *Clostridium* and by blue-green algae.

Animals obtain the nitrogen they need by eating plants or other animals and directly absorbing nitrogen-containing organic compounds from the digested plant or animal material (step 3 of Figure 24-2). No animals are hosts of symbiotic nitrogen-fixing bacteria, and few can use simple compounds such as nitrates (inorganic compounds containing the chemical group —NO$_3^-$) or ammonia in significant quantities as their nitrogen source. Both these substances are toxic to most animal life.

The body tissues of dead animals and plants and animal excretions serve as the source of nitrogen compounds for another part of the nitrogen cycle (step 4 of Figure 24-2). Complex nitrogen-containing organic compounds in the dead tissues and excretions are broken down to ammonia by detritus feeders. At this point, several kinds of soil bacteria then carry out two sequential steps that convert ammonia into nitrates. In the first step, ammonia is oxidized to nitrite (NO$_2^-$) by the removal of electrons and hydrogen and the addition of molecular oxygen:

$$2NH_4 + 3O_2 \rightarrow 2NO_2 + 2H_2O + 4H^+ + \text{energy} \quad (24\text{-}2)$$

This reaction, which yields usable free energy, is carried out almost exclusively by the soil bacteria of the genus *Nitrosomonas*. The inorganic nitrite produced, excreted as a waste material by *Nitrosomonas*, is then used as an energy source by another group of bacterial oxidizers, primarily *Nitrobacter*. The *Nitrobacter* oxidize nitrite to nitrate (—NO$_3^-$) by a second removal of electrons and addition of molecular oxygen:

$$2NO_2^- + O_2 \rightarrow 2NO_3^- + \text{energy} \quad (24\text{-}3)$$

This reaction also yields free energy for the bacteria. The net effect of the two reactions is the conversion of ammonia to nitrate. The two steps together are called **nitrification**.

Nitrification is an essential part of the nitrogen cycle. The inorganic nitrate compounds excreted by the soil bacteria are the primary source of nitrogen for plants that do not have symbiotic nitrogen-fixers in their roots (step 5 of Figure 24-2). Some nitrates, representing about 10 percent of the total atmospheric nitrogen converted into more complex compounds, are also produced by the action of lightning on the atmosphere during thunderstorms. Plants absorb the nitrates from the soil, convert them to ammonia, and use the ammonia to produce nucleic acids and proteins. These complex organic nitrogen compounds are then used in turn by animals feeding on plants.

Some of the ammonia released by detritus feeders is also used directly as a nitrogen source by plants. However, relatively little nitrogen makes its way to plants by this pathway (dotted lines in Figure 24-2) because much of the soil ammonia is converted into gaseous nitrogen by denitrifying bacteria (see below) or lost through runoff of groundwater.

The amount of nitrogen available as nitrate in soils is often a limiting factor for the growth of plants (other than the legumes that can use gaseous nitrogen fixed by their symbiotic *Rhizobium* bacteria). If soil nitrates are low, plant growth may be poor even though water and sunlight are available in abundance. This factor is so important that, in many parts of the world, low food supplies and famine result directly from poor crop yields caused by insufficient soil nitrates. In countries with more highly developed agricultural systems, the condition is corrected by adding fertilizers containing nitrates or ammonia to the soil.

The nitrogen cycle is closed (step 6 of Figure 24-2) by other soil bacteria that convert the ammonia excreted by decomposers into gaseous nitrogen in a process known as **denitrification**. The reactions of this series are essentially the reverse of nitrogen fixation.

In overall terms, nitrogen flows from its reservoir in the atmosphere to bacteria, which fix nitrogen into ammonia in the roots of legumes. The ammonia is then used by plants as an inorganic nitrogen source for the synthesis of nucleic acids and proteins. Animals, in turn, use the organic nitrogen compounds of plants as their nitrogen source. After death and decomposition of the body tissues of plants and animals, much of the nitrogen is returned to the soil as ammonia, which can either be converted into nitrate through nitrification (to enter food chains again through nonleguminous plants) or be returned to the atmosphere as gaseous nitrogen by denitrifiers.

The Carbon Cycle All organic compounds are based on chains or rings of carbon atoms. In addition to this structural role, carbon chains are oxidized by almost all organisms as an energy source. The major reservoirs of carbon, in the form of carbon dioxide, are the atmosphere and the sea. The atmosphere contains slightly more than 300 parts per million (or slightly more than 0.03 percent) carbon dioxide by volume. Much greater amounts of CO_2, about fifty times as much, are dissolved in the oceans and other bodies of water.

Atmospheric CO_2 enters the carbon cycle through CO_2 fixation in photosynthesis (Figure 24-4, step 1), carried out by the primary producers of ecosystems. In CO_2 fixation, the primary producers use light energy to add electrons and hydrogens to CO_2 and produce carbohydrates (for details, see Chapter 6):

$$CO_2 + H_2O \rightarrow \text{carbohydrates} + O_2 \qquad (24\text{-}4)$$

The carbohydrates are then used by plants to synthesize more complex carbohydrates, polysaccharides, and lipids or, after the addition of nitrogen, to synthesize amino acids, proteins, nucleotides, and nucleic acids.

The carbon cycle is completed through respiration and combustion of the organic compounds synthesized by plants. Some of this material is used by the plants themselves as an energy source (step 2, Figure 24-4) in a reaction that releases free energy and returns carbon to the atmosphere as CO_2 (for details, see Chapter 7):

$$\text{Carbohydrates} + O_2 \rightarrow CO_2 + H_2O + \text{energy} \qquad (24\text{-}5)$$

Animals and detritus feeders also use the organic carbon compounds of the plants as an energy source, eventually returning much of the remainder of the carbon fixed by plants to the atmosphere as CO_2 (step 3, Figure 24-4). The complex carbon compounds of plants are also converted directly into CO_2 and returned to the atmosphere through combustion of plant material by forest or grassland fires.

A fraction of the fixed, organic carbon compounds of plants, and to a lesser extent those of animals, becomes part of the sediment without immediate oxidation by detritus feeders or fire. This organic material (step 4, Figure 24-4) is slowly converted into deposits of peat, oil, coal, and other fossil fuels of our environment. Eventually this deposited carbon too is returned to the atmosphere as CO_2, either by natural fires or volcanic activity or through its use by humans for heating, light, or power sources of various kinds.

Some carbon is also deposited as sediment in the form of carbonates derived from the shells of marine and freshwater animals (step 5, Figure 24-4). This material is gradually compacted into carbonate rock; eventually, through uplift and weathering, the carbon of this rock material is returned to the atmosphere as CO_2.

Two parts of the carbon cycle are presently of great concern to the human community. One is the sedimentary portion including the deposition and gradual conversion of organic carbon compounds to coal and oil. Only a small fraction of organic matter is actually deposited to enter this part of the cycle; the process, moreover, requires millions of years to run its course. Since the beginning of the industrial revolution we have been burning these fossil fuels

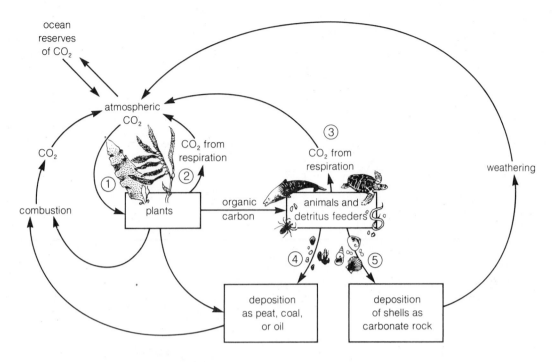

Figure 24-4 The carbon cycle (see text).

much faster than they are formed—so fast, in fact, that many authorities estimate that the available reserves of oil,

---

Only a small fraction of the matter flowing through the carbon cycle enters the sedimentary portion leading to formation of coal and oil, in a process that takes millions of years to complete.

---

the most widely used fossil fuel, may be exhausted in 50 to 100 years. Severe oil shortages already exist, as we are all aware, and both the world economy and the economic health of individual nations is increasingly dependent on access to the remaining oil resources of the earth. Whether oil and coal reserves run out 50, 100, or more years from now, it is certain that our demands for light, heat, and power will eventually have to be met by other energy sources, such as solar or fusion energy, if our present living standards are to be preserved and extended to underdeveloped parts of the world. (This problem and its possible solutions are discussed in greater detail in the final section of this chapter.)

The second part of the carbon cycle of concern to the human community is related to the use of fossil fuels as sources of light, heat, and power. Combustion of coal and oil has gradually raised the concentration of CO_2 in the atmosphere from levels estimated at 270 parts per million (by volume) in 1850 to 327 parts per million by 1975—an increase of about 20 percent. The increase would be even higher except for the oceans, which regulate the level of atmospheric CO_2 by absorbing excess CO_2 from the air. Absorption by the oceans, which takes up about 0.5 percent of the excess CO_2 of the atmosphere per year, has removed about half the extra CO_2 generated since 1850. Otherwise, present-day CO_2 levels would approach 400 parts per million in the atmosphere.

Atmospheric CO_2 concentration has increased steadily since the industrial revolution.

The possible effects of the extra CO_2 are presently the subject of extensive discussion among ecologists. Increases in CO_2 concentration are expected to raise atmospheric temperatures because CO_2 absorbs light energy and converts it to heat. Some ecologists have predicted that average world temperatures will rise as a consequence of the increased atmospheric CO_2 concentration, with results as dire as melting of the polar ice caps and inundation of coastal cities through the extra water released to the oceans. Records reveal that average temperatures have in fact risen by about 0.6°C from 1880 to 1940, supporting this contention to some extent. Average world temperatures have cooled again by about 0.2°C since the peak in

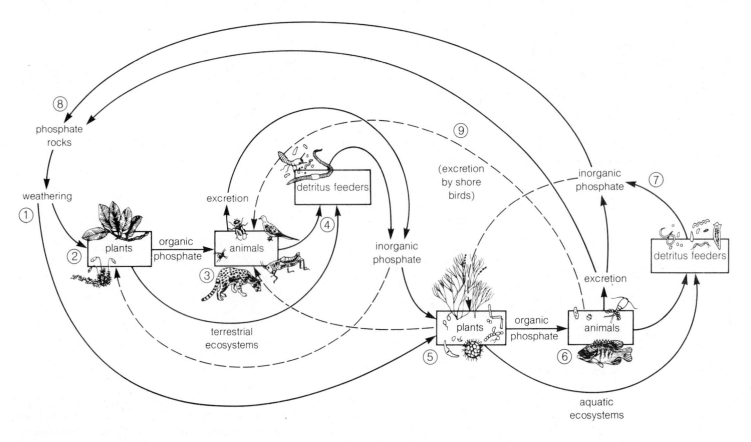

Figure 24-5 The phosphorus cycle (see text). In most ecosystems the greatest quantities of phosphorus follow the pathways indicated by the solid lines.

1940, however, making the atmospheric effects of the extra CO_2 uncertain. It is possible that dust particles in the atmosphere, which have also increased steadily in concentration since the industrial revolution, counterbalance the effects of the excess CO_2 by reflecting greater quantities of sunlight into outer space.

The Phosphorus Cycle Phosphorus is an essential element in nucleotides, nucleic acids, phospholipids, and many proteins. Because it is essential for life, and because many environments are deficient in phosphorus, a shortage in the supply of this element is frequently a significant factor limiting the plant growth of an area. In countries with highly developed agriculture millions of tons of fertilizer containing phosphorus (as phosphates) are added to soils yearly to correct this condition.

In contrast to nitrogen and carbon, the earth's major reservoirs of phosphorus are stored as deposits of phosphates in sedimentary rocks. These phosphates are slowly released by erosion and weathering to enter the phosphate

cycle (step 1, Figure 24-5). The phosphates, carried to the soil by water runoff, are taken up and used directly by plants to synthesize their organic phosphate compounds (step 2, Figure 24-5). The organic phosphates contained in plant tissues are then used by the primary and secondary consumers of ecosystems (step 3, Figure 24-5). The terrestrial cycle is closed by detritus feeders (step 4, Figure 24-5), which break down the organic phosphate compounds of dead plants and animals and release the phosphorus to the environment as inorganic phosphate. Many animals also regularly excrete large quantities of inorganic phosphate as a waste material.

The aquatic phosphate cycle follows a similar course. Almost all of the inorganic phosphates eroded from sedimentary rocks drains into streams, lakes, and seas, either directly or after movement through the terrestrial phosphate cycle. The phosphorus entering the aquatic environment is used there by primary producers (step 5, Figure 24-5) and flows as organic phosphates to the consumers of aquatic ecosystems (step 6, Figure 24-5). After death and decomposition of aquatic plants and animals, phosphorus is released as inorganic phosphate that settles to the bottom as sediment (step 7, Figure 24-5). In marine environments, which eventually receive most of the phosphate eroded from terrestrial rocks, the phosphate settles to the ocean floor in large quantity. Gradually, over millions of years, the phosphate sediments harden into rock, completing the phosphate cycle (step 8, Figure 24-5). The phosphate rock of the ocean floor may again enter terrestrial and aquatic phosphate cycles if it is raised and exposed to the air and weather by movements of the earth's crust.

Some of the phosphate entering marine environments is returned to the land in a side cycle carried out by shorebirds that feed on marine organisms (step 9, Figure 24-5). These birds, among them pelicans, cormorants, gannets, and other species, excrete phosphate wastes from digestion of fish on their island and shore nesting grounds. The phosphates are excreted in surprising quantities: between 300,000 and 400,000 tons annually. The excreted material, called *guano*, is mined as a phosphate source for fertilizers and other chemical needs. Some phosphate is also returned to terrestrial ecosystems in the excretions of terrestrial organisms, including humans, that feed on marine life.

Many ecologists are concerned that the balance of the phosphate cycle is being upset by our increased use of phosphates for fertilizers and detergents. The demand for phosphates for these purposes has expanded so greatly that mining of phosphate rock for human use has become a major enterprise. This mined phosphate, amounting to millions of tons annually, is added to the natural flow reaching lakes and seas, some through runoff from fertilized fields and some through the discharge of detergent wastes into streams and rivers. This inflow of phosphate has caused ocean phosphate sediments to build at a much greater rate than the return of phosphate to land through geological uplifting of the ocean floor, a process that requires millions of years to complete. As a result, the natural phosphate cycle has been converted largely to a one-way flow of phosphorus from land to sea. Although our terrestrial reserves of phosphate rock are extensive, continued one-way flow of phosphate to the ocean bottoms may so deplete these reserves that it may become necessary to mine the oceans to obtain our needed phosphates.

The increased phosphate inflow into bodies of water has also had some undesirable side effects on freshwater lakes and ponds. The phosphate entering these lakes from fertilizer runoff and sewage discharge greatly improves conditions for algal growth, causing "blooms" of algae so dense that the water turns a murky green. Although the

increased photosynthesis and resultant oxygen production by the algae might be expected to improve conditions in the lake, the opposite actually happens in an algal bloom. The algae foul the shorelines with green scum; at the end of the growing season, they die and sink in rotting masses to the bottom, where bacterial decomposition and respiration become so extensive that the oxygen of the lake is depleted. This condition may cause the death of most animal life of the lake, including both insect and crustacean larvae and the desirable game and commercial fish that feed on them. Many lakes and ponds have been turned into dead bodies of water in this way; only bacteria, a few species of worms, and undesirable fish such as carp can survive the low-oxygen conditions created by the phosphate inflow and algal blooms.

Inflow of phosphates to freshwater lakes and ponds from fertilizer runoff and detergents in sewage discharge often causes destructive algal blooms.

The nitrogen, carbon, and phosphorus cycles described in this section share several features common with most of the biogeochemical cycles of the ecosphere. In these cycles, an element is converted from inorganic to organic form by primary producers. The element then flows from primary producers to primary and secondary consumers in organic form; the animals feeding at these levels are often unable to use the element directly in inorganic form. Detritus feeders usually convert most of the element back to inorganic form. Thus detritus feeders often serve the important function of closing biogeochemical cycles and returning elements to their natural reservoirs.

ECOLOGY AND THE HUMAN POPULATION

We are inclined to regard our own population as an independent entity set apart from nature. But our population and its growth are subject to the same ecological laws governing the growth of any population of living organisms. Moreover, we fit into the structure of a community in which we take our place as both primary and secondary consumers in community food chains and webs, and our activities depend on the energy sources and nutrient cycles of the world ecosystem.

When the human population was relatively small, in times as recent as a few hundred years ago, the impact of human activities on our own and other communities and ecosystems was relatively slight. It was possible to ignore the fact that humans are part of nature and dependent on

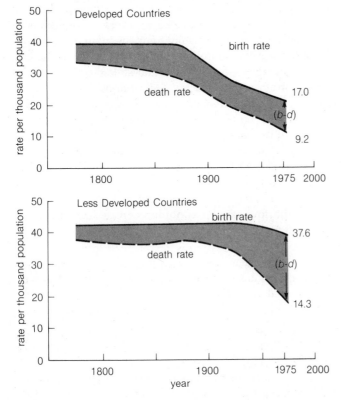

Figure 24-6 Difference between birth and death rates per thousand persons (shaded area) in developed and less developed countries between 1700 and 1975. Since the difference between births and deaths (*b* − *d*) is a positive number, the human population is growing exponentially in both cases. Data from United Nations *Demographic Yearbooks* and Population Reference Bureau.

nature for survival. But with the growth of the human population to levels that begin to approach the carrying capacity of the environment, and the expansion of our demands to the point that our sources of energy and inorganic nutrients are in danger of exhaustion, we can continue to exist in our present and future numbers only if we can find solutions to five major problems that face humanity: population overgrowth, food shortages, environmental pollution, depletion of inorganic raw materials, and the shortage of energy.

The Growth Problem

The Magnitude of the Problem Populations of any kind grow in size if the difference between births and deaths (*b* − *d* in Equation 22-4) is a positive number. As long as this number is positive, population growth will follow an exponential curve. The human population has grown exponentially for the same reason: for many centuries,

and especially since the industrial revolution, the quantity (*b* − *d*) for the world population has been a large and positive number. In recent times, since about 1750, the quantity· (*b* − *d*) has not only been positive but has increased in value, so that exponential growth has proceeded at an ever faster rate. This is not because the birth rate has increased, as many people believe; birth rates have actually fallen slightly over the world as a whole during the last 200 years. The exponential rate of growth has increased instead because death rates have fallen significantly since 1750 (Figure 24-6) due to increased efficiency in utilization of natural resources, a general improvement in living conditions, and better medical care.

The time required for the human world population to double gives an idea of the magnitude of our rate of population growth. Before the beginning of the Christian era, it is estimated that it took 2000 years or more for the human population to double in size. By the period 1650 to 1750, doubling time had dropped to 200 years; by 1800, to 100 years. After the industrial revolution in the mid-1800s, doubling time dropped at an ever-increasing rate. By 1910 doubling time dropped to 50 years; by 1950, to 40 years. The most recent estimates, for the period 1975–1979, indicate the doubling time has slowed again to about 41 years after peaking in the early 1970s. At this rate growth is still exponential, however, and the size of our own population is increasing rapidly. The human population cannot continue to grow indefinitely at this rate; like any population, it must either crash or level off and stabilize as it reaches the carrying capacity of the environment. Whether it crashes, with attendant disaster for the human race, or stabilizes near the carrying capacity depends on our ability to control our future growth.

Whether the human population crashes or stabilizes near the carrying capacity of the world environment depends on our ability to control our population growth.

A crash, if it comes, will result from an abrupt increase in the death rate through widespread famine, disease, nuclear war, or a combination of these factors. Stabilization of the population, if death rates are to remain steady or even decrease further, can come only through a reduction in the birth rate. These alternatives between crash and stabilization are often stated as the difference between birth control and death control for humanity.

The more preferable of these alternatives, a reduction in the birth rate, is possibly the most difficult joint task ever faced by the human race. We are all familiar with the religious, moral, and political objections raised against birth control, which make it difficult to institute organized programs to reduce the rate of births. Coupled with these objections is a worldwide attitude that the birth rate, no matter what the country, is a problem only in other parts of the world. (In some countries, particularly in Africa and South America, population growth is even promoted as a means for economic and political development.) Beyond these objections are additional problems that result from the age structure of the human world population. More than a third of the people in the world, about 36 percent, are under 15 years of age. Whenever a large proportion of a population is under reproductive age, the population will continue to grow even if births are reduced to the replacement level because, for many years, the number of people of childbearing age continues to increase.

The human population would continue to grow until 2070 even if births could be reduced to the replacement level today.

This means that even if our present attitudes could somehow be reversed, and a worldwide program of birth control developed today to stabilize births at the replacement level of 2.5 births per couple, the world population would continue to grow until stabilizing in 2070 at 6.3 billion persons. (The replacement level is estimated at 2.5 births per couple instead of 2.0 because, on the average, 0.5 children per couple do not become parents or do not survive to reach reproductive age.) At 6.3 billion persons, the world population would be 50 percent larger than the present population of about 4 billion. Since an immediate reduction of the birth rate to 2.5 children per couple from its present level of 4.7 is obviously impossible, the population will likely continue to grow past 2070. If the replacement rate of 2.5 births per couple is not reached until the year 2000, the population will increase to 8.2 billion before stabilizing, or about twice the size of the present world population. Further delays in reaching the replacement level, which is perhaps a more realistic outcome in view of our present objections and attitudes, would result in additional growth before the population stabilizes—10.5 billion if the replacement level is not attained until 2025 and 15.5 billion if not until 2045 (see Figure 24-7).

Thus the longer world programs of birth control are delayed, the more quickly our population will approach the carrying capacity of the world environment. Just where the carrying capacity of the world actually lies is a matter of some debate. Different authorities place the ultimate number somewhere between 30 and 50 billion; at this point the great majority of people in the world would exist in a chronic state of near-starvation. At present rates of growth, this number will be reached and passed within the next 100 years, with the inevitable crash soon to fol-

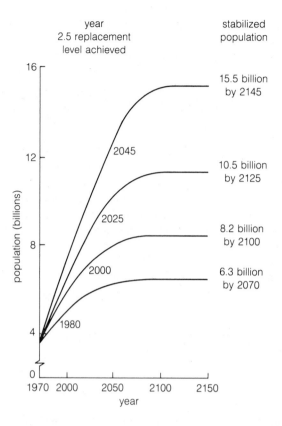

Figure 24-7 Projected growth of the human world population if replacement level of 2.5 births per couple is reached in 1980, 2000, 2025, or 2045.

low. If stabilization is achieved through regulation of births before the carrying capacity is reached, it is possible that a population crash may be avoided.

The Food and Pollution Problems

Our human population forms part of a larger community of organisms into which we fit as consumers in a complex food web. Like other consumers, we are ultimately dependent on a group of primary producers: the green plants that capture the energy of sunlight and convert it into the organic molecules serving as our energy source. And, like other organisms, we produce wastes that are broken down by the detritus feeders of our community. (Fortunately, the day has passed when humans also served as a significant food source for further levels of consumers.) The continued existence of this complex community, and our existence within it, depend on our ability to continue using the resources supplied by the community without destroying it.

Two questions are paramount concerning the position and role of our population in its community. One is related to the almost inevitable growth of the human population to levels at least twice as high as the present world population. Can we obtain enough food to feed the present and future human population without utterly destroying the community on which we depend? The second question concerns the increased levels of wastes we dump into our community. Can the detritus feeders and other populations of the community, including our own population, survive the onslaught of inorganic and organic chemicals, some of them highly toxic, that we release as wastes to our environment?

The Food Problem Humans can act by choice as either primary or secondary consumers in the food chains and webs of our community. As already noted (see p. 527), we can tap energy from our food chains most effectively by using primary producers as our food source, since, on the average, only 10 percent or less of the energy available at one trophic level is captured and made available at the next. Therefore the present and future needs of the human population can best be met by consumption of the plants acting as primary producers rather than the meat of animals such as beef cattle or fish.

But will enough food be available for future populations, even if humans learn to live on a vegetable rather than meat diet? The prospects seem doubtful if projections for human population growth prove accurate. Almost all the best farmland is now under cultivation; attempts to utilize the presently unfarmed land of the deserts, far north, mountain slopes, and tropics are likely to be unsuccessful or so limited in productivity that little food could be added to present levels. The tropics deserve special mention in this regard. Although the tropics would seem at first glance to have high potential as agricultural land, the opposite has turned out to be true. The soil of the tropics is typically thin, with few mineral nutrients and little organic matter. In many tropical regions, the soils contain large quantities of iron and aluminum oxides that bake to a rocklike consistency and become unworkable when the forest is cleared and the surface is exposed to sunlight. The only techniques developed so far for farming the land of the tropics, which has proved too unproductive to constitute a major food source, consist of clearing the tropical forest, farming for a few years until soil nutrients are exhausted or the surface becomes too hard to work, and abandoning the land to the wilderness.

Thus it is not likely that enough agricultural potential

remains available to feed the additional humans expected from the population increases projected for coming years. The problem is compounded by the fact that, if future trends echo present practice, much of the housing built to accommodate the future human population will be built on prime agricultural land, not on the unusable land of the deserts, mountains, far north, or tropics. If present trends continue, therefore, the amount of land available for agriculture is likely to decrease rather than increase in the years to come.

World production of food from agriculture and fishing probably cannot be increased significantly beyond present levels.

What about the oceans? It is often claimed that by harvesting the productivity of the oceans, many more humans could be adequately fed. The prospects here are just as gloomy, however. No methods have ever been developed, or seem feasible, for harvesting the primary producers of the ocean. Filtering the microscopic organisms of the phytoplankton and processing them for consumption would require so large a fleet of ships constantly plying the seas that the economic and energy costs would be prohibitive. Similar problems surround attempts to farm the seaweed growing along the shorelines of the continents: no workable methods have ever been developed. As a result, we are limited for practical reasons to acting as secondary, tertiary, or even higher-level consumers by eating the fish of the seas—but at these trophic levels only an estimated 1 percent of the original primary productivity of the oceans remains available. Thus our present and future problems resemble the situation on land: we are already harvesting our maximum possible share of the herbivores and carnivores of the oceans. In fact, our harvest of ocean fish has actually been decreasing because we have over-fished the present ocean populations to the extent that their reproductive capacity has been seriously diminished.

Unless there are unexpected breakthroughs in agriculture, finding food for the expected increases in the human population presents one of the most serious problems facing present and future generations. Moreover, the prospect of widespread human famine threatens not only our own population but also the survival of the community to which we belong. If the populations we use as a food source are to survive, we can harvest only a fraction of the individuals in these populations—a base level of reproducers must be retained to replace the individuals consumed as food. Maintaining this necessary reserve of reproducers may become impossible if the human demand for food becomes desperate. The result, if these reproducers are consumed in desperation, would mean elimination of much of the community on which humans depend for food as well as a crash of the human population.

The Pollution Problem The wastes dumped into the environment by the relatively small human populations of centuries past were so limited in quantity that they could easily be broken down by the detritus feeders of the community. Then, too, the wastes released were almost exclusively organic matter, not the toxic materials flowing to the environment as a result of our present-day industrial and agricultural activity. Many of the chemicals now deposited as industrial wastes, or sprayed on the environment as part of agricultural pest control, are so poisonous that the detritus feeders of our community, on which we depend for waste elimination and closure of the major biogeochemical cycles, are in danger of being wiped out. Many of the toxic materials being released also have detrimental effects on other populations in our community: birds, fish, plants, and our own population.

The toxic wastes dumped in the environment through human activities include radioactive substances, heavy metals, pesticides and herbicides, and air pollutants. Radioactive wastes are released from two primary sources: from atomic power plants and from weapons testing. Although radioactive pollution has received more publicity than any other source of toxic substances, the contamination of food chains and the environment by radioactive materials has remained slight up to this time. In fact, pollution from this source has declined to some degree since the cessation of aerial weapons testing in 1963. A major problem remains, however, in the form of radioactive wastes from power plants that are presently stored underground in large quantities. Although these deposits pose little danger to our present population and community, they are almost permanently radioactive and may prove lethal to populations as yet unborn if significant leakage begins from the accumulated wastes. The danger to future human and other populations is great enough to make the accumulation of radioactive wastes in our time a highly questionable practice.

The danger of future leakage and contamination of the environment makes the present accumulation and storage of radioactive wastes a highly questionable practice.

Heavy metals released as wastes to our community and its environment pose an immediate danger. The substances now released in quantity, including compounds of lead, mercury, arsenic, and cadmium, are toxic to our own population as well as other organisms of our community. All these substances tend to enter watersheds and accumulate in lakes and the oceans. Mercury, in particular, has already reached concentrations sufficiently high in some aquatic environments, such as Lake Erie and certain ocean estuaries, to make the fish in these regions unfit for human consumption. Lead, released principally through combustion of leaded gasolines in automobiles, has also reached high concentration in aquatic environments.

Pesticides are applied primarily to control insects that feed on crops used for human food. The substances used as insecticides are toxic to a wide variety of organisms besides insects, however, including the detritus feeders we depend on for waste breakdown and vertebrates such as fish and birds as well as humans. Herbicides, used widely to control the growth of pest weed species, also have detrimental effects on plant and animal life well beyond the weeds targeted by the treatment. Many of these pesticides and herbicides persist in the environment for as long as 50 years or more after application. Some, moreover, tend to accumulate in the organisms feeding at successively higher trophic levels in food chains. Although the use of DDT and other persistent and toxic pesticides is decreasing in the United States, application of these substances is still on the increase in other parts of the world.

Air pollution results primarily from the combustion of fuels of various kinds, which releases toxic substances such as nitrogen oxides (NO and NO_2), sulfur dioxide (SO_2), and carbon monoxide (CO) to the atmosphere. Air pollution in the United States, primarily from automobile exhaust, is at its highest levels in the Los Angeles Basin and the megalopolis extending from New York City north to Boston and south to Washington on the East Coast. In the Los Angeles region, air pollution has killed much of the pine forests in the mountains ringing the city and reduced the quality of life for the human population. The smog blanket of the East Coast has been even more destructive, at least for the human population. In 1953, from 175 to 260 persons died in the New York area from the effects of air pollution over the city. Air pollution also has secondary effects through toxic substances removed from the air and carried into streams, lakes, and the oceans in rain droplets as "acid rains." These rains have caused a marked increase in the acidity of lakes and streams in many parts of the world and are believed to have been responsible for the elimination of fish from several hundred lakes in New York alone.

Each of these sources of toxic wastes—radioactive substances, heavy metals, pesticides, and air pollutants— must be brought under control if the human population, and the community of living organisms on which we depend, are to survive. For our own population, the accumulating wastes spell out an increased incidence of cancer

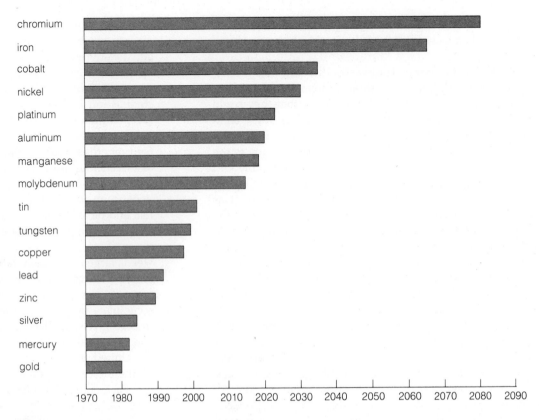

Figure 24-8 Projected times for 80 percent depletion of world reserves for 16 key metals if present consumption increases 2.5 percent per year. Data from U.S. Geological Survey 1973.

and circulatory and respiratory diseases. For our community, the toxic substances dumped on our environment result in a reduction of the populations of living organisms filling vital roles in food chains and webs, including the species we ourselves depend upon for food and breakdown of wastes.

The Raw Material and Energy Problems

The human population is so widespread that the ecosystem to which we belong is essentially the same as the ecosphere: the ecosystem of the entire world. Ecosystems include the sources of energy and inorganic substances as well as the living community, and it is these two areas, energy sources and mineral requirements, that pose some of the most serious problems for present and future generations of the human population.

Raw Materials Shortages already exist for some of the minerals needed for our technical culture, including tin, tungsten, and mercury; supplies of these substances are

predicted to run out by the year 2000 or soon thereafter. Many others, like zinc, silver, lead, and copper, will probably be exhausted by 2050—that is, at about the time the world population's needs will become most acute (Figure 24-8). One answer to these problems is obvious from the fact that mercury, one of the metals already in short supply, is presently dumped into the environment as a major pollutant: as many of these nonrenewable resources as possible must be recycled instead of dumped as wastes. It is still more economical to mine virgin resources than to recycle used materials, however, and little is being done to develop national programs for recycling metals.

One answer to the problem of resource shortages is recycling.

Energy Finding the energy needed to power our ecosystem presents an immediate and future problem that is no less serious than the others facing the human population. In this case, however, there are solutions in the form of conservation and alternative energy sources that, if developed, could supply our energy needs for a long time to come. Development of these alternative sources in time to avoid an energy crisis would mean the institution of immediate, worldwide programs of research and conversion from oil use, though, an effort that is not likely in view of the fact that oil, even today, is still the cheapest source of energy.

Avoidance of a future energy crisis depends on present-day development of alternative energy sources.

Much of the energy used in the world, particularly in the United States, is simply wasted. It has been estimated that per capita use of energy in the United States could be cut by 25 to 50 percent without a serious decline in the quality of life. Conservation, then, stands as an immediate means of reducing energy consumption and postponing the energy crisis until the world has sufficient time to develop alternative energy sources. This conservation could be accomplished in many ways: by using smaller and lighter cars; using cars longer before replacement; using carpools and vanpools for longer trips and relying on bicycles and walking for shorter trips when possible; shifting commercial transportation from the use of trucks and aircraft to the more efficient trains, ships, and pipelines; increasing insulation and reducing cooling, heating, and lighting levels in homes and commercial buildings; limiting the use of appliances such as dishwashers and clothes dryers; and recycling.

Conservation provides an immediate and workable means for postponing an energy crisis until alternative sources can be developed.

What are the alternative energy sources that, if developed in time, could circumvent an energy crisis and supply our present and future needs? These sources include coal, nuclear energy, oil shale, and tar sands as nonrenewable resources and solar energy, geothermal energy, water power, and wind as major renewable resources.

Nonrenewable Energy Sources Coal is probably the cheapest and most easily developed of the nonrenewable energy sources. The present coal reserves of the world represent more than 90 percent of the remaining fossil fuel supplies. The use of this coal would stretch our fossil fuels long enough to allow transition to nonfossil fuels. The energy of coal can be used directly for heating or transformed into electrical energy or liquid fuels to power automobiles. Use of this energy source, however, would result in consider-

able increases in atmospheric CO_2 concentration. Air pollution, particularly from sulfur-containing compounds that occur in coals of various kinds, would also increase if coal is used as a major energy source.

Oil shales and tar sands too offer the possibility of an interim energy supply to meet the human population's energy needs until nonfossil sources can be developed. Oil shales consist of sedimentary rocks that contain mixtures of rubberlike hydrocarbons that can be used as fuels. Tar sands contain solid, tarry deposits of hydrocarbons mixed with sand. As with coal, though, combustion of these fossil fuels as major energy sources would add significantly to atmospheric CO_2 levels and air pollution.

The possibilities for the remaining nonrenewable energy source, nuclear power, depend on two methods for releasing atomic energy: fission and fusion. Atomic fission provides the energy for the nuclear power plants now in use. In these plants, a heavy atom such as uranium-235 is split into smaller atoms and the energy released by the fission is used to heat water into steam. Electricity is then produced by generators driven by steam turbines. Nuclear fission plants were once hailed as the solution to our energy problems. The high cost of the electrical energy they produce, however, combined with the radioactive wastes released and the danger of nuclear accidents, have forced many countries to reassess the practicality of this potential energy source.

The high cost of nuclear fission plants and the danger of radioactive contamination have forced many countries to reassess the practicality of this energy source.

Fusion power, in contrast, offers the promise of practically unlimited energy supplies without the danger of extensive radioactive contamination. In atomic fusion, atoms of light elements such as hydrogen are forced under high temperature and pressure to fuse into heavier atoms such as helium. As in fission plants, the energy released in the reaction, which mimics the power source of the sun, would be used to heat water into steam and generate electricity. Depending on the atoms used for fusion, virtually no radioactive material is produced in the reaction.

Atomic fusion could meet all conceivable energy needs without danger of extensive radioactive contamination if it can be developed for human use.

There is still some doubt as to whether the energy of atomic fusion can be harnessed to generate energy for human use. If it can, the energy potentially available is almost limitless. One of the fuels usable for fusion reactions is deuterium or hydrogen-2, a hydrogen isotope containing one proton and one neutron in its nucleus. There is enough deuterium in the oceans to generate energy from atomic fusion amounting to many times the present per capita consumption for the next 100 billion years!

Renewable Energy Sources The renewable energy sources of importance include solar, water, wind, and geothermal energy. Solar energy is an immediately available energy source that could potentially meet 20 to 25 percent of our energy needs in the present and near future. The energy of sunlight can be utilized in two ways: directly, as an energy source for home and hot water heating and air conditioning, and indirectly, through electricity generated in photoelectric cells. Water power, which has been used as an energy source for many centuries, is limited in potential but could perhaps be further developed to supply as much as 3 percent of our future energy needs. This source of energy has the advantage of low environmental impact and operating costs. Wind too has been used as a power source for centuries. Energy from this source, developed through wind-driven generators, could perhaps supply as much as 13 times as much electrical energy as is now produced in the entire world if all the potential sites are used. Wind and water power are really indirect forms of solar energy, since the major air flows and water cycles of the world result from solar heating of the earth's surface. Geothermal power, developed through heat generated beneath the earth's surface by the decay of radioactive elements and subsurface volcanic activity, could also be developed to supply a significant fraction of our energy needs. Estimates of the potential of this source vary from as little as 5 percent to as much as the entire energy requirement of the United States.

Thus of the five major problems facing humanity—population overgrowth, food shortages, environmental pollution, exhaustion of raw materials, and energy shortages—only the problem of obtaining energy for our present and future needs shows promise of ready solution. Even in this case, it is obvious that solution will require an immediate, worldwide program of conservation and development of alternative resources if a crisis is to be avoided. Solving the remaining problems will depend not only on a continued study of the ecology of populations, communities, and ecosystems but also the application of the concepts of ecology to our own population. In essence, we must understand our place in the world community and ecosystem . . . and learn to keep it.

Questions

1. What is an ecosystem? What interactions are of primary interest in ecosystem ecology? How do ecosystems differ from communities? What is the ecosphere?

2. Why is heat an unusable form of energy in an ecosystem? How do the first and second laws of thermodynamics apply to ecosystems? In what way is entropy related to the one-way flow of energy through ecosystems?

3. What happens to the sunlight entering the earth's atmosphere and striking the surface of the earth?

4. What is the ozone layer? What effect does it have on the light passing through the atmosphere?

5. What are biogeochemical cycles? What natural reservoirs serve as sources for the nutrients cycling within ecosystems?

6. What features do the nitrogen, carbon, and phosphorus cycles have in common? In what ways are they different?

7. What important roles do detritus feeders have in biogeochemical cycles?

8. Why will the human population continue to grow even if the birth rate is reduced to the replacement level of 2.5 births per couple? What factors, acting singly or together, could produce a crash in the human population? What is meant by birth control versus death control in reference to human population growth?

9. Why is it more efficient for humans to act as primary rather than secondary consumers in the food chains of our community? What are the chances that future agricultural production or food from the sea will feed the human population adequately if our population doubles in size?

10. What major forms of toxic wastes are released by the human population to the community? Do you think that present needs justify the accumulation and storage of radioactive wastes from nuclear power plants? Why?

11. What conservation methods could be used to reduce per capita energy consumption in the United States? Do you think these conservation methods are necessary now? Why?

Suggestions for Further Reading

Krebs, C. J. 1978. *Ecology: The Experimental Analysis of Distribution and Abundance.* Harper & Row, New York.

McNaughton, S. J., and L. L. Wolf. 1979. *General Ecology.* 2nd ed. Holt, Rinehart & Winston, New York.

Miller, G. T., Jr. 1979. *Living in the Environment.* 2nd ed. Wadsworth, Belmont, California.

Pianka, E. R. 1978. *Evolutionary Ecology.* 2nd ed. Harper & Row, New York.

Ricklefs, R. E. 1979. *Ecology.* 2nd ed. Chiron Press, New York.

Whittaker, R. H. 1975. *Communities and Ecosystems.* 2nd ed. Macmillan, New York.

Chapter 12 (p. 256)

1. In the $CC \times Cc$ cross, the CC parent produces all C gametes, and the Cc parent produces ½C and ½c gametes. All offspring would have colored seeds—one-half homozygous (½CC) and one-half heterozygous (½Cc).

In the $Cc \times Cc$ cross, both parents produce ½C and ½c gametes. Of the offspring, three-fourths would have colored seeds (¼CC + ¾Cc) and one-fourth colorless (¼cc).

In the $Cc \times cc$ cross, the Cc parent produces ½C and ½c gametes, and the cc parent produces all c gametes. One-half of the offspring are colored (½Cc) and one-half are colorless (½cc).

2. The genotypes of the parents are Tt and tt.

3. Yes, the brown-eyed parents can have a blue-eyed child, but the blue-eyed parents *cannot* have a brown-eyed child. The chance of a blue-eyed child being born to the brown-eyed couple described is ¼. Since each combination of gametes is an independent event, the chance of the second child (and any child) having blue eyes is also ¼.

4. The chance that the fifth child will have brown eyes is ¾ and that it will have blue eyes is ¼.

5. Use a backcross; that is, cross the guinea pig having rough black fur with a double recessive individual, $rrbb$ (smooth white fur). If your animal is homozygous ($RRBB$) you would expect all of the offspring to have rough, black fur.

6. One gene probably controls fur color, with the alleles G (green pods) and g (yellow pods). The G allele for green pods is dominant to the g allele.

7. The tongue-rolling parents are both heterozygotes; that is, they are both carriers of the recessive trait. If T is the dominant allele, and t is the recessive, then both parents are Tt and the child is the homozygous recessive tt.

8. The cross $RR \times RR$ will produce ½RR and ½Rr offspring. The cross $Rr \times Rr$ will produce ⅓RR and ⅔Rr offspring, since rr is lethal and does not appear among the progeny.

9. The parental cross is $GGTTRR \times ggttrr$. All offspring of this cross are expected to be tall plants with green pods and round seeds, or $GgTtRr$. This heterozygous F_1 generation, when crossed, is expected to produce eight types of offspring, green-tall-round : green-dwarf-round : yellow-tall-round : green-tall-wrinkled : yellow-dwarf-round : green-dwarf-wrinkled : yellow-tall-wrinkled : yellow-dwarf-wrinkled in a 27:9:9:9:3:3:3:1 ratio.

Chapter 13 (p. 277)

1. Thirty-two different kinds of gametes can be produced.

2. In the first instance, the children would be expected to be: ¼ brown-eyed tasters, ¼ brown-eyed nontasters, ¼ blue-eyed tasters, and ¼ blue-eyed nontasters.

In the second instance, the children would be expected to be: 9/16 brown-eyed tasters, 3/16 brown-eyed nontasters, 3/16 blue-eyed tasters, and 1/16 blue-eyed nontasters.

ANSWERS TO GENETICS PROBLEMS

3. In the *RR* × *Rr* cross, one-half of the plants will have red flowers (½*RR*) and one-half will have pink flowers (½*Rr*).

In the *RR* × *rr* cross, all of the plants will have pink flowers (*Rr*).

In the *Rr* × *Rr* cross, one-fourth of the plants will have red flowers (¼*RR*), one-half will have pink flowers (½*Rr*), and one-fourth will have white flowers (¼*rr*).

In the *Rr* × *rr* cross, one-half of the plants will have pink flowers (½*Rr*) and one-half will have white flowers (½*rr*).

4. All sons would be color-blind (100 percent chance), but no daughters would be color-blind.

5. If the woman marries a normal male, the chance that her son would be color-blind is ½. If she marries a color-blind male, the chance that her son would be color-blind is also ½.

6. Polydactyly is caused by a dominant allele, and the trait is not sex-linked. Thus,

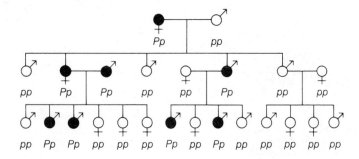

7. The genotypes are: bird 1, *FfPp*; bird 2, *FFPP*; bird 3, *FfPP*; bird 4, *FfPp*.

8. The genotype of the brown rabbit is *Cc^w*; the genotype of the chinchilla rabbit is *c^{ch}c^w*.

9. Yes. The child cannot be hers.

10. The sequence of the genes is ADBC.

11. Let the allele for normal body color = *B*, and the allele for black body = *b*. Let the allele for normal eye color = *P*, and the allele for purple eyes = *p*. Then the parents are

The F₁ flies with normal eye color and black bodies are

and the F₁ flies with purple eyes and normal body color are

12. Examples a, b, and c can be accepted as proving the hypothesis. Example d does not match results closely enough, and the hypothesis that it follows a 9:3:3:1 ratio is probably wrong.

13. A recessive allele *l* carried on one of the two X chromosomes of the female parental type used in the cross is lethal when present in males. Thus in the cross $X^LX^l \times X^LY = \frac{1}{4}X^LX^L + \frac{1}{4}X^LX^l + \frac{1}{4}X^LY + \frac{1}{4}X^lY$ half the males are lethals and die.

14. This cross is expected to produce white, tabby, and black kittens in a 12:3:1 ratio.

GLOSSARY

Abiotic Inanimate, not living.

Abscisic acid A plant hormone that inhibits plant growth by counteracting the effects of auxin and the cytokinins and gibberellins.

Acid A substance that releases hydrogen ions (H^+) when dissolved in water.

Acidity A measure of the relative concentrations of H^+ and OH^- ions in water. If the concentration of H^+ ions is higher, the solution is said to be acid. *See also* pH.

Acoelomate Having no internal body cavity.

Acrosome A vesicle in the sperm head containing enzymes that, when released, catalyze the breakdown of the egg surface coats.

Activation energy The energy that must be added to a spontaneous system to begin the reaction.

Active site The region of an enzyme that binds to the molecules being catalyzed in a reaction.

Active transport Movement of substances across a membrane against a concentration gradient—that is, from a region of low concentration to a region of high concentration; requires energy.

Adaptation Any structural or behavioral modification resulting from natural selection that increases the reproductive potential of an individual.

Adaptive radiation Branching evolution from a single ancestral species to several descendant species, each of which is adapted to a different environment.

Adhesion, selective The ability of embryonic cells to make and break attachments to other cells.

ADP (adenosine diphosphate) *See* ATP.

Aerobe A cell or organism that can use oxygen as final acceptor for the electrons removed in its oxidative reactions.

Aerobe, strict A cell or organism unable to live unless oxygen is present for use as final electron acceptor.

Age distribution (age structure) The number or percentage of individuals at each age level in a population.

Albino An individual with an inherited inability to produce body pigmentation.

Alcohol An organic compound with the reactive group

$$-\overset{\displaystyle H}{\underset{\displaystyle |}{\overset{\displaystyle |}{C}}}-OH.$$

Aldehyde An organic compound with the reactive group —CHO.

Allele One of the alternate forms of a gene having a unique DNA sequence.

Allen's rule An evolutionary tendency toward reduction in the size of body appendages of warm-blooded animals living in cold regions.

Alpha helix An arrangement of the amino acid backbone chain of a protein in a regular spiral held together by hydrogen bonds.

Amino acid A single unit of protein molecule consisting of a central carbon to which are attached an amino ($-NH_2$) group, a carboxyl ($-COOH$) group, and a characteristic side chain. Twenty different amino acids occur in living organisms.

Amino acid activation A two-part sequence of reactions that links an amino acid to its corresponding tRNA molecule to form a high-energy amino acid–tRNA complex.

Amniocentesis Extraction and testing of a sample of the fluid surrounding a human embryo for the presence of genetic defects or other disease.

Amniotic cavity The fluid-filled cavity surrounding the embryos of reptiles, birds, and mammals; formed from embryonic tissues that grow outward and surround the embryo.

AMP (adenosine monophosphate) *See* ATP.

Amyloplast In plant cells, a form of plastid in which starch is stored.

Anaerobe A cell or organism that is unable to use oxygen as final acceptor for electrons removed in cellular oxidations.

Anaerobe, facultative A cell or organism that is able to live aerobically or anaerobically.

Anaerobe, strict A cell or organism that is unable to use oxygen as final electron acceptor at any time.

Anaphase (mitosis) The division stage at which the two chromatids of each chromosome separate and move to opposite poles of the spindle.

Anastral spindle A spindle without centrioles or the starlike array of microtubules associated with the centrioles.

Angstrom (A) A unit of measurement equivalent to 0.1 nanometer or 0.0001 micrometer.

Animal pole The region or end of an animal egg containing the nucleus.

Annulus A ringlike plug of material that fills in the pores of the nuclear envelope.

Anterior Placed toward the front.

Anther The part of the male structures of a flower (the stamens) in which pollen is formed.

Antibiotic A substance produced by a living organism that is toxic to individuals of another species.

Antibody A protein, produced by a vertebrate organism, that is capable of combining with and inactivating proteins originating from other individuals of the same or different species. Each antibody inactivates a specific protein called its antigen.

Anticodon A triplet of nucleotides on a transfer RNA molecule that is complementary to the codon for an amino acid.

Antigen A foreign protein that stimulates production of antibodies when it enters the body of a vertebrate. Antigens are inactivated by combination with an antibody.

Archenteron The fluid-filled cavity of a gastrula-stage embryo.

Asexual (vegetative) reproduction Reproduction without fusion of gametes by budding or release of cells produced by mitotic division. The cells or buds grow separately into complete individuals.

Aster A starlike array of microtubules surrounding the centrioles at the spindle poles.

Atmospheric pressure Pressure exerted by the weight of the air over a given point; under standard conditions, a pressure equivalent to the weight of a column of mercury 760 millimeters high.

Atoll A ring-shaped island built up from coral surrounding a central lagoon.

Atom The smallest unit into which an element can be divided without loss of its chemical or physical properties; consists of an atomic nucleus surrounded by one or more electrons moving in orbitals.

Atomic number A number equivalent to the number of protons in the nucleus of an atom.

ATP (adenosine triphosphate) A compound containing adenine, ribose, and three phosphates. With ADP and AMP, the molecular system carrying cellular energy from reactions that release energy to reactions that require energy.

Autolysis Destruction of cell structures through the release of enzymes contained in lysosomes.

Autoradiography The detection of radioactive sites in tissue sections by covering them with a thin layer of photographic emulsion, which is exposed where there is radioactivity.

Autosomes Chromosomes with no differences in number or form in either sex of a species.

Autotroph A cell or organism capable of making all its required organic molecules from inorganic substances. Most autotrophs use light as an energy source for this synthesis. *Compare* Heterotroph.

Auxins A plant hormone that controls the elongation rate of embryonic cells.

Backcross *See* Testcross.

Bacteriochlorophyll The light-absorbing pigment of photosynthetic bacteria; closely related structurally to the chlorophylls of blue-green algae and the higher plants.

Bacteriophage A group of viruses that infects bacterial cells.

Barr body In the cell nuclei of female mammals, a block of chromatin produced by one of the X chromosomes, which remains tightly coiled and genetically inactive.

Basal body The centriole giving rise to the 9 + 2 system of microtubules in a flagellum.

Basal cell One of the two cells produced by the first division of the zygote in flowering plants; gives rise to the suspensor that nourishes the embryo.

Base A substance that increases the concentration of hydroxyl (OH^-) ions or reduces the concentration of hydrogen (H^+) ions in a water solution.

Basicity A measure of the relative concentrations of H^+ and OH^- ions in water. If the concentration of OH^- ions is higher, the solution is said to be basic. *See also* pH.

Bergman's rule An evolutionary tendency toward greater body size in warm-blooded animals living in colder regions.

Beta particle A high-energy electron released in the radioactive breakdown of an atomic nucleus.

Bias (of the fossil record) The limitation of the fossil record to those animals actually preserved, which tend to include only those with bony parts or shells that lived in wet environments.

Bilateral symmetry A form of animal body organization with two sides and front (anterior) and back (posterior) ends. *Compare* Radial symmetry.

Bilayer Fundamental membrane structure consisting of a double layer of lipid molecules; polar regions of the molecules are directed toward the membrane surface and nonpolar regions are associated in the membrane interior.

Binomial nomenclature The practice of giving a double name to all organisms: the genus and species names.

Biogeochemical cycle The flow of an element through the earth's ecosystem between its inorganic form in the environment and its organic form in living organisms.

Biological control The regulation of pest populations through the introduction of a natural predator, parasite, or disease organism attacking the pest.

Biology The study of life.

Biomass The total weight of organic matter in organisms living in an ecosystem or at any specified trophic level in an ecosystem. Usually expressed as dry weight.

Biome The major community of animals and plants typical for a terrestrial region.

Biosphere *See* Ecosphere.

Biotic Alive or living.

Birth rate The number of individuals born in a population per unit of time.

Bivalent A homologous chromosome pair at prophase of meiosis. *See also* Tetrad.

Blastocoel The fluid-filled cavity of a blastula-stage embryo.

Blastocyst An early mammalian embryo consisting of a hollow, fluid-filled ball of cells; equivalent to the blastula of sea urchin and amphibian embryos.

Blastodisc An early bird embryo consisting of a disklike layer of cells formed at the surface of the yolk; equivalent to the blastula of sea urchin and frog embryos.

Blastopore An opening leading from the exterior to the fluid-filled interior cavity (archenteron) of a gastrula-stage embryo.

Blastula An early embryo consisting of a hollow, fluid-filled ball of cells.

Bond, chemical *See* Covalent bond, Electrostatic bond, Hydrogen bond, Nonpolar bond, Polar bond.

Boundary membranes In mitochondria and chloroplasts, the outer limiting membranes.

C_4 cycle (Hatch–Slack cycle) A cycle of reactions in the photosynthetic dark reactions of some plants in which CO_2 is first incorporated into a series of four-carbon acids, including an amino acid.

calorie (with a lowercase *c*) The amount of energy required to raise 1 gram of water from 14.5 to 15.5°C at a pressure of 1 atmosphere.

Calvin cycle The cycle of dark reactions in chloroplasts in which CO_2 is reduced to units of carbohydrate (CH_2O).

Cambium A group of cells that retains the capacity to divide and gives rise to stem and root tissues of the adult plant.

Cancer Uncontrolled division of cells producing malignant growths that invade other tissues and interfere with normal body functions.

Canine tooth A conical, pointed tooth in mammals on either side of the upper and lower jaws; located between the incisors in the front and the molars at the sides.

Capsule A thick, jellylike layer surrounding the cell wall in many prokaryotes.

Carbohydrates Organic molecules containing carbon, hydrogen, and oxygen in the ratios 1C:2H:1O or CH_2O.

Carbon cycle The major biogeochemical cycle of carbon between its inorganic reservoirs in the atmosphere and bodies of water (as CO_2) and its organic form in living organisms.

Carcinogenic Capable of causing cancer.

Carcinoma A malignant tumor derived from epithelial tissue.

Carnivore An organism that lives by eating the flesh of animals.

Carotenoids A class of photosynthetic pigments that absorbs blue light and transmits yellow.

Carpel The female parts of a flower, consisting of the ovary, stigma, and style.

Carrier (of a recessive allele) An individual with one recessive and one dominant allele of a gene who exhibits only the characteristics of the dominant trait.

Carrying capacity (K) The maximum number of individuals of a population that an ecosystem can support.

Cartilage An elastic, supporting tissue consisting primarily of extracellular protein fibers with no bony elements.

Catalyst A substance that increases the rate of a chemical reaction without changing chemically itself as a result of the reaction.

Cell cycle The series of events carrying cells through one round of growth and division, including interphase, mitosis, and cytokinesis.

Cell elongation A basic embryonic growth process of plant cells in which the cells elongate in one direction.

Cell-free system A collection of organic molecules and enzymes isolated from a cell that is capable of carrying out a major cell reaction such as protein synthesis.

Cell line A group of cells with the same mitotic origins.

Cell matrix, cytoskeleton A network of microtubules, microfilaments, or intermediate filaments providing support to the nucleus and cytoplasm.

Cell plate The new wall separating the two products of a cell division in plants.

Cellulose A polysaccharide made up of a series of glucose molecules linked end to end.

Cell wall An extracellular coat of organic matter closely surrounding a cell just outside the surface (plasma) membrane.

Central singlets The central pair of microtubules in the 9 + 2 system of a flagellum.

Centrifugal force A force, developed in an object rotating around a center, that tends to propel the object away from the center.

Centrifuge A machine that increases the weight of cells, cell structures, or molecules by spinning them rapidly around a center; used to separate cell parts or molecules into purified fractions according to weight or density.

Centriole A barrel-shaped structure, consisting of nine sets of microtubules arranged in a circle, that gives rise to the microtubule system of flagella.

Centrolecithal egg An egg in which yolk is concentrated in a central region of the egg cytoplasm.

Centromere *See* Kinetochore.

Cervix The narrow opening of the uterus into the vagina in humans and other mammals.

Chemiosmotic hypothesis The hypothesis proposing that ATP synthesis in mitochondria and chloroplasts is driven by an H^+ ion gradient set up in turn by energy released by electrons flowing through an electron transport system.

Chemosynthesis Derivation of energy for cellular activity by breaking down inorganic molecules such as hydrogen sulfide and ammonia.

Chemotherapy The use of chemicals to treat cancer and other diseases.

Chiasma *See* Crossover.

Chitin A hard polysaccharide material that forms an outer, protective surface covering in insects and other animals.

Chlorophylls A group of photosynthetic pigments that strongly absorb red light and transmit green.

Chloroplast A membrane-bound organelle in the cytoplasm of plants that converts light to chemical energy and uses it to convert CO_2 and water to carbohydrates and other organic compounds.

Chromatid One of the two duplicate parts of a replicated chromosome.

Chromatin The hereditary material of the nucleus, consisting of DNA and its associated histone and nonhistone proteins.

Chromoplast A plastid containing pigments of various colors.

Chromosome A subunit of the hereditary material of a cell nucleus, consisting of a single DNA molecule with its associated histone and nonhistone proteins.

Chromosome mutation A rearrangement of DNA sequences involving several genes or large segments of chromosomes.

Cilium (pl. cilia) A short flagellum.

Circulatory system The system that delivers oxygen and nutrients and carries away wastes in the tissues of many-celled animals.

Cisterna A single unit of the endoplasmic reticulum, consisting of a flattened sac formed by a single, continuous membrane.

Citric acid cycle *See* Krebs cycle.

Class A taxonomic group above an order and below a phylum; usually includes several orders.

Cleavage Division of the fertilized egg into progressively smaller cells.

Climate The prevailing weather conditions (temperature, wind, moisture) of a region.

Climax community A stable community of plants and animals that is typical of a region as the final community established by successions.

Clone A line of genetically identical cells, tissues, or organisms derived from a single parental cell or individual by mitotic divisions or asexual reproduction.

Coacervate An aggregation of protein molecules in solution held together by attractions between charged groups on the proteins.

Coagulate To thicken to a more viscous state.

Codon A "word" in the nucleic acid code consisting of three adjacent nucleotides.

Coelomate Having a body cavity lined entirely with tissue of mesodermal origin.

Coenzyme An organic cofactor.

Coevolution The linkage of the evolution of two species through reciprocal, back-and-forth selection, as in the mutual selection of greater specialization in predator and prey species.

Cofactor A nonprotein group linked to an enzyme that contributes to its catalytic activity.

Cohesion A force holding the molecules of a substance together, developed through intermolecular attractions or links such as hydrogen bonds.

Colonial organism, colony An association of cells of the same species in which the individual cells function independently.

Combustion Rapid oxidation producing heat and light.

Commensalism A close relationship between organisms of two different species that benefits one but brings no advantage or disadvantage to the other.

Community All the populations of an ecosystem.

Community diversity The number of different species living in a community.

Community stability The resistance to change in the density and types of species populations in a community.

Competition The effort by different individuals or populations to use the same limited resource.

Competition, interspecific Competition between individuals of different species.

Competition, intraspecific Competition between members of the same species.

Competitive exclusion Extinction of the members of a species or population through competition by another species or population for a limited resource.

Complementarity The fixed relationship between the bases of the two nucleotide chains of a double helix established by base pairing.

Compound A substance produced by the chemical combination of two or more different kinds of atoms.

Concentration gradient A difference in the number of molecules per unit volume of a substance between two or more parts of a space or solution.

Condensation reaction A chemical reaction in which the elements of a molecule of water are derived as a by-product from the reacting groups.

Condensation stage of meiosis (leptotene) The initial stage of meiotic prophase in which the

chromosomes fold into structures large enough to be visible under the light microscope.

Cone The reproductive structure of coniferous plants.

Conformation The three-dimensional arrangement taken on by the atoms of a molecule.

Conformation, folding The three-dimensional arrangement of the amino acid chain of a protein molecule.

Conjugation A form of sexual reproduction in which cells join and genetic material is passed from one cell to another.

Conservative replication A hypothetical form of DNA replication in which the two nucleotide chains of the original DNA molecule reassociate to form a double helix consisting entirely of "old" chains, and the two newly synthesized chains associate to form a completely "new" double helix.

Constriction, primary and secondary Narrow regions in condensed chromosomes visible at metaphase of mitosis and meiosis. Primary constrictions contain the kinetochores, the points at which the chromosomes attach to microtubules of the spindle. Secondary constrictions are other narrow regions of the chromosomes; some of these contain nucleolar genes.

Consumer In ecosystems, an organism that survives by ingesting organic matter.

Contraception Prevention of fertilization or pregnancy.

Convergence, convergent evolution The tendency of distantly related organisms to evolve similar structures in similar environments.

Core particle, nucleosome The combination of eight histone protein molecules that forms the central mass of a nucleosome.

Cornea The transparent outer covering at the front of the eye that admits light to the interior.

Corpus luteum An enlarged, scarlike structure formed by follicle cells remaining at the surface of the ovary following ovulation in mammals.

Cortex In higher plants, relatively unspecialized cells that fill the space between the outer epidermis and the vascular tissue.

Cotyledon A leaflike nutritive structure formed by the embryos of higher plants.

Covalent bond A chemical bond formed by shared electrons traveling in orbitals between two atoms.

Crista In mitochondria, a fold of the inner membrane extending into the central mitochondrial matrix.

Crossover A physical breakage and exchange of corresponding segments between homologous chromosomes that produces new combinations of alleles in meiosis.

Crossover unit A relative unit of distance between genes on a chromosome based on the percentage of crossing over between them.

Cuticle A waxy coating on the stems and leaves of plants that retards water loss by evaporation.

Cyclic photosynthesis A photosynthetic pathway in which electrons removed from chlorophyll are returned to chlorophyll after passing through an electron transport system.

Cytokinesis Partition of the cytoplasm into daughter halves during cell division.

Cytokinins A group of plant hormones that regulate the rate of cell division.

Cytoplasm The portion of the cell outside the nucleus or nucleoid.

Cytoplasmic inheritance Transmission of genetic information through genes located in the cytoplasm rather than the nucleus.

Cytoplasmic streaming Active flowing movement of the cytoplasm.

Cytoskeleton *See* Cell matrix.

Dark reactions Reactions that are not directly dependent on light to produce organic molecules in chloroplasts.

Death rate The number of deaths in a population per unit of time.

Decomposer (detritus feeder) An organism that feeds on dead organic matter.

Decondensation Partial or complete unfolding of chromosomes to the interphase state.

Dedifferentiation Partial or complete return of a specialized cell to an embryonic state.

Degeneracy Use of more than one code word or codon for the same amino acid in the nucleic acid code.

Deletion (of nucleotides) A mutation in which one or more nucleotides are lost from a DNA sequence.

Denaturation An alteration in the three-dimensional structure of a protein that causes loss of functional activity.

Denitrification Conversion of nitrates to gaseous nitrogen by soil bacteria.

Density-dependent factor Any factor controlling population size that increases in intensity as the number of individuals per unit area or volume increases.

Density gradient A difference in the number of molecules of a substance occurring per unit of volume in different parts of a solution, producing a difference in the physical density (weight per unit volume) of the solution between the parts.

Density-independent factor Any factor controlling population size that has equal effects in populations with high or low numbers of individuals per unit area or volume.

Dependency load The ratio of dependents to the number of persons that are self-supporting in a human population.

Detritus The organic matter of dead organisms.

Deuterostome The group of phyla, including echinoderms and chordates, in which the mouth develops from a region of the embryo other than the blastopore.

Diakinesis *See* Recondensation stage.

Dicot A plant with embryos that contain two cotyledons in the seed.

Dictyosome *See* Golgi complex.

Differentiation The developmental process by which cells that were originally similar become specialized in function and structure.

Diffusion The net or average movement of molecules from regions of higher to lower concentrations. Diffusion is accomplished by the random motion of individual molecules.

Digestion Breakdown of complex organic matter to more simple substances, usually catalyzed by enzymes.

Dihybrid cross A genetic cross involving alleles of two separate genes.

Diploid Having two copies of each chromosome in the nucleus.

Diplotene *See* Transcription stage.

Directional evolution A change in the distribution of alleles in the gene pool of a population toward a different average.

Disaccharide A sugar molecule consisting of two monosaccharide units linked together.

Disulfide linkage A covalent bond between the sulfur atoms of two amino acids (cysteines) located at different points in the amino acid chain of a protein.

Divergent evolution Splitting of a population or species into two or more groups with different average characteristics or average distributions of alleles.

Division The highest and most inclusive taxonomic group in the plant kingdom; equivalent to a phylum in the animal and other kingdoms.

DNA (deoxyribonucleic acid) A nucleic acid molecule composed of two helically wound chains of nucleotides containing the sugar deoxyribose and the four bases adenine, thymine, guanine, and cytosine; the fundamental hereditary material of all organisms.

Dominant An allele that produces the same effect whether it is present on one or both chromosomes carrying it in a diploid organism.

Dorsal The top or upper side in a bilaterally or radially symmetric organism.

Double fertilization The process of fertilization in flowering plants in which one of the two sperm nuclei of the pollen fuses with the egg nucleus to form the zygote, and the other fuses with the polar cells to form the endosperm.

Double helix A regular double spiral formed by two nucleic acid chains wound together and held in place by hydrogen bonds between the paired bases of the two chains.

Doublet A fused microtubule pair located at the periphery of the microtubule circle in the 9 + 2 system of a flagellum.

Ecocline Gradual changes in plant characteristics or types along a gradient of physical characteristics such as climate or topography.

Ecological niche *See* Niche.

Ecosphere (biosphere) The largest ecosystem; includes the entire earth.

Ecosystem The community of organisms occupying a defined region, their interactions with each other and with the physical features of their environment, and the flow of matter and energy through them.

Ectoderm The outermost layer of cells in an embryo; gives rise to the skin and nervous system.

Egg A female gamete or germ cell.

Electron A negatively charged subatomic particle occurring outside the nucleus of an atom; its mass is equivalent to $1/1857$ that of a proton.

Electron microscope A microscope using a beam of electrons as an illumination source.

Electron transport system A sequence of electron carriers, each capable of accepting and releasing electrons. Each carrier accepts electrons at a lower energy level than the carrier preceding it in the chain.

Electrostatic (or ionic) bond A binding force created by the attraction between positively and negatively charged atoms.

Element A pure substance containing a single type of atom.

Embryo An organism developing from the fertilized egg. In plants, the embryo is contained in the developing or mature seed; in animals, the embryo is contained in surface coats derived from the egg or within the body of the mother. Release from these locations marks the end of the embryonic period.

Embryonic disc An early mammalian embryo with two cell layers at a stage just before differentiation of ectoderm, mesoderm, and endoderm.

Endocytosis The movement of substances into cells through the formation of pockets in the plasma membrane that subsequently pinch off to form cytoplasmic vesicles.

Endoderm The innermost layer of cells of an embryo; gives rise to the lining of the digestive tract.

Endometrium The spongy, blood-filled tissue lining the uterus that forms and breaks down each month as part of the menstrual cycle.

Endoplasmic reticulum (ER) A system of interconnected membranous sacs in the cytoplasm of eukaryotic cells. The system is called rough ER when ribosomes are attached to the membranes and smooth ER if no ribosomes are attached.

Endosperm A nutritive tissue surrounding the embryo of seed plants.

Energy The ability to accomplish work.

Energy efficiency In any system, the ratio of energy that is used to do work to the total energy input.

Entropy A measure of the energy unavailable to do work in a system; also considered as the degree of randomness or disorder of a system.

Environment The collective objects and conditions surrounding an individual or population, including other organisms, climate, light, and physical objects.

Enzyme A protein capable of increasing the rate of a chemical reaction without being used up in the reaction.

Enzyme, inducible An enzyme that can be increased or decreased in concentration inside a cell by changes in the chemical medium surrounding the cell.

Epidermis The outermost cell layer of a plant or animal.

Epistasis A pattern of inheritance in which one gene overrides the effects of another.

Epoch A division of geological time between ages and eras in length.

Equilibrium A state in which no net change occurs.

Era A division of geological time greater than an epoch.

Estrogens A group of hormones that stimulate the growth and activity of female reproductive organs.

Ethylene A plant hormone that stimulates fruits to ripen and increases the diameter of stems and roots in response to wounding.

Eugenics The improvement of races and breeds by selective breeding.

Eukaryote A cell or organism that contains nuclei surrounded by a nuclear envelope and has cytoplasm containing mitochondria and chloroplasts (in eukaryotes capable of photosynthesis).

Eutrophication A process of extensive accumulation of organic matter in bodies of water, leading to depletion of oxygen.

Evaporation The escape of molecules of a liquid into a gas or vapor.

Evolution A change in the average distribution of alleles in a population with time.

Excited state A state in which an electron occupies a higher-energy orbital lying at a greater distance from its atomic nucleus than its normal, ground-state orbital.

Excretion Elimination of the waste products of metabolism.

Exploitation competition A form of competition in which one individual or population reduces the availability of a limited resource to another by using it more efficiently.

Exponential growth Increases in population size that follow an exponential curve if plotted against time.

F_1, F_2 The first and second generation, respectively, following a mating cross between genetic types under study.

Facilitated diffusion A form of transport in which the diffusion of polar substances through cellular membranes is enhanced by membrane proteins.

Family A taxonomic group above a genus and below an order; usually includes several genera.

Fat A neutral lipid that is solid or semisolid at biological temperatures.

Fatty acid A class of molecules containing an unbranched chain of carbon atoms with attached hydrogens and a terminal —COOH group that gives the class its acidic properties.

Fermentation A glycolytic sequence in which an organic molecule is the final acceptor of electrons removed in oxidation.

Fertilization The fusion of egg and sperm or of male and female gametes; restores the chromosomes to the diploid number and initiates embryonic development.

Fetus The human embryo after 8 weeks of development and until birth.

First law of thermodynamics The statement that matter-energy can neither be created nor destroyed but merely transformed from one form to another.

Fitness A measure of the reproductive success of a population.

Flagellum A long, thin appendage of the cell surface containing an axial system of microtubules, capable of whiplike movement.

Flower The reproductive structure of a flowering plant (angiosperm), consisting of stamens, which bear the male reproductive cells, and carpels, which produce the female reproductive cells.

Fluid mosaic model The hypothesis for membrane structure proposing that proteins float as individual units on or within a framework formed by a lipid bilayer.

Fluorescence The release of energy as light by an electron falling from an excited to a ground-state orbital.

Follicle (1) A layer of accessory cells surrounding the egg in some animals. (2) A fluid-filled blisterlike outgrowth on the surface of the mammalian ovary containing the egg.

Follicle-stimulating hormone (FSH) A hormone that stimulates an egg in the ovary to complete meiosis and promotes the growth of the follicle surrounding the egg.

Food chain A sequence of organisms through which energy in the form of organic matter flows in an ecosystem.

Food web A branching and interconnected food chain.

Fossil The preserved remains of all or part of an organism or the impression of an organism left in mineral matter.

Founder effect The change in frequencies of alleles from the averages of a parent population created when a new population is founded by a few individuals colonizing a different region.

Free energy Energy available to do chemical or physical work.

Furrowing Cytoplasmic division in animals by progressive constriction of the cell periphery.

G_1, G_2 phases of interphase The periods of interphase during which no DNA synthesis takes place.

Gamete A haploid reproductive cell; an egg or sperm.

Gametogenesis The formation of reproductive cells (gametes).

Gametophyte The haploid phase of the life cycle in plants, which produces gametes following mitosis.

Ganglion A bundle of nerve cells that serves as a coordinating or association center.

Gastrula The stage of embryonic development in which the primary cell layers (ectoderm, mesoderm, and endoderm) are formed.

Gene A sequence of nucleotides in a DNA molecule coding for a unit of function, such as synthesis of a protein, rRNA, or tRNA molecule.

Gene flow A change in the frequency of alleles in a population due to immigration or emigration.

Gene (or allelic) frequency The number of individuals in a population carrying a gene or allele divided by the total number of individuals in the population.

Gene map A map showing the relative positions of genes on a chromosome or DNA molecule.

Gene mutation A change in the sequences of nucleotides within the boundaries of a gene.

Gene pool The total collection of genes and alleles in a population.

Generative cell One of the two cells of a pollen grain that divides to form the sperm nuclei.

Genetic code A code for the synthesis of proteins and RNA, and for regulatory functions, spelled out by the sequence of nucleotides in a nucleic acid molecule.

Genetic drift A change in the frequency of alleles in a population due to chance.

Genetic engineering Modification of the DNA sequences of one individual by introducing chromosomes or DNA from another.

Genotype The genetic constitution of an organism. *Compare* Phenotype.

Genus A classification unit containing several species with similar characteristics and common evolutionary origins.

Geological time scale A time scale derived by estimating the relative ages of rock strata.

Geographic or allopatric speciation The appearance of new species through the selective effects of the separation of a parent species or population into two or more populations occupying different geographic regions.

Germinate In plant or fungal reproductive cells, or in seeds, to resume growth after a period of arrest.

Germ line A line of cells and their descendants that gives rise to gametes, or eggs and sperm.

Gibberellins A group of plant hormones that control the rate of cell elongation.

Gloger's rule The tendency in warm-blooded animals for black surface pigments to be reduced in warm, dry climates and for brown surface pigments to be reduced in cold, moist climates.

Glycogen An animal starch consisting of branched chains of glucose units.

Glycolipid A molecular type formed by combined lipid and carbohydrate subunits.

Glycolysis Conversion and oxidation of glucose into pyruvic acid.

Glycoprotein A molecular type formed by combined protein and carbohydrate subunits.

Golgi complex (dictyosome) A system of ribosome-free sacs in the cytoplasm of eukaryotic cells that modifies proteins after their synthesis; also concerned with the assembly of polysaccharide units in plant cells.

Gradual speciation The appearance of new species through the slow accumulation of new mutations and adaptations over long periods of time.

Granum A stack of flattened membranous sacs inside the chloroplasts of higher plants associated with the light reactions of photosynthesis.

Gross primary production The total organic matter produced by the primary producers of an ecosystem. *Compare* Net primary production.

Ground meristem An embryonic cell layer that gives rise to parenchymal cells in the higher plants.

Ground state The orbital around an atomic nucleus representing the lowest energy level of an electron.

Group selection The hypothesis that altruistic or selfless behavior arises through the selective advantage brought to an entire population by the behavior of selfless individuals within the population.

Growth rate The change in number of individuals in a population per unit of time.

Habitat The place or region occupied by an individual or population.

Half-life The time required for half the atoms or molecules of a substance to be converted to another type.

Haploid or monoploid A cell or organism having one copy of each chromosome in the nucleus.

Hardy–Weinberg principle The unchanging distribution of alleles in a population that undergoes no evolutionary change.

Hatch–Slack cycle *See* C_4 cycle.

Herbivore An animal that lives by ingesting plant matter.

Heterotroph A cell or organism that must ingest organic matter to survive. *Compare* Autotroph.

Heterozygote An individual possessing two different alleles of a gene. *Compare* Homozygote.

Hexose A carbohydrate molecule formed by a carbon chain containing six carbon atoms.

Histone protein A basic (positively charged) protein that combines with DNA in the chromatin of eukaryotes.

Homolog, homologous chromosome A member of a chromosome pair having the same genes in the same sequence.

Homozygote An individual carrying identical alleles of a gene on both chromosomes of a homologous pair. *Compare* Heterozygote.

Hormone An organic substance capable of controlling the rate of a biological function, usually exerting an effect at a distance from the cell that produces it.

Hybrid An offspring of two parents genetically unlike or of two different species.

Hydrocarbon An organic molecule containing carbon and hydrogen.

Hydrogen bond An attraction set up between a hydrogen atom that has become positively charged through partial loss of unequally shared electrons and an adjacent atomic nucleus (usually oxygen or nitrogen) that has become negatively charged through partial gain of unequally shared electrons.

Hydrolysis reaction A chemical breakdown in which the elements of a molecule of water are added to the reacting groups.

Hydrophilic Tending to associate with polar substances such as water.

Hydrophobic Tending to associate with nonpolar substances.

Hypha A filamentous, threadlike unit that forms the body structure of most fungi.

Hypothesis, scientific A proposed explanation of the relationship of observed facts that can be tested by experiment.

Incomplete dominance *See* Lack of dominance.

Incompleteness (of the fossil record) The fact that the number of organisms preserved as fossils represents only a small fraction of the total living on the earth in the past.

Independent assortment The independent separation and delivery of the genes and alleles carried on one chromosome pair from the genes carried on any other chromosome pair during meiosis and fertilization.

Induction A developmental process in which one group of cells (the inducer) influences a second group of cells (the reactants) to develop along certain pathways.

Inflection point The point on a curve at which the rate of increase in slope becomes zero; in population curves, the point at which the rate of increase in growth first begins to slow.

Inhibition A reduction in the ability of an enzyme to catalyze a chemical reaction, detectable as a decrease in the rate of the reaction catalyzed by the enzyme.

Inorganic A term describing all molecules except the carbon or carbon-chain molecules that occur in living organisms.

Instantaneous growth rate The growth rate of a population at a vanishingly small instant of time.

Instructive hypothesis of induction The hypothesis that an inducing tissue "tells" a reactant tissue how to develop. *Compare* Permissive hypothesis of induction.

Interference competition A form of competition in which one individual or population directly blocks another's access to a resource.

Intermediate filament A fiber formed from nonmotile proteins that provides mechanical support to regions of the cytoplasm.

Internal skeleton A framework of hard or supportive elements located in the body interior.

Interphase The period of the cell cycle between divisions when most cell molecules are synthesized.

Intragenic recombination Recombination between two points located within the DNA of a single gene.

Intrinsic rate of population increase (r) The quantity obtained by subtracting deaths from births in a population.

Invagination An inpocketing of the surface of a cell, cell structure, or embryo.

Inversion A mutation caused by the removal and reversed reinsertion of a sequence of nucleotides in a DNA molecule.

Invertebrate An animal without a backbone (vertebral column) or notochord at any stage of development.

Ion An atom or molecule bearing a charge in solution due to loss or gain of electrons.

Ionic bond *See* Electrostatic bond.

Isolecithal egg An egg in which yolk is distributed uniformly throughout the cytoplasm.

Isotopes Atoms of the same element with the same number of protons but different numbers of neutrons.

J-shaped curve A growth curve representing rapid, exponential increase in population size.

K selection Selection of stable, highly adapted species by environmental conditions that remain constant for relatively long periods of time.

Karyotype The complete chromosome set of a cell or organism as seen at metaphase.

Keratin A fibrous protein forming the main constituent of horny surface structures in animals, such as skin, hair, claws, hooves, and feathers.

Ketone A carbon chain containing a carbonyl (C=O) group attached to one of the internal carbons of the chain.

Kinetic energy The energy associated with the motion of molecules or particles.

Kinetochore (centromere) A disklike structure forming the point at which chromosomes attach to spindle microtubules during cell division.

Kingdom The most inclusive unit used in classification.

Kin selection A hypothesis proposing that selfless behavior persists in evolution through the benefits brought to closely related individuals bearing the same alleles.

Krebs cycle (citric acid cycle) A cycle of reactions in which two-carbon acetyl units are oxidized to CO_2.

Label A radioactive or heavy isotope used to identify and trace a molecule or molecular group.

Lack of dominance (incomplete dominance) Incomplete masking of a recessive allele by a dominant allele when both are present in an individual. The external appearance of such a heterozygote is different from either homozygote.

Lamella A layer or membrane.

Land bridge Command given to pilot of a bridge.

Larva A developmental stage in which the young animal hatches from the egg coats to feed while completing transition to its adult form.

Law A fundamental scientific theory so extensively supported by repeated experiment that it is regarded as truth.

Leptotene *See* Condensation stage.

Leukemia Malignant growth arising from blood cells.

Leukoplast In plant cells, a colorless plastid filled with stored starch, lipid, or protein.

Lichen A composite life form consisting of a symbiotic association between an alga and a fungus.

Life expectancy (average) The age that at least half the individuals of a species will be expected to reach.

Light Energy traveling in waves at lengths (for visible light) between about 400 and 700 nanometers.

Light reactions The light-dependent reactions of photosynthesis in which light energy is converted to chemical energy.

Lignins A group of hard organic substances that occur with cellulose in the cell walls of higher plants.

Linkage The presence of two or more different genes on the same chromosome.

Linker, nucleosome The portion of DNA linking one nucleosome to the next in a chain of nucleosomes.

Lipids Biological molecules that are more soluble in nonpolar organic solvents such as acetone or ether than in water.

Lipoprotein A protein containing a chemically bound lipid group.

Logarithmic growth *See* Exponential growth.

Logistic growth An *S*-shaped population growth curve in which the curve is symmetric on either side of the inflection point.

Lumen The bore or central cavity of a tube.

Luteinizing hormone A hormone that stimulates the release of the egg from the ovary in mammals.

Lymphoma A malignant growth arising from or within the tissues of lymph glands.

Lysosome A cytoplasmic vesicle containing hydrolytic enzymes capable of breaking down all classes of cellular macromolecules.

Macroclimate The climate existing over extended geographical regions.

Macronucleus A large nucleus in ciliate protozoans that is the site of most of the RNA synthesis required for cell growth and activity.

Macronutrient A nutritive substance required by an organism in large quantity.

Map unit *See* Crossover unit.

Mass number The number of protons and neutrons in the nucleus of an atom.

Maternal chromosome The chromosome of a homologous pair that originates from the egg or maternal parent of an individual.

Matrix The substance filling the innermost space of a mitochondrion, bounded by the two mitochondrial membranes.

Medium The solution or substance in which reacting chemicals or cells are suspended.

Megaspore The haploid product of a meiotic division in the female reproductive structures of plants, which divides mitotically to form the female gametophyte.

Meiosis A sequence of two nuclear divisions without intervening DNA replication that reduces the chromosomes to the haploid number.

Melanoma A malignant growth arising from pigment cells.

Membrane A thin surface layer of lipid and protein molecules that acts as a partial or complete barrier to the passage of molecules.

Menstrual cycle The monthly cycle of growth and breakdown of the endometrial tissue lining the uterus in human and other primate females.

Meristem A localized region where cell divisions take place in a plant.

Mesenchyme Embryonic mesoderm cells that give rise to bone, cartilage, tendons, and other connective tissues.

Mesoderm The primary tissue layer of embryos, located between ectoderm and endoderm, that gives rise to the skeleton, circulatory system, and muscles.

Messenger RNA (mRNA) An RNA molecule that carries the coded information required to synthesize a complete protein or a polypeptide chain of a complex protein containing more than one chain.

Metabolism All the biochemical processes of an organism.

Metaphase The division stage during which chromosomes make attachments to spindle microtubules.

Metastasis Invasive growth or release of cells from a malignant tumor into other body regions.

Microclimate The climate in a restricted region such as the underside of a leaf or the surface of the soil.

Microfibril (cellulose) A fiber formed through the alignment and hydrogen bonding of cellulose molecules.

Microfilament An extremely fine cytoplasmic fiber of protein that is capable of producing movement.

Micrometer (μm) A unit of measurement equivalent to $1/1000$ of a millimeter, formerly called a micron.

Micron (μ) *See* Micrometer.

Micronucleus A small nucleus in ciliate protozoans that acts as the storage site for genetic information and is active only in cell division and reproduction.

Micronutrient A nutrient required by an organism in small or trace quantities.

Microorganism Any organism of microscopic dimensions, particularly bacteria, microscopic algae and fungi, and protozoans.

Micropyle The opening in the tissues of the ovule through which a pollen tube grows to reach the female gametophyte in flowering plants.

Microsphere A microscopic particle containing proteins or proteinoids condensed from solution.

Microspore A haploid product of meiosis in male reproductive structures of plants, which divides mitotically to form the male gametophyte.

Microtubule A fine tubular structure consisting of protein and capable of producing movement.

Midbody A residual structure left at the midpoint of the spindle in animals after chromosome division is complete. Cytoplasmic division proceeds through the region occupied by the midbody.

Mimicry An evolved similarity of one species to another.

Mineral An inorganic substance.

Missense chain The nucleotide chain of a DNA double helix that is complementary to the chain containing the code for a protein, RNA molecule, or other unit of function.

Mitochondrion A cytoplasmic membrane-bound organelle that contains (1) the oxidative reactions that break down pyruvic acid and acetyl units to CO_2 and H_2O and (2) the electron transport system that uses the energy released by these reactions to synthesize ATP.

Mitosis A nuclear division sequence in which the replicated chromosomes are precisely and equally divided and placed in two daughter nuclei that, as a result, are genetically identical.

Modified amino acids Additional amino acids produced by chemical changes in the 20 original amino acids incorporated into proteins during their synthesis.

Modified bases Additional nucleotides produced by chemical changes in the nitrogenous bases of the four original nucleotides incorporated into DNA and RNA during their synthesis.

Mole 6.022×10^{23} molecules of a compound; equivalent to the gram molecular weight (the atomic weight of a substance expressed as grams).

Molecular weight, gram molecular weight The sum of the atomic weights of the atoms of a molecule. Each atom is given a weight relative to a carbon atom (the isotope ^{12}C), which is assigned an atomic weight of 12. A gram molecular weight is the molecular weight expressed as grams.

Molecule A chemical combination of two or more like or unlike atoms; the smallest unit into which a compound can be divided without loss of its chemical properties.

Monocot A plant with embryos that contain one cotyledon in the seed.

Monohybrid cross A genetic cross involving the alleles of a single gene.

Monomer A molecule forming an individual unit of macromolecules consisting of a chain of repeated, identical molecular units.

Monoploid *See* Haploid.

Monosaccharide A carbohydrate molecule consisting of a chain of three to seven carbons linked to —OH groups and a single oxygen; the simplest carbohydrate unit.

Morphogenesis The growth and cell movements that produce the structures of an embryo.

Morphology The structure of an organism or subpart of an organism.

Mosaic development, mosaic eggs A pattern of development in which parts of the egg cytoplasm are programmed to develop into specific parts of the embryo.

Motility The ability to move.

Multiple alleles The presence of more than two alleles of a gene in the gene pool of a population or species.

Mutagen A chemical substance or physical agent that causes a change in the DNA sequence of a gene.

Mutation Any change in the nucleotide sequence of a gene.

Mutualism A close association between two species from which both derive benefit.

Mycelium The body of a fungus; formed by the growth of hyphae into a network.

NAD (nicotinamide adenine dinucleotide) A substance that accepts and carries high-energy electrons removed in cellular oxidations.

NADP (nicotinamide adenine dinucleotide phosphate) A substance that accepts and carries high-energy electrons generated in photosynthesis and cellular oxidations.

Nanometer (nm) A unit of measurement equivalent to $\frac{1}{1000}$ micrometer.

Natural increase (decrease) of populations The difference between the crude birth and death rates.

Natural selection Nonrandom reproduction due to environmental effects leading to a shift in the frequency or average distribution of alleles in a population.

Net primary production The total organic matter produced by the primary producers of an ecosystem minus the amounts used by the primary producers in respiration. *Compare* Gross primary production.

Neutral lipid (fats) Lipids containing no charged groups at pH 7. Typically, these lipids consist of a glycerol molecule linked to three fatty acids (a triglyceride).

Neutral solution A solution with pH equal to 7.

Neutron An uncharged particle in the nucleus of an atom having a mass similar to that of a proton.

Niche The total relationship of an organism or population to its environment.

Niche, fundamental The niche occupied by a population in the absence of competition.

Niche refuge A portion of the fundamental niche of a population in which competition with other populations is reduced or absent.

Niche specialization Evolution of the niche of a population toward a refuge that reduces or eliminates competition with other populations.

9 + 2 system The system of microtubules occurring in the axis of a flagellum, consisting of nine double microtubules (the doublets) arranged in a circle around two central single microtubules (the singlets).

Nitrification The conversion of ammonia to nitrates by soil bacteria.

Nitrogen cycle The major biogeochemical cycle of nitrogen between the major inorganic reservoir of N_2 in the atmosphere and the organic form in living organisms.

Nitrogen fixation The conversion of atmospheric nitrogen into nitrates and other compounds by soil bacteria.

Noncyclic photosynthesis A photosynthetic pathway in which electrons flow from water through the chlorophylls to an electron acceptor.

Nondisjunction The failure of the chromatids of one or more chromosomes to separate during meiosis.

Nonhistone protein Primarily an acidic (negatively charged) or neutral protein that associates with DNA and the histones in the chromatin of eukaryotes.

Nonpolar The characteristic of a molecule or molecular region in which all electrons are shared equally and no portions of the molecule or molecular region take on a relative charge.

Nonpolar bond A covalent bond in which electrons are shared equally, so that neither atom sharing the bond carries a charge.

Notochord An axial rod of stiff supportive tissue lying between the gut and nerve cord; present at some developmental stage in all chordates.

Nuclear envelope A double membrane system that separates the nucleus and cytoplasm in eukaryotic cells.

Nuclear region The part of a cell containing the DNA molecules storing and transmitting hereditary information.

Nucleic acid A chainlike molecule built up from a series of nucleotides linked end to end. Each nucleotide in the chain consists of a phosphate group, a five-carbon sugar, and a nitrogenous base.

Nucleoid The nucleuslike body containing the DNA of prokaryotes.

Nucleolar organizer The segment of a chromosome that contains genes coding for ribosomal RNA.

Nucleolus An irregularly shaped body in the nucleus, attached to the nucleolar organizer sites of the chromosomes, in which the subunits of ribosomes are synthesized.

Nucleoplasm All the substances within the cell nucleus enclosed by the nuclear envelope.

Nucleoprotein The combination of DNA with histone and nonhistone proteins.

Nucleosome A beadlike structure consisting of a length of DNA wrapped around a core of histones; forms the basic structural unit of chromatin.

Nucleotide One unit of a nucleic acid chain, consisting of a phosphate group, a five-carbon sugar, and a nitrogenous (purine or pyrimidine) base.

Nucleus, atomic The central, positively charged body of an atom containing neutrons and protons.

Nucleus, cell The central membrane-bound body of a eukaryotic cell; contains the chromosomes and the nucleolus.

Oil A neutral lipid that is liquid at biological temperatures.

Oocyte A reproductive cell that undergoes meiosis and develops into a mature egg.

Oogenesis The process, including meiosis and cytoplasmic maturation, by which an oocyte develops into a mature egg.

Operator A DNA sequence that controls an adjacent series of genes coding for proteins. Activation of the operator turns the adjacent series of genes "on" and they are transcribed; repression of the operator turns them "off."

Operon A group of adjacent genes in prokaryotic DNA that is transcribed into RNA as a single unit.

Orbital The average path around an atomic nucleus or nuclei followed by electrons, either singly or in pairs.

Order A unit of classification that includes several closely related families with common evolutionary origins.

Organ A many-celled structure of an organism, usually composed of several different kinds of tissues, that forms a functional unit.

Organelle A structural and functional unit inside a cell.

Organic chemical A compound containing carbon or carbon chains occurring in living organisms.

Organism Any living being.

Osmosis Net movement of water through a semipermeable membrane in response to a concentration gradient.

Osmotic pressure A force developed by the osmotic movement of water.

Ovary In animals, an organ that produces oocytes. In flowering plants, a part of the flower that houses the ovule.

Oviduct A tube through which eggs pass from the ovary to the outside, or, in mammals, to the uterus.

Ovulation The process by which a mature or maturing egg cell is released from the ovary.

Ovule A reproductive structure of seed plants, including the female gametophyte and a protective covering of tissue, that develops into the seed after fertilization.

Oxidation Removal or loss of electrons from an atom, group of atoms, or molecule.

Ozone Oxygen molecules formed through the combination of three oxygen atoms to form O_3.

Pachytene See Recombination stage.

Pairing stage of meiosis (zygotene) The stage of meiotic prophase during which homologous chromosomes come together and pair closely.

Parallelism, parallel evolution The tendency of closely related organisms to develop along similar pathways in similar environments.

Parasitism A close relationship between two species in which one, the parasite, lives at the expense of another, the host. Usually the host is injured through its association with the parasite.

Parenchyma A plant tissue containing thin-walled storage or photosynthetic cells; forms much of the interior bulk of stems, leaves, and roots and the pulp of fruits.

Parental chromosome One of the two chromosomes left unchanged by a breakage and exchange event in recombination.

Parthenogenesis The development of an egg into an embryo and adult without fertilization.

Partial diploid A haploid cell containing an extra piece of DNA of the same species that provides an additional copy of one or more genes.

Passive transport The transport of ions or molecules across membranes along concentration gradients.

Paternal chromosome The chromosome of a homologous pair originating from the sperm or paternal parent of an individual.

Pectins A family of complex polysaccharides that is present in plant cell walls as a filler substance between cellulose fibers.

Pedigree A family tree showing all marriages and offspring for several generations, distinguishing between males and females and showing the presence or absence of inherited traits.

Pentose A carbohydrate molecule formed by a carbon chain containing five carbon atoms.

Peptide An amino acid chain.

Peptide linkage, peptide bond A chemical linkage binding amino acids together in a peptide chain. The linkage includes four atoms in the form —N—C—.
$$\begin{matrix} | & \| \\ H & O \end{matrix}$$

Perinuclear compartment The space enclosed between the two concentric membranes of the nuclear envelope.

Permeable Allowing molecules to pass.

Permissive hypothesis of induction The hypothesis that in induction the developmental pathway triggered by the inducer is already programmed into the reactant tissue. Compare Instructive hypothesis of induction.

pH A scale used to describe the acidity or basicity of a solution. Solutions at pH 7 are neutral (the concentrations of H^+ and OH^- ions are equal); solutions at values below pH 7 are acidic (the concentration of H^+ ions is greater than that of OH^- ions); solutions at values above pH 7 are basic (the concentration of OH^- ions is greater). Technically, pH is the negative of the logarithm (base 10) of the H^+ ion concentration.

Phage See Bacteriophage.

Phagocytosis Cellular ingestion by a process in which the cytoplasm flows around and engulfs a particle.

Phenotype The physical and biochemical characteristics of an organism. Compare Genotype.

Pheromone A chemical used as a means of communication between individuals of the same species.

Phloem A tissue of vascular plants that conducts the organic products of photosynthesis to all parts of the plant.

Phospholipid A lipid molecule containing a phosphate group. Natural phospholipids consist of a glycerol unit bound to two fatty acids and a phosphate group; the phosphate is in turn linked to a nitrogenous alcohol.

Phosphorus cycle The biogeochemical flow of phosphorus between its natural reservoirs in rocks of the earth's crust and living organisms.

Phosphorylation The addition of phosphate groups to a substance.

Photophosphorylation The synthesis of ATP through electron transport in the light reactions of photosynthesis.

Photosynthesis A process in which light energy is absorbed and converted to chemical energy and used to synthesize complex organic substances from simple inorganic raw materials.

Photosystem A combination of chlorophyll, carotene, and protein molecules forming a light-absorbing unit in chloroplast membranes.

Phragmoplast A structure formed at the spindle midpoint of dividing plant cells that develops into the cell plate, the new wall separating the cytoplasm between daughter cells.

Phylum A classification unit that includes several closely similar classes with common evolutionary origins.

Phytoplankton Floating, microscopic plants.

Pigment molecule A molecule that is pigmented or colored in solution because it absorbs light at some wavelengths and transmits light at other wavelengths.

Pinocytosis Cellular ingestion of molecules by a process in which the cell surface membrane attaches to the molecules and then folds inward or invaginates, forming a vesicle that pinches off and sinks into the underlying cytoplasm.

Placenta An organ found in most mammals that is formed by tissues of both the mother and embryo and through which the embryo receives nutrients and eliminates wastes.

Plankton Floating, microscopic organisms suspended in a body of fresh or salt water.

Plaque A hole in a solid layer of bacteria growing in a culture dish, opened by the action of viruses infecting the bacteria.

Plasma membrane The surface membrane of a cell.

Plasmid A small, extra DNA circle occurring in bacteria in addition to the DNA of the nucleoid.

Plasmodesmata Minute pores or openings in the walls that separate plant cells, through which the cells make cytoplasmic connections.

Plastids A family of cytoplasmic organelles in eukaryotic plant cells, carrying out photosynthesis, containing pigments, or storing the products of photosynthesis.

Point mutation A change in a single nucleotide in the DNA sequence of a gene.

Polar The characteristic of a molecule or molecular part in which unequal electron sharing produces regions with relative charges.

Polar body A small nonfunctional cell produced by unequal cytoplasmic division during meiosis in oocytes.

Polar bond A covalent bond in which electrons are unequally shared by two atomic nuclei, giving the nuclei a relative charge.

Polar nuclei The two nuclei of a female gametophyte in flowering plants that fuse with a sperm nucleus to form the triploid endosperm.

Polarity (of eggs) The unequal distribution of yolk bodies, pigment granules, and other components of mature eggs.

Polarity (of a molecule) The degree to which a molecule develops charged regions due to unequal electron sharing by its atomic nuclei.

Pole, spindle One of the ends of the spindle.

Pollen, pollen grain The immature male gametophyte of a seed plant, enclosed within a hard, impermeable protective coat.

Pollen tube A tubelike outgrowth of a pollen grain, through which the sperm nucleus or nuclei are carried to the female gametophyte.

Pollination The delivery of pollen to the stigma of a flower of the same species.

Polygenic inheritance A pattern of inheritance resulting from the interaction of two or more different genes.

Polymer A molecule consisting of a series of repeating, identical molecular subunits linked into a chain.

Polymerization The chemical linkage of identical molecular subunits into a chain.

Polypeptide Three or more amino acids linked together by peptide linkages.

Polyploidy A condition in which more than the diploid number of chromosomes is present in the nuclei of an organism.

Polysaccharide A molecule consisting of three or more monosaccharide units linked end to end.

Polytene chromosome A chromosome containing hundreds or thousands of identical DNA molecules produced by repeated DNA replication without mitosis or cytoplasmic division.

Polyunsaturated The condition in which a fatty acid or neutral lipid contains multiple carbons sharing double bonds.

Population A group of interbreeding individuals.

Pore complex A pore in the nuclear envelope and its annulus. The pore serves as a channel through the nuclear envelope between nucleus and cytoplasm; the annulus fills the pore and controls diffusion through it.

Posterior The rear end of bilaterally symmetric organisms.

Potential A measure of the energy level or voltage of an electron.

Potential energy Energy that a molecule or object possesses because of its position or the degree of complexity or organization of its component parts.

Precursor A chemical substance from which another substance is formed.

Predation The act of feeding on another organism.

Preprophase band A band of microtubules that appears in some plants during interphase; the preprophase band marks the position at which the cell plate will form between daughter cells at the next cell division.

Prey An organism eaten or ingested by a predator.

Primary consumer The consumers feeding on the primary producers of an ecosystem or community.

Primary mesenchyme In embryos, loosely organized or scattered early mesoderm.

Primary producers The photosynthetic organisms of an ecosystem or community.

Primary spindle The initial bundle of spindle microtubules formed before the nucleus breaks down in animal mitosis or meiosis.

Primary structure (of proteins) The sequence of amino acids in a polypeptide chain.

Primate A member of the order Primates to which humans, monkeys, and apes belong.

Primer (DNA) A short length of DNA that is necessary as a starting point for enzymes (DNA polymerases) catalyzing the synthesis of DNA.

Primitive gut A tubelike endodermal structure in animal embryos that gives rise to parts of the digestive system of the adult.

Procambium A primary tissue of plant embryos that divides to give rise to vascular tissues.

Procentriole A developing centriole.

Progesterone A hormone in mammals that promotes growth of the uterine lining and stimulates the development of milk glands.

Progestin A synthetic hormone that mimics the effects of progesterone.

Prokaryotes Cells (bacteria and blue-green algae) that have a nucleoid instead of a nucleus. Membranous structures of the cytoplasm are limited to the plasma membrane and structures derived from it (no mitochondria or chloroplasts).

Prophase (mitosis) The initial stage of mitosis: the chromosomes fold or condense from the interphase to the compact metaphase state, and the spindle forms.

Proplastid A developing plastid.

Protamines Small, highly basic proteins that replace the histone chromosomal proteins during the development of sperm cells.

Protease An enzyme that catalyzes the breakdown of peptide linkages.

Protein An organic molecule consisting of a chain of amino acids linked by peptide bonds, held in a three-dimensional folding arrangement by hydrogen bonds and disulfide linkages.

Proteinoid A proteinlike molecule consisting of a random sequence of amino acids.

Proteoglycan A complex cell-surface molecule consisting of a protein combined with a carbohydrate unit derived from glucose.

Protoderm The cell layer giving rise to the epidermis in early plant embryos.

Proton A positively charged particle in the nucleus of an atom, assigned a relative mass of 1.

Protoplast The living part of a cell; in plants, the cell except for the cell wall.

Protostome A group of phyla (annelids, arthropods, and molluscs) in which the embryonic blastopore develops into the mouth.

Pseudocoelomate Having a body cavity incompletely lined by mesoderm.

Pseudopod A lobe or footlike extension of the cytoplasm associated with movement or feeding.

Purine A nitrogenous base that occurs in DNA; consists of two carbon–nitrogen rings.

Pyrimidine A nitrogenous base that occurs in DNA and RNA; consists of a single carbon–nitrogen ring.

Pyruvic acid oxidation A series of reactions in mitochondria in which pyruvic acid is oxidized and CO_2 is released, producing a two-carbon acetyl unit that forms the immediate fuel for the Krebs cycle.

Quantum speciation The sudden appearance of new species through rapid and extensive genetic changes.

Quaternary structure (of proteins) The three-dimensional structure of complex proteins containing two or more polypeptide chains.

r Selection Selection of species with short life span and high reproductive capacity by environments so unstable that favorable conditions persist only for brief periods.

Race A recognizable subgroup of individuals within a species, sharing similar genetic characteristics.

Radial symmetry An arrangement of body parts in a circle around a central axis running through the mouth. Organisms with radial symmetry have top (dorsal) and bottom (ventral) sides but no right or left sides. *Compare* Bilateral symmetry.

Radioactive dating A method for dating fossils by measuring the amount of decay of radioactive substances deposited with the fossil.

Radioactive decay *See* Radioactivity.

Radioactivity Spontaneous release of nuclear particles from an unstable isotope of an element.

Random coil In protein structure, an arrangement of amino acids with no regular repeating or periodic structure.

Range (of a population) The total surface area or volume of space occupied by the individuals of a population.

Reading frame The starting and continuing points from which bases are read three at a time in a nucleic acid coding sequence.

Realized niche The niche actually occupied by a population.

Recessive allele An allele of a gene whose effects are masked or compensated for by a dominant allele of the same gene.

Recombinant chromosome A chromosome with segments originating from opposite homologs of a chromosome pair, resulting from physical breakage and exchange between homologous pairs at prophase of meiosis.

Recombination The process by which alleles of homologous chromosomes are exchanged and mixed during prophase of meiosis, producing new combinations of alleles in the chromosomes of gametes.

Recombination stage of meiosis (pachytene) The stage of meiotic prophase during which homologous chromosomes exchange segments and allelic recombination occurs.

Recondensation stage of meiosis (diakinesis) The final stage of meiotic prophase, during which chromosomes fold down into compact form in preparation for the subsequent metaphase I. The recondensation stage is recognizable only if extensive unfolding of chromosomes has taken place during the previous synthesis (diplotene) stage of meiotic prophase.

Redox couple A pair of substances capable of interacting in electron transfer in an oxidation-reduction reaction. As the first substance is oxidized (loses electrons) the second substance is reduced (accepts the electrons).

Redox potential The characteristic energy associated with the electrons released by a substance.

Reduction Gain or acceptance of electrons by an atom, group of atoms, or molecule.

Regulation The control of the types and quantities of proteins synthesized in a developing or mature individual.

Regulation, transcriptional Regulation of protein synthesis by controlling the types and quantities of messenger RNAs synthesized in the cell nucleus.

Regulation, translational Regulation of protein synthesis by controlling the types and quantities of messenger RNAs translated into proteins on ribosomes.

Regulative development A form of development in which cells of the early embryo are flexible in developmental pattern; individual cells, if removed from the early embryo, can develop into complete individuals.

Regulative eggs Eggs in which subregions of the cytoplasm can give rise to complete embryos if compartmented into completely separate cells.

Regulator gene In the operon mechanism, a gene that codes for the synthesis of a repressor protein, a protein that determines whether the operator is active or inactive. *See* Operator.

Replication The semiconservative synthesis of DNA on a DNA template, producing an exact copy of the template molecule.

Repressor In the operon mechanism, a protein coded for by a regulator gene that controls the activity of an operator gene.

Reproductive barrier Any barrier to reproduction between members of different species.

Reproductive isolation A condition in which a population or species is unable to interbreed successfully with another population or species.

Resolution A measure of the ability of a microscope to distinguish specimen details of small size.

Resource Any substance or factor in the environment that can be used by the individuals of a population.

Respiration Cellular oxidations in which oxygen atoms are the final acceptors of the electrons removed.

Rhizoid A rootlike holdfast anchoring a plant to the soil.

Ribosomal RNA (rRNA) An RNA that forms an integral part of ribosomes.

Ribosome A complex cytoplasmic particle containing RNA and protein that synthesizes proteins.

RNA (ribonucleic acid) A nucleic acid consisting of a chain of nucleotides containing the five-carbon sugar ribose and the purine and pyrimidine bases adenine, uracil, guanine, and cytosine.

Root apical meristem A tissue of plant embryos that divides to give rise to roots in the adult plant.

Rough ER, rough endoplasmic reticulum *See* Endoplasmic reticulum.

S phase of interphase The period of interphase during which nuclear DNA is replicated.

S-shaped curve *See* Sigmoid curve.

Sarcoma A malignant tumor derived from connective tissue.

Saturated fat A fat (neutral lipid) in which all the possible bonds on the carbon chains of the fatty-acid units are linked to hydrogens (no double bonds in the fatty-acid chains).

Saturation, enzyme A condition in which substrate concentration is high enough to occupy all the available enzyme molecules in a solution; further increases in substrate concentration produce no increase in reaction rate.

Scientific method A technique for discovering truth through observation, hypothesis, and experimental test.

Second law of thermodynamics The statement that all interacting systems proceed toward a total condition of greater disorder.

Secondary consumer In a food chain, an organism that feeds on primary consumers.

Secondary structure (of proteins) The pattern of folding or twisting of the amino acid chain of a protein into random coil, alpha-helical, or other conformations.

Secretion The release of any cell product to the cell exterior.

Seed A plant embryo, with nutritive tissues, enclosed within protective coats derived from tissues of its parent plant. The embryo within the seed is arrested at an intermediate stage of development.

Segmentation The division of an animal body into repeating parts with similar or identical structure.

Semiconservative replication A pattern of DNA replication in which the two nucleotide chains of the parent DNA molecule separate and, after acting as templates for replication, remain with their newly synthesized copies. As a result, each replicated molecule consists of one old and one new nucleotide chain.

Semipermeable membrane A natural or artificial membrane that allows some molecules to pass but not others.

Sense chain The nucleotide chain of a DNA molecule that contains the code for a protein molecule, RNA molecule, or other unit of function and acts as a template for RNA transcription.

Serum, blood The fluid or watery part of blood.

Sessile Permanently attached to a surface or substrate.

Sex chromosomes Chromosomes that are structurally and genetically different in males and females of the same species.

Sex-influenced trait A genetic trait that is expressed differently in males and females of a species, usually through the influence of sex hormones.

Sex linkage (sex-linked inheritance) The pattern of inheritance of genes carried on the sex chromosomes of a species.

Sexual reproduction Reproduction involving the fusion of haploid gametes (usually egg and sperm).

Shoot apical meristem A tissue of plant embryos that gives rise to the stems, leaves, flowers, and other parts of the adult plant (except the roots).

Sigmoid curve A growth curve that follows an exponential pattern initially and then tapers off to an equilibrium level.

Smooth ER, smooth endoplasmic reticulum *See* Endoplasmic reticulum.

Solution A homogeneous mixture of two substances in which the molecules of the substances are individually suspended.

Solvent A substance (usually a liquid) capable of dissolving or dispersing the molecules of another substance, forming a homogeneous mixture of the two.

Somatic cell A body cell that does not give rise to gametes.

Specialization A change in the structure and function of a cell toward a particular type.

Speciation The formation of new species.

Species A group of organisms that resemble each other in structure or function more closely than the members of any other group and differ from the organisms of any other group in at least one clearly defined characteristic. In sexually reproducing organisms, the members of a species are all potentially capable of interbreeding to produce viable offspring like their parents in characteristics.

Specificity, enzyme The limitation of enzyme activity to catalysis of a single reaction or group of closely similar reactions or to recognition of a single substrate or group of closely similar substrates.

Sperm, spermatozoon A male gamete, usually flagellated and motile, that is different in morphology from the egg.

Spermatocyte A cell that undergoes meiosis and maturation to produce sperm.

Spermatogenesis The process by which spermatocytes develop into mature sperm cells.

Spicule A minute calcium or silica-containing supportive element in some sponges.

Spindle A bundle of parallel microtubules that divides the chromosomes during mitosis and meiosis.

Spindle midpoint The widest part of the spindle, halfway between the spindle poles.

Spindle pole One of the two ends of the spindle.

Spore In plants, a haploid reproductive cell produced by meiosis in sporophytes; gives rise to the gametophyte phase of the life cycle.

Sporophyte The diploid phase of plant life cycles. The sporophyte produces haploid spores by meiosis.

Stabilizing evolution Evolution toward more uniform average characteristics in a population, through elimination of individuals with characteristics falling near the extremes for the population.

Stamen The male reproductive parts of a flower that give rise to pollen.

Starch A polysaccharide storage molecule consisting of glucose units linked end to end.

Steroids Lipids based on a complex structure containing four interlocking carbon rings.

Sterol A type of steroid with the basic ring structure of the steroids plus a long carbon side chain.

Stigma The sticky tip of the carpel in flowers to which pollen grains adhere.

Stoma (plural = stomata) An opening in the epidermis of plant leaves and stems through which gases exchange between the plant interior and the atmosphere.

Stratum, rock A layer of rock solidified from sediment.

Stroma The inner fluid substance of a chloroplast enclosed by both boundary membranes.

Stromatolite A characteristic, moundlike mineral formation deposited by certain blue-green algae.

Substrate (1) The reactant molecule or molecules in a chemical reaction catalyzed by an enzyme. (2) The underlying solid substance on which an organism moves or is attached.

Succession In ecosystems, a sequence of communities that continues to change until the final climax community is established. Each successive community produces conditions that are favorable for the appearance of the next, until the stable climax community appears.

Succession, primary The sequence of communities occupying a previously barren region.

Succession, secondary The sequence of communities appearing in a region after the climax community of the region has been destroyed.

Supergene A group of alleles, closely linked on the same chromosome, that is inherited as a unit.

Surface tension An intermolecular force developed by molecules at the surface of a liquid, tending to hold the molecules together. As a result, the surface molecules form a semirigid layer.

Survivorship curve A curve produced by plotting the percentage of surviving individuals against their age.

Suspensor In seed plants, a stalklike column of cells anchoring the embryo to the tissues of the parent plant.

Symbiosis A close association between individuals of two different species.

Sympatric speciation The appearance of new species through evolutionary changes in populations occupying the same geographic region.

Synapsis The pairing of homologous chromosomes during meiotic prophase.

Synaptonemal complex A structure that fills in the narrow space between synapsed chromosomes during the recombination stage of meiosis.

Synthesis The chemical assembly of complex substances from simple substances.

System A defined or delimited part of the chemical or physical universe under study.

Systematics (taxonomy) The science describing and arranging organisms into a scheme of classification.

Taxonomy *See* Systematics.

Telolecithal egg An egg with yolk concentrated at one end of the cytoplasm.

Telophase (mitosis) The final stage of mitosis, during which the divided chromosomes return to the interphase state and become enclosed in newly formed nuclear envelopes.

Template In DNA replication, the "old" or parental nucleotide chain from which a new, complementary chain is copied.

Teratoma A malignant tumor derived from reproductive cells.

Terminator codon One of three codons (UAG, UAA, or UGA) in the nucleic acid code that specifies "stop" in the code for synthesis of a protein.

Territoriality The defense of a region used for feeding or mating by an individual, mating pair, or population.

Tertiary consumer An organism in a food chain that feeds on secondary consumers.

Tertiary structure (of proteins) The three-dimensional structure of a protein containing a single polypeptide chain.

Testcross (backcross) A genetic cross in which an individual being tested is crossed with a homozygous recessive (an individual that carries only the recessive alleles of a gene being studied).

Testis The organ of animals that produces sperm cells.

Tetrad A pair of homologous chromosomes at prophase of meiosis, consisting of four chromatids.

Theory A scientific hypothesis that has been repeatedly and completely supported by experimental tests.

Thermodynamics The scientific study of energy changes in reacting systems.

Thylakoid A structure containing the molecules that carry out the light reactions of photosynthesis; consists of a closed sac formed by a single, continuous membrane.

Tissue A group of cells having similar structure and function within a many-celled organism.

Tornaria larva The characteristic larva of the echinoderm–chordate phyla that (when present) has ciliary bands distributed in a ring encircling the mouth. *Compare* Trochophore larva.

Tracer *See* Label.

Transcription The synthesis of RNA on a DNA template, catalyzed by an RNA polymerase enzyme.

Transcription stage of meiosis (diplotene) A stage of meiotic prophase, following recombination, during which the chromosomes partially or completely uncoil and RNA transcription and protein synthesis occur.

Transduction The transfer of genetic material from one bacterial cell to another by a virus particle.

Transfer RNA (tRNA) A type of RNA that combines with amino acids and pairs with coding triplets on a messenger RNA molecule.

Transformation Genetic changes induced in a bacterial cell by DNA absorbed from the surrounding medium.

Transition A point mutation in which a different purine is substituted for the correct purine base, or a different pyrimidine for the correct pyrimidine base, in a DNA nucleotide sequence.

Translation The process of protein synthesis at the ribosome.

Translocation A mutation in which a chromosome segment breaks loose and attaches at a new location on the same or a different chromosome.

Transversion A point mutation in which a purine base is substituted for a pyrimidine base, or a pyrimidine for a purine base, in a DNA nucleotide sequence.

Triglyceride A neutral lipid consisting of a glycerol unit linked to three fatty-acid units.

Triose A carbohydrate molecule formed by a carbon chain containing three carbon atoms.

Triplet A code word or codon in the nucleic acid code, consisting of three adjacent nucleotides.

Triploid Having three copies of each chromosome in the nucleus.

Tritium A radioactive isotope (^{3}H) with one proton and two neutrons in its atomic nucleus.

Trochophore larva The characteristic larva of the annelid–arthropod–mollusc group of phyla that (when present) has cilia restricted to bands located in front of the mouth. *Compare* Tornaria larva.

Trophic level A group of producers or consumers of a food chain that uses the same source of energy or organic nutrients.

Trophoblast A tissue originating from the embryo that forms the membranes surrounding the embryo and the embryonic part of the placenta.

Tumor A growth produced by uncontrolled cell division.

Ultrastructure Cell structures too small to be visible in the light microscope.

Universality The use of the same codons to designate the same amino acids in all organisms.

Unsaturated fat A fat (neutral lipid) in which one or more double bonds occur in the carbon chains of the fatty-acid units.

Urethra The duct leading from the bladder to the exterior; also serves as a genital duct in male mammals.

Uterus A muscular sac in mammals in which embryonic development takes place.

Vacuole, vesicle A fluid-filled sac in the cell cytoplasm formed by a single, continuous membrane.

Vagina A muscular tube leading from the exterior to the uterus.

Variability The occurrence of different genetically determined characteristics in the individuals of a population.

Variegation Differences in color in segments of a plant leaf, stem, or flower.

Vascular plant A plant with true vascular tissues—that is, with xylem and phloem.

Vascular system, of plants The elements that conduct water and the products of photosynthesis in plants.

Vegetal pole The end of a mature egg opposite the animal pole.

Vegetative cell The cell of a pollen grain that regulates growth of the pollen tube during fertilization.

Vegetative reproduction *See* Asexual reproduction.

Ventral The bottom side of a bilaterally or radially symmetric organism.

Vertebrate An organism with an axial body support consisting of a chain of bony subunits, the vertebrae.

Vesicle *See* Vacuole.

Virus A particle consisting of a core of nucleic acid surrounded by a protein coat; capable of infecting a cell and directing the synthesis of virus particles of the same kind.

Vitamin A substance required in minute amounts by an organism.

Voltage *See* Potential.

Wavelength The length of the wave path, from crest to crest, followed by energy traveling in waves.

Wild-type allele An allele occurring as the most common type in natural populations of a species.

X radiation A form of energy traveling at very short wavelengths, between 0.00000001 and 0.00001 nanometer.

Xylem A tissue of vascular plants that conducts water and dissolved minerals from the roots to all parts of the plant.

Yolk Nutrient material, including proteins, carbohydrates, and lipids, stored in the cytoplasm of eggs.

Yolk body, yolk platelet A cytoplasmic vesicle containing yolk.

Z-pathway The Z-shaped pathway followed by electrons through the two photosystems and the electron transport chains of the light reactions in blue-green algae and the eukaryotic plants.

Zero population growth (ZPG) The equilibrium level of population growth, in which population size remains constant.

Zooplankton Floating microscopic animals in aquatic ecosystems.

Zygote The fertilized egg.

Zygotene *See* Pairing stage.

INDEX

photosynthesis (continued)
  chlorophyll in, 96–98
  cytochromes in, 100, *101*
  dark reactions, 138
  definition, 92–93
  electron transport in, 100–105
  in eukaryotes, 92–114
  in evolution, 481–483
  light absorption in, 96
  light reactions, 96–106
  photophosphorylation, 105–106
  and photosystem I, 100
  and photosystem II, 100, 103
  in prokaryotes, 114–116
  reaction centers, 98–99
  Z-pathway, 100ff, *101, 103*
photosynthetic bacteria, 114–116
photosystem I, 100
photosystem II, 100
phototropism, 354
phylum, 382
phytoplankton, 517, 519, *520*
pill, the, 325
pinna, 395
pinocytosis, 87–89
placenta, 319–321, 323–324
plankton, 517, 519, *520.* 544
plant cell, 65–68
Plantae, 382, 388–396
plants
  cells, 65–68
  classification, 388ff
  development, 393
  evolution, 396
  nonvascular, 393
  vascular, 396
plasma membrane, 49
plasmid, 245
plasmodesmata, 68, 204
plastid, 65
plastocyanin, 102
plastoquinone, 102
Platyhelminthes, 403
Pliohippus, 418
Pliocene, 418
ploidy, 206
  See also diploid, haploid
point mutation, 432
polar body, 234, 296
polar bond, 28
polar granules, 326–327
polar nuclei, 331, 335
polarity, 14
polarity, egg, 295–296
pole, of spindle, 217ff
pole plasm, 326–327
pollen, 334ff, 391
pollen tube, 335–336
pollution, 557, 559–560
polydactyly, 280
polygenic inheritance, 272
polymerase
  DNA, 192, 216–217
  RNA, 150, 151, 216, 217
polypeptide, 28
polyploid, 434–435
  and quantum speciation, 461–462
polysaccharide, 20
polyspermy, 301
polytene chromosome, 373
polyunsaturated fat, 22
populations, 381, 439–446, 491ff
  and carrying capacity, 494ff
  and competition, 509–510, 513
  definition, 490
  growth, 491–500, 509–510, 513
  human, 555ff

and predation, 510–513
pore complex, 57–58, *59,* 155ff
Porifera, 401
Porter, K. R., 362
postzygotic barriers, 457ff
potential energy, 36
predation, 438ff, 503, 510–513
  and species diversity, 29ff
predator, 439
prey, 439
prezygotic barrier, 456–457
Priestley, J., 94
primary acceptor, in photosynthesis,
    103
primary cell wall, 68
primary constriction, 196
primary consumer, 519–521
primary germ cell, 376
primary germ layers, 321ff
  derivatives, 304
primary meristem, 348
primary mesenchyme, 305
primary producer, 517–521
primary production, 517–521
primary structure, protein, 28
primary succession, 530
primary tissues, plant, 344ff
primase, 216–217
primer, 193, 216–217
primitive atmosphere, 475
primitive pit, 321
primitive streak, 308, 321
procambium, 346ff
procentriole, 217–218
producer, primary, 517–521
progesterone, 316–318, 319, 325
progestin, 325
projector lens, of electron
    microscope, 76
prokaryote, 54–69
  bacteria, 54ff
  blue-green algae, 54ff
  capsule, 54
  definition, 54
  division, 208–209
  evolution of, 481–483
  genetics, 284–286
  inducible enzymes, 178–181
  nucleoid, 54–55, *56,* 157
  oxidation in, 140, 141
  photosynthesis, 114–116
  ribosomal proteins, 172–173
  ribosomes, 172–173
  structure, 54–56
  transcriptional regulation in,
    178–181
prokaryotic cell, 54–56, 68–69
prophase, 194–198
prophase I, of meiosis, 228–232
  condensation stage, 228
  pairing stage, 228
  recombination stage, 230–231
  recondensation stage, 231–232
  transcription stage, 231
prophase II, of meiosis, 232
proplastid, 65, 333
prostaglandin, 300, 319
prostate gland, 315, *316*
protamine, 297
protein
  alpha helix, 29–30
  chromosomal, 191
  iron-sulfur, 100–101
  modification, 175–177
  oxidation, 134
  primary structure, 28, 29
  quaternary structure, 29

secondary structure, 29
  sequencing, 421
  structure, 27–30
  tertiary structure, 29
protein synthesis. *See* translation
proteinoid, 478
Protista, 382, 385–387
protoderm, 337, 338, 344ff, 345
proton, 8
protostome, 400
Protozoa, 385
pseudocoelomate, 399
pseudoextinction, 468–470
pseudopodia, 387
Pterophyta, 395
puff, chromosome, 373, *374*
Punnett, R. C., 252, 255
Punnett square, 252, 256
purine, 30, 32, 145ff
purple photosynthetic bacteria, 112
pyrimidine, 30, 32, 145ff
Pyrrophyta, 385–387
pyruvic acid, 112, 123ff, 138, 139
  oxidation, 124–127

quantum speciation, 459ff
quaternary structure, of protein, 29

r selection, 496
radial symmetry, 399, *401*
radioactive dating, 414
radioactive isotope, 8–9, 106
random coil, in proteins, 30
reaction center, 99
realized niche, 505
receptor, cell, 366–367
recessive trait, 250ff
recombinant chromatid, 261
recombinant DNA, 243–244
recombination, 221, 228, 230–231,
    236–238, 242–243, 256–263
  bacterial, 284–287
  in evolution, 437
  in viruses, 286–287
red algae, 394
redox couple, 97
redox potential, 97
reduction, 97
regulation
  of transcription, 178–183
  of translation, 177–178, 183
regulative egg, 296
regulator gene, 179ff
replication, DNA, 149, 190, 191–194,
    211–217, 243
  conservative, 212–215
  and meiosis, 222
  semiconservative, 192ff, 211–215
repressor, 181ff
reproductive isolation, in speciation,
    453ff
Reptilia, 405
respiration, 93, 124
restriction endonuclease, 243–245
reverse repeat, 152
Rh factor, 280–281
Rhizobium, 551
rhizoid, 395
Rhodophyta, 393–394
rhythm method, of birth control,
    324–325
ribonuclease, 217
ribonucleic acid (RNA), 30–33, 55,
    143–167ff
  in development, 304
  evolution, 478
  in meiosis, 231

messenger, 143
  polymerase, 150, 151, 216–217
  precursors, 152
  ribosomal, 144, 153–154, 159–160,
    167ff, 172–173, 231, 291
  synthesis, 159–160, 178–183, 334,
    356
  transfer, 143
ribose, 37, *38*
ribosomal protein, 172–173
ribosomal RNA (rRNA), 144,
    153–154, 167ff, 172–173, 231
  in oogenesis, 291
  synthesis, 159–160
ribosome, 160, 168ff
  chloroplast, 184–186
  in development, 304
  eukaryotic, 55, 61ff
  mitochondrial, 119, 184–186
  prokaryotic, 55, *56*
  rRNA and, 144
  structure, 172–173
  synthesis, 172, 173
  and translation, 144ff
ribulose 1,5-diphosphate (RuDP), 111
ribulose 1,5-diphosphate carboxylase
    (RuDP carboxylase), 111, 113
RNA. *See* ribonucleic acid
RNA synthesis. *See* RNA and
    transcription
Romalea, 242
root apical meristem, 339
rough ER. *See* endoplasmic reticulum
RuDP. *See* ribulose 1,5-diphosphate
RuDP, carboxylase. *See* ribulose 1,5-
    diphosphate carboxylase

S phase, of interphase, 191ff
sand, 502
Sarcodina, 387
sarcoma, 376
saturated fat, 22
saturation, of enzymes, 43–44
savanna biome, 544
scanning electron microscopy, 78–79
Schleiden, M. J., 49
Schroeder, T. E., 359, 362
Schwann, T., 49
scientific method, 2–5
scrotum, *316,* 318
second law of thermodynamics, 35,
    37, 547
secondary cell wall, 68
secondary consumer, 521–522
secondary structure, of protein, 29
secondary succession, 530
secondary tissues, 347
seed, 330, 343–344, 355, 391
  formation of, 339–341
  germination, 343–344
seed plants, 390
segmentation, 398, 399
segregation, 250
Selander, R. K., 425
selection
  K, 496, 505
  natural, 429ff, 443ff, 446ff, 510–513
  r, 496
selective cell adhesion, 305, 365–367
SEM. *See* scanning electron
    microscopy
semen, 315
semiconservative replication, 192ff,
    211–215, 243
seminal fluid, 315, 319
semipermeable membrane, 82
Sendai virus, 52